World-Systems Evolution and Global Futures

Series Editors

This series seeks to promote understanding of large-scale and long-term processes of social change, in particular the many facets and implications of globalization. It critically explores the factors that affect the historical formation and current evolution of social systems, on both the regional and global level. Processes and factors that are examined include economies, technologies, geopolitics, institutions, conflicts, demographic trends, climate change, global culture, social movements, global inequalities, etc. Building on world-systems analysis, the series addresses topics such as globalization from historical and comparative perspectives, trends in global inequalities, core-periphery relations and the rise and fall of hegemonic core states, transnational institutions, and the long-term energy transition. This ambitious interdisciplinary and international series presents cutting-edge research by social scientists who study whole human systems and is relevant for all readers interested in systems approaches to the emerging world society, especially historians, political scientists, economists, sociologists, geographers and anthropologists.

This book series is indexed in Scopus.

All titles in this series are peer-reviewed.

David J. LePoire • Leonid Grinin
Andrey Korotayev
Editors

Complexity in Universal Evolution

A Big History Perspective

Springer

Editors
David J. LePoire
Argonne National Laboratory
Argonne, IL, USA

Leonid Grinin
HSE University
Moscow, Russia

Andrey Korotayev
HSE University
Moscow, Russia

ISSN 2522-0985 ISSN 2522-0993 (electronic)
World-Systems Evolution and Global Futures
ISBN 978-3-032-20516-2 ISBN 978-3-032-20517-9 (eBook)
https://doi.org/10.1007/978-3-032-20517-9

This Springer imprint is published by the registered company Springer Nature Switzerland AG
The registered company address is: Gewerbestrasse 11, 6330 Cham, Switzerland

Foreword

For generations, the grandest intellectual challenge has been to weave the disparate threads of scientific knowledge into a single, coherent narrative, an evidence-based account of our shared cosmic, planetary, and human origins. This quest for a unified story, the core project of Big History, demands a conceptual framework robust enough to bridge the scales from quantum fluctuations to galactic superclusters, and from microbial evolution to global civilization. The present volume, *Complexity in Universal Evolution: a Big History Perspective*, edited by David J. LePoire, Leonid Grinin, and Andrey Korotayev, is not merely another contribution to this burgeoning field. It is, I contend, a landmark synthesis that marks a crucial maturation of the discipline. By systematically applying the analytical tools of complexity science, this collection of essays moves Big History beyond compelling narrative into a more rigorous, explanatory paradigm. It seeks to transform the field by correcting long-standing omissions to forge a new and more powerful grammar for understanding the universe.

The core originality of this volume lies in its steadfast commitment to a unified thesis: that the principles of complexity—specifically self-organization, energy flow, and information processing—are not just useful metaphors, but constitute the fundamental toolkit for analyzing the entire sweep of universal evolution. The essays here move decisively from describing what happened to explaining how it happened, tracing the common dynamics that govern the emergence of structure in physical, biological, and social systems alike. This shift from a primarily narrative account to an organized analytical framework is a profound intellectual achievement, in that it provides a flexible and empirically grounded model capable of unifying phenomena as diverse as stellar nucleosynthesis, the origins of life, and the growth of sociopolitical institutions. In doing so, this volume elevates the Big History project, offering a model that is at once comprehensive in its scope and precise in its application. This fundamental work not only refines our "maps of time," but also forges a theoretical compass for navigating the complex and interconnected landscapes they reveal.

The intellectual power of this collection of thematic articles stems from its clear and coherent central argument: that the 13.8-billion-year history of the cosmos is best understood as an accelerating story of rising complexity. This is a trajectory sustained by a dynamic triad of forces—the flow of energy, the growth of organization, and the processing of information. Moving beyond mere analogy, the analysis proceeds through concrete metrics, comparative approaches, and unifying theoretical models. From the formation of galaxies to the emergence of global civilization, we are shown the same fundamental processes at work, albeit manifesting in vastly different substrates and scales.

To build a comprehensive understanding of universal evolution, one must first establish a firm theoretical foundation. This volume achieves this by dedicating its opening sections to first providing the conceptual tools of complexity science and then applying them to the universe's earliest and grandest epochs. Parts I and II work in tandem, charting a course from abstract principles to their magnificent manifestation in the cosmos. *Part I: Complexity Theory for Big History* grounds the entire analysis in modern complexity science. The essays here articulate how concepts once seen as barriers to development, such as the second law of thermodynamics, are in fact its fundamental drivers. Under far-from-equilibrium conditions—demonstrated by tangible examples like Bénard cells forming in heated fluids—the relentless drive toward entropy fosters, rather than inhibits, the emergence of order. The chapters in Part I explain precisely how self-organized structures arise not in defiance of physical laws, but as a direct consequence of them, allowing systems to dissipate energy gradients more effectively.

Building on this foundation, *Part II: Cosmic Complexity* applies these theoretical tools to the vast scale of cosmic evolution. The chapters trace the emergence of structure from the primordial plasma, detailing the formation of galaxies and the first Population III stars. This section highlights the pivotal role of "creative destruction," particularly in the cataclysmic supernovae that forged and dispersed heavier elements. By tracing the stellar lifecycles from these massive first stars to later Population II and Population I stars, the book demonstrates the progressive enrichment of the cosmos with the carbon, oxygen, and iron that became the essential building blocks for every subsequent stage of complexity.

From these cosmic origins, the stage was set for the emergence of new, more localized forms of order on planetary scales. Since, in the grand narrative of Big History, the immense spans of time between the formation of a planet and the first spark of life can sometimes be treated as a passive interlude, this volume corrects this oversight by framing this phase as a critical and dynamic bridge, without which life as we know it would be impossible. The contributions in *Part III: Geological and Chemical Complexity* elevate geological and chemical evolution from a mere backdrop to an active and indispensable co-evolutionary process. It reframes geological evolution as a dynamic system in its own right. Processes such as plate tectonics and atmosphere formation were not just prerequisites for life but part of a complex feedback loop that shaped both the planet and the organisms that would eventually inhabit it. The chapters in Part III analyze Earth's unique geological trajectory, providing a powerful comparative basis for understanding the specific conditions that made life possible.

With these planetary foundations established, the narrative turns to the explosive and adaptive complexity of living systems. As the universal trend toward complexity manifests itself on Earth, it has produced new forms of organization in which information processing has taken on a central role, along with flows of energy.

The essays in the central portion of this book delve into the uniquely adaptive systems found in biology and human society, exploring how the storage, transmission, and interpretation of information became a primary driver of evolution. *Part IV: Biological Complexity* explores the intricate web of life, offering novel perspectives on long-standing evolutionary questions. Viruses, often viewed solely as pathogens, are reframed as fundamental "information agents" that transmit genetic material and fuel diversity.

Synthesizing these themes, *Part V: Human and Social Complexity* examines the recent and dramatic acceleration of complexity driven by human culture. This section explores artificial intelligence as both a product of human brain evolution and a potential future driver of cosmic evolution, and also reframes the Big History narrative itself as a modern creation myth—a scientifically grounded story that can provide a "shared map" to help us navigate the challenges of our time. Additionally, the section demonstrates how even conflict can be understood as a driver of complexity, analyzing war as a reflexive force where, as Charles Tilly famously noted, "War made the state and the state made war," revealing it as a process that is both built by and a builder of social organization."

Having established the foundational principles of social organization, the volume next demonstrates the power of complexity science as a diagnostic tool for understanding the macro-historical patterns that have shaped human history. This section moves from the micro-dynamics of human interaction to the grand lifecycles of civilizations, applying a rigorous, systems-based approach to the past. *Part VI: Civilization Complexity* masterfully applies thermodynamic and systems-based principles to analyze the rise, flourishing, and decline of large-scale societies. For example, the section explores how the dynamics of resource bubbles and historical dynasties can be understood as emergent dissipative structures, following predictable patterns of growth and decay. The analysis throughout this section is particularly insightful, presenting social phenomena not as arbitrary events but as processes governed by underlying rules of energy and organization.

In its final thematic section, *Part VII: Future Complexity*, the analytical framework is boldly applied to speculate on the future of complexity, both on Earth and in the cosmos. A rigorous, model-based approach to calculating the potential for intelligent life in the Milky Way is presented, with careful inclusion of uncertainty propagation serving as a model of scientific humility and highlighting the parameters most in need of refinement for future research. The exploration of a "Galactic Information Field" provides a speculative but powerful conclusion, challenging the reader to consider that intelligence as we know it may not be the final stage of evolution, but merely a precursor to an "over-civilization"—a distributed, superintelligent cosmic entity shaped by, but not reducible to, individual intelligences. Taken together, these contributions provide a uniquely comprehensive and analytically coherent exploration of complexity's role across the entire timeline of the universe.

Complexity in Universal Evolution: a Big History Perspective succeeds brilliantly in its ambitious goal. It offers not a rigid, deterministic law of history, but a flexible and empirically grounded framework to deepen our understanding of the Universe, its self-development and our place within it. It demonstrates that the ancient quest for a unifying story of origins need not be relegated to the realm of faith or metaphysics alone. Big History can indeed serve as a modern creation myth, and this volume provides the analytical depth and intellectual substance needed to make such a myth not only compelling but also credible. More than that, it makes it useful—a practical tool for navigating the profound uncertainties of our globalized, technologically saturated, and increasingly fragile future.

For this landmark achievement, we owe a great debt to the editors—D. J. LePoire, L. Grinin, and A. Korotayev—and to the international assembly of scholars they have brought together. They have not only advanced a field of study; they have provided a new and more powerful lens through which to view our world, our past, and our collective destiny.

International Big History Association
Rio de Janeiro, Brazil

Daniel Barreiros

Acknowledgments

First and foremost, we would like to express our gratitude to the International Big History Association and its members for continuing to foster and encourage communication concerning Big History. This continues the early efforts by David Christian, Fred Spier, and Cynthia Stokes-Brown in organizing and Bill Gates in funding this organization. We also thank Lowell Gustafson who helped organize the kick-off complexity in Big History workshop at Villanova University in Pennsylvania and is also the editor of the Journal of Big History, which continues to openly share early works in this field. Special thanks go to the Journal of Big History for permission to use its publications. We thank the referees, most of whom were also from this group, which reviewed the drafts and contributed greatly to improving their clarity and integration into this book.

About This Book

Complexity in Universal Evolution offers a comprehensive and interdisciplinary examination of how complexity emerges, transforms, and accelerates across the entire 13.8-billion-year history of the Universe. As the second volume in a series on Big History, this book brings together 22 chapters from leading researchers to explore the deep structure of complexity—from cosmic origins and planetary evolution to biological innovation, human development, civilizational dynamics, and possible future trajectories.

Despite the centrality of complexity in Big History, its meaning, measurement, and mechanisms remain fragmented across disciplines. This volume confronts these gaps by integrating advances in complexity science, thermodynamics, systems theory, social science, futures studies, evolutionary theory (including evolutionary biology), and universal evolutionism, in general, into a unified, though pluralistic, framework. Contributors examine how complexity arises through energy flows, self-organizing dynamics and self-regulating development, information processing, hierarchical organization, and how these processes unfold in distinct but interconnected phases of universal evolution.

The book is organized into seven thematic parts. Foundational chapters develop theoretical tools for understanding complexity across scales. Subsequent sections trace the rise of structural complexity in the cosmos; the crucial and often underappreciated role of chemical and geological evolution; the emergence of biological and biosocial complexity; and the evolution of human societies, technologies, and institutions. Additional chapters analyze civilizational transformations through thermodynamics, urban evolution, and conflict dynamics. The final chapters extend Big History into the future, modeling the potential development of intelligent life in the galaxy and exploring new forms of complexity that may surpass traditional conceptions of intelligence.

Together, these contributions reveal complexity not as a static state but as a dynamic, multilevel process marked by thresholds, feedbacks, crises, and innovations. By synthesizing perspectives across scientific and humanistic disciplines, the volume highlights complexity growth as a central trend in Big History—a trend that illuminates the past, clarifies the present, and shapes our understanding of humanity's possible futures.

Contents

Contributors

Nicolò Antonietti Institute of Radioastronomy – National Institute of Astrophysics (INAF-IRA), Bologna, Italy

Ken Baskin Philadelphia, PA, USA

Mark Ciotola San Francisco State University, San Francisco, CA, USA

Martin van Duin Sittard, The Netherlands

Leonid Grinin HSE University, Moscow, Russia

Eurasian Center for Big History and System Forecasting, Institute of Oriental Studies of the Russian Academy of Sciences, Moscow, Russia

Anton Grinin HSE University, Moscow, Russia

Lomonosov Moscow State University, Moscow, Russia

Lowell Gustafson Villanova University, Villanova, PA, USA

Antony Harper Lake Zurich, IL, USA

Andrey Korotayev HSE University, Moscow, Russia

Eurasian Center for Big History and System Forecasting, Institute of Oriental Studies of the Russian Academy of Sciences, Moscow, Russia

David J. LePoire Argonne National Laboratory, Argonne, IL, USA

Alexander Markov Haifa, Israel

Paolo Musso University of Insubria, Varese, VA, Italy

Paul Narguizian California State University, Los Angeles, CA, USA

J. N. Nielsen Portland, OR, USA

Alexander Panov Skobeltsyn Institute of Nuclear Physics (MSU SINP), Lomonosov Moscow State University, Moscow, Russia

Stephen Satkiewicz Rochester Hills, MI, USA

Ekaterina Sazhienko International University of Humanity and Nature, Dubna, Russia

Teófilo Vargas Universidad Nacional Mayor, de San Marcos, Peru

Complexity in Universal Evolution: A Big History Perspective—An Introduction

1

David J. LePoire, Leonid Grinin, and Andrey Korotayev

Abstract

Complexity is widely acknowledged as a foundational and pivotal concept in Big History, offering a unifying lens through which to examine the emergence and development of systems—from particles and galaxies to life, civilizations, and beyond. Yet, despite its centrality, major gaps remain in how we define, measure, and interpret complexity across different phases and scales. These gaps include challenges in theoretical focus, consistency, relevance across domains, and methodological integration. This volume was planned to provide a picture of complexity rising in Big History. By examining how complexity emerges, is sustained, and transforms across thresholds, as well as in different directions and dimensions, we seek to unify seemingly disparate phenomena under shared principles of self-organization, energy flow, and information processing. While the chapters in this volume do not claim to resolve all of these issues, they provide meaningful contributions to the broader scholarly conversation by offering fresh perspectives and exploratory approaches. In this volume, the authors also demonstrate complexity growth as a leading trend of Big History.

Keywords

Complexity · Universal evolution · Emergence · Cosmic evolution · Self-organization · Biological evolution · Science · Humanities · Sociocultural evolution · Big History · Universal evolution

D. J. LePoire
Argonne National Laboratory, Argonne, IL, USA
e-mail: dlepoire@anl.gov

L. Grinin · A. Korotayev (✉)
HSE University, Moscow, Russia

Eurasian Center for Big History and System Forecasting, Institute of Oriental Studies of the Russian Academy of Sciences, Moscow, Russia

D. J. LePoire et al. (eds.), *Complexity in Universal Evolution*, World-Systems Evolution and Global Futures, https://doi.org/10.1007/978-3-032-20517-9_1

This book is the second volume in our three-volume work dedicated to various issues and dimensions of Big History and universal evolution. The first was *Navigating Complexity in Big History: Exploring Periodization Across Cosmic and Biosocial Dimensions* (LePoire et al., 2025). It investigates the possibilities of different Big History periodizations and numerous issues connected with this topic.

This second book is named *Complexity in Universal Evolution: A Big History Perspective*. It brings together 22 chapters that examine the role of complexity across its key phases: cosmic, chemical, geological, biological, human, and civilizational evolution. The chapters reflect an open-ended and interdisciplinary approach to complexity, emphasizing the need for continued dialogue and synthesis across the sciences and humanities.

For an outline of the third book of our series (dedicated to phase transitions in the Big History and universal evolution), see chapter "A Big History Perspective on Complexity in Universal Evolution: Conclusions" (LePoire et al., 2026b) in this volume.

1.1 Complexity in Big History

Complexity is a foundational theme in Big History, offering a framework for understanding the dynamic of universal evolution, including evolution of the cosmos, the Earth, life, and society. Yet, despite its centrality, the concept remains fragmented across disciplines. This was a reason to name this volume *Complexity in Universal Evolution: A Big History Perspective* and organize it with the growth of complexity being a through line of research. The next chapter ("Introduction to Complexity Theory for Big History" [LePoire et al., 2026a, this volume]) develops a unifying perspective that integrates advances in complexity science, thermodynamics, and systems thinking with the macro-historical narrative of Big History, which is previewed briefly here.

Through the lens of complexity, Big History becomes much more than a linear chronicle. It is a multiscale, nonlinear process shaped by flows of energy, information dynamics, and emergent organization. From the formation of stars and galaxies to the rise of civilizations and Artificial Intelligence (AI), the Universe can be viewed as a nested series of phase transitions and attractor shifts, each marking a reorganization of matter, energy, and information.

Complexity in Big History is not static but dynamic, unfolding as systems interact, adapt, and transform over time. In the early Universe, simplicity gave way to structure through gravitational collapse and energy dissipation, setting the stage for stars, galaxies, and chemical diversity. Life on Earth emerged by capitalizing on these processes, forming adaptive biochemical networks that transformed energy dissipation into information-processing systems. Evolutionary mechanisms—genetic variation, selection, and cultural innovation—then fueled accelerating growth in complexity.

Human evolution exemplifies this trajectory. Cognitive and cooperative advances enabled the transition from simple tool use to symbolic culture, institutions, and global networks. This shift from biological to cultural and technological evolution created nested hierarchies that integrate physical, biological, cultural, and technological elements. Across domains, complexity emerges as an interplay of energy flow, organization, and information processing.

Beyond theory, complexity offers a practical framework for interpreting the past and anticipating the future. Every transition—galaxy formation, biological evolution, or societal transformation—involves trade-offs, feedback loops, and adaptive challenges. In today's rapidly evolving global system, the interplay of energy, information, and organizational change defines both unprecedented opportunities and profound uncertainties.

1.1.1 Core Features of Complex Systems

Despite varying definitions, most theorists agree that complex systems share key traits (Mitchell, 2009; Ladyman et al., 2013):

- **Emergence:** Macro-level properties arise from micro-level interactions.
- **Self-organization:** Order emerges without central control.
- **Nested hierarchies:** Systems exist within systems across scales.
- **Nonlinearity:** Small changes can trigger large effects.
- **Adaptivity:** Systems reconfigure in response to environmental change.
- **Integration:** Cooperation among components yields new properties, or "aromorphosis" (Maynard Smith & Szathmáry, 1995).

These traits appear in phenomena as varied as weather systems, ecosystems, language, markets, and civilizations.

1.1.2 Complexity, Thermodynamics, and Emergence

From a physical perspective, the rise of life and civilization might seem improbable, given the Second Law of Thermodynamics and destabilizing planetary conditions. Yet research shows that entropy can foster, not inhibit, complexity. Far-from-equilibrium conditions drive self-organization, allowing systems to dissipate energy gradients more effectively (Schneider & Sagan, 2006). Examples include Bénard cells forming in heated fluids or gravitational collapse concentrating matter to form stars and planets. Complexity thus emerges not in defiance of physical law but because of it, under specific conditions.

Evolutionary trajectories of complex systems often follow cycles of destabilization, reorganization, and metastable order (Grinin, 2013). These patterns appear across cosmology, biology, and society, producing recurring phases of growth, crisis, and transformation.

1.1.3 Complex Systems in Big History

Complexity in Big History can be broadly categorized into:

- **Physical complex systems:** Self-organizing but non-adaptive systems such as galaxies, stars, hurricanes, and vortices. Governed by physical laws, they stabilize into dynamic patterns without processing information.
- **Complex adaptive systems (CAS):** Systems capable of learning and adaptation, such as organisms, ecosystems, and societies. CAS evolve through cycles of variation, selection, and memory, adjusting their internal structures to remain viable.

Some researchers also suggest a third category—**quantum complex systems**—characterized by rich probabilistic dynamics and structured interactions, even without ongoing energy flow.

These systems demonstrate that complexity emerges when openness, interaction, and energy flows combine. In human history, increased energy capture and information processing underpinned successive transitions, from genetic evolution to cultural and technological revolutions. Each step integrated smaller units into larger, more complex wholes, though often at the cost of greater fragility and systemic risk (Tainter, 1988, 1996).

1.1.4 Interdisciplinary Insights

The growth of complexity science has been fueled by institutions like the Santa Fe Institute, the New England Complex Systems Institute, and the Complexity Science Hub Vienna, along with influential thinkers such as Page (2015), Bar-Yam (1997), Krakauer (2019), as well as Thurner and his colleagues (2018). Their work spans theoretical, computational, and applied domains, highlighting universal dynamics of self-organization and adaptation. Within Big History, Chaisson (2011) and Spier (1996) have emphasized energy flow and information processing as key metrics for tracking complexity across cosmic, biological, and social scales.

Despite disciplinary differences—from algorithmic measures in computer science to resilience studies in ecology—these perspectives converge on a pluralistic understanding: complexity is a dynamic interplay of many interacting parts operating across time and scale.

1.2 Overview of Book Parts and Chapters

Complexity in Big History reveals a universe in constant transformation. From the emergence of structure in the early cosmos to the adaptive networks of life and the accelerating interconnections of human society, complexity offers a unifying framework to explore both resilience and fragility. Understanding these principles not only deepens our grasp of the past but also sharpens our insight into possible futures—where energy, information, and organization will continue to shape the evolving tapestry of the Universe.

In the complexity introduction section (**Part I: Complexity Theory for Big History** of this volume), three chapters examine how general principles of complexity apply across Big History periods. David LePoire, Leonid Grinin, and Andrey Korotayev offer an expanded perspective on complexity theory and its relevance to Big History and universal evolution,[1] Lowell Gustafson advocates for complexity as a unifying framework across disciplines, including its role in rethinking the purpose and structure of the modern university,[2] while Martin van Duin's chapter highlights the role of energy flows within Big History's complex systems at different levels of hierarchy.[3]

In the cosmic evolution section (**Part II: Cosmic Complexity** of this volume), papers explore the symmetry between early cosmic cooling and the accelerating complexity of life on Earth, the dissipation of gravitational and nuclear energy in cosmic structures, stellar and galactic modeling to estimate conditions for habitable planets, and the risks of contact with extraterrestrial intelligence (ETI). These studies highlight the dynamic interplay of energy, matter, and structure in the formation of complexity on a universal scale.[4]

Bridging cosmic and terrestrial phases, geological and chemical evolution plays a crucial yet underappreciated role in Big History. This is analyzed in **Part III: Geological and Chemical Complexity** of this volume. Chapters by Leonid Grinin and David LePoire emphasize Earth's transformation through plate tectonics, atmosphere formation, and biosphere interaction.[5] These processes not only made the planet habitable but also initiated co-evolutionary feedback loops that shaped both geology and life. Recognizing this phase as more than background to biological evolution enriches our understanding of planetary complexity and resilience. Another chapter by Leonid Grinin presents an overview of chemical evolution through various phases of Big History.[6]

The section on biological evolution (**Part IV: Biological Complexity** of this volume) expands on this by addressing less-explored dimensions of life's

[1] See chapter "Introduction to Complexity Theory for Big History" (LePoire et al., 2026a) in this volume.

[2] See chapter "Complexity as a Unifying Framework: From the Big Bang to Human Societies" (Gustafson, 2026, this volume).

[3] See chapter "Scale dependence of FERD in Big History's Hierarchical Systems" (Van Duin, 2026) in this volume.

[4] See chapter "The Evolution of Complexity in the Stelliferous Era: Observations and Patterns" (Korotayev & LePoire, 2026, this volume) and chapter "Energy Flow and Emergence of Complexity in the Universe's Structure" (LePoire, 2026a) in this volume.

[5] See chapter "Geological Evolution: A Bridge Between Cosmic and Earthly Development in Big History" (Grinin, 2026c, this volume) and chapter "The Earth's Pivotal Role in Big History: From Cosmic Evolution to Life's Emergence" (LePoire, 2026b, this volume),

[6] See chapter "Chemical Evolution in Cosmos and Earth" (Grinin, 2026b) in this volume.

complexity. Anton Grinin discusses the evolutionary role of viruses as information agents and ecological influencers.[7] Paul Narguizian highlights how complexity science can transform undergraduate biology education, fostering systems thinking.[8] Meanwhile, Leonid Grinin, Alexander Markov, and Andrey Korotayev compare biological and social evolution, emphasizing shared mechanisms such as selection, adaptation, and information transmission—thereby bridging biological and cultural complexity.[9]

Within the domain of human and social evolution, several contributions explore how emerging technologies and long-term human patterns can inform our understanding of complexity (see **Part V: Human and Social Complexity** of this volume). Andrey Korotayev undertakes an attempt to propose a periodization of the human evolution on the basis the most recent scientific data, taking into account cognitive, technological, cultural dimensions, and biological dimensions of complexity growth during the human evolution phase of the Big History.[10] David LePoire and Ekaterina Sazhienko examine artificial intelligence as both a product and driver of complex evolution, comparing AI development to brain evolution and exploring future human–AI symbioses.[11] Nick Nielsen introduces the concept of "peer complexity" in the Stelliferous Era, placing Earth-bound complexity within a wider cosmological framework.[12]

The section on civilizational evolution (see **Part VI: Civilization Complexity** in this volume) investigates how complexity unfolds in societies, resource systems, and historical dynamics. Ken Baskin reframes Big History as a modern creation myth, proposing that integrating complexity science can guide societal transformation in the face of global challenges.[13] Mark Ciotola applies thermodynamic principles to model the rise and fall of dynasties and resource bubbles.[14] Antony Harper examines urban size distributions over millennia, revealing punctuated complexity trends.[15]

[7] See chapter "Viruses in Big History: Evolutionary Roles and Complex Systems" (Grinin, 2026a, this volume).

[8] See chapter "Biology Education Through Big History's Complex Systems Thinking" (Narguizian, 2026, this volume).

[9] See chapter "Biological and Social Phases of Big History and Universal Evolution: Similarities, Differences, and the Growth of Complexity" (Grinin et al., 2026, this volume).

[10] See chapter "Human Evolution in the Complexity Growth Perspective: Toward Periodization of the Big History Biosocial Era" (Korotayev, 2026, this volume). See also Korotayev, 2025b.

[11] See chapter "The Role of Artificial Intelligence in Big History: Understanding Human Evolution and Future Scenarios" (Sazhienko & LePoire, 2026, this volume).

[12] See chapter "Rethinking Complexity in the Universe: Implications for SETI and the Fermi Paradox" (Nielsen, 2026, this volume).

[13] See chapter "Myth, Complexity, and Big History: Rethinking Our Shared Map for the Future" (Baskin, 2026, this volume).

[14] See chapter "Emergence, Complexity, and Thermodynamics in Resource Bubbles and Historical Dynasties" (Ciotola, 2026, this volume).

[15] See chapter "Punctuated Growth and Complexity in Urban Evolution: Insights from Historical City Size Distributions" (Harper, 2026, this volume).

Stephen Satkiewicz explores the co-evolution of war and social organization, reinforcing the link between organized conflict and state formation.[16]

The final section on future complexity (see **Part VII: Future Complexity** in this volume) considers possible processes in the Universe including the location, probability of formation of planets suitable for life,[17] and the possible future development of complexity that intelligent life on these planets might form.[18]

Importantly, many of these studies transcend individual phases of Big History, illustrating how complexity evolves across nested systems and timescales. Some draw parallels between cosmic and terrestrial transitions, compare energy dissipation in physical and adaptive systems, or reflect on the human lifespan relative to the Universe's age. This integrative approach is vital for advancing a more coherent and comprehensive understanding of complexity in Big History.

While consensus remains elusive, the contributions in this issue reflect the vitality of ongoing inquiry. By confronting complexity's unresolved questions and proposing new directions, these chapters strengthen the foundation for future research. In doing so, they reaffirm complexity's place at the heart of Big History and encourage continued exploration of how complex systems emerge, evolve, and shape the unfolding story of the Universe and humanity.

We have presented a very brief overview of the contents of all the parts of the present volume. Now, we will spell out the contents of all the chapters of this monograph in more detail.

1.2.1 Introduction to Complexity in Universal Evolution

The theoretical issues of complexity dynamics in universal evolution are considered in **Part I** of this volume—**Complexity Theory for Big History—**that consists of three chapters.

Chapter "Introduction to Complexity Theory for Big History" (LePoire et al., 2026a, this volume) by David LePoire, Leonid Grinin, and Andrey Korotayev explores the role of complexity dynamics in Big History through an integrative theoretical framework that spans physical, chemical, biological, social, cognitive, and civilizational domains. Building on foundational work in systems theory, thermodynamics, and evolutionary theory, the chapter argues that complexity can serve as a conceptual bridge across disciplines. By examining how complexity emerges, is sustained, and transforms across thresholds, as well as in different directions and dimensions, we seek to unify seemingly disparate phenomena under

[16] See chapter "War and the Rise of Social Complexity: A Big History Perspective" (Satkiewicz, 2026, this volume).

[17] See chapter "Modeling Galactic Evolution and the Potential for Intelligent Life in the Milky Way" (Vargas et al., 2026, this volume).

[18] See chapter "Beyond Intelligence: The Galactic Information Field and the Evolution of Complexity" (Panov, 2026, this volume).

shared principles of self-organization, energy flow, and information processing. The aim is not to impose a rigid definition, let alone paradigm, but to offer a flexible, empirically grounded definition and understanding of complexity, which may pave way to creation of a model that deepens our understanding of the Universe, its self-development, and our place within it.

Lowell Gustafson's chapter "Complexity as a Unifying Framework: From the Big Bang to Human Societies" (Gustafson, 2026, this volume) emphasizes that complexity serves as a unifying theme that addresses fundamental questions about both the structure of the Universe and human nature. It provides an intellectual framework that bridges disciplines across the natural sciences, social sciences, and humanities—connecting everything from quarks to global societies and even humanity's curiosity about intergalactic relations. The concept of complexity begins with its unidirectional emergence, tracing a path from the Big Bang to the present. It then expands to a multidirectional perspective, encompassing the formation of planets, galaxies, and potential extraterrestrial life. However, complexity does not always progress; it can remain static, regress to simpler forms, or even collapse entirely. Speculations on the existence of multiple universes further extend this idea, suggesting the possibility of infinite complexity.

Martin van Duin in his chapter "Complexity Pyramids and Energy Rate Density: A Quantified System Hierarchy for Big History" (Van Duin, 2026, this volume) notes that in their position paper for a Big History 2.0, Henriques and Volk (2023)[19] have advocated to distinguish between (i) emergence in realms vs. that when moving from one realm to another, and (ii) the complexification path from quarks to culture vs. that of astronomical objects, which are not part of that sequence. In his re-examination of energy rate density (ERD; proposed by Chaisson [2001] as a metric for complexity), Solis (2024) has concluded that there is a need to further explore systems at different levels of hierarchy to see if there is a general correlation between ERD and complexity. Martin van Duin's chapter follows up these suggestions by applying the "complexity pyramid" concept, proposed by Reeves (1991), as a framework to represent the hierarchical organization of (complex) systems. Applying a rigorous part/system approach, a series of complexity pyramids is constructed for systems from the material, astronomical, biological, technological, and social realms. Next, ERD is used to quantify the variation of horizontal complexity on a pyramid level (size and diversity) and the change in vertical complexity in a pyramid (emergence). The main observation is that ERD of a system is typically smaller than the largest ERD of its parts on the level below. This results in a narrowing of the ERD range of systems on a next level and a decrease of ERD of system's "engines," when going upward in a complexity pyramid. This is rather counterintuitive, since complexity and thus ERD are supposed to increase, when parts combine in a larger system with emergent properties at the next-level. However usually, both active, energy-processing (ERD > 0) and inactive, structural parts (ERD $= 0$) converge to the larger system. ERD of that larger system is the

[19] See also Volk and Henriques (2025).

mass-average of the ERDs of the (in)active parts and, consequently, is typically smaller than the largest ERD of its parts. Finally, the complexity pyramids for the various realms are integrated in an overlay, representing "all" systems from the physical reality of our world. The ERD quantification of the systems in the individual pyramids allows the quantitative positioning of the systems from the various pyramids in this overlay. A comparison of ERD values of systems from two different complexity pyramids results in pars-pro-toto issues, because there is no objective way to establish which levels from different pyramids may be compared. Chaisson's (2001) sequence of galaxies → stars → planets → plants → animals → brains → society is not only a rather tortuous path through the pyramid overlay, but the observed increase of ERD over big time is also somewhat problematic.

1.2.2 Cosmic Complexity

As we explore the vast narrative of Big History—from the Big Bang to the rise of human civilization—a key question emerges: How did complexity increase across such different phases of the Universe? In recent years, scholars have begun to focus more closely on the transition between cosmic and terrestrial complexity. This shift—from the formation of stars and galaxies to the emergence of life and intelligence on Earth—marks a critical turning point in the story. Understanding this transition helps us see how energy, matter, and organization came together to create the conditions necessary for life, and later, for technological societies like ours.

Several recent studies have taken up this challenge from different angles. Some focus on the buildup of elements through stellar processes, while others explore how energy flow patterns shaped conditions for life. New models estimate when and where intelligent life might arise in our galaxy, and some even question whether intelligence, as we know it, is the final stage of evolution. Together, these works expand our understanding of complexity in Big History—linking the deep past to possible futures and opening new perspectives on humanity's role in the cosmos.

These topics are considered in **Part II** of this volume—**Cosmic Complexity**. It comprises two chapters that discuss various aspects of complexity growth at the cosmic phases of the universal evolution. Chapters of this part explore the symmetry between early cosmic cooling and the accelerating complexity of life on Earth, the dissipation of gravitational and nuclear energy in cosmic structures, stellar and galactic modeling to estimate conditions for habitable planets, and the risks of contact with extraterrestrial intelligence (ETI). These chapters highlight the dynamic interplay of energy, matter, and structure in the formation of complexity on a universal scale.

Andrey Korotayev and David LePoire explore the interesting transition between the cooling after the Big Bang and the acceleration of complexity increase on Earth in their chapter "The Evolution of Complexity in the Stelliferous Era: Observations and Patterns" (Korotayev & LePoire, 2026) in this volume. Their previous studies (Korotayev, 2024, 2025a; LePoire, 2016, 2020a, 2020b, 2025) compared the symmetries of the cooling after the Big Bang and the development of life, humans, and civilization on Earth. A simple model of the relative rates of change depicts two cones joined at their bases representing the Big Bang expansion and the acceleration of complexity development on Earth through life evolution, human evolution, and civilization development. This indicates points of rapid change at both ends corresponding to the Big Bang and our current technological and demographic change. However, this depiction oversimplifies complexity growth, specifically at the transitions from the Universe's decelerating physical complexity development and Earth's accelerating evolution of complex adaptive systems. In their previous work, Korotayev and LePoire found that for about 150 million years after the Big Bang Singularity, one could observe a hyperbolic deceleration of the universal complexity growth driven by hyperbolically decelerating cooling of the Universe. However, around this time, with the end of the cosmological Dark Ages and the start of the Stelliferous Era, the further cooling of the Universe stopped playing any significant role in the further growth of the universal complexity. Against this background, a new driver of the universal complexity growth emerged. There are some grounds to maintain that this was the nucleosynthesis in the emerging stars that produced elements more complex than hydrogen and helium that appeared in the Epoch of Big Bang Nucleosynthesis (between 10 and 1000 seconds after the Big Bang Singularity). Korotayev and LePoire propose possible proxy measures of complexity that could help us to estimate the universal complexity dynamics in the Stelliferous Era of the Big History in more detail than was done earlier. One such measure is what the astronomers call "metallicity." Astronomers define as "metals" any elements that are heavier than hydrogen and helium (thus, according to them, oxygen, or carbon are also "metals"). Then, the "metallicity" turns out to be a measure indicating the ratio of "metals" (all the elements that are heavier [and, thus, more complex] than helium [let alone hydrogen]) to the "non-metals." Korotayev and LePoire suggest that the mean metallicity of the Universe can be regarded as a measure of universal chemical–physical complexity (reflecting the universal complexity of the element base for the chemical evolution). Their analysis also suggests that, since 150 million years after the Big Bang Singularity, the universal metallicity experienced exponential acceleration with subsequent quasi-logistic acceleration getting to the maximum level around 10–11 billion years before present (BP), which was followed by a new wave of hyperbolic deceleration.

David LePoire's chapter "Energy Flow and the Emergence of Complexity in the Universe's Structure" (LePoire, 2026a, this volume) compares the physical complex system of the developing structure in the Universe with the later complex adaptive systems evolving on Earth. The structural development of the Universe differs significantly from the initial rapid cooling after the Big Bang and the subsequent development of life on Earth. This development spans various spatial structures, from the cosmic web to galaxies, stars, and planets. Traditionally, it was believed that the free energy rate density (FERD) indicated a trend of increasing energy density over time, implying that galactic structures were less complex than stellar structures based on estimated energy flows. However, recent findings in cosmic evolution suggest that large energy flows during the formation of the cosmic web and galaxies contributed to these structures. By incorporating energy flows from the cosmic web's gravitational collapse, active galactic nuclei (AGN) in younger systems, stellar energy, and Earth's cooling from gravitational collapse and radioactive decay, it is suggested that FERD might actually decrease during this cosmic phase. This decrease in energy gradients and cooling improved conditions for life. For later stages of life's evolution on Earth, Chaisson's FERD calculations might apply in a limited scope, particularly in the evolving atmosphere and ocean influenced by sunlight. However, it is believed that the cascade of energy flow seen in turbulent fluid dynamics is a more fitting analogy for cosmic structural development.

1.2.3 Geological and Chemical Complexity

In telling the grand story of the Universe—from the Big Bang to the rise of human civilization—Big History connects many fields of science and humanities. Yet one important phase in this story often receives less attention than it deserves: the geological evolution of Earth. While cosmic evolution and biological evolution are usually emphasized, the transformation of Earth's physical structure and surface over billions of years is a critical bridge between the formation of the planet and the rise of life. Understanding Earth's geological changes not only helps explain how life began and developed but also provides key insights into the planet's long-term stability—and potential future. Transition from cosmic to Earthly evolution meant that the scope of evolution narrows. As a result, the possibilities for the emergence of its new levels and the speed of its changes increased dramatically. This is due to the fact that the evolution of smaller objects requires less energy, and consequently, the opportunities for evolution increase. This rule is confirmed by the transition from the Universe to the Solar System, from the Solar System to individual planets, and from geological and biological evolution to social.

Recent work is beginning to highlight the importance of integrating geological evolution more fully into Big History. By examining Earth's dynamic processes—plate tectonics, volcanic activity, atmospheric changes, and more—we can better understand how the planet became habitable and how it continues to support life. These interactions also reveal important feedback loops between geology and biology, including those that regulate the climate. Besides, examination of geological evolution gives new insight on planetary evolution.

Part III of the present volume—**Chemical and Geological Complexity**—consists of three chapters that address this vital but underexplored topic. Leonid Grinin offers a broad conceptual framework for why chemical and geological evolution should play a more central role in Big History, while David LePoire explores the specific complexities that shaped Earth's development and its eventual transition into a life-bearing world.

Leonid Grinin's chapter "Chemical Evolution in Cosmos and Earth" (Grinin, 2026b, this volume) discusses the limited research on chemical evolution within both Big History and evolutionary studies, despite its central role in the universal evolution. Chemical evolution has often served as a leading process, particularly during the formation of pre-life on Earth around five billion years ago. This paper traces the trajectory of chemical evolution from the emergence of the first atoms in the early Universe to abiogenesis on Earth, laying the groundwork for the appearance of living organisms. Three broad stages can be distinguished: (1) the formation of atoms (pre-evolution), (2) processes leading up to abiogenesis on Earth, and (3) abiogenic chemical evolution itself. Building on this framework, the chapter proposes a more detailed periodization of chemical evolution prior to life's origin. A distinctive feature of this process is its inherently co-evolutionary character: unfolding first in tandem with cosmic and stellar-galactic evolution, then as part of planetary evolution where conditions favored chemical reactions, and subsequently within geochemical and biochemical contexts on Earth. Chemical evolution, therefore, emerges as a continuous, co-evolutionary thread linking the cosmic, planetary, geological (including mineralogical), and biological domains of Big History.

A brief look at complexity in the formation and development of Earth is explored in David LePoire's chapter "The Earth's Pivotal Role in Big History: From Cosmic Evolution to Life's Emergence" (LePoire, 2026b, this volume). This chapter includes references to other Big History writers' perspectives. Earth's formation was a complex and unpredictable process shaped by gravitational, thermal, and chemical interactions, along with numerous astronomical events. From the collapse of gas and dust into a disk around the Sun to the Moon-forming collision, the early Solar System's chaos ultimately created a stable, life-supporting planet. Key features like Earth's large moon, active core, and magnetic field have been crucial for its habitability. As Earth evolved from a barren rock to a thriving world, geology and biology began to influence each other. Processes like ocean currents, atmospheric changes, and carbon dioxide absorption shaped environments where life could emerge and, in turn, transform the planet. Feedback loops, such as oxygenation of the atmosphere and erosion driven by plant life, show how Earth and its biosphere co-evolved. The Gaia hypothesis suggests life itself helps stabilize Earth's conditions, keeping it habitable. However, human activities now pose new challenges to this balance.

Leonid Grinin in his chapter "Geological Evolution: A Bridge Between Cosmic and Earthly Development in Big History" (Grinin, 2026c, this volume) analyzes how Earth's physical features and processes have changed over time. He notes that this is much better understood than cosmic evolution, which deals with the Universe as a whole. However, geological evolution often gets overlooked in the broader context of Big History—a field that connects the history of the Universe, Earth, life, and humanity. In Big History courses, geological evolution is often treated as a minor backdrop to biological evolution, rather than a subject deserving detailed study. This gap is surprising because geological evolution has unique characteristics that set it apart from cosmic and planetary evolution. While it is well-studied in connection with how life has evolved on Earth, it has not been fully integrated into the larger framework of how all forms of evolution—from the Universe to life—are interconnected. By making comparisons between Earth's geological changes and similar processes on other planets, we can deepen our understanding of Earth's role in the grand story of evolution across the cosmos. In geological terms, there are quite a lot of similarities between the evolution of Earth and that of other planets in the Solar System. This is particularly evident in the early phases of their development (see the end of the chapter and its appendix). Yet, we can also note differences in the evolution of Earth and other celestial bodies. At the beginning of the Earth history, these differences seemed superficial (we are not considering chemical and biological evolution here), but then comes formation of hydrosphere which, in addition to the Earth, only occurred on one planet (Mars), and later the tectonics of lithospheric plates, which radically changes the situation. Understanding how a hot, dead planet without a crust became one that continuously renews itself and is alive is extremely important. This still remains an intriguing mystery, but it is impossible to solve this mystery without delving into geological evolution. This chapter and especially its appendix reveal many interesting facts on the subject of similarities and differences between the evolution of the Earth and the other planets in the Solar System.

1.2.4 Biological Complexity

Understanding evolution through the lens of Big History requires a broader and more integrated perspective—one that includes biological, informational, and social complexity. While much attention has been given to the evolution of life and human societies, new research highlights the importance of underexplored contributors and frameworks that shape our understanding of these processes.

These topics are considered in **Part IV** of this volume—**Biological Complexity**. It comprises three chapters that offer diverse but interconnected insights into evolutionary complexity.

In chapter "Viruses in Big History: Evolutionary Roles and Complex Systems" (Grinin, 2026, this volume), Anton Grinin explores the crucial role of viruses for understanding evolution, though it remains under-researched due to their microscopic nature and limited physical traces in the geological record. This chapter explores the role of viruses in evolutionary development within the framework of Big History, providing an overview of current perspectives and challenges in understanding their position in evolutionary history. The origins and evolution of viruses are examined through hypotheses such as the regressive, protobiont, and symbiotic theories. It highlights the unique position of viruses on the edge of disorder, displaying both instability and adaptability, sparking debate on whether they can be considered alive. From an energetic perspective, viruses significantly influence information and energy dynamics within ecosystems. They can be seen as basic information agents transmitting genetic material, serving as a fundamental tool of evolution. Lastly, viruses are examined as complex self-regulating systems, drawing parallels to computer viruses, cancer, and languages. The report emphasizes the similarities between these systems and explores potential explanations for these resemblances.

Paul Narguizian explores how to teach biology through the lens of complex systems in his chapter "Teaching Biology Through Complexity: Integrating Systems Thinking in Undergraduate Education" (Narguizian, 2026, this volume). This chapter explores the integration of Complex Systems Thinking (CST) into undergraduate biology education, advocating for a more holistic approach to understanding living systems. Traditional reductionist methods, while valuable for dissecting specific biological components, often fail to capture the complexity, interconnectedness, and emergent properties inherent in biological systems. By embracing CST, biology educators can provide students with the tools needed to comprehend and navigate the ever-growing complexity of life. The article delves into the significance of CST in biological education, discussing key concepts such as complexity, emergence, and systems thinking, and highlighting the pedagogical implications for both educators and students. It also offers practical strategies for incorporating these interdisciplinary approaches into the curriculum. Ultimately, the goal is to equip future biologists and scientifically literate citizens with a deeper, more integrated understanding of biological systems, preparing them to address the multifaceted challenges of the natural world.

Building on their earlier work,[20] Leonid Grinin, Alexander Markov, and Andrey Korotayev aim to better understand the connection between two phases in chapter "Biological and Social Phases of Big History and Complexity Growth" (Grinin et al., 2026, this volume). The comparison between biological and social evolution is an important but often overlooked area in Big History. Although these two phases of evolution are usually studied separately, recent research suggests that examining their similarities and differences can offer valuable insights into the larger patterns and transitions in evolution. This chapter explores how biological and social evolution compare across several key areas: the structure of organisms, how selection works, how major changes occur, how information is passed on, and how certain traits are preserved. Grinin et al. also look at how some biological traits may have helped prepare the way for the rise of social evolution. While there is still debate about whether Darwinian ideas apply to social and cultural change, the authors argue that many useful comparisons can still be made, helping us see both continuity and transformation in the history of life and society.

1.2.5 Human and Social Complexity

As Big History expands to encompass not just the deep past but also humanity's emerging future, new questions arise about how complexity continues to unfold—both on Earth and across the cosmos. Increasingly, these questions are being addressed through interdisciplinary approaches that blend science, technology, philosophy, and cultural analysis. At the heart of these inquiries is a desire to understand our evolving place within a universe marked by transformation, uncertainty, and emergent possibilities.

The three chapters of **Part V** of this volume—**Human and Social Complexity**—offer distinct but complementary perspectives on these themes. Andrey Korotayev's

[20] Grinin et al. (2009, 2011, 2012, 2013, 2020).

chapter "Human Evolution in the Complexity Growth Perspective: Toward Periodization of the Big History Biosocial Era" (Korotayev, 2026, this volume) proposes a new periodization of the Big History Biosocial (Anthropogenesis) Era grounded in the latest scientific evidence. The framework is complexity-based, with each epoch defined by a phase transition that produced significant increases in both biological and sociocultural complexity within evolving planetary biosocial systems. Five major epochs are identified:

1. **Australopithecine Epoch (c. 4.2–2.9/2.4 million BP):** Emergence of australopithecines with early, though limited, signs of sociocultural complexity.
2. **Oldowan–Habilis Epoch (c. 2.9/2.4–2.0 million BP):** Appearance of the first Homo species and the Oldowan stone tool industry.
3. **Early Acheulean–Erectus Epoch (c. 2.0–0.9/0.5 million BP):** Rise of *Homo erectus* and Early Acheulean technologies.
4. **Late Acheulean–Heidelbergensis Epoch (c. 0.9/0.5–0.3 million BP):** Development of *Homo heidelbergensis*, Late Acheulean tools, composite implements, systematic fire use, and construction of dwellings.
5. **Middle Paleolithic Epoch (c. 300,000–50,000 BP):** Emergence of Neanderthals, Denisovans, and early *Homo sapiens*, accompanied by major advances in sociocultural complexity, including Levallois stone technologies, symbolic burials, ornaments, pigments, engravings, and structured spatial use.

Within this scheme, the Upper Paleolithic Revolution (c. 60–40 BP) marks the conclusion of both the Middle Paleolithic Epoch and the Biosocial Era itself. The study also demonstrates that widely cited complexity-based periodizations of Big History (e.g., Modis, Panov) conflict with current evidence of human evolution and require significant revision. However, updating the dates of biosocial complexity jumps does not undermine previous mathematical analyses of planetary complexity growth: recalculations continue to reveal the same hyperbolic acceleration patterns when corrected dates are applied.

Ekaterina Sazhienko and David LePoire in their chapter "The Role of Artificial Intelligence in Big History: Understanding Human Evolution and Future Scenarios" (Sazhienko & LePoire, 2026, this volume) explore how AI can shape both the future and our understanding of the past. Artificial Intelligence (AI) is rapidly becoming an integral part of our daily lives. This chapter explores how AI can enhance Big History in two significant ways: by providing insights into the complexity of human evolution, and by offering a better understanding of future scenarios as the relationship between humans and AI evolves. Additionally, it discusses how AI tools can enhance scientific discovery and facilitate access to information through improved search and communication capabilities. The chapter examines AI and brain understanding by reviewing the latest developments in AI, comparing them to brain evolution and current brain architecture. The integration of AI and humans is explored through various future scenarios, including an ongoing relationship resembling that of a parent and child. The future development of this complex human–AI system is projected to follow multiple pathways, including immersing humans in virtual and augmented realities created by AI, integrating AI systems with human activities and organizational processes, and embedding AI directly within human minds.

Nick Nielsen's chapter "Rethinking Complexity in the Universe: Implications for SETI and the Fermi Paradox" (Nielsen, 2026, this volume) focuses on the types of emergent complexity that characterize the time period when stars are actively forming, living, and dying, that is, the Stelliferous Era, using the framework from Adams and Laughlin's *The Five Ages of the Universe* (1999) as an extension of Big History. The paper defines the concept of peer emergent complexity and contextualizes familiar human phenomena—life, consciousness, intelligence, technology, and civilization—within this framework. By situating Earth-originating emergent complexities within a broader cosmological class of peer emergent complexities, Nielsen aims to develop a less anthropocentric approach to the search for biosignatures and technosignatures, suggesting that these categories may be too narrow. The implications of these considerations for SETI (Search for Extra=Terrestrial Intelligence) are explored, including stronger and weaker interpretations of the Fermi paradox.

1.2.6 Civilization Complexity

As Big History continues to grow as an integrative framework for understanding the evolution of the Universe, life, and human society, the concept of complexity has emerged as a central unifying theme. From thermodynamic systems to urban networks and civilizations, complexity helps us trace patterns of emergence, transformation, and collapse across vastly different domains. The four chapters of **Part VI** of this volume—**Civilization Complexity**—demonstrate how diverse forms of complexity—physical, social, political, and intellectual—can be analyzed using tools from thermodynamics, systems theory, and historical modeling. Together, they offer a multiscale view of how complexity develops, interacts with energy and entropy, and underpins major transitions in both natural and human history.

Ken Baskin in his chapter "Myth, Complexity, and Big History: Rethinking Our Shared Map for the Future" (Baskin, 2026, this volume) notes that Big History, as envisioned by David Christian, is more than a story of the Universe—it is a modern creation myth. Like the traditional myths that helped past societies make sense of their world, Big History can offer a shared map to help us understand our place in a vast and changing Universe. As E.O. Wilson put it, this "evolutionary epic" gives us a unifying story grounded in science. In today's world, where we face serious challenges like climate change and social instability, such a shared story may be more important than ever. This chapter suggests that combining Big History with complexity science—a field that studies how systems grow, adapt, and evolve—can deepen our understanding of this modern myth and help us better respond to today's crises. By looking at how past societies, such as those during the Axial Age and Modernity, overcame periods of great change by rewriting their core beliefs, we can see how new stories can guide social transformation. This chapter explores four key ideas: Christian's concept of a shared myth, what complexity science tells us about the Universe, how complex systems evolve, and what happens when we bring these insights into Big History.

Mark Ciotola in his chapter "Emergence, Complexity, and Thermodynamics in Resource Bubbles and Historical Dynasties" (Ciotola, 2026, this volume) explores how the concepts of emergence and complexity can be applied to analyze physical resource bubbles and historical dynasties. It introduces key terms such as energy, thermodynamic laws, efficiency, and power, and discusses relevant extensions of the Second Law of Thermodynamics. Various definitions of complexity, including the energy rate density approach, are examined and critiqued. Comparisons are made between energy rate density and entropy-based approaches, highlighting their divergence from complexity artifacts. The principles of extended thermodynamics are then applied to understand the emergence of medium-term dissipative structures, such as the rise and fall of resource bubbles and historical dynasties. The chapter discusses how complexity changes as these structures progress through their stages of emergence. Patterns in complexity change are identified, leading to the articulation of a characteristic complexity profile for resource bubbles and historical dynasties.

Antony Harper's chapter "Punctuated Growth and Complexity in Urban Evolution: Insights from Historical City Size Distributions" (Harper, 2026, this volume) shows that, at the macro level, the maximum population size of urban areas throughout history follows a hyperbolic growth pattern. However, when the data is log-transformed, a more nuanced structure is revealed: periods of stability or stasis, punctuated by phases of rapid expansion. This study investigates the implications of this historical trend on the size-frequency distribution and complexity of urban areas relative to their maximum population size. Several foundational assumptions frame this research. Firstly, it is posited that urban area size-frequency distribution mirrors a power-law (Zipf-Pareto) distribution. Secondly, all maximum urban area sizes are considered historical upper limits, contingent upon the type and access to energy sources, such as solar energy for agrarian systems and fossil fuels for industrial systems. By fitting historical city size distributions for each century to extract the power-law's steepness, a pattern of punctuated equilibria emerges. In this study, "complexity" is defined as the diversity of urban area sizes within an interconnected system. Analyzing this proxy for complexity over centuries reveals a consistent trend of increasing complexity, marked by alternating phases of positive and negative slopes. Further exploration is ongoing regarding the relationship between this complexity and entropy. Noteworthy periods of increased complexity coincide with epochs of system reorganization, including the Dark Ages as delineated by Sing C. Chew (2006). Additionally, the correlation between power-law steepness and complexity highlights distinct phases in societal development, corresponding to shifts from agrarian societies and Bronze Age empires to industrial and post-industrial eras.

Stephen Satkiewicz in his chapter "War and the Rise of Social Complexity: A Big History Perspective" (Satkiewicz, 2026, this volume) maintains that, from a Big History perspective, the relationship between the emergence of social complexity and war is reflexive, with each influencing the other as both cause and effect, depending on the context. This dynamic is evident from the origins of organized violence in the Neolithic period to the present day, particularly in the formation of early states,[21] empires, and various socio-political systems. Historical sociologist Charles Tilly (1929–2008) famously summarized this relationship with the phrase, "War made the state and the state made war." This study builds upon interdisciplinary work within civilizational analysis to examine the thresholds of emergent social complexity and the processes of social macroevolution. The goal is to help ground Big History within the human and social sciences where applicable. War and organized violence do not occur in a vacuum; their nature is a reflection of the complex contexts in which they arise.

1.2.7 Future Complexity

The two chapters of **Part VII** of this volume—**Future Complexity**—analyze possible future developments.

Teófilo Vargas, Paolo Musso, and Nicolò Antonietti extend their previously present model to calculate the average age of potential intelligent life in the Milky Way in their chapter "Modeling Galactic Evolution and the Potential for Intelligent Life in the Milky Way" (Vargas et al., 2026, in this volume). The model relies on various inputs, including gas surface density, total star mass, initial mass function, supernovae explosions, and metallicity. Each input carries an uncertainty due to the diverse and complex nature of measurements. These uncertainties can be propagated to estimate the overall uncertainty in the average age of potential intelligent life. This process will help identify which inputs contribute most to the final uncertainty and which measurements need refinement for a more precise estimate. Additionally, the model will be enhanced to account for variance relative to a two-dimensional (2D) radial symmetric model.

Alexander Panov considers a potentially wider Big History in his chapter "Beyond Intelligence: The Galactic Information Field and the Evolution of Complexity" (Panov, 2026, this volume). This study challenges the conventional perspective of Big History, proposing that intelligence, as traditionally defined, may not represent the peak of cosmic evolution. Instead, the increasing complexity of the Universe could give rise to novel forms of matter organization that transcend our current understanding of intelligence. Building on the concept of "superintelligence," this study explores a scenario in which intelligent civilizations across the Galaxy engage in continuous interstellar communication, gradually forming a vast, interconnected Galactic information field. Over time, this distributed network accumulates and influences planetary civilizations, evolving into an over-civilization—an entity shaped by but not reducible to individual intelligences. Due to the

[21] On state formation see also, e.g., Feinman (2023); Jennings and Earle (2023); Kradin (2023, 2025); Popov (2023); Grinin (2025); Askerov et al. (2025); Bondarenko (2025).

finite speed of light and the vastness of space, no single civilization can fully grasp or control this emergent structure. The Galactic information field is thus proposed as a superintelligent entity, distinct from traditional intelligence and beyond the comprehension of any one planetary mind. This perspective challenges the assumption that intelligence is the ultimate evolutionary outcome, suggesting instead that the Universe may be progressing toward forms of complexity that defy human-centric definitions of cognition and agency.

In concluding chapter "A Big History Perspective on Complexity in Universal Evolution: Conclusions" (LePoire et al., 2026b, this volume), David LePoire, Leonid Grinin, and Andrey Korotayev summarize the main findings of the present volume.

1.3 Concluding Remarks

Complexity growth as a Big History trend is inextricably linked with universal evolution. It is extremely important to consider more closely the growth of complexity as a certain trend of self-development of the Universe and its individual parts; as a set of general movements of various systems and objects, as well as forms of their interaction. This trend in some senses can be considered leading in Big History. Understanding, explaining, and analyzing this trend actually constitutes one of the main objects of Big History research.

By examining how complexity emerges, is sustained, and transforms across thresholds, as well as in different directions and dimensions, we seek to unify seemingly disparate phenomena under shared principles of self-organization, energy flow, and information processing. The aim is not to impose a rigid definition, let alone paradigm, but to offer a flexible, empirically grounded definition and understanding of complexity, which may pave way to creation of a model that deepens our understanding of the Universe, its self-development, and our place within it.

The study of Big History invites us to explore the vast continuum of complexity—from the explosive beginnings of the Universe to the intricate networks of human culture and society. The complexity lens allows us to interpret Big History not as a linear chronicle but as a multiscale, multidirectional, nonlinear process shaped by energy flows, information dynamics, and emergent and developed organization. The history of the Universe—from the formation of stars and galaxies to the rise of civilizations and AI—can be reframed as a nested series of phase transitions and attractor shifts, each marking a significant reorganization of matter, energy, information, and self-management (see, e.g., Grinin et al., 2024).

In chapter "Introduction to Complexity Theory for Big History" (LePoire et al., 2026a, this volume), we have collected some definitions of complexity. Mostly these definitions characterize only some (one or two) fields of science and complexity remains a concept marked by fragmentation across disciplines. So, it seems that complexity resists a single definition. For the scope of this book, it is even more important that there are no unified definitions of complexity in Big History and Evolution. Perhaps, it is impossible. However, each new capacious definition will allow us to see the development of complexity in a different aspect. That is why we suggest our own definition.

Complexity is a category used to describe, evaluate, and compare the degree or level of organization and functionality within systems and objects. This applies across a wide spectrum, from individual systems to a conceptually defined level of evolutionary efficiency. As a scientific tool, complexity is crucial for understanding and ranking the evolutionary efficiency and position of objects within their specific domains or "kingdoms of complexity." Naturally, higher levels of complexity have sublevels, perhaps it would make sense to identify such sublevels, to make some kind of systematization of complexity.

Evolutionary efficiency is the ability to adapt and generate breakthrough innovations (aromorphoses) that open up new evolutionary space for development. It is an ability to go beyond the limits set by complexity niches. Within these niches and levels, both their filling and horizontal evolution occur, which, however, prepares pre-adaptations and other mechanisms for a breakthrough. There are also parallel lines of evolution (such as the planetary one, for example), where the movement toward greater complexity takes place, but it leads to dead ends until a certain point. The more parallel, lateral, and additional lines, the higher the possibility for an evolutionary breakthrough toward higher complexity. Since all objects differ in their ability to self-preservation and change, one of the most important dimensions of complexity is the balance between self-preservation and the ability to change, evolve, and increase in complexity.

For Big History, the ability to find criteria for measuring the level of complexity is especially important. It was said above that it is impossible to talk about any one criterion expressed in a formula or quantitatively measurable variable, but only about a group of criteria (and their list can be open). However, identifying each new criterion moves us toward understanding complexity. These points will be discussed systematically throughout this volume.

References

Adams, F., & Laughlin, G. (1999). *The five ages of the universe: Inside the physics of eternity*. The Free Press.

Askerov, M.-A., Issaev, L., & Korotayev, A. (2025). Islamist state formation in Somalia. *Social Evolution & History, 24*(2), 78–102. https://doi.org/10.30884/seh/2025.02.04

Bar-Yam, Y. (1997). *Dynamics of complex systems*. Perseus Books.

Baskin, K. (2026). Myth, complexity, and Big History: Rethinking our shared map for the future. In: D. J. LePoire, L. Grinin, & A. Korotayev (Eds.), *Complexity in universal evolution: A Big History perspective* (469–486). Springer Nature (in this volume). https://doi.org/10.1007/978-3-032-20516-2_16

Bondarenko, D. M. (2025). Evolution and forms of socio-political organization: Historical and anthropological approaches (introductory remarks). *Social Evolution & History, 24*(2), 118–126. https://doi.org/10.30884/seh/2025.02.06

Chaisson, E. J. (2001). *Cosmic evolution; The rise of complexity in nature*. Harvard University Press.

Chaisson, E. (2011). Energy rate density as a complexity metric and evolutionary driver. *Complexity, 16*(3), 27–40. https://doi.org/10.1002/cplx.20323

Chew, S. (2006). *The recurring dark ages: Ecological stress, climate changes, and system transformation*. Altamira Press.

Ciotola, M. (2026). Emergence, complexity, and thermodynamics in resource bubbles and historical dynasties. In: D. J. LePoire, L. Grinin, & A. Korotayev (Eds.), *Complexity in universal evolution: A Big History perspective* (487–518). Springer Nature (in this volume). https://doi.org/10.1007/978-3-032-20516-2_17

Feinman, G. M. (2023). Rethinking premodern governance: Passing the torch. *Social Evolution & History, 22*(2), 4–17. https://doi.org/10.30884/seh/2023.02.01

Grinin, L. (2013). *Macrohistory and globalization*. Uchitel Publishing House.

Grinin, L. E. (2025). On the stages of the evolution of statehood. *Social Evolution & History, 24*(2), 157–161. https://doi.org/10.30884/seh/2025.02.12

Grinin, A. (2026a). Viruses in Big History: Evolutionary roles and complex systems. In: D. J. LePoire, L. Grinin, & A. Korotayev (Eds.), *Complexity in universal evolution: A Big History perspective* (247–256). Springer Nature (in this volume). https://doi.org/10.1007/978-3-032-20516-2_10

Grinin, L. (2026b). Chemical evolution in cosmos and earth. In: D. J. LePoire, L. Grinin, & A. Korotayev (Eds.), *Complexity in universal evolution: A Big History perspective* (193–210). Springer Nature (in this volume). https://doi.org/10.1007/978-3-032-20516-2_7

Grinin, L. (2026c). Geological evolution: A bridge between cosmic and earthly development in Big History. In: D. J. LePoire, L. Grinin, & A. Korotayev (Eds.), *Complexity in universal evolution: A Big History perspective* (221–244). Springer Nature (in this volume). https://doi.org/10.1007/978-3-032-20516-2_9

Grinin, L., Markov, A., & Korotayev, A. (2009). Aromorphoses in biological and social evolution: Some general rules for biological and social forms of macroevolution. *Social Evolution & History, 8*(2), 6–50.

Grinin, L., Markov, A., & Korotayev, A. (2011). Biological and social aromorphoses: A comparison between two forms of macroevolution. In L. Grinin, R. L. Carneiro, A. Korotayev, & F. Spier (Eds.), *Evolution: Cosmic, biological, and social* (pp. 162–211). Uchitel.

Grinin, L., Markov, A., & Korotayev, A. (2012). Biological and social phases of Big History: Similarities and differences of evolutionary principles and mechanisms. *Evolution, 2*, 158–198.

Grinin, L., Markov, A., & Korotayev, A. (2013). On similarities between biological and social evolutionary mechanisms: Mathematical modeling. *Cliodynamics, 4*(2), 185–228. https://doi.org/10.21237/C7clio4221334

Grinin, L., Korotayev, A., & Markov, A. (2020). Big History and macro-evolution. Evolutionary principles and mechanisms at biological and social phases of the big history. In C. Benjamin, E. Quaedackers, & D. Baker (Eds.), *The Routledge companion to Big History* (pp. 72–105). Routledge.

Grinin, L., Grinin, A., & Korotayev, A. (2024). *Cybernetic revolution and global aging. Humankind on the way to cybernetic society, or the next hundred years*. Springer Nature. https://doi.org/10.1007/978-3-031-56764-3

Grinin, L., Markov, A., & Korotayev, A. (2026). Biological and social phases of Big History and complexity growth. In: D. J. LePoire, L. Grinin, & A. Korotayev (Eds.), *Complexity in universal evolution: A Big History perspective* (pp. 283–356). Springer Nature (in this volume). https://doi.org/10.1007/978-3-032-20516-2_12

Gustafson, L. (2026). Complexity as a unifying framework: From the Big Bang to human societies. In: D. J. LePoire, L. Grinin, & A. Korotayev (Eds.), *Complexity in universal evolution: A Big History perspective* (pp. 91–118). Springer Nature (in this volume). https://doi.org/10.1007/978-3-032-20516-2_3

Harper, A. (2026). Punctuated growth and complexity in urban evolution: Insights from historical city size distributions. In: D. J. LePoire, L. Grinin, & A. Korotayev (Eds.), *Complexity in universal evolution: A Big History perspective* (pp. 519–536). Springer Nature (in this volume). https://doi.org/10.1007/978-3-032-20516-2_18

Jennings, J., & Earle, T. (2023). Social stratification, genealogical distance, and state formation in early Hawaii and the central Andes: Steps, slopes, and off-ramps in social evolution. *Social Evolution & History, 22*(2), 44–66. https://doi.org/10.30884/seh/2023.02.04

Korotayev, A. (2024). Complexity growth patterns in the Big History. Preliminary results of a quantitative analysis. *Journal of Big History, 7*(2), 9–27. https://doi.org/10.22339/jbh.v7i2.7203

Korotayev, A. (2025a). Comparing and contrasting Big History singularity trends of the Big Bang and terrestrial evolution. In D. J. LePoire, L. Grinin, & A. Korotayev (Eds.), *Navigating complexity in Big History. Exploring periodization across cosmic and biosocial dimensions* (pp. 101–157). Springer Nature. https://doi.org/10.1007/978-3-031-85410-1_5

Korotayev, A. (2025b). Toward periodization of the biosocial era of the Big History: A reanalysis. *Journal of Big History, 8*(2), 1–37. https://doi.org/10.22339/jbh.v8i2.8201

Korotayev, A. (2026). Human evolution in the complexity growth perspective. Toward periodization of the Big History biosocial era. In: D. J. LePoire, L. Grinin, & A. Korotayev (Eds.), *Complexity in universal evolution: A Big History perspective* (pp. 539–410). Springer Nature (in this volume). https://doi.org/10.1007/978-3-032-20516-2_13

Korotayev, A., & LePoire, D. J. (2026). The evolution of complexity in the Stelliferous Era: Observations and patterns. In: D. J. LePoire, L. Grinin, & A. Korotayev (Eds.), *Complexity in universal evolution: A Big History perspective* (pp. 151–172). Springer Nature (in this volume). https://doi.org/10.1007/978-3-032-20516-2_5

Kradin, N. N. (2023). Origins of the early Russian state: Anthropological perspectives. *Social Evolution & History, 22*(2), 67–93. https://doi.org/10.30884/seh/2023.02.05

Kradin, N. N. (2025). Debates on the origin of the state: Some results and perspectives. *Social Evolution & History, 24*(2), 147–156. https://doi.org/10.30884/seh/2025.02.11

Krakauer, D. (2019). *Worlds hidden in plain sight: The evolving idea of complexity at the Santa Fe Institute, 1984–2019*. SFI Press.

Ladyman, J., Lambert, J., & Wiesner, K. (2013). What is a complex system? *European Journal for Philosophy of Science, 3*(1), 33–67. https://doi.org/10.1007/s13194-012-0056-8

LePoire, D. J. (2016). Exploring temporal patterns in Big History dynamics. *KronoScope, 16*(2), 229–249.

LePoire, D. J. (2020a). Exploring the singularity concept within Big History. In A. V. Korotayev & D. J. LePoire (Eds.), *The 21st century singularity and global futures. A Big History perspective* (pp. 77–97). Springer. https://doi.org/10.1007/978-3-030-33730-8_3

LePoire, D. J. (2020b). *Time patterns in Big History. Cycles, fractals, waves, transitions, and singularities*. David LePoire Consulting, LLC.

LePoire, D. J. (2025). A simple, compatible, and extensible Big History framework and periodization based on previous findings. In D. J. LePoire, L. Grinin, & A. Korotayev (Eds.), *Navigating complexity in Big History—Exploring periodization across cosmic and biosocial dimensions* (pp. 159–193). Springer Nature. https://doi.org/10.1007/978-3-031-85410-1_6

LePoire, D. J. (2026a). Energy flow and the emergence of complexity in the universe's structure. In: D. J. LePoire, L. Grinin, & A. Korotayev (Eds.), *Complexity in universal evolution: A Big History perspective* (pp. 173–190). Springer Nature (in this volume). https://doi.org/10.1007/978-3-032-20516-2_6

LePoire, D. J. (2026b). The Earth's pivotal role in Big History: From cosmic evolution to life's emergence. In: D. J. LePoire, L. Grinin, & A. Korotayev (Eds.), *Complexity in universal evolution: A Big History perspective* (pp. 211–220). Springer Nature (in this volume). https://doi.org/10.1007/978-3-032-20516-2_8

LePoire, D. J., Grinin, L., & Korotayev, A. (Eds.). (2025). *Navigating complexity in Big History—Exploring periodization across cosmic and biosocial dimensions*. Springer Nature. https://doi.org/10.1007/978-3-031-85410-1

LePoire, D. J., Grinin, L., & Korotayev, A. (2026a). Introduction to complexity theory for Big History. In: D. J. LePoire, L. Grinin, & A. Korotayev (Eds.), *Complexity in universal evolution–*

A Big History perspective (pp. 29–90). Springer Nature (in this volume). https://doi.org/10.1007/978-3-032-20516-2_2

LePoire, D. J., Grinin, L., & Korotayev, A. (2026b). A Big History perspective on complexity in universal evolution. Conclusions. In: D. J. LePoire, L. Grinin, & A. Korotayev (Eds.), *Complexity in universal evolution: A Big History perspective* (pp. 585–608). Springer Nature (in this volume). https://doi.org/10.1007/978-3-032-20516-2_22

Maynard Smith, J., & Szathmáry, E. (1995). *The major transitions in evolution*. Oxford University Press.

Mitchell, M. (2009). *Complexity: A guided tour*. Oxford University Press.

Narguizian, P. (2026). Biology education through Big History's complex systems thinking. In: D. J. LePoire, L. Grinin, & A. Korotayev (Eds.), *Complexity in universal evolution: A Big History perspective* (pp. 257–282). Springer Nature (in this volume). https://doi.org/10.1007/978-3-032-20516-2_11

Nielsen, N. (2026). Rethinking complexity in the universe: Implications for SETI and the Fermi paradox. In: D. J. LePoire, L. Grinin, & A. Korotayev (Eds.), *Complexity in universal evolution: A Big History perspective* (pp. 431–468). Springer Nature (in this volume). https://doi.org/10.1007/978-3-032-20516-2_15

Page, S. E. (2015). What sociologists should know about complexity. *Annual Review of Sociology, 41*(1), 21–41. https://doi.org/10.1146/annurev-soc-073014-112230

Panov, A. (2026). Beyond intelligence: The galactic information field and the evolution of complexity. In: D. J. LePoire, L. Grinin, & A. Korotayev (Eds.), *Complexity in universal evolution: A Big History perspective* (pp. 575–582). Springer Nature (in this volume). https://doi.org/10.1007/978-3-032-20516-2_21

Popov, V. A. (2023). Annual festivals and potlatch-like forms of redistribution in early political formations of Upper Guinea. *Social Evolution & History, 22*(2), 18–31. https://doi.org/10.30884/seh/2023.02.02

Reeves, H. (1991). *The hour of our delight—Cosmic evolution, order and complexity*. W.H. Freeman and Company.

Satkiewicz, S. (2026). War and the rise of social complexity: A Big History perspective. In: D. J. LePoire, L. Grinin, & A. Korotayev (Eds.), *Complexity in universal evolution: A Big History perspective* (pp. 537–560). Springer Nature (in this volume). https://doi.org/10.1007/978-3-032-20516-2_19

Sazhienko, E., & LePoire, D. J. (2026). The role of artificial intelligence in Big History: Understanding human evolution and future scenarios. In: D. J. LePoire, L. Grinin, & A. Korotayev (Eds.), *Complexity in universal evolution: A Big History perspective* (pp. 411–430). Springer Nature (in this volume). https://doi.org/10.1007/978-3-032-20516-2_14

Schneider, E., & Sagan, D. (2006). *Into the cool: Energy flow, thermodynamics, and life*. University of Chicago Press.

Solis, K. (2024). Reexamining free energy rate density as a complexity metric. *Journal of Big History, 7*(1), 19–28. https://doi.org/10.22339/jbh.v7i1.7102

Spier, F. (1996). *The structure of Big History: From the Big Bang until today*. Amsterdam University Press.

Tainter, J. A. (1988). *The collapse of complex societies*. Cambridge University Press.

Tainter, J. A. (1996). Complexity, problem solving, and sustainable societies. In R. Costanza, O. Segura, & J. Martinez-Alier (Eds.), *Getting down to earth: Practical applications of ecological economics* (pp. 61–76). Island Press.

Thurner, S., Klimek, P., & Hanel, R. (2018). *Introduction to the theory of complex systems*. Oxford University Press.

Van Duin, M. (2026). Complexity pyramids and energy rate density: A quantified system hierarchy for Big History. In: D. J. LePoire, L. Grinin, & A. Korotayev (Eds.), *Complexity in universal evolution: A Big History perspective* (pp. 119–148). Springer Nature (in this volume). https://doi.org/10.1007/978-3-032-20516-2_4

Vargas, T., Musso, P., & Antonietti, N. (2026). Modeling galactic evolution and the potential for intelligent life in the milky way. In: D. J. LePoire, L. Grinin, & A. Korotayev (Eds.), *Complexity in universal evolution: A Big History perspective* (pp. 563–574). Springer Nature (in this volume). https://doi.org/10.1007/978-3-032-20516-2_20

Volk, T., & Henriques, G. (2025). Integrating the tree of knowledge and combogenesis approaches to Big History. In D. J. LePoire, L. Grinin, & A. Korotayev (Eds.), *Navigating complexity in Big History—Exploring periodization across cosmic and biosocial dimensions* (pp. 195–215). Springer Nature. https://doi.org/10.1007/978-3-031-85410-1_7

David J. LePoire researches, develops, and applies science principles in environmental issues, Big History evolutionary trends, and particle scattering. He has a BS in physics from CalTech, a PhD in computer science from DePaul University, and over 30 years of experience at the Argonne National Laboratory in the development of scientific analyses, software, training, and modeling. His research includes Big History synergistic trends among energy, environment, organization, and information. Currently, he is the President of the International Big History Association (IBHA).

Leonid Grinin is Senior Research Professor at the Center for Stability and Risk Analysis at the HSE University in Moscow, as well as Director of Eurasian Center for Big History & System Forecasting at the Institute of Oriental Studies, Russian Academy of Sciences. He is co-editor of the international journals *Social Evolution & History* and the *Journal of Globalization Studies*, as well as co-editor of the international almanacs *Evolution, History & Mathematics*, and *Kondratieff Waves*. He is author of more than 600 publications, including 30 monographs in Russian, English, and Chinese. His current research focuses on comparative political studies, theory of revolution, political anthropology, global economy, global history, historical sociology, and futurology. Among his monographs are *The Cybernetic Revolution and the Forthcoming Epoch of Self-Regulating Systems* (Uchitel, 2016, with Anton Grinin), *Big History of Globalization: The Emergence of a Global World System* (Springer 2019, in co-authorship with Julia Zinkina, David Christian, and others), and *Islamism, Arab Spring, and the Future of Democracy: World System and World Values Perspectives* (Springer, 2019; with Andrey Korotayev and Arno Tausch). In 2004, he was awarded as a winner of the best publication of *Socis* journal during the year for the article: "People of Publicity—a New Social Layer?" In 2012, he was awarded the Gold Kondratieff Medal by the International N. D. Kondratieff Foundation. In 2016, he received the Belyaev award for the book *From Bifaces to Nanorobots* (the book published in Russian in co-authorship with Anton Grinin). 12, Rozhdestvenka, Moscow, 107031, Russian Federation; email: leonid.grinin@gmail.com; ORCID 0000-0003-0278-2619.

Andrey Korotayev heads the Center for Stability and Risk Analysis at the HSE University in Moscow, Russia. He is also Senior Research Professor at Eurasian Center for Big History & System Forecasting at Institute of Oriental Studies, Russian Academy of Sciences and Institute for African Studies, Russian Academy of Sciences, as well as at the Faculty of Global Studies of the Lomonosov Moscow State University. Beginning as an historian of Arabia, he made focused studies, such as seen in "Two Social Ecological Crises and Genesis of Tribal Organization in the Yemeni North-East" (1996). Korotayev researches topics in quantitative cross-cultural anthropology and seeks to understand biological and social macroevolution, as in "A Compact Macromodel of World System Evolution" (2005) and "Mathematical Modeling of Biological and Social Phases of Big History" (2014). He has authored or co-authored over 650 scholarly publications, including monographs such as *Ancient Yemen* (Oxford University Press, 1995), *World Religions and Social Evolution of the Old World Oikumene Civilizations: A Cross-Cultural Perspective* (The Edwin Mellen Press, 2004), *Introduction to Social Macrodynamics: Compact Macromodels of the World System Growth* (URSS, 2006; with Daria Khaltorina and Artemy Malkov), *Introduction to Social Macrodynamics: Secular Cycles and Millennial Trends* (URSS, 2006; with Daria Khaltorina and Artemy Malkov), *Great Divergence and Great Convergence: A Global Perspective* (Springer,

2015; with Leonid Grinin), *Economic Cycles, Crises, and the Global Periphery* (Springer, 2016; with Leonid Grinin), and *Islamism, Arab Spring, and the Future of Democracy: World System and World Values Perspectives* (Springer, 2019; with Leonid Grinin and Arno Tausch). He is a founding member of the Eurasian Center for Megahistory & System Forecasting and founder of *Evolution*, an almanac dedicated to the study of Universal Evolution, and thus is closely linked with Big History. In 2020, he produced, with David LePoire, a collective monograph, *The 21st Century Singularity and Global Futures: A Big History Perspective* (Springer Nature, 2020). Recently, he has produced, with David LePoire and Leonid Grinin, another Big History collective monograph, *Navigating Complexity in Big History: Exploring Periodization Across Cosmic and Biosocial Dimensions* (Springer Nature, 2025). He is a laureate of a Russian Science Support Foundation in "The Best Economists of the Russian Academy of Sciences" Nomination (2006); in 2012, he was awarded the Gold Kondratieff Medal by the International N. D. Kondratieff Foundation. He serves as a board member of the International Big History Association. 20, Myasnitskaya, Moscow 101000, Russian Federation; email: akorotayev@gmail.com; ORCID 0000-0003-3014-2037.

Part I

Complexity Theory for Big History

Introduction to Complexity Theory for Big History

2

David J. LePoire, Leonid Grinin, and Andrey Korotayev

Abstract

This chapter explores the role of complexity dynamics in Big History through an integrative theoretical framework that spans physical, chemical, biological, social, cognitive, and civilizational domains. Building on foundational work in systems theory, thermodynamics, and evolutionary theory, the chapter argues that complexity can serve as a conceptual bridge across disciplines. By examining how complexity emerges, is sustained, and transforms across thresholds, as well as in different directions and dimensions, we seek to unify seemingly disparate phenomena under shared principles of self-organization, energy flow, and information processing. The aim is not to impose a rigid definition, let alone paradigm, but to offer a flexible, empirically grounded definition and understanding of complexity, which may pave way to creation of a model that deepens our understanding of the Universe, its self-development, and our place within it.

Keywords

Complexity · Complex adaptive system · Entropy · Dissipation · Self-organization · Big History · Universal evolution

This chapter is an extensively re-worked version of the text originally published in Journal of Big History (LePoire et al., 2025d).

D. J. LePoire
Argonne National Laboratory, Argonne, IL, USA
e-mail: dlepoire@anl.gov

L. Grinin · A. Korotayev (✉)
HSE University, Moscow, Russia

Eurasian Center for Big History and System Forecasting, Institute of Oriental Studies of the Russian Academy of Sciences, Moscow, Russia

D. J. LePoire et al. (eds.), *Complexity in Universal Evolution*, World-Systems Evolution and Global Futures, https://doi.org/10.1007/978-3-032-20517-9_2

2.1 Introduction: Complexity as a Unifying Theme in Big History

Complexity is a foundational and organizing theme in Big History, offering a framework through which we can understand the dynamic evolution of the cosmos, life, and society. Yet despite its centrality, complexity remains a concept marked by fragmentation across disciplines (see for detail below). This chapter aims to develop a unifying perspective that brings together recent advances in complexity science, thermodynamics, and systems thinking with the macro-historical narrative of Big History.

The complexity lens allows us to interpret Big History not as a linear chronicle but as a multiscale, multidirectional, nonlinear process shaped by energy flows, information dynamics, and emergent and developed organization. The history of the Universe—from the formation of stars and galaxies to the rise of civilizations and artificial intelligence (AI)—can be reframed as a nested series of phase transitions and attractor shifts, each marking a significant reorganization of matter, energy, and information.

The study of Big History invites us to explore the vast continuum of complexity—from the explosive beginnings of the Universe to the intricate networks of human culture and society. Throughout this chapter, we have traced the emergence and evolution of complex systems, revealing how fundamental physical laws, self-organizing processes, and feedback mechanisms shape the Universe at every scale.

At its core, complexity in Big History is not a static or singular attribute but *a dynamic process that unfolds as systems interact, adapt, and transform over time*. In the early Universe, simplicity gave way to structure through gravitational collapse and energy dissipation—a process that laid the groundwork for the formation of stars, galaxies, and, ultimately, the chemical diversity required for life. This initial phase of physical complexity, governed by principles of thermodynamics and conservation, set into motion a series of transitions that have progressively increased in order and information content.

As life emerged on Earth, it did so by capitalizing on these physical processes. Biochemical networks and autocatalytic sets, as described by pioneering thinkers like Kauffman (1993) and Prigogine (1980), illustrate how molecular systems can organize into self-sustaining, adaptive networks. The emergence of life signified a shift from static, energy-dissipative systems to adaptive, information-processing systems—where the interplay of genetic variation, selection, and cultural innovation fuels *an ever-accelerating growth in complexity*.

Human evolution, with its hallmark developments in cognition and cooperation, stands as a vivid testimony to this trajectory. The transformation from simple tool use and symbolic artifacts to the sophisticated institutions and global networks of the modern era exemplifies the shift from biological evolution to cultural and technological evolution.[1] This progression is characterized by the integration of diverse

[1] See chapter "Human Evolution in the Complexity Growth Perspective: Toward Periodization of the Big History Biosocial Era" (Korotayev, 2026) in this volume; see also Korotayev, 2025b.

elements—physical, chemical, biological, cultural, and technological—into increasingly nested hierarchies. Whether observed through the lenses of neural plasticity, gene-culture coevolution, or economic modeling, the evolution of human societies confirms that *complexity is best understood as an emergent property of myriad interconnected processes*.

Beyond merely an academic concept, complexity in Big History serves as a practical framework for understanding our past and anticipating our future. It reminds us that every transition—from the formation of galaxies to the organization of civilizations—involves trade-offs, feedback loops, and adaptive challenges. In our rapidly evolving global system, the interplay of energy flow, information processing, and organizational change suggests that our future will likely be defined by both unprecedented opportunities and profound uncertainties.

In synthesizing insights from physics, chemistry, biology, cognitive science, and social theory, this chapter has laid out a unified narrative *where complexity is the thread that connects the cosmos to human culture*. It calls for an integrative approach that bridges disciplines and scales, emphasizing that the history of our Universe is a story of constant transformation—a dynamic tapestry woven from the same fundamental principles that continue to shape the evolution of complexity today.

While complexity resists a single definition, most theorists agree that *it involves a high degree of interdependence among system components, nonlinearity, feedback, and emergent behavior* (Mitchell, 2009; Ladyman et al., 2013). Complex systems are neither fully ordered nor chaotic; rather, they inhabit the edge of chaos where adaptive capacity and innovation are maximized (Langton, 1990).

Several key attributes characterize complex systems:

- Emergence: Properties arise at the macro level that are not present at the micro level.
- Self-organization: Systems can spontaneously order themselves without external direction.
- Nested hierarchies: Systems exist within systems, across spatial and temporal scales.
- Nonlinearity: Small causes can have disproportionately large effects.
- Adaptivity: Complex systems respond to changing environments through internal reconfiguration.
- Integration ability: Complex systems are able to unite and cooperate for their benefits and to gain as a result new emergent properties sometimes resulting in new aromorphoses (both in biological and, especially, in social world—see Sect. 2.3.3.2 "The Evolution of Cooperation: From Cells to Civilizations" below in this chapter; see also Maynard Smith & Szathmáry, 1995).

These features are found in phenomena as diverse as weather systems, cellular metabolism, biological symbioses, languages, ecosystems, markets, politics, and civilizations.

From many traditional scientific perspectives, the emergence of life—and particularly complex technological civilization—appears improbable. Physics reminds us that the second law of thermodynamics drives closed systems toward disorder, reducing the likelihood of spontaneously forming ordered states. Planetary science reveals a range of destabilizing conditions: solar winds strip atmospheres, planetary magnetic fields can collapse, and rotational instabilities threaten surface conditions. Biologically, ecosystems can overshoot resource use, leading to collapse and extinction. Against these odds, Earth has sustained the conditions for complex life and civilization.

Several twentieth and twenty-first-century thinkers speculated that humanity's accelerated development might catalyze a major evolutionary or societal transformation (Teilhard de Chardin, 1987; Vernadsky, 1997; Tsiolkovsky, 1964; Kurzweil, 2005; Fuchs-Kittowski & Krüger, 1997; see also about this Grinin et al., 2024: Chap. 1). Others warned of collapse due to limits to growth or crises of complexity (Tainter, 1996; Moiseyev, 2000; Ayres, 2006; Rodrigue, 2023).

Yet recent developments in complexity science suggest that many of the supposed barriers to complexity—such as entropy—instead play essential roles in fostering it. Schneider and Sagan (2006) argue *that complexity can flourish under thermodynamic constraints when systems operate far from equilibrium, enabling self-organization to dissipate energy gradients more effectively*. This helps explain how gravitational forces, though entropy-increasing overall, enable localized decreases in entropy by concentrating matter to form stars, planets, and, eventually, life.

Emergent order arises not despite the second law of thermodynamics, *but because of it—under very specific conditions* (see below). Benard cells, for example, form spontaneously in heated fluids as they adopt more efficient heat transfer patterns. These self-organized structures exemplify how fluctuations within a system can generate novel order when energy gradients are sufficiently high (Haken, 1988).

Self-organization operates as both culmination and beginning in evolutionary sequences. As systems transition from chaos to order and again into new forms of instability, they traverse a cycle of destabilization, reorganization, and metastable complexity (Gunderson & Holling, 2002; Grinin, 2014; Grinin et al., 2016; Grinin & Korotayev, 2009a, 2012; Korotayev et al., 2016; Korotayev & LePoire, 2025). The history of the Universe is thus marked by cycles of emergent order punctuated by crises and transitions to new attractor states (see chapter "Myth, Complexity, and Big History: Rethinking Our Shared Map for the Future" [Baskin, 2026] in this volume; see also Korotayev, 2006a, 2006b; Korotayev et al., 2006b).

2.1.1 Some Ideas About Complexity and Complex Systems

Complex systems are defined not only by their structure but also by their dynamics (Thurner et al., 2018). They operate across multiple time and spatial scales: metabolism sustains them in the short term; learning adapts them to their environment; evolution enables long-term transformation. The coordination of interacting components at various levels—molecular, individual, institutional—constitutes the fabric of adaptive complexity.

In Big History, we can categorize systems into equilibrium (non-complex), dissipative, and complex adaptive systems. While simple molecules like H_2O exhibit emergent properties, hurricanes and chemical reaction-diffusion systems exemplify dissipative structures; *higher complexity arises when systems not only dissipate energy but also process information and revise their behavior.* Complex adaptive systems (CAS), such as organisms and civilizations, learn from their environment, adjust their models, and evolve.

Jantsch (1980) described this process as a recursive cycle involving energy, information, and structure—across physical, biological, and sociocultural domains. Modern CAS research, including Friston's (2010) Bayesian models, builds on this by showing how systems actively experiment, predict, and revise their behavior to remain viable in uncertain environments.

Big History adds a comparative and evolutionary perspective. Researchers have traced energy flows across evolutionary transitions (Fox, 1988; Chaisson, 2001; LePoire & Chandrankunnel, 2020; Smil, 2008), highlighting how increasingly complex systems manage greater resource throughput. Evolution proceeds through aggregation—first through physical bonds, then biological symbiosis, and finally social coordination. As complexity rises, cycles shorten, coordination costs rise, and systemic fragility may increase (Tainter, 1996); see Fig. 2.1.

Crucially, complexity depends on variation, selection, and memory. Yet the growing global integration of human society raises questions: What happens when there is only one global system? Can evolutionary experimentation continue? Do fossil-fuel-dependent trajectories foreclose alternatives? Counterfactual thinking and comparative planetary models can help illuminate these questions.

In recent decades, a number of interdisciplinary research centers dedicated to the study of complexity have emerged, producing a wealth of academic literature and advancing our understanding of complex systems. Prominent among these are institutions such as the Santa Fe Institute, the New England Complex Systems Institute, and the Complexity Science Hub in Vienna. These institutes have supported foundational and applied research in complexity science and have contributed to a growing corpus of influential works by scholars such as Scott Page (2011), Yaneer Bar-Yam (1997, 2004), Stefan Thurner and his colleagues (2018), David Krakauer (2019), and many others. Their work spans topics from theoretical frameworks to empirical studies of emergent behavior in social, biological, technological, and ecological systems.

In addition to scholarly research, several authors have produced accessible and pedagogical books designed to serve as textbooks or computational guides for

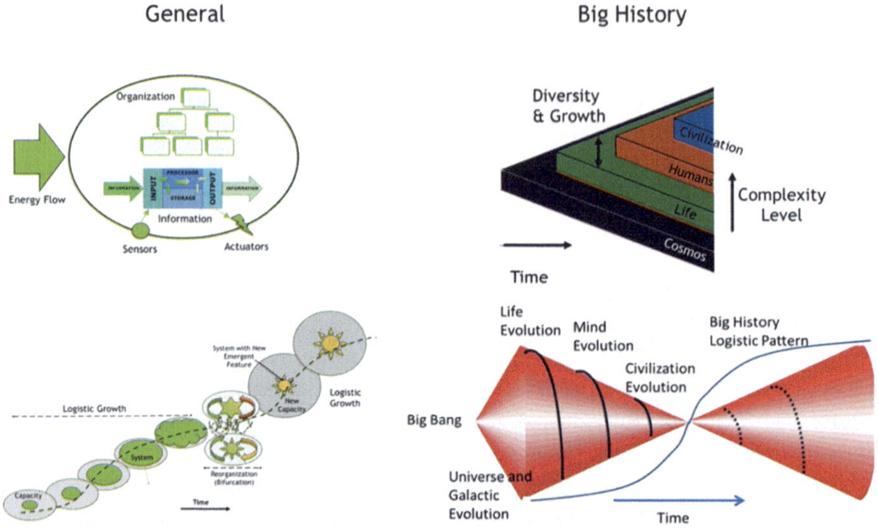

Fig. 2.1 Diagrams showing general systems processes (left) and applications in Big History (right). The upper left shows the integration of energy flow, organization, and information processing in a general complex adaptive system. The upper right shows the building up of complexity layers in Big History. The lower left shows a general growth of a complex adaptive system. It starts with the identification of new energy resources which can be extracted with new organization and information processing. The growth continues (in a logistic growth pattern) until limited by the available resources. Then the system can either continue by identifying a new energy resource and strategy or become sustainable (no growth) or collapse (if the resources are diminished beyond regeneration). The lower right shows the large-level logistic-like growth in Big History. It starts with the slowing down and expansion after the Big Bang to the formation of Earth-like planets. Then complex adaptive systems (life) develop multiple layers of complexity by evolving through genetics, a combination of genetics and culture (human evolution), and then by dominance of cultural evolution. The current situation seems to be in the middle at the point where the two cones meet. A possible future is the reflective logistic growth pattern into the future, which indicates a possible slowdown in the rate of evolution

students and practitioners. For example, Paul Fieguth's (2017) *An Introduction to Complex Systems* offers a structured approach to computational exploration and modeling, making complexity science more accessible to those entering the field.

Complexity theory has also been incorporated into broader historical and cosmological narratives. In particular, Eric Chaisson (2001, 2011) and Fred Spier (2010, 2011) have examined complexity within the framework of Big History, tracing the emergence and evolution of increasingly complex systems from the Big Bang through to human society. Their contributions highlight the role of energy flow and information processing as central metrics in understanding long-term evolutionary trends across different domains of existence.

Many complexity institutes have also embraced data-intensive methods, leveraging artificial intelligence and machine learning to analyze vast datasets and uncover underlying patterns in social, economic, and biological systems (Mefteh &

Mejri, 2020; Roman & Bertolotti, 2023). These efforts aim to identify universal principles of self-organization, adaptation, and network dynamics.

It is important to note that different disciplines interpret and apply the concept of complexity in diverse ways (see below). In computer science, complexity is often associated with algorithmic efficiency and computational resource requirements. In economics, it may refer to the adaptive behavior of markets and agents, nonlinear feedback loops, or the resilience of financial systems. In cybernetics, they refer to the growth of control of various systems with complex regulation (e.g., Wiener 1948; Ashby 1956; see also Grinin et al., 2024: Chap. 3). These varying definitions reflect the multifaceted nature of complexity and underscore the need for a pluralistic, cross-disciplinary approach to its study.

In short, complexity is both resilient and fragile. Its history is contingent, its future uncertain. But by understanding the principles that underlie its emergence and transformation, we gain insight not only into our past, but also into our possible futures.

2.1.2 Many Interacting Parts

While large-scale systems such as galaxies or civilizations may seem too vast or abstract to intuitively grasp, smaller-scale self-organizing systems provide more accessible illustrations of complexity. Examples include weather phenomena like tornadoes and hurricanes, where local energy differentials drive spontaneous structure formation. Even the familiar vortex of a draining bathtub illustrates how rotational symmetry and fluid dynamics generate order from what initially seems like a uniform flow (Feynman et al., 2010).

Another well-studied example is the formation of Bénard cells (Fig. 2.2)—geometric convection patterns in a fluid heated from below—demonstrating how thermal gradients can produce coherent structures through self-organization (Koschmieder, 1993). In chemical systems, the Belousov–Zhabotinsky (BZ) reaction shows how dynamic, oscillating patterns can emerge spontaneously in far-from-equilibrium conditions (Winfree, 1984). These are classic instances of dissipative structures, as described by Prigogine (1980), where energy throughput maintains complex order.

In digital realms, Conway's Game of Life simulates emergent behavior from simple rules applied to grid-based agents (Gardner, 1970). Building on such examples, Stephen Wolfram (2002) argued that the Universe itself might be reducible to computational rules among discrete interacting elements. These systems, though varied, share fundamental properties: many interacting parts, localized rules, multilevel organization, and emergent behaviors driven by energy or information flow.

Complex systems typically exhibit diverse components operating across spatial and temporal scales. These components may be organized into recursive or nested hierarchies, often forming networks with both local and global interactions (Barabási, 2002). Through feedback—both negative (stabilizing) and positive

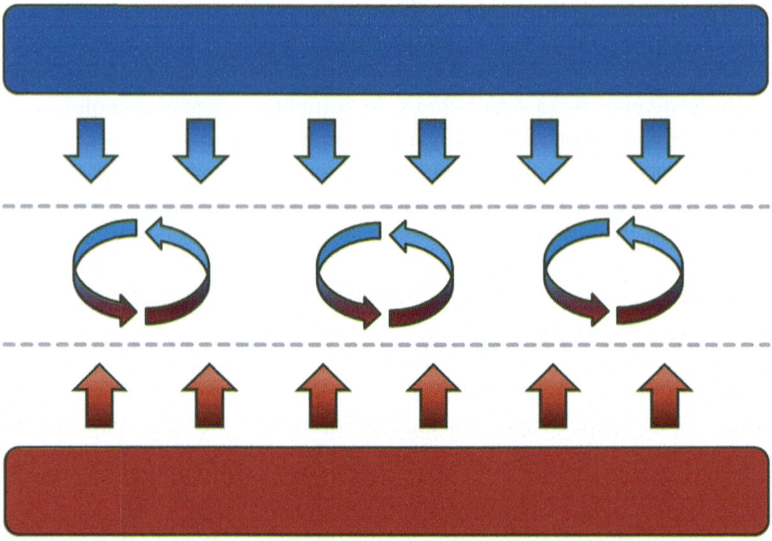

Fig. 2.2 Formation of patterns when the entropy gradient is large. The upper and lower plates are kept at different temperatures. If the temperature difference is quite small, the fluid between them conducts heat from the hotter to the colder plates. However, when the temperature difference becomes large enough, the fluid between them starts to form patterns of convective cells

(amplifying)—systems can maintain coherence while also enabling growth and learning. When such systems incorporate sensing, memory, and decision-making, they become *complex adaptive systems* (CAS), capable of evolving and responding to changing environments (Holland, 1992).

In biological and cultural systems, adaptation is often organized around cycles of growth, breakdown, reorganization, and renewal. This adaptive cycle, also known as the panarchy cycle, has been widely used in resilience studies and systems ecology (Gunderson & Holling, 2002). It illustrates how systems oscillate between stability and transformation as they confront new challenges or opportunities. We also can point out technological and economic cycles, which also usually lead to growth of adaptiveness and complexity (see, e.g., Grinin et al., 2016, 2024; Grinin & Grinin, 2020).

Beyond descriptive traits, researchers have explored whether deeper physical laws might underlie the dynamics of complex systems. For instance, Stuart Kauffman (2000), Lee Smolin (1997), and others (e.g., Cortês et al., 2024) have proposed a "fourth law of thermodynamics" governing far-from-equilibrium systems—suggesting that such systems evolve to maximize entropy production while maintaining local order.

The Constructal Law, introduced by Adrian Bejan (2000), similarly argues that flow systems evolve to facilitate greater access to currents—whether of air, water, heat, or information. These principles may help explain the scaling behaviors observed in nature, from vascular networks to urban infrastructure.

Schneider and Kay (1994) emphasized that rather than contradicting the second law of thermodynamics, life and self-organizing systems enhance entropy production by accelerating the dissipation of energy gradients. In this view, biological order does not defy physics—it fulfills it. Systems are stabilized not by equilibrium but by dynamic steady states maintained by continuous energy throughput.

Other theoretical frameworks, such as the Principle of Least Action, have been invoked to understand complexity's path-dependent evolution (Prokopenko et al., 2009). Though a unified physical law of complexity remains elusive, these proposals suggest that complexity arises not in defiance of thermodynamics but as a natural consequence of open systems operating under specific constraints.

Taken together, these examples and theoretical models demonstrate that *complexity emerges when systems are sufficiently open, interconnected, and energized*. They also show that complexity is not confined to life or society—it is a general principle manifesting across scales, from fluids to galaxies to algorithms.

2.1.3 Complex Systems Types

Natural complexity emerges in two general categories (Fig. 2.3): physical complex systems and complex adaptive systems (CAS). Physical complex systems exhibit

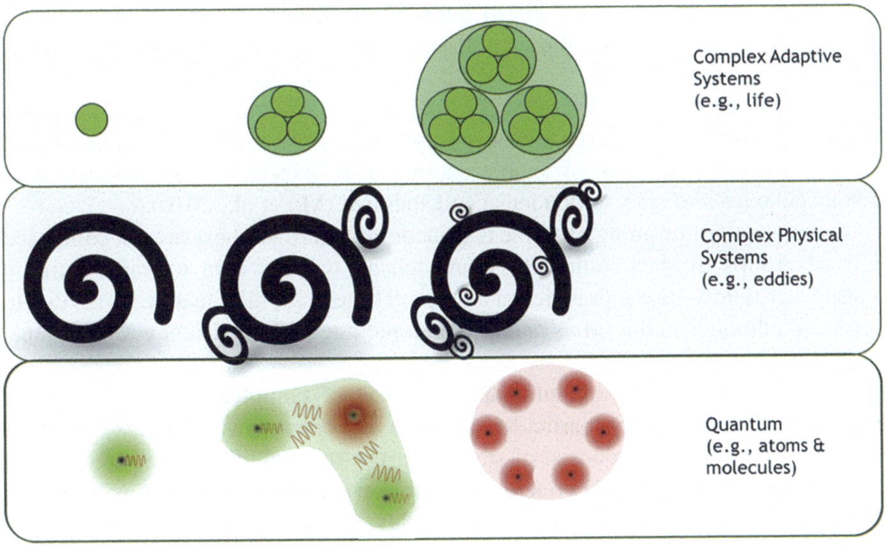

Fig. 2.3 Three possible forms of complex systems in Big History. The first two are well established: (1) the complex adaptive systems of life and social organization and (2) the complex physical systems driven by entropy gradients but without information processing. The third system on the bottom, quantum systems, are not traditional complex systems since they do not require energy flow to maintain their structure and so in a sense are static as opposed to dynamic structures. However, the detailed theories require the exchange of virtual quanta of various fundamental fields to establish the energy levels of the structures

self-organization and energy dissipation but lack information processing capacity. Examples include the cosmic web, spiral galaxies, hurricanes, and vortices in draining fluids. These systems are governed primarily by physical laws—such as gravity, thermodynamics, and fluid dynamics—and tend to stabilize into recurring, dynamic patterns (Prigogine, 1980; Bejan, 2000).

In contrast, CAS are systems that gather, store, and process information to adapt over time. These include living organisms, ecosystems, and human societies. CAS are characterized by learning, memory, and purposeful behavior through feedback, replication, and selection processes (Holland, 1992).

Importantly, these two system types exhibit distinct evolutionary pathways. CAS often evolve from small, modular units that aggregate into higher-level structures—for example, cells to tissues, to organs, to organisms, and to social networks (West et al., 1997). Physical complex systems, however, tend to evolve through large-scale structure formation and cascading interactions. In turbulence, for instance, large eddies initiated by an energy input (such as a paddle) decay into smaller eddies through a process known as the energy cascade, until dissipated by viscosity at the microscale (Kolmogorov, 1941).

In astronomy, many structures assume either spherical or disk-like geometries. These forms emerge from conservation of angular momentum and gravitational collapse. During the formation of solar systems, for instance, matter flattens into a protoplanetary disk perpendicular to the net angular momentum vector (Montmerle et al., 2006; Lin, 2008; Grinin, 2018). Planetary orbits carry most of the system's angular momentum, leaving the central star with relatively little spin (Armitage, 2011).

Cooling is another critical process for maintaining structure. As matter collapses inward, it heats up; unless excess energy is radiated away, structure formation stalls. Radiative cooling mechanisms enable entropy to be exported, allowing structures such as galaxies and stars to condense and stabilize (Mo et al., 2010).

One especially intriguing example is galactic spiral arms. These are not composed of fixed groups of stars; rather, they are density waves—akin to traffic jams in rotating star fields—through which stars pass. These wave-like features arise due to correlated elliptical stellar orbits and gravitational interactions (Binney & Tremaine, 2008).

When considering the evolutionary sequence of complexity, a key question arises: should elementary particles or atoms be considered complex systems? While protons and atoms consist of smaller parts—quarks, gluons, and electrons—they are typically equilibrium systems without internal energy flows. Thus, they may fall outside the traditional definition of complexity used in open, far-from-equilibrium systems (Gell-Mann, 1994).

Yet quantum systems are anything but simple. Describing the helium nucleus, for instance, involves modeling the interactions of 12 quarks (within 2 protons and 2 neutrons), and countless virtual particles exchanging gluons according to quantum chromodynamics (Wilczek, 2008). Electrons in atoms obey quantum rules such as the Pauli exclusion principle, leading to emergent properties in electron shells and chemical bonding.

These quantum behaviors suggest a possible third category: **quantum complex systems**, characterized by probabilistic dynamics, entanglement, and virtual energy exchanges. Though typically in equilibrium, these systems display richly structured internal dynamics and offer the basis for all higher levels of complexity. Future research may further illuminate how quantum behavior connects with macroscopic emergent structures across cosmology and biology.

2.1.4 Definitions of Complexity in Different Scientific Fields

Complexity is a concept that appears in various disciplines, including mathematics, computer science, physics, biology, and social sciences. Its definition often depends on the context in which it is used. Below are some common definitions of complexity, along with references:

- **Complexity in Systems Theory:** In systems theory, complexity refers to the degree of interconnectivity and interdependence among components of a system. A system is considered complex if it exhibits emergent behavior that cannot be easily predicted from the behavior of its individual components (Bar-Yam, 2004).
- **Structural Complexity:** Structural complexity refers to the complexity of an object or system based on its structure, such as the number of components, their arrangement, and the relationships between them. For example, the complexity of a graph can be measured by metrics like the number of nodes, edges, and connectivity (Simon, 1962).
- **Hierarchical Complexity:** Hierarchical complexity measures the number of distinct levels of organization within a system, where each level is composed of subsystems or components that interact to form higher-order structures. A system is considered more complex if it has more levels of hierarchy and if the interactions between these levels are intricate (Simon, 1962).
- **Behavioral Complexity:** Behavioral complexity refers to the range of behaviors or states a system can exhibit. For example, in dynamical systems, complexity can be associated with chaotic behavior, where small changes in initial conditions lead to vastly different outcomes (Gleick, 1987).
- **Social Complexity:** In social sciences, complexity often refers to the interactions and relationships within social systems, such as organizations, societies, or economies. Social complexity is characterized by the presence of multiple agents, feedback loops, and emergent phenomena (Miller & Page, 2007).
- **Kolmogorov Complexity:** Kolmogorov complexity, also known as algorithmic complexity, measures the complexity of an object (e.g., a string) as the length of the shortest possible description (program) that can produce the object on a universal Turing machine (Kolmogorov, 1965).
- **Computational Complexity:** In computer science, computational complexity refers to the amount of resources (time, space, etc.) required to solve a problem as a function of the size of the input. For example, the time complexity of an algorithm is often expressed using Big-O notation (Sipser, 2012).

- **Thermodynamic Depth Definition:** Thermodynamic depth is a measure of the complexity of a system based on the amount of information and thermodynamic work required to describe the system's history and construct it from a disordered state. It reflects the system's "causal history" and the number of intermediate steps involved in its formation (Lloyd & Pagels, 1988).

All these definitions might be useful; however, they mostly characterize only some (one or two) fields of science and confirm what we said above—that complexity remains a concept marked by fragmentation across disciplines.

2.2 Basics

2.2.1 Components: Energy, Organization, and Information

In the study of complexity within Big History, three fundamental concepts—energy, organization, and information—emerge as critical drivers shaping the Universe's evolution (Fig. 2.4). These elements are not isolated but intricately interlinked, forming a dynamic interplay that underpins the development of all complex systems, from cosmic structures to human civilizations.

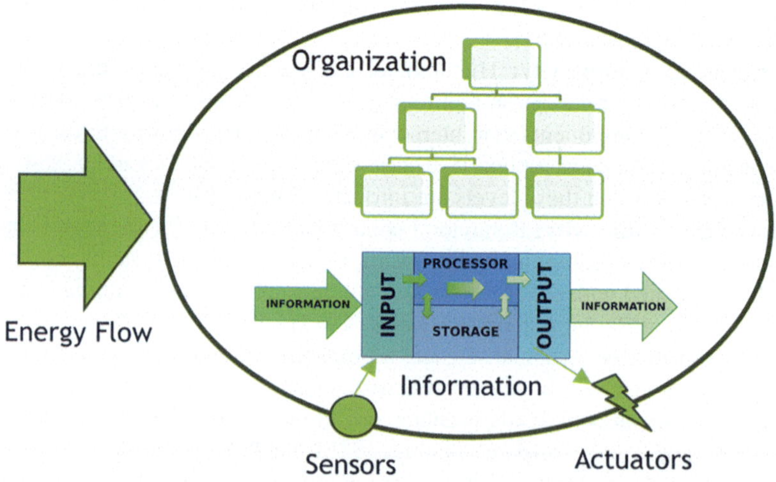

Fig. 2.4 The integration of energy flow, organization, and information processing in a general complex adaptive system is shown. The energy flow is derived from an entropy gradient such as objects with two different temperatures (e.g., light from the Sun versus light radiating from the Earth at night). The CAS builds an organization based on the blocks of the previous level of complexity (e.g., molecules from atoms, tissues from cells) and maintains information processing to monitor both the internal state of organization and the external environment to adjust accordingly to maintain the organization by extracting resources from the environment and avoiding environmental threats

Energy serves as the vital fuel for the maintenance and development of complexity. The flow of free energy enables systems to dissipate gradients and maximize entropy production. Understanding energy's role requires considering the number of participants, energy utilization efficiency, coordination among components, and the availability of surplus energy for new organizational layers.

Organization refers to the hierarchical structures and self-organizing principles that characterize complex systems. In adaptive systems like biological organisms and societies, complexity arises through the aggregation and layering of smaller, semi-autonomous units, forming sophisticated networks with feedback and memory. This contrasts with physical systems, where large structures often dissipate into smaller components. As systems grow, complexity may intensify internally through reorganization and integration rather than just spatial expansion, reflecting cyclical patterns of centralization and decentralization in human organizations.

Information represents the capacity of systems to acquire, store, interpret, and communicate data, evolving from basic molecular recognition to advanced cognitive functions, language, and collective intelligence. The evolution of life is deeply intertwined with increasingly complex modes of information processing, leading to neural networks, brains, and symbolic reasoning. Crucially, information is not static; its meaning and potential are revealed through interpretive frameworks and models. These models, whether quantitative or qualitative, guide our understanding, demonstrating how even abstract concepts can catalyze scientific breakthroughs and shape our perception of reality, illustrating the dynamic relationship between data and interpretive systems.

Together, energy, organization, and information provide an integrative framework for comprehending the emergent order, adaptive capacity, and evolutionary trajectories of complex systems throughout Big History.

2.2.1.1 Energy: Flow, Efficiency, and the Limits of Metrics

Free energy flow is essential for maintaining and developing complexity in both physical and adaptive systems. This insight underpins the proposed "fourth law" of thermodynamics, which posits that systems naturally evolve to dissipate energy gradients and maximize entropy production (e.g., maximize heat flow through combination of convection and conduction) at the fastest possible rate (Schneider & Sagan, 2006). Adrian Bejan's Constructal Law supports this perspective, showing that systems self-organize to optimize flow access, particularly when comparing modes of heat transfer such as conduction and convection between plates (Bejan, 2000; Camazine et al., 2001).

However, quantifying complexity solely through energy flow is problematic. Eric Chaisson (2001) introduced the metric of energy rate density (ERD), measured in watts per kilogram, as a proxy for complexity across cosmic, biological, and cultural systems. While useful, ERD oversimplifies the nuanced dynamics of complex adaptive systems (CAS), which rely on additional factors including information processing, cooperation, and innovation (Chaisson, 2011).[2]

To better characterize complex systems, several additional variables must be considered:

1. Number of participants or total system mass
2. Efficiency of energy utilization (output vs. input)
3. Degree of coordination and cooperation among components
4. Availability of surplus energy for new layers of organization
5. Presence of energy disruptors, feedback loops, or sinks

In physical systems, complexity often emerges from the dissipation of large energy gradients, such as turbulence cascading from large to small eddies (Kolmogorov, 1941). In contrast, CAS grow through modular aggregation and internal differentiation, forming layered networks of memory and control (West et al., 1997). These fundamental differences limit the applicability of a universal ERD metric across all system types (see in particular Van Duin, 2025b).

Chaisson's approach has led to surprising conclusions—for example, that stars are more "complex" than galaxies based on ERD alone (Chaisson, 2001, 2011). This counterintuitive result underscores the need for a more multidimensional framework when evaluating system complexity (Van Duin, 2025b).

The Jevons paradox further complicates the narrative. Greater energy efficiency often leads to increased total energy consumption, not less (Alcott, 2005). Historical shifts in energy regimes illustrate this dynamic:

- In the Late Roman Empire, hydraulic technologies like water wheels improved efficiency and possibly displaced labor in some sectors of economy (Widdowson, 2023).
- Medieval Europe adopted technological improvements such as horse collars and heavy plows, boosting agricultural output and population (Gimpel, 1976; Grinin et al., 2024: Chap. 2).
- Expansion of trade networks like the Silk Road facilitated the diffusion of technologies across Eurasia, catalyzing regional development (Frankopan, 2017).

Marc Widdowson (2023) proposed that population thresholds may influence technological caps. Following Rome's decline, reconnected Eurasian trade routes potentially reinvigorated cultural and technical exchange, with China's innovations spreading rapidly once they reached medieval Europe.

In conclusion, free energy flow is vital for sustaining order and fostering innovation in complex systems. However, the intricate interplay between energy efficiency, system organization, and historical context demands a multifaceted

[2]On very serious problems with the use of energy rate density as a universal measure of complexity, see Van Duin (2025a, 2025b).

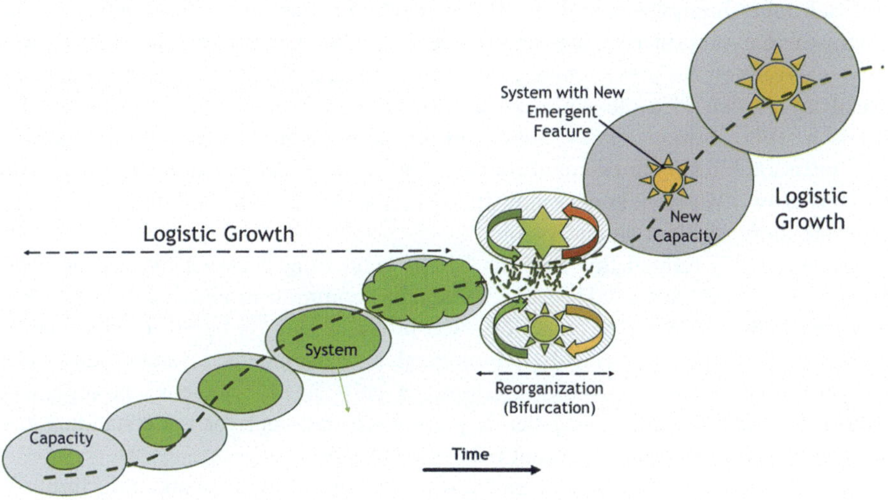

Fig. 2.5 A general growth of a complex adaptive system. It starts with the identification of new energy resources which can be extracted with new organization and information processing. The growth continues (in a logistic growth pattern) until limited by the available resources. Then the system can either continue by identifying a new energy resource and strategy (as shown in the reorganization or bifurcation phase) or become sustainable (no growth) or collapse (if the resources are diminished beyond regeneration). Often there are many systems exploring the combinations of environment, organization, and information processing but only a few are successful at continuing the growth pattern

approach. Energy rate density alone cannot capture the full story. Instead, an integrative model—one that includes energy flow, network coordination, feedback, and adaptive capacity—is required to understand the evolution and sustainability of complex systems.

2.2.1.2 Organization

The development of hierarchical structures varies significantly between physical and adaptive complex systems. In complex adaptive systems (CAS) such as biological organisms or human societies, complexity emerges through the aggregation of smaller, semi-autonomous units—cells into tissues, tissues into organs, and so forth (Jantsch, 1980). These systems demonstrate increasing sophistication by layering organizational levels, often supported by feedback mechanisms and memory processes (Fig. 2.5).

In contrast, physical complex systems—like atmospheric eddies or galactic formations—typically begin with large-scale structures that dissipate into smaller components as energy gradients are smoothed (Prigogine, 1980; Kolmogorov, 1941). In systems like Bénard cells or hurricanes, hierarchical organization is present but tends not to evolve toward increasing complexity. Rather, such systems maintain dynamic equilibrium until external conditions change.

Erich Jantsch emphasized that self-organization occurs across multiple scales in CAS, where evolutionary processes operate in parallel across hierarchical levels. He viewed evolution not just as change over time, but as the synthesis of new organizational layers—an idea closely aligned with Herbert Simon's theory of hierarchy (Jantsch, 1980; Simon, 1962). Simon proposed that stable intermediate structures (or "subassemblies") are essential for building higher-order systems efficiently. His parable of two watchmakers demonstrated that systems with modular, semi-stable units are more resilient and adaptable.

Salthe (1985) refined this analysis by distinguishing between two hierarchical forms: "is-a" and "has-a." The "is-a" hierarchy, common in classification systems like Linnaean taxonomy or object-oriented programming, is based on inheritance. The "has-a" hierarchy, typical of biological and engineered systems, describes compositional relationships—for example, a cell "has-a" nucleus, an organism "has-a" nervous system. The latter is particularly relevant for understanding the structural complexity and emergent behavior in CAS.

As human systems become increasingly global, questions arise about the limits of hierarchical expansion. Rather than expanding spatially to interplanetary or interstellar scales, complexity may intensify internally through reorganization, greater coordination, and integration of existing layers (Arrighi, 1994; Grinin & Korotayev, 2015). This aligns with the logistic growth model in evolutionary theory, where early exponential growth is followed by plateau phases marked by consolidation and refinement (Odum, 2001).

Scientific discovery itself mirrors this pattern. Early Modern physics saw rapid conceptual expansion—defining energy, mass, and force—followed by slower advances within quantum and relativistic domains due to growing experimental complexity and theoretical abstraction (LePoire, 2005). Recurrent anomalies now demand new paradigms, echoing the need for restructuring in other complex systems.

Human organizations also oscillate between phases of centralization and decentralization. New hierarchical layers may simplify governance at lower levels, but sometimes creative destruction—dismantling outdated structures—is needed to renew systemic coherence (Tainter, 1988). Such transformations reflect an underlying dynamic between expanding and refining organizational complexity.

*In sum, the insights of Jantsch (*1980*), Simon (*1962*), and Salthe (*1985, 1993*) offer a robust theoretical scaffold for understanding how hierarchical structures emerge, stabilize, and adapt.* Recognizing the interplay between modularity, classification, and compositional layering enables us to map complexity across biological, technological, and social domains—critical for building integrated Big History models of systemic transformation.

2.2.1.3 Information

2.2.1.3.1 Information: Processing and Evolution

The evolution of life is intricately tied to increasingly complex modes of information processing. Beginning with simple molecular recognition mechanisms and genetic

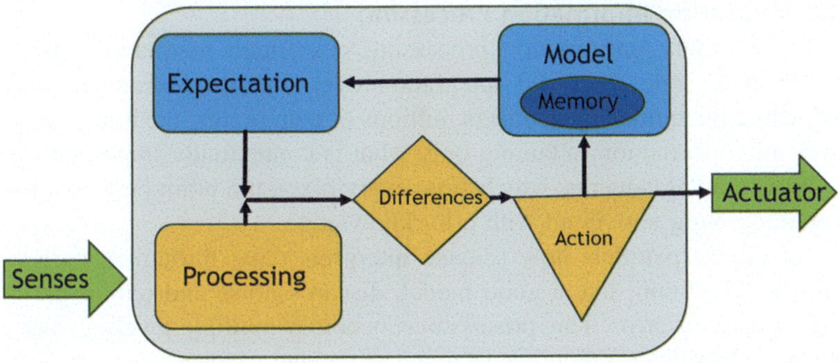

Fig. 2.6 A highly abstract version of the information process with learning. The CAS needs both sensors of the environment (and self) and actuators (e.g., muscles) to respond to the environmental signals. The information is processed to extract meaningful signals (or symbols) which are compared with the expectations it has based on the previous activities, models, and memories. If the senses do not match the model, then some learning can occur by adjusting the models and memories. This learning action is then coupled with some response to the signal on the environment to gather resources, avoid threats, or test the model (similar to an experiment)

systems, this evolutionary trajectory advances to the formation of neurons, brains, languages, and civilizations. These developments enable organisms—especially humans—to acquire, store, interpret, and communicate information with growing sophistication.

In earlier evolutionary stages, basic sensory systems such as chemoreception and photoreception allowed organisms to respond to their environment. Over time, neural networks evolved to facilitate rapid information transmission and integration. The emergence of brains enabled the development of higher cognitive functions, such as planning, symbolic reasoning, and theory of mind—capabilities foundational to complex social behavior and culture.

Karl Friston's Free Energy Principle (FEP) offers a unifying theory for brain function, framing it as a predictive engine that minimizes the difference between expected and actual sensory input. This continuous hypothesis-testing process drives learning and decision-making (Friston, 2010). It mirrors the scientific method: noticing anomalies, generating hypotheses, and revising beliefs based on evidence (Fig. 2.6).

The trajectory from genes to societies also includes epigenetic regulation and the activation of dormant genes under certain environmental pressures. These mechanisms add adaptability and nuance to classical evolutionary models, emphasizing interaction between genetic inheritance and environmental context.

Human collaboration relies increasingly on collective cognitive structures—institutions, shared myths, and symbolic communication. These systems externalize memory and distribute cognition across individuals and generations (Donald, 1991). Storytelling and religious narratives help encode worldviews and coordinate group behavior, enabling cultural evolution alongside biological evolution (see, e.g., Korotayev, 2004).

2.2.1.3.2 Measuring Information Processing

Despite its centrality, information processing is difficult to quantify. Biological systems prioritize relevance and integration over the raw storage of data. For example, while the human eye collects millions of data points, the brain compresses and filters this information, retaining only what is contextually important (Tononi, 2008). Unlike digital cameras, which store every pixel, the brain performs selective attention, integrating new input with prior knowledge.

This selectivity parallels how science interprets data: through paradigms and expectations. The brain, like a good model, discards noise and refines predictions based on surprise or error. This process occurs across multiple levels—from neural pathways to scientific communities testing and validating ideas.

2.2.1.3.3 Where Is the Information?

The case of zircon crystals illustrates the contextual nature of information. On its own, a zircon grain holds no inherent "bit" of data. But when studied through the lens of radiometric dating, its uranium content and decay rates become a reliable chronometer for Earth's early history (Valley et al., 2005). This informational potential is only revealed through the application of theories, technologies, and interdisciplinary frameworks.

Information, therefore, is not a static substance but a dynamic relationship between data and interpretive systems. Without the right theoretical models, even the most data-rich systems may appear meaningless. As knowledge evolves, previously inert objects become meaningful sources of insight.

In sum, the evolution of complexity in life and society is fundamentally entwined with the capacity to process information. From genetic coding to digital communication, systems that better anticipate, interpret, and respond to change tend to survive and thrive. Measuring and modeling these capacities remains a frontier of interdisciplinary science—very important to understanding Big History.

2.2.1.3.4 Information, Models, and Uncertainty

Information processing is inseparable from the frameworks or models used to interpret data. These models guide the organization and analysis of information, shaping expectations and directing inquiry. Depending on context, models can vary significantly in their level of abstraction—from precise mathematical simulations to conceptual, qualitative theories.

Quantitative models, such as those in physics or engineering, require high precision and predictive accuracy. They often seek close alignment between theoretical predictions and empirical observations. In contrast, qualitative models emphasize general patterns, relationships, or causal structures. These are especially useful in fields like evolutionary biology, where mechanisms may be inferred despite incomplete or noisy data.

A classic example is Charles Darwin's theory of evolution by natural selection. Despite lacking genetic theory or radiometric dating, Darwin's conceptual framework provided a powerful explanatory model, unifying disparate observations into a coherent, testable theory (Mayr, 1991). His success illustrates how abstract models can catalyze transformative science even in the absence of precise data. Scientific models thus serve as cognitive scaffolding. They help scientists reason about uncertainty, anticipate anomalies, and pursue discoveries—shaping what data are considered meaningful and how they are interpreted (Kuhn, 1962).

In the social sciences, model evaluation relies on robust frameworks for validation. One influential contribution is the work of Cook and Campbell (1979), who distinguish between internal and external validity. Internal validity assesses whether a study accurately identifies causal relationships within its specific context—ensuring that observed effects result from the hypothesized mechanisms. External validity examines whether findings can be generalized beyond the study's immediate sample or setting. This dual framework ensures that theoretical models are not only consistent with empirical observations but also applicable in broader, real-world scenarios. Their approach underscores the importance of both rigorous design and practical relevance—essential for theory-building in complex, open systems.

Douglas Hofstadter (2008) introduced the concept of "strange loops"—self-referential feedback structures that emerge across hierarchical levels of cognition. These loops create recursive dynamics where symbols refer to other symbols, producing the illusion (or reality) of consciousness. In his seminal work *Gödel, Escher, Bach*, Hofstadter explores how abstraction, analogy, and recursion generate complex cognitive architectures. He argues that consciousness arises from a recursive pattern of symbols reflecting upon themselves, akin to feedback in formal systems or neural networks. This model bridges low-level neural processing with higher-order cognition. Although still debated, Hofstadter's ideas have influenced research in artificial intelligence, philosophy of mind, and systems theory. By situating information processing within recursive, symbolic, and emergent systems, Hofstadter expands the conceptual toolkit for understanding how brains and cultures generate meaning—a core concern of Big History's integrative narrative.

2.2.2 Complex Systems Dynamics

2.2.2.1 Change and Adaptive Dynamics in Complex Systems

Complex systems evolve by adapting to, modifying, or constructing their environments. This process of change occurs through various modes: passive environmental response, active manipulation, and the construction of entirely new environmental niches (Odling-Smee et al., 2003). In complex adaptive systems (CAS), these dynamics are particularly prominent, manifesting as cycles of growth, stability, collapse, and renewal.

A central framework for understanding such transitions is the adaptive cycle, which describes four key phases: exploitation (rapid exponential growth), conservation (slowdown or saturation, often modeled by logistic growth), release (collapse or crisis), and reorganization (innovation and restructuring) (Gunderson & Holling, 2002). When scaled across long historical periods, this cycle can resemble hyperbolic growth, as seen in models of civilizational development (Korotayev et al., 2006b); a similar interplay of cyclical and hyperbolic dynamics can be observed as regards global biodiversity growth in the Phanerozoic Eon (Grinin et al., 2013, 2025; Markov & Korotayev, 2007, 2008; Korotayev & Markov, 2014; see also chapter "Biological and Social Phases of Big History: Similarities, Differences, and the Growth of Complexity" [Grinin et al., 2026] in this volume).

2.2.2.2 Key Behavioral Patterns of Complex Systems

Researchers have identified several hallmark behaviors that characterize the dynamics of complex systems:

1. **Network Structures:** Complex systems are defined by interconnected nodes with nontrivial topologies, often following power-law distributions (Barabási, 2002).
2. **Fractal Geometry:** Many systems exhibit self-similarity across scales, reflecting recursive growth and hierarchical structuring (Mandelbrot, 1982).
3. **Feedback Loops:** Both positive and negative feedback loops govern adaptation and stability. Positive loops amplify change, while negative loops promote regulation (Meadows, 2008).
4. **Sensitivity to Initial Conditions:** Known as the "butterfly effect," small changes can produce vastly different outcomes—a hallmark of chaotic systems (Lorenz, 1963).
5. **Scale-Free Fluctuations:** Fluctuation dynamics often conform to power laws, suggesting persistent, nonrandom correlations across time (Bak et al., 1988).
6. **Nonlinear Responses:** Complex systems exhibit disproportionate reactions to stimuli, which can lead to emergent properties (Holland, 1992).
7. **Bistability and Path Dependence:** Systems may stabilize in one of several states, depending on historical pathways, making transitions contingent and path-dependent.
8. **Self-Organized Criticality:** Many systems evolve toward a critical point where minor perturbations can cause cascading changes (Bak et al., 1988).

Earth's climate demonstrates many of the properties listed above (Fig. 2.7). It exhibits bistability under similar solar inputs—oscillating between ice-covered and ice-free states due to albedo-driven feedback loops. Ice reflects sunlight, promoting cooling; conversely, open water absorbs heat, promoting warming. The same external energy can thus support multiple stable states, depending on prior conditions (Lenton & van Oijen, 2002).

2.2.2.3 Growth Trajectories and Transitions

Growth within complex systems follows various mathematical patterns (Fig. 2.8).

- Exponential Growth occurs when the rate of growth is proportional to the system's size.
- Logistic Growth introduces a capacity limit, slowing expansion as the system nears saturation.

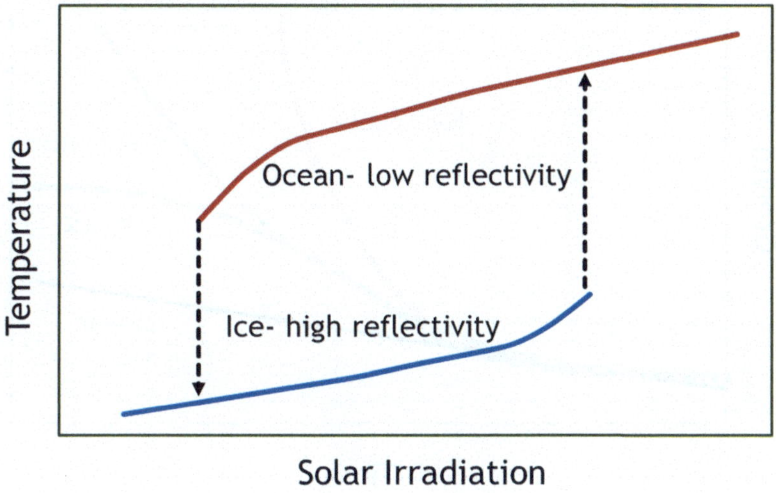

Fig. 2.7 A complex physical system like the atmosphere demonstrates path dependence, where the system's current state is influenced by its history. For example, there are multiple temperatures associated with a given level of solar irradiation, depending on the system's prior conditions. As solar input gradually changes, the atmosphere may reach critical thresholds where the temperature suddenly shifts—exhibiting a nonlinear response to what seems like a continuous input. Such abrupt transitions are characteristic of complex adaptive systems (CAS), where outputs are not directly proportional to inputs. Instead, the system's response depends on the sequence of prior states, meaning the range of possible outcomes at any moment is shaped by its historical trajectory

- Hyperbolic Growth, where growth accelerates super-exponentially, has been proposed to describe certain historical and evolutionary transitions characterized by second-order positive feedback loops (von Foerster et al., 1960; LePoire, 2020a, 2020b; Korotayev, 2020, 2025a).

However, no real-world system can sustain indefinite growth. As such, periods of collapse, bifurcation, or transition often follow unsustainable trajectories. These transitions may be:

- Marginal or Critical, requiring a tipping point to shift.
- Trap-based, where the system cannot escape suboptimal states.
- Cyclic, with recurring adaptive phases.
- Path-dependent, where earlier choices constrain future options.

Understanding these behavioral dynamics helps anticipate the tipping points and potential reorganization paths within biological, ecological, and social systems.

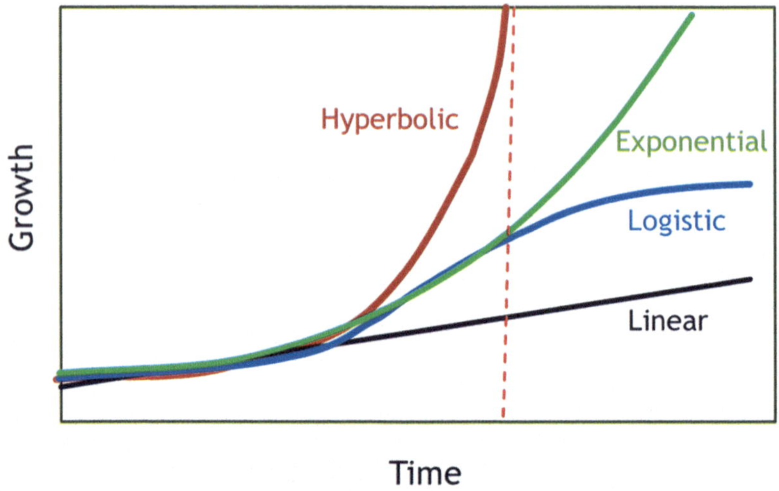

Fig. 2.8 Comparison of different trend patterns. The linear trend (black, bottom) grows by the same amount for a given time period. An exponential growth (green) grows by the same factor for each given time period. The logistic trend (blue) transitions from a low value to a higher value over time. The hyperbolic (red, top) is derived from a differential equation very similar to the exponential trend but has a very different behavior. Instead of continuously growing over time, it has a trend toward infinity at a specific time. The hyperbolic growth is seen when the growth rate coefficient (similar to interest rate) is proportional to the size of the growing entity (the larger the size, the higher the relative growth rate)

2.2.3 Measures

2.2.3.1 Measuring Complexity: Indices, Layers, and Abstractions

Attempting to measure complexity is akin to measuring an economy—particularly its health. Like Gross Domestic Product (GDP), a single number may provide a crude proxy, but it inevitably fails to capture the rich, multidimensional nature of the underlying system. GDP, for instance, overlooks unpaid labor (e.g., housework, volunteering), includes economically harmful activities (e.g., pollution), and treats expenditures on social dysfunction (e.g., policing) as indicators of positive growth (Stiglitz et al., 2009).

Similarly, measuring complexity requires a pluralistic approach. There is no single metric that can account for the interwoven dimensions of energy flow, hierarchical organization, and information processing. As in economics, where multiple indices (e.g., inflation, unemployment, productivity) supplement GDP, complex systems must be evaluated across several proxies.

Adam Smith's *The Wealth of Nations* (1776) recognized the emergent properties of market economies as outcomes of decentralized decisions by many interacting agents. This framework set the stage for understanding economies as complex adaptive systems, characterized by local rules, risk-sharing institutions, and emergent behaviors such as inflation, cycles, and innovation.

However, early economic models often treated systems as static equilibria, similar to mechanical systems like bridges. These simplifications collapsed under the reality of innovation, behavioral shifts, and systemic disruptions. Schumpeter (1942) later described capitalist economies as subject to "creative destruction," where innovation destabilizes existing structures, leading to continual renewal.

Modern developments—such as the Internet and mobile computing—exemplify complexity layered over simplicity. These technologies make everyday tasks (shopping, communication, navigation) appear simple to the user while masking immense structural complexity handled by specialists. As complexity increases, functionality often improves, but so does dependency, reducing users' autonomy.

2.2.3.2 Proposed Complexity Measures and Analogies

- Thermodynamic Depth (Lloyd, 2001): A system's complexity reflects how much energy has flowed through it over time in non-redundant, structured ways.
- Hierarchical Depth (Simon, 1962): The number of nested levels in a system's structure.
- Kolmogorov Complexity of an Object (e.g., Kolmogorov, 1998 [1963]): The length of the shortest possible description of that object in a fixed language.
- Economic Complexity Index (ECI; Hidalgo & Hausmann, 2009): Measures a country's productive knowledge based on the diversity and sophistication of its exports.
- Cognitive Complexity (Misra et al., 2018): In software, metrics evaluate the depth and modularity of object-oriented designs, offering insights into adaptability and scalability.
- Molecular Complexity (Hendrickson et al., 1987): Looks at the network of bonds between atoms in a molecule to estimate complexity.
- World System Complexity: The KOF Index of Globalization (Dreher, 2006) is a composite index that measures the degree of globalization in countries across economic, social, and political dimensions (the adequacy of this index is, however, strongly criticized [e.g., Zinkina et al., 2013]).

These perspectives underscore that complexity is contextual and multidimensional. Some complexity may be "deep" and meaningful, supporting resilience or adaptability; other complexity may be "shallow" or unnecessarily convoluted. Furthermore, systems may evolve toward apparent simplicity—delegating complexity into abstracted modules or specialized domains. For example, smartphones simplify daily life while embedding users in vast digital ecosystems that most do not understand or control. Evaluating complexity requires both quantitative proxies and qualitative judgment, guided by the purpose of measurement: understanding, optimization, resilience, or innovation.

2.2.3.3 Toward a Composite Complexity Measure

While energy flow, organizational hierarchy, and information processing each provide valuable perspectives on complexity, integrating these dimensions into a unified metric remains a challenge. Each component offers a proxy: energy flow might be measured in terms of net free energy dissipation (Chaisson, 2001), organization by the number of hierarchical levels (Simon, 1962), and information by bits processed or the abstraction level of processing (Tononi, 2008).

A potential way to synthesize these proxies into a composite complexity score is to use a weighted geometric mean:

$$C = E^{a*}\, O^{b*}\, I^{c},$$

where:

- E = proxy for energy flow
- O = proxy for organizational depth
- I = proxy for information processing
- a, b, c = weighting coefficients summing to 1

This structure parallels economic models that combine labor, capital, and land, and allows tuning of emphasis based on the system under study. For example, biological systems may prioritize energy and information, while technological systems might emphasize information and organization.

Absolute measures of complexity are difficult, but relative complexity can be estimated more easily by comparing systems with known structural differences. For instance, if one system includes three additional levels of hierarchy over its component parts, its relative complexity increases accordingly. Copying a system does not double its complexity—it simply adds the minor complexity of replication mechanisms (Bar-Yam, 1997).

In evolutionary systems, complexity tends to increase via discrete transitions involving new organizational levels, energy intensification, and enhanced information processing. These transitions often follow geometric or hyperbolic trends in complexity over time (Korotayev & LePoire, 2020; Korotayev, 2018, 2020; LePoire, 2020a; LePoire et al., 2025a, 2025b, 2025c, 2026). If the evolving system's energy flow increases by a fixed geometric factor with each level, the number of levels (n) can be derived using logarithmic transformation:

$$\log(En/Ei) = n - i.$$

Assuming this pattern holds, the relative complexity difference between great apes and early agricultural societies is comparable to the difference between those early societies and modern industrial civilization—a progression of approximately six hierarchical layers. These assumptions require more rigorous empirical validation, including better estimates of energy consumption and organizational architecture in historical periods such as post-Roman Europe (Tainter, 1988).

This integrated model opens the door to formalizing complexity not merely as an abstract concept, but as a measurable, comparative quantity across different systems and epochs. Further work should refine the scaling coefficients and develop robust metrics for each domain to advance this approach.

2.3 Big History Complexity

The history of the Universe is a grand narrative of increasing complexity, unfolding from the profound simplicity of its earliest moments to the intricate biological and social systems we observe today. This journey of escalating organization and information processing can be broadly understood through four major phases (Fig. 2.9): the emergence of cosmic structures in the early Universe, the multifaceted evolution of life from molecular networks to planetary ecosystems, the unique trajectory of human evolution marked by advanced cognition, and civilization history marked by the advance of cooperation and technology.

2.3.1 The Early Universe: Complexity from Simplicity

The physics of the first few minutes after the Big Bang has been a major focus of cosmological research for decades. This period, often referred to as the "first three minutes" following Steven Weinberg's seminal 1977 book *The First Three Minutes* (Weinberg, 1977), is understood primarily through three lines of observational evidence: the cosmic microwave background (CMB), the abundances of light elements from Big Bang nucleosynthesis, and the observed redshift of galaxies

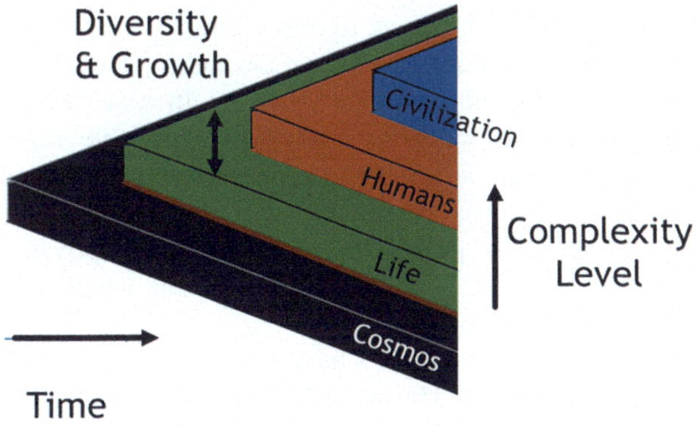

Fig. 2.9 The three-dimensional (3D) diagram depicts conceptual development of complexity layers in Big History. The base layer is the cosmic development of galactic, stellar, and planetary structures along with the fundamental building blocks of elements and some minerals and molecules. The layer shows the development in time as moving left to right. Life originates (first green point from the left) and develops from these elements of the cosmic layer in the environment of a terrestrial planet like Earth and the energy from a star like the Sun. As time proceeds, the diversity of life and its ecosystems grows. Human evolution and further social evolution in civilizations are later complexity layers. (Note, that two authors of this chapter (Grinin and Korotayev) regard human and civilization phases as one—social—phase that can be subdivided into subphases. Thus, the human evolution ("anthropogenesis") period can be classified as a transitional phase between biological and social main phases. See Grinin and Grinin (2025); Grinin and Korotayev (2025); see also chapter "Human Evolution in the Complexity Growth Perspective: Toward Periodization of the Big History Biosocial Era" (Korotayev, 2026, this volume))

indicating cosmic expansion. By the 1970s, these pillars of modern cosmology were well established, painting a remarkably coherent picture of a Universe expanding and cooling from a hot, dense origin.

Yet, despite these successes, profound mysteries remain. We still do not know why there is something rather than nothing, why the Universe has more matter than antimatter, what dark matter and dark energy are, how supermassive black holes formed so quickly after the Big Bang.[3] These are not merely gaps in data—they are challenges to our fundamental understanding of physics and cosmology.

Physicist Peter Atkins has provocatively described the emergence of the Universe as the result of "nothing," arguing that given the laws of physics—particularly the principle of least action and the symmetries underpinning quantum field theory—it might be expected that a Universe could emerge from a state of zero total energy. In this view, gravity's negative energy balances out the positive energy of matter, suggesting the net energy of the Universe could be zero, allowing something to arise from nothing without violating conservation laws (Atkins, 2018).

Still, this minimalist elegance confronts a complex reality: the precise values of fundamental constants—such as the strength of gravity, the masses of elementary particles, and the charge of the electron—appear finely tuned for the emergence of structure, life, and consciousness. Some theorists, such as Don Page (2003), have attempted to derive constraints on these parameters based on internal consistency or the existence of observers. This connects to broader anthropic reasoning, which posits that the Universe must have properties that allow for conscious observers, because otherwise we would not be here to observe it.

Such arguments have sparked intense debate. Do anthropic explanations have real predictive power? Are we forced to assume a vast "multiverse" where all parameter values exist somewhere, and we merely inhabit a habitable region? Or is the Universe fine-tuned by design, or through a self-consistent logic, as John Wheeler's "participatory universe" suggests? Clement Vidal (2014) offers a nuanced view: without knowing the possible probability distributions of physical constants—assuming they are variable at all—we cannot yet evaluate whether our Universe is "special" in any rigorous statistical sense.

At a more empirical level, simulations have become essential tools in exploring how structure emerged in the early Universe. Even small gravitational instabilities in the primordial plasma were amplified over time, especially with the presence of cold dark matter, eventually forming stars, galaxies, and galaxy clusters. The development of powerful computational models has enabled researchers to explore these nonlinear processes with increasingly high resolution.

Interestingly, the gravitational collapse of matter has been likened to origami folding, where sheets of dark matter collapse into filaments and nodes along preferred directions, giving rise to the cosmic web (Shandarin et al., 2005). Some simulations track millions or even billions of particles, solving Newton's equations of motion across large volumes of space, accounting for gravity, gas dynamics,

[3] There are even doubts whether there was the Big Bang itself (see, e.g., Grinin, 2019).

cooling, star formation, and feedback from supernovae and black holes. While the underlying equations are conceptually simple, the sheer scale and complexity of the calculations demand cutting-edge computing resources and clever approximations (Springel et al., 2005).

Moreover, the role of magnetic fields, radiative transfer, and turbulence introduces even more layers of complexity. As in other complex systems, emergent behavior—spiral arms in galaxies, global correlations in spin alignment, or feedback-regulated star formation—arises from simple local rules played out across vast scales.

2.3.2 Evolving Views of Life's Evolution: From Molecular Networks to Planetary Ecosystem

The study of evolution has expanded dramatically beyond traditional Darwinian frameworks that emphasize natural selection acting on random variation. While natural selection remains central, particularly in biological and social contexts, contemporary research increasingly incorporates concepts from complex systems theory, thermodynamics, network science, and nonlinear dynamics. This broader approach captures how life, and the systems in which it is embedded, evolve through feedback loops, emergent properties, and phase transitions that reshape evolutionary landscapes.

Together, these frameworks—Eigen's singularities, Kauffman's autocatalytic sets, Lane's thermodynamic constraints, and Lovelock's Gaia system—offer a multiscale perspective on evolution. They move beyond the traditional selectionist paradigm by emphasizing:

- Feedback loops that drive nonlinear growth and change
- Self-organization in chemical and biological networks
- Energy flow and disequilibrium as the engines of complexity
- Planetary-scale interactions between life and environment

These ideas align with the broader concept of *major evolutionary transitions* proposed by Maynard Smith & Szathmáry (1995), where *complexity increases through cooperation, integration, and information-sharing among evolving units—whether molecules, cells, or societies.*

2.3.2.1 Evolution as Accelerating Change

A pivotal shift in evolutionary modeling came in the 1970s with Manfred Eigen's groundbreaking work on the mathematics of molecular evolution. Eigen developed the concept of *quasispecies* and introduced models where molecular replication and mutation drive a hypercycle of autocatalytic growth (Eigen & Schuster, 1979). His work revealed that, under certain conditions, evolution itself accelerates—an effect that leads not to simple exponential but *hyperbolic* or *singular* growth. In such systems, the rate of change increases over time until reaching a critical threshold or

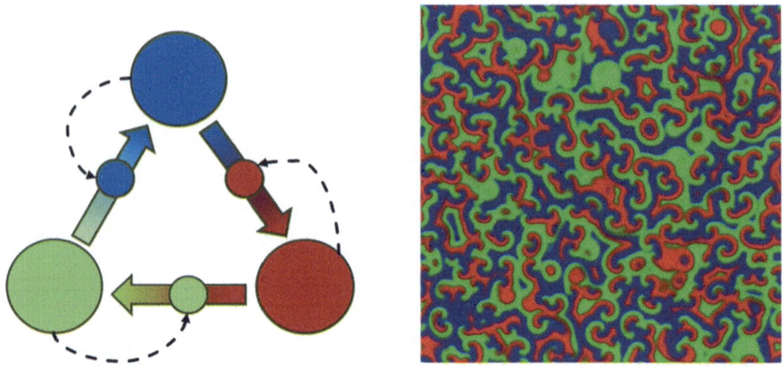

Fig. 2.10 Simplified BZ reactive-diffusion model. The schematic reactions on the left represent autocatalytic feedback. When diffusion is introduced, dynamic spatial patterns form, exemplified in the static image on the right. (*Source*: https://en.wikipedia.org/wiki/Belousov%E2%80%93 Zhabotinsky_reaction#/media/File:The_Belousov-Zhabotinsky_Reaction.gif)

"singularity," a moment of qualitative transformation but this transformation is connected usually with sharp changing trends and very often with decline, crises, and collapse (as it was in the late Roman Empire).

This concept of evolutionary singularity was not confined to the molecular domain. Researcher Widdowson extended these ideas to historical and cultural evolution. He proposed that civilizations like the Roman Empire underwent similar dynamics: initial rapid growth, followed by instability, plateau, and eventual reorganization or collapse (Widdowson, 2023). These historical inflection points resemble biological singularities—moments when the existing system cannot sustain its growth trajectory and must either evolve new strategies or fragment. The repetition of such patterns across different scales suggests a meta-evolutionary process, where systems evolve mechanisms for adapting to growth limits, whether through institutional reforms, technological innovations, or ecological adaptations.

2.3.2.2 Autocatalytic Networks and the Origins of Life

The question of how life emerged from non-life has prompted models that go beyond random assembly. Stuart Kauffman's theory of *autocatalytic sets* provides a foundational framework for understanding life as a self-sustaining network of interacting components (Fig. 2.10). In his view, life began not with a singular "miracle molecule" but with a collective of molecules that catalyzed each other's formation (Kauffman, 1993). As the diversity of molecular species increased, the likelihood of forming such autocatalytic networks approached certainty. This probabilistic emergence of order from chaos represented a new paradigm, where metabolism could precede replication—a reversal of the classical view.

Kauffman's ideas built upon the work of Ilya Prigogine, who explored *dissipative structures* in nonequilibrium systems. Prigogine demonstrated that energy flows through chemical systems could lead to spontaneous pattern formation, such as the oscillating waves in the Belousov-Zhabotinsky (BZ) reaction (Prigogine & Stengers,

1984). These dynamic patterns reflect self-organization driven by thermodynamic disequilibrium, a hallmark of life itself.[4]

2.3.2.3 Hydrothermal Vents and Metabolic Emergence

Nick Lane introduced an energetically grounded model of the origin of life, emphasizing the geochemical conditions present in alkaline hydrothermal vents. These vents offer a unique combination of factors: steep proton gradients, porous FeS structures, and abundant hydrogen—all of which may have facilitated the formation of primitive metabolic pathways (Lane, 2009; Lane & Martin, 2012). Unlike surface environments, these undersea systems maintain stable energy flow from the Earth's interior, providing long-term disequilibrium conditions ideal for life's emergence.

Lane's "metabolism-first" hypothesis posits that early life began as networks of energy-coupled reactions, gradually evolving the machinery to process information (i.e., genetic material). Initially, compartmentalization may have occurred within the mineral pores of vent structures, with lipid membranes later evolving to encapsulate and regulate these systems. This bottom-up view of life's origins has profound implications for understanding how energy flow constrains and guides evolutionary complexity (Lane, 2009).

2.3.2.4 Hierarchy: Bottom-Up and Top-Down Constraints

As Stanley N. Salthe has emphasized in his work on hierarchy theory, complex systems evolve through nested levels of organization, each constrained and enabled by processes at lower and higher levels. His developmental approach to evolution frames complexity not merely as a product of selection, but as a *dynamic tension between bottom-up emergence and top-down constraint* (Salthe, 1985, 1993). This perspective aligns well with Big History's cross-scale narratives, where new levels of structure—from molecules to civilizations—emerge through the interplay of energetic, informational, and structural dynamics.

2.3.2.5 The Earth as a Living System: Lovelock's Gaia Theory

While models such as Eigen's and Lane's focus on the emergence and dynamics of life at micro and meso scales, James Lovelock proposed a macro-scale evolutionary system: the Earth itself. The *Gaia Hypothesis*, developed in the 1970s by Lovelock and microbiologist Lynn Margulis, posits that life and the environment co-evolve in a feedback-stabilized system. Biological processes influence atmospheric composition, ocean salinity, and planetary temperature, forming a planetary-scale homeostasis (Lovelock, 1979; Margulis & Lovelock, 1989).

Initially met with skepticism, Gaia has evolved from metaphor to scientific theory, integrated into Earth system science, geophysiology, and systems ecology (Lenton & van Oijen, 2002). For instance, the transition from a CO_2-dominated early

[4]On some modern ideas how life emerged, see Grinin (2024), as well as chapter "Chemical Evolution in Cosmos and Earth" (Grinin, 2026b) in this volume.

atmosphere to today's oxygen-rich environment is attributed to photosynthetic organisms, particularly cyanobacteria. Gaia theory frames this planetary transformation not as a passive result of biology but as an active, cybernetic feedback loop driven by life itself, maintaining conditions suitable for its own persistence.

2.3.3 Human Evolution

2.3.3.1 The Most Intricate Puzzles in Evolutionary Science

The evolution of human cognition—especially the emergence of symbolic reasoning, language, and theory of mind—represents one of the most intricate puzzles in evolutionary science. Despite an abundance of physical artifacts and fossil evidence, the underlying neurological and cognitive mechanisms remain only partially understood. Stone tools, symbolic carvings, and cave art signal the presence of abstract thinking and cultural transmission, but these objects cannot fully reveal how the brain evolved to support such functions (Tattersall, 2012).

Key physical adaptations laid the groundwork for advanced cognition, including:

1. Bipedalism, which freed the hands for tool use
2. Fine motor control, enabling delicate manipulation and symbolic representation
3. Laryngeal descent, which facilitated a broader range of vocalizations critical for complex speech (Lieberman, 2007)
4. Reduction in body hair, possibly linked to thermoregulation and extended social bonding via touch and grooming (Jablonski, 2006)

The study of human cognition has moved beyond behavioral archaeology to integrate insights from neuroscience, artificial intelligence, and evolutionary theory. While symbolic artificial intelligence focused on rule-based reasoning systems in the mid-twentieth century, the rise of deep neural networks has shifted attention to distributed, emergent representations that more closely mimic biological learning processes (Lake et al., 2017).

One particularly ambitious theory in neuroscience is Karl Friston's Free Energy Principle (FEP), which posits that cognitive systems strive to minimize surprise (or "free energy") by maintaining predictive internal models of the world (Friston, 2010). This framework integrates perception, action, and learning through a continuous process of hypothesis testing:

- The brain generates predictions about sensory inputs
- Compares them with actual inputs
- Updates internal models to reduce prediction error
- Reorganizes itself when persistent discrepancies arise

This theory presents cognition as a dynamic, inference-based process that emphasizes adaptability and self-organization—properties crucial for the emergence of symbolic reasoning and problem-solving.

Neuroscientific studies support the idea of co-evolving neural structures and cognitive functions. For instance, the maturation of pyramidal neurons, especially in the prefrontal cortex, is associated with symbolic capacity, working memory, and language development (Elston, 2003). Comparative studies across primates show that human brain evolution involved selective expansion and reorganization of cortical and subcortical areas, such as the temporoparietal junction and Broca's area, involved in language and social cognition (Smaers et al., 2017).

However, recent findings challenge some longstanding assumptions. Research by Fedorenko and colleagues suggests that language processing may be more distinct from symbolic reasoning than previously thought, with specialized neural circuits dedicated exclusively to language, independent of domain-general symbolic logic (Fedorenko & Varley, 2016).

Understanding human cognition remains uniquely difficult, as the brain is both the instrument and object of study—a recursive situation that complicates efforts to trace the origins of abstract thought. While the archaeological "creative explosion" of symbolic artifacts between 40,000 and 10,000 years ago signals mature symbolic cognition, the neurocognitive foundation of this shift may have evolved much earlier (d'Errico et al., 2003; see also chapter "Human Evolution in the Complexity Growth Perspective: Toward Periodization of the Big History Biosocial Era" [Korotayev, 2026, this volume] and chapter "Biological and Social Phases of Big History and Universal Evolution: Similarities, Differences, and the Growth of Complexity" [Grinin et al., 2026] in this volume).

2.3.3.2 The Evolution of Cooperation: From Cells to Civilizations

The capacity for cooperation represents a major evolutionary innovation in both biology and society. In humans, Robin Dunbar's research has highlighted a strong correlation between neocortex size and group size, suggesting that human brains evolved, in part, to manage the complexities of social relationships. The concept of "Dunbar's number"—a cognitive limit of roughly 150 stable relationships—has been observed in various social and organizational settings (Dunbar, 1993).

Beyond human societies, cooperation has deep evolutionary roots. One of the most profound examples is the endosymbiotic origin of mitochondria, where ancestral eukaryotic cells incorporated formerly free-living bacteria. Over time, these symbionts lost their independence and transferred most of their DNA to the host nucleus, giving rise to complex multicellular life (Margulis, 1970; Lane, 2009). Such symbiosis exemplifies how entities may sacrifice autonomy for collective stability—a recurring theme across evolutionary levels.

This dynamic also plays out in the social domain. Peter Turchin has examined how cooperation scaled up from tribes to "ultrasocieties"—large, hierarchical civilizations held together by cultural institutions, ideological narratives, and collective defense mechanisms. In his Secular Cycles model, Turchin demonstrates how elite overproduction and inequality lead to oscillations between growth and collapse, underscoring the fragile balance in large-scale cooperative systems (Turchin, 2016; see also Grinin, 2013; Grinin & Korotayev, 2011; Korotayev et al., 2006b, 2016).

Modern humans have developed symbolic tools—like money and law—to facilitate cooperation across large, anonymous populations. Money, in particular, represents a symbolic medium of trust and time that can substitute for direct reciprocity, allowing complex trade and credit systems to flourish. These innovations mark a transition from immediate, face-to-face cooperation to abstract, institutionalized social contracts (Graeber, 2011; Grinin, 2013; Grinin et al., 2009).

2.3.3.3 Co-Evolution and Dual Inheritance Theory

Human evolution cannot be fully understood without recognizing the interaction between genes and culture. The theory of gene-culture coevolution, also known as dual inheritance theory, holds that cultural traits influence genetic selection and vice versa. This process has enabled rapid adaptation beyond the pace of genetic mutation alone (Richerson & Boyd, 2005).

Key coevolutionary dynamics include:

1. Prosocial emotions such as empathy and shame, which evolved to stabilize cooperation
2. Physiological changes like vocal tract reconfiguration and fine motor control to support speech and tool use
3. Cognitive mechanisms for cumulative cultural transmission, such as imitation and norm enforcement
4. Complex social structures, including hierarchies and institutions, which are sustained by culturally transmitted norms and beliefs

Coevolution has given humans a social brain—equipped with a preference for fairness, revenge, and cooperation—that functions not just for survival, but for managing reputation and alliances in complex groups. Importantly, this process continues today: our cultural environment influences genetic selection pressures, especially in areas like diet, immunity, and even cognitive traits (Henrich, 2016).

2.3.4 Social and Economic Complexity

Mathematical modeling has become indispensable in the study of historical and economic evolution. The earliest such model, Thomas Malthus's population theory, posited that unchecked population growth would outstrip food supply, leading to collapse. This framework helped illuminate the pressures behind societal transformations and the eventual escape from the "Malthusian trap" (North & Thomas, 1973).

Modern approaches use System Dynamics and Agent-Based Modeling (ABM) to simulate complex historical processes. Jay Forrester's system dynamics models emphasized feedback loops in ecological and economic systems, exemplified by the Club of Rome's "Limits to Growth" (Meadows et al., 1972; see also Sadovnichy et al., 2023). ABMs, on the other hand, tracks micro-level agents to reveal emergent macro-patterns, such as in Turchin's work on elite dynamics and civilizational collapse.

Other modeling efforts include:

- Korotayev and Widdowson's analyses of technological growth and social singularities (Widdowson, 2023; Korotayev, 2005, 2007; Korotayev & Zinkina, 2017; Korotayev et al., 2006a)
- Bettencourt and West's urban scaling laws, showing how infrastructure and innovation scale nonlinearly with population (Bettencourt et al., 2007)
- Sornette's "Dragon Kings," rare outliers in city growth and innovation that signify system-level phase changes (Sornette, 2009; Malkov et al., 2012)
- The Economic Complexity Index (ECI), developed at MIT, which uses network theory to assess the sophistication of national economies (Hausmann et al., 2014)

Some ancient societies incorporated reset mechanisms, such as the Jewish Jubilee system, which forgave debts every 50 years—a social innovation that parallels modern calls for wealth redistribution to avoid systemic instability.

Human evolution is a multilayered process shaped by genetic, cultural, social, and ecological interactions. From the neural reorganization enabling symbolic thought to the cooperative dynamics of mitochondria and cities, humanity's development is a testament to the power of self-organization, adaptation, and innovation across scales. The integration of models from biology, cognitive science, and economics continues to deepen our understanding of the forces that shaped our past—and that will define our future.

2.3.5 Common Patterns

Patterns that are common to the various stages of Big History have been developed. Three main groups of patterns are:

1. Those that consider the relationship of sizes of Big History objects and how the size of the primary drivers of evolution change with time
2. Patterns of evolution of Earth with the evolution-development relationship, that is, the order of evolutionary innovations is often reflected in the order of major events during the development of an individual
3. Various energy flow relationships, although as mentioned before the complex physical systems show a different pattern than the evolution of life on Earth

Energy Several researchers have discussed energy extraction from the environment, storage, and use throughout evolution (Fox, 1988; Niele, 2005; Smil, 2008). However, Eric Chaisson (2001) focused on energy flow. In the course of evolution from galaxies, stars, planets, to forms of life and civilization, the energy flow density (W/kg) has increased (Fig. 2.11). The system, however, had to be defined spatially and temporally in order to make these estimates. As an example, when a galaxy's

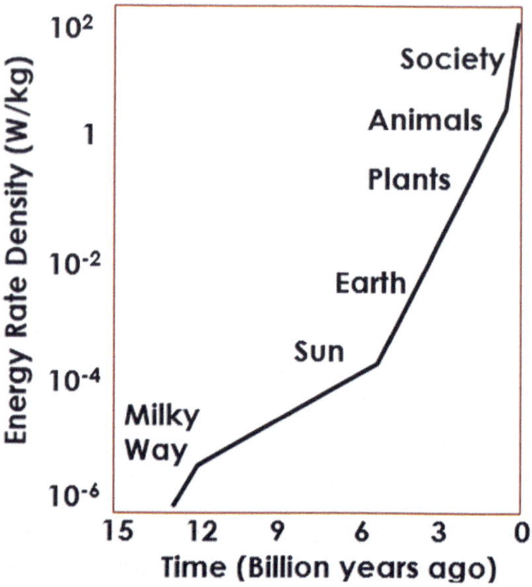

Fig. 2.11 Relationship identified by Chaisson (2011) (modified) showing how the energy flow per mass changes with the time at which big history objects first form. This indicates that energy flow might be important for considering periodization. Note that the energy rate density for society is based on the energy usage per person and the mass of a person (not all the equipment and infrastructure). The energy rate density of Earth is for a thin layer of water and atmosphere (not considering the whole planet mass)

black hole is active, energy flow through it can increase quite a bit. Based on these assumptions, galaxies are less complex than stars, which is contrary to a Big History textbook (Christian et al., 2014).[5] Structure and dynamics of galaxies appear more complex with many interconnected pieces, though they are somewhat dynamically stable. There are several parts to a galaxy including the central black hole, the central bulge, arms, and a halo of dark matter.

The energy efficiency of natural and engineered systems differs as well. Computers require much more energy flow per mass and are not as complex as the human brain, which works with only 20W. Furthermore, by mitigating increasing entropy, the total energy flow (not just density) through an expanding system should increase as it evolves (LePoire, 2020a, 2020b).

Organization Organizational change is another aspect of evolving complex adaptive systems. Many transitions occurred when previously independent entities merged, according to Volk (2017). Initially, this was due to the binding energies in the physical world. A combination of these entities (e.g., an electron and proton) would not be stable if the temperature exceeded the binding energy. In the course of expansion, the Universe cooled, resulting in a sequence of merging. This occurred during the first 5 of the 12 steps identified by Volk, namely, fundamental quanta (e.g., quarks), nucleons, atomic nuclei, atoms, and molecules. The four steps in biological evolution are simple (prokaryote) cells, eukaryote cells, multicellular organisms, and social animal groups. Combination steps required previously independent entities to come together to achieve some environmental advantage.

[5] On the critique of Chaisson's claims, see also Van Duin (2025a, 2025b).

According to Volk, the most recent three steps focus on human civilization in terms of tribal groups, agro-villages, and geopolitical states. Again, the entities had to loosen their independence in order to reap the benefits of combining. Possibly this can be seen as a social cooling process, in which disagreements between groups are resolved enough to facilitate sharing common goals, such as economy of scales, a reduction of barriers, or defense.

Several ways were possible to combine the elements of the previous step to create the elements of the next step, resulting in diversity between steps. The combination of nucleons to form nuclei of elements is an example in the physical realm. It was then possible to combine the different elements to form many different molecules. These combinations undergo the evolutionary dynamics of reproduction, variation, and selection. Through this, they are able to explore new spaces in order to maintain the stability of the inherited information.

Information is clearly emphasized in this view of organization, but little is said about energy flow, environmental interactions, and challenges as the systems approach their limits. Apart from combining elements, systems can also develop complexity by growing and specializing. Animal organ evolution demonstrates this.

One information processing mechanism in social systems is the transfer of ideas from smaller systems that have reached their limits to larger systems that offer additional management for further development. One example of this can be seen in the series of leading capitalist countries, which change about once every century (LePoire 2010). The new leading country had twice the population of the previous one at the time of the transition. Consequently, as capitalism grew, a larger population base was required to support the necessary evolving organizations.

Environment/Limits The role of the environment is illustrated, in Fig. 2.12, by the inverse relationship between sequential evolutionary objects and their environments (Jantsch, 1980). This relationship was extended to include modern technological civilizations (LePoire, 2004, 2020a, 2020b). For example, as a result of gravitational instability, the tiny quantum fluctuations in the early Universe later evolved into cosmic structures such as the cosmic web and eventually into galaxies (see chapter "Energy Flow and Emergence of Complexity in the Universe's Structure" [LePoire, 2026] in this volume; see also Grinin, 2019).

Examples of this relationship continued in terrestrial evolution. These examples include single-celled organisms that had to evolve aerobic respiration, as the buildup of oxygen in the planetary atmosphere was required for further evolution. A higher level of oxygen in the air allowed multicellular organisms to diffuse oxygen more easily. Also, at about 540 million years ago, the Cambrian Explosion, which occurred after about 88% of Earth's current age, brought about the dominance and diversification of multicellular organisms, which could compete in large ecosystems. Both humans and some of the tools that formed their evolving environment were similar in size during human evolution. However, as human societies grew in scope, their tools depended on technology that is more precise. For example, an environment of microchips is used to evolve a global system that exists today.

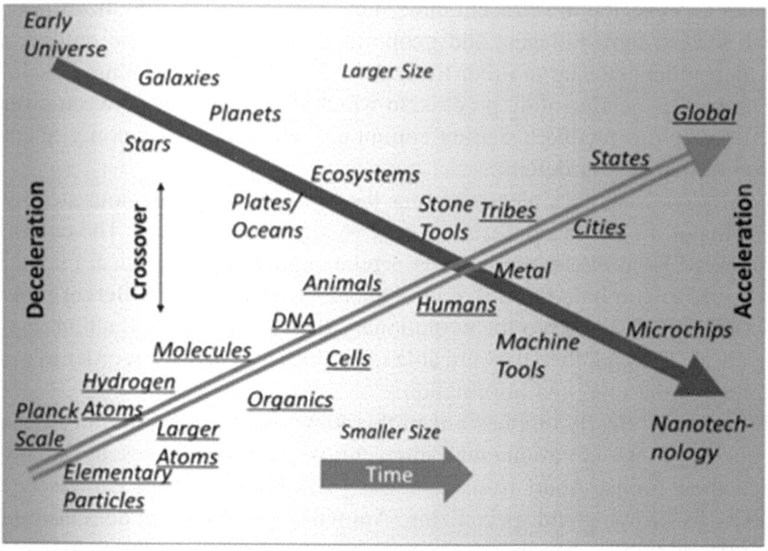

Fig. 2.12 Relationship first identified by Jantsch of the relationship between the relative sizes of evolving systems and the environment in which the evolution takes place. This indicates that hierarchy of organization and considerations of environmental scale might be important for considering periodization

Evolutionary Dynamics The great polymath John von Neumann had identified the acceleration of technological change in the 1950s (Ulam 1958). Von Foerster (1960) identified evidence that global population was growing faster than exponentially. Kremer (1993) and Kapitza (1996) extended this concept far into the past. In 1971, Eigen discovered that evolving systems might follow a hyperbolic trajectory (Eigen, 1971). These growth patterns follow increasingly quicker exponential growth as the system adapts to its environment. A singularity trend was proposed by Modis (2002) and Panov (2005, 2020) as an explanation for cumulative learning acceleration. It is Panov's event set (Fig. 2.13) that comes closest to the extended Jantsch approach, due to the geometric factor of 3 that he discovered (independently with Snooks [2005]). Compared to the acceleration scheme with a factor of 3 across 19 events, Modis' list provides 28 events.

Although the difference between hyperbolic and exponential growth equations is small, it makes an enormous qualitative difference in the evolutionary process (see, e.g., LePoire, 2020a; Korotayev, 2018, 2020, 2025a). A simple exponential growth, such as money collecting interest in a bank, will double in the same amount of time. The value of money invested with a 7% annual interest rate doubles every 10 years, for example. In the simple hyperbolic evolution model, however, progress begins very slowly. Even so, as it approaches a specific time, the singularity time, it accelerates rapidly. (For example, if a population doubles within 10 years and then doubles again in half the time it took before, then the singularity time, when the population would become infinite, will only take 20 years from the beginning of the

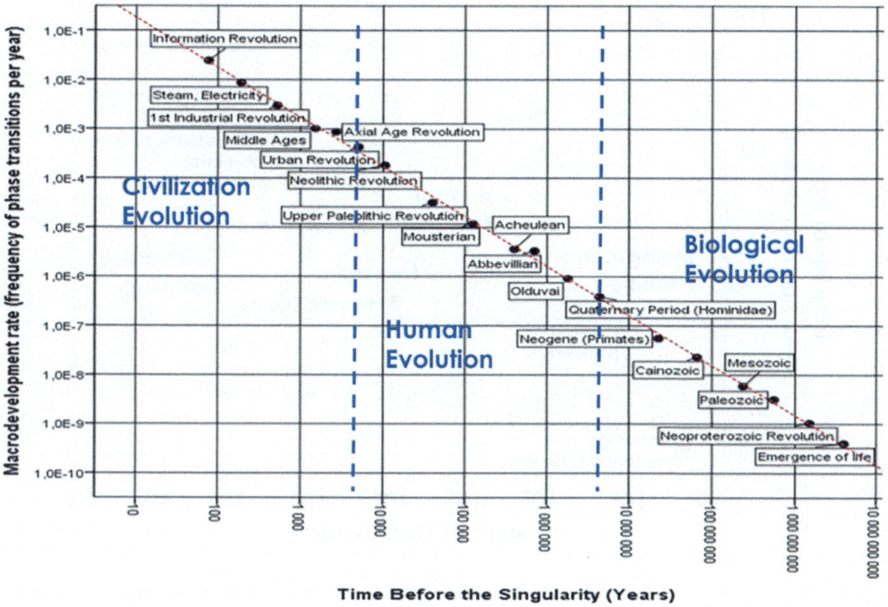

Fig. 2.13 Relationship of phase transition points identified by Panov (2005, 2020) showing a singularity trend with the singularity date of 2027 CE. This indicates that cumulative learning, and its consequential hyperbolic trend, might be important for considering periodization. Note that the transitions from 5 billion years ago to 5 million years ago cover biological evolution; from 5 million to 5000 years ago cover human evolution; and from 50 to 5000 years ago cover civilization history

process.) Even though the trend suggests continued acceleration to ever-higher (indefinite) levels, due to finite resource limits, this growth pattern will shift to another. A delay between the growth and the impact of that growth, for example, increases the chances of the system overshooting and collapsing. There are, however, more optimistic scenarios in which growth is restrained before this condition is achieved.

A link was discovered by Ekstig (1994) between the time of an emergent feature's emergence and its expression in a developing organism (Fig. 2.14). In general, the development of a characteristic follows the evolution of that characteristic. The range of characteristics includes animal tissue formation, humans' ability to walk and speak, and cultural learning levels such as reading, writing, and math. The results of this analysis later were incorporated into the identification of a super-exponential growth in complexity (Ekstig 2012, 2015, 2017).

As discussed in Schneider and Kay (1994) and Azarian (2022), the second law of thermodynamics can be applied to the origin and evolution of life. Physical dissipative systems will self-organize when there is a gradient of entropy (e.g., temperature differences). The formation of Benard cells is a result of fluctuations around the normal conditions that lead to increased energy flows reducing this entropy gradient. In some environments, such as alkaline hydrothermal vents, this principle could

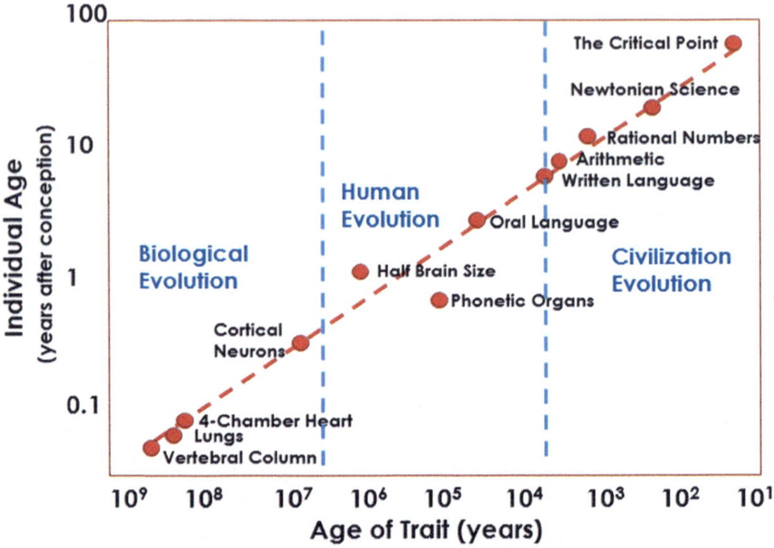

Fig. 2.14 Relationship of evolutionary time (age of traits) and development time (individual time) identified by Ekstig (2017) along with the critical point. This indicates that information processing and reproduction might be important for considering periodization

even be taken further, suggesting that life originated to reduce entropy gradients (Lane 2009).

Finally, the hope is that these trends may indicate that the full scope of evolution can be considered at some level to be from similar processes that share some characteristics. The fact the spatial scale of Big History objects from the Universe to the atomic nucleus (Fig. 2.15) can be estimated with relatively simple physical relationships and fundamental constants, provides one example of a straightforward relationship between the scales.

2.3.6 Evolutionary Rules and Principles Associated with the Formation of Complexity

The Big History narrative charts an accelerating trend of complexity, from cosmic structures to life and culture (Spier, 1996, 2010, 2011). But the trend is not inevitable nor indefinite. Critical transitions—such as the emergence of life or the emergence of humans—might easily not have occurred. As complexity builds, marginal returns may decline, raising questions about sustainability, measurement, and alternative developmental pathways.

Thus, the transition to new levels or attractors of complexity is not programmed. Development can stop or go in a way that does not lead to the emergence of a new level of complexity. On the other hand, a breakthrough to new levels, although closely connected with chance, is at the same time prepared by preceding

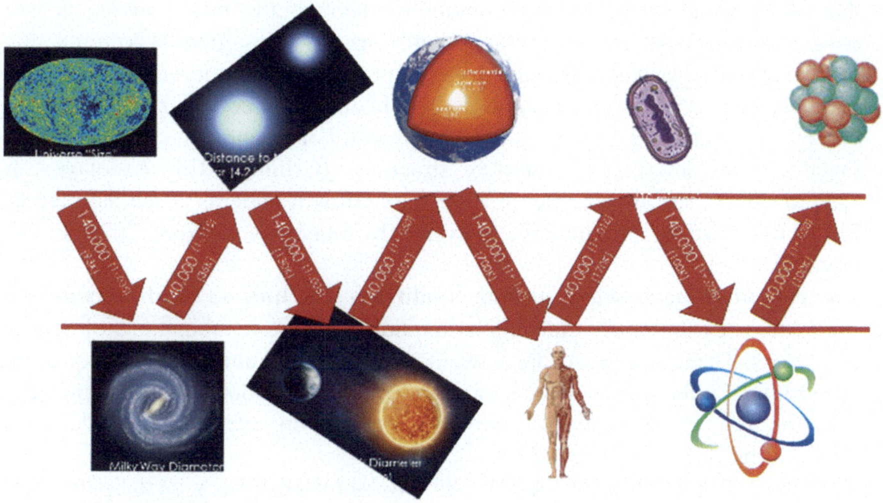

Fig. 2.15 Relationship of sizes of Big History objects from the Universe to an atomic nucleus based on fundamental physics considerations and constants identified in part by Press and Lightman (1983). This indicates that spatial scales seem to be related, giving hope for considering integrated periodization across cosmic, biological, and cultural systems. (Figure from LePoire, 2019)

development; the potential for a breakthrough to be realized grows with each attempt of evolution to move to new frontiers of complexity.

All this is largely explained by the fact that the transition to something fundamentally new, is always not an inevitable result, but the result of the development of exceptional conditions. Obviously, such conditions develop very rarely and in places that are quite far from the main line of development of evolution and Big History.

The transition to higher levels of complexity follows a number of evolutionary rules which can be also formulated as principles (Grinin et al., 2009, 2013, 2020; Korotayev et al., 2015). These include in particular:

- **The rule of paying for evolutionary progress (the cost of complexity principle):** Evolutionary progress comes at a price. It means that an evolutionary breakthrough usually follows numerous failed paths that accumulate pre-adaptations, resources, and favorable conditions. Each major transition is typically preceded by several dead-end developments, reflecting the trial-and-error nature of reaching greater complexity. It is also important to note here that along with the main direction, lateral, additional ones also appear such as planetary, chemical, mineralogical evolution. The more such additional lines, the higher the possibility to enhance the chances of a breakthrough by increasing diversity (planets are an additional line of development; chemical evolution; mineralogical evolution; etc.).
- **The rule of the preparatory work of evolution (the preparation principle):** Evolutionary leaps are not random accidents but the result of long-term processes

that build up the necessary preconditions. Though the final convergence of conditions may involve chance, the groundwork is laid by gradual accumulation of matter, energy, and directional development. Preparatory conditions also manifest themselves in the creation of a certain density of concentration of matter, energy, elements, and directions of development. For example, increasing cosmic metallicity helped enable complex structures (Grinin, 2014; Korotayev & LePoire, 2025; see also chapter "Complexity Growth Patterns in the Stelliferous Era of Big History: Preliminary Analysis" [Korotayev & LePoire, 2026] in this volume).
- **The rule of coincidence of unique conditions (the unique conditions principle):** Major evolutionary innovations require very rare combinations of special conditions. The greater the change, the more specific and unique the prerequisites. Accordingly, coincidence of such rare conditions and events requires a number of breakthrough attempts, which provide a successful attempt with additional opportunities (e.g., pre-adaptations).[6]
- **The rule of transition to new levels (the semi-periphery principle):** New evolutionary levels tend to emerge not in the core or margins, but in transitional zones with diverse and boundary-crossing conditions—such as Earth's surface.
- **The rule of the localization of evolutionary breakthrough (the localization principle):** Evolutionary breakthroughs require not only the convergence of multiple rare and unique conditions but also coincidence of them within a confined space and time. These conditions must interact within a single or closely linked system during a limited period to trigger a transition.

2.4 Applied to a Big History Framework

The evolution of complex adaptive systems on Earth, from the emergence of life to the rapid progression of human societies and technologies, represents a remarkable and accelerating trajectory. This pattern of hyperbolic acceleration underscores how life has continuously adapted to new challenges and opportunities, creating increasingly intricate systems of information processing, energy flow, and organization. By understanding these patterns, we can better navigate the accelerating pace of change and consider how humanity's future can contribute to—or disrupt—this extraordinary trend of growing complexity.

This section presents one way a framework can be constructed based on some of the theories and approaches presented earlier.

A consensus approach includes both considering the traditional approach of physical and complex adaptive systems and the nuanced intricacies of evolution. The following figures gather some of these concepts into the following areas of Physical and Complex Adaptive Systems:

[6] However, uniqueness applies only to certain critical traits, not the entire system, which may otherwise remain ordinary. It is often unclear which factors were essential versus incidental.

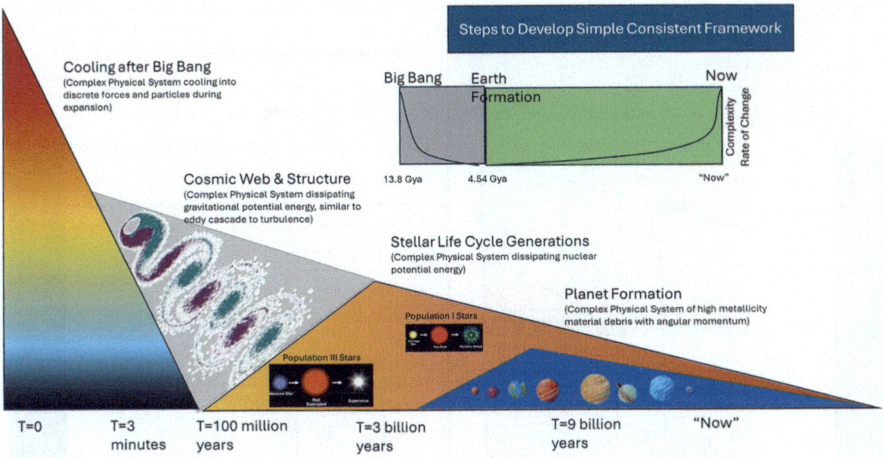

Fig. 2.16 The development of physical complexity in the Universe from the Big Bang until today. The insert in the upper right shows this development starting at the left (gray area). The continuation of complexity growth due to life (shown in green in the insert) is detailed in Fig. 2.17. The simplified phases of the development of physical complexity begin with the Big Bang where the expansion resulted in cooling and the laws and particles of physics unfolded (Atkins, 2018). This resulted in a rather smooth density-distribution of mostly hydrogen and helium atoms which held large gravitational and nuclear potential energies. The gravitational potential energy was dissipated through gravitational collapse with the assistance of dark matter as the cosmic web formed. This gravitational potential was also reduced when the matter cooled enough to form stars, which then ignited the nuclear fusion reactions resulting in reducing the nuclear potential. As the heavier elements accumulated through stellar explosions, further star formation took place with smaller masses and also the heavier elements in orbit that could form planets. The Sun and Solar System formed about 4.5 billion years ago (approximately 9 billion years after the Big Bang), that is, the Sun and Earth are approximately a third of the Universe's age now

- Outline of physical complex system development from the Big Bang as the gravitational and nuclear potential energy left from the Big Bang expansion and cooling were decreased with gravitational collapse and stellar nuclear ignition (Fig. 2.16).
- Outline of complex adaptive systems on Earth through the evolution of life, humans, and civilization. A geometric acceleration pattern is identified with factors of 1000 on the large scale of stages and factors of 3 on the smaller nested substeps. The factor of 3 has some supporting evidence besides being identified by others (Fig. 2.17).

2.4.1 Step 1: Physical and CAS

The development of physical complexity in the Universe is a story of transformation spanning from the Big Bang to the present day, shaped by fundamental processes and forces. This progression is outlined in the main figure, with the inset in the upper

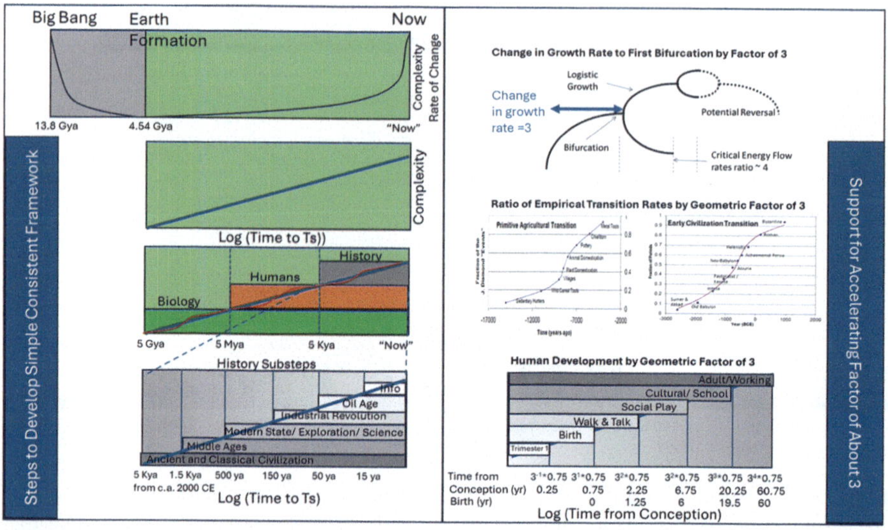

Fig. 2.17 The development of complex adaptive agents (life) on the Earth. The insert in the upper left shows this development starting at the right (green area). The earlier development of physical complexity (shown in gray in the insert) is detailed in Fig. 2.16. The simplified phases of the evolution of complex adaptive systems are shown on the left with three levels of detail. Various supporting evidences for the accelerating (scaling) factor of 3 are shown on the left. On the left, the three levels of detail include the following. (1) Top: the large-scale hyperbolic growth of complexity (on log–log scale with the time until the singularity on the x-axis). (2) The next detailed structure of three phases based on the information processing which correspond to the traditional disciplines of biological evolution, human evolution from the branching of other great apes, and the evolution of technology in written history. Each phase has an accelerating factor of about 1000 with beginning times of the phases at approximately 5 billion years ago (ya), 5 million ya, and 5 thousand ya. (3) The further detail within each phase of geometrically accelerating substeps. The History phase is chosen as an example where the substeps correspond closely to the traditional periods of history: Ancient and Classical, Middle Ages, modern state/exploration, industrial revolution, fossil fuel age, and the information age. These six substeps with geometrical decreasing durations by a factor of 3, combine to give the factor of 1000 within the phases (i.e., 3^6 is approximately 1000). On the right, three supporting cases for the geometrically accelerating factor of 3 are shown: (1) (top) the cascade toward chaos has its first bifurcation transition after the initial growth rate increases by a factor of 3; (2) (middle) empirical evidences of the rate of events during two of the historical substeps (agricultural transition, and ancient and classical civilization) are shown to have durations about two-thirds of the time to singularity from when they start, leaving the last third for all the remaining substeps; and (3) human development from conception shows that major steps occur about a factor of 3 from the previous

right providing a focused view of key stages. It all began approximately 13.8 billion years ago with the Big Bang, a singular event that set the Universe into rapid expansion. This expansion caused cooling, which allowed fundamental physical laws to manifest and the formation of elementary particles such as quarks, electrons, and eventually protons and neutrons (Atkins, 2018). These particles combined to form the simplest elements, hydrogen and helium, which dominated the early Universe.

Initially, the Universe was a smooth, nearly uniform expanse of matter and energy, but small density fluctuations, magnified by gravitational forces, began to grow. The gravitational potential energy stored in these regions began to dissipate as matter clumped together, aided by the unseen influence of dark matter. This process led to the formation of the cosmic web—a vast interconnected structure of filaments and voids that became the scaffolding for galaxies. As regions of matter collapsed under gravity, they cooled and condensed, eventually forming stars, the first sites of nuclear fusion. These stars transformed hydrogen into helium and other light elements, releasing vast amounts of energy and reducing the nuclear potential energy stored in their cores.

As stars aged and died, many underwent cataclysmic explosions known as supernovae. These events were critical for the next stage of complexity: they dispersed heavier elements like carbon, oxygen, silicon, and iron into space. These elements, forged in the cores of stars, were the building blocks for planets and life. Over billions of years, this enrichment of the interstellar medium allowed for the formation of second- and third-generation stars surrounded by planetary systems. One such system, our Solar System, formed about 4.5 billion years ago from a dense molecular cloud enriched by earlier stellar generations. This marked a pivotal point in the Universe's evolution, as the Sun and Earth emerged approximately 9 billion years after the Big Bang—about two-thirds of the Universe's current age.

The Earth, with its stable environment and abundance of heavy elements, provided the conditions necessary for the next leap in complexity: the emergence of life. This phase, detailed in Fig. 2.17, represents a profound continuation of the Universe's trend toward increasing complexity. The interactions of physical and chemical processes on Earth gave rise to living systems, which, through biological evolution, introduced entirely new forms of complexity. As life evolved, it began to influence its environment in unprecedented ways, eventually giving rise to human culture and civilization (e.g., LePoire, 2023)—adding new layers of complexity to the cosmic story.

Thus, the journey of physical complexity—from the simple hydrogen atoms of the early Universe to the complex systems of planets, ecosystems, and intelligent life—illustrates the Universe's remarkable capacity for transformation and innovation, driven by the interplay of fundamental forces, matter, and energy over billions of years.

The evolution of complex adaptive systems, particularly life on Earth, is a pivotal chapter in the broader narrative of increasing complexity in the Universe. This development, illustrated in the upper-left inset (green area), builds upon the earlier stages of physical complexity (gray area, detailed in Fig. 2.16). The phases of life's evolution and the subsequent emergence of human society and technology are marked by accelerating rates of complexity growth, governed by patterns of information processing, energy flows, and organizational innovation. These patterns reveal the accelerating progression of life and its interconnected systems through three levels of detail, underscoring how the evolution of complexity has followed a geometric acceleration across successive stages of development.

2.4.2 Steps 2–4: Hierarchy of CAS on Earth

- **Hyperbolic Growth:** The large-scale hyperbolic growth of complexity is depicted on a log-log scale (Fig. 2.17: second from top left), with the x-axis representing the time remaining until a hypothetical singularity—a point of infinitely rapid change. This hyperbolic curve reflects how complexity has increased explosively across vast timescales, suggesting a trend of accelerating transformations in the Universe's history. While this trend is evident, the curve also hints at a profound challenge: understanding and managing the accelerating pace of change as systems grow more complex.
- **Major Steps:** This level breaks the evolution of complexity into three primary phases, each corresponding to major shifts in information processing capabilities (Fig. 2.17: third from top left). Each phase progresses approximately 1000 times faster than the one before, reflecting a hyperbolic acceleration in the pace of innovation and adaptation:
  - *Biological Evolution (c. 5 billion years ago)*: Beginning with the origins of life, this phase encompasses the emergence of self-replicating molecules, the development of cells, multicellular organisms, and ecosystems. Key milestones include the Cambrian explosion, the rise of plants and animals, and the diversification of species.
  - *Human/biosocial Evolution (c. 5 million years ago)*: This phase marks the divergence of early hominins from other great apes, leading to the development of *Homo sapiens*. It encompasses the rise of tool use, language, culture, and social structures, which allowed humans to adapt and thrive in diverse environments.
  - *History and Technological Evolution (c. 5000 years ago)*: Commencing with the advent of written records, this phase represents the rapid advancement of human societies through agriculture, science, industrialization, and the digital age. Innovations in information storage and processing (e.g., writing, printing, computers) have hyperbolically increased the capacity for knowledge accumulation and dissemination.
- **Focus on One Substep Sequence:** Within each of the three phases, smaller substeps exhibit a geometric pattern of acceleration, with each successive step occurring approximately one-third the duration of the previous one (Fig. 2.17: left, bottom). For example, within the History Phase, six substeps correspond to traditional historical periods:
  - *Ancient and Classical Era*: The emergence of early civilizations, such as Mesopotamia, Egypt, Greece, and Rome, marked by innovations in governance, writing, and trade.
  - *Middle Ages*: A period of feudalism, religious dominance, and gradual recovery after the fall of classical empires.
  - *Modern State and Exploration*: The Renaissance and Age of Exploration saw the rise of centralized nation-states and global trade networks.
  - *Industrial Revolution*: Marked by mechanization, steam power, and transformative economic systems.

- *Fossil Fuel Age*: Powered by coal, oil, and gas, enabling rapid technological and societal development.
- *Information Age*: Defined by digital technologies, global communication, and artificial intelligence. Each substep decreases in duration by a factor of 3, culminating in a geometric acceleration of change. Together, these substeps demonstrate how complexity builds on prior advancements, creating a hyperbolic trajectory.

2.5 Complexity Growth as a Leading Trend of Big History

2.5.1 On Possible Periodization of Complexity Growth in Big History

Some views on complexity and its evolutionary development, as well as its basic components (energy, organization, information, etc.) have been discussed above. The main focus was on the analysis of complex physical and adaptive systems (CAS). Physical complex systems exhibit self-organization and energy dissipation but lack information processing capacity. Examples include the cosmic web, spiral galaxies, hurricanes, and vortices in draining fluids. These systems are governed primarily by physical laws—such as gravity, thermodynamics, and fluid dynamics—and tend to stabilize into recurring, dynamic patterns (Prigogine, 1980; Bejan, 2000). In contrast, CAS are systems that gather, store, and process information to adapt over time. These include living organisms, ecosystems, and human societies. CAS are characterized by learning, memory, and purposeful behavior through feedback, replication, and selection processes (Holland, 1992).

However, it is extremely important to consider more closely the growth of complexity as a certain trend of self-development of the Universe and its separate parts; as a set of general movements of various systems and objects, as well as forms of their interaction. This trend in some senses can be considered leading in Big History. Understanding, explaining, and analyzing this trend actually constitutes one of the main objects of Big History research.

To begin with it makes sense to propose a periodization of the development of complexity in Big History based on division of systems into complex physical, complex adaptive systems, and complex integrative systems, which connected with higher stages of biological evolution and especially with social evolution (see below).

The periodization below can be presented as a through line sequence of development of complexity in a large-scale periodization of the complexity growth (whereas in each stage, a whole series of sub-stages of development of complexity can be distinguished).

1. Creation of conditions for the growth of large-scale physical complexity, or more precisely for the growth of large-scale physical complex systems (the emergence of hadrons, atomic nuclei, atoms, molecules, molecular clouds, etc.)
2. The emergence and development of large-scale physical complex systems, that is, stars, galaxies, cosmic web, etc.
3. The emergence and development of complex adaptive systems (starting with abiogenic evolution)
4. The emergence of complex integrative systems within the framework of social evolution

At each new stage, the forces of development of complexity of all previous stages continue to operate. Thus, complex integrative systems are naturally adaptive (thus, in some respects, complex integrative systems can be regarded as a form of CAS).

2.5.2 Complex Integrative Systems Within the Framework of Social Evolution

We can see that the main line of biological evolution was connected with a growth of integration of biological organisms into more complex systems. These are colonies of unicellular organisms, transition from unicellular organisms to multicellular ones, different types of symbiosis, for example, mitochondria in eukaryotic cells. Other types of integration examples can be found among social insects as well as gregarious animals. However, although biosocial evolution played a significant role in biological evolution, resulting in the development of social insects and animals, the biosocial direction was not the leading one. It was lateral (and according to the above-described rules of evolution, a new form of complexity grew out of it). Unlike biological evolution, where the formation of integrative systems uniting biological organisms is more noticeably represented at lower levels (bacteria, colonies of unicellular organisms, etc.), social evolution can be represented as a constant growth of the size of systems from tens of individuals to billions of people in one system (e.g., Grinin & Korotayev, 2023). One can imagine the following sequence of growth of integrative systems: independent community—simple chiefdom and its analogues—complex chiefdom and its analogues—simple state and its analogues—complex state and its analogues—empire and its analogues.[7] But there are great many integrative systems in economics, religion, and other spheres of life. At the same time, the density of connections is constantly growing. Today we can talk about the World System. The trend toward strengthening integration (with fluctuations) can be traced throughout history.

[7] On chiefdom and state analogues, see, e.g., Bondarenko et al. (2002); Bondarenko and Korotayev (2003); Grinin et al. (2004); Grinin and Korotayev (2009b, 2011); Korotayev et al. (2000); Popov (2023).

2.5.3 About Information, Adaptability, and Organization

It was shown above in the chapter that one of the main differences between physical complex systems and complex adaptive systems is the processing of information by the latter. Indeed, the breakthrough in terms of information with the advent of life was very significant. However, information does not appear with the advent of life. Information as one of the functions of physical interactions and forces exists at all levels. The four main interactions carry information, otherwise there would be no interaction; light carries information, otherwise we would not be able to learn anything about the Universe (dark matter does not give us such an opportunity; there, information is reduced only to gravity). It is important that information interaction can only occur with a certain mutual correspondence of the properties of objects (Yankovsky 2000). Many molecules already have a kind of selectivity, or "preferences," in general—the ability to recognize "related" or necessary components (see, e.g., Eglinton, 1969).[8] At the nano level, fine-tuning of forms and fine mechanics are amazing forms of information verification, which clearly allows us to implement the friend–foe principle (see Melikhov, 2018).

Already at the level of chemical interactions, information begins to separate from the main function (Grinin, 2024; see also chapter "Chemical Evolution in Cosmos and Earth" [Grinin, 2026b] in this volume). Here the greater difference between the energetic and informational sides is already more clearly visible. It can be seen in such an example as catalytic interaction, which is used in the general theory of information (Yankovsky, 2000). Crystals also give certain signals (Lima-de-Faria, 1988). At the level of even not very complex biochemistry, information noticeably separates from the reactions themselves. Information, in essence, already becomes a code for the formation of proteins, etc. Robert M. Hazen and Michael L. Wong (2024) explore mineralogical evolution. They formulate the law of increasing functional information. They explain that the functional information of a system will increase (i.e., the system will evolve) if many different configurations of the system are subjected to selection for one or more functions. In this regard, they define evolution as a process of increasing information that is driven by selective pressures on a system of many interacting components. A key consequence is that selection for function causes an increase in system information—a temporally asymmetric arrow of increasing order that is not derivable from the second law of thermodynamics' increase in entropy. They then emphasize that mineral evolution, which considers Earth's changing diversity and distribution of minerals at a planetary scale through deep time, provides an important quantitative test of these ideas. These views are consistent with our understanding of information growth in

[8] There is a peculiar filter in crystals, which was discovered by V. M. Goldshmidt. Since different ions have different sizes (in particular, positively charged ions, anions, are much larger than negatively charged ones—cations), their crystallochemical structure forms a framework that plays the role of a peculiar geochemical filter, facilitating the differentiation of chemical elements by the size of their ions. Not all elements with the necessary valence can enter a specific crystallochemical structure, but only those whose ions have the corresponding radius size (Dobrovolsky, 2003: 20).

evolution. After all, if evolution is a process of information growth, then information cannot arise only at some phase, but must be there initially, but in an undeveloped form. However, it is with the emergence of life that the role of information increases qualitatively and colossally.

With the emergence of life, we can say that information becomes specialized. Viruses are almost pure information (see Grinin, 2025; as well as chapter "Viruses in Big History: Evolutionary Roles and Complex Systems" [Grinin, 2026a in this volume]), and if we imagine that protoviruses existed, then with them the development of information went very strongly. At the same time, their information exists in two forms—biochemical and physical. Let us also note: we are not talking about the emergence of information, but about its functional allocation.

Adaptability of behavior also exists in the physical world (gravity requires it). We should talk about the growth of multi-aspect adaptation of behavior. At the same time, the number of aspects is growing on the planets.

The organization (structure) includes such parameters as: (a) The number of basic elements (the emergence of new chemical elements during the life cycle of stars is indicative). (b) The concentration of scarce components—this is usually underestimated; however, only at a certain concentration (of different parameters, including energy) can breakthroughs be realized. (c) The number of parts of the structure—the density of their interaction and the number, strength, and density of direct and feedback connections in the system (here we are approaching control). (d) The level of basic elements (see above). But it seems that to these basic points it is worth adding control, since control is not covered by the information dimension (the concept of control also includes self-organization, self-regulation, the ability to learn, self-development, etc.). Control can be traced back to the emergence of large objects, when gravity began to control many objects in a certain dimension. However, each new level of complexity brings innovations to control.

2.5.4 Complexity and Evolution

Complexity growth as a trend is inextricably linked with evolution. However, only exceptionally rare evolutionary changes lead to increase in complexity—such evolutionary changes can be denoted as aromorphoses (Grinin et al., 2009, 2020; see also chapter "Biological and Social Phases of Big History: Similarities, Differences, and the Growth of Complexity" [Grinin et al., 2026] in this volume).

Such changes may take place only when evolutionary niches are filled, abrupt changes in the environment are observed, and exceptional conditions, which we discussed above, emerge. But without evolution, evolutionary search, and numerous trials and errors (see above on the rule of paying for evolutionary progress or the cost of complexity principle), there can be no movement to new levels and combinations of complexity (for detail, see chapter "Biological and Social Phases of Big History: Similarities, Differences, and the Growth of Complexity" [Grinin et al., 2026] in this volume).

Evolution actually means the destruction of stability and identification. From this point of view, at any stage and in any sphere of evolution the matter can be divided into two types: the one that is capable of self-preservation and the one that is capable of self-transformation (of course, these characteristics are manifested in different objects and systems in different proportions). In other words, one may speak about evolutionary and non-evolutionary matters. Within human society there also exist rather conservative elements and there still exist societies which are not quite prone to changes, and this phenomenon was even more strongly pronounced in the previous epochs. An average lifespan of a biological species is less than 10 million years. At the same time there are species which have endured for 200–300 million years. Thus, the presumable age of cyanobacteria is several billions of years, and they have not changed significantly since the Archean Eon. Thus, in biology one can observe species that have existed for hundreds of millions of years without radical changes as well as species that have given impetus to powerful typogenesis (i.e., the formation of new taxa), or species that are disappearing rapidly in biological terms within hundreds of thousands of years. One of the most important discoveries of the second half of the twentieth century was the discovery of the so-called dark matter whose abundance in the Universe far exceeds by mass the baryonic matter visible to us. But at the same time, it seems that dark matter is hardly able to evolve in comparison with baryonic matter.

At any phase, the evolving matter makes up the minority; thus, the baryonic matter according to some current views amounts for only 3–5% of the total mass of the Universe. It is amazing that this proportion is relevant even to human society in which, according to some reports, the number of innovators is also 3–5%. Actually, any object, system, or any form of matter can evolve, but this ability differs so much among various types and objects that it is reasonable to talk about the evolutionary rule of inability of some objects to evolutionary changes. In addition, evolutionary changes require a certain time rate of change of external conditions (or special conditions), which is far from always available. At the same time the inability to evolve means the ability of the matter to self-preservation. And in some cases, this turns into a clear advantage, while in others it becomes a disadvantage. Thus, one can see that the diversity of forms of existence (and development) in our Universe is also manifested in a hugely varying ability of different objects and forms of matter to change, evolve, and increase in complexity.[9] In short, existence fluctuates between stability and variability over a huge continuum.

Both characteristics—stability and variability—have great advantages, as well as disadvantages; they are both necessary for the existence of objects, species, and the world in general. This can also be observed in social evolution. There are more stable institutions which remain fundamentally unchanged when undergoing transformations; there are nations that have adapted to their way of life, so that

[9] One can assume that dark matter is not completely devoid of the ability to change; it only requires much more time than the baryonic matter for such changes. The stars also used to be considered unchanging.

they can exist without radical changes for a long time (millennia); and in some societies and situations the evident rapid changes lead to considerable qualitative transformations. We believe that such an inability is not genetic or race-related (although for the period of anthropogenesis it is quite possible), but depends on the certain societies' circumstances including natural and social environment, the role of factors, like the emergence of outstanding personalities, etc.

2.5.5 On Evolutionary Balance Between Breakthroughs and Stability

Evolutionary efficiency refers to a system's capacity for adaptation and its ability to generate breakthrough innovations (aromorphoses) that unlock new evolutionary pathways. It is about transcending the boundaries of established complexity niches. Within these niches, systems undergo both internal development and horizontal evolution, which in turn prepare them for breakthroughs. Parallel evolutionary paths, such as planetary evolution, also contribute to increased complexity, though some may lead to dead ends. The presence of more parallel, lateral, and additional evolutionary lines increases the potential for significant evolutionary breakthroughs toward higher complexity. Given that all objects vary in their capacity for self-preservation and change, a critical dimension of complexity lies in the balance between maintaining stability and the ability to evolve and increase in complexity.

For the field of Big History, establishing criteria to measure complexity levels is particularly vital. As previously noted, no single criterion—including one that can be expressed in a formula or quantitatively measured—is sufficient. Instead, we must consider a set of criteria, with the understanding that this list can remain open-ended. Nevertheless, identifying each new criterion advances our comprehension of complexity. Each successive level of complexity facilitates the formation of more developed organizational structures in relation to the fundamental levels of matter, energy, organization, and information. For instance, if we consider a quality like the breadth of distribution, we observe that less organized objects tend to be far more widespread than highly organized ones.

Some views on complexity and its evolutionary development, as well as its basic components (energy, organization, information, etc.) were discussed above. The main focus was on the development of complex physical and adaptive systems. It was also emphasized that the development of physical complexity in the Universe is a story of transformation spanning from the Big Bang to the present day, shaped by fundamental processes and forces; and also that higher complexity arises when systems not only dissipate energy but also process information and adapt behavior.

It is extremely important to consider more closely the growth of complexity as a certain trend of self-development of the Universe and its individual parts; as a set of general movements of various systems and objects, as well as forms of their interaction. This trend in some senses can be considered leading in Big History. Understanding, explaining, and analyzing this trend actually constitutes one of the main objects of Big History research.

As has been mentioned above, natural complexity emerges in two general categories: physical complex systems and complex adaptive systems (CAS). Physical complex systems exhibit self-organization and energy dissipation but lack information processing capacity. Examples include the cosmic web, spiral galaxies, hurricanes, and vortices in draining fluids. These systems are governed primarily by physical laws—such as gravity, thermodynamics, and fluid dynamics—and tend to stabilize into recurring, dynamic patterns (Prigogine, 1980; Bejan, 2000). In contrast, CAS are systems that gather, store, and process information to adapt over time. These include living organisms, ecosystems, and human societies. CAS are characterized by learning, memory, and purposeful behavior through feedback, replication, and selection processes (Holland, 1992).

2.6 Conclusion: Some Ideas for Further Study of Complexity

This chapter has identified complexity as a foundational and unifying theme within Big History, offering a comprehensive framework for understanding the dynamic evolution of the cosmos, life, and society. It emphasizes that complexity is not a static attribute but a dynamic process shaped by energy flows, information dynamics, and emergent organization across various scales and dimensions.

We began this chapter by exploring various definitions of complexity, yet this exploration is ongoing. Each new, comprehensive definition offers fresh perspectives on the development of complexity. Below, we propose our own definition.

Complexity is a category used to describe, evaluate, and compare the degree or level of organization and functionality within systems and objects. This applies across a wide spectrum, from individual systems to a conceptually defined level of evolutionary efficiency. As a scientific tool, complexity is crucial for understanding and ranking the evolutionary efficiency and position of objects within their specific domains or "kingdoms of complexity." Naturally, higher levels of complexity comprise sub-levels, and systematizing these could be a valuable endeavor.

The Universe's history can be viewed as a series of phase transitions and attractor shifts, each involving a significant reorganization of matter, energy, and information. From the simplicity of the early Universe, structure emerged through gravitational collapse and energy dissipation, laying the groundwork for stars, galaxies, and chemical diversity. The emergence of life on Earth further propelled this trajectory, shifting from static, energy-dissipative systems to adaptive, information-processing systems, with genetic variation, selection, and cultural innovation driving an accelerating growth in complexity.

Human evolution, characterized by advancements in cognition and cooperation, demonstrates this progression from biological to cultural and technological evolution, integrating diverse elements into increasingly nested hierarchies. This suggests that complexity is an emergent property of interconnected processes.

Beyond being an academic concept, complexity in Big History provides a practical framework for understanding our past and anticipating our future. It highlights that transitions involve trade-offs, feedback loops, and adaptive challenges, implying that our future will be shaped by both opportunities and uncertainties.

While a single definition for complexity remains elusive, it generally involves a high degree of interdependence among system components, nonlinearity, feedback, and emergent behavior. Key attributes include emergence, self-organization, nested hierarchies, nonlinearity, and adaptivity, which are evident in diverse phenomena from weather systems to civilizations.

The chapter also discusses the "fourth law of thermodynamics," suggesting that systems evolve to maximize entropy production while maintaining local order, and the Constructal Law, which posits that flow systems evolve to facilitate greater access to currents. These principles help explain how complexity can flourish under thermodynamic constraints and how emergent order arises not despite but because of the second law of thermodynamics under specific conditions.

Ultimately, this unified narrative, synthesizing insights from various scientific disciplines, positions complexity as the connective thread from the cosmos to human culture. It advocates for an integrative approach that bridges disciplines and scales, underscoring that the Universe's history is a story of constant transformation, driven by fundamental principles that continue to shape the evolution of complexity today.

References

Alcott, B. (2005). Jevons' paradox. *Ecological Economics, 54*(1), 9–21. https://doi.org/10.1016/j.ecolecon.2005.03.020

Armitage, P. J. (2011). *Astrophysics of planetary formation*. Cambridge University Press.

Arrighi, G. (1994). *The long twentieth century: Money, power, and the origins of our times*. Verso.

Ashby, R. (1956). *An introduction to cybernetics*. Chapman and Hall.

Atkins, P. (2018). *Conjuring the universe: The origins of the laws of nature*. Oxford University Press.

Ayres, R. U. (2006). *Turning point: The end of the growth paradigm*. Earthscan.

Azarian, B. (2022). *The romance of reality: How the universe organizes itself to create life, consciousness, and cosmic complexity*. BenBella Books.

Bak, P., Tang, C., & Wiesenfeld, K. (1988). Self-organized criticality. *Physical Review A, 38*(1), 364–374. https://doi.org/10.1103/PhysRevA.38.364

Barabási, A. L. (2002). *Linked: The new science of networks*. Perseus Publishing.

Bar-Yam, Y. (1997). *Dynamics of complex systems*. Perseus Books.

Bar-Yam, Y. (2004). *Making things work: Solving complex problems in a complex world*. NECSI Knowledge Press.

Baskin, K. (2026). Myth, complexity, and Big History: Rethinking our shared map for the future. In: D. J. LePoire, L. Grinin, & A. Korotayev (Eds.), *Complexity in universal evolution: A Big History perspective* (pp. 469–486). Springer Nature (in this volume). https://doi.org/10.1007/978-3-032-20516-2_16

Bejan, A. (2000). *Shape and structure, from engineering to nature*. Cambridge University Press.

Bettencourt, L. M., Lobo, J., Helbing, D., Kühnert, C., & West, G. B. (2007). Growth, innovation, scaling, and the pace of life in cities. *Proceedings of the National Academy of Sciences, 104*(17), 7301–7306. https://doi.org/10.1073/pnas.0610172104

Binney, J., & Tremaine, S. (2008). *Galactic dynamics* (2nd ed.). Princeton University Press.

Bondarenko, D., & Korotayev, A. (2003). Early state in cross-cultural perspective: A statistical re-analysis of Henri. J. M. Claessen's Database. *Cross-Cultural Research, 37*(1), 104–131. https://doi.org/10.1177/1069397102238924

Bondarenko, D. M., Grinin, L. E., & Korotayev, A. V. (2002). Alternative pathways of social evolution. *Social Evolution & History, 1*(1), 54–79.

Camazine, S., Deneubourg, J. L., Franks, N. R., Sneyd, J., Theraulaz, G., & Bonabeau, E. (2001). *Self-organization in biological systems*. Princeton University Press.

Chaisson, E. J. (2001). *Cosmic evolution: The rise of complexity in nature*. Harvard University Press.

Chaisson, E. J. (2011). Energy rate density as a complexity metric and evolutionary driver. *Complexity, 16*(3), 27–40. https://doi.org/10.1002/cplx.20323

Christian, D., Stokes Brown, C., & Benjamin, C. (2014). *Big History: Between nothing and everything*. McGraw Hill.

Cook, T. D., & Campbell, D. T. (1979). *Quasi-experimentation: Design & analysis issues for field settings*. Houghton Mifflin.

Cortês, M., Kauffman, S. A., Liddle, A. R., & Smolin, L. (2024). Biocosmology: Towards the birth of a new science. *arXiv:2204.09378*.

d'Errico, F., Henshilwood, C., Lawson, G., Vanhaeren, M., Tillier, A. M., Soressi, M., et al. (2003). Archaeological evidence for the emergence of language, symbolism, and music—An alternative multidisciplinary perspective. *Journal of World Prehistory, 17*, 1–70. https://doi.org/10.1023/A:1023980201043

Dobrovolsky, V. V. (2003). *Fundamentals of biogeochemistry*. Academy.

Donald, M. (1991). *Origins of the modern mind: Three stages in the evolution of culture and cognition*. Harvard University Press.

Dreher, A. (2006). Does globalization affect growth? Evidence from a new index of globalization. *Applied Economics, 38*(10), 1091–1110. https://doi.org/10.1080/00036840500392078

Dunbar, R. I. M. (1993). Coevolution of neocortical size, group size and language in humans. *Behavioral and Brain Sciences, 16*(4), 681–694.

Eglinton, G. (1969). Organic geochemistry. The organic chemist's approach. In G. Eglinton & M. T. J. Murphy (Eds.), *Organic geochemistry* (pp. 20–73). Springer. https://doi.org/10.1007/978-3-642-87734-6_2

Eigen, M. (1971). Selforganization of matter and the evolution of biological macromolecules. *Naturwissenschaften, 58*(10), 465–523.

Eigen, M., & Schuster, P. (1979). *The hypercycle: A principle of natural self-organization*. Springer.

Ekstig, B. (1994). Condensation of developmental stages and evolution. *BioScience, 44*(3), 158–164.

Ekstig, B. (2012). Superexponentially accelerating evolution. *World Futures, 68*(1), 40–48. https://doi.org/10.1080/02604027.2011.653263

Ekstig, B. (2015). Complexity, natural selection and the evolution of life and humans. *Foundations of Science, 20*, 175–187. https://doi.org/10.1007/s10699-014-9358-y

Ekstig, B. (2017). Complexity, progress, and hierarchy in evolution. *World Futures, 73*(7), 457–472. https://doi.org/10.1080/02604027.2017.1357967

Elston, G. N. (2003). Cortex, cognition and the cell: New insights into the pyramidal neuron and prefrontal function. *Cerebral Cortex, 13*(11), 1124–1138.

Fedorenko, E., & Varley, R. (2016). Language and thought are not the same thing: Evidence from neuroimaging and neurological patients. *Annals of the NY Academy of Sciences, 1369*(1), 132–153. https://doi.org/10.1111/nyas.13046

Feynman, R. P., Leighton, R. B., & Sands, M. (2010). *The Feynman lectures on physics* (Vol. 2). Basic Books.

Fieguth, P. (2017). *An introduction to complex systems*. Springer.

Fox, R. F. (1988). *Energy and the evolution of life*. W.H. Freeman.

Frankopan, P. (2017). *The silk roads: A new history of the world*. Vintage.

Friston, K. (2010). The free-energy principle: A unified brain theory? *Nature Reviews Neuroscience, 11*(2), 127–138. https://doi.org/10.1038/nrn2787

Fuchs-Kittowski, K., & Krüger, K. (1997). Evolution and complexity: Some philosophical and scientific considerations. *World Futures, 49*(1–4), 153–166.

Gardner, M. (1970). Mathematical games: The fantastic combinations of John Conway's new solitaire game "life". *Scientific American, 223*(4), 120–123.

Gell-Mann, M. (1994). *The quark and the jaguar: Adventures in the simple and the complex*. W. H. Freeman.

Gimpel, J. (1976). *The medieval machine: The industrial revolution of the Middle Ages*. Holt, Rinehart and Winston.

Gleick, J. (1987). *Chaos: Making a new science*. Viking.

Graeber, D. (2011). *Debt: The first 5000 years*. Melville House.

Grinin, L. (2013). *Macrohistory and globalization*. Uchitel Publishing House.

Grinin, L. (2014). The star-galaxy era of Big History in the light of universal evolutionary principles. In L. Grinin, D. Baker, E. Quaedackers, & A. Korotayev (Eds.), *Teaching & researching Big History: Exploring a new scholarly field* (pp. 163–187). Uchitel.

Grinin, L. E. (2018). Evolution of the early solar system in terms of Big History and universal evolution. *Journal of Big History, 2*(1), 15–26.

Grinin, L. E. (2019). Was there a Big Bang? *Journal of Big History, 3*(1), 51–59. https://doi.org/10.22339/jbh.v3i1.3130

Grinin, L. (2024). Chemical evolution in Big History. *Journal of Big History, 7*(2), 97–105. https://doi.org/10.22339/jbh.v7i2.7208

Grinin, A. (2025). Viruses and evolution: The role of viruses in Big History. *Journal of Big History, 8*(1), 51–56. https://doi.org/10.22339/jbh.v8i1.8108

Grinin, A. (2026a). Viruses in Big History: Evolutionary roles and complex systems. In: D. LePoire, L. Grinin, & A. Korotayev (Eds.), *Complexity in universal evolution: A Big History perspective* (pp. 247–256). Springer Nature (in this volume). https://doi.org/10.1007/978-3-032-20516-2_8

Grinin, L. (2026b). Chemical evolution in cosmos and Earth. In: D. LePoire, L. Grinin, & A. Korotayev (Eds.), *Complexity in universal evolution: A Big History perspective* (pp. 193–210). Springer Nature (in this volume). https://doi.org/10.1007/978-3-032-20516-2_7

Grinin, L., & Grinin, A. (2020). The cybernetic revolution and the future of technologies. In A. V. Korotayev & D. LePoire (Eds.), *The 21st century singularity and global futures. A Big History perspective* (pp. 377–396). Springer Nature. https://doi.org/10.1007/978-3-030-33730-8_17

Grinin, L., & Grinin, A. (2025). Big History periodization. Complexity, directions and phase characteristics. In D. J. LePoire, L. Grinin, & A. Korotayev (Eds.), *Navigating complexity in Big History—Exploring periodization across cosmic and biosocial dimensions* (pp. 319–342). Springer Nature. https://doi.org/10.1007/978-3-031-85410-1_10

Grinin, L., & Korotayev, A. (2009a). Social macroevolution: Growth of the world system integrity and a system of phase transitions. *World Futures, 65*(7), 477–506. https://doi.org/10.1080/02604020902733348

Grinin, L., & Korotayev, A. (2009b). The epoch of the initial politogenesis. *Social Evolution and History, 8*(1), 52–91.

Grinin, L., & Korotayev, A. (2011). Chiefdoms and their analogues: Alternatives of social evolution at the societal level of medium cultural complexity. *Social Evolution & History, 10*(1), 276–335.

Grinin, L., & Korotayev, A. (2012). Does "Arab Spring" mean the beginning of world system reconfiguration? *World Futures, 68*(7), 471–505. https://doi.org/10.1080/02604027.2012.697836

Grinin, L., & Korotayev, A. (2015). *Great divergence and great convergence. A global perspective*. Springer Nature. https://doi.org/10.1007/978-3-319-17780-9

Grinin, L., & Korotayev, A. (2023). Social evolution: Theoretical aspects and general outlines. *Social Evolution & History, 22*(2), 111–158. https://doi.org/10.30884/seh/2023.02.07

Grinin, L., & Korotayev, A. (2025). Possible methodological ideas for periodizing Big History. In D. J. LePoire, L. Grinin, & A. Korotayev (Eds.), *Navigating complexity in Big History—Exploring periodization across cosmic and biosocial dimensions* (pp. 49–71). Springer Nature. https://doi.org/10.1007/978-3-031-85410-1_3

Grinin, L., Carneiro, R., Bondarenko, D., Kradin, N., & Korotayev, A. (Eds.). (2004). *The early state, its alternatives and analogues*. Uchitel.

Grinin, L., Markov, A., & Korotayev, A. (2009). Aromorphoses in biological and social evolution: Some general rules for biological and social forms of macroevolution. *Social Evolution & History, 8*(2), 6–50.

Grinin, L., Markov, A., & Korotayev, A. (2013). On similarities between biological and social evolutionary mechanisms: mathematical modeling. *Cliodynamics, 4*(2), 185–228. https://doi.org/10.21237/C7clio4221334

Grinin, L., Korotayev, A., & Tausch, A. (2016). *Economic cycles, crises, and the global periphery*. Springer Nature. https://doi.org/10.1007/978-3-319-41262-7

Grinin, L., Korotayev, A., & Markov, A. (2020). Big History and macro-evolution. Evolutionary principles and mechanisms at biological and social phases of the Big History. In C. Benjamin, E. Quaedackers, & D. Baker (Eds.), *The Routledge companion to Big History* (pp. 72–105). Routledge.

Grinin, L., Grinin, A., & Korotayev, A. (2024). *Cybernetic revolution and global aging. Humankind on the way to cybernetic society, or the next hundred years*. Springer Nature. https://doi.org/10.1007/978-3-031-56764-3

Grinin, L. E., Markov, A., & Korotayev, A. V. (2025). Biological and social phases of Big History and universal evolution: Similarities, differences, and the growth of complexity. *Journal of Big History, 9*(3), 1–51. https://doi.org/10.22339/jbh.v8i3.8301

Grinin, L., Markov, A., & Korotayev, A. (2026). Biological and social phases of Big History and complexity growth. In: D. LePoire, L. Grinin, & A. Korotayev (Eds.), *Complexity in universal evolution: A Big History perspective* (pp. 283–356). Springer Nature. https://doi.org/10.1007/978-3-032-20516-2_12

Gunderson, L. H., & Holling, C. S. (2002). *Panarchy: Understanding transformations in human and natural systems*. Island Press.

Haken, H. (1988). *Information and self-organization: A macroscopic approach to complex systems*. Springer.

Hausmann, R., Hidalgo, C. A., Bustos, S., Coscia, M., & Simoes, A. (2014). *The atlas of economic complexity: Mapping paths to prosperity*. MIT Press.

Hazen, R. M., & Wong, M. L. (2024). Open-ended versus bounded evolution: Mineral evolution as a case study. *PNAS Nexus, 3*(7), pgae248. https://doi.org/10.1093/pnasnexus/pgae248

Hendrickson, J. B., Huang, P., & Toczko, A. G. (1987). Molecular complexity: A simplified formula adapted to individual atoms. *Journal of Chemical Engineering & Computer Sciences, 27*, 63–67.

Henrich, J. (2016). *The secret of our success: How culture is driving human evolution, domesticating our species, and making us smarter*. Princeton University Press.

Hidalgo, C. A., & Hausmann, R. (2009). The building blocks of economic complexity. *Proceedings of the National Academy of Sciences, 106*(26), 10570–10575.

Hofstadter, D. R. (2008). *I am a strange loop*. Basic Books.

Holland, J. H. (1992). *Adaptation in natural and artificial systems*. MIT Press.

Jablonski, N. G. (2006). *Skin: A natural history*. University of California Press.

Jantsch, E. (1980). *The self-organizing universe: Scientific and human implications of the emerging paradigm of evolution*. Pergamon.

Kapitza, S. (1996). The phenomenological theory of world population growth. *Physics-uspekhi, 39*(1), 57–71. https://doi.org/10.1070/PU1996v039n01ABEH000127

Kauffman, S. A. (1993). *The origins of order: Self-organization and selection in evolution*. Oxford University Press.

Kauffman, S. A. (2000). *Investigations*. Oxford University Press.

Kolmogorov, A. N. (1941). The local structure of turbulence in incompressible viscous fluid for very large Reynolds numbers. *Doklady Akademii Nauk SSSR, 30*(4), 301–305.

Kolmogorov, A. N. (1965). Three approaches to the quantitative definition of information. *Problems of Information Transmission, 1*(1), 1–7.

Kolmogorov, A. (1998 [1963]). On tables of random numbers. *Theoretical Computer Science, 207*(2), 387–395. https://doi.org/10.1016/S0304-3975(98)00075-9

Korotayev, A. (2004). *World religions and social evolution of the old world Oikumene civilizations: A cross-cultural perspective*. The Edwin Mellen Press.

Korotayev, A. V. (2005). A compact macromodel of world system evolution. *Journal of World-Systems Research, 11*(1), 79–93. https://doi.org/10.5195/jwsr.2005.401

Korotayev, A. (2006a). The World System history periodization and mathematical models of socio-historical processes. *History & Mathematics, 1*, 39–98.

Korotayev, A. (2006b). The World System urbanization dynamics: A quantitative analysis. *History & Mathematics, 2*, 44–62.

Korotayev, A. (2007). Compact mathematical models of world system development, and how they can help us to clarify our understanding of globalization processes. In G. Modelski, T. Devezas, & W. R. Thompson (Eds.), *Globalization as evolutionary process: Modeling global change* (pp. 133–160). Routledge. https://doi.org/10.4324/9780203937297

Korotayev, A. (2018). The 21st century singularity and its Big History implications: A re-analysis. *Journal of Big History, 2*(3), 73–119. https://doi.org/10.22339/jbh.v2i3.2320

Korotayev, A. (2020). The 21st century Singularity in the Big History perspective. A re-analysis. In A. V. Korotayev & D. LePoire (Eds.), *The 21st century Singularity and global futures. A Big History perspective* (pp. 19–75). Springer Nature. https://doi.org/10.1007/978-3-030-33730-8_2

Korotayev, A. (2025a). Comparing and contrasting Big History Singularity trends of the Big Bang and terrestrial evolution. In D. J. LePoire, L. Grinin, & A. Korotayev (Eds.), *Navigating complexity in Big History. Exploring periodization across cosmic and biosocial dimensions* (pp. 101–157). Springer Nature. https://doi.org/10.1007/978-3-031-85410-1_5

Korotayev, A. (2025b). Toward periodization of the biosocial era of the Big History: A reanalysis. *Journal of Big History, 8*(2), 1–37. https://doi.org/10.22339/jbh.v8i2.8201

Korotayev, A. (2026). Human evolution in the complexity growth perspective. Toward periodization of the Big History biosocial era. In: D. LePoire, L. Grinin, & A. Korotayev (Eds.), *Complexity in universal evolution: A Big History perspective* (pp. 359–410). Springer Nature (in this volume). https://doi.org/10.1007/978-3-032-20516-2_13

Korotayev, A., & LePoire, D. J. (Eds.). (2020). *The 21st century singularity and global futures: A big history perspective*. Springer Nature. https://doi.org/10.1007/978-3-030-33730-8

Korotayev, A. V., & LePoire, D. J. (2025). Complexity growth patterns in the stelliferous era of Big History. Preliminary observations. *Journal of Big History, 8*(1), 1–12. https://doi.org/10.22339/jbh.v8i1.8102

Korotayev, A. V., & LePoire, D. J. (2026). Complexity growth patterns in the stelliferous era of Big History. Preliminary analysis. In: D. LePoire, L. Grinin, & A. Korotayev (Eds.), In: D. LePoire, L. Grinin, & A. Korotayev (Eds.), *Complexity in universal evolution: A Big History perspective* (pp. 151–172). Springer Nature. https://doi.org/10.1007/978-3-032-20516-2_3

Korotayev, A., & Markov, A. (2014). Mathematical modeling of biological and social phases of Big History. In L. Grinin et al. (Eds.), *Teaching & researching Big History: Exploring a new scholarly field* (pp. 188–219). Uchitel.

Korotayev, A., & Zinkina, J. (2017). Systemic boundary issues in the light of mathematical modeling of world-system evolution. *Journal of Globalization Studies, 8*(1), 78–96.

Korotayev, A., Kradin, N., de Munck, V., & Lynsha, V. (2000). Alternativity of social evolution: Introductory notes. In N. N. Kradin, A. V. Korotayev, D. M. Bondarenko, V. de Munck, & P. K. Wason (Eds.), *Alternatives of social evolution* (pp. 12–51). Far Eastern Branch of the Russian Academy of Sciences.

Korotayev, A., Malkov, A., & Khaltourina, D. (2006a). *Introduction to social macrodynamics. Compact macromodels of the world system growth.* KomKniga/URSS.
Korotayev, A., Malkov, A., & Khaltourina, D. (2006b). *Introduction to social macrodynamics. Secular cycles and millennial trends.* KomKniga/URSS.
Korotayev, A. V., Markov, A., & Grinin, L. E. (2015). Modeling of biological and social phases of Big History. In L. E. Grinin & A. V. Korotayev (Eds.), *Evolution: From Big Bang to nanorobots.* Uchitel. www.sociostudies.org/almanac/articles/modeling_of_biological_and_social_phases_of_big_history/
Korotayev, A., Zinkina, J., & Andreev, A. (2016). Secular cycles and millennial trends. *Cliodynamics, 7*(2), 204–216. https://doi.org/10.21237/C7clio7230484
Koschmieder, E. L. (1993). *Bénard cells and Taylor vortices.* Cambridge University Press.
Krakauer, D. (2019). *Worlds hidden in plain sight: The evolving idea of complexity at the Santa Fe Institute, 1984–2019.* SFI Press.
Kuhn, T. S. (1962). *The structure of scientific revolutions.* University of Chicago Press.
Kurzweil, R. (2005). *The singularity is near: When humans transcend biology.* Viking.
Ladyman, J., Lambert, J., & Wiesner, K. (2013). What is a complex system? *European Journal for Philosophy of Science, 3*(1), 33–67. https://doi.org/10.1007/s13194-012-0056-8
Lake, B. M., Ullman, T. D., Tenenbaum, J. B., & Gershman, S. J. (2017). Building machines that learn and think like people. *Behavioral and Brain Sciences, 40,* e253. https://doi.org/10.1017/S0140525X16001837
Lane, N. (2009). *Life ascending: The ten great inventions of evolution.* W. W. Norton.
Lane, N., & Martin, W. (2012). The origin of membrane bioenergetics. *Cell, 151*(7), 1406–1416. https://doi.org/10.1016/j.cell.2012.11.050
Langton, C. G. (1990). Computation at the edge of chaos. *Physica D: Nonlinear Phenomena, 42*(1–3), 12–37.
Lenton, T. M., & van Oijen, M. (2002). Gaia as a complex adaptive system. *Philosophical Transactions of the Royal Society of London.* Series B: *Biological Sciences, 357*(1421), 683–695. https://doi.org/10.1098/rstb.2001.1014
LePoire, D. J. (2004). A 'perfect storm' of social and technological transitions? *Futures Research Quarterly, 20*(3), 25–40.
LePoire, D. J. (2005). Application of logistic analysis to the history of physics. *Technological Forecasting and Social Change, 72*(4), 471–479. https://doi.org/10.1016/S0040-1625(03)00044-1
LePoire, D. J. (2010). Long-term population, productivity, and energy use trends in the sequence of leading capitalist nations. *Technological Forecasting and Social Change.* https://doi.org/10.1016/j.techfore.2009.07.003
LePoire, D. J. (2019). Size relationships of Big History objects: From the universe to the atomic nucleus. *Journal of Big History, 4*(1), 44–48. https://doi.org/10.22339/jbh.v4i1.4160
LePoire, D. J. (2020a). Exploring the singularity concept within Big History. In A. V. Korotayev & D. LePoire (Eds.), *The 21st century Singularity and global futures. A Big History perspective* (pp. 77–97). Springer Nature. https://doi.org/10.1007/978-3-030-33730-8_3
LePoire, D. J. (2020b). *Time patterns in Big History. Cycles, fractals, waves, transitions, and singularities.* David LePoire Consulting, LLC.
LePoire, D. J. (2023). Exploring phenomenological models for societal and technological transitions of the neolithic revolution and early civilization formation. *Cliodynamics, 14*(1), 5–26. https://doi.org/10.21237/C7clio14162728
LePoire, D. J. (2026). Energy flow and emergence of complexity in the Universe's structure. In: D. LePoire, L. Grinin, & A. Korotayev (Eds.), In: D. LePoire, L. Grinin, & A. Korotayev (Eds.), *Complexity in universal evolution – A Big History perspective* (pp. 173–190). Springer Nature. https://doi.org/10.1007/978-3-032-20516-2_4
LePoire, D. J., & Chandrankunnel, M. (2020). Energy flow trends in Big History. In A. V. Korotayev & D. J. LePoire (Eds.), *The 21st century singularity and global futures: A Big*

History perspective (pp. 185–200). Springer Nature. https://doi.org/10.1007/978-3-030-33730-8_9

LePoire, D. J., Grinin, L., & Korotayev, A. (2025a). Navigating complexity in Big History—Exploring periodization across cosmic and biosocial dimensions: An introduction. In D. J. LePoire, L. Grinin, & A. Korotayev (Eds.), *Navigating complexity in Big History. Exploring periodization across cosmic and biosocial dimensions* (pp. 1–27). Springer Nature. https://doi.org/10.1007/978-3-031-85410-1_1

LePoire, D. J., Grinin, L., & Korotayev, A. (2025b). Discussion: Cross-cutting topics and a variety of periodizations. In D. J. LePoire, L. Grinin, & A. Korotayev (Eds.), *Navigating complexity in Big History. Exploring periodization across cosmic and biosocial dimensions* (pp. 471–494). Springer Nature. https://doi.org/10.1007/978-3-031-85410-1_15

LePoire, D. J., Grinin, L., & Korotayev, A. (2025c). Conclusions: Exploring Big History periodizations across cosmic and biosocial dimensions with ideas for further research. In D. J. LePoire, L. Grinin, & A. Korotayev (Eds.), *Navigating complexity in Big History. Exploring periodization across cosmic and biosocial dimensions* (pp. 495–525). Springer Nature. https://doi.org/10.1007/978-3-031-85410-1_16

LePoire, D. J., Grinin, L., & Korotayev, A. (2025d). Complexity in Big History. An introductory exploration. *Journal of Big History, 8*(3), 98–139. https://doi.org/10.22339/jbh.v8i3.8304

LePoire, D. J., Grinin, L., & Korotayev, A. (2026). A Big History perspective on complexity in universal evolution. Conclusions. In: D. J. LePoire, L. Grinin, & A. Korotayev (Eds.), *Complexity in universal evolution: A Big History perspective* (pp. 585–608). Springer Nature (in this volume). https://doi.org/10.1007/978-3-032-20516-2_22

Lieberman, P. (2007). The evolution of human speech: Its anatomical and neural bases. *Current Anthropology, 48*(1), 39–66. https://doi.org/10.1086/509092

Lima-de-Faria, A. (1988). *Evolution without selection: Form and function by autoevolution*. Elsevier.

Lin, D. N. C. (2008). The genesis of planets. *Scientific American, 298*(5), 50–59. https://doi.org/10.1038/scientificamerican0508-50

Lloyd, S. (2001). Measures of complexity: A nonexhaustive list. *IEEE Control Systems Magazine, 21*(4), 7–8.

Lloyd, S., & Pagels, H. (1988). Complexity as thermodynamic depth. *Annals of Physics, 188*(1), 186–213. https://doi.org/10.1016/0003-4916(88)90094-2

Lorenz, E. N. (1963). Deterministic nonperiodic flow. *Journal of the Atmospheric Sciences, 20*(2), 130–141. https://doi.org/10.1175/1520-0469(1963)020

Lovelock, J. E. (1979). *Gaia: A new look at life on Earth*. Oxford University Press.

Malkov, A., Zinkina, J., & Korotayev, A. (2012). The origins of dragon-kings and their occurrence in society. *Physica A: Statistical Mechanics and Its Applications, 391*(21), 5215–5229. https://doi.org/10.1016/j.physa.2012.05.045

Mandelbrot, B. B. (1982). *The fractal geometry of nature*. W. H. Freeman and Company.

Margulis, L. (1970). *Origin of eukaryotic cells*. Yale University Press.

Margulis, L., & Lovelock, J. E. (1989). Gaia and geognosy. *Global Ecology: towards a science of the biosphere*, 1–29.

Markov, A., & Korotayev, A. (2007). Phanerozoic marine biodiversity follows a hyperbolic trend. *Palaeoworld, 16*(4), 311–318. https://doi.org/10.1016/j.palwor.2007.01.002

Markov, A., & Korotayev, A. (2008). Hyperbolic growth of marine and continental biodiversity through the Phanerozoic and community evolution. *Zhurnal Obshchei Biologii, 69*(3), 175–194.

Maynard Smith, J., & Szathmáry, E. (1995). *The major transitions in evolution*. Oxford University Press.

Mayr, E. (1991). *One long argument: Charles Darwin and the genesis of modern evolutionary thought*. Harvard University Press.

Meadows, D. (2008). *Thinking in systems*. Chelsea Green Publishing.

Meadows, D. H., Meadows, D. L., Randers, J., & Behrens, W. W., III. (1972). *The limits to growth*. Universe Books.

Mefteh, W., & Mejri, M.-A. (2020). Complex systems modeling overview about techniques and models and the evolution of artificial intelligence. In Á. Rocha, H. Adeli, L. P. Reis, S. Costanzo, I. Orovic, & F. Moreira (Eds.), *Trends and innovations in information systems and technologies* (pp. 678–688). Springer.

Melikhov, I. V. (2018). *Fiziko-khimicheskaya evolyutsiya tverdogo veshchestva*. BINOM. Laboratoriya znaniy.

Miller, J. H., & Page, S. E. (2007). *Complex adaptive systems: An introduction to computational models of social life*. Princeton University Press.

Misra, S., Adewumi, A., Fernandez-Sanz, L., & Damasevicius, R. (2018). A suite of object oriented cognitive complexity metrics. *IEEE Access, 6*, 8782–8796.

Mitchell, M. (2009). *Complexity: A guided tour*. Oxford University Press.

Mo, H., van den Bosch, F. C., & White, S. (2010). *Galaxy formation and evolution*. Cambridge University Press.

Modis, T. (2002). Forecasting the growth of complexity and change. *Technological Forecasting and Social Change, 69*, 377–404.

Moiseyev, N. (2000). *Man and the noosphere*. Progress.

Montmerle, T., Augereau, J.-C., Chaussidon, M., Gounelle, M., Marty, B., & Morbidelli, A. (2006). Solar system formation and early evolution: The first 100 million years. *Earth, Moon, and Planets, 98*(1–4), 39–95. https://doi.org/10.1007/s11038-006-9087-5

Niele, F. (2005). *Energy: Engine of evolution*. Elsevier.

North, D. C., & Thomas, R. P. (1973). *The rise of the western world: A new economic history*. Cambridge University Press.

Odling-Smee, J., Laland, K. N., & Feldman, M. W. (2003). *Niche construction: The neglected process in evolution*. Princeton University Press.

Odum, H. T. (2001). Energy, hierarchy and money. In *Understanding complexity* (pp. 139–148). Springer US.

Page, D. N. (2003). Anthropic estimates of the charge and mass of the proton. *arXiv preprint*.

Page, S. E. (2011). *Diversity and complexity: Primers in complex systems*. Princeton University Press.

Panov, A. (2005). Scaling law of the biological evolution and the hypothesis of the self-consistent Galaxy origin of life. *Advances in Space Research, 36*(2), 220–225. https://doi.org/10.1016/j.asr.2005.03.001

Panov, A. (2020). Singularity of evolution and post-singular development in the Big History perspective. In A. V. Korotayev & D. J. LePoire (Eds.), *The 21st century singularity and global futures* (pp. 439–465). Springer Nature. https://doi.org/10.1007/978-3-030-33730-8_20

Popov, V. A. (2023). Annual festivals and potlatch-like forms of redistribution in early political formations of Upper Guinea. *Social Evolution & History, 22*(2), 18–31. https://doi.org/10.30884/seh/2023.02.02

Press, W. H., & Lightman, A. P. (1983). Dependence of macrophysical phenomena on the values of the fundamental constants. *Philosophical Transactions of the Royal Society of London. Series A, Mathematical and Physical Sciences, 310*(1512), 323–336.

Prigogine, I. (1980). *From being to becoming: Time and complexity in the physical sciences*. W. H. Freeman.

Prigogine, I., & Stengers, I. (1984). *Order out of chaos: Man's new dialogue with nature*. Bantam Books.

Prokopenko, M., Boschetti, F., & Ryan, A. (2009). An information-theoretic primer on complexity, self-organization, and emergence. *Complexity, 15*(1), 11–28. https://doi.org/10.1002/cplx.20249

Richerson, P. J., & Boyd, R. (2005). *Not by genes alone: How culture transformed human evolution*. University of Chicago Press.

Rodrigue, B. H. (2023). The bedrock of survival: Humanity, history, and adaptation. *Journal of Globalization Studies, 14*(1), 137–169. https://doi.org/10.30884/jogs/2023.01.09

Roman, S., & Bertolotti, F. (2023). Global history, the emergence of chaos and inducing sustainability in networks of socio-ecological systems. *PLoS One, 18*(11), e0293391. https://doi.org/10.1371/journal.pone.0293391

Sadovnichy, V., Akaev, A., Ilyin, I., Malkov, S., Grinin, L., Aleshkovski, I., & Korotayev, A. (2023). Macrohistorical approach. In V. Sadovnichy et al. (Eds.), *Reconsidering the limits to growth. A report to the Russian Association of the Club of Rome* (pp. 17–34). Springer. https://doi.org/10.1007/978-3-031-34999-7_2

Salthe, S. N. (1985). *Evolving hierarchical systems: Their structure and representation*. Columbia University Press.

Salthe, S. N. (1993). *Development and evolution: Complexity and change in biology*. MIT Press.

Schneider, E. D., & Kay, J. J. (1994). Life as a manifestation of the second law of thermodynamics. *Mathematical and Computer Modelling, 19*(6–8), 25–48.

Schneider, E. D., & Sagan, D. (2006). *Into the cool: Energy flow, thermodynamics, and life*. University of Chicago Press.

Schumpeter, J. A. (1942). *Capitalism, socialism and democracy*. Harper & Brothers.

Shandarin, S. F., et al. (2005). The shape of the dark matter cosmic web. *arXiv preprint*.

Simon, H. A. (1962). The architecture of complexity. *Proceedings of the American Philosophical Society, 106*, 62–76.

Sipser, M. (2012). *Introduction to the theory of computation*. Cengage Learning.

Smaers, J. B., Turner, A. H., Gómez-Robles, A., & Sherwood, C. C. (2017). A cerebellar substrate for cognition evolved multiple times independently in mammals. *eLife, 6*, e29265. https://doi.org/10.7554/eLife.35696

Smil, V. (2008). *Energy in nature and society: General energetics of complex systems*. The MIT Press.

Smith, A. (1776). *An inquiry into the nature and causes of the wealth of nations*. W. Strahan and T. Cadell.

Smolin, L. (1997). *The life of the cosmos*. Oxford University Press.

Snooks, G. D. (2005). Big History or big theory? Uncovering the laws of life. *Social Evolution and History, 4*(1), 160–188.

Sornette, D. (2009). Dragon-kings, black swans and the prediction of crises. *International Journal of Terraspace Science and Engineering, 2*(1), 1–18.

Spier, F. (1996). *The structure of Big History from the big bang until today*. Amsterdam University Press.

Spier, F. (2010). *Big History and the future of humanity*. Wiley-Blackwell.

Spier, F. (2011). Complexity in Big History. *Cliodynamics, 2*(1), 146–166. https://doi.org/10.21237/C7clio21213

Springel, V., White, S. D., Jenkins, A., Frenk, C. S., Yoshida, N., Gao, L., et al. (2005). Simulations of the formation, evolution and clustering of galaxies and quasars. *Nature, 435*, 629–636. https://doi.org/10.1038/nature03597

Stiglitz, J. E., Sen, A., & Fitoussi, J.-P. (2009). *Report by the commission on the measurement of economic performance and social progress*. Retrieved from https://ec.europa.eu/eurostat/documents/8131721/8131772/Stiglitz-Sen-Fitoussi-Commission-report.pdf

Tainter, J. A. (1988). *The collapse of complex societies*. Cambridge University Press.

Tainter, J. A. (1996). Complexity, problem solving, and sustainable societies. In R. Costanza, O. Segura, & J. Martinez-Alier (Eds.), *Getting down to earth: Practical applications of ecological economics* (pp. 61–76). Island Press.

Tattersall, I. (2012). *Masters of the planet: The search for our human origins*. Palgrave Macmillan.

Teilhard de Chardin, P. (1987). *The phenomenon of man*. Harper & Row. (Original work published 1955).

Thurner, S., Hanel, R., & Klimek, P. (2018). *Introduction to the theory of complex systems*. Oxford University Press.

Tononi, G. (2008). Consciousness as integrated information: A provisional manifesto. *The Biological Bulletin, 215*(3), 216–242. https://doi.org/10.2307/25470707

Tsiolkovsky, K. (1964). *The will of the universe: Unknown intelligent forces*. NASA translation (original Russian publication 1928).
Turchin, P. (2016). *Ultrasociety: How 10,000 years of war made humans the greatest cooperators on Earth*. Beresta Books.
Ulam, S. (1958). Tribute to John von Neumann. *Bulletin of the American Mathematical Society, 64*(3), 1–49.
Valley, J. W., Peck, W. H., King, E. M., & Wilde, S. A. (2005). Early Archean crust and the evolution of the Earth's surface. *Geology, 33*(9), 729–732. https://doi.org/10.1130/G21634.1
Van Duin, M. (2025a). Development of mass and energy rate (density) of dissipative systems over their lifetimes: A comparison of a low-mass star, like our sun, a human and the Roman Empire. In D. J. LePoire, L. Grinin, & A. Korotayev (Eds.), *Navigating complexity in Big History—Exploring periodization across cosmic and biosocial dimensions* (pp. 345–385). Springer Nature. https://doi.org/10.1007/978-3-031-85410-1_11
Van Duin, M. (2025b). What more can we learn from a master plot of energy rate versus mass for a wide variety of (complex) systems? *Journal of Big History, 8*(2), 38–53. https://doi.org/10.22339/jbh.v8i2.8202
Vernadsky, V. I. (1997). *Scientific thought as a planetary phenomenon*. Synergetic Press.
Vidal, C. (2014). *The beginning and the end: The meaning of life in a cosmological perspective*. Springer.
Volk, T. (2017). *Quarks to culture, how we came to be*. Columbia University Press.
von Foerster, H., Mora, P. M., & Amiot, L. W. (1960). Doomsday: Friday, 13 November, A.D. 2026. *Science, 132*(3435), 1291–1295. https://doi.org/10.1126/science.132.3436.1291
Weinberg, S. (1977). *The first three minutes: A modern view of the origin of the universe*. Basic Books.
West, G. B., Brown, J. H., & Enquist, B. J. (1997). A general model for the origin of allometric scaling laws in biology. *Science, 276*(5309), 122–126.
Widdowson, M. (2023). *Complexity and collapse in historical systems: Evolutionary singularities revisited*. [Manuscript in preparation].
Wiener, N. (1948). *Cybernetics, or control and communication in the animal and the machine*. MIT Press.
Wilczek, F. (2008). *The lightness of being: Mass, ether, and the unification of forces*. Basic Books.
Winfree, A. T. (1984). The prehistory of the Belousov-Zhabotinsky oscillator. *Journal of Chemical Education, 61*(8), 661–663.
Wolfram, S. (2002). *A new kind of science*. Wolfram Media.
Yankovsky, S. Y. (2000). *The concept of the general theory of information*. Beta-Izdat (In Russian).
Zinkina, J., Korotayev, A., & Andreev, A. (2013). Measuring globalization: Existing methods and their implications for teaching global studies and forecasting. *Campus-Wide Information Systems, 30*(5), 321–339. https://doi.org/10.1108/CWIS-08-2013-0036

David LePoire researches, develops, and applies science principles in environmental issues, Big History evolutionary trends, and particle scattering. He has a BS in physics from CalTech, a PhD in computer science from DePaul University, and over 30 years of experience at the Argonne National Laboratory in the development of scientific analyses, software, training, and modeling. His research includes Big History synergistic trends among energy, environment, organization, and information. Currently, he is the President of the International Big History Association (IBHA).

Leonid Grinin is Senior Research Professor at the Center for Stability and Risk Analysis at the HSE University in Moscow, as well as Director of Eurasian Center for Big History & System Forecasting at the Institute of Oriental Studies, Russian Academy of Sciences. He is co-editor of the international journals *Social Evolution & History* and the *Journal of Globalization Studies*, as well as co-editor of the international almanacs *Evolution, History & Mathematics*, and *Kondratieff Waves*. He is author of more than 600 publications, including 30 monographs in Russian, English, and Chinese. His current research focuses on comparative political studies, theory of revolution, political anthropology, global economy, global history, historical sociology, and futurology. Among his monographs are *The Cybernetic Revolution and the Forthcoming Epoch of Self-Regulating Systems* (Uchitel, 2016, with Anton Grinin), *Big History of Globalization: The Emergence of a Global World System* (Springer 2019, in co-authorship with Julia Zinkina, David Christian, and others), and *Islamism, Arab Spring, and the Future of Democracy: World System and World Values Perspectives* (Springer, 2019; with Andrey Korotayev and Arno Tausch). In 2004, he was awarded as a winner of the best publication of *Socis* journal during the year for the article: "People of Publicity—a New Social Layer?" In 2012, he was awarded the Gold Kondratieff Medal by the International N. D. Kondratieff Foundation. In 2016, he received the Belyaev award for the book *From Bifaces to Nanorobots* (the book published in Russian in co-authorship with Anton Grinin). 12, Rozhdestvenka, Moscow, 107031, Russian Federation; email: leonid.grinin@gmail.com; ORCID 0000-0003-0278-2619.

Andrey Korotayev heads the Center for Stability and Risk Analysis at the HSE University in Moscow, Russia. He is also Senior Research Professor at Eurasian Center for Big History & System Forecasting at Institute of Oriental Studies, Russian Academy of Sciences and Institute for African Studies, Russian Academy of Sciences, as well as at the Faculty of Global Studies of the Lomonosov Moscow State University. Beginning as an historian of Arabia, he made focused studies, such as seen in "Two Social Ecological Crises and Genesis of Tribal Organization in the Yemeni North-East" (1996). Korotayev researches topics in quantitative cross-cultural anthropology and seeks to understand biological and social macroevolution, as in "A Compact Macromodel of World System Evolution" (2005) and "Mathematical Modeling of Biological and Social Phases of Big History" (2014). He has authored or co-authored over 650 scholarly publications, including monographs such as *Ancient Yemen* (Oxford University Press, 1995), *World Religions and Social Evolution of the Old World Oikumene Civilizations: A Cross-Cultural Perspective* (The Edwin Mellen Press, 2004), *Introduction to Social Macrodynamics: Compact Macromodels of the World System Growth* (URSS, 2006; with Daria Khaltorina and Artemy Malkov), *Great Divergence and Great Convergence: A Global Perspective* (Springer Nature, 2015; with Leonid Grinin), and *Islamism, Arab Spring, and the Future of Democracy: World System and World Values Perspectives* (Springer Nature, 2019; with Leonid Grinin and Arno Tausch). He is a founding member of the Eurasian Center for Megahistory & System Forecasting and founder of *Evolution*, an almanac dedicated to the study of Universal Evolution, and thus is closely linked with Big History. In 2020, he produced, with David LePoire, a collective monograph, *The 21st Century Singularity and Global Futures: A Big History Perspective* (Springer Nature, 2020). Recently, he has produced, with David LePoire and Leonid Grinin, another Big History collective monograph, *Navigating Complexity in Big History: Exploring Periodization Across Cosmic and Biosocial Dimensions* (Springer Nature, 2025). He is a laureate of a Russian Science Support Foundation in "The Best Economists of the Russian Academy of Sciences" Nomination (2006); in 2012, he was awarded the Gold Kondratieff Medal by the International N. D. Kondratieff Foundation. He serves as a board member of the International Big History Association. 20, Myasnitskaya, Moscow 101000, Russian Federation; email: akorotayev@gmail.com; ORCID 0000-0003-3014-2037.

3 Complexity as a Unifying Framework: From the Big Bang to Human Societies

Lowell Gustafson

Abstract

Complexity serves as a unifying theme that addresses fundamental questions about both the structure of the universe and human nature. It provides an intellectual framework that bridges disciplines across the natural sciences, social sciences, and humanities—connecting everything from quarks to global societies and even humanity's curiosity about intergalactic relations. The concept of complexity begins with its unidirectional emergence, tracing a path from the Big Bang to the present. It then expands to a multidirectional perspective, encompassing the formation of planets, galaxies, and potential extraterrestrial life. However, complexity does not always progress; it can remain static, regress to simpler forms, or even collapse entirely. Speculations on the existence of multiple universes further extend this idea, suggesting the possibility of infinite complexity.

Keywords

Complexity · Transdisciplinary · Emergence · University

This chapter is an extensively reworked version of the text originally published in Journal of Big History (Gustafson, 2024).

L. Gustafson (✉)
Villanova University, Villanova, PA, USA

D. J. LePoire et al. (eds.), *Complexity in Universal Evolution*, World-Systems Evolution and Global Futures, https://doi.org/10.1007/978-3-032-20517-9_3

3.1 Introduction

The topic of complexity draws on a universe of knowledge and presents an intellectual rationale for the contemporary university.[1] The topic, like the contemporary university, is predictably complex. Universities house multiple schools, colleges, departments, centers, and other administrative units. Unifying key evidence from each of its disciplines to substantiate an account of a universe of knowledge is the intellectual rationale rather than just the administrative management of a university. Complexity is one theme that lends itself to that rationale. It includes the ideas of unidirectional and multidirectional emergence, stasis, devolution, and infinity.

Emergent complexity responds to age-old fundamental questions. "Like our ancestors, we look up at the heavens and wonder. What is the structure of the universe? How significant are we? Are we alone?" (Library of Congress, n.d.) How did we get here? What does that mean for who we are now and what we can reasonably expect in the future? Complexity is one theme that permits all the disciplines within universities to contribute to a coherent set of responses to such questions. As Mitchell observed, "As science probes the nature of life, society, and technology ever more closely, what it finds there is complexity. The sophisticated group behavior of social insects, the unexpected intricacies of the genome, the dynamics of population growth, and the self-organized structure of the World Wide Web—these are just a few examples of complex systems . . ." (Mitchell, 2009).

Disciplines from the natural sciences, social sciences, and humanities contribute evidence towards a substantiated account of development from quarks, protons and neutrons, atoms with 1 or 2 protons and electrons, stars and galaxies, atoms with more than 2 protons, chemicals, second and third generation stars, terrestrial planets like Earth, the origin and increasing complexity of life forms, one of these forms being hominins and then humans, and increasingly complex relationships among humans in kinship groups, villages, cities, nations and empires, and global systems. Humans, each with a hundred billion neurons and maybe a trillion synapses, now number some 8 billion, with more humans connected digitally to each other than ever before. And throughout the human experience, our species has looked to the skies and wondered what our place is in the cosmos. "Complex systems research is becoming ever more important in both the natural and social sciences. It is commonly implied that there is such a thing as a complex system, different examples of which are studied across many disciplines" (Ladyman et al., 2013).

Emergent complexity does not only go from the Big Bang to humans, although it does do that as well. Rather, it is multidirectional. It leads to different types of stars, galaxies, planets, and life forms. The known life forms are still those on Earth, but astrobiology suggests that they may well be evolving elsewhere as well.

Also, complexity does not always continue to emerge. Those units within each new stage of complexity often remain in stasis; they do not become more complex.

[1] For various definitions of complexity, see chapter "Introduction to Complexity Theory for Big History" (LePoire et al., 2026) in this volume.—*Editors.*

Additionally, complexity includes reversal as well as emergence and stasis. What complexity has existed often breaks down into simpler levels. The emergence to increasingly simple levels of complexity eventually leads to endings or death. There is informed speculation at the universal level about this leading to rebirth or a new cycles, as well as about the greatest complexity being an infinite number of types of universes.

3.2 The University, the Universe, and Unidirectional Emergence

Unidirectional emergent complexity, the traditional big history account, which leads us from the Big Bang to humanity, offers an important rationale for contemporary universities. The International Big History Association defines the field of big history like this: "Big History seeks to understand the integrated history of the Cosmos, Earth, Life, and Humanity, using the best available empirical evidence and scholarly methods." Some major sources on big history include books, presentations, and textbooks (Stokes-Brown, 2007; Chaisson, 2006; Christian, 2004, 2008, 2010, 2011, 2015; Christian et al., 2014; Spier, 2015).

Contemporary universities very often cover portions of this account in their departments and disciplines. Evidence about the cosmos is especially included in physics and astronomy. Material from chemistry is included in all four areas. Geology and geography are especially important in the history of the Earth. Biology is central to the discussion of life's origin and evolution. Physical anthropology presents the origins and evolution of hominins to humans. Once we get to human studies, the range of social sciences, humanities, law, nursing, medical, business, and engineering provides material about collective learning that big historians examine. The many departments and colleges within universities have been able to provide so much important knowledge because of their disciplined foci. This often comes at the expense of synthesizing knowledge from different disciplines. The tremendous work that has been done within disciplines has overwhelmed universities' attempts to provide an intellectual rationale for themselves within a universe of knowledge. Directionless programs in "General Studies" rarely make an effort to synthesize knowledge. Universities rarely seek to succinctly offer students a single course or major which value synthesis of all disciplines. Expertise and intellectual skills from the dissection of knowledge are the best many have to offer.

Universities have sometimes come to operate as holding companies for a variety of colleges. A college is often a holding company for a variety of disciplines. Even disciplines are sometimes uneasy collections of sub-fields without clearly stated definitions. Specialization and expertise are often seen as the key intellectual virtues and the only justification for research support. Hiring, prestige, tenure, and promotion are generally tied to recognition by disciplinary programs and journals. College and university administrators struggle to meet budgets, allocate available funds and hiring lines to departments and colleges, and generally keep the trains running on time. Deans and provosts do not always have the luxury to reflect on intellectual mission.

The study of unidirectional emergent complexity seeks to overcome this difficulty and provide an intellectual rationale—a complex universe of knowledge—to which all disciplines of the university contribute. Each part of this account has been made possible only through disciplinary focus. The account as a whole would be possible through the synthesis of each of these parts all within a university; this needs to be done within a course or a series of courses far more often than it is.

3.3 Disciplines and Unidirectional Emergent Complexity

3.3.1 Physics and Astrophysics

3.3.1.1 Big Bang

The beginning of our story is the Big Bang of 13.82 billion years ago, sometimes thought to have emerged from an infinitely hot and dense point without mass, or nothing (Sing, 2004). Or not. It may be that nothing is really something; it is always pulsating and foaming and regularly turning into a variety of forms of something. Perhaps we live in a popcorn multiverse with an infinite number of Big Bangs going off all the time in ways we cannot detect or imagine (Greene, 2011). Other universes with different fundamental laws and forces may be sharing our space or off in other locales. Or maybe our own yo-yo universe has an infinite set of cycles of trillions of years (Steinhardt & Turok, 2007).

We used to think there was only one galaxy that the Milky Way was the universe with a few clouds of something or nebulae circling around it. Then we wondered if there were other inhabitable planets. We now know there are great numbers of both, many or most of which we cannot currently see as they are now. Why should ours be the only universe? However, for now we will prosaically restrict our attention to our own universe.

The extraordinarily hot and dense radiation and plasma immediately after the Big Bang was dramatic. A second after the Big Bang, the temperature is said to have been 10^{32} Kelvin degrees, or 18 billion degrees Fahrenheit (Hooper, 2019). It was as largely as uniform a situation as has existed in our universe's history.

All but immediately after our own universe's Big Bang, when energy first congealed into normal or baryonic matter, six types of quarks appeared. They can appear again if protons and neutrons are smashed into each other at sufficient energy levels, which is done rather routinely at Fermilab and CERN. Four of these quarks lead extraordinarily brief lives before returning to energy; they do not go on to form more complex forms of matter. However, two of them—the up and down quarks—did form relationships as they appeared. This will be a pattern. Some things go on to participate in emergent complexity. Many do not.

For a billion and one bits of matter that appeared, a billion bits of anti-matter did as well. "Antimatter particles share the same mass as their matter counterparts, but qualities such as electric charge are opposite. The positively charged positron, for example, is the antiparticle to the negatively charged electron" (CERN, n.d.). Rather than playing well together, matter and anti-matter annihilate each other. This mayhem is a rather good thing from our point of view, since if all the matter that appeared survived, the universe would have been just too crowded to ever have developed into us. And plenty remained. Enough matter to eventually make a hundred billion galaxies each with an average of a hundred billion stars all have been formed by the leftovers of the great annihilation. Those quarks that survived formed relationships that have lasted billions of years. Destruction can be very creative.

The lucky surviving quarks did not exist in isolation; they formed pairs or threesomes. Their relationship is structured by the strong force that is mediated by the exchange of the charmingly named gluons. Two quark pairs are mesons (often very short-lived); threesomes are often very long-lived baryons, whether protons or neutrons. Two up quarks and a down one form a positively charged proton; two downs and an up form a neutron. All that were formed were single or double protons, what would become later the nuclei of hydrogen and helium with a very small smattering of deuterium and others. It was too hot and dense for protons to form relationships with electrons, so there were no atoms yet.

Why is the strong force exactly as strong as it is and not weaker or stronger? Is it different in other universes? Who knows? It is just the way we do things in our universe. But if it differed at all, we would not be here and neither would anything else that we know of.

The quarks do not merge into one undifferentiated blob. Each proton and neutron is constituted by two different types of quarks. Relationship includes individuality. The one and the many exist together. The quarks relate to each other through the strong force, but they keep their distance as well. Relative to their own size, quarks have a rather pronounced need for personal space. Each of these three moves in a constant dance around the others. They are always related, always moving, always distinct. Nature at rest is hard to find. Nature is spinning, moving, and restless.

The protons and neutrons that were formed quickly after the Big Bang are with us still after almost 14 billion years. In fact, they are us, and everything else that we can see or feel. The structured relationships among individual quarks have been remarkably sustained. There are a lot of them that make each of us. As inventive and creative as nature is, it also keeps certain things around for a long time. And as we will see, lots of very new and more complex types of relationships keep forming. If liberalism is about change and conservatism about keeping things the way they are, we can answer an interesting question. Something came from nothing (or something else) at the Big Bang. That is change. Quarks can maintain their relationships for tens of billions of years. Can't get much more of a status quo than that. So is the universe liberal or conservative? And the answer is—yes.

About 380,000 years after the Big Bang, when the universe had expanded enough to cool to around 3000 degrees Kelvin, the electromagnetic force mediated by the exchange of photons could structure a sustained relationship between protons and electrons. Atoms appeared. Hydrogen, with one proton and one electron, appeared in the greatest numbers. If you add up their mass, about three quarters of all atoms in the universe are still hydrogen. If you count atoms by number, they constitute about

90% of all atoms. They also constitute 63% of the number of atoms in your body (10% by mass). As has been said, hydrogen is an odorless, colorless gas that, given enough time, becomes you. And me.

Helium, with two protons and two electrons each, formed about a quarter of all atoms' mass that then existed (9% by number). There was also a dash of deuterium, or heavy hydrogen (one proton, one neutron, and an electron), helium isotopes, and lithium (three protons and electrons). It's possible that a handful of other atoms existed as well. Vast primal clouds of hydrogen and helium atoms, millions of light years across, still majestically float in certain areas of space nearly 14 billion years later. Some have gone on to form greater complexity; many have not. As Eric Chaisson has pointed out, "Far many more atoms are alone and isolated; only ~0.4% of the universe comprises bound atoms within complex, structured systems, roughly ten times that is loose baryonic (yet still normal) matter, which floats amidst the intergalactic beyond (all else, i.e., the remaining ~96% is 'dark', at least to our senses)." Sometimes complexity emerges; more often it doesn't.

Once formed, and left on their own, these atoms tended to keep their distance. While the strong force bound quarks together and protons and neutrons together within atoms, these atoms left to themselves generally liked their own company. They might approach each other as they moved about, but usually swerved off, avoiding connections with each other.

We sometimes hear about an "atomistic society." This usually refers to a rather asocial condition in which individuals have little to do with each other. The analogy might be a billiard table, with hard billiard balls usually sitting by themselves, but occasionally knocking into each other, sending each other off in various directions. Atoms may be the basic building blocks; but in our experience, blocks usually just sit there by themselves. We are each made of about $6.7 * 10^{27}$ atoms. What are we then like at our most constitutive level? Are we like the individuals discussed by Hobbes in the *Leviathan*? Do we live lives largely isolated from others? The only natural relationship is hostility; a war of all against all. By nature, are we as asocial as atoms? Should Libertarians seek out new sympathizers among the universe's vast majority of unaffiliated atoms? If we seek to form relationships, do we need to find ways to overcome our natural proclivity for individualism? Are atoms the ultimate existentialists, destined to live lives of lonely desperation and then die alone? On a dark, rainy night. And since we are built from atoms, is that what we are really like, all niceties aside?

But what if the story of which we are a part is one of emergent communitarianism rather than the individualism so celebrated in Western Liberalism? Recall that even the simplest of atoms—those that have only one or two protons and are still the most abundant in the universe—are each a set of sustained, structured relationships. Quarks which just moments before had not existed, started to be related through the exchange of gluons mediating the strong force. Atoms, which had not existed before the Big Bang plus 380,000 years, added a relationship between protons and electrons. Atoms are sets of sustained, structured relationships.

At our most constitutive core, we are built more from relationships than from building blocks. Quarks and electrons are more fuzzy than blocky. Their "hardness" comes from forces defining their relationships. What exists between things is as real as the things themselves.

3.3.1.2 Stars

But what about atoms naturally avoiding each other? Relationships within atoms are fine, but beyond that, they naturally stay at a distance. Well, atoms are not left to their own devices. They exist within a larger framework that acts upon them.

When they did form, atoms were not perfectly distributed, if by perfect you mean absolutely equally. They were a little more densely distributed here, a little less there. This asymmetry, unequal distribution, or imperfection was another very fortunate occurrence. Inequality can contribute a lot. Gravity has no force at the relatively small distances between quarks. However, the space between slightly densely packed atoms can be just enough to let it start operating. A larger clump of atoms here can exert gravitational attraction on a smaller clump there. If all atoms had been equally distributed, their gravitational attraction on each other would have canceled it all out, and they would never have been drawn to each other. However, with the asymmetry, the denser regions could start drawing in the slightly less densely packed atoms. Gravity kept pulling them together, increasing their density and heat. (It may be that there just is not enough normal matter to have created enough gravity to have had the effects that we will soon see. It seems to be that there is "dark matter" that actually creates the additional gravity necessary to form the universe's structure that developed.)

As atoms were pulled closer together, they began to spin faster like a figure skater drawing in her arms. Once sufficient density and heat developed, with atoms moving about more and more quickly, the atoms overcame their preference to stay away from each other. More accurately, the Coulomb barrier refers to the electrostatic interaction that two nuclei need to overcome so they can get close enough to fuse. The protons needed to "tunnel" through this barrier to overcome the Coulomb barrier and fuse into one heavier element. The result of the quantum tunneling was to produce a two-proton helium nucleus, all held together by the strong force. Most protons remained separate, but enough fused to maintain a star's fusion reaction.

Each newly fused atom was less than the sum of its parts. Each new helium atom weighed slightly less than the two hydrogen atoms which had combined to form it. The missing matter had turned into energy. The fusion caused energy to burst out. Gravity kept trying to draw the atoms in. The uneasy equilibrium between these two forces resulted in the formation of stars. The black sky began twinkling. The dark age was over.

As the helium was formed, gravity drew it in more, until it heated up enough for it to start fusing into heavier elements, such as nitrogen. This released energy and permitted gravity to draw the newly formed elements further in, until they too began to fuse, forming carbon and neon. This was repeated as oxygen, magnesium, silicon, and sulfur were each fused. The largest stars with enough mass to permit gravity to keep drawing the newly fused elements further in developed an onion like structure, with the lighter elements on the periphery; the heavier ones successively formed layers closer to the core. Not only can there be new things under the stars, the stars themselves were something new. The strong force, electromagnetism, gravity, and fusion formed relationships between atoms within the structure of a star.

Gravitational attraction between stars and dark matter formed galaxies or groupings of stars in distinct patterns. Galaxies formed relationships due to gravity in local groups and even larger patterns. The theoretical work of Fr. Georges Lemaître, confirmed by the evidence collected by Edwin Hubble, demonstrated that not only were there more galaxies than our own Milky Way, but that once they got to be further away from each other than those in the local group, they are racing away from each other. It may be that dark energy or anti-gravity is causing the galaxies to keep falling out, with space and the universe expanding at ever faster speeds the further from each other they are. It may also be that the relative amount of dark energy has been changing over the course of universal history.

When the largest of the stars began to make iron with its 26 protons, energy was consumed rather than released. The equilibrium between gravity and fusion was broken. Almost immediately, the star exploded in a supernova. The sudden increase in temperatures during the explosion permitted the almost instantaneous formation of all of the elements with more than 26 protons per atom, all sent streaming into space at incredible speeds, often mixing with pre-existing clouds of hydrogen and helium that had been floating since the Big Bang.

3.3.1.3 LIGO and NANOGrav

Another way of forming all the elements in the universe could come from the collision of neutron stars, which could be detected a billion or more years later as having also produced gravitational waves. In 2016, the Laser Interferometer Gravitational-Wave Observatory (LIGO) detected the gravitational waves predicted by Einstein a century before. Two observatories used mirrors that were placed four kilometers apart and could detect a wave of less than one ten-thousandth the diameter of a proton that had been traveling across the universe for a billion years since two neutron stars collided. Adam Frank (2023) lyrically writes of observations of the North American Nanohertz Observatory for Gravitational Waves (NANOGrav):

> The whole universe is humming. Actually, the whole universe is Mongolian throat singing. Every star, every planet, every continent, every building, every person is vibrating along to the slow cosmic beat.
>
> That's the takeaway from yesterday's remarkable announcement that scientists have detected a "cosmic background" of ripples in the structure of space and time.

In addition producing waves, colliding neutron stars are, along with fusion within stars, another way of fusing elements with fewer protons into element with great numbers of them, or virtually all the elements. So your gold ring could have been made in a supernova or in a neutron star collision. In one way or the other, we get the full range of atoms.

3.3.2 Chemistry and Molecules

Atoms form in such a way that electrons orbit protons in shells. The innermost shell is full with two electrons, the second with eight, the third with eighteen, the fourth with thirty-two, the fifth with fifty. Hydrogen, with its one electron, has a vacancy sign out in its only electron shell. That shell seems to want one more electron to form a full house. Oxygen, with its eight electrons, has two in its first shell and six in its second. This leaves two vacancies in its second shell. This is a match made in the heavens. If two hydrogen atoms hook up with an oxygen atom, each sharing their electrons, each hydrogen atom can have two electrons in its only shell and oxygen can have 8 in its second shell. Everybody is happy because a new relationship between atoms is formed: H_2O—water. This molecule has a new property. At the right temperature, it has the property of wetness, which did not exist before. Water, which is abundant throughout space, is not the only molecule that forms. Dozens of molecules with 2, 3, 4, 5, or more atoms evolve naturally. Many atoms due to the way electron shells work led to the formation of these new relationships called molecules.

Not all atoms are anxious to form molecules. Helium has two electrons in its only shell and has a No Vacancy sign well lit. It is called a noble gas. Having all they need; nobility does not require additional relationships with the lesser types that are needy. Relationship added to relationship is not much part of helium's story. While hydrogen becomes us, helium often just goes floating off into space. Not everything is social. Not everything forms polity, or sustained, ordered relationships. We saw that same aloofness with four of the six quarks. A subatomic particle formed in nuclear fusion, neutrinos, is much the same. Like photons, they go shooting from stars off into space, but almost never interact with anything. They can sail through 20 miles of lead and never hit anything. It has taken extraordinary measures to detect them at all. History and polity are not built on the backs of two-thirds of quarks, neutrinos, helium, or other asocial phenomena. They are indeed the rugged individualists of the universe. The story of emergent complexity is not uniform.

3.3.3 Geology, Earth

After a nearby supernova shot its star dust out into neighboring space, disturbing pre-existing clouds of hydrogen and helium, gravity again began pulling together the mixture of elements and molecules. A second-generation star not only with mostly hydrogen and helium but also with traces of heavier elements in it—including oxygen, carbon, neon and iron—eventually began shining. This process may have been in its third round when our sun formed 4.6 billion years ago. It is not big enough to permit gravity to create densities high enough to fuse elements heavier than helium. This is good for us, since huge stars live fast and die young. Our sun goes along at a nice leisurely pace of fusing 600 million tons of hydrogen each second, turning it into 596 million tons of helium and more energy than mankind has ever produced in our species' entire history. It is because of all their mass that stars like our sun produce so much heat and light. Surprisingly, once you get down to the energy released bit by bit, the energy density flow is about the same as a reptile's metabolism (Chaisson, 2001).

The sun's rate of consuming its stock of hydrogen will permit it to continue shining for a total of about 10 billion years, meaning it is at mid-life now. Its 5-billion-year history has provided energy and the time for Earth to develop. We've got billions more years before the sun turns into a Red Giant, evaporates the oceans, and engulfs the Earth. There is time before anyone needs to get tickets for a trip to another solar system.

While gravity drew together 99.86% of the total mass of the Solar System to make the sun, the leftover debris was put to good use. On the outskirts of the spinning disk that eventually ignited as the sun, these leftovers from part of the supernova started accreting through the power of gravity. Gases and chunks of iron, nickel, silicon, and bits or gold, silver, uranium, and other elements and molecules bumped into each other and stuck together. The planets, planetoids, comets, and asteroids were formed. On the emerging new Earth, all this knocking together that created kinetic energy, not to mention the radioactive decay of uranium and other such elements. This made for a molten, hot planet that formed its own structure from thousands of molecules and the minerals they produced. Heavier iron and nickel sank into a dense core that is still as hot as the surface of the sun. Silicon and other lighter elements rose to the top. Eventually, a thin layer made of basalt made for oceanic floors and the frothy granite cooled enough to permit land to form. Lighter, cooler outer layers spinning around denser iron and nickel produced a magnetic shield around the planet that protected it from solar winds that might otherwise blow away Earth's atmosphere.

3.3.4 Biology and the Emergence of Life

The process of chemical evolution that had begun in space continued on Earth. The most common elements on the surface of the Earth continued to combine in many ways. Hydrogen, carbon, nitrogen, oxygen, sodium, magnesium, phosphorus, sulfur, chlorine, potassium, calcium, iron, and other elements on Earth interacted to form over 4700 minerals. Around black smokers at the bottom of the oceans where tectonic plates separated and mineral rich heated waters bellowed up, on the relatively cooler white (alkaline) smokers, or on sun-soaked pools of water on rocky beaches, the process of chemical evolution continued. Lipids that created films formed, eventually forming membranes. Carbon, with its four electrons in its second orbit and a total of six overall, was able to combine with many other elements, and was central to the Krebs cycle which spins off amino acids. These molecules continued to combine until they integrated membranes, metabolism or access to energy, and RNA and DNA that permitted reproduction with variation in response to environmental changes. The Last Common Universal Ancestor—LUCA—was combined in the most complex relationship in universal history to date—that we know of. The first prokaryote cells were earthlings, formed of the commonly available chemicals and elements on Earth's surface. They were also children of the universe, with elements forged in stars that had died long before.

3.3.4.1 Biological Evolution

It has been said that the dream of every bacterium, the simplest of cells, is to become two bacteria. Reproduction has to be important for any species that plans on surviving, since the death of any given individual is part of the way life works. Sustained relationship is not eternal relationship. The nice thing about being a bacterium is that your dreams can come true about every 20 min. Reproduction with variation in response to environmental changes is a skill perfected by prokaryote cells. You just can't argue with success. They live in virtually any setting, however extreme the condition on Earth can be. From deep underground to thermal waters, prokaryotes are there. There are more bacterial cells in and on your body than there are cells that constitute your body. They help you digest food. And when you die, they will digest you. These types of cells have survived for almost 4 billion years. They will be on Earth long after humans have vanished. Many prokaryote cells follow a plan that isn't broken and doesn't need fixing, although they do keep adjusting to new conditions such as antibiotics. They evolve quickly, but as a group, they have not become fundamentally more complex.

However, after a couple billion years of happily reproducing at their same level of complexity, some did become more complex. A prokaryote cell may have tried to eat and digest a mitochondrial, but instead somehow managed instead to form a long-lasting and mutually beneficial relationship with it instead in a new, more complex type of cell. About 2 billion years ago, eukaryote cells appeared with a membrane covered kernel in which more complex DNA was kept. Hosting the mitochondrial cell created a new way of obtaining energy; it was now able to burn carbohydrates and eventually permit us to enjoy eating donuts.

A more complex set of relationships within the cell led to more complex relationships among cells. Films of bacteria on the surface of the ocean or accretions of them in rock like formations of stromatolites in tidal pools were steps towards multicellular life forms. One generation of cells died off, only to be covered by future layers of descendants. Another step in multicellular cooperation came with creatures like sponges. These are formed by the same type of cells that could still specialize in serving different functions. Some cells drew in nutrient rich water; others expelled nutrient drained water. Same type of cells; different tasks. Push these cells through a sieve so that they are separated as they fall to the bottom of a tank, and they scoot back together to form another new sponge. These are cooperative cells, not hardy individualists.

Relationships among increasingly complex body structures formed by different types of cells are seen in such examples as cnidarians, or jellyfish, first seen about 800 million years ago. They have little harpoons that can inject prey with poison, have such structures as a mouth/anus, and have two layers of tissue. Their nervous system is pretty uniformly spread out throughout the animal. Jellyfish are still around and doing fine. The Scarecrow in the *Wizard of Oz* seemed to get along pretty well without a brain, and so have the cnidarians. They have existed 4000 times longer than *Homo sapiens* have. They see no reason to develop more complexity.

Still, there were additional mutations that worked out in the environment of the time. Flatworms introduced a body plan about 590 million years ago with a right and a left side, an up and down, and a front and a back. Sense organs were put up front, along with a ganglia of nerve cells to interpret the incoming data. Chordates like the currently existing hagfish put a cord along its back to protect the flow of information from the ganglia to the rest of the body, as well as putting the mouth up front and an anus in the rear. About 525 million years ago, vertebrates started breaking that cord into bony segments, offering better protection and definition. The first animals to venture out from the seas onto land, such as Tiktalik, had wrists to help scoot on land and a neck to help look around. About 360 million years ago, the first amniotes could recreate the watery world in which reproduction had originally taken place, and start producing eggs with a protective shell and watery interior. About 360 million years ago, mammals first appeared, which had, among other things, a more complex auditory system with more parts that helped them hear better. The story of evolution is in part a story of increasing complexity of body structures, with more complex relationships among greater numbers of parts.

3.3.4.2 Relations Among Animals and Plants

The development of increasingly complex relationships among quarks, protons and electrons, atoms, molecules, cells, and body parts were followed by increasingly complex relations among and between species. This process of natural evolution becomes a process of pre-human social evolution Wilson (2012). From quorum sensing of bacteria to schools of fish, bee hives, ant colonies, flocks of birds, herds of bison, troops of chimpanzees, and many other examples, animals often live in groups and groups often form ecosystems.

Not all animals live in groups. Many seem to exist in splendid isolation for most of their lives, coming together just long enough for reproduction without any care for offspring after birth. Mother guppies and sharks would just as soon eat their babies. Sea turtles lay their eggs on the beach, return to the sea, and may hope for the best for their offspring, but likely don't think about them. Crocodiles help their offspring out of their eggshell and out of the nest; after that, the kids are on their own. Childcare is of course more of an issue for various lengths of time for many species. From weeks of care to a couple years is common. Mothers, fathers, and others are involved in different ways, depending on the species.

3.3.5 Physical Anthropology and Hominins

By the time we get to hominins, our ancestors' survival strategy and increasingly complex sociability went hand in hand. *Australopithecus* and its ancestors were the hunted rather than the hunters. They may have scavenged, eating bone marrow of leftover carcasses, but gathering fruits, nuts, tubers, and leaves likely provided a main stay of their diet. Other than that, they tried to stay out of the way of predators.

They had few natural weapons. Their teeth and fingernails were no match for lions. Their speed was no match for cheetahs. They had no shells for defense nor wings for flight. No wonder that there do not seem to have been huge numbers of hominids, that most species went extinct, and that our own ancestors came close to extinction. They just did not have that much going for them.

Bipedalism may have been an advantage when a drier climate led to more savannah grasslands and fewer forests with tree to swing from. Standing on two feet exposed less of the body to the hot sun, made it easier to see over tall grasses, and freed the use of the arms, hands, and opposable thumbs. A parent could hold a child and pick fruit all at once. But every benefit comes with a cost. It also altered the skeleton, restricting the birth canal, making child birth more painful and dangerous.

This problem was aggravated as hominids' greatest advantage developed. Brain size from *Australopithecus* to *Homo sapiens* tripled, with Neanderthals winning the brain size competition. (Brain size for *Australopithecus* averaged between 375 and 550 cm^3, *Homo habilis* from 500 to 800, *Homo erectus* 750–1225, *Homo sapiens* 1200–1750, and Neanderthals 900–1880.) Hominids couldn't outfight competing species, but they could start to outthink them. Brains rather than brawn would eventually win the day.

The development of the hardware enabling life forms to think reaches back to bacteria using their flagella to scoot towards light and away from toxins. From there to hominin brain development is a long process. Brains gave species from jellyfish to humans all kinds of abilities. The eighteenth-century naturalist, Carl Linnaeus, first placed us as *Homo sapiens* within the Latin binomial nomenclature he developed for species. There are other types of men, but we are wise men. Our brains are what we most identify with. They grew in size and complexity; but why? Maybe it was originally because earlier species were *Homo habilis*—handy men who developed and used tools. It is the technological prowess that our brains gave us that is our central advantage. Or maybe it was a positive feedback loop between greater brain sized and complexity that permitted and was selected by more complex social relationships and cooperation that provided what little advantage we had early on.

Even with only partial brain development and soft skulls at birth, delivering children had become highly risky. To permit time for the brain to develop to maturity, grow a fused, bony skull, and learn all that they required to survive, childhood for hominids took years. Breastfeeding and childcare-giving mothers developed close relations with offspring over long childhoods.

Child mortality was still likely high. For a handful of children to reach sexual maturity, birth would need to be given to a number more. Especially for those with lifespans in the 30s or so for adults who got through childhood, this meant that most or all of a female's adult life was involved with pregnancy and childcare—and more. Working mothers were the norm. They likely provided the bulk of the calories through gathering and carried out many other important tasks. Still, they would have needed support as they did the primarily important work of getting children to adulthood so the species could survive. Long-term relations between mothers and children and between childcare-taking females and males were necessary for the fat-headed hominids to survive.

It is one thing to get together briefly to copulate. That is all sharks need to do since childcare is not a problem. Once they give birth, offspring on their own. A sea turtle female lays its eggs on a beach and never sees her offspring even hatch. Hominins faced a wholly other set of problems. Long childhoods required care-takers to work together for many years to raise children, a problem that hominins had to figure out if the species was going to survive. Resolving the issues of food, shelter, and other necessities for a kinship group over years takes problem solving and relationships to a whole different level Haraway (2016). The increased demands of a long childhood and the long-term adult relations it required selected for an increased ability to figure out how to live together for many years at a time. The gender relations made necessary by being a big brained bipedal species is a root of hominin polity. Sexual politics has changed markedly recently with longer lifespans and lower mortality rates. Mothers no longer spend their entire adult lives dealing with pregnancy and childcare and have the time and energy to do much else.

As Michael Duffy, who writes within the Montessori tradition, notes that as we go through evolution, "organisms produce fewer and fewer offspring and require longer and longer periods of care, leading to more important and deeper relationships. Fish produce thousands of eggs and rarely care for their young, reptiles produce hundreds of eggs and have only limited contact with their offspring, most mammals produce only a litter of a half dozen young and care for them for a long time through nursing, and humans have one or maybe two babies at a time and produce the most parent dependent creatures on Earth!" (Gustafson, 2013).

Many species have long developed their own ways of developing and maintaining relationships. Baboons groom each other, checking for parasites in the fur. Frans de Waal discusses how bonobos use sex for much the same purposes. Social primates, who were not genetically identical like ants within a colony are, developed a "theory of mind;" they could understand each other's reactions. They could even sometimes "feel for each other," or empathize. The law of the jungle, as de Waal argues, includes the social practices and understandings that would later be self-consciously developed into ethics.

Picking lice out of children's hair and having sexual relations have forever been part of hominid mothers' lives as well. Hominids' survival strategy led to developed abilities to relate to each other. For their relations to develop, they would need to exchange a lot more than just gluons and photons. If you thought physics was hard to grasp, just try politics as previous types of hominins evolved in *Homo sapiens* with all the complexities of human memories, imagination, symbolic thinking, trade, and culture.

3.3.5.1 Migration

In addition to the relationships among care-givers for the purpose of child raising have been the relationships of hominins and other species with the land that supports them and on which they make their nests of various sorts Wilson (2012). When land can no longer support the number of people living on it, that is one common reason for some of them to move to new territory. Hominins were not the first group of species who intentionally migrated. Queen bees will look for a new place to nest.

Matriarchs will guide elephant herds to water over long distances, often making adjustments to age old routes in response to changed circumstances. As noted earlier, prokaryotes with flagellum seem to move intentionally. Movement in response to environmental changes requires complex thinking at various levels. It is a form of creative activity. *Homo erectus* migrated by about 2 million years ago from Africa to as far as Asia. *Homo sapiens* left Africa about 70,000 years ago. From the time that some humans left Africa, perhaps due to climate changes that were making available food sources more limited, people gradually spread out to inhabit the entire world, except for Antarctica, over the next 50,000 years. Humans reached the Americas via Beringia by 15,000–20,000 years ago. The many cultural adaptations that made survival in varied geographical and climatic variations are made possible by cognitive abilities. Piecing together now through the use of archaeology and genetic analysis, we are able to reconstruct the movements of people into new territories. The current story is one of very early human creativity, imagination, and courage.

3.3.6 Cultural Anthropology: Memory, Imagination, Symbolic Thinking, and Exchange

Defining what humans are is notoriously difficult. But somehow it includes a collection of physical characteristics such as bipedalism, opposable thumbs, a large brain, and smaller teeth. It also entails some combination of behaviors and cultural characteristics, such as memory, imagination, symbolic thinking, exchange, and empathy or having a "theory of mind."

3.3.6.1 Memory

Memory is an incredibly complicated topic. Virtually all species remember, although in very different ways. The long childhoods in which each person remembers their period of dependency create long-term memories of caretakers. Hominid adults still remember their own childhoods and their caretakers. They remember how these important experiences were carried out by those who are now old or dead. What was so important is now gone, but remains important in memory. Memories of what is no longer may be pondered while going about present tasks.

Child bearing for hominids also entails the expectation of repeating a long-term set of relationships. I am going to have to do for my children what was done for me. This baby will require years of nurture to get it to sexual maturity. What is a baby now will in a number of years become an adult if I do what I need to do to help it survive. I can imagine a long-term future which does not yet exist, but which I can help create. Memory, imagination, planning, and execution go hand in hand.

3.3.6.2 Symbols and Language

Being able to remember what no longer is—and imagine what is not yet—is facilitated by symbolic thinking and language. Vervet monkeys will make one call for threats from above such as an eagle, another for threats in trees such as snakes, or those on the ground such as big cats. When one monkey makes such a call, others in the troop look in the right direction. One screech signifying eagle causes other monkeys to look up. A sound and an expressed / perceived meaning is linked correctly, helping the group's survival. However, the monkey does not make the sound in the absence of the threat. It does not discuss how to better prepare for a future threat. Vervet monkeys do not sit around at night discussing that day's eagle attack. They do not draw pictures of eagles. They do not intellectually manipulate or exchange symbols.

The development of syntax or grammar and vocabulary went along with that of symbolic thought. Being able to consider words and meaning in the absence of immediately present referents, adjust them, move them around and think of alternative arrangements, was facilitated by language. Being able to communicate these ideas in novel yet understandable ways meant that new meanings could be created.

Remembering and imagining in the absence of the referent is a source of symbolic thinking, planning, and eventually realizing possibilities. The road from the communication of monkeys to the symbolic thinking of hominids is long, complex, and still not exactly understood. But that it took place seems clear. By over two and a half million years ago at the Gona River in Ethiopia, *Australopithecus* or *Homo habilis* was making stone tools. Other species use tools as well. Crows, wolves, chimps, and others will use stones and sticks to achieve various purposes. However, the Gona River chipped tools were fashioned by toolmakers. They had to first select which type of rock they wanted to alter. Some types of rock are too soft to make good tools. Then they had to be able to imagine the tool that was in the right kind of rock, to imagine how it could be made into a cutting, scraping, or digging tool. Then they had to carry out a series of steps to create the tool. This was probably done with others looking on and learning how to do this as well. And remember, all of this was going on over 2 million years before *Homo sapiens* appeared.

Tool-making was added to older tool-using skills when symbolic thinking and imagination were possible due to eye—hand and brain development, relative to earlier species. Those who had emerged from nature now began to adjust what they found in nature. Nature in these complex pockets called hominids could begin to select what helped them survive and live better. Evolution could begin to be not only in response to environment, but determinative of it. Nature became partially self-selecting in hominids.

Nature had long exhibited how creative it is. There was nothing and then there was something. There were not protons and then there were. Same with atoms, molecules, stars, terrestrial planets, and life. The transition from one to the next are times of change and natural creativity, but there were long periods of stasis in between each one. Relative to these periods, the time it took for hominids to develop their tool making was rather quick, even if it seems to be agonizingly slow to us. By the Oldowan period from about 2.6–1.7 million years ago, *Australopithecus* and/or *Homo habilis* had developed more sophisticated tools. By the Acheulean period about 1,650,000–100,000 years ago, tools had become bifacial, larger, and more varied. The oval or pear-shaped tools were not only functional; they also have shapes that are pleasing to us and perhaps to their makers. Natural emergence had become hominids' creativity. The road from physics to art was being paved.

Adjusting nature was done in various ways. Eating meat and tough tubers was hard on the digestive track of early hominids. Cooking them made them easier to digest and taste better. Exactly when this began is not certain, although it seems to have started between 150,000 and 790,000 years ago with the fire altered stones at Gesherbenot-Ya'aqov in Israel. The transition from scavenging to hunting had been made at least by a half million years ago, as indicated by spear points and skeletal wounds in prey found at Boxgrove, England, and Kathu Pan 1 in South Africa (Donald, 1991).

3.3.6.3 Imagination

Burials indicate a new level of relationship. Other species such as elephants will clearly mourn dead members of the group. But the careful burial of the dead is a human activity. Again, exactly when this began is not clear, but there are burials from 80,000 to 120,000 years ago in Qafzeh, Israel. Here, we have living members of the group remembering the people who had died and imagining they have an obligation to them even after they die. Burial is a relationship with the dead, requiring memory of what is no longer. What is real in the present is only part of what matters. Memories of the past—kept in the electrical/chemical relationships among neurons—can be more important than the hard stuff that one can feel now in the present.

Hunters had long understood the difference between life and death. Causing an animal to bleed from wounds transformed the beast from one running through the woods to one lying on the ground. Did the hunters begin to think symbolically about the "life" being in the blood that sank into the ground? Does the life of the body go into the Earth looking for a new form to inhabit? Is the spirit of the dead animal believed to be angry at the hunter, planning to return to the surface world to make trouble if proper steps of propitiation are not taken by the hunter?

Once grave goods become included in the burials, we seem to also have imagination of the future added to memory of the past. Burial goods suggest that people thought they could indeed take it with them. Everything had a spirit: people, mountains, rivers, pots, weapons, etc. The life or spirit of the dead person will need the spirits of various tools or weapons in the next life. Members of the group were socially close to those now dead. They remembered them and valued these memories. They wanted to imagine that their beloved would live on, and that proper actions by the living could help the dead live well. Ancestor worship may be one origin of religion. This seems to indicate the powerful social attachments our ancestors had with each other (Tomasello, 1999).

3.3.7 Art

The discoveries at Blombos Cave in South Africa from about 75,000 years ago include an etched, rectangular rock. A net or diamond like design is scratched, with diagonal and parallel sets of lines. This is not just aimless doodling. This is done by a person interested in perceiving and creating patterns. What other patterns were being perceived and analyzed? Seasons? Plant growth? Movements of animals? Behaviors of fellow members of the group? Did the patterned lines have symbolic meaning of some sort in a way that etched crosses, six pointed stars, or crescents often have for us?

Shells with drilled holes were also found at Blombos. The cave is near the coast, and a diet of sea food sustained them. Did they wear the shells as a way to offer the spirits of the dead animals a place to live after their bodies had been ingested? Did they wear necklaces of shells out of a sense of beauty made possible by using or improving on what nature offers? What do these artifacts indicate about their symbolic thinking? By perhaps 48,000 years ago, at the El Castillo Cave in Spain, an artist painted animals and designs from dots and lines on the walls. This was the case later as well at Chauvet, Lascaux, and elsewhere. The animals that were painted were not modeling for them. The artists worked from memory. What purposes did they have in painting these animals and designs underground? What were the artists thinking about the animals and designs they painted? It is hard not to speculate. Was the cave where the spirits of dead animals went to live after their blood drained from their bodies? Were these spirits looking for new bodies to inhabit? What was the meaning of the paintings for those who drew or first viewed them? The artists also spit painted the outline of their hand multiple times. Were they leaving their signature, wanting those who would view the painting in the future to know who painted them? Were they touching the rock behind which the spirits of the animals they painted lived?

The importance of reproduction and fertility is made explicit by the so-called Venus figures found at Hohle Fels in Germany from the Upper Paleolithic period, the Woman of Willendorf from about 24,000 years ago, the Woman of Laussel from about 20,000 years ago, and many others. These palm size statuettes of women with exaggerated breasts and hips may have offered comfort to mothers going through pregnancy or delivery, or had any number of other possible meanings. Whoever made the statues did so while thinking about fertility and sexuality rather than engaging in sex. These statues demonstrate symbolic thinking about sex in the immediate absence of sexual behavior.

3.3.7.1 Music

The emergence from sound to detecting sounds to creating sounds for communication and expression too is a complex topic. The hardware necessary to transforming the waves through a medium such as air into perceived sounds in the brain began with early land dwellers feeling vibrations in their bones. Sight is great, but you can't see around the bend or over the hill. Sound provides crucially important information. The patterns and tones of sound provide important information about the environment.

Many species produce sounds as well as perceive them. Some birds will sing to announce territorial claims or attract mates. Whales and others too will sing to communicate over long distances. Sounds can convey information to others.

With the malleus, incus, and stapes as part of their auditory system, mammals became able to hear in ways that reptiles cannot. Listening to the sound waves caused by ocean waves, lion roars, chirping crickets, and howling winds all had important meanings for hominids. Hearing and responding to a dependent babies cry, parting the lips and calling "Ma" with various inflections of tone elicited powerful responses among caretakers. Different sounds would have elicited other profound emotional responses, such as fear or sexual desire. Rhythmic music and drumming would have enhanced group identity during kinship groups' dances. Eventually, fife and drums communicated information and bolstered courage during battle. Campaign theme songs would identify candidates. National anthems would stir patriotism. Perceiving and making music has a long history of the relationships between animals and their environments, and animals such as humans with each other.

3.3.7.2 Creativity

Symbolic thinking and imagination made combination beyond natural referents possible. A wonderful example of this is the Löwenmesch or Lion Man from Germany from about 30,000 years ago. A bipedal man's body with a lion's head was not something the artist had ever seen. This was work not from memory alone but from imagination and combination. This indicates the ability to manipulate symbols separate from natural perception. It also indicates a crucially important political ability of combining what had not yet been combined in nature.

Nature had combined much in the past through increasingly complex relationships. Quarks, atoms, molecules, minerals, cells, body parts, animal groups, and ecosystems all kept putting thing together in larger and novel combinations. Now, humans could do this at a faster pace and self-consciously.

Placing value on symbols for their own sake was exhibited by early artists as well. For example, there is a beautiful ivory horse sculpture from Vogelherd, Germany, from about 32,000 years ago. The artist did not try to include all the musculature of a real horse. Instead, it is an idealized shape with a series of flowing curves. This is not so much a representation of a physical horse as an ideal one expressing a sense of beauty. The artist took delight in abstraction. Plato was a bit of a Johnny-come-lately with his theory of the forms. Relationships through the exchange of words, music, and symbols developed human relationships. Exchange of goods did too. This too has a long history, going back to sharing food to enhance group relations. Specialized tool production homo habilis sites relatively far from sources of rock that were used indicates trade as much as 2 million years ago. Trading routes become increasingly extensive and established, until by 14,000 years ago the obsidian trade in the Near East and then the famous Silk Road establish what some see as a central core political system.

3.3.8 The Social Sciences and Development

Economics, sociology, and political science examine the emergence and structure of various ways of providing the material goods needed to sustain and enjoy life, groups of persons, decision-making, and implementation—the development of human organization over time. Humanity does not begin with the individual. It begins with the social structure required for our reproductive strategy to sustain the species—kinship.

3.3.8.1 Kinship

The growth of symbolic thinking and exchange of goods, words, glances, gestures, musical sounds, and artistic images facilitated political development. We have discussed the importance of kinship groups. Long-term bonding of childcare givers required sophisticated relationships demanding lots of exchanges. Kinship groups within a scavenger/gatherer and then hunter/gatherer economy likely became complex, but were still limited in size to perhaps fifty or a hundred persons. Larger trading routes would have permitted development of complexity of relationship. Family groups needed to exchange offspring for mating in the next generation. This led over generations to complex sets of inter-kinship relations. Terms such as "second cousin once removed" start to indicate such complexity.

In kinship relationships, lineage is important. Loyalties are to caretakers and common ancestors. Family and kinship remain important in our own day. The powerful resonances are indicated by larger groups attempting to appropriate kinship relations. Nationalists sometimes have referred to their country as a Motherland. In the United States, George Washington is referred to as the "Father of the Country." Members of the Roman Catholic church call their priests "Father." *Pope* is derived from the Greek *pappas*: father. Larger, non-lineage groups often seek to call upon the powerful forces of kinship. One of the values of Big History is its scientific story of the real lineage of all persons, going back to a small group in Africa about 200,000 years ago; of all life to LUCA, and the universe to a single point. It turns out that we really do all have a common background. Big History is the scientific story for a period of Human Politics.

3.3.8.2 Agriculture and Villages

One of the major thresholds of Big History is the Agricultural Revolution. The transition from hunting and gathering to growing crops and raising certain animals is of crucial importance. It also entails a stage of political development. Hunting/gathering went along with kinship polities. With agriculture came settled villages of increasing size, beginning to include different kinship lines. This presented the village with an enormous political problem: how to establish a sustained, structured set of relationships beyond kinship.

One way to do this was to create dynasties; village lineages that all could be persuaded or forced to adopt. Lineage now became a symbolic political category rather than a biological one. In many regions of the world, mounds and other monumental burial sites enshrined the lineage of the village. Those within one lineage might still have the right to rule, but all needed to exchange the symbols that helped nurture loyalty to it.

The political leaders of these settlements or villages during the early agricultural era were sometimes those who had access and control over the best growing areas. We start to see increased social stratification and inequalities in wealth as the agricultural era proceeded. Some residences and some graves are noticeably grander than others. Hierarchy in the hunter/gatherer era was more likely based on strength, size, or cunning. In each period, leadership could also be exercised by those we call shamans, or those who could impress their fellows with their special insights and relationships. When some went through fasting, whether by choice or necessity, carried out rhythmic dancing while listening to repetitive rhythmic music, added various hallucinogens, and perhaps inflicted self-flagellation, they likely could report any number of special insights and experiences. Shapes would have shifted, experienced as traveling in other realms. These were similar to dream like states. Dreams while sleeping and trances while awake offered symbolic connections with what was beyond normal referents. Imagined relationships with abstract designs, ancestors, and the supernormal by some could have impressed others and established a claim to leadership.

Village identity could be developed and expressed through styles of clothing, certain verbal expressions, or other identifiers. Stories about the village could be told at gatherings. It took enormous effort and creativity to incorporate loyalty to the family within loyalty to the village.

3.3.8.3 Cities and Empires

Monumental, ceremonial architecture reinforced the claim by some of symbolic leadership that legitimized claims to leadership. Standing in awe not directly of the universe but of some people's special connections with it were impressive. From Watson Brake in Ouachita Parish in Louisiana from about 5400 hundred years ago to Imhotep's Saqarra in Egypt about 4700 years ago, grand burial sites began to announce the emergence of full-time leading families. Large, stylized burial mounds called attention if not of the gods, at least of the humbled onlookers who stood before them during ceremonies. Equivalents in modern America are the tall, stiff obelisk in honor to the Father of the Country, or the Jefferson or Lincoln Memorials in which political pilgrims can stand reverently in front of larger-than-life leaders who have mythical meaning and personify the presidential succession that leads to the current national leader.

Large, monumental architecture also announces the emergence of new political units of cities with larger populations and relations of cities within regional associations and nations or empires. Eridu, Uruk, Ur, Çatalhöyük, Jericho, Damascus, Mohenjo-daro, Tenochtitlán, Teotihuacan, Xi'an, and other great cities represent a transition to larger, more complex political units. Sometimes these became the hubs of empires; sometimes they were combined with other cities within empires such as the Akkadian Empire of Sargon the Great from 2400 BCE, the fifteenth century BCE New Kingdom of Ancient Egypt ruled by Thutmose III, the Assyrian empire of 2000–612 BCE, the Median Empire in Persia by the sixth century BCE, the Achaemenid Empire from 550–330 BCE, the Mauryan Empire from 321 to 185 BCE, the Roman, Han, Byzantine, Qing, Mongol, Arabian, Ottoman, Ashanti, and Mughal empires.

The modern European empires were transformative through their incorporation of the Industrial Revolution. The British, French, Dutch, German, and Japanese empires were built from steel, oil powered ships, railroads, gasoline powered vehicles. The Russian and American empires combined these in the Information Age with nuclear power and nuclear weapons. A critique of the defense of the national state as a principal purpose of universities has been made the 1996, "The University in Ruins" (Readings, 1996).

Empires have survived for various lengths of time, sometimes lasting for a number of centuries. Imperial overstretch often exhausted them. This happened most recently with the Soviet empire, which broke up as many of its satellite states gained independence. It may be happening now with the American empire, with a state that is quickly becoming hopelessly indebted. Hundreds of US military bases add to a military budget that is equivalent to those of the next twenty states combined—and to US budget deficits that, along with entitlements and the interest on previous borrowing, add to the skyrocketing of American borrowing.

The struggles for power within empires and between some of them are the stuff of traditional history. The endless lists of battles and army flanks can make for a depressing account of the human past. Homer's account of the Trojan War is heroic enough, but it is also just another deadly battle scene. And things don't seem to have improved much. We started the twentieth century with a war to end all wars, followed by a horrific Second World War 20 years later. Since the end of WWII, there have been about 250 wars with over 50 million people killed, tens of millions more wounded, and countless made homeless. Emergence is at best a very mixed bag.

3.4 Multidirectional Emergence

The substantiated account from the Big Bang to ourselves today is a phenomenal accomplishment of humanity that can be told if the evidence offered by the disciplines in a university is integrated. How nations and civilizations emerged and developed are meaningful to many; placing humanity within our common universal context needs to be as well. The answer to the question of where we came from and how we got here is fantastic and as true as we can know it now. This account, told briefly above, is marvelous in itself. It also may well have profound effects on its readers and listeners. Perhaps this story teaches us to be not only national citizens, but in a way, global and universal ones as well. All of us have common origins in the Big Bang, LUCA, and small bands in Africa.

However, complexity is not only a universal Great Chain of Being from the Big Bang to ourselves now. The Big Bang leads not only to our Milky Way, our sun, our planet, and our species—although it does do that as well. It also leads to many types of stars: main sequence, red giants, white dwarfs, neutron, and others (NASA, n.d.-b). Stars are often categorized by how hot to cool from O, B, A, F, G, K, and M. Each star within one of the letter classes is placed along a spectrum numeric spectrum from zero being hottest and nine being coolest. It leads to a trillion other galaxies of various types: spiral, elliptical, lenticular, "irregular, active, seyfert, quasars, and blazars" (NASA, n.d.-c). It leads to many other types of planets than Earth. Within our own solar system, there are gaseous and terrestrial planets of different sizes and structures. And in recent years we have increasingly learned about the range of plants in other solar systems about five thousand light years from us or less. These nearby Milky Way exoplanets are divided into four types: gas giants, Neptune-like, super-Earth, and terrestrial (NASA, n.d.-d). And then there is the matter of life forms in addition to humanity. There is a dizzying range of life forms on Earth, with astrobiology and habitable planets suggesting that there may well be an even greater number elsewhere. The universe has by no means developed in only one direction.

It is important to be well aware of how the universe has led to so much more than ourselves. If we think that we are the most important outcome, or the only one worth studying, we might wonder what is the point of the rest of such vast universe? Who needs galaxies billions of light years away? This sort of species solipsism is similar to the unfortunate existence of the pathological and narcissistic individualist who sees everything only in relationship to him/herself. Humanism is a valuable antidote to various forms of groupism; but it can lead to its own dismissal of the value of non-humans. A mature person is aware of, and values, other persons for themselves. A mature species cares about other species and other worlds. Reveling in a universal emergence that leads to so much other than ourselves is a joy. Studying and caring about the emergence of other species on Earth and other types of planets, stars, and perhaps universes change and enrich our own perspective here and now.

3.4.1 The Humanities

The account of emergence of many different types of galaxies, stars, planets, life forms—and perhaps even other universes—creates a recognition of, and appreciation for, all kinds of diversity. There is no one privileged story in our story. When we measure the red shifts of other galaxies, it appears to us that we are at the center of the universe and everything else is racing away from us. However, if we were sitting in another galaxy, it would look the same. We are the center of the universe. So is every other place. We live within a magnificently large and complex and diverse whole.

This makes the universe story of complexity more consistent with post-modern humanities disciplines than is often thought. Post-modern scholars often see science and modernity as going together. The fear is that modern science is an attempt to establish claims to objective truth that are used to develop a grand narrative in the service of power. Modernity comes out of a mostly Western European process from the Italian Renaissance, French Enlightenment, English scientists and mathematicians as Isaac Newton and Charles Darwin, the German Albert Einstein, and many others. The intellectual hegemony of Western Europe dismisses non-Western cultures, a form of neo-colonialism.

As a rule, the sciences carry out experiments in laboratories in order to analyze nature. The humanities as a rule study and produce written texts in libraries and archives in order to examine and imagine humanity. As Carl Sagan (1980) said:

> What an astonishing thing a book is. It's a flat object made from a tree with flexible parts on which are imprinted lots of funny dark squiggles. But one glance at it and you're inside the mind of another person, maybe somebody dead for thousands of years. Across the millennia, an author is speaking clearly and silently inside your head, directly to you. Writing is perhaps the greatest of human inventions, binding together people who never knew each other, citizens of distant epochs. Books break the shackles of time. A book is proof that humans are capable of working magic.

Among the key teachings of the study of the humanities, of reading books and other materials that have been produced throughout the world, is the stunning variety not only of the natural world but also of the human experience. The Big Bang develops into many varieties of nature and human nature.

The masterful reflection on the *Pale Blue Dot* by Carl Sagan has rightly been frequently repeated as a powerful humanistic piece by a revered scientist (Sagan, 1994). It was inspired by the picture taken by Voyager 1 in 1990 at Saga's suggestion NASA (n.d.-a). It turned around so its camera could take a picture of Earth from 4 billion miles away. The Earth appeared as a tiny dot. Sagan wrote,

> Look again at that dot. That's here. That's home. That's us. On it everyone you love, everyone you know, everyone you ever heard of, every human being who ever was, lived out their lives. The aggregate of our joy and suffering, thousands of confident religions, ideologies, and economic doctrines, every hunter and forager, every hero and coward, every creator and destroyer of civilization, every king and peasant, every young couple in love, every mother and father, hopeful child, inventor and explorer, every teacher of morals, every corrupt politician, every "superstar," every "supreme leader," every saint and sinner in the history of our species lived there—on a mote of dust suspended in a sunbeam.
>
> The Earth is a very small stage in a vast cosmic arena. Think of the rivers of blood spilled by all those generals and emperors so that, in glory and triumph, they could become the momentary masters of a fraction of a dot. Think of the endless cruelties visited by the inhabitants of one corner of this pixel on the scarcely distinguishable inhabitants of some other corner, how frequent their misunderstandings, how eager they are to kill one another, how fervent their hatreds.
>
> Our posturings, our imagined self-importance, the delusion that we have some privileged position in the universe, are challenged by this point of pale light. Our planet is a lonely speck in the great enveloping cosmic dark. In our obscurity, in all this vastness, there is no hint that help will come from elsewhere to save us from ourselves.
>
> The Earth is the only world known so far to harbor life. There is nowhere else, at least in the near future, to which our species could migrate. Visit, yes. Settle, not yet. Like it or not, for the moment the Earth is where we make our stand.
>
> It has been said that astronomy is a humbling and character-building experience. There is perhaps no better demonstration of the folly of human conceits than this distant image of our tiny world. To me, it underscores our responsibility to deal more kindly with one another, and to preserve and cherish the pale blue dot, the only home we've ever known.

3.4.2 Static Levels of Complexity

So far, we have been tracing some of the thresholds or transitions between increasingly complex sets of relationships within relationships. However, it is usually more common for each level of complexity to remain more or less as it is. There are still vast clouds of unattached atoms of hydrogen and helium in the interstellar medium. They have not become anything more than that. There are vast numbers of planets that do not have the tectonic plates nor the layers that Earth has, and no live has emerged on them. On Earth, there are huge numbers of single cell organisms that have not become more complex life forms. The "Boring Billion" between 1.8 and 0.8 billion years ago showed minor tectonic developments, climatic change, and or biological evolution. Things can stay pretty much the same many times and places for a very long time.

Once we move into biological evolution, it is important to recall that there was no steady rise from simplicity to complexity. LUCA presumably was a prokaryote cell nearly 4 billion years ago. There are still enormous number of prokaryote cells that evolve without ever becoming eukaryote cells or multicellular life forms. With each new level of biological complexity, many species stay with their level without change. Sponges, cndarians, colelacanths, and many life forms that are millions and hundreds of millions of years old remain at their level of complexity.

Within humanity, there are many people who might be considered tribal or national and who have no interest in more complex forms of relationships. They count how many medals their nation's athletes win at the Olympics and care little for vague notions of the three values of Olympism: excellence, friendship and respect, which "constitute the foundation on which the Olympic Movement builds its activities to promote sport, culture and education with a view to building a better world" (International Olympics Committee, n.d.). Try finding a Human Passport that will get you through customs at the airport. There is no legal global citizenship.

At each level of complexity, from a hydrogen proton to nations, stasis is the more common part of the theme of complexity. There is often considerable time between levels of complexity if there is any emergence. In order to transition to a more complex relationship, it can at times take considerable energy and effort. For example, fusing hydrogen and helium into heavier elements takes enormous heat produced by gravity's creating greater densities. Moving from villages and cities into nations and empires has often required wars.

3.4.3 Greater Simplicity

In addition to the emergence of more complex relationships within relationships, multidirectional emergence, and statis, there is also a common experience of greater simplification. Relationships break down. Stars burn out and "die" or blow themselves up in super novae. Cells fail to relate as well as before and life forms age and die.

Various reasons for extinction have often killed off many complex life forms and sent the story back to simpler times. Five major extinction periods between 450 mya and 65 mya caused huge interruptions. And those followed the even more destructive Oxidation event. This is only part of the reason why over 99% of the species that have ever existed are now extinct. We may be going through a sixth (self-induced) extinction period that we hope does not conclude with our own species' disappearance. It would be a shame to be a mere 300,000 year-long flash in a pan. If we are, we may well have left an Earth with lots of extinct life forms and a host of simpler ones.

Dark energy seems to be pulling galaxies apart, perhaps followed by pulling apart even elements, and may pull everything apart into an end of existence as we know it. Origins and emergent complexity are more fun to discuss, but decline, disarticulation, collapse, endings, and death are just as real—and may have the last word.

Our sun will become increasingly hot until life on Earth may be impossible within a few billion years. It will become a Red Giant within 5 billion years, expanding until it evaporates Earth's oceans and fries any creature still hanging onto life. There will be a long universal future after the Earth is gone. However, eventually, dark energy may pull all the galaxies in our universe apart. Many keep vanishing beyond an event horizon, never to be seen again by us. Given enough time, most of the galaxies in our universe will have sped out of our view, leaving us with a mostly black sky. And then, our own local galaxies and even matter might come apart. William Butler Yeats (1919) and Chinua Achebe (2010) were indeed right; things do indeed fall apart. Or things get ripped apart. In the long run, everything. In this view, the Big Rip follows the Big Bang. It is not only we as persons and as a species that will end, it is our solar system and our entire universe—perhaps. And the Big Chill follows the Big Rip, with a return to absolute dark and cold. Or perhaps that is not the end.

3.4.4 Emergent Complexity from Conflict and Destruction?

There can mistakenly be a comforting feeling about emergent complexity, where it does exist. Optimists have long hoped that there is progress, that the arc of the universe is long, but it bends toward what is good, that it is getting better all the time. Emergence is often associated with violence of some sort. The beginning of the universe is probably misnamed, but the term Big Bang does not get the story off to a peaceful start. Gravity creating the densities that force protons together produce the fusion and energy of hug number of nuclear bombs going off every second just in our own sun. Exploding stars in their death throes, or neutron stars colliding, are needed to produce elements heavier than iron. Once we have biology, we have a never ending arms race. Creatures are endlessly creative in devising toxic chemical assaults, infections, harpoons, fangs, talons, wings, shells, brains, and any number of defensive and offensive weapons. Once we get to human politics, there is a bellicist theory of the state that sees that war has often made the state and the state often makes war. If humanity ever does get itself structured globally, it may well be an outcome of violent struggle.

3.4.5 Infinite Complexity

The end of our universe may lead to the birth of new ones. And new universes, some still born and others different from anything we can imagine, or others almost identical to ours, may be already be out there in infinite numbers. We have increasingly seen our own universe as so much more complex than we had imagined it to be. Now many suspect that our notions of complexity need to become infinitely more complex.

So, there we have it. Existence includes emergent complexity of relationships within relationships, the multidirection of emergence, stasis, emergent simplicity and the breakdown of relationships, or an infinite multiverse that is beyond our imagination. This draws on all the natural science, social science, and humanities disciplines of contemporary universities. It offers a theme that synthesizes knowledge rather than dissecting it (Morin, 2001). It points to a universe of knowledge that provides a rationale for a university. Every person from every location, every life form, every planet, every galaxy is included in the account of complexity. This is a story of unity and diversity, fact and imagination, and relationship and individual uniqueness that could reanimate a new idea of the university in our time.

References

Achebe, C. (2010). *The African trilogy: Things fall apart; No longer at ease; Arrow of god*. Everyman's Library.

Brown, C. B. (2007). *Big History: From the Big Bang to the present*. New Press.

CERN. (n.d.). *The Big Bang should have created equal amounts of matter and antimatter. So why is there far more matter than antimatter in the universe?* Retrieved from https://home.cern/science/physics/matter-antimatter-asymmetry-problem

Chaisson, E. (2001). *Cosmic evolution: The rise of complexity in nature*. Harvard University Press.

Chaisson, E. (2006). *Epic of evolution: Seven ages of the cosmos*. Columbia University Press.

Christian, D. (2004). *Maps of time: An introduction to Big History*. University of California Press.

Christian, D. (2008). *Big History, the Big Bang, life on Earth, and the rise of humanity*. The Great Courses.

Christian, D. (2010). *Big History: From the Big Bang to us and the future*. Lecture at Villanova University.

Christian, D. (2011). *The history of our world in 18 minutes*. TED talk. Retrieved from http://www.ted.com/talks/david_christian_big_history

Christian, D. (2015). *How did we get here? Big History 101*. World Economic Forum Annual Meeting 2015. Retrieved from http://www.weforum.org/sessions/summary/how-did-we-get-here-big-history-101

Christian, D., Cynthia, S. B., & Craig, B. (2014). *Big History: Between nothing and everything*. McGraw Hill Education.

Donald, M. (1991). *Origins of the modern mind: Three stages in the evolution of culture and cognition*. Harvard University Press.

Frank, A. (2023, June 29). Scientists found ripples in space and time. And you have to buy groceries. *The Atlantic*. Retrieved from https://www.theatlantic.com/science/archive/2023/06/universe-gravitational-waves-nanograv-discovery/674570/

Greene, B. (2011). *The hidden reality: Parallel universes and the deep laws of the cosmos*. Alfred A. Knopf. https://doi.org/10.1002/9781118080835

Gustafson, L. (2013, May 30). Big History and big politics. *Metanexus*. Retrieved from https://metanexus.net/big-history-and-big-politics/

Gustafson, L. (2024). Complexity: A Rationale for the University. *Journal of Big History, 7*(2), 106–124. https://doi.org/10.22339/jbh.v7i2.7209

Haraway, D. J. (2016). *Staying with the trouble: Making Kin in the Chthulucene*. Duke University Press. https://doi.org/10.1215/9780822373780
Hooper, D. (2019). *At the edge of time: Exploring the mysteries of our universe's first seconds*. Princeton University Press.
International Olympics Committee. (n.d.). *Olympic values*. Retrieved from https://olympics.com/ioc/olympic-values
Library of Congress. (n.d.). Seth MacFarlane Collection of the Carl Sagan and Ann Druyan Archive (Collection Identifier: MSS85590). Manuscript Division, Library of Congress, Washington, D.C. Retrieved March 18, 2026, from https://findingaids.loc.gov/repositories/19/resources/5583. Also see the LOC page, Retrieved March 18, 2026, from https://www.loc.gov/static/collections/finding-our-place-in-the-cosmos-with-carl-sagan/about-this-collection/overview.html
Ladyman, J., Lambert, J., & Wiesner, K. (2013). What is a complex system? *European Journal for Philosophy of Science, 3*, 33–67. https://doi.org/10.1007/s13194-012-0056-8
LePoire, D. J., Grinin, L., & Korotayev, A. (2026). Introduction to complexity theory for Big History. In D. J. LePoire, L. Grinin, & A. Korotayev (Eds.), *Complexity in universal evolution: A Big History perspective* (pp. 29–90). Springer Nature. (In this volume). https://doi.org/10.1007/978-3-032-20516-2_2
Mitchell, M. (2009). *Complexity: A guided tour*. Oxford Academic. https://doi.org/10.1093/oso/9780195124415.001.0001
Morin, E. (2001). *Seven complex lessons in education for the future*. UNESCO.
NASA. (n.d.-a). *Voyager 1's Pale Blue Dot*. https://science.nasa.gov/mission/voyager/voyager-1s-pale-blue-dot/
NASA. (n.d.-b). *Types of stars*. Retrieved from https://universe.nasa.gov/stars/types/
NASA. (n.d.-c). *Types of galaxies*. Retrieved from https://universe.nasa.gov/galaxies/types/
NASA. (n.d.-d). *What is an Exoplanet?* Retrieved from https://exoplanets.nasa.gov/what-is-an-exoplanet/planet-types/overview/
Readings, B. (1996). *The university in ruins*. Harvard University Press.
Sagan, C. (1980). What an astonishing thing a book is. *Cosmos: A Personal Voyage*, Ep.11, The Persistence of Memory [Online video].
Sagan, C. (1994). *Pale Blue Dot: A vision of the human future in space*. Random House.
Sing, S. (2004). *The origin of the universe*. Fourth Estate. https://doi.org/10.1017/CBO9780511756599
Spier, F. (2015). *Big History and the future of humanity*. John Wiley & Sons. https://doi.org/10.1002/9781118949807
Steinhardt, P. J., & Turok, N. (2007). *Endless universe: Beyond the Big Bang*. Doubleday.
Tomasello, M. (1999). *The cultural origins of human cognition*. Harvard University Press.
Wilson, E. O. (2012). *The social conquest of the Earth*. Liveright.
Yeats, W. B. (1919). *The second coming*. Retrieved from Poetry Foundation. https://www.poetryfoundation.org/poems/43290/the-second-coming

Lowell Gustafson is Professor of Political Science at Villanova University in Pennsylvania (USA). His course on "Our Social Nature" uses a Big History approach, and he has also taught it at the Graterford maximum security prison near Villanova. His research has included how science explains the origin and development of polity and how emergent complexity provides an intellectual rationale for universities. He has served as secretary, vice-president, and president of the International Big History Association (IBHA), and as editor of *Origins: The Bulletin of the IBHA*, and the *Journal of Big History*.

Complexity Pyramids and Energy Rate Density: A Quantified System Hierarchy for Big History

4

Martin van Duin

Abstract

In their position paper for a Big History 2.0, Henriques & Volk (*Journal of Big History*, 6(3), 1–4, 2023) have advocated to distinguish between (1) emergence in realms *versus* that when moving from one realm to another, and (2) the complexification path from quarks to culture *versus* that of astronomical objects, which are not part of that sequence. In his re-examination of energy rate density (ERD; proposed by Chaisson [*Cosmic evolution; The rise of complexity in nature*. Harvard University Press, 2001] as a metric for complexity), Solis (*Journal of Big History,* 7(1), 19–28, 2024) has concluded that there is a need to further explore systems at different levels of hierarchy to see if there is a general correlation between ERD and complexity. This study follows up these suggestions by applying the "complexity pyramid" concept, proposed by Reeves (*The hour of our delight—Cosmic evolution, order and complexity.* W. H. Freeman and Company, 1991), as a framework to represent the hierarchical organization of (complex) systems. Applying a rigorous part/system approach, a series of complexity pyramids is constructed for systems from the material, astronomical, biological, technological, and social realms. Next, ERD is used to quantify the variation in horizontal complexity on a pyramid level (size and diversity) and the change in vertical complexity in a pyramid (emergence). The main observation is that ERD of a system is typically smaller than the largest ERD of its parts on the level below. This results in a narrowing of the ERD range of systems on a next level and a decrease in ERD of system's "engines," when moving upward in a complexity pyramid. This is rather counterintuitive, since complexity and thus ERD are supposed to increase, when parts combine in a larger system with emergent properties at the next level. However usually, both active, energy-processing parts (ERD > 0) and inactive, structural parts (ERD = 0) converge

M. van Duin (✉)
Sittard, The Netherlands

D. J. LePoire et al. (eds.), *Complexity in Universal Evolution*, World-Systems Evolution and Global Futures, https://doi.org/10.1007/978-3-032-20517-9_4

to the larger system. ERD of that larger system is the mass-average of the ERDs of the (in)active parts and, consequently, is typically smaller than the largest ERD of its parts. Finally, the complexity pyramids for the various realms are integrated in an overlay, representing "all" systems from the physical reality of our world. The ERD quantification of the systems in the individual pyramids allows the quantitative positioning of the systems in this overlay. A comparison of ERD values of systems from two different complexity pyramids results in *pars-pro-toto* issues, because there is no objective way to establish which levels from different pyramids may be compared. Chaisson's sequence of galaxies → stars → planets → plants → animals → brains → society is not only a rather tortuous path through the pyramid overlay, but the observed increase in ERD over big time is also somewhat problematic.

Keywords

Big history · Universal evolution · Energy rate density · Complexity · Complex adaptive system · Complexity pyramid

4.1 Introduction

Big history (BH) is the narrative and the study of the origin and development of our world.[1] BH starts at the Big Bang and proceeds *via* galaxies, stars, and planets, as well as living organisms (including hominins), machines, and social systems up to the human society on Earth today (Christian, 2005; Stokes Brown, 2012; Spier, 2015; Fewster, 2016). BH is not just multidisciplinary, but interdisciplinary, and often transdisciplinary.[2] Typically, BH is explored in the context of three realms,[3] namely the cosmological (material and astronomical), biological, and cultural (technological and social) realms (Chaisson, 2001; Volk, 2020). Various frameworks have been proposed to structure BH, each emphasizing different aspects, such as:

- the distinction between micro- *versus* macroevolution (Jantsch, 1980);
- the increase in free-energy rate density (free-energy flow, normalized to mass) of dissipative systems over deep time (Chaisson, 2001);
- the thresholds between major eras (Christian, 2005);
- the goldilocks conditions enabling major transitions (Spier, 2015);

[1] "Our world" refers to everything that we encounter in our daily lives and anything that exists in the universe, namely the whole of our reality.

[2] Multidisciplinary: combining knowledge from various disciplines, such as cosmology, astronomy, geology, geography, biology, history, economy, and sociology; interdisciplinary: crossing interdisciplinary boundaries and integrating (insights from) different perspectives; transdisciplinary: creating a holistic approach.

[3] A "realm" is characterized by a particular set of system parts and structures, interactions and forces, as well as mechanisms for matter, energy, and information processing.

- the coevolution of systems in the various realms with top-down and bottom-up causation (Walker et al., 2012; Quaedackers, 2024) ;
- the importance of information processing in addition to matter and energy processing (Jantsch, 1980; Christian, 2005; Smart, 2009; Solis, 2018; Grassie, 2018; Azarian, 2022).

Today, BH still shows the diverging characteristics of a young discipline with authors using different terminologies and definitions, as well as proposing various "theories of everything" in a BH context (cf. article archive of *Journal of Big History*[4]). It appears to be difficult to align and integrate the various frameworks and converge to a single, theoretical basis for BH, probably because of the enormous width and complicatedness of the field, the continuing, scientific progress in all the underlying disciplines, and, possibly, also the rigidity of the experts involved.

Probably the most striking trend throughout BH is the apparent increase in the complexity of systems, originating and evolving locally over time and, on top of that, at an apparently increasing rate (Jantsch, 1980; Reeves, 1991; Chaisson, 2001; Christian, 2005; Kurzweil, 2005; Spier, 2015; Fewster, 2016).[5] The term "apparent (ly)" is used here on purpose, because complexity is typically judged qualitatively and intuitively. There is not yet a generally accepted definition of a complex system and a single, practical measure for complexity is lacking (Mitchell, 2011; Ladyman *et al.*, 2013; Siegenfeld & Bar-Yam, 2020; Lloyd, n.d.).[6] This is probably the result of the very large diversity of systems considered, especially in a BH context, with different structures, as well as different matter, energy, and information processes. Typically, complex systems are described as ensembles of interacting parts with new properties emerging at the system level. Systems themselves are often parts of a next-level system, resulting in a nestedness and hierarchy of systems. Systems at the higher end of the complexity scale, especially biological and social systems, have attributes, such as self-organisation, robustness, and adaptivity. In their position paper for a BH 2.0, Henriques & Volk (2023)[7] have advocated to distinguish between (1) the emergence in realms *versus* the emergence when moving from one realm to another, and (2) the specific path of complexification in the grand sequence from quarks to culture *versus* the evolution of aggregates which are not part of that sequence, such as stars, planetary systems, and galaxies.

Man-made pyramids have a broad base that supports the bulk up to the top. In a more abstract fashion, pyramids are often used as metaphors to visualize the hierarchical organization of systems with different levels, as for example for feudalism, public governance, company, army, ecosystems, capitalism, culture, human needs

[4]https://jbh.journals.villanova.edu/issue/archive

[5]See also chapter "The Evolution of Complexity in the Stelliferous Era: Observations and Patterns" (Korotayev & LePoire, 2026) in this volume.

[6]For a review of various operationalizations of complexity, see chapter "Introduction to Complexity Theory for Big History" (LePoire *et al.*, 2026) in this volume.

[7]See also Volk & Henriques (2025).

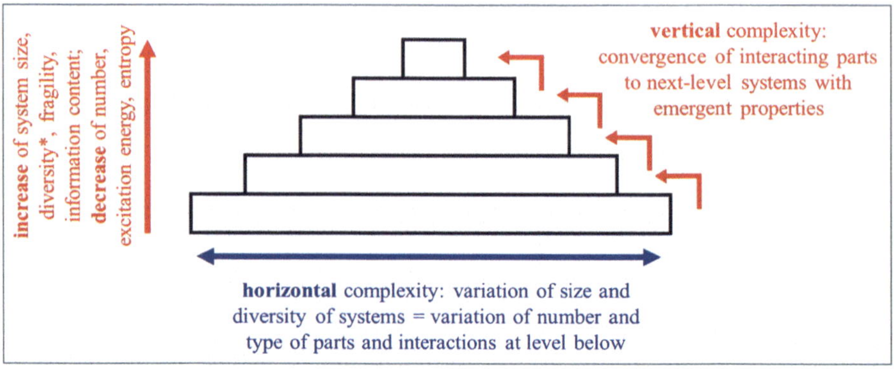

* at higher pyramid levels, diversity decreases again: see text

Fig. 4.1 Horizontal and vertical complexity in a complexity pyramid, following part/system hierarchy; * at higher pyramid levels, diversity decreases again: see text

and motivation, safety, energy saving, information, and network theory, to name just a few. Often the hierarchy in these pyramids represents differences in importance and status. In this study, the term "hierarchy" will be used as a neutral term though, referring to the systematics of the part/system organization. Reeves (1991) has coined the term "complexity pyramid" for a complex system hierarchy in such a strict part/system context. He has organized systems in a pyramid with quarks and electrons at the basis, nucleons, atoms, simple molecules, biomolecules, and cells at subsequently higher levels, and, finally, organisms at the highest level. This sequence of systems follows the direction of unfolding time, since first smaller parts had to be present before larger, next-level systems could be formed. Reeves has discussed the change of several characteristics, when moving upward in this complexity pyramid (Fig. 4.1):

- the size of the systems becomes larger at each successive level, while the number of systems becomes smaller, which is illustrated by the narrowing of the pyramid to the top;
- the systems become more diverse with new properties emerging at each next level;
- the local entropy of the systems increases, which is compensated by a larger increase in the global entropy, and the information content of the systems increases; both are interpreted as an increase in complexity;
- the excitation energy for transitions between ground and excited states of the material systems decreases; as a result, the systems become increasingly more fragile.[8]

Similar, part/system pyramids are used, for example, in:

[8]The magnitude of the binding energy of material systems more or less parallels their excitation energy and, thus, also decreases when going upward in the material complexity pyramid.

- the astronomical realm: hydrogen/helium (H/He) plasma, stars, stellar clusters, galaxies (groups and [super]clusters), cosmic web, (observable) universe, and possibly multiversum;
- the biological realm: biomolecules, organelles, cells, tissues, organs, organ systems, and organisms;
- the technological realm: materials, parts, machines, and chemical and power plants;
- the social realm: traditionally: human individual, family, clan, tribe, and people; modern: human, family, city, nation, and global society ;
- non-physical realm: consciousness, mind, and spirit.

The complexity pyramid resembles the iceberg metaphor in system theory, not only in its inverted shape but also in its reference to emergence, that is, the visible reality of observable events is a manifestation of patterns and structure that are below the surface and cannot be directly observed (Vandenbroeck, 2015). The complexity pyramid provides a clear distinction between horizontal and vertical complexity (Fig. 4.1):

- horizontal complexity: systems consist of various numbers and types of parts and, thus, vary themselves in size and type on a particular pyramid level; they have similar structural features as well as exploit similar energy and information processing mechanisms and
- vertical complexity: Follows the part/system hierarchy with new systems with different structures, different energy, and information processing, as well as new, emergent functions at the next level.

In organization theory, such a distinction between horizontal and vertical complexity is visualized in an organization tree. The different departments at a single level represent the horizontal complexity, while the hierarchical line represents the vertical complexity (Daft, 2010; Siegenfeld & Bar-Yam, 2020). A similar distinction is made for the evolution of biological complexity between the number of different part types at a given hierarchical level *versus* the number of levels of nestedness (McShea, 2021). Anderson has made the same distinction with "more is different" at the horizontal level and symmetry breaking in the vertical direction (Anderson, 1972). The complexity pyramid framework is very well suited to illustrate this distinction between horizontal and vertical complexity. It also clarifies similar points made in a BH context by Spier (2015) on more and more varied interactions among increasing numbers of more varied building blocks (horizontal complexity), while emergence of new complexity (vertical complexity) is a key element in most BH frameworks.

There are issues with some of these different versions of the complexity pyramid. First, most of these pyramids follow a pattern with only one type of parts converging to the larger system. In reality, in most pyramids, different types of parts converge to the larger system. For example, cells are not just composed of organelles but also of plasma (water with dissolved molecules and salts), as well as membraneless structures and other functional components. Sometimes, systems from different realms are combined in one complexity pyramid, which is somewhat awkward, because of step-changes in material structure and mechanisms of matter, energy, and information processing. For example, atoms and molecules from the material realm are combined with cells and organisms of the biological realm, and sometimes even with cities and nations from the social realm. It is more elegant to construct multiple complexity pyramids for different realms, which aligns with Baskin's (2023) view on the universe as a collection of interconnected, nested networks. In addition, sometimes a temporal or evolutionary link is shown, for example, prokaryotic cell → eukaryotic cell in the biological pyramid (Jagers op Akkerhuis, 2010; McShea, 2021). Indeed, endosymbiosis of prokaryotic cells has resulted in the origination of energy-converting organelles in the first eukaryotic cells, such as mitochondria and chloroplasts. However, prokaryotes are not parts of eukaryotes and, thus, should not be included in the biological complexity pyramid. Admittedly, (proto)galaxies are the nurseries of stars, but stars are the parts forming a galaxy. Finally, hierarchic links, such as employee → manager → chief executive officer, do not represent a part/system organization either.

Dissipative systems are a particular class of complex systems, which are defined as open, out-of-equilibrium systems with energy converted by the systems to maintain their energy gradients (Chaisson, 2001; Niele, 2005). Illustrative examples of dissipative systems (their corresponding energy processes between brackets) from the cosmological, biological, and cultural realms are stars (nuclear fusion), black holes (matter accretion), plants (photosynthesis), animals (aerobic respiration), cars (fuel combustion), and cities (oil, natural gas, electricity). Mass is a measure for the size of the system. Energy rate (ER = power in J/s = W) is a measure for the distance of the system from equilibrium and its effort to maintain its complexity. Chaisson (2001, 2010, 2014) has proposed energy rate density (ERD in W/kg), defined as ER normalized to mass (in kg), as a single metric for the complexity of a dissipative system. ERD is easy to calculate and applicable to systems from all realms. For example, ERD of a star is calculated from its luminosity, ERD of a black hole from its matter accretion rate, ERD of a plant by the amount of carbohydrate formed, ERD of an animal by its oxygen consumption, ERD of a driving car by its fuel consumption, and ERD of a city by its total energy consumption. Chaisson has shown that ERD has increased super-exponentially over deep time, thus quantitatively corroborating the intuitive notion that complexity has increased during cosmic evolution and BH. Quite some issues have been identified with respect to the calculation of ERD and its application in a BH context though (Spier, 2005, 2015; Solis, 2024; van Duin, 2024a, 2025a, 2025b).[9] Amongst others:

[9] An issue that seems to have not been raised so far, is that Chaisson's sequence of systems with increasing ERD over time resembles that of a list of records, for example, for speed. Not just the speed records on the marathon are compared over time, but also those on other track athletic distances (10 km, 1000 m, 100 m) and other sports (skating, cycling). And also speed records outside the human regime are compared, such as those of other animals (birds, gazelles, jaguars), machines (cars, airplanes, rockets), and even astronomical objects (asteroids, planets, stars). This

- ERD is not applicable to inactive systems without energy processing (matter, parts);
- ERD varies with the activity level of the system;
- ERD decreases when the efficiency of energy processing increases;
- ERD underestimates the complexity of biological and social systems for which information processing is dominant;
- Chaisson applies a limited definition for the mass of the Earth and social systems (Sect. 4.2.2).

In addition, Chaisson (2001) has compared the ERD values of the human brain and body with those of chips and computers, as well as those of fighter jets and the human society. Dissipative systems are open systems with matter and/or energy being exchanged with their environment through their permeable boundaries. Consequently, they are always parts of a larger, next-level system. As a result of this nestedness of complex and dissipative systems, one runs into a selection issue when one wants to compare systems from different complexity pyramids. There is no objective criterion to establish which systems from which levels of two different complexity pyramids may be compared. Referring to this *pars-pro-toto* (part for the whole) issue, Solis (2024) has concluded that there is a need to further explore many examples of systems at different levels of hierarchy, to see if there is a general correlation between ERD and complexity.

This study will make preliminary steps to achieve some convergence in the BH discipline and, in a way, follows the suggestions by:

- Henriques & Volk (2023): to distinguish between the emergence in realms and that when moving between realms, as well as between the complexification path from quarks to culture and that of astronomical objects and
- Solis (2024): to further explore systems at different levels of hierarchy to establish a general correlation between ERD and complexity.

This study will do so by constructing a series of complexity pyramids for different realms and using ERD to quantify the systems at the various levels, thus combining two approaches discussed above. In more detail:

1. the complexity-pyramid concept will be applied as a qualitative framework to structure system hierarchy in the various realms, decoupled from evolution history; next, these pyramids will be combined in an overlay to cover "all" systems in our world and

appears like an apples-and-pears comparison, but actually speed is a sound metric for how fast objects are moving. And in the same way, ERD is an appropriate metric for how far away dissipative systems are from equilibrium and how much effort systems have to make to maintain their energy gradients.

2. ERD will be used to quantify the systems at the various levels in the different complexity pyramids and, subsequently, to quantitatively connect the individual pyramids in the pyramid overlay.

The first step will help in clarifying the confusing situation when dealing with and comparing "all" (complex) systems of our world by making a careful distinction between horizontal (variation of size and diversity) and vertical complexity (next-level emergence) within a particular complexity pyramid. In addition, separate complexity pyramids will be constructed for different realms. The second step will demonstrate that despite the issues mentioned above, ERD can be used as a practical and universal measure in a complexity context. It will be shown that ERD decreases when moving upward in a complexity pyramid, which is a somewhat intriguing observation considering Chaisson's notion that ERD quantifies the complexity of a dissipative system. However, this trend is easily explained by understanding how parts converge into a next-level system with ERD of that larger system being the mass-averaged ERD of the parts. Finally, it will be shown that the typical BH timeline follows a rather tortuous path through the overlay of complexity pyramids.

4.2 Complexity Pyramids and Energy Rate Density

4.2.1 Complexity Pyramids

Hierarchies of complexity pyramids for the material, astronomical, biological, technological, and social realms are shown in Table 4.1a–e, respectively, and will first be discussed qualitatively. For each pyramid level, representative system groups are listed together with a specific system, typically the "engine" (see Sect. 4.2.4). The latter will be used to construct the pyramid overlay in Sect. 4.2.5. The numbers[10] between brackets in Table 4.1 correspond to (ranges of) ERD values, which will be discussed in detail in the next section. The key criterion for identifying a next-level system has been the convergence of parts to a new system, moving upward in the complexity pyramid. This is accompanied by the emergence of new properties and functions, resulting from the collective interactions of the parts at the previous level. These emergent characteristics are not shown by the interacting parts ("the whole is more than the sum of its parts") and are unexpected, unpredictable, and even surprising for higher-end complex systems.[11]

[10] The American number notation is used with a comma to separate thousands and a decimal point to separate whole numbers from fractions; for very large and small numbers the scientific notation is used.

[11] Unexpected, unpredictable, and surprising properties are often used as characteristics to define the emergence of complex systems at a next level. However, these terms do not provide a sufficient and absolute characterization, because of progressive scientific insight and understanding over time. Complexity is viewed here not as a black-to-white step change, but on a gradual scale with many shades of gray in between (more than 50 ☺). Admittedly, simple complex systems, such as

Table 4.1 Complexity hierarchies representative for (a) material, (b) astronomical, (c) biological, (d) technological, and (e) social realms; for each level, first a system group is listed with a range of ERD values (in W/kg), followed by a specific system (typically an "engine": see text) with a single ERD value; material systems have ERD = 0 (values omitted; except for a: order of magnitude of excitation energy [in J] in italic); for rationale behind choices of levels and systems, see text; data from (van Duin 2024a, 2025b)

realm	hierarchy level	system group	specific system
a: material	6	**materials**	polymers, minerals, metals
	5	molecules (~10^{-24}); salts	H_2, H_2O, DNA; NaCl, $CaCO_3$, SiO_2
	4	atoms (~10^{-19}); ions	H, He, C; Na^+, Cl^-, CO_3^{2-}
	3	nuclei (~10^{-12})	^1H, ^2He, ^6C
	2	nucleons (~10^{-10})	proton, neutron
	1	fundamental particles	up & down quarks, electron
b: astronomical	4	**universe**	observable universe (10^{-5})
	3	galaxies (10^{-5} - $4x10^{-2}$)	Milky Way ($2x10^{-5}$)
	2	stars ($2x10^{-8}$ - 150), planets (10^{-11} - 10^{-6}), moons (10^{-10} - 10^{-7}), black holes (10^{-19} - 10^{-6})	Sun ($2x10^{-4}$)
	1	materials	plasma, gas, dust, ice, minerals
	3	**stars** ($2x10^{-8}$ - 150)	Sun ($2x10^{-4}$)
	2	stellar cores, shells, zones	Solar core ($4x10^{-4}$)
	1	plasma	H, He, "metals"
	4	**planets** (10^{-11} - 10^{-6})	Earth ($2x10^{-8}$)
	3	climaspheres, geospheres	Earth's climasphere (0.0075)
	2	atmospheres, hydrospheres, biospheres	Earth's biosphere (0.09)
	1	materials	water, air, minerals, organics
c: biological	6	**animals** (0.1 – 20)	human (2.0)
	5	organs (0.2 – 50)	brain (11)
	4	cells (1 - 30)	neuron (26)
	3	intracellular structures	mitochondrion (300)
	2	materials	water, lipid bilayers, respiratory complex (1,200)
	1	molecules	H_2O, lipids, cytochrome oxidase (1,700)
d: technological	4	**(super)computers** (2 - 250)	personal computer (55)
	3	keyboards, screens, batteries, CPUs (12 - $3x10^4$)	Intel Core i7 CPU ($2.6x10^4$)
	2	ICs ($6x10^3$ - $6x10^5$)	Intel Core i7 IC ($6x10^5$)
	1	materials	silicon, copper, plastics
	3	**windmills** (1 - 15)	Enercon E-126 (1.3)
	2	towers, rotor blades, foundations, generators	generator (13)
	1	materials	concrete, steel, composites
	3	**Solar panels** (19 - 27)	reidential Solar panel (24)
	2	frame, foil, cables, Solar cells (380 - 490)	Solar cell (489)
	1	materials	silicon, cupper, steel, plastic
	3	**cars** (1 - 130)	1994 car (130)
	2	chassis, chairs, fuel tanks, engines (5 - 820)	engine (820)
	1	materials	steel, plastics, rubber, glass, textile
	3	**airplanes** (500 - 2,000)	Boeing 747 (140)
	2	fuselages, radars, engines (10^4 - $3x10^4$)	engine (720)
	1	materials	aluminum, steel, plastics, composites
	3	**fighter jets** (2,000 - 9,000)	Lockheed Martin F-35 A (4,600)
	2	fuselages, guns, engines (9,000 - $1.4x10^5$)	engine ($3.6x10^4$)
	1	materials	aluminum, titanium, composites, plastics
	3	**space rockets** (10^4 - 10^6)	Space Shuttle ($1.4x10^4$)
	2	frameworks, engines, fuel tanks	engine ($1.5x10^5$)
	1	materials	aluminum, titanium, steel, composites, ceramics
	3	**ships** (0.07 - 23)	Titanic (0.85)
	2	hulls, ankers, engines (14 - 140)	main engine (14)
	1	materials	steel, aluminum
	3	**refrigerators** (4 - 7)	refigerator (5)
	2	shelfs, thermocouples, compressors (30 - 60)	compressor (45)
	1	materials	steel, plastics, freon
e: social	3	**ant colony** (0.2)	ant colony (0.2)
	2	ant hills (0.15 - 0.5), ants (2 - 4)	ants (3)
	1	materials	organic molecules & polymers
	5	**human society** (0.017 - 1.9)	today's global society (0.017)
	4	cities (0.01 – 0.03), plants (0.3 - 15), trains (0.4 - 40)	Barcelona (0.015)
	3	homes; cars (100 - 150)	single-family home (0.014)
	2	humans (1 - 3); boilers (80 - 800); lighting (40 - 300); appliances (0.7 - 4,000)	refrigerator (5)
	1	materials	bricks, concrete, glass, metals, organics

Rigorous application of the part/system hierarchy, while excluding evolutionary effects, has enabled the construction of the complexity hierarchies shown in Table 4.1 in a straightforward and natural fashion. Following a reductionist approach as a useful check, larger systems have been analyzed going downward in the hierarchy and diverged in smaller components. In principle, these bottom-up and top-down approaches are complementary and should provide the same overall hierarchy. The hierarchies shown do not provide a comprehensive overview of all systems in the whole world, but present a selection of illustrative and representative systems from all realms. Only systems with known ERD values have been included in order to enable quantitative discussions and avoid the presentation of just another qualitative framework to structure complex systems. The hierarchies shown in Table 4.1 are sufficient for the purpose of this study, enabling a sound discussion of general trends and making additional, comparable hierarchies redundant. Finally, they have been constructed in such a way that they overlap with each other, which will facilitate the overlay of pyramids, as presented in Sect. 4.2.5.

It is beyond the scope of this chapter to discuss all systems at all levels from all pyramids from Table 4.1 in detail. The reader has probably sufficient knowledge to understand their positions in the pyramids. More detailed information can easily be found in textbooks, encyclopedias, and Wikipedia. Material systems (Table 4.1a) are often viewed as not complex at all, but they are included here because they also follow the part/system hierarchy. Fundamental particles (here just quarks and electrons as building blocks of matter) form the basis of all other systems.[12] The hierarchy shown is adapted to Reeves (1991) with ions as well as salts and metals added next to atoms and molecules, respectively. Molecules, salts, and metals represent the three types of bonding: covalent with shared electron pairs, Coulombic with oppositely charged ions, and metallic with cations in an electron cloud, respectively. Molecules and ions converge to materials in different physical states (solid, liquid, gas, and plasma) with typical emergent properties, such as modulus, viscosity, pressure, and temperature. Note the distinction between N_2/O_2, H_2O, and $Si^{4+}O^{2-}{}_2$ as molecules and salt versus air, water, and silica, respectively, as materials.

Materials are not only presented as the upper level of the material pyramid, but also form the bases of many other pyramids. For example, various materials converge to a variety of astronomical systems (Table 4.1b) as next level, that is, stars,

molecules, stars, and planetary systems, can be viewed as deterministic and their properties can be fully described from first principles. Even the development of the cosmic web and the formation of galaxies can be simulated today on supercomputers (von Däniken, 2025). Higher-end complex systems, such as living organisms and social systems, have unexpected and unpredictable properties, including self-organization and adaptivity, which cannot (yet) be explained from their parts and interactions.

[12] Preons are hypothetical particles that may be components of fundamental particles like quarks and leptons, but direct experimental evidence is lacking. Alternatively, fundamental particles may be one-dimensional, vibrating strings with the vibration modes determining how the strings are perceived.

planets, moons, and (supermassive) black holes. In principle, asteroids and stellar remnants (white dwarfs, neutron stars) are also present at this level, but these are systems with (very) low ERD merely fading away. Galaxies used to be viewed as just a bunch of stars, but today it is becoming more and more clear that they are complex systems by themselves in steady states for billions of years. They have complicated structures and magnetic fields, as well as matter and energy flows, mediated by a central, supermassive black hole (Grete et al., 2025). Planetary systems (planets orbiting a star) and stellar systems (two or more stars orbiting each other), as well as galaxy groups and (super)clusters have not been included as separate levels. These systems are viewed as just gravitationally bound without any additional energy conversion and dissipation on the system level. This is substantiated by their ERD ranges fully overlapping with those of single stars and galaxies, respectively. The universe forms the top of the astronomical pyramid. Planets may have intricate mass and heat flow patterns in their molten interiors, gas atmospheres, and liquid hydrospheres, which are driven by temperature gradients derived from their original, molten states, radioactive decay, and stellar radiation. In addition, there may be life on planets resulting in biospheres, which are coupled to the planetary geosystems (Boston, 2008). They are thus viewed as dissipative systems, warranting a separate complexity hierarchy in Table 4.1b.

The complexity hierarchy for animals (Table 4.1c) represents the biological realm. Unfortunately, ERD data for active proteins, intercellular structures, and cells are very scarce. For tissues, no ERD values have been found at all[13] and, thus, they have been excluded. At the base, material level of the biological pyramid, energetically inactive matter, such as water, deoxyribonucleic acid (DNA), lipids, minerals, and structural proteins, complement active proteins, such as ATP-ase, DNA polymerase, and kinesin. Organelles have traditionally been viewed as parts of eukaryotic cells, while they were believed to be absent in prokaryotic cells. However, eukaryotic cells also contain membraneless structures, while many prokaryotes cells also contain intracellular structures that perform cellular metabolic functions and, thus, can be seen as primitive organelles, (Murat et al., 2020; Grandpre, 2024). Hence, the use of the more general term "intracellular structures" in the biological pyramid. The number of levels in the complexity pyramid of unicellular microbes is smaller than that of multicellular, eukaryotic animals and plants.

Various complexity hierarchies for the technological realm (Table 4.1d) represent the straightforward part/system organization of man-designed and -made systems. The hierarchy for computers shows four levels, starting at base materials via integrated circuits (ICs) and central processing units (CPUs) to the (super)computer itself. Chemical and power plants have also four different levels, starting at materials and inactive parts (tubes, housing) *via* devices (pumps, thermocouples), and unit operations (reactors, heat exchangers) to the complete plant. Unfortunately, ERD

[13] For adipose tissue, an ERD value of 0.2 W/kg has been found, but adipose tissue is now widely recognized as a metabolically active, endocrine organ, not just a passive tissue for fat storage.

values have not been found for all these levels. For most energy-producing systems (wind turbines, solar cells) and energy-consuming systems (devices, machines, vehicles), three levels are distinguished, namely materials, the core generator or engine, and the technological system itself. For such technological systems, a lot of ERD data are available, as listed in Table 4.1d. Man-powered tools are comparable with inactive parts: both have ER(D) = 0.

ERD data for social systems are very scarce, but sufficient data are available to construct two representative complexity hierarchies (Table 4.1e). The first is for an ant colony composed of ants living in an ant hill. The thermal stability of the colony is high, as a result of heat carrier processes, such as ventilation and insulation, and a dominating contribution of microbial heating (Coenen-Stass et al., 1980). The second shows our human society, which combines active systems from the biological (humans, pets) and technological pyramids (devices, machines, production plants). The inactive parts with ER = 0 (ant hill, house) dominate the mass of social systems, resulting in rather small ERD values. Unfortunately, there are no ERD data for nations in between cities and the human society.

Finally, classification schemes of material systems often follow the part/system hierarchy. The Eightfold Way presents the combination of elementary particles in hadrons, including the proton and neutron nucleons. The Periodic System shows how nuclei and electrons make up the atoms of the various elements. Mineral classification is based on chemical composition and crystalline organization. In contrast, classification of stars, according to their magnitude or luminosity and color or temperature (Morgan–Keenan system and Hertzsprung–Russell diagram), does not follow the part/system hierarchy. The same holds for the classification of planets, based on their mass, composition, and orbital regime, as well as black holes, based on their mass (black hole, massive black hole, and supermassive black hole). Biological taxonomy does not reflect a part/system hierarchy either (organisms in a domain are not composed of organisms in kingdoms, *etc.*), but one based on evolutionary relationships. Social classification schemes, such as organizational flow and socioeconomic stratification charts, emphasize the formal hierarchy of social systems.

4.2.2 Energy Rate Density

A database with mass (in kg), ER (in J/s = W), and corresponding ERD (in W/kg) values of all sorts of dissipative systems from the cosmological, biological, and cultural realms has been set up (van Duin, 2024a, 2025b). Currently, it consists of more than 5000 entrees and is continuously updated with new data for systems of interest. For full details of data sources and calculations, see the original papers and their supplementary materials. The elusive dark matter and dark energy have been excluded from the ERD calculations for galaxies and the universe. Since ER reflects the energy converted by the dissipative system in action, preferably the mass of the complete system is used, that is, bacteria with water, trees with wood, and machines with fuel. Without the latter material parts, the dissipative systems would not be living nor in operation. For social systems, such as ant colonies and the global society not only the mass of the living organisms, but also the mass of the social constructs (ant hill, human infrastructure and architecture) is included. Chaisson (2001) and Barton (2020) have calculated ERD of various human societies over time, using the mass of humans only. However, humans do not eat coal and drink oil, nor do they inhale natural gas, while electricity may be harmful. Devices, machines, as well as power and production plants facilitate the energy flow maintaining human social systems and, thus, their masses should be included. As a result, the ERD values for social systems presented here are smaller than those from Chaisson (2001). Similarly, Chaisson has used just the mass of the Earth's climasphere (30 km atmosphere and 30 m ocean layer). Here the mass of the whole planet is used to calculate its ERD, because the climasphere is supported by the rest of the planet.

Free energy flowing in a dissipative system (ER_{in}) is partly converted to useful work (ER_{out}),[14] may be stored inside the system (ER_{st}), and is also dissipated to the environment (ER_{dis}). The energy efficiency of a system, defined as (ER_{out} + ER_{st})/ ER_{in}, typically varies between 1 and 99%. Unfortunately, the energy efficiency of many systems is not known. All four ER types are found in the database:

- ER_{in}: food and molecular oxygen consumption by living, aerobic organisms, fossil fuel and electricity consumption by machines and cities, as well as matter accretion by black holes;
- ER_{out}: power of athletes, machines, and energy-generating devices;
- ER_{st}: carbohydrates produced and stored in plants and trees; ER_{st} may also be converted, as in nuclear fusion of H/He plasma in stars
- ER_{dis}: typically heat, but also as light and sound.[15]

In addition, for many systems, ER fluctuates over time. The luminosities of stars and galaxies vary strongly over their lifetimes. The luminosities of black holes depend on the matter accretion rates. For living organisms, metabolic rates in rest (basal metabolic rates) or in the field (total energy expenditures) have been used. Animals have peaks in ER_{in}, when eating and digesting food, and their activity levels vary over the day ($ER_{out,day} > ER_{out,night}$). Photosynthesizing plants have a large ER_{in} in light, but none in darkness. The power of machines can be adjusted and even shut off. Typically, for machines maximum power data and for cosmological objects today's luminosities are listed. As a result of these issues, the ERD data of the various systems are not fully comparable.

---

[14] Interpretation of the term "useful" depends on the viewpoint of the beholder. Heat, light, and sound are typical forms of dissipated energy, but may also be useful work, as produced by heaters, lamps, and radios, respectively. Is Solar luminosity ER_{dis} or ER_{out}, considering that only a tiny fraction of the heat radiation of our Sun is used for photosynthesis by cyanobacteria and green plants on Earth?

[15] See footnote 14.

The systems in the database range from tiny active proteins and nano-machines *via* living organisms and machines, as well as stars and black holes to the observable universe and span roughly 70 orders of magnitude of mass and ER (van Duin, 2024a). A double logarithmic plot of ER *versus* mass shows that most data points fall in a diagonal band width of 13 orders of ERD magnitude from the lower, left corner (small mass and ER) to the upper, right corner (large mass and ER). Only stable, dissipative systems, which are characterized by ERD below 10^5 W/kg, have been included in the complexity hierarchies in Table 4.1. Unstable, explosive systems (technical explosives, nuclear weapons, supernovae, and cosmological transients) with ERD larger than 10^5 W/kg, as well as dying and dead systems (very small ERD and ERD = 0, respectively) have been excluded from Table 4.1 (van Duin, 2024b).

Systems from the material pyramid form the basis of all other pyramids. However, quarks and electrons, nucleons, nuclides, atoms and ions, molecules, and salts, as well as materials are not dissipative systems, that is, they do not process energy just by themselves.[16] As a result, ER(D) = 0, preventing a comparison with the systems in the other complexity pyramids. Reeves (1991) has provided quantitative data on energy levels (in eV = 1.60×10^{-19} J) of the ground and excited states for most of the matter systems. The numbers in italic in Table 4.1a represent the typical excitation energies (in J).

4.2.3 Horizontal and Vertical Complexity

A good way to start the discussion on the ERD-quantified complexity hierarchies in Table 4.1, explained in the previous two sections, is by distinguishing horizontal and vertical complexity.

Horizontal Complexity

Systems at a particular level of a complexity pyramid typically consist of various numbers and types of parts from the previous level. These range from:

- a relatively small number of the same parts:[17] eight S atoms → S_8 elemental sulfur molecule;

[16] Material systems as such do not process energy. However, they do contain chemical and nuclear binding energy, which is characterized by their energy density (in J/kg). Such materials may be used as "fuel" for powering dissipative systems. For example, wood for a burning fire in a stove, petrol in an internal combustion engine of a car, and H plasma for nuclear fusion in a star. In addition, material systems may be transformed into the tools and devices that can process energy, such as a quartz crystal in an old-fashioned radio receiving radio waves, the glass bulb and metal wire of an incandescent light bulb emitting light, as well as an engine, fuel tank, and many other parts of a driving car.

[17] A note on "sameness." At the lowest level of the material pyramid, parts are identical. All up quarks, down quarks, and electrons are the identical, and the same holds for protons and neutrons. However, nuclei of a particular element with the same number of protons may have different number of neutrons and, thus, formally are not the same, namely different isotopes with different

- a small number of dissimilar parts: three up and down quarks → proton or neutron; eight protons and eight neutrons → O-16 nucleus; one N nucleus plus seven electrons → N atom; one C atom plus four H atoms → CH_4 natural gas molecule; 79 organs (heart, lungs, kidneys, brain, *etc.*) → human body; engine, chassis, bumpers, battery, window wipers, and other parts → car;
- a relatively large number of the same parts: 3.3×10^{22} H_2O molecules → one liter of water or one kilo of ice;
- a large number of dissimilar parts: 2.1×10^{11} C, H, O, N, and P atoms in human DNA; 4×10^9 muscle, epithelial, connective, and nervous tissue cells[18] → heart; 8.5×10^9 humans, 2.1×10^6 cars, 6600 subway cars, and 200 movie theatres → New York city; 8.5×10^{56} protons, 8.2×10^{55} He nuclei, and 10^{57} electrons → Sun; ~3×10^{11} stars of all types and sizes as well as ~10^{11} white dwarfs, neutron stars, and stellar black holes → Milky Way.

All the systems of various sizes and compositions at a pyramid level may be formed by convergence of different numbers and types of parts from the previous level. In addition, systems formed in such a way may change in number and type of parts on the same pyramid level over time, either developing as single systems during their lifetimes or as a group of systems during its evolution[19] (van Duin, 2025b). For example, in the material realm nuclei may undergo nuclear reactions (fusion and fission), changing into other nuclei with different atomic numbers and masses. Atoms, ions, and molecules may undergo chemical reactions by exchanging atoms and electrons, changing into other chemical species with different physical and chemical properties. Materials are shaped for their application as well as wear and age over time. In the astronomical realm, the birth, main sequence, and aging of stars are accompanied by changes in their elemental composition, size, and mass, as well as in their structure of burning core and shells, non-burning envelope, as well as convective and radiative zones. Stellar ERD shows quite strong fluctuations in the initial proto-star and final giant stages. Eventually, stars die in dramatic explosions with huge ERD for short periods of time or simply fade away with decreasing ERD.

atomic masses (H-1, H-2, and H-3, as well as O-16, O-17, and O-18). Molecules of a compound have in principle the same bruto-equation (H_2O for water), but are composed of different isotopes (all combinations of the mentioned H and O isotopes correspond to water). Polymers of a thermoplastic material consist of the same monomer repeat units and have the same chemical composition, but have different molar masses. Higher up in the pyramid overlay (Fig. 4.3), "sameness" becomes even more fuzzier. Cells forming a particular tissue are similar, but not identical. For sure, all human individuals belong to the same *Homo sapiens* species, but one-by-one they are considered to be different. All stars are composed of plasma consisting of mainly H, some He, and traces of other elements ("metals"), but probably none of the 2×10^{24} stars in the observable universe is identical. It seems that for systems higher in the pyramid overlay, there are as many different types of systems as there are systems, themselves and, as a result, the system number and the number of system types (diversity) are equal.

[18] Amazingly, grown out of one fertilized egg cell!

[19] A strict distinction is made here between the evolution of a group of systems and the lifetime of a single system. For example, the evolution of a galaxy actually is the lifetime of a galaxy.

Stellar populations evolve over time with increasing metallicity as well as decreasing minimum and maximum mass, resulting in a decrease in their ERD range. The changes of planets during their lifetimes are governed by their compositions and the changes of their stellar companions, explaining the wide variety of the more than 6000 (exo)planets observed by today. Black holes grow via matter accretion, which is accompanied by an increase in ERD, and mergers. Galaxies change over their lifetimes, especially with respect to their star formation rates, with their ERD probably going through maxima.

In the biological realm, cells originate, grow, divide, and die with maximum ERD at maximum growth rate. Living organisms are born, grow, mature, age, and eventually die over their lifetimes with ERD of animals decreasing with age. Biological species have evolved into new species with different cellular structures and functions, as well as different cell number. Overall, the biological evolution of prokaryotes *via* unicellular eukaryotes to multicellular eukaryotes has resulted in an increase in mass with ERD going through a maximum. In the technological realm, innovation has resulted in a huge variety of tools, devices, machines, and plants. Size and mass as well as power (= ER) have been varied in all directions and combinations for improved performance and fitness-for-purpose, resulting in a large variation of ERD. For example, vehicles have been developed for transport of humans and goods on land, as well as in water, air, and space, ranging in size from nano-devices to megastructures and exceeding human capabilities. In the social realm, human history tells about the changes of human social systems during their lifetimes and evolution, involving changes in all their parts (humans themselves, animals, plants, infrastructure, and architecture), organization, and especially energy resources. ERD of the global society has decreased with time. Evolution in the broadest sense can be viewed as the process of groups of systems exploring the ER *versus* mass area below the ERD = 10^5 W/kg limit. Different groups of systems (stars, cells, living organisms, machines, cities) follow different ER *versus* mass trajectories during their evolution (van Duin, 2025b). Apparently, there is no universally preferred route for ER and mass evolution.

The variations in system size, diversity, and complexity for groups of systems at a particular pyramid level are often correlated, which substantiate the typical size/complexity rules (Bell & Mooers, 1997; Bonner, 2004; Bettencourt *et al.*, 2007; Brummitt *et al.*, 2012; Niklas *et al.*, 2014; Böttcher, 2018; Amado *et al.*, 2018; Butler & Georgiev, 2022) and diversity/complexity rules (Carneiro, 1967; Page, 2010; Ferguson-Gow *et al.*, 2014; Trible & Kronauer, 2017; Sharma *et al.*, 2023) in the astronomical, biological, and social realms. For dissipative systems with energy processing as key feature of complexity (Chaisson, 2001), all the empirical scaling laws for energy consumption *versus* mass, observed for many groups of systems from all realms (van Duin, 2024a), can be viewed as another set of size/complexity rules. All the variations in size, diversity, and complexity over the lifetimes and evolution (of groups) of systems have resulted in variations of their mass, ER, and corresponding ERD (van Duin, 2025a, 2025b). The latter is illustrated by the range of ERD values for groups of systems listed in Table 4.1.

Vertical Complexity

Different numbers and types of parts at a level may converge to systems of different sizes and compositions at the next level, as was already discussed at horizontal complexity. This is usually accompanied by the emergence of new properties and functions. Interestingly, systems of different pyramid levels may be combined. For example, nuclei (composed of nucleons) combine with electrons (fundamental particles) in atoms and ions. The latter ions combine with electrons in metals. Similarly, the biosphere comprises tiny, unicellular organisms (archaea, bacteria, unicellular eukaryotes), but also large multicellular organisms with specialized parts (organs). The global society consists of a large number of cities with numerous families and singles, but also of families and singles living isolated in the country side. In addition, systems of different complexity pyramids may be combined. For example, a city comprises humans, animals, and trees from the biological pyramid, as well as houses, cars, and buses from the technological pyramid. Human societies consist of cities, power plants, and trains, as well as agricultural fields and cattle from the technological and biological pyramids, respectively. Planet Earth contains all systems from the biological, social, and planetary spheres pyramids. Finally, the Milky Way and, ultimately, the observable universe contain all systems from all pyramids with mass and ER(D) being dominated by astronomical objects though.

Reeves (1991) has already identified several trends when going upward in the material complexity pyramid, namely the number, excitation energy, and local entropy of systems decrease when movoing upward, whereas the size (and thus mass), diversity, fragility, information content, and complexity of systems increase (Fig. 4.1). These changes are accompanied by longer time scales for the dynamics of the systems. In the material pyramid, the diversity of systems at a pyramid level increases indeed, when going upward. For example, today 17 different fundamental particles, 88 different hadrons, around 2000 different nuclides, and roughly 2×10^8 different molecules have been identified, showing an "exponential" increase in diversity. In the astronomical, biological, technological, and social pyramids, the diversity continues to increase in theory when moving to higher levels, because of the increased number of possible, different combinations of parts. However, in reality the diversity starts to decrease at some point, because it is limited by the total number of fundamental particles present in the observable universe. This is most clearly illustrated by the decreasing diversity in the astronomical realm, namely 2×10^{24} different stars, 10^{24} different stellar systems, 2×10^{12} different galaxies, 10^{11} different galaxy groups, 4×10^8 different galaxy clusters, and just one observable universe. Similarly, more than 400 different cell types make up 60 tissue types and 79 organs, which make up one human body. There is also just one global society at the top of the human social pyramid. The number of levels in the various complexity hierarchies (Table 4.1) ranges from 3 to 6, which is partly the result of some choices made, as explained in Sect. 4.2.1. The number of systems on a pyramid level is determined by the ratio of the total number of parts available and the typical number of parts in those systems. As a result, the number of levels in a pyramid is limited by the same ratio. This resembles the way how the number of trophic levels in a food pyramid in biology (by the way: not a part/system but an energy hierarchy)

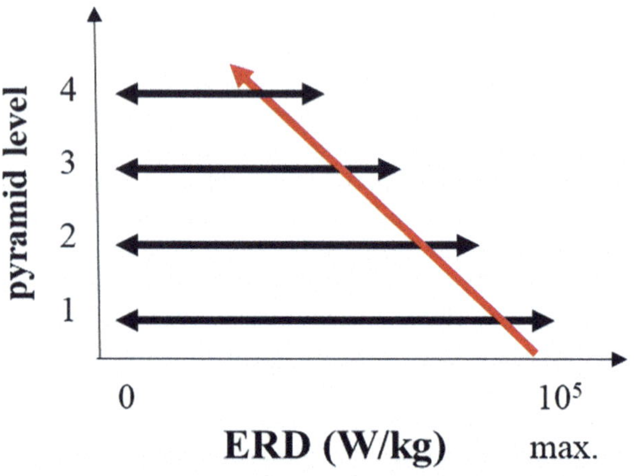

Fig. 4.2 ERD of a next-level system is always smaller than largest ERD of its parts at level below; as a result, ERD range narrows (black horizontal arrows) and ERD of "engine" (see text) of systems decreases (red arrow), when moving to higher levels of a complexity pyramid

is determined by the amount of the primary producers at the bottom and the efficiency of energy transfer to each higher level. Typically, this results in five trophic layers in an ecosystem.[20]

The correlation of diversity with size applies to horizontal complexity on all pyramid levels. As explained, it also applies to vertical complexity in the material pyramid, but not for the other pyramids. The evolution of groups of systems occurs in both horizontal and vertical directions in a complexity pyramid with the former representing "more of different" (Anderson, 1972) and the latter representing "more of new." The evolution of complexity in a horizontal direction proceeds gradually, while that in the vertical direction results in step-changes in complexity and may be viewed as revolution.

4.2.4 ERD Decreasing Upward in Complexity Pyramid

The most intriguing observation from the complexity hierarchies with their quantified ERD levels in Table 4.1 is that ERD values of systems at a next level are typically smaller than the largest ERD value of its parts at the level below. As a result, the ERD range of system groups narrows (Fig. 4.2: black arrows) and ERD of the "engines" of systems decreases (red arrow), when moving upward in a complexity pyramid. The term "engine" is used here for that part of a system that has the largest ERD. Typically, the engine is the part that drives the system at the next level, such as the engine of a machine, the H-fusing core of a star, as well as the ATP-ase

[20] The largest number of trophic layers (9) has been found for marine Paja Formation biota from the Early Cretaceous of Colombia (Cortés & Larsson, 2024). Interestingly, a preliminary analysis has shown that broad ERD ranges are overlapping and, thus, ERD does not significantly change, when moving upward in a food pyramid.

protein and mitochondrion in a cell. For other systems, it is more difficult to identify the "engine." For example, is it the heart, lungs, or brain of an animal? Or the stars or (supermassive) black holes in a galaxy? The specific systems listed in Table 4.1 are typically the engines or have comparable importance.

Admittedly, for system groups the changes in ERD ranges from one pyramid level to a next level are not always that large or clear, which is probably related to the variety of systems groups and the corresponding, broad ranges of ERD values at a pyramid level (Sect. 4.2.3). However, for the specific systems the trend is quite clear: ERD of a specific system on a particular level is always smaller than that of the system at the level below without any overlap. For example, the ERD range for animals from 0.1 to 20 W/kg is somewhat smaller though with some overlap than that for organs from 0.2 to 50 W/kg, which in its turn overlaps with that of cells from 0.5 to 30 W/kg. In contrast, ERD of a human body is significantly smaller than that of a brain, which in its turn is smaller than that of a neuron ($2 < 11 < 26$ W/kg). The decrease in ERD for the specific systems from the six levels of the biological pyramid is very consistent and shows an average ERD decrease of 4.5 per step. ERD values are available for brain regions, such as the cerebellar and temporal cortices (15 and 18 W/kg). Although these regions are typically not considered as a separate level in the biological complexity pyramid, the corresponding ERD values fit in quite nicely between those of a neuron and the brain as a whole. Another example: the ERD range for cars from the 1896 to 1994 period (1–130 W/kg) is somewhat below that for the corresponding car engines (5–820 W/kg), but with quite some overlap. However, ERD of a 1994 car is significantly smaller than that of the corresponding engine (130 *versus*. 820 W/kg). Similar observations can be made for the other machine-plus-engine examples in Table 4.1d. The ERD ratio for engine *versus* machine varies between 5 and 20. The lack of ERD data for parts of chemical and power plants was already mentioned in Sect. 4.2.2. But the fact that ERD values of power plants (1–7 W/kg), synthetic rubber plant (0.3 W/kg), and chips production platforms (0.9–11 W/kg) are relatively small compared to those for tools and machines seems to confirm the decrease in ERD in the corresponding plant pyramids. In the social realm, ERD of an ant colony is also smaller than that of the ants themselves. ERD in the human society pyramid shows a significant decrease from refrigerator (5 W/kg) to single-family home, but then it does not change anymore to Barcelona and the global society (0.014, 0.015, and 0.017 W/kg are same within error). Still, these data may be interpreted as a corroboration of the definition of the mass of human social systems to include the mass of all infrastructure and architecture, as done in this study, instead of just the mass of humans, as done by Chaisson (2001). ERD of the global society in 2019 calculated in the former way is 0.017 W/kg, which fits the trend. In contrast, ERD calculated in Chaisson's way of 54 W/kg is much larger and does not fit the trend.

On first sight, the observation of a decrease in ERD, when going upward in a complexity pyramid, is somewhat counterintuitive. Chaisson (2001) has proposed and used ERD as a metric for complexity with a smaller ERD representing a dissipative system more close to equilibrium, with a smaller energy gradient and, thus, with a lower complexity. Therefore, a decrease in ERD upward in a complexity pyramid would correspond to a decrease in complexity. However, vertical complexity is supposed to increase, when going to new systems with emergent functions on higher pyramid levels. The observation that ERD decreases when going upward in the complexity pyramids for all realms is the simple and direct result of convergence of smaller parts on a particular level to a larger system at the next level. All the parts have a typical mass, which varies in the case of dissimilar parts. ER of different active parts also varies and is often correlated with their mass according to a power law. The energy flow through inactive, structural parts is zero. The total mass and ER of the system as a whole correspond to the sum of the mass and ER of all parts, respectively, that is, these quantities are additive. As a result, the mass and ER of the system are (much) larger than those of the single parts. However, ERD of the system is not the sum of the ERDs of all parts, that is, ERD is not additive. ERD of the system is the ratio of the ER and mass of the whole system, that is, it is the mass-averaged value of the ERD values of its parts. As a result and in contrast to mass and ER, in theory ERD of the system must be equal or smaller than the largest ERD of its parts, while in practice it is typically (much) smaller. A simple example will illustrate this point. Starting with a system composed of just one type of active parts, results in $ERD_{system} = ERD_{part}$, independent of the number of parts. Next adding other active parts with smaller ERD_{part} and/or inactive parts with $ERD_{part} = 0$ to the system, results in $ERD_{system} < ERD_{part}$. More detailed examples for real parts/systems are presented in Table 4.2, listing the mass and ER(D) values of (parts of) the Milky Way, the body of an 18 year old male, and a typical Dutch, single-family home, representative for the cosmological, biological, and social realms, respectively. The data and parts of the discussion below are from van Duin (2024a, 2025b).

The Milky Way galaxy consists of both active and inactive parts in terms of energy processing (Table 4.2 top). Stars powered by nuclear fusion form the majority of the active parts of a galaxy. Their number decreases, but their luminosity (= ER) increases with their mass. An active supermassive black hole with matter accretion, supernovae explosions, and planets with flow patterns in their geo-, hydro-, and atmospheres are other active parts. Interstellar gas and dust are the inactive parts of a galaxy. Stellar remnants (stellar black holes, neutron stars, white dwarfs) take up intermediary positions, because in isolation they are just cooling and fading away. The mass and ER values of the individual parts do not add up to those of the Milky Way as a whole (last row in Table 4.2 top), because the former values still have to be multiplied by their corresponding numbers. In addition, the various types of supernovae and cosmological transients (gamma-ray bursts, fast-radio bursts, tidal disruption events) do have huge peak powers, exceeding all other astronomical phenomena, but only for short periods of time (milliseconds to months). Therefore, time-averaging is needed to obtain their effective ER contributions. Sagittarius A*, that is, the supermassive black hole in the center of the Milky Way with a rather low matter accretion today, hardly contributes to the mass and ER(D) of the galaxy as a whole. Overall, interstellar gas and dust, as well as dwarf stars are the main contributors to the total mass of the Milky Way, while stars dominate its ER. As a result, ERD of the Milky Way (1.7×10^{-4} W/kg) falls somewhere in the middle of

Table 4.2 Mass, ER, and ERD (of parts) of (top) Milky Way, (center) human body, and (bottom) typical Dutch single-family home, representing systems in the cosmological, biological, and cultural realms, respectively

typical Dutch, single-family home			
part*	**mass (kg)**	**ER (W)#**	**ERD (W/kg)#**
active			
father (40 yr)	82	150	1.8
mother (40 yr)	75	115	1.5
boy (10 yr)	40	105	2.6
girl (5 yr)	20	65	3.3
dog	10	35	3.6
cat	4	10	2.4
gas central heater	35	220	6.2
electric cooker	80	3000	38
washing machine	80	500	6.3
television	13	60	4.6
portable computer	2	100	50
living room lamp	3.5	100	29
food mixer	2.7	400	150
hair drier	0.6	2500	4200
inactive			
house	$2x10^{5}$	0	0
living room table	100	0	0
fauteuil	75	0	0
basic chair	5	0	0
parents bed	150	0	0
bed table	1	0	0
total			
family and pets	230	480	2.1
house	$2.1x10^{5}$$	400&	0.0019
total	**$2.1x10^{5}$**	**880**	**0.0043**

human body			
part*	**mass (kg)**	**ER (W)**	**ERD (W/kg)**
active (organs)			
brain	1.3	15	11
heart	0.3	9.7	32
kidneys	0.3	7.0	23
liver	1.4	17	12
gastrointestinal tract	1.1	13	12
skeletal muscles	27	14	0.50
lungs	0.6	4	6.7
skin	5.0	1.5	0.30
adipose tissue	21	4.6	0.22
inactive			
skeleton	2.5	0	0
blood fluid	2.5	0	0
total#	**74**	**91**	**1.2**

* non-comprehensive lists of both active and inactive parts; # mass and ER of body are average mass and basal metabolic rate of 18 yr old male.

Milky Way			
part	**mass (kg)***	**ER (W)***	**ERD (W/kg)***
stars			
hypergiants	$9x10^{31}$	$6x10^{32}$	7.4
supergiants	$4x10^{31}$	$7x10^{31}$	1.8
red giants	$3x10^{30}$	$1x10^{30}$	0.32
subgiants	$1x10^{31}$	$3x10^{30}$	0.12
blue dwarfs (O)	$1x10^{32}$	$2x10^{32}$	2.0
blue dwarfs (B)	$1x10^{31}$	$2x10^{30}$	0.09
A-type stars	$4x10^{30}$	$7x10^{27}$	0.0017
F-type stars	$3x10^{30}$	$2x10^{27}$	$5.4x10^{-4}$
yellow dwarfs (G)	$2x10^{30}$	$3x102^{6}$	$1.8x10^{-4}$
orange dwarfs (K)	$1x10^{30}$	$9x10^{25}$	$5.7x10^{-5}$
red dwarfs (M)	$5x10^{29}$	$6x10^{24}$	$7.3x10^{-6}$
brown dwarfs	$9x10^{28}$	$2x10^{23}$	$5.7x10^{-6}$
stellar remnants in isolation			
stellar black holes	$6x10^{31}$	$1x10^{25}$	$1.8x10^{-7}$
neutron stars	$3x10^{30}$	$7x10^{30}$	2.0
white dwarfs	$1x10^{30}$	$2x10^{24}$	$1.1x10^{-6}$
other objects			
supermassive black hole	$8x10^{36}$	$1x10^{29}$	$1.2x10^{-8}$
supernovae#	$3x10^{31}$	$2x10^{36}$	$3.7x10^{5}$
planets	$3x10^{26}$	$1x10^{18}$	$2.3x10^{-7}$
moons	$7x10^{22}$	$1x10^{15}$	$1.4x10^{-8}$
asteroids	$<10^{21}$	0	0
interstellar gas and dust	$2x10^{41}$	0	0
total	**$3x10^{41}$$**	**$5x10^{37}$**	**$1.7x10^{-4}$**

Data from van Duin (2024a, 2025b)

the ERD range of its active parts and is comparable with ERD of a low-mass, yellow dwarf (1.8×10^{-4} W/kg) like our Sun.

The human body also consists of active and inactive parts (Table 4.2 center), which together enable it to live and function. The list of parts is just illustrative and not comprehensive. As a result, the total mass and basal metabolic rate (= ER in rest) of the human body are larger than the sum of the mass and ER of the (in)active parts listed. The inactive parts of the human body with ER(D) = 0 represent just a minor part of its mass. ERD of the body of a human male of 18 year in rest (1.2 W/kg) falls at the low side of the ERD range of the active organs (0.22–32 W/kg). The brain is often considered as the most complex organ of the human body, but its ERD is smaller than those of the heart and the kidneys (11 *versus* 32 and 23 W/kg). This shows that physical activities, such as pumping and filtering of blood, are more energy demanding than body control and cognition. During physical activities, not only ER(D) of the skeletal muscles but also those of other organs, such as the heart and lungs, will increase, resulting in a larger ERD of the human body. The average total energy expenditure (= typical ER) of an 18 years old male is 2.2 W/kg, while during intense physical and sport activities it may be as high as 6 W/kg. Interestingly, intense thinking and high creativity, such as during a chess game, do not affect ERD.

As a third example, a typical Dutch single-family home also comprises active and inactive parts (Table 4.2 bottom). The active parts consist of the human residents and their pets, powered by food and oxygen, as well as the tools and devices used by the family, powered by electricity and natural gas. The inactive parts are structural, such as the house itself, or functional, such as the furniture, curtains, and decoration. Although these inactive parts do not consume any energy, they also facilitate the family living in the home. Again, the overview is not comprehensive, that is, there are more active parts used and more inactive parts present in the home than listed. As a consequence, the total mass and ER of the home are larger than the sum of the mass of the (in)active parts listed. ER of the tools and devices relates to their maximum power, but they are not continuously in operation and, thus, their time-averaged ER values are (much) smaller than those listed. ER of the home as a whole is simply calculated from the average yearly consumption of natural gas and electricity by a Dutch family. The mass of the home is dominated by the mass of the structural components, such as the bricks and concrete of the house, which have ER = 0. The family living in the house as well as the tools and devices hardly contributes to the total mass, but govern the total ER with a ~ 55/45 split. Because of the relatively large mass of the house, ERD of the single-family home is rather low and far below the ERD range of the active parts (0.0043 *versus* 1.5–4200 W/kg).

The decrease in the excitation energy, when moving upward in the material pyramid (Table 4.1a), parallels the decrease in ERD when moving upward in the other pyramids (Table 4.1b to e). However, these two trends cannot be directly compared, since:

- excitation energy (in J) relates to an event in a single particle species without a time specification; it can be converted to specific excitation energy (in J/kg) for a collection of particles and
- ERD relates to the energy conversion (in J/s), which maintains the out-of-equilibrium structure (in kg) and energy gradient of a dissipative system.

In this respect, Reeves' original pyramid is somewhat a mix-up, because quarks up to biomolecules are matter particles, whereas cells and organisms are dissipative systems. This explains why Reeves did not provide excitation energies for the latter. Most forms of matter contain chemical and nuclear binding energy, have a corresponding energy density, and sometimes can be used as fuel. Some preliminary calculations (van Duin, 2024a) have yielded ERD values for:

- molecules: approximately 10^7 W/kg for a hypothetical, downscaled system of just one C atom reacting with one O_2 molecule at the same rate as coal burning in a power plant and
- nuclei: around 10^{15} W/kg for a downscaled system of just four H nuclei fusing to one He nucleus at the same rate as nuclear fusion in the Sun.

These values suggest a continuation of the increasing ERD scale downward in the complexity pyramids for the cosmological, biological, and technological realms into the material realm. In a way, such a continued ERD increase might be expected, since moving downward in the material pyramid corresponds not only to increased excitation energies, but also to increased binding energies combined with decreased mass.

4.2.5 Overlay of Complexity Pyramids

For most complexity hierarchies in Table 4.1, the systems at the top levels correspond to systems at the bottom levels of other pyramids. These connections allow the qualitative integration of the complexity pyramids from the different realms in one pyramid overlay, comprising "all" systems from our world. In addition, the ERD quantification of the systems in the individual complexity pyramids (Table 4.1b–e) enables the quantitative positioning of these systems in the overlay on a single ERD scale. Figure 4.3 presents this overlay with the logarithm of ERD increasing downward[21] as the y-axis and the logarithm of time as a provisional x-axis. The material systems at the bottom of the overlay have ERD $= 0$ and are quantified with the logarithm of the excitation energy (in gray), also decreasing downward. The specific systems (mainly "engines") from Table 4.1 are used in Fig. 4.3 to represent "all" systems relevant for BH. The material pyramid (Table 4.1a) forms the basis of most other pyramids, explaining why it extends to the full width of the pyramid

[21] In contrast to the convention that a quantity on the y-axis of a plot increases when moving upward.

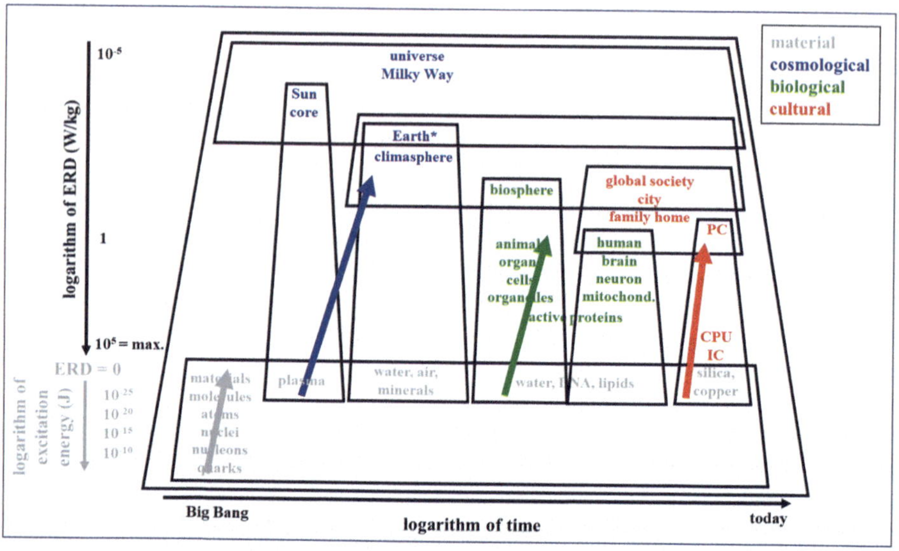

* the position of Earth in the overlay does not match with its relatively low ERD value: see text

Fig. 4.3 Overlay of complexity pyramids of various realms from Table 4.1, positioned with respect to logarithm of ERD on y-axis increasing downward and logarithm of time as provisional *x*-axis; for material systems, ERD = 0 and logarithm of excitation energy at lower *y*-axis decreases downward (in gray); systems from different realms are presented in different colors; * the position of Earth in the overlay does not match with its relatively low ERD value: see text

overlay. H/He plasma constitutes stars. Water, air, and minerals make up the various Earth spheres. Water, organic chemicals, and minerals make up living cells and organisms. Note that active proteins are borderline systems, because they cannot only be inactive molecules (material) but also dissipative systems (biological). A large variety of materials, ranging from concrete, bricks, wood, metals, glass, and polymers, are used for the construction of simple tools, as well as (parts of) machines and production plants. Humans, animals (cattle), and plants (agriculture), as well as machines and vehicles converge in the human society. The latter in its turn together with the biosphere, the Earth's atmosphere, and the ocean surface layer converge in the climasphere. This in its turn converges with the geosphere in the Earth planet. Finally, the latter converges with the Sun in the Milky Way and, eventually, in the observable universe, which is the ultimate convergent system at the top of the overlay. Typically, ERD decreases when going upward in this pyramid overlay, in the same way as it does in the smaller, individual pyramids. Thus, ERD and complexity not only change in a consistent way in the horizontal and vertical directions in a single, complexity pyramid but also when shifting from one to another complexity pyramid in the overlay. The position of Earth in the overlay conflicts with its low ERD value (2×10^{-8} W/kg), probably because it is a rather insignificant part of the Milky Way and not its "engine."

The four bold arrows in Fig. 4.3 represent the directions of the evolutions of systems in the material, cosmological, biological, and cultural realms. The x-axis represents a provisional, logarithmic time axis, starting at the Big Bang and pointing toward today. The individual pyramids have been positioned in such a way that the transitions between the different matter types in the expanding and cooling, primordial universe (gray arrow) are on the left side. Next, astronomical objects, including the Milky Way, Sun, and Earth with its various spheres, were formed (blue arrow) with new stars and planets continue being formed today and in the future. Later on, uni- and multicellular life including hominins originated and evolved on Earth (green arrow). More recently, cultural systems originated and developed, including social and technological systems (red arrow on the right side). The overlay shows the nestedness of the various systems across the four realms. (Groups of) systems change during their lifetime and evolution at all levels of all pyramids. In addition, changes in parts affect a next-level system and changes in a system affect its parts at a lower level. The combined result of these horizontal and vertical dynamics is that systems in the four realms have not evolved independently over time, but have coevolved in parallel. BH does not follow a simple nor single part/system trajectory, but several part/system trajectories in parallel. The sequence of galaxies → stars → planets → plants → animals → brains → society, as followed by Chaisson (2001), becomes a rather tortuous path in this overlay. This sequence of cosmic evolution corresponds to steps downward in the astronomical pyramid (galaxies → stars → planets), next to the biological pyramid (first on the same level: plants → animals; but then downward: animals → brains), and, finally, to the social pyramid (society). When materials are included preceding astronomical objects, the path becomes even more tortuous.

ERD values of systems at a particular pyramid level can be compared. For example, in astronomy stars, black holes, and galaxies are studied and compared on the basis of their luminosity-to-mass ratios (= ERD) and in luminosity/mass planes. In biology, living organisms are compared using their specific basal metabolic rates and specific total energy expenditures (= ERD). Allometric scaling laws between ERD and mass have been established. In sports, athletes are compared *via* their specific power (= ERD) and the same applies in technology to machines. The comparison of ERD values of systems from different levels in a complexity pyramid has shown that ERD decreases when moving upward, as a result of convergence of parts to a system on the next level (Sect. 4.2.4). As a result, ERD values of systems in different complexity pyramids cannot be compared, because there is no objective way to establish which levels of different pyramids should be compared. It is not clear whether:

- stellar cores, stars, galaxies, or the universe should be compared with
- biospheres, climaspheres, or a planets
- active proteins, intracellular structures, cells, organs, or living organisms
- ICs, CPUs, or computers, and engines or vehicles
- homes, cities, or the global society.

This corresponds to the *pars-pro-toto* issue, discussed in the introduction when raising the question whether ERD values of brains, chips, and fighter jets should be used to represent humans, computers, and the human society, respectively. A consequence is that the increase in ERD over deep time in the sequence of galaxies → stars → planets → plants → animals → brains → society (Chaisson, 2001) becomes problematic, because systems from different pyramids cannot compared. In addition, the different definition of the mass of Earth (not just climasphere, but whole planet) and social systems (not just humans, but including social constructs) results in lower ERD values for Earth and human society compared to Chaisson's values. This disrupts the continuous ERD increase in Chaisson's sequence.

4.3 Conclusions

Following a rigorous part/system approach, a series of complexity pyramids has been constructed for systems from the material, astronomical, biological, technological, and social realms. ERD has been used to quantify the variation in horizontal complexity on a pyramid level (size and diversity) and the change in vertical complexity in a pyramid (emergence). The main observation is that ERD of a system is typically smaller than the largest ERD of its parts on the level below. This results in a narrowing of the ERD range of systems on a next level and a decrease in ERD of system's "engines," when going upward in a complexity pyramid. This is rather counterintuitive, since complexity and thus ERD are supposed to increase, when parts combine in a larger system with emergent properties at the next level. However usually, both active, energy-processing (ERD > 0) and inactive, structural parts (ERD $= 0$) converge to the larger system. ERD of that larger system is the mass average of the ERDs of all (in)active parts and, consequently, is typically smaller than the largest ERD of its parts. Finally, the complexity pyramids for the various realms have been integrated in an overlay, representing "all" systems from the physical reality of our world. The ERD quantification of the systems in the individual pyramids allows the quantitative positioning of the various pyramids in this pyramid overlay. A comparison of ERD values of systems from different complexity pyramids results in *pars-pro-toto* issues, because there is no objective way to establish which levels from two different pyramids may be compared. Chaisson's (2001) sequence of galaxies → stars → planets → plants → animals → brains → society is not only a rather tortuous path through the overlay, but the observed increase in ERD over big time is also somewhat problematic.

Acknowledgments Georgi Georgiev (Assumption University, USA) and Esther Quaedackers (University of Amsterdam, NL) are acknowledged for critical and constructive discussions on the use of ERD in complexity pyramids and BH.

References

Amado, A., Batista, C., & Campos, P. R. (2018). A theoretical approach to the size-complexity rule. *Evolution, 72*(1), 18–29. https://doi.org/10.1111/evo.13392

Anderson, P. W. (1972). More is different. *Science, 177*(4047), 393–396. https://doi.org/10.1126/science.177.4047.393

Azarian, S. B. (2022). *The romance of reality*. BenBella Books.

Barton, S. (2020). *A proxy for progress: Energy rate density*. https://medium.com/@samhbarton/a-proxy-for-progress-energy-rate-density-c1e724903aec

Baskin, K. (2023). *A practical guide to complexity science in big history*. Lecture during International Big History Association Conference.

Bell, G., & Mooers, A. (1997). Size and complexity among multicellular organisms. *Biological Journal of Linnean Society, 60*, 345–363. https://doi.org/10.1006/bijl.1996.0108

Bettencourt, L. M. A., *et al.* (2007). Growth, innovation, scaling, and the pace of life in cities. *Proceedings of the National Academy of Sciences of the United States of America, 104*(17), 7301–7306. https://doi.org/10.1073/pnas.0610172104

Bonner, J. T. (2004). Perspective – The size-complexity rule. *Evolution, 58*(9), 1883–1890. https://www.jstor.org/stable/3449441

Boston, P. J. (2008). Gaia hypothesis. In *Encyclopedia of ecology*. Elsevier. https://doi.org/10.1016/B978-008045405-4.00735-7

Böttcher, T. (2018). From molecules to life: Quantifying the complexity of chemical and biological systems in the universe. *Journal of Molecular Evolution, 86*, 1–10. https://doi.org/10.1007/s00239-017-9824-6

Brummitt, C.D., Gomez-Lievano, A., Goudemand, N., & Haslam, G. (2012). Hunting for keys to innovation: The diversity and mixing of occupations do not explain a city's patent and economic productivity. *Complex Systems Summer School Proceedings.*

Butler, T.H., & Georgiev, Y.G. (2022). *Self-organization in stellar evolution: Size-complexity rule.* In G.Y. Georgiev & M. Shokrollahi-Far (Eds.), *Efficiency in complex systems: Self-organization towards increased efficiency*. https://doi.org/10.1007/978-3-030-69288-9_4

Carneiro, R. L. (1967). On the relationship between size of population and complexity of social organization. *Southwestern Journal of Anthropology, 23*(3), 234–243.

Chaisson, E. J. (2001). *Cosmic evolution; the rise of complexity in nature*. Harvard University Press.

Chaisson, E. J. (2010). Energy rate density as a complexity metric and evolutionary driver. *Complexity, 16*(3), 27–40. https://doi.org/10.1002/cplx.20323

Chaisson, E. J. (2014). The natural science underlying big history. *The Scientific World Journal., 2014*, 384912. https://doi.org/10.1155/2014/384912

Christian, D. (2005). *Maps of time: An introduction to big history*. University of California Press.

Coenen-Stass, D., Schaarschmidt, B., & Lamprecht, I. (1980). Temperature distribution and calorimetric determination of heat production in the nest of the wood ant, *Formica Polyctena. Ecology, 61*(2), 238–244. https://doi.org/10.2307/1935180

Cortés, D., & Larsson, H. C. E. (2024). Top of the food chains - an ecological network of the marine Paja formation biota from the early cretaceous of Colombia reveals the highest trophic levels ever estimated. *Zoological Journal of the Linnean Society, 202*, 1–19. https://doi.org/10.1093/zoolinnean/zlad092

Daft, R. L. (2010). *Organization theory and design* (10th ed.). South-Western, Cengage Learning.

Ferguson-Gow, H., Sumner, S., Bourke, A. F. G., & Jones, K. E. (2014). Colony size predicts division of labour in attine ants. *Proceedings of Royal Society, B281*(2014), 1411. https://doi.org/10.1098/rspb.2014.1411

Fewster, H. (2016). *Big history examines our past, explains our present, imagines our future*. Dorling Kindersley Ltd..

Grandpre, T. (2024). Mystery droplets inside cells may play vital roles in life. *Scientific American*. https://www.scientificamerican.com/article/mystery-droplets-inside-cells-may-play-vital-roles-in-life/
Grassie, W. (2018). The great matrix of all beings. *Journal of Big History, 2*(2), 37–52. https://doi.org/10.22339/jbh.v2i2.2301
Grete, P., *et al.* (2025). The XMAGNET exascale MHD simulations of SMBH feedback in galaxy groups and clusters: Overview and preliminary cluster results. *The Astrophysical Journal, 988*, 155. https://doi.org/10.3847/1538-4357/adde45
Henriques, G., & Volk, T. (2023). Toward a big history 2.0. *Journal of Big History, 6*(3), 1–4. https://doi.org/10.22339/jbh.v6i3.6301
Jagers op Akkerhuis, G. (2010). *The operator hierarchy: A chain of closures linking matter, life and artificial intelligence*. PhD thesis. Radboud University Nijmegen.
Jantsch, E. (1980). *The self-organizing universe: Scientific and human implications of the emerging paradigm of evolution*. Pergamon.
Korotayev, A., & LePoire, D. J. (2026). The evolution of complexity in the stelliferous era: Observations and patterns. In D. J. LePoire, L. Grinin, & A. Korotayev (Eds.), *Complexity in universal evolution – A Big History perspective* (pp. 151–172). Springer Nature (in this volume). https://doi.org/10.1007/978-3-032-20516-2_5
Kurzweil, R. (2005). *The singularity is near: When humans transcend biology*. Penguin Books.
Ladyman, J., Lambert, J., & Wiesner, K. (2013). What is a complex system? *European Journal for Philosophy of Science, 3*, 33–67. https://doi.org/10.1007/s13194-012-0056-8
LePoire, D. J., Grinin, L., & Korotayev, A. (2026). Introduction to complexity theory for Big History. In D. J. LePoire, L. Grinin, & A. Korotayev (Eds.), *Complexity in universal evolution – A big history perspective* (pp. 29–90). Springer Nature (in this volume). https://doi.org/10.1007/978-3-032-20516-2_2
Lloyd, S. (n.d.). *Measures of complexity a non-exhaustive list*. https://web.mit.edu/esd.83/www/notebook/Complexity.pdf
McShea, D. W. (2021). Evolution of complexity. In L. Nuño de la Rosa & G. B. Müller (Eds.), *Evolutionary developmental biology*. Springer.
Mitchell, M. (2011). *Complexity – A guided tour*. Oxford University Press.
Murat, D., Byrne, M. & Komeili, A. (2020). Cell biology of prokaryotic organelles. *Cold Spring Harbor Perspectives in Biology*. https://doi.org/10.1101/cshperspect.a000422
Niele, F. (2005). *Energy: Engine of evolution*. Elsevier.
Niklas, K. J., Cobb, E. D., & Dunker, A. K. (2014). The number of cell types, information content, and the evolution of complex multicellularity. *Acta Societatis Botanicorum Polonia, 83*(4), 337–347. https://doi.org/10.5586/asbp.2014.034
Page, S. E. (2010). *Diversity and complexity*. Princeton University Press.
Quaedackers, E. (2024). *A little big history of Tiananmen: An example of a new type of long-term history*. PhD thesis. University of Amsterdam.
Reeves, H. (1991). *The hour of our delight – Cosmic evolution, order and complexity*. W. H. Freeman and Company.
Sharma, A., Czégel, D., Lachmann, M., Kempes, C. P., Walker, S. I., & Cronin, L. (2023). Assembly theory explains and quantifies selection and evolution. *Nature, 622*, 321–328. https://doi.org/10.1038/s41586-023-06600-9
Siegenfeld, A. F., & Bar-Yam, Y. (2020). An introduction to complex systems science and its application. *Complexity, 2020*, 6105872. https://doi.org/10.1155/2020/6105872
Smart, J. M. (2009). Evo devo universe? A framework for speculations on cosmic culture. In S. J. Dick & M. L. Lupisella (Eds.), *Cosmos and culture: Cultural evolution in a cosmic context*. Govt. Printing Office.
Solis, K. (2018). The unfolding of information in big history. *Journal of Big History, 2*(1), 43–62. https://doi.org/10.22339/jbh.v2i1.2254
Solis, K. (2024). Reexamining free energy rate density as a complexity metric. *Journal of Big History, 7*(1), 19–28. https://doi.org/10.22339/jbh.v7i1.7102

Spier, F. (2005). How big history works: Energy flows and the rise and demise of complexity. *Social Evolution and History, 4*(1), 87–135. https://www.researchgate.net/publication/240623637

Spier, F. (2015). *Big history & the future of humanity*. Wiley.

Stokes Brown, C. (2012). *Big history: From the Big Bang to the present*. The New Press.

Trible, W., & Kronauer, D. J. C. (2017). Caste development and evolution in ants: It's all about size. *Journal of Experimental Biology, 220*, 53–62. https://doi.org/10.1242/jeb.145292

van Duin, M. (2024a). What can we learn from a master plot of energy rate *versus* mass for a very wide variety of (complex) systems? *Journal of Big History, 7*(1), 29–78. https://doi.org/10.22339/jbh.v7i1.7103

van Duin, M. (2024b). Dissipative systems have a maximum energy rate density of 10^5 W/kg. *The European Physical Journal, B97*(11), 171. https://doi.org/10.1140/epjb/s10051-024-00785-2

van Duin, M. (2025a). Development of mass and energy rate (density) of dissipative systems over their lifetimes: A comparison of a low-mass star, like our sun, a human and the Roman empire. In D. J. LePoire, L. Grinin, & A. Korotayev (Eds.), *Navigating complexity in big history—Exploring periodization across cosmic and biosocial dimensions* (pp. 345–385). Springer Nature. https://doi.org/10.1007/978-3-031-85410-1_11

van Duin, M. (2025b). What more can we learn from a master plot of energy rate *versus* mass for a very wide variety of (complex) systems? *Journal of Big History, 8*(2), 38–52. https://doi.org/10.22339/jbh.v8i2.8202

Vandenbroeck, P. (2015). System thinking and four forms of complexity. *Shift^n*. https://shiftn.com/_uploads_pdf/shiftN_WP_Systems-Thinking_Web-kopie.pdf

Volk, T. (2020). The metapattern of general evolutionary dynamics and the three dynamical realms of big history. *Journal of Big History, 4*(3), 31–53. https://doi.org/10.22339/jbh.v4i3.4320

Volk, T., & Henriques, G. (2025). Integrating the tree of knowledge and combogenesis approaches to big history. In D. J. LePoire et al. (Eds.), *Navigating complexity in big history—Exploring periodization across cosmic and biosocial dimensions* (pp. 195–215). Springer Nature. https://doi.org/10.1007/978-3-031-85410-1_7

von Däniken, T. (2025). *Largest-ever cosmological simulation created using algorithms for Euclid mission*. https://phys.org/news/2025-09-largest-cosmological-simulation-algorithms-euclid.html

Walker, S. I., Cisneros, L., & Davies, P. C. W. (2012). Evolutionary transitions and top-down causation. *Artificial Life, 13*, 283–290. https://doi.org/10.7551/978-0-262-31050-5-ch038

Martin van Duin has completed a PhD study in physical, organic chemistry at the Technical University of Delft (NL) in 1986. Next, he has worked for 35 years in research, development, and innovation in the international polymer and rubber industry and academia. Main theme of his work has always been the understanding of structure/reactivity/properties relationships as the starting point for solving problems, as well as developing new product and application solutions. He is a (co-)author of 190 scientific and technical publications and (co-)inventor of 10 patents. After his retirement, he started a studies in big history with emphasis on understanding main trends using quantitative approaches, such as the use of energy rate density.

Part II

Cosmic Complexity

The Evolution of Complexity in the Stelliferous Era: Observations and Patterns

5

Andrey Korotayev and David J. LePoire

Abstract

A previous study compared the symmetries of the cooling after the Big Bang and the development of life, humans, and civilization on Earth. A simple model of the relative rates of change depicts two cones joined at their bases representing the Big Bang expansion and the acceleration of complexity development on Earth through life evolution, human evolution, and civilization development. This indicates points of rapid change at both ends corresponding to at the Big Bang and our current technological and demographic change. However, this depiction oversimplifies complexity growth patterns. In their previous work, Korotayev and LePoire found that for about 150 million years after the Big Bang Singularity, one could observe a hyperbolic deceleration of the universal complexity growth driven by hyperbolically decelerating cooling of the Universe. However, around this time, with the end of the cosmological Dark Ages and the start of the Stelliferous Era, the further cooling of the Universe stopped playing any significant role in the further growth of the universal complexity. Against this background, a new driver of the universal complexity growth emerged. There are some grounds to maintain that this was the nucleosynthesis in the emerging stars that produced elements more complex than hydrogen and helium that appeared in

This chapter is an extensively re-worked version of the text originally published in *Journal of Big History* (Korotayev & LePoire, 2025).

A. Korotayev (✉)
HSE University, Moscow, Russia

Eurasian Center for Big History and System Forecasting at Institute of Oriental Studies, Russian Academy of Sciences, Moscow, Russia

D. J. LePoire
Argonne National Laboratory, Argonne, IL, USA
e-mail: dlepoire@anl.gov

D. J. LePoire et al. (eds.), *Complexity in Universal Evolution*, World-Systems Evolution and Global Futures, https://doi.org/10.1007/978-3-032-20517-9_5

the Epoch of Big Bang Nucleosynthesis (between 10 and 1000 s after the Big Bang Singularity). Korotayev and LePoire propose possible proxy measures of complexity that could help us to estimate the universal complexity dynamics in the Stelliferous Era of the Big History in more detail then was done this earlier. One such measure is what the astronomers call "metallicity." Astronomers define as "metals" any elements that are heavier than hydrogen and helium (thus, according to them, oxygen, or carbon are also "metals"). Then, the "metallicity" turns out to be a measure indicating the ratio of "metals" (all the elements that are heavier [and, thus, more complex] than helium [let alone hydrogen]) to the "non-metals." Korotayev and LePoire suggest that the mean metallicity of the Universe can be regarded as a measure of universal chemical–physical complexity (reflecting the universal complexity of the element base for the chemical evolution). Their analysis also suggests that, since 150 million years after the Big Bang Singularity, the universal metallicity experienced exponential acceleration with subsequent quasi-logistic acceleration getting to the maximum level around 10–11 billion years BP, which was followed by a new wave of hyperbolic deceleration.

Keywords

Complexity growth · Big Bang · Chemical–physical complexity · Stelliferous Era · Metallicity · Universal complexity · Big History

5.1 Introduction

In our prior research (Korotayev, 2024, 2025; LePoire, 2016, 2025), we conducted a quantitative examination of two distinct trajectories of complexity expansion within Big History. The first trajectory entails a decelerating universal (cosmic) evolutionary progression, which has manifested in the cosmos over several billion years following the Big Bang (approximately 13.8 billion years ago). The second trajectory encompasses an accelerating global (biosocial) evolutionary advancement, observed over roughly 4 billion years on Earth since the advent of life and continuing until the early 1970s.

Our analysis revealed that the initial pattern can be succinctly represented by the equation: $y = C_1/(t—t_1*)$, where y signifies the rate of universal complexity growth (measured as the frequency of phase transitions), C_1 represents a constant, and $t - t_1*$ denotes the time elapsed since the Big Bang Singularity (t_1*–13.8 billion years ago).

Meanwhile, our prior research (Korotayev, 2018, 2020) has demonstrated that the second trajectory can be effectively characterized with the following equation: $y = C_2/(t_2* - t)$. In this equation, y represents the rate of accelerating global (biosocial) evolutionary development, C_2 stands for another constant, and $t_2* - t$ signifies the time remaining until the twenty-first century Singularity (t_2*, estimated

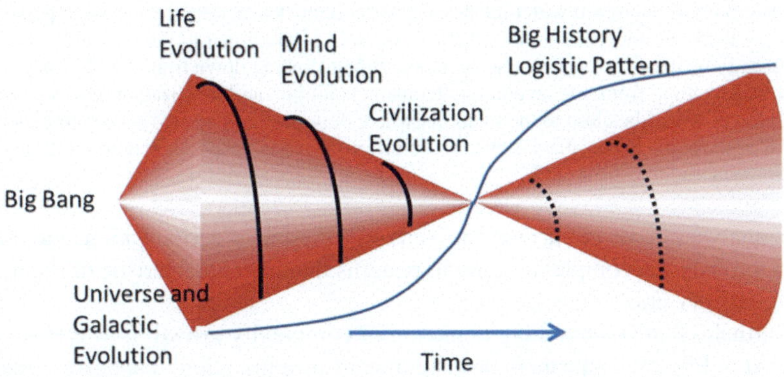

Fig. 5.1 Patterning of the Big History suggested earlier by David LePoire (2020a: 89)

to occur around 2027 or 2029 CE). Thus, the hyperbolic decline in universal complexity growth rate post-Big Bang and the hyperbolic rise in the growth rate of global complexity over the last 4 billion years have both followed the same pattern.

Here, we encounter a striking symmetry. The rate of universal (cosmic) complexity growth diminishes as we progress from the Big Bang Singularity, while the rate of global complexity growth escalates as we approach the twenty-first century Singularity. More precisely, as the time since the Big Bang Singularity increases by a factor of *n*, the universal (cosmic) complexity growth rate decreases by the same factor of *n*, whereas as the time until the twenty-first century Singularity decreases by a factor of *n*, the global complexity growth rate increases by the same factor of *n*. Additionally, a somewhat more intricate symmetry emerges concerning the interplay between energy dynamics and complexity growth within both processes.

Note that the quantitative analysis we conducted on complexity growth patterns in Big History suggests the delineation of the following major Big History eons:

1. The eon characterized by hyperbolic deceleration of universal complexity growth, spanning from the Big Bang Singularity to 4 billion years before present (YBP).
2. The eon marked by hyperbolic acceleration of global complexity growth, extending from 4 billion YBP to the early 1970s.
3. A new eon demonstrating potentially hyperbolic deceleration of global complexity growth, commencing from the early 1970s.

Note that this periodization is very congruent with the patterning of the Big History suggested earlier by David LePoire (2016, 2020a, 2020b) (see Fig. 5.1):

LePoire provides the following rather useful explication to a similar figure published in an earlier work of his (LePoire, 2016: 231):

The Big Bang is on the left with time flowing to the right. The height of the curve represents the time between important events (that is, inversely proportional to the rate of change or complexity increase) as the universe develops, at first slowing down from the Big Bang and then accelerating from life evolution, human evolution, and civilization development. The present would be close to the midpoint where the two cones meet. This would indicate the highest rate of change, that is, when an inflection occurs in the progress represented in the logistic pattern.

Our analyses have confirmed the veracity of such a periodization, identifying more rigorously the complexity growth patterns that are characteristic of the respective Big History eras.

Nevertheless, the overarching depiction of complexity growth in Big History, as illustrated in Fig. 5.1, appears to be a significant simplification; it appears somewhat too tidy to be entirely accurate. A closer scrutiny reveals that the patterns of hyperbolic deceleration and hyperbolic acceleration of universal/global complexity growth were particularly well-observed around a few hundred million years surrounding the singularities. Concerning the hyperbolic acceleration of global complexity, there are substantial doubts regarding whether such a pattern was consistently observed during the initial 3 billion years of life's existence on Earth (see, for instance, Tsirel, 2020, or Glötzl, 2025). Conversely, a distinctly neat pattern of hyperbolic deceleration of universal complexity growth seems to have been observed only until approximately 150 million years post the Big Bang Singularity—up to the end of the cosmological Dark Ages.

As we suggested earlier (Korotayev, 2024), after the Big Bang, the rate at which complexity in the Universe grew slowed down as the Universe cooled. This cooling process allowed for the formation of increasingly complex structures, from quarks to atoms and molecules (see, e.g., Baumann, 2022; Gorbunov & Rubakov, 2018; Grinin, 2013; Hawking, 2009; LePoire, 2016, 2020b; Mukhanov, 2005; Ryden, 2017). Initially, the Universe cooled rapidly, following a hyperbolic pattern, leading to a rapid increase in complexity. So, it does not appear to be of any surprise that as the cooling slowed hyperbolically over time, so did the growth of complexity.

However, around 150 million years after the Big Bang, with the end of the cosmological Dark Ages and the start of the Stelliferous Era, the further cooling of the Universe stopped playing any significant role in the further growth of the universal complexity. Against this background, a new driver of the universal complexity growth emerged.

5.2 Metallicity as a Measure of Universal Complexity

The further growth of complexity in the Universe was achieved mostly through the chemical evolution that produced more and more complex molecules (up to the living matter). In fact, the chemical evolution began well before the end of the Dark Ages (Palla et al., 1995; Lepp et al., 2002; Galli & Palla, 2013; Güsten et al., 2019), with the first chemical compound, the helium hydride ion HeH^+, emerging already around 100 thousand years after the Big Bang (Engel et al., 2005; Güsten et al., 2019), that is even before the end of the Recombination.

However, before the onset of the Stelliferous Era, the chemical evolution had to rely on a very narrow element base, as the primordial Big Bang nucleosynthesis (that took place from 10 s to 20 min after the Big Bang Singularity) only produced atom nuclei of just three elements—hydrogen, helium, and a very small amount of lithium[1] (see, e.g., Coc, 2016; Gorbunov & Rubakov, 2018; Ishkhanov et al., 2019; Mathews et al., 2017; Mukhanov, 2005; Steigman, 2006), from which many thousands years later respective atoms emerged in the process of the so-called recombination (Gorbunov & Rubakov, 2018; Mukhanov, 2005; Ryden, 2017; Switzer & Hirata, 2008; Zel'dovich et al., 1969).

Actually, a few students of "the chemistry of the early universe" have demonstrated that, even before the onset of the Stelliferous Era, the chemical evolution could produce an unexpectedly large number of possible compounds (Palla et al., 1995; Lepp et al., 2002; Galli & Palla, 2013, etc.). However, it is quite clear that, with such a limited element base, the chemical evolution had very strong limitations, being totally unable to produce any really complex molecules.

This possibility arose only when the Stelliferous Era began. The emergence and evolution of stars dramatically expanded the elemental foundation for chemical evolution. It's important to highlight that the formation of the first stars significantly boosted universal complexity, as stars were more complex than the gas clouds from which they originated, as they self-organized into complex physical systems that released the potential nuclear energy through hydrogen fusion.

> In fact, a miracle occurs: from a completely amorphous, structureless, opaque and cold mass, from a gas fog, a huge shining hot body with a complex structure grows, capable of existing for billions of years. In a word, we have before us a vivid example of self-organization under the influence of gravity and the laws characteristic of a gaseous environment[2] (Grinin 2013: 128).

However, the first stars (the so-called population III stars) only consisted of hydrogen, helium, and extremely small quantities of lithium and, hence, they were very massive and evolved (and died) very quickly (Bromm & Larson, 2004; Ohkubo et al., 2009; Sobral et al., 2015; Tominaga et al., 2007; Vink, 2014; Volpato et al., 2023). Already during their life the nuclear fusion within them produced numerous new elements, and as their life typically ended with explosions in pair-instability

[1]It also produced an even smaller amount of beryllium radioactive isotope beryllium-7 (^{7}Be), which, however, was very unstable and very soon decayed into lithium.

[2]Note that this suggests that the growth of the overall complexity of the Universe can well be estimated through the average stellar density of its galaxies (a proxy of the total number of living stars in the observable Universe), whereas the data on the dynamics of this indicator in the past 13 billion of years are more or less available (Madau & Dickinson, 2014: 463). However, in this paper we will avoid trying to estimate the universal complexity growth dynamics through the estimates of the dynamics of the average stellar density of galaxies (though we plan to do this in the future).

supernovae all those elements created during the life and death of the star[3] were usually ejected into the interstellar medium (Belkus et al., 2007; Fraley, 1968; Fryer et al., 2001; Thielemann et al., 2018; Volpato et al., 2023).[4] On one hand, the dramatic demise of the star led to an abrupt and catastrophic loss of complexity. The intricate structure of the star swiftly transformed into an amorphous gas-dust cloud, potentially with a black hole or neutron star at its center (Bandyopadhyay & Kar, 2022; Branch & Wheeler, 2017; Graur, 2022). However, this was essentially a form of creative destruction.

The crucial aspect to consider is that the pair-instability supernova explosions not only obliterated population III stars but also generated numerous complex chemical elements (Belkus et al., 2007; Fraley, 1968; Fryer et al., 2001; Thielemann et al., 2018; Volpato et al., 2023). Consequently, this process contributed to the increasing complexity of the Universe's chemical composition, laying the foundation for the emergence of progressively intricate molecules during the course of chemical evolution [see, e.g., chapter "Chemical Evolution in Cosmos and Earth" (Grinin, 2026) in this volume; see also Grinin, 2024].

In fact, this suggests possible proxy measures of complexity that could help us to estimate the universal complexity dynamics in the Stelliferous Era of the Big History in more detail then we did this earlier (LePoire, 2016, 2020a, 2020b; Korotayev, 2024). One such measure is what the astronomers call "metallicity." Astronomers define as "metals" any elements that are heavier than hydrogen and helium (thus, according to them, oxygen, or carbon are also "metals") (Acharyya, 2022; Kewley & Kobulnicky, 2007; Maiolino & Mannucci, 2019). Then, the "metallicity" turns out to be a measure indicating the ratio of "metals" (all the elements that are heavier [and, thus, more complex] than helium [let alone hydrogen]) to the "non-metals"[5] (Magrini et al., 2012; Matteucci, 2001; Nagao et al., 2006; Wuyts et al., 2014).

While this measure may seem unconventional, when considering the metallicity of a specific galaxy of a particular type, it can serve as a proxy for the complexity of matter within that galaxy. For instance, if we observe that the metallicity of galaxy A is one order of magnitude (one dex) higher than that of galaxy B of the same type and size, we can reasonably infer that the "element base" for chemical evolution in galaxy A is, on average, an order of magnitude richer than that in galaxy B. Consequently, there is a correspondingly higher probability of finding more complex molecules (including those of a "living" nature) in galaxy A compared to galaxy B.

It makes sense to stress that the average metallicity of galaxies at any moment of time can be regarded as a proxy of the *universal* complexity in the respective

[3]This included some heavier than iron elements (up to rubidium) produced by neutrino-driven wind (NDW) *r*-process (see, e.g., Wanajo, 2013; Yamazaki et al., 2022; Arcones & Thielemann, 2023).

[4]Note that the life of Population III stars typically ended with explosions in pair-instability supernovae that left no stellar remnant behind; thus, such explosions enriched the interstellar medium with complex elements in the most effective way.

[5]That is, just H + He.

moment. Actually, we do not know any other measures (except for the star formation rate,[6] to which we will return a bit later) that can be realistically used to trace the dynamics of the universal complexity after the onset of the Stelliferous Era.

At the moment, the only measure used to study the complexity in the Big History is the one proposed by Eric Chaisson (2001)—the free energy rate density/FERD (Chaisson, 2003, 2011; Aunger, 2007; Baker, 2015, 2019, 2020; Grinin et al., 2020; Spier, 2005, 2008, 2010, 2011; Vidal, 2010, 2016). However, this measure has very serious problems[7] (Panov, 2012; Solis, 2024; van Duin, 2024a, 2024b, 2025a, 2025b[8]); and, anyway, there do not seem to be any realistic way to apply it to measure the universal complexity.

On the other hand, the measures that we have just discussed—the mean metallicity of the Universe,[9] mean star formation rate of the Universe, and mean stellar mass density within the galaxies of the Universe (a proxy of the total mass of the living stars in the Universe)—can be used to study the dynamics of the truly *universal* complexity. To be more exact, the mean metallicity of the Universe can be regarded as a measure of universal chemical–physical complexity (reflecting the complexity of the element base for the chemical evolution[10]).

Consider now what drives the growth of the universal metallicity. Quite clearly this is firstly the supernova explosions.[11] We do not know any collected empirical

---

[6] And, possibly, mean stellar mass density within the galaxies of the Universe.

[7] Note also that, as references above indicate, FERD is used as a universal measure of complexity almost exclusively by big historians, and is hardly used to measure the complexity by anyone outside the Big History studies (with a very few exceptions [first of all, Georgi Georgiev])—this could be easily checked by printing "free energy rate density/FERD" in the search window of Google Scholar (https://scholar.google.ru/scholar?hl=en&as_sdt=0%2C5&q=%22free+energy+rate+density%22+ferd&btnG=).

[8] See also chapter "Complexity Pyramids and Energy Rate Density: A Quantified System Hierarchy for Big History" (van Duin, 2026) in this volume.

[9] Note that Madau and Dickinson (2014: 471) denote this variable just this way—"mean metallicity of the Universe".

[10] Note that, within the chemical evolution, the growth of complexity occurs through the emergence of more and more complex molecules. Note also that it is difficult to imagine the ways to estimate more or less exactly the mean chemical complexity of the Universe (or even a given galaxy). For example, the chemical complexity of a given galaxy will clearly depend on whether life is present in this galaxy or not, whereas the present-day science does not possess any way to find this with a slightest degree of confidence; on the other hand, the modern science has developed rather sophisticated and reliable methods allowing to identify the mean degree of chemical–physical complexity (~ metallicity) of both the Universe and a given galaxy in a given moment of time (e.g., Acharyya, 2022; Kewley & Kobulnicky, 2007; Madau & Dickinson, 2014). Note also that the proposed differentiation between chemical-physical and chemical complexity corresponds quite well to the present-day differentiation between chemistry, understood as "the science that deals with the properties, composition, and structure of substances (defined as elements and compounds) and the transformations they undergo [that is the transformations of certain molecules into other molecules]" and chemical physics [as a branch of physics dealing with physical foundations of chemistry (e.g., Fitts, 1999)].

[11] Supernova explosions eject in the interstellar medium the main mass of "metals" with atomic numbers of less than 38. As regards the elements with atomic numbers of 38 and more, according to

evidence on anything like dynamics of the "supernova explosion rate," but we have very good evidence [mainly due to a colossal effort of Madau and Dickenson (2014)] on the long-term universal dynamics of what could be considered a proxy of the universal metallicity/complexity growth rate—the star formation rate in the Universe.

Indeed, supernova explosions mark the end of life of massive stars (more than 8 solar masses) (Bandyopadhyay & Kar, 2022; Graur, 2022; Vink, 2014). However, the life of those stars is quite short—tens of millions, or even just millions of years (Belkus et al., 2007; Branch & Wheeler, 2017; Bromm & Larson, 2004; Fryer et al., 2001; Ohkubo et al., 2009). Hence, the star formation rate now suggests the "supernova explosion rate"/chemical–physical complexity growth rate just tens of millions of years later. This suggests that the increase (or decrease) in the star formation rate now should be accompanied with the increase (or decrease) in the metallicity/chemical–physical complexity growth rate with a lag of tens of millions of years. Note that such a lag is not very large for cosmological scale of many billions of years. Hence, indeed, there are grounds to maintain that the dynamics of the universal chemical–physical complexity growth should correlate quite closely (albeit with a lag of a few dozen million years [still not quite visible on the scale of many billions of years of the universal history]) with the universal dynamics of the star formation.

Consider now the results obtained by Madau and Dickinson (2014) as regards the star formation rate (see Fig. 5.2):

The Madau–Dickinson reconstruction suggests that the universal star formation rate (and, hence, the rate of the growth of universal chemical–physical complexity measured through the growth rate of the mean metallicity of the Universe [this is confirmed by Madau–Dickinson estimates of the universal metallicity growth, see Fig. 5.3]) was increasing up to 11–10 billion YBP ($\approx$ 3–4 billion years after the Big Bang Singularity) when it reached its maximum.

Note that this billions of years can be regarded as the period of the fastest growth of the chemical–physical complexity of the Universe.

In fact, in view that by the end of the cosmological Dark Ages (about 150 million years after the Big Bang Singularity $\approx$ 13.6 billion YBP $\approx$ redshift $z = 20$) the star formation rate/SFR was just zero, Madau–Dickinson reconstruction suggests that for a few hundred million years the SFR growth rate was characterized by exponential acceleration. This appears quite logical, as the increased metallicity of the interstellar medium facilitated the formation of new stars (quite a clear case of positive feedback).

the current views, they are mostly produced through rapid neutron capture (*r*-process) in neutron star mergers/kilonova explosions and collapsars (single massive stars collapsing to black holes) as well as through slow neutron capture (*s*-process) in asymptotic giant branch (AGB) stars (Arcones & Thielemann, 2023; Johnson, 2019; Kasen et al., 2017; Yamazaki et al., 2022). However, elements with atomic numbers of 38 and more only account for a very minor part of the universal metallicity, in comparison to elements with atomic numbers between 3 and 37 mostly ejected into the interstellar medium with supernova explosions (see, e.g., Arcones & Thielemann, 2023: 3–7).

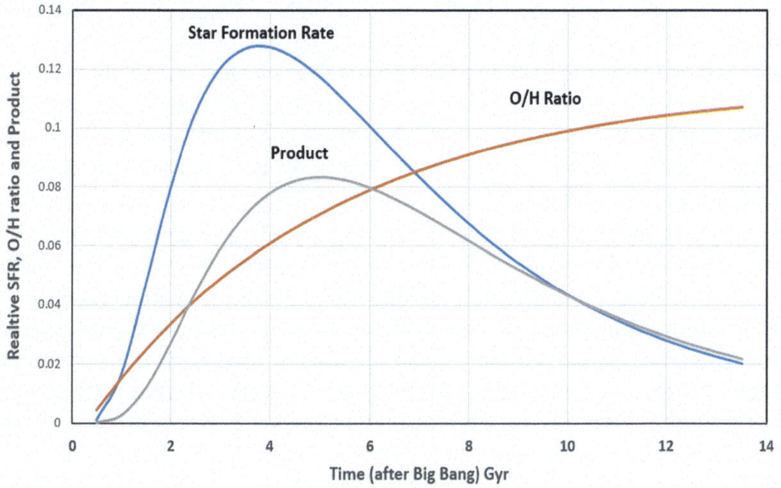

Fig. 5.2 Universal star formation rate dynamics. Data source Madau and Dickinson (2014: 473)

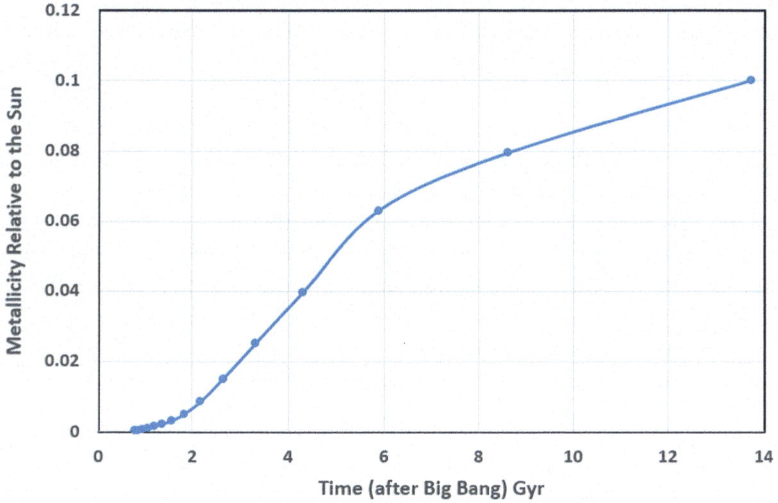

Fig. 5.3 Universal metallicity dynamics. Data source Madau and Dickinson (2014: 471)

However, after a few hundred million years the acceleration of SFR growth, it started changing towards deceleration. Between one and three billion after the Big Bang Singularity, the universal SFR continued to grow, but the speed of this growth was declining. As has been mentioned above, the universal star formation rate (and, hence, the rate of the growth of universal chemical–physical complexity measured through the growth rate of the mean metallicity of the Universe) reached its maximum around 11–10 billion YBP (≈ 3–4 billion years after the Big Bang Singularity).

After that, the universal star formation rate (and, consequently, the universal metallicity growth rate) began declining. This is explained by the following.

These where only those first population III stars that ended their lives with explosions in pair-instability supernovae that returned all their material (including the metals that had been created during their lives and deaths) back into interstellar medium where it could be used for the formation of new significantly more metal-rich stars. As new generations of stars were significantly more metal-rich, they did not have to be as massive as the first population III stars (Kippenhahn et al., 2012; Lequeux, 2013; Pinsonneault & Ryden, 2023; Ryan & Norton, 2010; Salaris & Cassisi, 2005). Hence, more massive of them were most likely to end their lives not as pair-instability, but rather as core collapse type II supernovae explosions, as a result of which only a part of their material was returned back into the interstellar medium in a form accessible for the future star formation, whereas the rest went out of star formation—star death—star formation cycle as black holes and neutron stars[12] (Bandyopadhyay & Kar, 2022; Graur, 2022; Vink, 2014). On the other hand, the growth of the metallicity of the interstellar medium led to the formation of huge numbers of intermediate mass stars.

The stars whose mass is between 0.5–0.6 and *c.* 8 solar masses ($M_\odot$) play a significant part in the enrichment of the Universe with carbon, nitrogen, and oxygen produced at the final phases of their lives.[13] Part of it returns back into the interstellar medium in a form accessible for the future star formation initially as “planetary nebulae” (e.g., Gurzadyan, 1997); yet, a very substantial part of this material goes out of star formation—star death—star formation cycle as white dwarfs[14] (Kippenhahn et al., 2012; Lequeux, 2013; Masevich & Tutukov, 2019; Pinsonneault & Ryden, 2023; Ryan & Norton, 2010; Salaris & Cassisi, 2005).

Stars with masses below 0.5–0.6 $M_\odot$ are classified as “red dwarfs.” At present they are the most common stars in the Universe (about 75% of stars in our galaxy are estimated to belong to this class). Due to their very low mass they restrict themselves to the thermonuclear fusion of hydrogen into helium and are unable to produce any elements that are more complex than helium; hence, the evolution of these stars does not lead to any increase in the universal metallicity.[15] Red dwarfs in the 0.35–0.6 $M_\odot$

[12] Though the latter still could contribute substantially later to the growth of the universal metallicity through neutron star mergers and accompanying *r*-processes (see, e.g., Arcones & Thielemann, 2023; Berger et al., 2013; Johnson, 2019; Kasen et al., 2017; Yamazaki et al., 2022).

[13] And, during an asymptotic giant branch (AGB) phase of their life, some of them can produce some quantities of more complex elements up to the lead (Pb) through the slow neutron-capture process (s-process) (Boothroyd, 2006; Busso et al., 1999; Ishkhanov et al., 2019)

[14] Yet, some part of the material of some of those white dwarfs might be returned back into this cycle, if a white dwarf accretes matter from another star, thus trespassing the Chandrasekhar limit, which could result in a type Ia supernova explosion; however, this is relevant for a small minority of white dwarfs, which still have contributed substantial amount of such elements as chromium or nickel to a total mass of those elements available in the present-day Universe (e.g., Arcones & Thielemann, 2023; Johnson, 2019; Poludnenko et al., 2019).

[15] Note that the red dwarfs are not only the smallest, but also the longest living stars in the Universe. Their lifespans far exceed the age of the Universe; hence, not a single red dwarf appears to have

range are still expected to be able to move to the red giant phase, and, hence, a part of their material would be returned (albeit without any enrichment with metals) to the interstellar medium (with a possibility to be used for the formation of new stars). On the other hand, red dwarfs in the 0.08–0.35 $M_\odot$ range (and these stars are most numerous of all the red dwarfs, in particular, and all the stars in general) are fully convective; thus, during their very long lives (that, according to the current astrophysical models may last for trillions of years) they can fuse all their hydrogen into helium, at the end transforming themselves into helium blue dwarfs (and later black dwarfs) (Adams & Laughlin, 1997; Adams et al., 2004; Kippenhahn et al., 2012; Lequeux, 2013; Padmanabhan, 2001; Pinsonneault & Ryden, 2023; Reiners & Basri, 2009; Ryan & Norton, 2010; Salaris & Cassisi, 2005; Stevenson, 2019). Thus, the material used for the formation of these most numerous stars goes totally out of the process of the further growth of universal chemical–physical complexity, and, in general, out of star formation—star death—star formation cycle.

Hence, as we see, in every star formation—star death—star formation cycle some part of material returns back into the process (and that is very important as it brings an increase in metallicity in the next generation of stars); however, a very substantial part of the material goes out of the process as stellar remnants (black holes, neutron stars, white dwarfs) and the smallest red dwarfs whose matter mostly can never be used for the formation of new stars.

In fact, Madau and Dickinson (2014: 464) estimate the return fraction[16] (R) to be as low as 0.27. That is after the death of the given generation of stars only 27% of their materials returns back to interstellar medium and can be used for the formation of new stars. Hence, the exhaustion of the material for the formation of new stars was bound to be felt more and more acutely. By about 10 billion YBP its amount turned out to be insufficient to support that very high universal star formation rate/SFR that was achieved around 11–10 billion YBP, and the universal SFR started declining, which was bound to lead to the decline of the universal chemical–physical complexity/metallicity growth rate.

What was the pattern of decline of the universal SFR after 10 billion YBP? Figure 5.2 suggests its logarithmic decline, which allows to expect the hyperbolic decline of the universal metallicity growth rate. Figure 5.3 supports this expectation. Kewley and Kobulnicky (2007) estimate the metallicity growth rate of disk galaxies between $0 < z < 3$ as 0.15 dex/z, which is illustrated with Fig. 5.4:

Recalculating this to natural (rather than logarithmic scale), we arrive at the following picture (see Fig. 5.5):

As we see, Fig. 5.5 suggests that if estimates of Kewley and Kobulnicky are correct, the universal metallicity in the recent billions of years grew logarithmically, and, hence, the universal metallicity growth rate declined hyperbolically.

"died" by now, and all what we say here about the later phases of the evolution and death of the red dwarfs has been obtained through theoretical astrophysical modeling; however, at present those models look very well founded indeed.

[16] Madau and Dickinson (2014: 418) define it as "the mass fraction of each generation of stars that is put back into the interstellar medium (ISM) and intergalactic medium (IGM)"

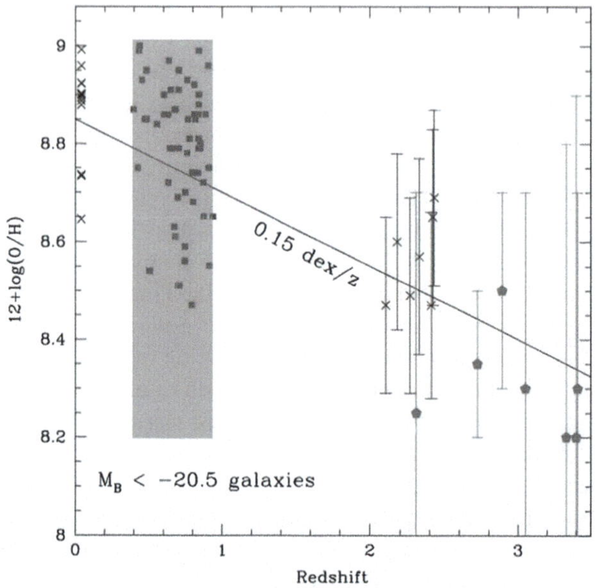

Fig. 5.4 Metallicity evolution for star-forming (emission-line) galaxies between $0 < z < 3$ (= Kewley & Kobulnicky, 2007: Fig. 5.3)

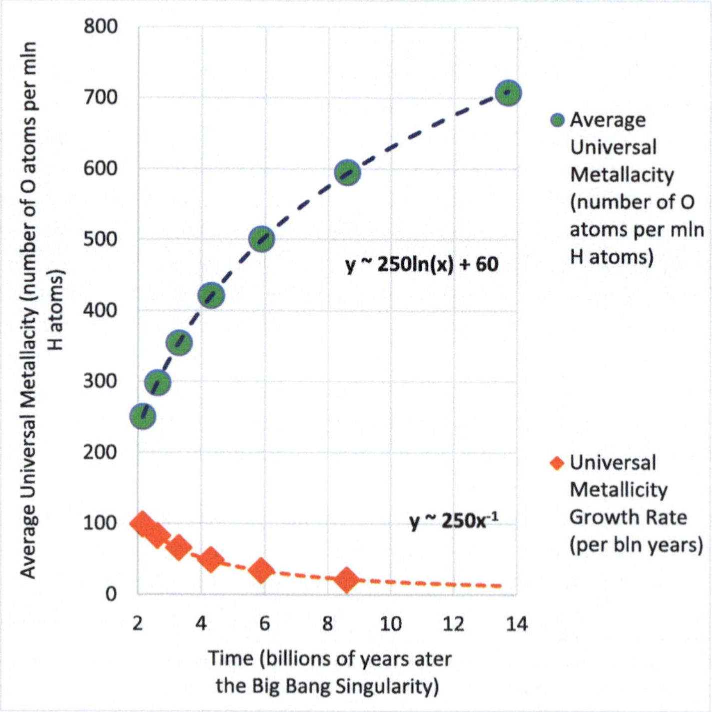

Fig. 5.5 Dynamics of universal metallicity and universal metallicity growth rate

Note that, as has been mentioned above, our prior research (Korotayev, 2018, 2020) has demonstrated that the accelerating pattern of global (biosocial) complexity growth, observed over roughly 4 billion years on Earth since the advent of life and continuing until the early 1970s can be effectively characterized with the following equation:

$$y = dn/dt = C_2/(t_2^* - t). \tag{5.1}$$

In this equation, y (= dn/dt) represents the rate of accelerating global (biosocial) complexity (n) growth, C_2 stands for a constant, and $t_2^* - t$ signifies the time remaining until the twenty-first century Singularity (with t^*, estimated to occur around 2027 or 2029 CE).

Note that the solution of differential Eq. (5.1) looks as follows:

$$n_t = A_2 - C_2 \cdot ln(t_2^* - t), \tag{5.2}$$

where A_2 is constant (e.g., Korotayev, 2020: 56–57). This shows that at this phase of the Big History the accelerating logarithmic increase in global biosocial complexity described by Eq. (5.2)[17] was accompanied by the hyperbolic acceleration of global biosocial complexity growth rate described by Eq. (5.1).

Figure 5.5 suggests again a rather symmetric situation as regards the universal chemical–physical complexity growth since about 3–4 billion years after the Big Bang Singularity (= 10–11 billion YBP). Indeed, since X-axis at Fig. 5.5 depicts time after the Big Bang Singularity, the equation describing the universal metallicity (m) growth since about 3–4 billion years after the Big Bang Singularity (= 10–11 billion YBP) looks as follows:

$$m_t = A_1 + C_3 \cdot ln(t - t_1^*). \tag{5.3}$$

In this equation, m represents universal metallicity, C_3 and A_1 are constants, and $t - t_1^*$signifies the time elapsed since the Big Bang Singularity (t_1^*–13.8 billion years ago). Note that Eq. (5.3) can be regarded as the solution of the following differential equation:

$$dm/dt = C_3/(t - t_1^*). \tag{5.4}$$

This shows that since about 3–4 billion years after the Big Bang Singularity (= 10–11 billion YBP) the decelerating logarithmic increase in universal chemical–physical complexity described by Eq. (5.3) was accompanied by the hyperbolic

[17] Note that logarithmic Eq. (5.2) describes a pattern of accelerating growth of complexity, because, with the growth of value of t, the value of $ln(t_2^* - t)$ (log time till the Singularity) decreases at faster and faster speed, but as the term $C \cdot ln(t_2^* - t)$ is subtracted from constant A, the accelerating decline in the value of the term $C \cdot ln(t_2^* - t)$ results in the accelerating growth of the value of n (that is the global biosocial complexity).

deceleration of universal chemical–physical complexity growth rate described by Eq. (5.1).

As has been mentioned above, 11–10 billion YBP ($\approx$ 3–4 billion years after the Big Bang Singularity) can be regarded as the period of the fastest growth of the chemical–physical complexity of the Universe. But in no way was this period favorable for the growth of the chemical complexity per se—that is, for the emergence of more and more complex molecules. The point is that supernova explosions tended to destroy all complex molecules in their vast vicinities (see, e.g., Gowanlock et al., 2011). Hence, after each supernova explosion, the chemical evolution in the respective part of the Universe had to start almost from zero; however, every time, the chemical evolution had at its disposal a somehow richer element base (as, that very supernova, that has destroyed all complex molecules in its vicinity, enriched the interstellar medium with enormous quantities of complex [much more complex than hydrogen, or even helium] chemical elements). Thus, at the next cycle, the chemical evolution had rather good chances to achieve a somehow higher level (characterized by the synthesis of somehow more complex molecules) than at the previous cycle (whose end was marked by the explosion of the above-mentioned supernova). On the other hand, the chemical evolution is likely to have accelerated after the end of the period of the most intensive star formation and the fastest growth of chemical–physical complexity (that is after 10 billion YBP) against the background of the declining star formation rate and the concomitant decelerating growth of the universal metallicity.

The point is that the element base for chemical evolution still continued to grow (albeit at a declining rate); among other things this led to the formation of planetary systems that could only appear around stars with a certain level of metallicity,[18] whereas the planetary systems enhanced enormously possibilities for emergence of more and more complex molecules.

On the other hand, decline of the star formation rate in a given galaxy made it a much safer place for the growth of chemical complexity, as it increased the probability that in its given part this growth will not be extinguished by a supernova explosion in its close vicinity (cp. Gowanlock et al., 2011).

5.3 Conclusion

Taking into account what has been mentioned above, the patterns of the global complexity growth against the background of the universal complexity growth patterns can be described as follows (see Table 5.1):

[18] Not less than 10% of the metallicity of the Sun (Johnson & Li, 2012).

Table 5.1 Patterns of the global complexity growth against the background of the universal complexity growth patterns

Time since the Big Bang Singularity/ BBS (years)	Time BP (billion years)	Verbal description	Universal complexity growth pattern	Universal complexity growth driver	Global complexity growth pattern
0–150 million	c. 13,8–13,65 billion years BP	Since the start of the Planck Epoch till the end of the Dark Ages	Hyperbolic deceleration	Hyperbolically decelerating cooling of the Universe	Not relevant
150 million–*c.* 3.8 billion	13.65–*c.* 10 billion years BP	Since the end of the Dark Ages till the peak of the star formation epoch (= peak SFR values)	Exponential acceleration with subsequent quasi-logistic acceleration getting to the maximum level around 10–11 billion years BP	Exponentially accelerating universal star formation rate (SFR) with subsequent slowdown and getting to the maximum level around 10–11 billion years BP	Not relevant
3.8 billion–*c.* 13.8 billion and beyond	*c.* 10 billion years BP–present and beyond	Since the peak SFR values epoch until the present and beyond	Hyperbolic deceleration	Hyperbolically (?) decelerating star formation rate (SFR)	***c.* 10–*c.* 4 billion years BP:** exponential or hyperbolic (?) acceleration of universal pre-global chemical-protobiological complexity growth if the Sharov–Panov–Anissimov “pseudo-panspermia”[a] hypothesis is correct
					c. 4 billion years BP–early 1970s: hyperbolic acceleration of

(continued)

Table 5.1 (continued)

Time since the Big Bang Singularity/ BBS (years)	Time BP (billion years)	Verbal description	Universal complexity growth pattern	Universal complexity growth driver	Global complexity growth pattern
					the global sociobiological complexity growth
					Since the early 1970s and beyond: logistic (?) or hyperbolic (?) deceleration of the global sociocultural complexity growth

[a] See, e.g., Panov (2008), Markov et al. (2010), Sharov and Gordon (2018)

References

Acharyya, A. (2022). *Evolution of gas-phase metallicity across cosmic time*. Doctoral dissertation. The Australian National University.

Adams, F. C., & Laughlin, G. (1997). A dying universe: The long-term fate and evolution of astrophysical objects. *Reviews of Modern Physics, 69*(2), 337–372. https://doi.org/10.1103/RevModPhys.69.337

Adams, F. C., Laughlin, G., & Graves, G. J. (2004). Red dwarfs and the end of the main sequence. *Revista Mexicana de Astronomía y Astrofísica, 22*, 46–49.

Arcones, A., & Thielemann, F. K. (2023). Origin of the elements. *The Astronomy and Astrophysics Review, 31*(1), 1–109. https://doi.org/10.1007/s00159-022-00146-x

Aunger, R. (2007). A rigorous periodization of 'big' history. *Technological Forecasting and Social Change, 74*(8), 1164–1178. https://doi.org/10.1016/j.techfore.2007.01.007

Baker, D. (2015). Standing on the shoulders of giants: Collective learning as a key concept in Big History. *Globalistics and Globalization Studies, 4*, 301–318.

Baker, D. (2019). Scientific Doomsday scenarios: Foresight projections for the near and deep future. *Cahiers François Viète, 3*(7), 71–93. https://doi.org/10.4000/cahierscfv.647

Baker, D. (2020). Complexity in the future: Far-from-equilibrium systems and strategic foresight. In A. V. Korotayev & D. LePoire (Eds.), *The 21st century singularity and global futures. A Big History perspective* (pp. 397–417). Springer. https://doi.org/10.1007/978-3-030-33730-8_18

Bandyopadhyay, D., & Kar, K. (2022). *Supernovae, neutron star physics and nucleosynthesis*. Springer. https://doi.org/10.1007/978-3-030-95171-9

Baumann, D. (2022). *Cosmology*. Cambridge University Press.

Belkus, H., Van Bever, J., & Vanbeveren, D. (2007). The evolution of very massive stars. *The Astrophysical Journal, 659*(2), 1576–1581. https://doi.org/10.1086/512181

Berger, E., Fong, W., & Chornock, R. (2013). An r-process kilonova associated with the short-hard GRB 130603B. *The Astrophysical Journal Letters, 774*(2), 1–4. https://doi.org/10.1088/2041-8205/774/2/L23

Boothroyd, A. I. (2006). Heavy elements in stars. *Science, 314*(5806), 1690–1691. https://doi.org/10.1126/science.1136842

Branch, D., & Wheeler, J. C. (2017). *Supernova explosions*. Springer. https://doi.org/10.1007/978-3-662-55054-0

Bromm, V., & Larson, R. B. (2004). The first stars. *Annual Review of Astronomy and Astrophysics, 42*, 79–118. https://doi.org/10.1146/annurev.astro.42.053102.134034

Busso, M., Gallino, R., & Wasserburg, G. J. (1999). Nucleosynthesis in asymptotic giant branch stars: Relevance for galactic enrichment and solar system formation. *Annual Review of Astronomy and Astrophysics, 37*(1), 239–309. https://doi.org/10.1146/annurev.astro.37.1.239

Chaisson, E. (2001). *Cosmic evolution: The rise of complexity in nature*. Harvard University Press.

Chaisson, E. J. (2003). A unifying concept for astrobiology. *International Journal of Astrobiology, 2*(2), 91–101. https://doi.org/10.1017/S1473550403001484

Chaisson, E. J. (2011). Energy rate density. II. Probing further a new complexity metric. *Complexity, 17*(1), 44–63. https://doi.org/10.1002/cplx.20373

Coc, A. (2016). Primordial nucleosynthesis. *Journal of Physics: Conference Series, 665*(1), 012001–012016. https://doi.org/10.1088/1742-6596/665/1/012001

Engel, E. A., Doss, N., Harris, G. J., & Tennyson, J. (2005). Calculated spectra for HeH+ and its effect on the opacity of cool metal-poor stars. *Monthly Notices of the Royal Astronomical Society, 357*(2), 471–477. https://doi.org/10.1111/j.1365-2966.2005.08611.x

Fitts, D. D. (1999). *Principles of quantum mechanics: As applied to chemistry and chemical physics*. Cambridge University Press.

Fraley, G. S. (1968). Supernovae explosions induced by pair-production instability. *Astrophysics and Space Science, 2*(1), 96–114. https://doi.org/10.1007/BF00651498

Fryer, C. L., Woosley, S. E., & Heger, A. (2001). Pair-instability supernovae, gravity waves, and gamma-ray transients. *The Astrophysical Journal, 550*(1), 372–382. https://doi.org/10.1086/319719

Galli, D., & Palla, F. (2013). The dawn of chemistry. *Annual Review of Astronomy and Astrophysics, 51*, 163–206. https://doi.org/10.1146/annurev-astro-082812-141029

Glötzl, E. (2025). Unification of biological and cultural evolution through natural periodization. In D. J. LePoire, L. Grinin, & A. Korotayev (Eds.), *Navigating complexity in Big History—Exploring periodization across cosmic and biosocial dimensions* (pp. 265–318). Springer Nature. (in this volume. https://doi.org/10.1007/978-3-031-85410-1_9

Gorbunov, D. S., & Rubakov, V. A. (2018). *Introduction to the theory of the early universe: Hot big bang theory*. World Scientific Publishing Company.

Gowanlock, M. G., Patton, D. R., & McConnell, S. M. (2011). A model of habitability within the Milky Way galaxy. *Astrobiology, 11*(9), 855–873. https://doi.org/10.1089/ast.2010.0555

Graur, O. (2022). *Supernova*. The MIT Press.

Grinin, L. (2013). *Bolshaya istoriya razvitiya mira: Kosmicheskaya evolyutsiya*. Uchitel.

Grinin, L. (2024). Chemical evolution in Big History. *Journal of Big History, 7*(2), 97–105. https://doi.org/10.22339/jbh.v7i2.7208

Grinin, L. (2026). Chemical evolution in Cosmos and Earth. In D. J. LePoire, L. Grinin, & A. Korotayev (Eds.), *Complexity in universal evolution – A Big History perspective* (pp. 193–210). Springer Nature (in this volume). https://doi.org/10.1007/978-3-032-20516-2_7

Grinin, L. E., Korotayev, A. V., & Baker, D. (2020). Beyond global studies. An introductory lecture into a Big History course. *Evolution, 7*, 246–255.

Gurzadyan, G. A. (1997). *The physics and dynamics of planetary nebulae*. Springer.

Güsten, R., Wiesemeyer, H., Neufeld, D., Menten, K. M., Graf, U. U., Jacobs, K., Stutzki, J., et al. (2019). Astrophysical detection of the helium hydride ion HeH+. *Nature, 568*(7752), 357–359. https://doi.org/10.1038/s41586-019-1090-x

Hawking, S. (2009). *A brief history of time: From big bang to black holes*. Random House.

Ishkhanov, B. S., Kapitonov, I. M., & Tutyn, I. A. (2019). *Nukleosintez vo Vselennoy. Osnovnyye etapy razvitiya Vselennoy ot momenta Bolshogo Vzryva*. Librokom/URSS.

Johnson, J. A. (2019). Populating the periodic table: Nucleosynthesis of the elements. *Science, 363*(6426), 474–478. https://doi.org/10.1126/science.aau9540

Johnson, J. L., & Li, H. (2012). The first planets: The critical metallicity for planet formation. *The Astrophysical Journal, 751*(2), 1–11. https://doi.org/10.1088/0004-637X/751/2/81

Kasen, D., Metzger, B., Barnes, J., Quataert, E., & Ramirez-Ruiz, E. (2017). Origin of the heavy elements in binary neutron-star mergers from a gravitational-wave event. *Nature, 551*(7678), 80–84. https://doi.org/10.1038/nature24453

Kewley, L., & Kobulnicky, H. A. (2007). The metallicity history of disk galaxies: Evolution between 0< z< 3. In *Island universes: Structure and evolution of disk galaxies* (pp. 435–440). Springer Netherlands.

Kippenhahn, R., Weigert, A., & Weiss, A. (2012). *Stellar structure and evolution* (2nd ed.). Springer.

Korotayev, A. (2018). The 21st century singularity and its Big History implications: A re-analysis. *Journal of Big History, 2*(3), 71–118. https://doi.org/10.22339/jbh.v2i3.2320

Korotayev, A. (2020). The 21st century singularity in the Big History perspective. A re-analysis. In A. Korotayev & D. LePoire (Eds.), *The 21st century singularity and global futures. A Big History perspective* (pp. 19–75). Springer. https://doi.org/10.1007/978-3-030-33730-8_2

Korotayev, A. (2024). Complexity growth patterns in the Big History. Preliminary results of a quantitative analysis. *Journal of Big History, 7*(2), 9–27. https://doi.org/10.22339/jbh.v7i2.7203

Korotayev, A. (2025). Comparing and contrasting Big History singularity trends of the big bang and terrestrial evolution. In D. J. LePoire, L. Grinin, & A. Korotayev (Eds.), *Navigating complexity in Big History — Exploring periodization across cosmic and biosocial dimensions* (pp. 101–157). Springer Nature. https://doi.org/10.1007/978-3-031-85410-1_5

Korotayev, A. V., & LePoire, D. J. (2025). Complexity growth patterns in the Stelliferous era of Big History. Preliminary observations. *Journal of Big History, 8*(1), 1–12. https://doi.org/10.22339/jbh.v8i1.8102

LePoire, D. J. (2016). Exploring temporal patterns in Big History dynamics. *KronoScope, 16*(2), 229–249.

LePoire, D. J. (2020a). Exploring the singularity concept within Big History. In A. V. Korotayev & D. LePoire (Eds.), *The 21st century singularity and global futures. A Big History perspective* (pp. 77–97). Springer. https://doi.org/10.1007/978-3-030-33730-8_3

LePoire, D. J. (2020b). *Time patterns in Big History. Cycles, fractals, waves, transitions, and singularities.* David LePoire Consulting, LLC.

LePoire, D. J. (2025). A simple, compatible, and extensible Big History framework and periodization based on previous findings. In D. J. LePoire, L. Grinin, & A. Korotayev (Eds.), *Navigating complexity in Big History—Exploring periodization across cosmic and biosocial dimensions* (pp. 159–193). Springer Nature. https://doi.org/10.1007/978-3-031-85410-1_6

Lepp, S., Stancil, P. C., & Dalgarno, A. (2002). Atomic and molecular processes in the early Universe. *Journal of Physics B: Atomic, Molecular and Optical Physics, 35*(10), R57–R80. https://doi.org/10.1088/0953-4075/35/10/201

Lequeux, J. (2013). *Birth, evolution and death of stars*. World Scientific Publishing Company.

Madau, P., & Dickinson, M. (2014). Cosmic star-formation history. *Annual Review of Astronomy and Astrophysics, 52*, 415–486. https://doi.org/10.1146/annurev-astro-081811-125615

Magrini, L., Hunt, L., Galli, D., Schneider, R., Bianchi, S., Maiolino, R., Valiante, R., et al. (2012). Scaling relations of metallicity, stellar mass and star formation rate in metal-poor starbursts—II. Theoretical models. *Monthly Notices of the Royal Astronomical Society, 427*(2), 1075–1088. https://doi.org/10.1111/j.1365-2966.2012.22055.x

Maiolino, R., & Mannucci, F. (2019). De re metallica: The cosmic chemical evolution of galaxies. *The Astronomy and Astrophysics Review, 27*(1:3), 1–187. https://doi.org/10.1007/s00159-018-0112-2

Markov, A. V., Anisimov, V. A., & Korotayev, A. V. (2010). Relationship between genome size and organismal complexity in the lineage leading from prokaryotes to mammals. *Paleontological Journal, 44*(4), 363–373. https://doi.org/10.1134/S0031030110040015
Masevich, A., & Tutukov, A. (2019). *Evolutsiya zvezd: Teoriya i nablyudeniya* (2nd ed.). Lenand/URSS.
Mathews, G. J., Kusakabe, M., & Kajino, T. (2017). Introduction to big bang nucleosynthesis and modern cosmology. *International Journal of Modern Physics E, 26*(08), 1741001. https://doi.org/10.1142/S0218301317410014
Matteucci, F. (2001). *The chemical evolution of the Galaxy*. Springer. https://doi.org/10.1007/978-94-010-0967-6
Mukhanov, V. (2005). *Physical foundations of cosmology*. Cambridge University Press.
Nagao, T., Maiolino, R., & Marconi, A. (2006). Gas metallicity diagnostics in star-forming galaxies. *Astronomy and Astrophysics, 459*(1), 85–101. https://doi.org/10.1051/0004-6361:20065216
Ohkubo, T., Nomoto, K. I., Umeda, H., Yoshida, N., & Tsuruta, S. (2009). Evolution of very massive population III stars with mass accretion from pre-main sequence to collapse. *The Astrophysical Journal, 706*(2), 1184–1193. https://doi.org/10.1088/0004-637X/706/2/1184
Padmanabhan, T. (2001). *Theoretical astrophysics*. Cambridge University Press.
Palla, F., Galli, D., & Silk, J. (1995). Deuterium in the universe. *Astrophysical Journal, 451*, 44–50.
Panov, A. (2008). *Universal'naya evolyutsiya i problema poiska vnezemnogo razuma (SETI)*. LKI/URSS.
Panov, A. (2012). Comments on the Russian translation of "How Big History works: Energy flows and the rise and demise of complexity". *Evolyutsiya, 4*, 123–126.
Pinsonneault, M., & Ryden, B. (2023). *Stellar structure and evolution*. Cambridge University Press.
Poludnenko, A. Y., Chambers, J., Ahmed, K., Gamezo, V. N., & Taylor, B. D. (2019). A unified mechanism for unconfined deflagration-to-detonation transition in terrestrial chemical systems and type Ia supernovae. *Science, 366*(6465), eaau7365. https://doi.org/10.1126/science.aau7365
Reiners, A., & Basri, G. (2009). On the magnetic topology of partially and fully convective stars. *Astronomy and Astrophysics, 496*(3), 787–790. https://doi.org/10.1051/0004-6361:200811450
Ryan, S. G., & Norton, A. J. (2010). *Stellar evolution and nucleosynthesis*. Cambridge University Press.
Ryden, B. S. (2017). *Introduction to cosmology* (2nd ed.). Cambridge University Press.
Salaris, M., & Cassisi, S. (2005). *Evolution of stars and stellar populations*. Wiley.
Sharov, A. A., & Gordon, R. (2018). Life before earth. In R. Gordon & A. A. Sharov (Eds.), *Habitability of the universe before Earth* (pp. 265–296). Academic Press. https://doi.org/10.1016/B978-0-12-811940-2.00011-3
Sobral, D., Matthee, J., Darvish, B., Schaerer, D., Mobasher, B., Röttgering, H. J., Hemmati, S., et al. (2015). Evidence for PopIII-like stellar populations in the most luminous Lyα emitters at the epoch of reionization: Spectroscopic confirmation. *The Astrophysical Journal, 808*(2), 1–14. https://doi.org/10.1088/0004-637X/808/2/139
Solis, K. (2024). Reexamining free energy rate density as a complexity metric. *Journal of Big History, 7*(1), 19–28. https://doi.org/10.22339/jbh.v7i1.7102
Spier, F. (2005). How Big History works: Energy flows and the rise and demise of complexity. *Social Evolution & History, 4*(1), 87–135.
Spier, F. (2008). Big History: The emergence of a novel interdisciplinary approach. *Interdisciplinary Science Reviews, 33*(2), 141–152.
Spier, F. (2010). *Big History and the future of humanity*. Wiley-Blackwell.
Spier, F. (2011). Complexity in Big History. *Cliodynamics, 2*(1), 146–166. https://doi.org/10.21237/C7clio21213
Steigman, G. (2006). Primordial nucleosynthesis: Successes and challenges. *International Journal of Modern Physics E, 15*(01), 1–35. https://doi.org/10.1142/S0218301306004028

Stevenson, D. S. (2019). *Red Dwarfs: Their geological, chemical, and biological potential for life*. Springer.

Switzer, E. R., & Hirata, C. M. (2008). Primordial helium recombination. III. Thomson scattering, isotope shifts, and cumulative results. *Physical Review D, 77*(8), 083008-1–083008-23. https://doi.org/10.1103/PhysRevD.77.083008

Thielemann, F. K., Isern, J., Perego, A., & Von Ballmoos, P. (2018). Nucleosynthesis in supernovae. *Space Science Reviews, 214*(3), 1–62. https://doi.org/10.1007/s11214-018-0494-5

Tominaga, N., Umeda, H., & Nomoto, K. I. (2007). Supernova Nucleosynthesis in Population III 13–50 $M_{\odot}$ stars and abundance patterns of extremely metal-poor stars. *The Astrophysical Journal, 660*(1), 516–540. https://doi.org/10.1086/513063

Tsirel, S. (2020). Big History and singularity as metaphors, hypotheses and prediction. In A. Korotayev & D. LePoire (Eds.), *The 21st century singularity and global futures. A Big History perspective* (pp. 119–144). Springer. https://doi.org/10.1007/978-3-030-33730-8_6

Van Duin, M. (2024a). What can we learn from a master plot of energy rate *versus* mass for a very wide variety of (complex) systems? *Journal of Big History, 7*(1), 29–78. https://doi.org/10.22339/jbh.v7i1.7103

Van Duin, M. (2024b). Dissipative systems have a maximum energy rate density of 10^5 W/kg. *The European Physical Journal, B97*(11), 171. https://doi.org/10.1140/epjb/s10051-024-00785-2

Van Duin, M. (2025a). Development of mass and energy rate (density) of dissipative systems over their lifetimes: A comparison of a low-mass star, like our sun, a human and the Roman Empire. In D. J. LePoire, L. Grinin, & A. Korotayev (Eds.), *Navigating complexity in Big History – Exploring periodization across cosmic and biosocial dimensions* (pp. 345–385). Springer Nature. https://doi.org/10.1007/978-3-031-85410-1_11

Van Duin, M. (2025b). What more can we learn from a master plot of energy rate *versus* mass for a very wide variety of (complex) systems? *Journal of Big History, 8*(2), 38–52. https://doi.org/10.22339/jbh.v8i2.8202

Van Duin, M. (2026). Complexity pyramids and energy rate density: A quantified system hierarchy for Big History. In D. J. LePoire, L. Grinin, & A. Korotayev (Eds.), *Complexity in universal evolution – A Big History perspective* (pp. 119–148). Springer Nature (in this volume). https://doi.org/10.1007/978-3-032-20516-2_4

Vidal, C. (2010). *Big History and our future: Extension, evaluation and significance of a universal complexity metric*. Research proposal. http://evodevouniverse.com/uploads/9/98/Vidal_2010-Big_History_and_our_Future.Pdf

Vidal, C. (2016). Stellivore extraterrestrials? Binary stars as living systems. *Acta Astronautica, 128*, 251–256. https://doi.org/10.1016/j.actaastro.2016.06.038

Vink, J. S. (Ed.). (2014). *Very massive stars in the local universe*. Springer. https://doi.org/10.1007/978-3-319-09596-7

Volpato, G., Marigo, P., Costa, G., Bressan, A., Trabucchi, M., & Girardi, L. (2023). A study of primordial very massive star evolution. *The Astrophysical Journal, 944*(1), 1–14. https://doi.org/10.3847/1538-4357/acac91

Wanajo, S. (2013). The r-process in proto-neutron-star wind revisited. *The Astrophysical Journal Letters, 770*(2), L22. https://doi.org/10.1088/2041-8205/770/2/L22

Wuyts, E., Kurk, J., Schreiber, N. M. F., Genzel, R., Wisnioski, E., Bandara, K., Wilman, D. J. et al. (2014). A consistent study of metallicity evolution at $0.8 < z < 2.6$. *The Astrophysical Journal Letters, 789*(2), L40.

Yamazaki, Y., He, Z., Kajino, T., Mathews, G. J., Famiano, M. A., Tang, X., & Shi, J. (2022). Possibility to identify the contributions from collapsars, supernovae, and neutron star mergers from the evolution of the r-process mass abundance distribution. *The Astrophysical Journal, 933*(1), 112. https://doi.org/10.3847/1538-4357/ac721c

Zel'dovich, Y. B., Kurt, V. G., & Syunyaev, R. A. (1969). Recombination of hydrogen in the hot model of the universe. *Soviet Journal of Experimental and Theoretical Physics, 28*, 146–150.

Andrey Korotayev heads the Center for Stability and Risk Analysis at the HSE University in Moscow, Russia. He is also Senior Research Professor at Eurasian Center for Big History and System Forecasting at the Institute of Oriental Studies, Russian Academy of Sciences and Institute for African Studies, Russian Academy of Sciences, as well as at the Faculty of Global Studies of the Lomonosov Moscow State University. Beginning as an historian of Arabia, he made focused studies, such as seen in "Two Social Ecological Crises and Genesis of Tribal Organization in the Yemeni North-East" (1996). Korotayev researches topics in quantitative cross-cultural anthropology and seeks to understand biological and social macroevolution, as in "A Compact Macromodel of World System Evolution" (2005) and "Mathematical Modeling of Biological and Social Phases of Big History" (2014). He has authored or co-authored over 650 scholarly publications, including such monographs as *Ancient Yemen* (Oxford University Press, 1995), *World Religions and Social Evolution of the Old World Oikumene Civilizations: A Cross-Cultural Perspective* (The Edwin Mellen Press, 2004), *Introduction to Social Macrodynamics: Compact Macromodels of the World System Growth* (URSS, 2006; with Daria Khaltorina and Artemy Malkov), *Introduction to Social Macrodynamics: Secular Cycles and Millennial Trends* (URSS, 2006; with Daria Khaltorina and Artemy Malkov), *Great Divergence and Great Convergence: A Global Perspective* (Springer Nature, 2015; with Leonid Grinin), *Economic Cycles, Crises, and the Global Periphery* (Springer, 2016; with Leonid Grinin), and *Islamism, Arab Spring, and the Future of Democracy: World System and World Values Perspectives* (Springer Nature, 2019; with Leonid Grinin and Arno Tausch). He is a founding member of the Eurasian Center for Megahistory and System Forecasting and founder of *Evolution*, an almanac dedicated to the study of Universal Evolution, and thus is closely linked with Big History. In 2020, he produced, with David LePoire, a collective monograph, *The 21st Century Singularity and Global Futures: A Big History Perspective* (Springer, 2020). Recently, he has produced, with David LePoire and Leonid Grinin, another Big History collective monograph, *Navigating Complexity in Big History: Exploring Periodization Across Cosmic and Biosocial Dimensions* (Springer Nature, 2025). He is a laureate of a Russian Science Support Foundation in "The Best Economists of the Russian Academy of Sciences" Nomination (2006); in 2012, he was awarded the Gold Kondratieff Medal by the International N. D. Kondratieff Foundation. He serves as a board member of the International Big History Association.

David LePoire researches, develops, and applies science principles in environmental issues, Big History evolutionary trends, and particle scattering. He has a BS in physics from CalTech, a Ph.D. in computer science from DePaul University, and over 30 years of experience at the Argonne National Laboratory in the development of scientific analyses, software, training, and modeling. His research includes Big History synergistic trends among energy, environment, organization, and information.

Energy Flow and Emergence of Complexity in the Universe's Structure

6

David J. LePoire

Abstract

The Universe's structural development contrasts with the rapid cooling after the Big Bang and the later emergence of life on Earth, spanning scales from the cosmic web to planets. Earlier studies on free energy rate density (FERD) suggested increasing energy density over time, with galaxies appearing less complex than stars. However, recent findings indicate that large energy flows—such as gravitational collapse in the cosmic web and active galactic nuclei (AGN)—were crucial in forming these structures in successive stages. Including energy from cosmic web collapse, AGN phases, stellar processes, and Earth's cooling suggests FERD may decrease as energy gradients dissipate, stabilizing conditions for life. This structural evolution resembles a cascading energy flow, akin to turbulent fluid dynamics, offering a better framework for understanding complexity growth in the cosmos.

Keywords

Cosmic web · Energy flow · Cascade · Dissipate · Complexity

This chapter is an extensively re-worked version of the text originally published in Journal of Big History (LePoire, 2025).

D. J. LePoire (✉)
Argonne National Laboratory, Argonne, IL, USA
e-mail: dlepoire@anl.gov

D. J. LePoire et al. (eds.), *Complexity in Universal Evolution*, World-Systems Evolution and Global Futures, https://doi.org/10.1007/978-3-032-20517-9_6

6.1 Introduction

There are two mega phases in Big History: cosmic physical evolution and terrestrial evolution of life, humans, and civilization. Terrestrial phases occur at rapidly accelerating rates, forming an approximate geometric logistical transition (LePoire, 2015). A cosmic phase, however, differs from a terrestrial phase in that physical transitions occur very early and decrease with time as the Universe expands (Christian, 2011; Christian et al., 2014; Panov et al., 2020; Korotayev, 2024, 2025a). Additionally, it differs in other ways; for example, the evolution of structure in the Universe occurs as a result of dissipating the potential nuclear and gravitational energy that remains after the first 3 min of the Big Bang. In the course of this energy flow, dissipative structures such as the cosmic web, galaxies, stars, and eventually planets are formed. It differs from the bootstrapping of life on Earth where information storage and processing are key to continuously better adapting to the evolving environment.

According to the extended Jantsch model (1980), the cosmic phase/"eon" consists of the Big Bang, followed by about two-thirds of the Universe's age until the formation of our planet 4.5 billion years ago. That is, the Universe is about three times older than the Earth. During this time, the Big Bang, the cosmic web, galaxies, stars, and planets developed. It was during the Big Bang that the hot particles cooled down as they expanded. There was a factor of two decrease in temperature for each quadrupling of the Universe's age (during its first few minutes) in an inverse singularity growth pattern (Korotayev, 2024, 2025a). There is a Big Bang sequence of events and particles with a three-order reduction in binding energies of protons, nuclei, atoms, molecules, and solids (from 1000 MeV for protons to 25 meV for current terrestrial temperatures) followed by another decrease in temperature of two orders of magnitude to the cosmic microwave background temperature of 3 degrees K (0.25 meV) (LePoire, 2014).

Accordingly, there was a period of about 100 million years (1%) of cosmic phase up to the formation of the early stars, followed by 98% of the cosmic phase, ending with another 100 million years (1%) during which the solar system was formed. The mid 98% of the time was spent cooling galactic processes and creating elements for rocky planets (Binney & Tremaine, 2011; Korotayev & LePoire, 2025).[1] A similar pattern is seen in the wear and tear of mechanical systems (and also software systems) as most issues occur at the beginning (bad parts or mismatched components) or after some wear and tear, with a relatively smooth operational productive middle period.

Additionally, during this phase, the evolving environment gradually shrinks from the Universe to the cosmic web, galaxies, stars, and planets (Gaite, 2019). Fundamental particles, nuclei, atoms, and molecules are being simultaneously formed as the components for later evolution. These components, however, are not complex or

[1] See also chapter "Complexity growth patterns in the stelliferous era of Big History: Preliminary analysis" (Korotayev & LePoire, 2026) in this volume.

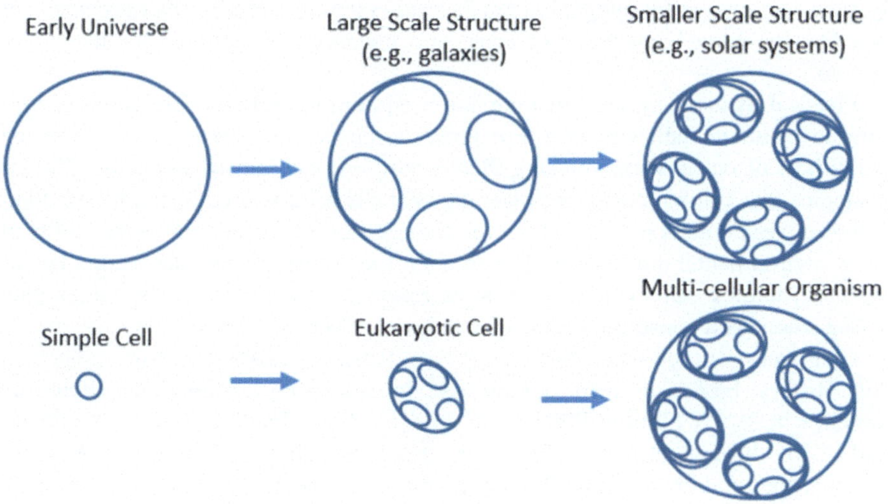

Fig. 6.1 Differences in the evolution of sizes in the cosmic and terrestrial life phases. The complex physical system starts from the largest size, the Universe, and develops into small structures (galaxies, stars, planets). CAS, like life, evolve from the bottom-up, that is, from simple cells, to larger multi-cellular organisms and social groups

dissipative systems since they are simply physically bound with any energy flow, and they only combine when the temperature drops below the binding energy. Even so, the larger structures are not complex adaptive systems (CAS), but they are dissipative systems, forming complex physical systems. In contrast, in terrestrial evolutionary phases, organisms do form CAS and tend to grow larger from cells to animals to people to societies and civilizations (Fig. 6.1). Nevertheless, the environment in which they are involved tends to decrease from the challenge of changing the oxygen content of the Earth through simple cellular life to microchips fueling globalization.

6.2 Multiple Forms of Complex Systems

Big History encompasses a variety of complexities across its systems. These include complex adaptive systems (CAS) with agency, dynamic complex physical systems, and arrangements of particles in intricate structures, such as organic molecules. Such systems extend beyond historical contexts, appearing in modern technological and social frameworks as well. This broader study of complexity spans multiple disciplines, as explored by Holland (2014), Mitchell (2009), and Waldrop (1992).

Complex adaptive systems (CAS), such as living systems with agency and energy flow (Niele, 2005; Kauffman, 1995), utilize environmental information to learn, grow, avoid threats, reproduce, and evolve. The continuous energy flow and agency prevent these systems from reaching equilibrium, which would signify death. These systems can evolve by integrating smaller subsystems, a process that provides net benefits, including economies of scale and enhanced capabilities for the larger system.

Physical dynamic systems that are out of equilibrium rely on continuous energy flow to maintain their structure. For instance, hurricanes derive energy from the rising heat of ocean water, making them examples of dissipative systems. Similar examples that can be studied in a laboratory include Bénard cells in a heated fluid and reactive chemical systems. However, these systems lack memory or the ability to store environmental information. Importantly, they illustrate how the second law of thermodynamics can facilitate self-organization in specific contexts, rather than acting solely as a constraint (Schneider & Kay, 1994).

Chaisson proposed using free energy rate density (FERD) to classify objects in Big History based on their complexity, incorporating assumptions about the system's mass. By plotting FERD over time, he showed an increase in complexity and FERD throughout history. However, this approach raises questions: Why not measure the total energy flow through a system instead? And why would stars be considered more complex than galaxies, which contain far more interacting components? Regarding FERD, for a planet, this metric would account for heat from gravitational collapse and radionuclide decay within its interior, averaged over the planet's mass—not just the sunlight it receives.[2]

This chapter aims to examine the early energy flows that shaped the Universe, focusing on the evolution of galaxies, stars, planets, and the cosmic web from the relatively uniform density of the post-Big Bang state. Unlike living complex systems, these cosmic structures are expected to behave differently, resembling eddy formations created by wind around buildings. Notably, the initial formation of these structures may not directly involve energy dissipation. Instead, the structures evolve, breaking down into smaller "eddies" that eventually dissipate energy. This process occurs across multiple spatial scales, with the distribution of these structures following a scaling power law (Gaite, 2019).

Particles like protons, nuclei, atoms, and molecules—formed from fundamental quarks and electrons—are sometimes described as "complex" due to their structured nature. However, unlike physical dissipative systems or complex adaptive systems with agency, these systems lack energy flow and remain in equilibrium. In equilibrium, their dynamics involve the exchange of virtual particles to calculate a "dynamic" equilibrium, but their quantum state remains fixed with no time variation or energy transfer in the ground state. Since there isn't a specific term for this type of system development, the concept of "virtual complexity" could be introduced. In contrast, larger systems can form through interactions among these units when a thermal gradient is present, creating a physical dissipative system.

[2]For more detail on problems with the use of FERD as the main complexity metrics in the Big History, see chapter "Complexity Pyramids and Energy Rate Density: A Quantified System Hierarchy for Big History" (van Duin, 2026) in this volume. See also Solis (2024) and van Duin (2024, 2025a, b).

6.3 Cosmic Structures and Processes

6.3.1 Big Bang

The events surrounding the Big Bang and their interpretations remain a topic of active debate. Peter Atkins (2018) argues that the unfolding of forces and particles during the Big Bang was uneventful, as it adhered strictly to the basic principles of physics. However, Gleiser (2010) highlights that the breaking of symmetries was crucial to making the Universe "interesting." For instance, an asymmetry between matter and antimatter ensured that a small fraction of matter (approximately one part of a billion) survived after temperatures dropped below the threshold for pair formation. To explain the remarkable coincidences that allowed for a life-supporting Universe, researchers like Hawking (Hertog 2023) and Davies (2006) have proposed the idea of a self-consistent, participatory Universe during this phase.

Following the Big Bang, physics manifested in the form of particles and forces governed by quantum fields. About 3 min after the Universe's inception, cosmic nucleosynthesis produced helium atoms, converting roughly a quarter of the Universe's mass into this element. Quantum fluctuations from the early Universe eventually evolved into cosmic structures such as galaxies and webs as the Universe cooled over an extended period (Loeb, 2006; Binney & Tremaine, 2011). For gravitational collapse and the formation of these structures to occur, the Universe's gas needed to cool sufficiently. Dark matter, which interacts only through gravity and does not collide to generate heat, played a critical role in this cooling. By contrast, ordinary matter converted half of its gravitational potential into heat, shaping the eventual structures of the Universe.

6.3.2 Cosmic Web

The formation of structure in the Universe began with gravitational forces acting on small density fluctuations in dark matter, which played a foundational role in shaping the cosmic web. Unlike normal ionized matter, dark matter could respond earlier in the Universe's evolution, initiating large-scale flows that reduced gravitational potential energy. This energy reduction allowed matter to move more efficiently toward the dense nodes of the cosmic web, where galaxies would eventually form (Hogan, 2001). These gravitational flows continue to channel matter through galaxies to this day, sustaining their dynamics (Walter et al., 2020). However, star formation within galaxies reached its peak roughly 10 billion years ago (Sobral et al., 2013; Madau & Dickinson, 2014), well before the formation of our solar system.

The Universe's early structures—cosmic web, galaxies, and stars—arose through processes that dissipated gravitational and nuclear potential energy. Over time, these once-dramatic energy flows began to settle. Meanwhile, successive generations of stars went through life cycles of formation and supernova explosions, enriching the interstellar medium with heavy elements forged in their cores. This gradual buildup of elements was critical for the formation of planets and, ultimately, the emergence of life on at least one of them—Earth (Chaisson, 2001; Korotayev & LePoire, 2025; see also chapter "Complexity Growth Patterns in the Stelliferous Era of Big History. Preliminary Analysis" [Korotayev & LePoire, 2026] in this volume).

6.3.3 Galaxies

The active galactic nuclei (AGN) of massive black holes in galaxies may have played a significant role in regulating stellar formation rates by influencing the dynamics and structure of galaxies. Walter et al. (2020) estimate that stellar formation peaked approximately 3 billion years after the Big Bang, a period often referred to as "cosmic noon." Since then, the rate of star formation has declined by a factor of about eight, with the Milky Way currently forming stars at an average rate of one per year. Over the next 5 billion years, the influx of material from the cosmic web necessary for star formation is projected to decline to half its current rate.

Throughout their history, galaxies have captured ionized hydrogen gas from the cosmic web. This gas subsequently cools into neutral hydrogen, eventually condensing into molecular hydrogen within galaxies, where it fuels star formation. Galaxies, being gravitationally bound systems, have frequently collided and merged over cosmic time. While the stars within colliding galaxies rarely collide directly, the massive black holes at their centers often merge, producing gravitational waves as a result. These galactic interactions also contributed to the formation of spiral galaxies. The gravitational dynamics concentrated material into the two-dimensional planes of spiral arms, creating regions conducive to further star formation.

Recent surveys of over 3000 galaxies reveal a wide range of properties, including variations in mass, rotation, bulge size, disk structure, spiral arms, and halos (ASTRO-3D, 2021). Some findings, such as the correlation between a galaxy's total mass and the much smaller mass of its central black hole, remain puzzling.

Other intriguing patterns, such as the alignment of galactic rotation axes among nearby galaxies, suggest connections to their formation within the cosmic web (Neyrinck, 2014). This alignment might be partially explained by the influence of the cosmic web during galaxy formation. Unlike regular matter, dark matter interacts only through gravity and cannot shed energy (it emits no light or heat), resulting in unique gravitational dynamics. A theoretical model likens the development of the cosmic web to the folding of origami paper, where constraints in folding create inherent correlations. Similar processes could underlie the observed alignment of rotation axes in neighboring galaxies.

6.3.4 Stars

Stars formed from collapsing clouds of gas and dust, a process that generated immense heat, reaching temperatures similar to those of the Universe just minutes after the Big Bang. This extreme heat triggered nuclear fusion, releasing vast amounts of energy as hydrogen atoms fused into helium. It is important to note

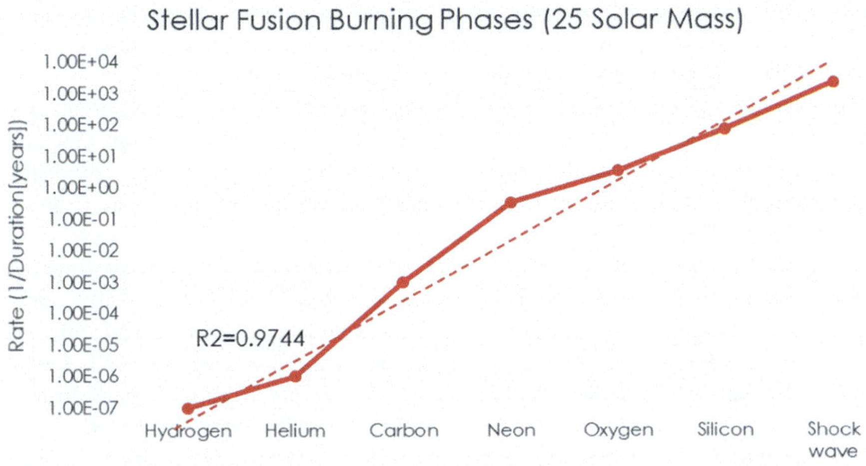

Fig. 6.2 The reciprocal of the stellar fusion phase for various elements

that this temperature is insufficient to cause the colliding hydrogen nuclei to fuse. Instead, a quantum tunneling effect facilitates the temperature at a much lower temperature, and consequently slower rate. About 100 million years after the Big Bang—roughly 1% of the 9 billion years it took to form the Earth and the Solar System—the first stars emerged.

The most massive of these early stars rapidly consumed their hydrogen fuel, continuing to fuse heavier atomic nuclei until reaching iron, the most stable nucleus (Fig. 6.2). Once nuclear fusion ceased to generate sufficient energy to counteract gravitational forces, these stars collapsed under their own weight. This collapse often triggered spectacular supernova explosions, ejecting material from the stars' layered interiors, enriched with newly formed elements.

These processes were critical for seeding the Universe with heavier elements, which later contributed to the formation of planets and life. Recent studies, including those by Kewley and Kobulnicky (2007), Madau and Dickinson (2014), and Korotayev and LePoire (2025), have refined estimates of stellar formation rates and the gradual enrichment of the Universe's metallicity over time, providing deeper insights into the role of stars in cosmic evolution.[3]

6.3.5 Planets

There are many phenomena that are still being investigated concerning gravitational collapse of various sizes and their complicated interactions. These include the role of dark matter in early gravitational structure formation, the fragmentation of

[3] See also chapter "Complexity Growth Patterns in the Stelliferous Era of Big History. Preliminary Analysis" (Korotayev & LePoire, 2026) in this volume.

gravitational cloud collapse, the transfer of angular momentum, the highly dynamic orbits of forming planets, the effect of galactic collisions, and the role of the galactic central massive black hole in regulating star formation.

The gravitational collapse occurs when the temperature of the cloud is sufficiently low, or the density is large. As the cloud collapses there might be conditions in which the cloud fragments, that is, the conditions for smaller collapses occur within the larger cloud. This often happens in star formation leading to multiple stars forming from the collapse of a larger cloud.

The cloud collapse leading to our sun and solar system might have been triggered by the collision of the Milky Way with a small galaxy. While stars would rarely directly collide, the gas and stars undergo gravitational forces. It is believed that such collisions helped form the disk-shaped galaxies from the originally ellipsoidal galaxies. In the disk galaxies, the material is more condensed as it goes from a three-dimensional object to one closer resembling the two-dimensional disk.

The gravitational collapse into planets, however, is a bit more subtle. The dust particles of heavier elements and ice particles are just a small fraction of the (mostly hydrogen) gas in the protoplanetary disk. The disk is also already moving in orbit around the star, that is, not isolated. In these circumstances, the mass of the dust and ice must build up to planetesimal size (about 1 km in diameter) before gravity starts its feedback loop in accelerating the aggregation process. Much has been learned recently about this process such as friction and static cling of particles, but much is still unknown (Blum & Wurm, 2008; Johansen, 2014).

Lineweaver (2001) used stellar formation rates by mass and time along with metallicity growth and dispersion within galaxies to estimate the "Earth formation rate", which is similar in approach to the metallicity and stellar formation rate product in Fig. 6.3. This "Earth formation rate" estimate could also be used as a measure of complexity. This rate shows an early exponential increase from 2 to 3 billion years ago, followed by a relatively stable peak from 3 to 7 billion years after the Big Bang, ending with an exponential decrease to about 1/3 the peak formation rate now. The solar system formed just after the end of the relatively stable peak formation period.

Besides the mass collapsing and heat transfer, there seems to be quite a bit of angular momentum that is transferred from the forming star (Ray, 2012). For example, in the solar system the sun has about 98.7% of the mass of the whole solar system but only less than 1% of the angular momentum. Jupiter has a large fraction of the angular momentum. As the sun formed some of its angular momentum might have been transferred to matter that was later ejected from the solar system. The mechanism of this transfer is still being investigated but it seems like the interaction of magnetic fields and the small amount of ionized particles in the solar system disk might be a major contributor to this transfer.

However, comparing our solar system to other's being found, leaves many mysteries concerning our solar system uniqueness in having a wide range of planetary masses (Choi, 2015). In the simple nebular hypothesis, the planets formed where they are currently located and the rocky, gaseous, and icy planets are separated by distances from the sun that will support these different phases.

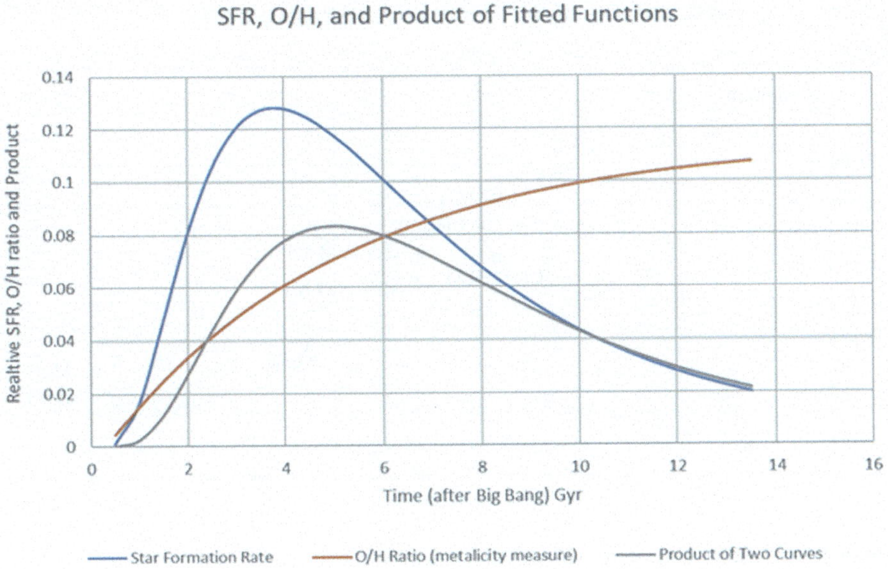

Fig. 6.3 The relative stellar formation rate (blue), metallicity buildup (O/H ratio in orange), and product (gray) which is an indication of planetary formation rate

However, more evidence is being accumulated to point to large protoplanetary movements especially that of Jupiter and Saturn as the rocky planets formed. This disturbed some of their development (e.g., Mars small mass) but also led to the availability of water to enter the atmospheres of the rocky planets.

Our solar system formation seems to have been catalyzed by a nearby supernova which released radioactive material such as Al-26. The age of this Al-26 seems to indicate that the supernova was about 25–100 million years before the sun and solar system completely formed (Tillman, 2021).

6.4 Analysis of Energy Flow

Hogan (2001) investigated the energy flow through the various mechanisms that have dissipated gravitational and nuclear potential in the Universe considering energy flow, duration, and frequency (Fig. 6.4). This energy was released through expansion of the Universe, the formation of galaxies and stars, the evolution of stars, including supernovae, pulsars and blazers, and active galactic nuclei which helped form quasars in the early Universe.

The results of phenomena dating back to early times are being explored as new detector systems, such as the James Webb telescope and gravitational wave LIGO instruments, report them. Although no one really knows what 95% of the Universe is as dark energy and dark matter, computer modeling continues. Some scientists speculate that some of the seeming independent current astronomical mysteries

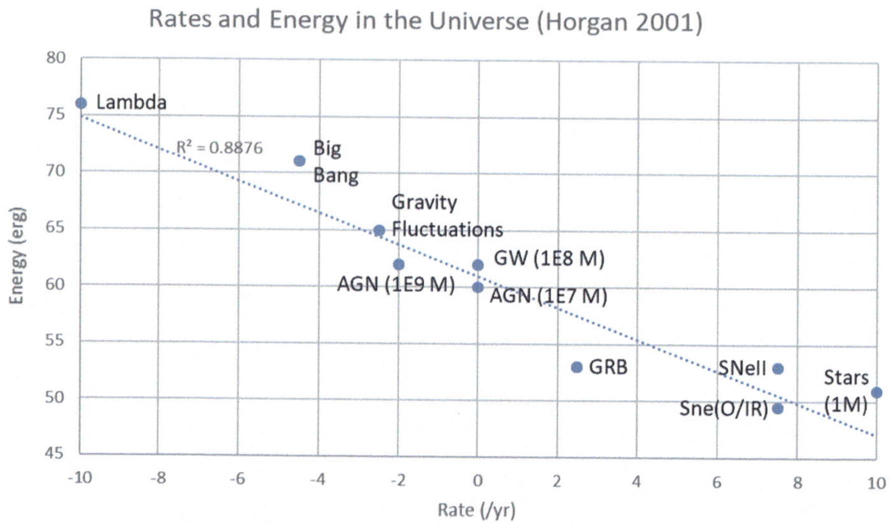

Fig. 6.4 Rates and energy in the universe (Modified Hogan 2001)

might be connected. According to a recent hypothesis from Brookhaven National Laboratory (Davoudiasl et al., 2022), the hypothetical axion particle could be responsible for dark matter, baryogenesis, and unusually large black holes. However, high energy physics experiments have crushed many similar recent hypotheses by not finding any new particles which combine with astronomical measures that restrict the acceptable properties of such particles. From Hogan (2001):

> A glance shows a remarkable coincidence: in spite of 20 orders of magnitude variation in mass and timescale (and N), the integrated power is comparable for all of these populations . . . This coincidence can be understood if there are feedback loops controlling the release of energy—a globally regulated choreography coupling the formation rate of stars, the events leading to their death and the transformation and ejection of the elements, the formation of galaxies and quasars and the feeding of their central engines. Although the scales of the individual sources all derive directly from fundamental physics, their frequency in the cosmos depends on this poorly understood "cosmo-ecology" of interacting systems.

6.5 Comparison to Dissipative Systems

The Big Bang's gravitational and nuclear potential is dissipated in new ways as the Universe develops. Despite the fact that large stars may have existed before galaxies, most of them exist now along with other material that includes massive central black holes, haloes from earlier stars, remnant stars from galactic collisions, galactic arms, pulsars, nebula, molecular clouds, and planets. These cosmic structures don't form from new combinations like life does; instead, gravitational collapses and nuclear reactions form from further collapse of larger previous structures.

Fig. 6.5 Energy cascade of an eddy, showing physical structure (top) and energy vs. size (bottom). Source: https://en.wikipedia.org/wiki/Energy_cascade

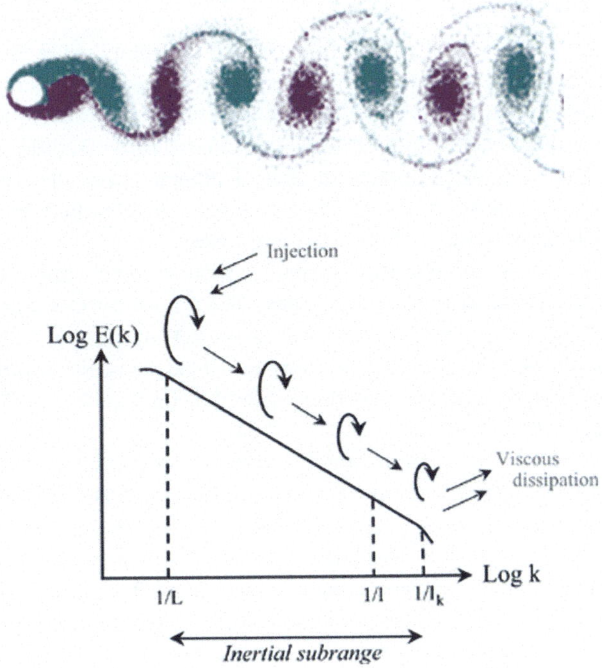

A cascade of energy dissipation in turbulent flow can be compared to this type of increase in entropy. When an obstructing body releases an eddy, the energy within it interacts with the surrounding fluids to form counter eddies that continue to form. As the eddy moves away from the obstructing body, it begins to develop structure until it finally reaches a size small enough to match the fluid's impedance and therefore transfer energy (Alexakis & Biferale, 2018). From the largest to the smallest, eddies in turbulent systems follow a fractal scaling law, which is called a cascade (Fig. 6.5).

In light of this, Chiasson's analysis of energy density flow (FERD), which includes astronomical structure formation is interesting. However, it fails to distinguish between the different behaviors of cosmic dissipative systems following cascades and complex adaptive systems that evolve through combination and energy extraction. Energy flow and complexity should be examined separately during these two main phases of evolution, according to this theory. In other words, by comparing the whole system to turbulence, it becomes clear that the Universe is more complex than a galaxy, and a galaxy is more complex than a solar system.

6.6 Patterns in the Complexity Levels

A complex system is often formed by many various objects at different scales. Therefore, we would expect to see the large numbers of galaxies, stars, and planets showing some intermediate substructures and variations. Table 6.1 shows a rough estimate of the number of galaxies in the Universe, the number of stars in the Milky Way, and an estimated number of planetesimals that formed a rocky planet in our solar system. It also lists a number of substructures and properties that cause variation with a galaxy, star, and planet.

This seeming scaling of about a trillion objects might continue to lower levels into the terrestrial phase of evolution. The reason for this number of objects in a complex system is unknown both from a bottom-up perspective, that is, why gravitational collapse and subsequent formation lead to these numbers, and from a top down perspective, that is, why there should be this number of objects to form a new level of complex objects.

However, relationships between the fundamental physics constants have shown estimates for the temporal and spatial scales of Big History objects from the Universe to atoms (Carr & Rees, 1979; Press & Lightman, 1983; Vidal, 2014). The relative size of a group of Big History length scales depends on a simple factor constructed from a few fundamental physics constants. The ratio is about 140,000 in going from nuclei, atoms, cells, humans, Earth's core, distance to the sun, distance between stars, galactic size, and size of the Universe (LePoire, 2020a). In this way the Big History objects of both the cosmic and terrestrial phases are linked by considering simple physical processes and how they relate to the physical constants. Further connections are being explored between the development of life and the cosmic structures (Panov, 2005).

Table 6.1 The number of objects and components in various cosmic scale systems

Scale	Number of smaller objects	Components
Universe	2 trillion galaxies ($2 \cdot 10^{12}$)	Galaxies, clusters, cosmic web, voids
Milky Way	100–400 billion stars ($2 \cdot 10^{11}$)	Massive central black hole, disks/arms, haloes, bulge, dark matter, satellite galaxies (50)
Solar System	Planets: 8 Number of planetesimals[a] to make a rocky planet: about $2 \cdot 10^{12}$	Star, planets, moons, asteroids, Oort objects, solar wind

[a] A minimal self-gravitational planetesimal might be 1 km in diameter. This gives the number of about $2 \cdot 10^{12}$ of these minimal planetesimals to make a planet (Earth) that is about 13,000 greater in diameter. However, the definition of self-gravity at this initial size of a planetesimal means that it can grow faster by accumulating gravitation attraction of smaller rocks, rather than collisions of two small planetesimals

6.7 Conclusion

There was a period of structure formation in the Universe between the rapid cooling following the Big Bang and the acceleration of life events on Earth about 9 billion years later that has been well studied but contains many mysteries that continue to attract great interest. In the formation of galactic and stellar structures, gravitational and nuclear energy potentials dissipated, creating their complexity. After the Big Bang, the Universe cooled and stabilized, resulting in fewer recent quasars, a decrease in star formation by about an order of magnitude, and a more stable environment that allowed stellar debris of heavy elements to form rocky planets like Earth. There were two major subphases of complexity growth during cosmic phase—the rapid cooling after the Big Bang and the structural formation subphase, during which gravitational and nuclear potentials dissipated, leaving materials and energy to develop solar systems like our own, in which at least one planet provided the conditions for long-term evolution of life. Various interpretations are possible for the Big-Bang sub-phase, including stating that nothing really happened (Atkins, 2018) or stating that quite a bit happened (Gleiser, 2010).

Complex adaptive systems (life) evolved on Earth, starting from simple cells, which, through a series of combinations and reorganizations, resulted in a wide variety of life forms. Humans, one of these species, developed during a phase where genetics and culture played a significant role in forming biological characteristics that would then contribute to later cultural evolutions. Cultural developments such as the use of tools and fire, communication and storytelling, and cooperation were developed along with the predominantly biological evolution of a larger brain, smaller stomach, ability to speak, ability to walk upright, and ability to run long distances (see chapter "Human Evolution in the Complexity Growth Perspective. Toward Periodization of the Big History Biosocial Era" [Korotayev, 2026] in this volume; see also Korotayev, 2025b).

The differences and similarities between the cosmic and terrestrial phases of complexity warrant further investigation. While the analysis of free energy rate density (FERD) may provide interesting insights, a broader perspective on the mechanisms of complexity development seems to yield deeper understanding. For example, cosmic complexity might resemble a dynamic cascade of turbulence formation, where length scales are connected by fundamental physical constants. Furthermore, organized energy flow could be more significant than energy density flow, and a combination of physical phenomena may act synchronistically in the development of structure. Observations and models of cosmic structure formation continue to evolve rapidly. As the nature of dark matter and dark energy remains a mystery, future discoveries are likely to expand our knowledge. These findings, combined with advances in understanding the origins and expansion of life complexity, will contribute to a more comprehensive understanding of their connections in Big History (LePoire, 2020b).

Earlier considerations of energy flow in the Universe continue to be found in strikingly different scientific disciplines. For example, the 2004 paper compares different energy considerations of form and structure based on ecological studies

(Brown et al., 2004). Geoffrey West's work on biological and social scaling laws provides a valuable theoretical foundation for interpreting patterns of complexity across cosmic structures. Although his analyses focus primarily on organisms, cities, and corporations, the underlying mechanisms—such as network constraints, energy flow, and fractal geometries—may extend to non-living systems as well. West (2017) argues that scaling laws reflect universal constraints that transcend domains, suggesting that the same principles governing metabolic efficiency in mammals or infrastructure scaling in cities could apply to self-organizing structures in the Universe. This perspective offers a conceptual link to the observed regularities in cosmic size scales, such as the roughly 10^5-fold transitions between nuclei, atoms, cells, humans, planets, and galactic structures. These discrete steps in spatial and temporal scales may reflect universal principles of hierarchical organization under energy and information constraints—an idea that resonates with West's proposal that all complex systems must balance the cost of energy distribution with the need to maintain coherence and adaptability.

The analysis in this chapter will need to be reviewed and updated based on new findings, for example, from the James Webb Telescope. Some indications, such as early galaxy bar formation and rotation curves of massive elliptical, point to structures forming in the Universe earlier than expected (Lea, 2024; Fernandez, 2025). While the analysis in this chapter has assumed the λCDM (i.e., Lambda cold dark matter) model with dark matter and energy, insightful research continues to be explored for the Modified Newtonian Dynamics, including deriving a relativistic version consistent with General Relativity from the consideration of the quantum transaction interpretation model (Schlatter & Kastner, 2023). This model also seems to have predicted earlier formation of structures in the Universe (McGaugh et al., 2024).

References

Alexakis, A., & Biferale, L. (2018). Cascades and transitions in turbulent flows. *Physics Reports, 767–769*, 1–101. https://doi.org/10.1016/j.physrep.2018.08.001

ASTRO 3D. (2021, February 2). Complex mechanics of the evolution of the universe: The secrets of 3000 galaxies laid bare. *SciTechDaily*.

Atkins, P. (2018). *Conjuring the universe: The origins of the laws of nature*. Oxford University Press.

Binney, J., & Tremaine, S. (2011). *Galactic dynamics*. Princeton University Press.

Blum, J., & Wurm, G. (2008). The growth mechanisms of macroscopic bodies in protoplanetary disks. *Annual Review of Astronomy and Astrophysics, 46*, 21–56. https://doi.org/10.1146/annurev.astro.46.060407.145152

Brown, M. T., Odum, H. T., & Jorgensen, S. E. (2004). Energy hierarchy and transformity in the universe. *Ecological Modelling, 178*(1–2), 17–28. https://doi.org/10.1016/j.ecolmodel.2003.12.002

Carr, B. J., & Rees, M. J. (1979). The anthropic principle and the structure of the physical world. *Nature, 278*(12). https://doi.org/10.1038/278605a0

Chaisson, E. J. (2001). *Cosmic evolution: The rise of complexity in nature*. Harvard University Press.

Choi, C. Q. (2015). Jupiter's 'smashing' migration may explain our oddball solar system. *Space.com*.
Christian, D. C. (2011). *Maps of time: An introduction to Big History*. University of California Press.
Christian, D. C., Brown, S., & Benjamin, C. (2014). *Big History: Between nothing and everything*. McGraw Hill Education.
Davies, P. (2006). *Cosmic jackpot: Why our universe is just right for life*. Penguin Press.
Davoudiasl, H., Denton, P. B., & Gehrlein, J. (2022). Supermassive black holes, ultralight dark matter, and gravitational waves from a first order phase. *Physical Review Letters, 128*, 081101. https://doi.org/10.1103/PhysRevLett.128.081101
Fernandez, S. (2025, June). Largest map of the universe announced revealing 800,000 galaxies, challenging early cosmos theories. *Phys.org*.
Gaite, J. (2019). The fractal geometry of the cosmic web and its formation. *Advances in Astronomy, 2019*, 6587138. https://doi.org/10.1155/2019/6587138
Gleiser, M. (2010). *A tear at the edge of creation: A radical new vision for life in an imperfect universe*. Free Press.
Hertog, T. (2023). *On the origin of time: Stephen Hawking's final theory*. Bantam Books.
Hogan, C. J. (2001). Energy flow in the universe. In R. G. Crittenden & N. G. Turok (Eds.), *Structure formation in the universe* (Vol. 565, NATO Science Series C). Springer. https://doi.org/10.1007/978-94-010-0540-1_13
Holland, J. H. (2014). *Complexity: A very short introduction*. Oxford University Press.
Jantsch, E. (1980). *The self-organizing universe: Scientific and human implications of the emerging paradigm of evolution*. Pergamon.
Johansen, A. (2014). Formation of planetesimals: The building blocks of planets. In R. Amils et al. (Eds.), *Encyclopedia of astrobiology*. Springer. https://doi.org/10.1007/978-3-642-27833-4_5251-1
Kauffman, S. A. (1995). *At home in the universe: The search for laws of self-organization and complexity*. Oxford University Press.
Kewley, L., & Kobulnicky, H. A. (2007). The metallicity history of disk galaxies: Evolution between $0 < z < 3$. In *Island universes: Structure and evolution of disk galaxies* (pp. 435–440). Springer. https://doi.org/10.1007/978-1-4020-5573-7_75
Korotayev, A. V. (2024). Complexity growth patterns in the Big History – Preliminary results of a quantitative analysis. *Journal of Big History, 7*(2), 9–27. https://doi.org/10.22339/jbh.v7i2.7203
Korotayev, A. (2025a). Comparing and contrasting Big History singularity trends of the Big Bang and terrestrial evolution. In D. J. LePoire, L. Grinin, & A. Korotayev (Eds.), *Navigating complexity in Big History: Exploring periodization across cosmic and biosocial dimensions* (pp. 101–157). Springer. https://doi.org/10.1007/978-3-031-85410-1_5
Korotayev, A. (2025b). Toward periodization of the biosocial era of the Big History: A reanalysis. *Journal of Big History, 8*(2), 1–37. https://doi.org/10.22339/jbh.v8i2.8201
Korotayev, A. (2026). Human evolution in the complexity growth perspective. Toward periodization of the Big History biosocial era. In D. J. LePoire, L. Grinin, & A. Korotayev (Eds.), *Complexity in universal evolution – A Big History perspective* (pp. 359–410). Springer Nature (in this volume). https://doi.org/10.1007/978-3-032-20516-2_13
Korotayev, A. V., & LePoire, D. J. (2025). Complexity growth patterns in the stelliferous era of Big History: Preliminary observations. *Journal of Big History, 8*(1), 1–12. https://doi.org/10.22339/jbh.v8i1.8102
Korotayev, A. V., & LePoire, D. J. (2026). The Evolution of complexity in the Stelliferous Era: Observations and patterns. In D. LePoire, L. Grinin, & A. Korotayev (Eds.), *Complexity in universal evolution: A Big History perspective* (pp. 151–172). Springer Nature. https://doi.org/10.1007/978-3-032-20516-2_5
Lea, R. (2024, April 26). James Webb Space Telescope discovers some early universe galaxies grew up surprisingly fast. *Space.com News*.

LePoire, D. J. (2014, August). *Two contrasting views of energy and time scales*. Paper presented at the International Big History Conference, Dominican College, San Rafael, CA.
LePoire, D. J. (2015). Interpreting Big History as complex adaptive system dynamics with nested logistic transitions in energy flow and organization. Emergence: Complexity & Organization, 17(1), 1–16. https://doi.org/10.emerg/10.17357.dbe7c8d6fae7f082c4f33c5b35df8287
LePoire, D. J. (2020a). Size relationships of Big History objects: From the universe to the atomic nucleus. *Journal of Big History, 4*(1). https://doi.org/10.22339/jbh.v4i1.4160
LePoire, D. J. (2020b). Exploring the singularity concept in Big History. In A. Korotayev & D. LePoire (Eds.), *The 21st century singularity and global futures: A Big History perspective*. Springer. https://doi.org/10.1007/978-3-030-33730-8_3
LePoire, D. J. (2025). Energy flow and complexity in the cosmic structure development phase. *Journal of Big History, 8*(1), 13–20. https://doi.org/10.22339/jbh.v8i1.8103
Lineweaver, C. H. (2001). An estimate of the age distribution of terrestrial planets in the universe: Quantifying metallicity as a selection effect. *Icarus, 151*(2), 307–313. https://doi.org/10.1006/icar.2001.6607
Loeb, A. (2006). The dark ages of the universe. *Scientific American, 295*(5), 46–53. https://doi.org/10.1038/scientificamerican112006-2jrkRVe32VGFRMBCEkBWCB
Madau, P., & Dickinson, M. (2014). Cosmic star-formation history. *Annual Review of Astronomy and Astrophysics, 52*, 415–486. https://doi.org/10.1146/annurev-astro-081811-125615
McGaugh, S. S., Schombert, J. M., Lelli, F., & Franck, J. (2024). Accelerated structure formation: The early emergence of massive galaxies and clusters of galaxies. *The Astrophysical Journal, 976*(1), 13. https://doi.org/10.3847/1538-4357/ad834d
Mitchell, M. (2009). *Complexity: A guided tour*. Oxford University Press.
Neyrinck, M. C. (2014). Cosmological origami: Properties of cosmic-web components when a non-stretchy dark-matter sheet folds. *arXiv*. https://doi.org/10.48550/arXiv.1408.2219
Niele, F. (2005). *Energy: Engine of evolution*. Elsevier.
Panov, A. (2005). Scaling law of the biological evolution and the hypothesis of the self-consistent galaxy origin of life. *Advances in Space Research, 36*(2), 220–225. https://doi.org/10.1016/j.asr.2005.03.001
Panov, A., LePoire, D., & Korotayev, A. (2020). The 21st century singularity in the Big History perspective: An overview. In A. Korotayev & D. LePoire (Eds.), *The 21st century singularity and global futures: A Big History perspective* (pp. 1–18). Springer Nature. https://doi.org/10.1007/978-3-030-33730-8_1
Press, W. H., & Lightman, A. P. (1983). Dependence of macrophysical phenomena on the values of the fundamental constants. *Philosophical Transactions of the Royal Society A, 310*, 323–336. https://doi.org/10.1098/rsta.1983.0094
Ray, T. (2012). Losing spin: The angular momentum problem. *Astronomy & Geophysics, 22*. https://doi.org/10.1111/j.1468-4004.2012.53519.x
Schlatter, A., & Kastner, R. E. (2023). Gravity from transactions: Fulfilling the entropic gravity program. *Journal of Physics Communications, 7*, 065009. https://doi.org/10.1088/2399-6528/acd6d7
Schneider, E. D., & Kay, J. J. (1994). Life as a manifestation of the second law of thermodynamics. *Mathematics, Computing, and Modeling, 19*(6–8), 25–48. https://doi.org/10.1016/0895-7177(94)90188-0
Sobral, D., et al. (2013). A large Hα survey at z = 2.23, 1.47, 0.84 and 0.40: The 11 Gyr evolution of star-forming galaxies from HiZELS. *Monthly Notices of the Royal Astronomical Society, 428*(2). https://doi.org/10.1093/mnras/sts096
Solis, K. (2024). Reexamining free energy rate density as a complexity metric. *Journal of Big History, 7*(1), 19–28. https://doi.org/10.22339/jbh.v7i1.7102
Tillman, N. T. (2021, December 13). Multiple supernovas may have implanted our solar system with the seeds of planets. *Space.com*.

van Duin, M. (2024). What can we learn from a master plot of energy rate *versus* mass for a very wide variety of (complex) systems? *Journal of Big History, 7*(1), 29–78. https://doi.org/10.22339/jbh.v7i1.7103

Van Duin, M. (2025a). Development of mass and energy rate (density) of dissipative systems over their lifetimes: A comparison of a low-mass star, like our sun, a human and the Roman empire. In D. J. LePoire, L. Grinin, & A. Korotayev (Eds.), *Navigating complexity in Big History – Exploring periodization across cosmic and biosocial dimensions* (pp. 345–385). Springer Nature. https://doi.org/10.1007/978-3-031-85410-1_11

Van Duin, M. (2025b). What more can we learn from a master plot of energy rate *versus* mass for a very wide variety of (complex) systems? *Journal of Big History, 8*(2), 38–52. https://doi.org/10.22339/jbh.v8i2.8202

Van Duin, M. (2026). Complexity pyramids and energy rate density: A quantified system hierarchy for Big History. In D. J. LePoire, L. Grinin, & A. Korotayev (Eds.), *Complexity in universal evolution – A Big History perspective* (pp. 119–148). Springer Nature (in this volume). https://doi.org/10.1007/978-3-032-20516-2_4

Vidal, C. (2014). *The beginning and the end: The meaning of life in a cosmological perspective*. Springer.

Waldrop, M. M. (1992). *Complexity: The emerging science at the edge of order and chaos*. Simon & Schuster.

Walter, F., et al. (2020). The evolution of the baryons associated with galaxies averaged over cosmic time and space. *The Astrophysical Journal, 902*(2). https://doi.org/10.3847/1538-4357/abb82e

West, G. B. (2017). *Scale: The universal laws of life, growth, and death in organisms, cities, and companies*. Penguin Press.

David LePoire researches, develops, and applies science principles in environmental issues, Big History evolutionary trends, and particle scattering. He has a BS in physics from Caltech, a Ph.D. in computer science from DePaul University, and over 30 years of experience at the Argonne National Laboratory in the development of scientific analyses, software, training, and modeling. His research includes Big History synergistic trends among energy, environment, organization, and information. Currently, he is the President of the International Big History Association (IBHA).

Part III

Chemical and Geological Complexity

Chemical Evolution in Cosmos and Earth

7

Leonid Grinin

Abstract

Chemical evolution is a very important part of Big History without which the emergence of life would have been impossible. Chemical evolution was the leading line in the formation of pre-life on the Earth 5 billion years ago. However, unfortunately, there is insufficient research on the course of chemical evolution within the framework of both Big History and evolutionary studies. The lack of attention to chemical evolution is all the more disappointing, and the reasons for such inattention are not clear. This chapter presents a brief history of chemical evolution based on Big History: from the formation of the first atoms in the Universe to abiogenesis on the Earth, that is, the stage of pre-life and the formation of prerequisites for the emergence of the first living organisms. The history of chemical evolution before life's origin can be divided into three stages: the formation of atoms (pre-evolution); history before the start of the abiogenic phase on the Earth; and abiogenic chemical evolution. However, the author aims to elaborate a more detailed periodization of chemical evolution before life's origin. One should also pay attention to an important feature of chemical evolution, which distinguishes it from other branches of evolution, namely, its co-evolutionary nature. The author demonstrates that chemical evolution at all its stages acted as a part of a co-evolutionary tandem: first, as a part of cosmic and stellar-galactic evolution, then as a part of planetary evolution since it is on planets (where temperature parameters are much more comfortable for chemical

This chapter is an extensively re-worked version of the text originally published in Journal of Big History (Grinin, 2024).

L. Grinin (✉)
HSE University, Moscow, Russia

Eurasian Center for Big History and System Forecasting, Institute of Oriental Studies, Russian Academy of Sciences, Moscow, Russia

D. J. LePoire et al. (eds.), *Complexity in Universal Evolution*, World-Systems Evolution and Global Futures, https://doi.org/10.1007/978-3-032-20517-9_7

reactions) that a new qualitative stage in the development of chemical evolution begins. Finally, on the Earth, it developed first as a part of geochemical evolution, and then as a part of biochemical evolution, and this development continues until now.

Keywords

Chemical evolution · Megaevolution · Big History · Cosmochemical evolution · Co-evolution · Biochemical evolution · Geochemical evolution · Star-Galaxy era

7.1 Introductory Notes

It is strange, but there is little, if any, research on chemical evolution in the framework of the study of Big History. Why? This is a difficult question. Perhaps, no one knows the answer. I do not know it either but I have an idea, which will be presented below. One way or another, considering the chemical line of Big History can significantly enrich our ideas and understanding about the general course of Big History and about the path to increasing complexity. Moreover, without understanding the history of chemical evolution, one can hardly grasp either the mystery of the origin of life and development of life in the early periods.[1]

The formation of hydrogen and helium atoms in the first period after the Big Bang and the accumulation of heavy element atoms as a result of the star collapses were the most important events in chemical evolution. However, the formation of atoms cannot be yet considered as a chemical evolution in the full sense of the word. Chemical evolution is *the emergence and development of different and more complex types of molecules and substances.* One should realize that such evolution could hardly begin in a very hot Universe, nor could it take place in the depths of stars.

Thus, it is important to realize that chemical reactions:

(a) Can occur when the temperature drops to 5000 degrees, but in fact the most favorable condition for them is at relatively low plus temperatures
(b) Take place constantly in space, even at deep sub-zero temperatures; some of the characteristics of such chemical reactions are known from the studies of gas and dust clouds

[1]See Bernal (1969), Betekhtin (2007), Galimov (2008), Glyantsev (2019), Guotmi and Cunningham (1960), Degens and Roiter (1967), Dickerson (1981), Dobretsov (2005), Zavarzin (2003), Zaguskin (2014), Kalvin (1971), Kamshilov (1970, 1979), Lima de Faria (1991), Rudenko (1969), Spiridonov (2019), Haldane (1949), Lyons et al. (2014), Grinin (2013, 2017, 2018, 2020), Grinin and Grinin (2019).

(c) Should be even more active within the framework of evolution of planets and other bodies (including comets), as can be inferred from studies of the bodies of the Solar system

Some chemical compounds could have been formed only during the collapse of stars or on the surface of cooling stars.

7.2 Chemical Evolution as a Peripheral and Parallel Line of Big History

Chemical evolution can be regarded as a peripheral and parallel line of Big History. Why? From the above said one can make the following important conclusions: first, the chemical evolution could only begin after the cooling of the Universe. Second, it always evolved not in the main sequence, that is, not in stars and galaxies, but at the periphery of the Universe. It developed mainly in gas and dust clouds and on peripheral celestial bodies, especially on planets. And consequently, for many billions of years, the "achievements" of chemical evolution have been somewhat invisible (see Figs. 7.1 and 7.2).

Let us now return to the question of why so little attention is paid to the chemical evolution. I believe that one of the main reasons is the parallel development of chemical evolution within Big History.[2] Other reasons are its co-evolution with geological, including mineralogical, and biological evolution and the fact that we know very little about the abiogenic chemical evolution. This, however, in no way diminishes the role of chemical evolution; but, on the contrary, it makes its study really relevant.

Chemical evolution began even before the Star-Galaxy era, that is, already in the first millions of years after the Big Bang, in gaseous hydrogen–helium clouds. This is where the first molecules were formed. But, of course, this evolution could not proceed actively without the formation of a sufficient variety of chemical elements. Thus, chemical evolution progressed in parallel with the star-galaxy evolution. At the same time, (in clouds, on planets, in comets and meteorites, etc.) there are many dozens of different not only inorganic but also many organic substances, including water, alcohols, acids, monosaccharides, and even amino acids, in particular glycine. The synthesis of simple organic substances constantly occurs in various cosmic environments.

We do not know when the first planets formed within ancient star-planetary systems, but with their emergence, the rate of chemical evolution increased

[2]One of the few who pointed out chemical evolution as a special trend in Big History was Eric Chaisson in his book *Epic of Evolution: Seven Ages of the Cosmos* (Chaisson, 2005). However, he did not take into account its parallel development. As a result, he considered the Chemical Epoch as a separate epoch, which is unjustified. Chemical evolutionary development can only be considered as a continuous line of growing chemical complexity in Big History and as an integral part of other phases (see Grinin, 2024).

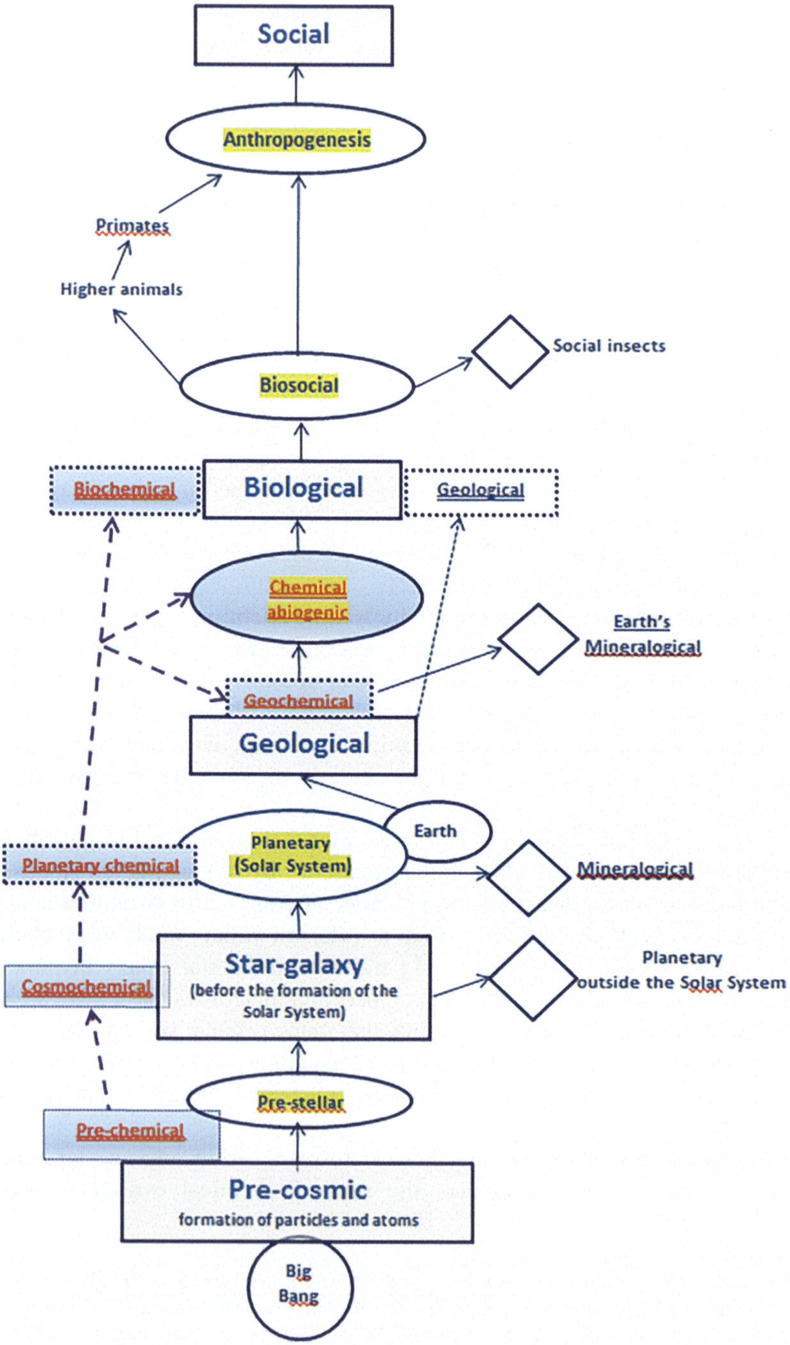

Fig. 7.1 Unfolding of Big History (Megaevolution). Phases and lines of Big History

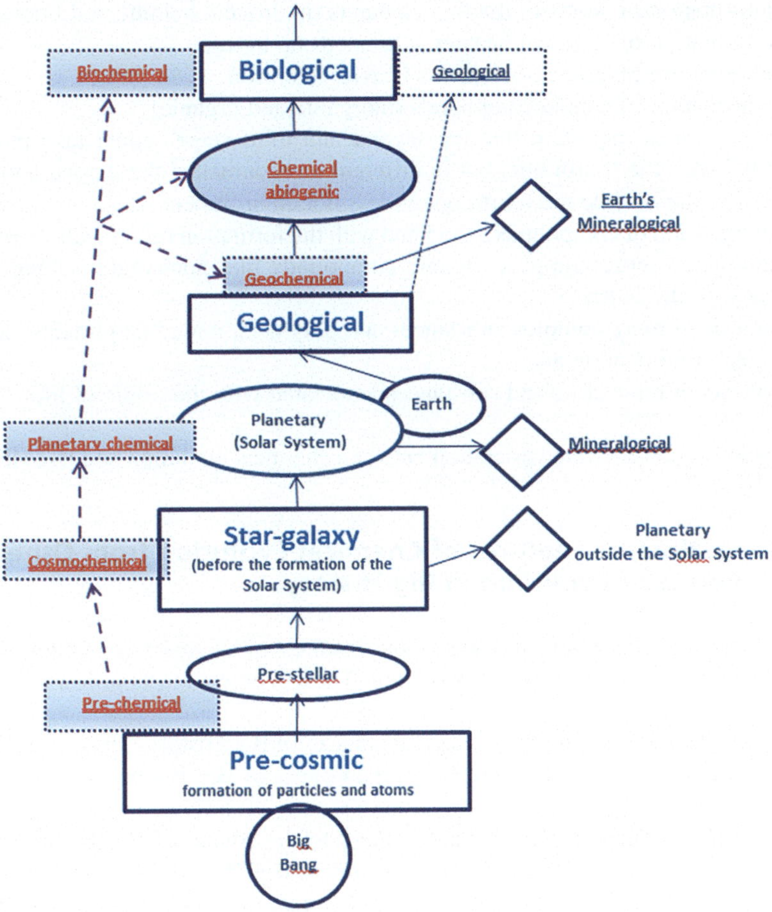

Fig. 7.2 Chemical evolution as a peripheral and parallel line of Big History

considerably due to the variety of chemical processes on the planets, including in different gaseous and liquid media.

The scheme (see Fig. 7.1) demonstrates the unfolding of Big History, which is structure into ten phases—five major phases and five transitional ones (for details about this scheme, see Grinin & Grinin, 2025). On the left one can see the line of chemical evolution.

7.3 Periodization of Chemical Evolution

We distinguish the following sequence of stages of chemical evolution before the origin of life (Grinin, 2020):

1. The formation of atoms of the first elements (hydrogen, helium, and lithium)
2. The formation of atoms of heavier substances up to iron
3. The formation of atoms of elements heavier than iron
4. The formation of simple compounds (inorganic and organic)
   However, it is important that the second and third stages could take place in parallel with the fourth one, but in different environments: the second and third stages in stars, while the fourth one in less hot environment
5. The formation of compounds associated with the formation of minerals on planets
6. Synthesis of more complex organic compounds like nucleotides taking place already on the Earth
7. Synthesis of more complex substances and polymers, including proteins, not yet capable of replication and
8. Synthesis of replicators and substances associated with the origin of life

Now let us consider the correlation between chemical evolution and Big History.

7.4 Distinctive Features of Chemical Evolution from Other Forms of Evolution in Big History

Figures 7.1 and 7.2 show the important features of chemical evolution which distinguish it from other forms of evolution. These features are as follows:

1. All other forms of evolution are separate phases of Big History. Thus, one form of evolution, having been realized at a certain phase of Big History, is replaced by another form.
2. However, chemical evolution goes parallel to the course of Big History. More precisely, it co-evolves with different phases of Big History as a constituent part of each of them. Thus, one can see that chemical evolution acted as a part of a co-evolutionary tandem at all phases of Big History (see Fig. 7.3).

Let us now briefly consider the development of chemical evolution in its relation to the phases of Big History.

Chemical evolution after the inflationary phase appears as a part of the pre-stellar phase. I have pointed out above that chemical evolution began in pre-stellar clouds. But it was still pre-chemical evolution.

The star-galaxy phase, which includes the formation of planets outside the Solar System, corresponds to Cosmochemical evolution. It is during this phase that the first chemicals are formed. Thus, a new qualitative stage in the development of chemical evolution (where the temperature was much more favorable for chemical reactions than in stars) began on ancient planets. However, we know practically nothing about this evolution.

With the formation of the Solar System and the beginning of the planetary phase, one can talk about the planetary chemical evolution since we know quite a lot about chemical processes and substances on the planets of the Solar System.

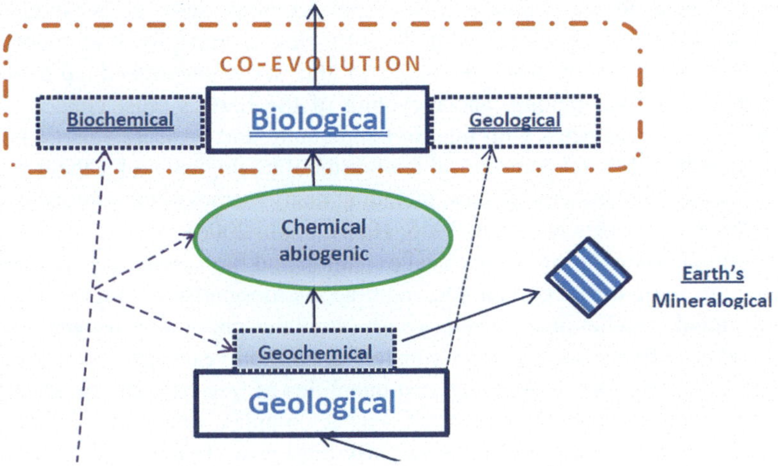

Fig. 7.3 Chemical evolution on the Earth and its increasing evolutionary role. Phenomenon of co-evolutionism

The formation of the Solar System means that the main line of Big History begins to focus on the Earth, where geological processes begin. Finally, on the Earth, chemical evolution developed first as a part of geological and then as a part of biological evolution. This development is still ongoing.

Thus, for the first time, chemical evolution moves to the center of evolutionary development at the level of chemical abiogenic phase (see Fig. 7.3). In this phase, the role of chemical evolution rapidly increases to the level of a transitional phase.

The period between the formation of the Earth and the emergence of life was pivotal for the whole Big History, and at the same time, the least known and the most obscure. During this period, chemical evolution was integral and interrelated with geological, mineralogical, and biological evolution. It was the co-evolutionary tandem mentioned above. Let me give some examples of such coevolution. Let's start with mineralogical evolution. The mineral kingdom is an example of a non-living system that has systematically increased in diversity and complexity through the Earth's 4.56-billion-year history (Hazen & Wong, 2024). Naturally, without the development of chemical evolution, this increase in diversity and complexity would be impossible. In the case of minerals, their combinatorial richness derives from possible combinations of chemical elements, coupled with permutations of their formula coefficients (Hazen & Wong, 2024). Due to systemic processes and the interaction between shells, the mineral composition on the Earth, ultimately became much richer than that on other planets (Hazen et al., 2008; Hazen & Wong, 2024; Grinin, 2020, 2026 in this volume).

With respect to geological evolution, we should make the following remarks. It is well known that geological processes, including those associated with the concentration of ores and minerals, and various chemical transformations resulting from magmatic and other processes, lead to complex migrations of chemical elements within and beneath the Earth's crust. Plate tectonics also leads to chemical transformations which, in turn, enable the formation of minerals. It is important to speak about the role of plate tectonics in the above-mentioned co-evolution. Repeated subduction, melting, and recycling of the Earth's crust created a wide range of new environments with intense temperature and pressure variations. As a result of geochemical and petrological reworking of geological rocks, there was rise to an estimated 1500 mineral species, including many metamorphic minerals that we know today (see Garrels & Christ, 1965; Hazen et al., 2008).

Concerning co-evolution of chemical evolution and biological one, we can point out numerous directions. For example, multiple microorganisms from the beginning affected global geochemistry. Then, a very dramatic phase was triggered by the Great Oxidation Event (GOE), approximately 2.4 billion years ago, when photosynthetic cyanobacteria began pumping vast quantities of free oxygen into the Earth's atmosphere and oceans. Of course, it was a complex process that had many peculiarities which are not understood clearly until now (Lyons et al., 2014). This newly available oxygen transformed existing minerals by oxidizing them and opened up thousands of new chemical pathways. Also, later mineral evolution is characterized by biomineralization, the direct creation of minerals by living organisms. Starting with the formation of simple phosphate and carbonate minerals in early life forms, this process began to explode during the Cambrian period with the evolution of complex skeletons, shells, and teeth. Such biogenic minerals as calcite, aragonite, and apatite are fundamental components of the Earth's crust cycle.

7.5 From the Abiogenic Phase to the Origin of Life

7.5.1 Strengths and Weaknesses of Evolutionary Hypotheses About the Origin of Life

Most researchers agree that without chemical evolution and its preparatory work, life could not have appeared.[3] According to the origin-of-life scenarios, a planet must go through at least some stages of chemical processing as a prerequisite for life (Hazen et al., 2008; Brack et al., 2010).

There are various hypotheses about abiogenic chemical evolution and the origin of life including the so-called RNA world.[4] Although some progress has been made

[3]Only the proponents of the panspermia theory consider abiogenic evolution to be less important (e.g., Anisimov, 2013).

[4]It emerged in the mid-1980s. The importance of this theory is that it presents RNA, not DNA, as the first molecule of life. Three types of molecules perform primary functions in the body. DNA is responsible for storing hereditary information, that is, it performs the information-replication function. RNA transfers information from DNA to proteins, that is, it performs the information-communication function. Proteins then perform various specialized functions, figuratively ensuring the functioning of the body (the working function). According to the RNA world theory, all functions in primordial abiotic organisms were performed by different RNA molecules. This

in many respects, especially in the last 20 years, none of them seems to be completely satisfactory yet. This is mainly due to the extreme complexity of the problem itself. But from the point of view of evolutionary theory, the weaknesses of these approaches are in the following points:

1. They deliberately or involuntarily reduce evolution to one of its lines
2. They take one evolutionary mechanism as the main one in all cases
3. The achievements of later periods, already related to the biological phase, are extrapolated to the abiogenic phase

In addition, models of the origin of pre-life were subject to simplification, as scientists attempted to show that life was possible using only four to five elements. Hazen and Sverjensky (2010) provide a list of such attempts. The simplest chemical models for life's origins use only four essential elements—C, H, O, and N (e.g., Oparin, 1938; Urey, 1951; Morowitz, 1992; Morowitz et al., 2000) with possible addition of S (Waächtershaäuser, 1990; de Duve, 1995) and/or P (Westheimer, 1987; Kornberg et al., 1999). Nevertheless, it can be assumed that the palette of chemical elements necessary for the emergence of life was greater.

We believe that the possibility of a major breakthrough exists only if there are a number of different development lines and paths. Moreover, each of these lines is limited and usually develops only one mechanism or innovation. But these lines compete and complement each other. As a result, there comes a time when the innovations of different lines are merged and formed into a fundamentally new system. This means the beginning of a powerful breakthrough to a new level of complexity. However, the beginning of such a breakthrough, after the formation and development of the new level, is difficult or even impossible to detect. This corresponds to the important idea of Pierre Teilhard de Chardin (1987) that transitional forms leave no visible material traces. We have also formulated the rule of archaic character of primary systems. Systems do not emerge in the mature form. They usually require several transformations to reach maturity and sustainability, including cycles of destruction and reforming. Primary systems as a rule look archaic and are unlikely to survive.

Therefore, the first pre-living systems (the so-called protobionts) should not be considered as direct ancestors of the first living organisms, but as their analogues.[5] These analogues were already comparable to the most primitive living systems in a

means that RNA may have appeared before proteins, and especially DNA. Biochemical discoveries in the late twentieth and early twenty-first centuries largely confirmed the validity of this hypothesis. [About possible role of RNA for emergence of life see Brack et al., 2010; Markov, 2014; Grinin, 2020; see also Chapter "Biological and Social Phases of Big History and Universal Evolution: Similarities, Differences, and the Growth of Complexity" (Grinin et al., 2026) in this volume].

[5] I also think that the same principle—at first simpler analogues—lies at the basis of its main chemical structure, that is, RNA. For example, Brack et al. (2010: 71) point out that "Because the structural units that make up RNA are quite complicated, it has been proposed that the first life contained simpler RNA analogues or surrogates that evolved to RNA."

number of functions. But in general they were organized differently (it is now extremely difficult to say how exactly). In addition, one should also take into account that the conditions on the young Earth were peculiar. Consequently, such structures could have formed, but modern scientists are unlikely to believe in their existence until concrete facts are available.

7.5.2 Evolutionary Directions of Abiogenic Organic Substances

We can assume that one of the most important directions of chemical evolution even before the emergence of life was the integration into geochemical processes and the development of the ability: (a) to accelerate these processes using catalysts; (b) to use these processes to one's advantage; and (c) to store energy and expend it purposefully. Here, we can assume a huge breakthrough in terms of the discovery of a developmental lineage involving cycles of energy storage and expenditure, consistent with the success and needs of proto-systems and proto-organisms. Of course, these were far less efficient energy mechanisms than those of archaea today, but the most important thing is that a fundamentally important step was taken. Further adaptation led to an increase in the efficiency of these energy mechanisms (through the development of catalysis, etc.), and through attempts to replicate to maintain the ability to expand. For billions of years, up until the present day, protozoa have been part of geochemical cycles of substances, transforming them into biogeochemical ones (see Zavarzin, 2007).

In any case, we should note that one of the fundamental differences of the abiogenic chemical evolution is the acquisition of the ability to store energy through chemical transformations during the life processes of a system (which did not previously exist), and then use this energy for its own needs. An internal source of energy for evolution has emerged.

In conditions requiring a sharp increase in the concentration of substances, efficient energy use, and so on, the main evolutionary advantages accrue to systems that develop on the basis of reactions with maximum useful work. Accordingly, exothermic reactions (i.e., reactions with an internal energy source), at sufficiently high temperatures capable of occurring at a noticeable rate in an isolated system, were of particular value (Rudenko, 1969: 120, 122).

Thus, one of the most important directions in chemical evolution was the creation of systems that best support the conditions for chemical reactions, utilizing their energy, and being capable of regularly reproducing them. It can be said, therefore, that the emergence of complex chemical systems with some form of separation from the external environment (and the development of this ability), is, to some extent, a rather logical result of previous evolution. We recall that the creation of shells to separate from the external environment and protect from uncontrolled environmental influences (beyond the optimal parameters required by the system) has been observed since the emergence of stars (see Baade, 2002; Surkova, 2005; Grinin, 2013, 2014; Grinin & Grinin, 2019). This has clearly advanced in the course of planetary evolution, and on deeper analysis, we see that this quality is also present at the level of basic elements (atoms and molecules) (it can even be discerned at the subatomic level and at levels of elementary particles capable of maintaining structure). *In other words, structural properties are always associated with some form of separation from the external environment.* Let us note that the ability to maintain conditions conducive to reactions (temperature and many others) has become not just a crucial, but even a decisive property of biological systems. This property developed until relatively recently (when warm-blooded animals emerged who raise their fetuses in placentas).

Mikhail Fedonkin (2003) believes, and I agree with him, that the most essential quality of life is the high speed of biochemical reactions, which are controlled and accelerated by enzymes. The active catalytic centers of most known enzymes are composed of metal atoms. It is possible that metal ions served as catalysts and sources of electrons during the initial, possibly pre-organic, phase of life's development.

Here, it is worth paying special attention to the very interesting formulation of the question: the pre-organic phase of life's development. In other words, a number of important innovations in chemical evolution could have occurred within the framework of inorganic reactions; more precisely, mixed inorganic and organic reactions. So the boundary between the two areas of chemical evolution was significantly narrower than is often imagined. Early biochemical reactions took place in a restorative environment saturated with free electrons and rich in ions of various metal catalysts, including heavy ones. The next stage was the emergence of life based on organic matter.

Thus, one can argue that the evolution of abiogenic organic substances occurred in the following different directions:

(a) Increasing complexity of chemical compounds and structures
(b) Increase in energy output and reaction rate
(c) Selection of elements and compounds according to certain parameters
(d) Concentration of substances
(e) The ability of complex compounds and proto-organisms to expand and grow fast
(f) The selectivity and recognition of some substances by others, according to the important evolutionary pattern for self and non-self-discrimination

7.6 The Most Important Pre-adaptations for the Beginning of Biological Evolution

The important pre-adaptations are worth special mentioning. The most important ones for the beginning of biological evolution are:

1. Creation of a system isolated from the external environment, in which constant cycles of chemical and biochemical reactions could take place

2. Constant maintaining of conditions, concentrations, energy balance, and the desired rate of reactions within this isolated environment, etc.
3. Effective responses to external conditions and stimuli
4. Replication (i.e., the ability to reproduce)
5. Preservation without major distortions of the initial code
6. Control of complex chemical processes through the use of increasingly advanced catalysts and substances
7. Autocatalysis and the ability to self-assemble[6]

These breakthroughs and pre-adaptations laid the foundation for biochemical synthesis and expansion. Especially important were the ability to store energy, and the ways to speed up reactions and to increase the concentration of a substance. Along with this, a new type of information (chemical and biochemical) emerged, which reached a very significant development later in biological evolution.

These and other achievements, of course, could not combine immediately and simultaneously. They combined much later when the basic mechanisms of life and the living cell were formed.

The primary conditions after the origin of the Earth were unique. Without them the transition to the emergence of pre-life and then life was impossible. Will these unique conditions ever be precisely known? Probably, they will not. But in any case, there must have been an abundance of available energy. Consequently, the fundamental difference between abiogenic chemical evolution and the previous stages of evolution was *the acquisition of the ability to store energy through chemical transformation during a system's lifespan and to use it for its own benefit.*

7.7 Protoviruses

There may have been one more intermediate phase between the abiogenic chemical and biological phases—the phase of protoviruses [see Fig. 7.4; about viruses, see Chapter "Viruses in Big History: Evolutionary Roles and Complex Systems" (Grinin, 2023) in this volume].

Below we will show the possible place of this phase in the megaevolutionary process. One should take into account that chemical reactions played a great role in the origin and development of protoviruses.

[6] About the role of autocatalysis as well as role of catalysts in the origin of life, see Rudenko (1969), Eigen (1971), Eigen and Schuster (1979), Kauffman (1993), Williams and Frausto da Silva (2002), Markov (2014). Lima de Faria (1991) also connected autocatalysis with autoevolution.

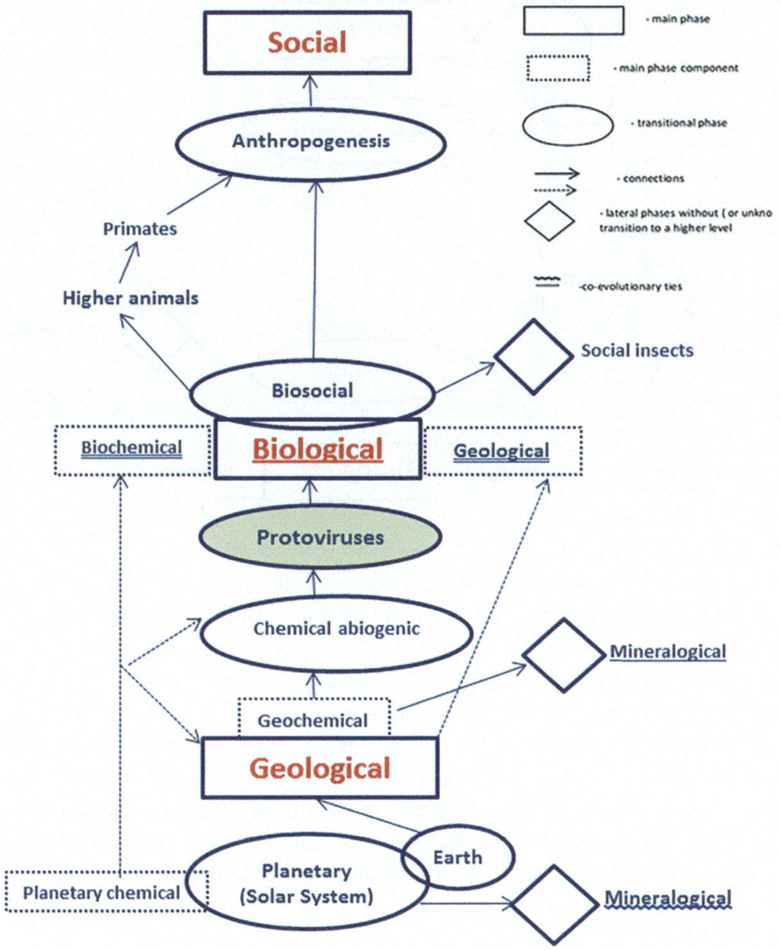

Fig. 7.4 Evolutionary phases of Big History including the phase of protoviruses

7.8 Concluding Remarks

For a long time, the abiogenic organic chemical evolution was only lateral and marginal in the general flow of inorganic chemical evolution. Then it was able to advance to a new level of evolution, that is, to life, taking place in a complex co-evolutionary movement. Abiogenic chemical evolution was involved in a whole bundle of evolutionary developments: geological, mineralogical, and geochemical. Thus, one can assume that initially one of the most important directions of chemical evolution was the integration of protobionts into geochemical processes,

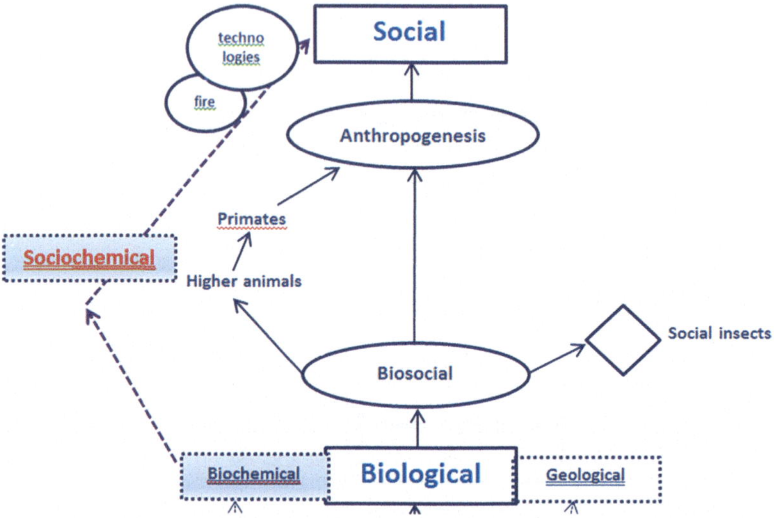

Fig. 7.5 Sociochemical evolution

such as sulfur springs, and the development of the ability to use these processes for one's own benefit.

Thus, gradually abiogenic chemical evolution gained momentum.

However, the role of chemical evolution remained very important. It again becomes a part of a larger—the biological—phase. In the scheme of the phases of Big History, we do not trace a further development of chemical evolution, but one should remember that it has also become an important component of social evolution (see about Grinin, Korotayev 2023), which can be called sociochemical. At the same time, its importance begins to appear already in the phase of anthropogenesis, from the moment when humans learned how to control fire. It is widely known that there is no point in talking about the further role of chemical evolution in the social phase of Big History; it is widely known. Nevertheless, one can argue that neither technology nor ordinary life would be possible without continuous efforts to master new chemical substances and reactions (Fig. 7.5). The complete line of chemical evolution from the Big Bang to social evolution you can see in Fig. 7.6.

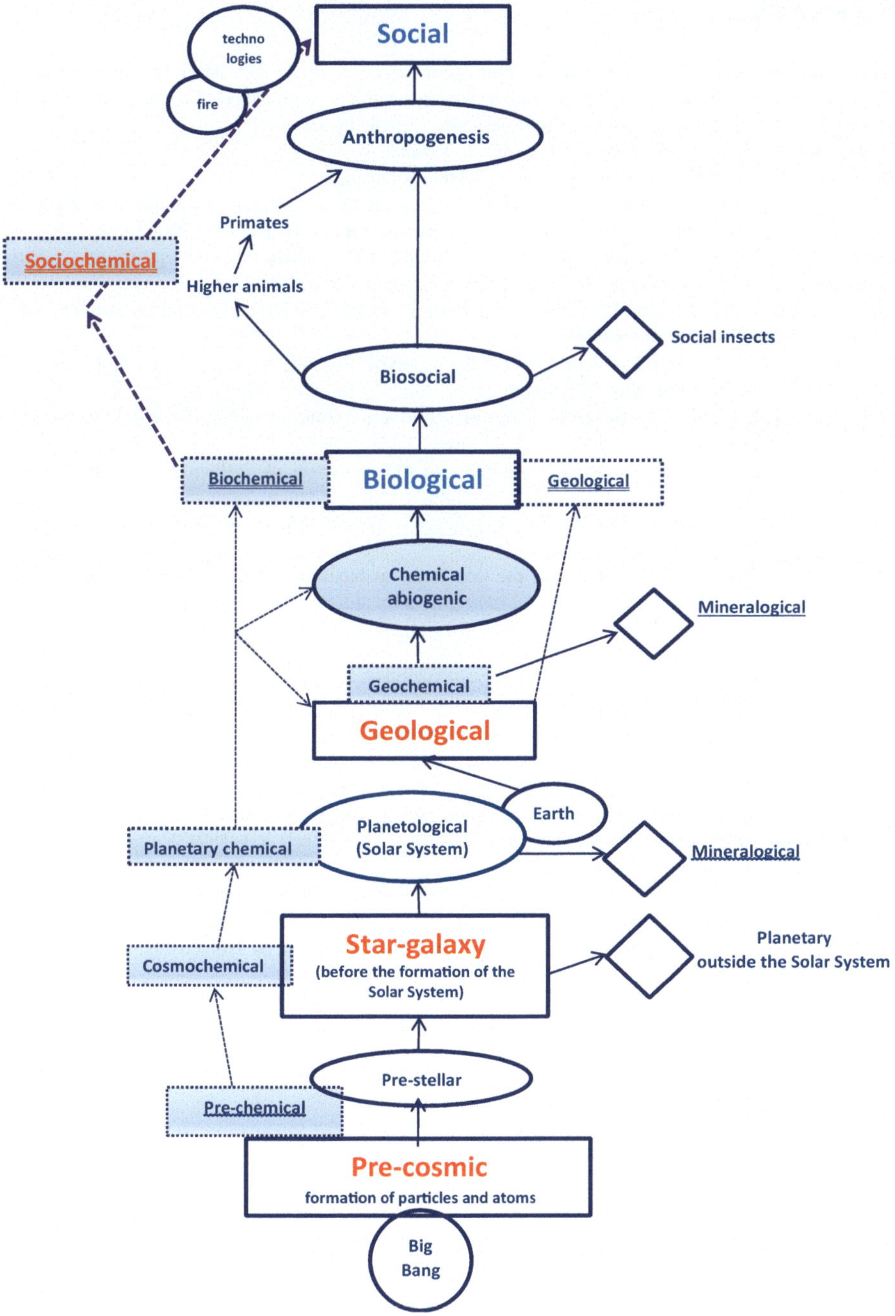

Fig. 7.6 The complete line of chemical evolution from the Big Bang to social evolution

References

Anisimov, V. A. (2013). Gipoteza emnogoabiogeneza v svete dannykh paleontologii, molekulyarnoy biologii i analiza khimicheskogo sostava molekul RNK. *Evolyutsiya, 5*, 14–31.
Baade, W. (2002). *Evolution of stars and galaxies*. URSS. In Russian.
Bernal, J. (1969). *The emergence of life*. Mir. In Russian.
Betekhtin, A. G. (2007). *Mineralogy course*. KDU. In Russian.
Brack, A., Horneck, G., Cockell, C. S., White, G. J., et al. (2010). Origin and evolution of life on terrestrial planets. *Astrobiology, 10*(1), 69–76. https://doi.org/10.1089/ast.2009.0374
Chaisson, E. (2005). *Seven ages of the cosmos*. Columbia University Press.
de Duve, C. (1995). *Vital dust: Life as a cosmic imperative*. Basic Books.
Degens, E., & Roiter, J. (1967). *Analytical methods of research in organic geochemistry. Organic geochemistry*. Nedra. In Russian.
Dickerson, R. (1981). Chemical evolution and the origin of life. In E. Mayr, F. Ayala, R. Dickerson, et al. (Eds.), *Evolution*. Mir. In Russian.
Dobretsov, N. L. (2005). On the early stages of origin and evolution of life. *Vestnik VOGiS, 9*(1), 43–54. In Russian.
Eigen, M. (1971). *Self-organization of matter and the evolution of biological macromolecules*. Springer.
Eigen, M., & Schuster, P. (1979). *The hypercycle. A principle of natural self-organization*. Springer.
Fedonkin, M. A. (2003). Narrowing of the geochemical basis of life and eukaryotization of the biosphere: A causal relationship. *Paleontologicheskiy zhurnal, 6*, 33–40. In Russian.
Galimov, E. M. (Ed.). (2008). *Problems of biosphere origin and evolution*. LIBROCOM. In Russian.
Garrels, R. M., & Christ, C. L. (1965). *Solutions, minerals, and equilibria*. Harper & Row.
Glyantsev, A. (2019). *A crucial 'molecule of life' has been discovered in the interstellar medium for the first time*. https://nauka.vesti.ru/article/1152892 (In Russian).
Grinin, L. E. (2013). *Big history of the world development: Cosmic evolution*. Uchitel. In Russian.
Grinin L. E. (2014). The star-galaxy era of big history in the light of universal evolutionary principles. In L. E. Grinin, D. Baker, E. Quaedackers, A. V. Korotayev (Eds.), *Teaching and Researching Big History: Exploring a New Scholarly Field* (pp. 263–187). Volgograd: Uchitel Publishing House.
Grinin, L. E. (2017). *Big History of the world development: History and evolution of the solar system*. Uchitel. In Russian.
Grinin, L. E. (2018). Evolution of the early solar system in terms of Big History and universal evolution. *Journal of Big History, 2*(1), 15–26.
Grinin, L. E. (2020). *Big History of the world development: Planets of the solar system. Their History and Evolution. Chemical Evolution in Space and on Earth*. Uchitel. In Russian.
Grinin, L. E. (2023). Chemical evolution in big history. In L. E. Grinin, A. V. Korotayev (Eds.), *Evolution: Complexity in Nature, Society, and Cognition* (p. 44–56). Volgograd: Uchitel Publishing House.
Grinin, L. (2024). Chemical evolution in Big History. *Journal of Big History, 7*(2), 97–105. https://doi.org/10.22339/jbh.v7i2.7208
Grinin, A. (2026). Viruses in Big History: Evolutionary roles and complex systems. In D. J. LePoire, L. Grinin, & A. Korotayev (Eds.), *Complexity in universal evolution – A big history perspective* (pp. 247–256). Springer Nature (in this volume). https://doi.org/10.1007/978-3-032-20516-2_10
Grinin, L., & Grinin, A. (2019). The star-galaxy era in terms of Big History and universal evolution. *Journal of Big History, 3*(4), 69–92.
Grinin, L., & Korotayev, A. (2023). Social evolution: Theoretical aspects and general outlines. *Social Evolution & History, 22*(2), 111–158. https://doi.org/10.30884/seh/2023.02.08

Grinin, L., & Grinin, A. (2025). Big History evolutionary phases. In D. J. LePoire, L. Grinin, & A. Korotayev (Eds.), *Navigating complexity in Big History – Exploring periodization across cosmic and biosocial dimensions* (pp. 319–342). Springer Nature. https://doi.org/10.1007/978-3-031-85410-1_10

Grinin, L., Markov, A., & Korotayev, A. (2026). Biological and social phases of Big History and complexity growth. In D. J. LePoire, L. Grinin, & A. Korotayev (Eds.), *Complexity in universal evolution – A Big History perspective* (pp. 283–356). Springer Nature (this volume). https://doi.org/10.1007/978-3-032-20516-2_12

Guotmi, A., & Cunningham, R. (1960). The action of crystal facet in catalysis. *In Catalysis. A study of the surface of catalysts* (pp. 74–117). IL. In Russian.

Haldane, J. B. S. (1949). *What is life?* Lindsay Drummond.

Hazen, R. M., & Sverjensky, D. A. (2010). Mineral surfaces, geochemical complexities, and the origins of life. *Cold Spring Harbor Perspectives in Biology, 2*, a002162.

Hazen, R. M., & Wong, M. L. (2024). Open-ended versus bounded evolution: Mineral evolution as a case study. *PNAS Nexus, 3*(7), 248. https://doi.org/10.1093/pnasnexus/pgae248

Hazen, R. M., Papineau, D., Bleeker, W., Downs, R. T., Ferry, F., VcCoy, T., Sverjensky, D., & Yang, H. (2008). Mineral evolution. *American Mineralogist, 93*, 1693–1720.

Kalvin, M. (1971). *Chemical evolution. Molecular evolution leading to the emergence of living systems on earth and other planets*. Mir. In Russian.

Kamshilov, M. M. (1970). *Biotic cycle*. Nauka. In Russian.

Kamshilov, M. M. (1979). *Evolution of the biosphere*. Nauka. In Russian.

Kauffman, S. A. (1993). *The origins of order: Self-organization and selection in evolution*. Oxford University Press.

Kornberg, A., Rao, N. N., & Ault-Riche, D. (1999). Inorganic polyphosphate: A molecule of many functions. *Annual Review of Biochemistry, 68*, 89–125.

Lima de Faria, A. (1991). *Evolution without selection: Autoevolution of form and function*. Mir. In Russian.

Lyons, T. W., Reinhard, C. T., & Planavsky, N. J. (2014). The rise of oxygen in earth's early ocean and atmosphere. *Nature, 506*, 307–315.

Markov, A. (2014). *Rozhdeniye slozhnosti*. Corpus (AST).

Morowitz, H. J. (1992). *The beginnings of cellular life: Metabolism recapitulates biogenesis*. Yale University Press.

Morowitz, H. J., Kostelnik, J. D., Yang, J., & Cody, G. D. (2000). The origins of intermediary metabolism. *Proceedings of the National Academy of Sciences, 97*, 7704–7708.

Oparin, A. I. (1938). *The origin of life*. Macmillan.

Rudenko, A. P. (1969). *Theory of self-development of open catalytic systems*. MSU. In Russian.

Spiridonov, V. (2019). *The origin of life on earth: A proven theory or an unsolved mystery*. https://ria.ru/20180518/1520873401.html

Surkova, L. P. (2005). *Stars and stellar clusters in our galaxy*. Zabaikalsky State University. In Russian.

Teilhard de Chardin, P. (1987). *The phenomenon of man*. Nauka. In Russian.

Urey, H. C. (1951). The origin and development of the earth and other terrestrial planets. *Geochimica et Cosmochimica Acta, 1*, 209–277.

Waächtershaäuser, G. (1990). Evolution of the first metabolic cycles. *Proceedings of the National Academy of Sciences, 87*, 200–204.

Westheimer, F. H. (1987). Why nature chose phosphates. *Science, 235*, 1173–1178.

Williams, R. J. P., & Frausto da Silva, J. J. R. (2002). The involvement of molybdenum in life. *Biochemical and Biophysical Research Communications, 292*(2), 293–299.

Zaguskin, S. L. (2014). Emergence and evolution of life from the position of chronobiology. *Prostranstvo i vremya, 3*(17), 275–282. In Russian.
Zavarzin, G. A. (2003). The formation of the system of biogeochemical cycles. *Paleontologicheskii Zhurnal, 6*, 16–24. In Russian.
Zavarzin, G. A. (2007). Anti-market in nature. *Vysshee obrazovanie v Rossii, 4*, 123–133. In Russian.

Leonid Grinin is Senior Research Professor at the Center for Stability and Risk Analysis at the HSE University in Moscow, as well as Director of Eurasian Center for Big History & System Forecasting at the Institute of Oriental Studies, Russian Academy of Sciences. He is co-editor of the international journals *Social Evolution & History* and the *Journal of Globalization Studies*, as well as co-editor of the international almanacs *Evolution, History & Mathematics*, and *Kondratieff Waves*. He is author of more than 700 publications, including 35 monographs in Russian, English, and Chinese. His current research focuses on comparative political studies, theory of revolution, political anthropology, global economy, global history, historical sociology, and futurology. Among his monographs are *The Cybernetic Revolution and the Forthcoming Epoch of Self-Regulating Systems* (Uchitel, 2016, with Anton Grinin), *Big History of Globalization: The Emergence of a Global World System* (Springer 2019, in co-authorship with Julia Zinkina, David Christian, and others), *Islamism, Arab Spring, and the Future of Democracy: World System and World Values Perspectives* (Springer, 2019; with Andrey Korotayev and Arno Tausch); and Cybernetic Revolution and Global Aging. Humankind on the Way to Cybernetic Society, or the Next Hundred Years (Springer 2024 with Anton Grinin and Andrey Korotayev). In 2004, he was awarded as a winner of the best publication of *Socis* journal during the year for the article "People of Publicity—a New Social Layer?". In 2012, he was awarded the Gold Kondratieff Medal by the International N. D. Kondratieff Foundation. In 2016, he received the Belyaev award for the book *From Bifaces to Nanorobots* (the book published in Russian in co-authorship with Anton Grinin).

The Earth's Pivotal Role in Big History: From Cosmic Evolution to Life's Emergence

8

David J. LePoire

Abstract

The Earth's evolution occupies a unique position in the grand narrative of Big History, bridging the gap between cosmic evolution and the rise of complex life forms. This chapter explores the Earth's formation and the dynamic interplay between geological and biological processes that enabled life to emerge and evolve. The process of planet formation is examined in terms of gravitational collapse, thermal gradients, and the impact of celestial events such as the Moon-forming collision. The co-evolution of Earth's geological systems and life is discussed, with emphasis on feedback mechanisms like carbon sequestration and the oxygenation of the atmosphere. Challenges such as the Snowball Earth event, mass extinctions, and the evolution of complexity are considered in light of the Gaia hypothesis. Ultimately, this work underscores the rare and fragile set of conditions that have allowed the Earth to sustain life, offering crucial insights for the search for life-bearing planets beyond our solar system.

Keywords

Complexity · Co-evolution · Gradients · Feedback · Geological evolution · The Earth · Biological evolution · The Gaia hypothesis

This chapter is an extensively re-worked version of the text originally published in *Journal of Big History* (LePoire, 2025).

D. J. LePoire (✉)
Argonne National Laboratory, Argonne, IL, USA
e-mail: dlepoire@anl.gov

D. J. LePoire et al. (eds.), *Complexity in Universal Evolution*, World-Systems Evolution and Global Futures, https://doi.org/10.1007/978-3-032-20517-9_8

8.1 Introduction

The Earth lies at a unique place in the context of Big History (Fig. 8.1). Earth-like planets are both one of the smallest objects ("eddies") in the cosmic cascade and also the platform from which complex adaptive systems (life) arise. However, Big History discussions often focus on the more dramatic Big Bang and stellar evolution, or the rise of life to civilizations, instead of this hinge at the cosmic and life developments (Curtis, 2017; Hazen, 2021). Figure 8.1 emphasizes the unique position by placing the Earth in the center with the cosmic developments starting with the Big Bang on the far left, leading to the expansion and development of structures, until some of the smallest objects in this sequence, planets, are formed (Lineweaver, 2001). The Earth then is the stage (or platform) where life evolves, but it is not passive, as life and the Earth co-evolve to overcome challenges over the billions of years with changes in the local solar system environment as the energy output from the Sun increases and the planets and asteroids shift (Lovelock, 2006).

Robert M. Hazen, a mineralogist and geophysicist at the Carnegie Institution for Science, has significantly reshaped modern understandings of mineral evolution, planetary complexity, and the co-evolution of geosphere and biosphere (Hazen, 2021). His work, rooted in the Earth and materials science but radiating outward to astrobiology and Big History, integrates mineralogy with systems thinking to describe the deep-time transformation of the Earth's surface and interior. Hazen introduced the concept of mineral evolution, a framework for understanding how the diversity and distribution of minerals on the Earth have changed over geological time (Hazen et al., 2008). Unlike the classical view of minerals as static substances, this approach posits that minerals are dynamic participants in the Earth's history, with their diversity increasing in discrete phases linked to planetary differentiation, plate tectonics, and the emergence of life. In this view, the Earth began with perhaps no more than a dozen types of minerals in the early solar system. Through processes such as planetary accretion, crustal formation, and biological activity (especially the Great Oxygenation Event), the number has since grown to over 5000 known mineral species today (Hazen, 2013). Notably, life itself is a mineralizing agent, responsible for at least two-thirds of the Earth's mineral diversity, primarily through biological

Fig. 8.1 A schematic image showing the Earth's position in Big History as the transition from the previous cosmic evolution (or development, on the left) to the terrestrial evolution of life (including humans and civilization, on the right)

oxygenation and biomineralization. Hazen's team has also pioneered the emerging field of mineral ecology, which applies statistical, network, and machine learning methods to predict undiscovered mineral species and patterns of occurrence. Just as ecologists use models to estimate undiscovered species in a forest, Hazen and collaborators use frequency-distribution and probabilistic methods to estimate how many mineral species remain undiscovered on the Earth—and where they might be found (Hazen et al., 2015).

To continue this perspective, the complexity of geology (Press & Siever, 1978; Westbroek & Westbroek, 1991) will be split into these two phases: the formation of planets and the possible dynamics leading to and incorporating life. The first phase reviews the various phenomena leading to planet formation including their distance from the star, mass, spin, composition, and moons (Ward & Brownlee, 2004). One question is what it takes and how probable are the conditions that lead to a long-term stable planet that might support appropriate conditions for life over billions of years (Ward, 2009; Mieli et al., 2023). The second phase focuses on various energy gradients leading to self-organizing convection cells (Lane, 2015). Besides this aspect of complexity, the diversity of environments it might support gradually changes over time, causing dynamic evolution with new challenges and opportunities for life to expand and explore.

8.2 Earth Formation

The formation of planets might seem to be straightforward. The stellar system formation starts with a cloud of gas and dust which cools sufficiently or is condensed at first by a nearby stellar explosion by pressure waves. Most of the energy in the system is gravitational potential, which was the starting point for larger level structure formation of the cosmic web and galaxies. As the cloud collapses under gravity into a star, part of the material is left in a rotating disk around the star. The rotation prevents the material from falling into the star. Over time, the light gases are pushed outward, leaving rocky material closer to the star. These then further undergo gravitational collapse to form planets with some remaining disk again constrained by rotation to stay in orbit as moons (Ward & Brownlee, 2004).

However, the process is much more complicated when considering heat, magnetic fields, phase changes of materials, and orbital mechanics of many interacting objects (Goldreich et al., 2004; Goldreich & Ward, 1973; Sari & Goldreich, 2004). This last factor of other moving objects seems to be a key ingredient in the formation of the Earth. The planets might not be fixed at a given distance from the Sun during early solar system, where the dust and gas might lead to drag forces that could lead Jupiter and Saturn to come nearer to the Sun before moving out again. This could clear some of the debris for planet formation, for example, Mars seems to be quite small compared to the Earth and Venus. Mars' smaller size led to it losing part of its atmosphere and cooling down faster. The weaker magnetic field led to cosmic rays stripping away another part of the atmospheric gases.

However, a different gravitational interaction seemed to help the Earth form a large moon to give the rotation its stability (Comins, 1993). The dominant hypothesis is that two planet size objects formed in the Earth orbit, a proto-Earth and Theia. Eventually, their paths crossed in a large planetary swiping (not head on) collision. The resulting heat and ejected material later reformed with most of the heavy metals settling on the Earth, while the lighter materials formed the moon. This is quite different than the process by which other planets' moons are captured instead of by collision. There is one planet, Uranus, which also seems to have a major collision, since its spin is very misaligned with its orbital motion.

The collision that formed the moon helped heat up the planets causing the heavier elements to move to the center and differentiate the materials on the Earth. The heat of the Earth's formation along with the heat of the radioactive decay of the material kept the center hot and active, forming a protective magnetic field, along with providing convective mantle cells that would sustain plate tectonics and heat spots. The source of water on the Earth is still being investigated but it seems to come from two sources—asteroids that hit the Earth later and release of bound water in the rocks.

Therefore, this first phase saw the gravitational potential energy of the gas and dust collapse into many interacting gravitational objects. The heat from the collapse caused chemical and magnetic effects. The Moon's formation helped stabilize these for the long-term.

Recently there has been great progress in reducing the uncertainty concerning planet formation with new observations and models, although no observations of life on other planets or solar systems have been discovered (Kaltenegger, 2024). A recent estimate of the probability of each of the four characteristics [(1) in the solar habitable zone with a mass to capture atmosphere, (2) stable spin because of large moon, (3) active heated liquid outer core to sustain magnetic field and geothermal activity, and (4) avoiding catastrophic astronomical events] was roughly about the same at 20% (Mieli et al., 2023). This would lead to an overall probability of a long-term stable Earth-like planet around an appropriate Sun-like star to be about roughly 0.2%. About 10% of the 100 billion disk stars in the Milky Way are Sun-like with from 0.45 to 1.4 solar masses (K, G, and F stars). Most stars in the galactic disk are either too small with insufficient mass and energy output or too massive which have relatively shorter lives.

8.3 Rise of Life

From a complexity perspective (Fig. 8.2), the gravitation collapse led to the conditions for many self-organizing convective cells (Curtis, 2017) (similar to the Benard cells in a heated pan). These cells include the convective mantle, the Hadley cells of the atmosphere, and ocean currents. This led to a diversity of environments (biomes) which life could later explore. These environments also had both quick changes (such as weather) and longer-term variations in climate, along with the formation and evolution of mountains and river systems. Life seems to evolve to

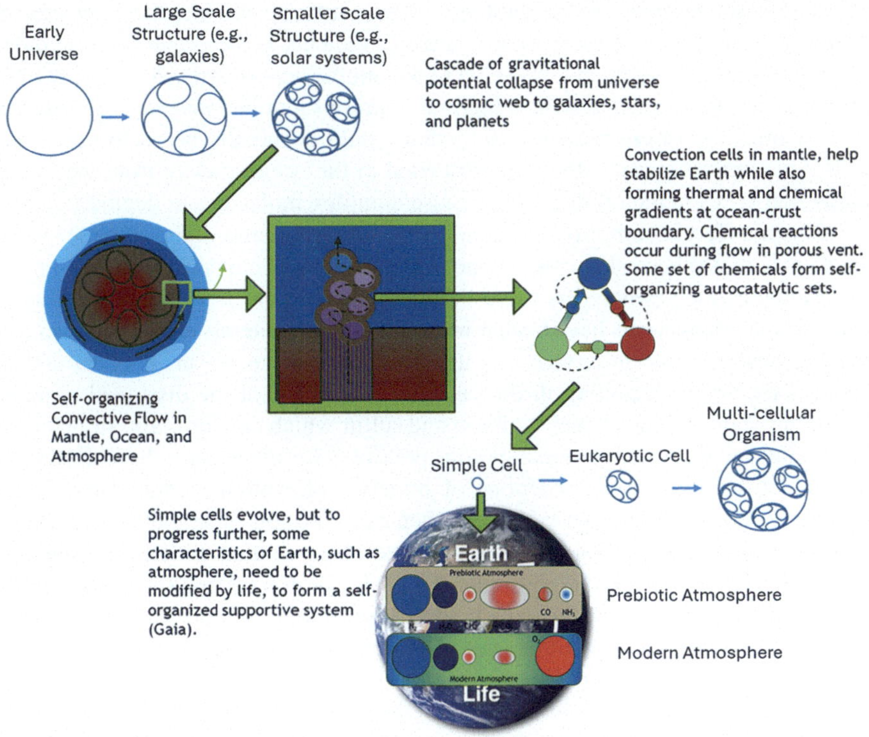

Fig. 8.2 A schematic diagram showing some of the complexity processes in the formation and development of life on the Earth

accommodate and use these gradients (Bejan & Lorente, 2011) and the dynamic landscapes they generate (Scheffer, 2009).

However, the stability of the weather system is not determined (Boschi et al., 2013; Lovelock, 2016). For the conditions on the Earth, it seems like for a given solar irradiation, there are two stable climates. One is warmer, with less ice and therefore more absorption of the solar energy. The other is cooler, with the ice reflecting the solar irradiation maintaining the cold. It has been hypothesized that the Earth did experience these two climates, with the cooler one leading to an Earth almost completely covered with ice, known as Snowball Earth. This is especially interesting since the Sun has been becoming more energetic as time progresses due to the production of Helium. There are significant questions about how the Earth managed to avoid the Snowball Earth soon after the magma cooled. This has led to the debated Gaia hypothesis that life self-organizes to maintain the chemical and temperature of the Earth within a manageable range. Another hypothesis is that the Earth trended towards cooler temperatures despite the Sun warming. Higher complexity might require cooler temperatures to dispense of the heat generated by the additional complexity. This might lead to an evolution on the Earth that may have similar patterns on other such planets (Schwartzman, 1999, 2020).

The interaction between the geology of the Earth and the first life indicates two processes that seem to be important: thermal gradients at the ocean floor boundary and introduction of sufficient oxygen into the atmosphere (Curtis, 2017). First in the origin of life, there needed to be a thermal (entropy) gradient necessary to drive the self-organization of chemical reactions that would eventually be enclosed in early cells. One major theory is that this happened in the oceans, away from the intense ultraviolet (UV) radiation that would cause complex molecules to degrade.

A major effort is currently investigating the characteristics of the alkaline "white-smokers" hydrothermal vents (Lane, 2015) which are not directly on the mid-Atlantic ridge (where the more thermally intense black smoker hydrothermal vents form). These vents have heated water filled with minerals enter the ocean from magma near the ocean surface. As the water enters the ocean, it cools and the chemistry changes, resulting in the deposition of some of the dissolved minerals. These minerals include elements of iron and sulfur which can form iron sulfide (FeS) in "chimneys" that maintain small porous pathways for the water to flow. Hydrogen, resulting from exothermic reactions of minerals like olivine with water, is also dissolved in the heated water which can then cause proton separation in the alkaline environment. An indication of the importance of this is that all life still maintains metabolism based on structures similar to FeS for proton separation. These thermal and chemical gradients might have held the conditions for early life formation of metabolism and cellular membranes as the way the information is passed through genetic means began to evolve.

Whatever path the origin of life took on the Earth, it was relatively soon after the late asteroid bombardment and the cooling of the surface to hospitable conditions. This rapid evolution has been investigated as a possible result of rapid fluctuations in exchanges of components in self-catalyzing networks. Growth with improvement is known to possibly lead to hyperbolic change as identified by (Eigen, 1971) who was puzzled by the rapid evolution.

However, life stayed at this level of simple prokaryote cell for billions of years. The mechanisms might have been refined but the cell did not develop any new organization. The new energy mechanism developed was the ability to extract energy from the solar light (photosynthesis) which allowed life to be independent from the hydrothermal vents (Lane, 2015).

8.4 Life's Challenges

What seemed to be happening is that there were no major challenges to life to cause it to explore new paths. What eventually led to the evolution of complex cells was the crisis caused by the generation of oxygen from the photosynthesis (Lane, 2016). This of course required the growth of the number of cells and the accumulation of oxygen, which first reacted with iron in the ocean to help form iron bands, and only after the ocean turned from green (with the iron) to blue, could the oxygen accumulate in the atmosphere. Since oxygen was a poison to early life, a challenge was how to handle it. Mechanisms evolved to form respiration for organisms that did not have photosynthesis. This solution led to a manageable level of oxygen in the atmosphere that then would lead to conditions where multicellular organism could form because enough oxygen could diffuse through the cells to supply each one.

At one point, as the Sun grew increasingly hotter and oxygen levels in the Earth's atmosphere rose, the Earth became too warm. To counteract this, carbon dioxide (CO_2) was sequestered as coal, long before organisms evolved the ability to digest plant material effectively. This process aligns with the Gaia hypothesis, which posits that the Earth's systems self-regulate to maintain conditions suitable for life. The formation of coal, and later oil, appeared to be a "gift" of energy resources for humanity during the Industrial Revolution. However, as James Lovelock argued, the storage of carbon in these forms was the Earth's response to an overly warm climate. He framed humanity's subsequent burning of fossil fuels—and the massive release of CO_2—as the "Revenge of Gaia," warning that by undoing this natural carbon storage, we are dismantling one of life's mechanisms for maintaining moderate global temperatures (Lovelock, 2006).

As life evolved in both the ocean and land, plate tectonics continued to reshuffle the light granite floats of land on the more dense basalt ocean beds (Press & Siever, 1978). This caused variations in the volcanic activity, mountain building, erosion, water distribution, and sedimentation. While natural erosion and weathering of rocks occur, the process was accelerated by life first with fungi and then by other plants (Smith, 2018). The feedback between life and plate tectonics itself is being pursued (Hall, 2017) with some hypothesizing that life needed plate tectonics and/or plate tectonics needed life. The disruptions caused by massive releases of carbon dioxide and ash into the atmosphere contributed to many of the great extinctions.

As mentioned before, the feedback mechanism of the Earth system may be diminishing in part due to the warming Sun and in part due to the evolution of complexity in life (e.g., humans), leading to greater use and extraction of its resources. Besides the greenhouse gases that might interfere with atmospheric processes, humans have taken control of much of the arable land for food and material production, stressing the habitats of other organisms. This might have been an early effect of humans as some point to the coincident demise of large mammals when early human populations arrived (Curtis, 2017).

8.5 Conclusion

In conclusion, the Earth evolution occupies a rare and pivotal position in the grand narrative of Big History, serving as a bridge between cosmic evolution and the rise of complex adaptive systems like life. While discussions in Big History often gravitate towards the monumental events of the Big Bang, stellar formation, and the evolution of civilizations, it is this intermediate phase—the formation of planets and the emergence of life—that underscores the intricate interplay of cosmic and geological processes.

The formation of the Earth was not a linear, predictable path but a dynamic and multifaceted sequence of gravitational, thermal, and chemical interactions, influenced by countless astronomical and planetary events. From the collapse of gas and dust into a rotating disk around the Sun, to the chaotic interactions that led to the Moon-forming collision, the early solar system's turbulence laid the foundation for a stable, life-sustaining planet. This phase highlights the significance of planetary features like the Earth's large moon, active core, and magnetic field—each of which contributes to the planet's long-term stability and habitability.

As the Earth transitioned from a barren rock to a living, breathing world, the interplay between geology and biology became increasingly pronounced. Convection cells, ocean currents, and atmospheric dynamics created diverse environments that life adapted to and transformed over time. The feedback loops between life and the Earth's systems—such as the sequestration of carbon dioxide, the oxygenation of the atmosphere, and the erosion driven by plant life—illustrate the co-evolution of the planet and its biosphere.

However, the Earth's journey has not been without its challenges. From Snowball Earth events to mass extinctions triggered by volcanic eruptions and asteroid impacts, the planet's history reflects the delicate balance required to sustain complex life. The Gaia hypothesis suggests that life itself may play a regulatory role, stabilizing the planet's conditions to remain within habitable bounds. Yet, as humanity exerts unprecedented influence on the Earth's systems, the stability of this feedback loop faces new uncertainties.

Understanding the Earth's unique role within Big History not only provides insights into our origins but also raises profound questions about the rarity and fragility of life-bearing planets elsewhere in the cosmos. As our exploration of exoplanets advances, the lessons learned from the Earth's formation and evolution may guide our search for other worlds capable of supporting life, reminding us of the extraordinary set of circumstances that made our existence possible.

References

Bejan, A., & Lorente, S. (2011). The constructal law and the evolution of design in nature. *Physics of Life Reviews, 8*(3), 209–240. https://doi.org/10.1016/j.plrev.2011.05.010

Boschi, R., Lucarini, V., & Pascale, S. (2013). Bistability of the climate around the habitable zone: A thermodynamic investigation. *Icarus, 226*(2), 1724–1742. https://doi.org/10.1016/j.icarus.2013.03.017

Comins, N. F. (1993). *What if the moon didn't exist? Voyages to Earths that might have been.* Harper Collins.

Curtis, W. (2017). *The biggest picture: From the big bang to the development of the big bang theory*. GeobookStudio.

Eigen, M. (1971). Self-organization of matter and the evolution of biological macromolecules. *Die Naturwissenschaften, 58, 465.*

Goldreich, P., & Ward, W. R. (1973). The formation of planetesimals. *Astrophysical Journal, 183*, 1051–1062. https://doi.org/10.1086/152291

Goldreich, P., Lithwick, Y., & Sari, R. (2004). Final stages of planet formation. *The Astrophysical Journal, 614*(1), 497–507. https://doi.org/10.1086/423612

Hall, S. (2017). Earth's tectonic activity may be crucial for life – And rare in our galaxy. *Scientific American*. https://www.scientificamerican.com/article/earths-tectonic-activity-may-be-crucial-for-life-and-rare-in-our-galaxy/

Hazen, R. M. (2013). *The story of earth: The first 4.5 billion years, from stardust to living planet*. Viking Press.
Hazen, R. M. (2021). Mineral evolution and the co-evolution of the geosphere and biosphere. In A. H. Knoll & A. J. Kaufman (Eds.), *Fundamentals of geobiology* (2nd ed., pp. 34–55). Wiley.
Hazen, R. M., Grew, E. S., et al. (2008). Mineral evolution. *American Mineralogist, 93*(11–12), 1693–1720. https://doi.org/10.2138/am.2008.2955
Hazen, R. M., Morrison, S. M., Downs, R. T., et al. (2015). Earth's "missing" minerals. *American mineralogist, 100*(11–12), 2344–2347. https://doi.org/10.2138/am-2015-5366
Kaltenegger, L. (2024). *Alien earths: The new science of planet hunting in the cosmos*. St. Martin's Press.
Lane, N. (2015). The vital question.
Lane, N. (2016). *Oxygen: The molecule that made the world* (Rev. impression). Oxford University Press.
LePoire, D. J. (2025). Geological complexity: Bridging cosmic evolution and the evolution of life. *Journal of Big History, 8*(1), 45–50.
Lineweaver, C. H. (2001). An estimate of the age distribution of terrestrial planets in the universe: Quantifying metallicity as a selection effect. *Icarus, 151*(2), 307–313. https://doi.org/10.1006/icar.2001.6607
Lovelock, J. (2006). *The revenge of Gaia: Earth's climate in crisis and the fate of humanity*. Basic Books.
Lovelock, J. (2016). *Gaia: A new look at life on Earth*. Oxford University Press.
Mieli, E., Valli, A. M. F., & Maccone, C. (2023). Astrobiology: Resolution of the statistical Drake equation by Maccone's lognormal method in 50 steps. *International Journal of Astrobiology, 22*(4), 428–537.
Press, F., & Siever, R. (1978). *Earth* (2nd ed.). W. H. Freeman.
Sari, R., & Goldreich, P. (2004). Planet-disk symbiosis. *The Astrophysical Journal, 606*(1), L77–L80. https://doi.org/10.1086/421080
Scheffer, M. (2009). *Critical transitions in nature and society*. Princeton University Press.
Schwartzman, D. W. (1999). *Life, temperature, and the earth: The self-organizing biosphere*. Columbia university press.
Schwartzman, D. (2020). Biospheric evolution is coarsely deterministic. *Journal of Big History, 4*(2), 60–66. https://doi.org/10.22339/jbh.v4i2.4230
Smith, N. (2018). The kingdom: How fungi made our world [Film. In *Smith & Nasht, real to reel productions, ZDF/Arte, CBC, and discovery science*].
Ward, P. D. (2009). *The Medea Hypothesis: Is life on earth ultimately self-destructive?* Princeton University Press.
Ward, P. D., & Brownlee, D. (2004). *Rare earth: Why complex life is uncommon in the universe*. Copernicus.
Westbroek, P., & Westbroek, P. (1991). *Life as a geological force: Dynamics of the earth*. Norton.

David LePoire researches, develops, and applies science principles in environmental issues, Big History evolutionary trends, and particle scattering. He has a BS in Physics from CalTech, a PhD in Computer Science from DePaul University, and over 30 years of experience at the Argonne National Laboratory in the development of scientific analyses, software, training, and modeling. His research includes Big History synergistic trends among energy, environment, organization, and information.

Geological Evolution: A Bridge Between Cosmic and Earthly Development in Big History

9

Leonid Grinin

Abstract

Geological evolution is a central but often overlooked aspect of Big History and megaevolution, serving as a crucial link between cosmic and terrestrial development. This study emphasizes the importance of geological evolution in understanding the history of Earth, particularly how it shares similarities with the evolution of other planets in the Solar System, especially during their early developmental stages. Yet, as Earth's geological evolution progressed, distinctive features such as plate tectonics, continental drift, and the formation of oceans have set it apart from other planetary bodies. While geological evolution is well documented, it remains insufficiently studied within Big History frameworks, often relegated to the background of biological evolution. This chapter explores key paradoxes, discrepancies between theory and empirical research, and important evolutionary patterns revealed by geological processes, such as competition for resources and the cyclical alternation between slow and abrupt changes. It also highlights significant gaps in our understanding, particularly in the deep Earth layers, and discusses how the geological features of Earth set it apart from other planets in the Solar System. Ultimately, this study calls for a more comprehensive integration of geological evolution into Big History, positing that a deeper exploration of Earth's geological past is essential for understanding the broader context of megaevolution.

This chapter is an extensively re-worked version of the text originally published in *Journal of Big History* (Grinin, 2025).

L. Grinin (✉)
HSE University, Moscow, Russia

Eurasian Center for Big History and System Forecasting, Institute of Oriental Studies of the Russian Academy of Sciences, Moscow, Russia

D. J. LePoire et al. (eds.), *Complexity in Universal Evolution*, World-Systems Evolution and Global Futures, https://doi.org/10.1007/978-3-032-20517-9_9

Keywords

Big History · Megaevolution · Universal evolution · Evolution · Geological evolution · The Kola super-deep borehole · The evolutionary pace · The Hadean Eon · The Archean Eon · The Proterozoic Eon · The Paleozoic Era · The Mesozoic Era · The Cenozoic Era

9.1 Introduction

The history of Earth is an extremely important part of Big History and megaevolution. The geological evolution is the central element of Earth's history. Geological evolution is to a certain extent a bridge between cosmic evolution and evolution of Earth. I mean that in geological terms there are quite a lot of similarities between the evolution of Earth and that of other planets in the Solar System. This is particularly evident with respect to the early phases of the development of Earth and of other planets (see Appendix). Yet, we can also note differences in the evolution of Earth and other celestial bodies. At first, these differences seem superficial (we are not considering chemical and biological evolution here), but then comes a division into continents and oceans, which only occurred on one planet (Mars), and a little later—the tectonics of the lithospheric plates, which radically changes the situation. Understanding of how a hot and dead planet without any crust became one that constantly renews itself and is alive is extremely important. This still remains an intriguing mystery, but it is impossible to solve it without delving into the geological evolution.

9.2 The Need for Greater Attention to Geological Evolution Within the Framework of Big History

For some time now, more than 3 years, I have been studying geological evolution as a part of megaevolution and Big History. I would like to share my thoughts on some important characteristics and features of geological evolution. Of course, due to the scope of this chapter, I can present only a few ideas and facts in this chapter.

Geological evolution is incomparably better known than cosmic evolution. However, geological evolution is clearly insufficiently studied within the framework of Big History. At best, it serves as a background to biological evolution. So, a rather interesting paradox has arisen within the framework of Big History. Although there are tens of times more facts about geological evolution than about cosmic evolution, the latter has far more researchers within the framework of Big History, although the analysis of many problems is extremely difficult due to insufficient facts and knowledge. Meanwhile, geological evolution is obviously understudied within the framework of Big History. In my opinion, this greatly distorts the idea of the course of Big History and mega-evolution; at least, it narrows their horizons. This interesting contradiction calls for more attention to be paid to geological evolution, which is either absent from Big History courses or is treated in a very cursory and superficial way. But it is worth noting, and I say this from my own experience, that for an evolutionist and a big historian the main problem in the study of geological evolution is to select and systematize the processes and facts that are most suitable for evolutionary analysis.

Geological evolution has many similarities with other evolutionary phases. On the other hand, geological evolution naturally has its own major characteristics. Below I will talk about both.

9.3 The Discrepancies Between Theory and Empirical Research

Despite the huge number of geologists, including those who study historical geology, and a vast geological literature, they have really studied only the structure of the evolution of the Earth's outer shell (the Earth's crust, lithosphere and oceanic crust), and partially the magnetosphere and the atmosphere. For the asthenosphere, mantle, and especially the outer and inner cores, there are only a few facts (or, rather, indirect evidence) and hypotheses about their evolution. But even ideas about the upper part of the Earth's crust are very different from what the real facts show.

Let me give you a very illustrative example. Ultra-deep drilling is one of the most important geological methods for exploring the upper part of the Earth's crust. The Kola super-deep borehole is considered to be the deepest vertical borehole. It was drilled intermittently for more than 20 years between 1970 and 1994. It was drilled primarily for scientific purposes, to confirm theoretical positions. But it turned out that the reality greatly differs from theory. Due to the discrepancy between expectations and reality, there were repeated accidents, falls, broken drills, etc. Now the structure of the borehole resembles a tree top rather than a straight borehole. The depth of the borehole was 12,262 meters; diameter of the upper part—92 cm, diameter of the lower part—21.5 cm (about this borehole, see Alekseyev et al., 1972; Smythe et al., 1994; Andrei, 2014; Ault, 2015).

Most importantly, the actual data from the Kola superdeep borehole showed that many of the predictions about the structure of the Earth's crust in this area were not confirmed, namely:

1. The pattern that was expected in the first 5 km of the borehole extended to 7 km.
2. The basalts predicted at a depth of 7 km were not found, even at a depth of 12 km.
3. Instead of the boundary between the granite and basalt layers of the Earth's crust predicted by the geophysical data (Conrad's section), there was a boundary between granites and less durable and fractured rocks—gneisses.
4. In other words, the most important position, that the granite layer of the crust is followed by a basalt layer, has not been confirmed in this area. *This does not mean that there cannot be basalt under the granite layer elsewhere, but it does mean*

that the theory must be modified. This is one of the important theoretical and cognitive aspects of drilling the Kola superdeep borehole.

5. There are many discrepancies in other predicted parameters. In particular, a higher than expected geothermal gradient was observed. The ambient temperature exceeded 70 °C at a depth of 5 km, 120 °C at 7 km, and sensors recorded 212 °C at a depth of 12 km.
6. It is worth noting that the drilling of a borehole has provided a lot of new information for estimating the time of origin of life on Earth. For example, traces of ancient microorganisms are present at a much greater depths than previously thought (Boyko & Prokaten, 2014, p. 59).
   If there are such large discrepancies between theories and practice only in relation to the Earth's crust, *what discrepancies with theory can we expect to reveal with the emergence of new methods for exploring the great depths?*

9.4 Geological Evolution Demonstrates Patterns and Regularities That Are Common to All Levels and Stages of Evolution

Geological evolution shows that many regularities and patterns are characteristic of all levels and stages of evolution.

Let us take, for example, the law of the competition for resources and habitats that we have formulated. This law is usually attributed only to biological and social phases of Big History (about this law in social evolution see Grinin & Korotayev, 2023). *However, the struggle for resources is a general mechanism of selection at all levels and stages of evolution.*

For example, it is known that large galaxies can "swallow" smaller ones, and that one star can capture the matter from another (Gibson & Ibata, 2007; Grinin & Grinin, 2019; see also Zwart, 2016). Similarly, during the formation of a planetary system, the *struggle for resources and living space can be defined as a general evolutionary law*. However, the selection mechanisms themselves vary considerably. Accordingly, certain advantages, including random ones, play a huge role since they can significantly influence the selection process.

What does geological evolution show? There are many examples, in particular, one tectonic plate can intercept the resources of another. I will give in more detail just one example (Armand, 1975, pp. 43, ff. 41–44).

The manifestation of the competition between river systems thus takes the form of river interception. As is known, the possibility of interception arises when the adjacent headwaters of two rivers are at different elevations. In the general case, the lower river has the advantage. It has a large capacity for intercepting groundwater and for faster back-up cutting, which determines the outcome of the competition. The absolute heights of the headwaters and their slopes depend (at the same distance from the mouths) on the heights of the corresponding erosional bases and the degree of approximation of the longitudinal profile to the developed, equilibrium profile.

The result of competition is not the destruction of one of the systems, but only its partial suppression.

According to the rather interesting opinion of the author of the book, the decisive conditions that determine the outcome of the competition between river systems are the same energy potential and the degree of perfection of the organization, that is, the level of entropy. This reflects the internal unity of the mechanism that determines both the geomorphological phenomenon of interceptions and the biological phenomenon of the struggle for existence. Of course, there are profound differences in the manifestation of competition in the first and second cases.

Let us show how other general evolutionary rules manifest themselves in geology.

Let us take *the rule* we formulated *about the cyclical alternation of abrupt and slow changes*. Its essence is that evolution organically combines processes of slow, almost imperceptible growth with explosive growth, the periods of rapid transformation, often associated with destruction or even collapse, can be followed by periods of a slow accumulation of changes. The result may be the emergence of objects with qualitatively new properties. And order may be replaced by disorder or chaos. *This rule organically combines gradualism and catastrophism, which remain controversial in geology and biology.*

In cosmic evolution, the combination of the two principles can be observed in many cases. Here, as nowhere else at other levels of evolution, these methods are organically combined, for example, in the individual fates of stars. The main sequence of stars where there is a very long process of hydrogen burning—an obligatory phase of any star—demonstrates the gradualness and importance of slow and long processes. However, there are similar catastrophes in the life of every star.

Geological evolution also has many examples of such alternations between abrupt and slow changes. Here's an interesting one.

India broke away from Antarctica about 100 million years ago. These are drastic changes. Then the Indian plate drifted for 70 million years at a rate of more than 9 cm/year. These are slow, imperceptible changes. As a result, in 70 million years it traveled a distance of more than 6.5 thousand km. Finally, 30 million years ago (i.e., already in the Cenozoic era), it collided with the southern edge of the Asian continent, causing dramatic changes, namely the formation of the Himalayan mountain system. These are again catastrophic changes (Sorokhtin & Ushakov, 2002: Ch. 8).

9.5 The Geological Evolution Has Many Peculiarities

On the other hand, the geological evolution shows many characteristics comparable to other phases of Big History. Let us point out the following (of many) features.

1. In geological evolution we deal with a single object, and not with a mass of objects, as in cosmic or planetary, biological, or social evolution. Of course, in

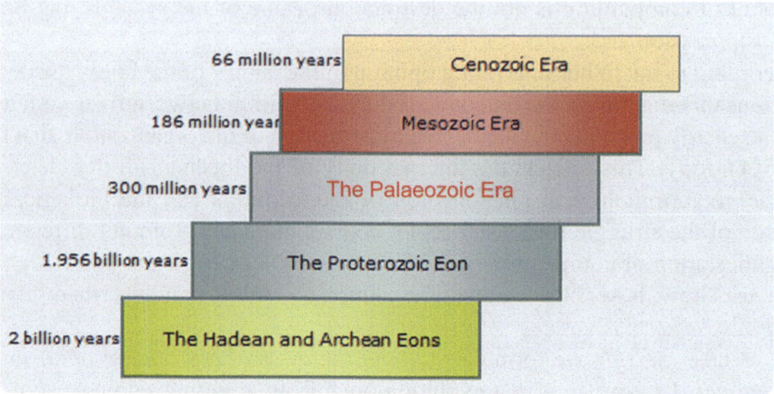

Fig. 9.1 The pace of geologic evolution

relation to the biosphere and humanity, we can in a sense speak about the evolution of a single object. But this unity is still rather (or at least largely) theoretical. And the unity of the Earth as a system is quite obvious (Khain, 2010; Khain & Koronovsky, 2007; Puchkov, 2016).

2. The pace of evolution does not have the same trends in all phases of Big History. It is known that cosmic evolution is initially rapid and then slows down. The planets of the Solar System also show fairly rapid rates of evolution in the early periods and an apparent slowing down or near-stalling in later periods. However, the evolutionary phases associated with the Earth show a clear tendency towards acceleration (Gumerova, 2010; Yanvarev, 1993). It should be taken into account that the pace and acceleration of biological evolution have been greatly influenced by the acceleration of geological evolution. Let's see how geological evolution accelerated (see Fig. 9.1).
   - The first 2 billion years: the Hadean and Archean eons (4.57–2.5 billion BP).
   - The Proterozoic Era lasted almost 2 billion years (1.956 billion years to be precise).
   - The Palaeozoic Era lasted slightly less than 300 million years, which is 8.5 times shorter than the preceding Proterozoic eon. Fantastic acceleration of speed!
   - The Mesozoic total duration, therefore, is 186 million years.
   - Cenozoic—66 million years.
3. *The uniqueness of geological evolution in terms of showing traces of past periods.* No other science, except geology, has materials similar to geological formations that represent the materialized time of the distant past itself. The Earth's crust can be considered as an almost unique system containing inexhaustible reserves of information about processes taken place over several billion years (Vinkovetsky, 1971, pp. 79, 91). Geology is time and processes frozen in time, which led to the result (Krut', 1978, p. 97).

9.6 Evolutionary and Other Features That Are Worth Paying Attention to

Finally, I will simply list the important evolutionary features in geological processes that I do not have opportunity to discuss.

1. Consistency is increasing. Perhaps, in no other area is it so clear how changes in some parameters (e.g., sea level) lead to changes in many others (e.g., climate), how positive and negative feedbacks are formed.
2. The increasingly complicated consistency and interactions are particularly evident at the junction of the Earth's shells.
3. Metamorphosis, that is, the transformation of matter, often multiple transformations as a result of geological processes in the evolution of the Earth, is remarkably clear.
4. Changes in geological regimes and cyclicity (transgressions and regressions; the formation of supercontinents and their disintegration, etc.) are a very important feature of geological evolution (e.g., Bozhko, 2009).
5. In geological evolution, we see a special combination of additive and replacement evolution.
6. The directional evolution of ore deposits can be traced. Empirical evidence shows that deposit types evolve from simpler and less organized to more complex and organized. At the same time, younger deposits become more complex in structure, composition, and also more individual in their characteristics. In younger epochs, the role of deposits of rare and trace elements, requiring increasingly higher concentrations, etc., increases progressively.
7. The result is a combination of stable and unstable zones, stable and renewed (on a large scale, because, of course, the surface of the plates is renewed). The plates are stable, and the oceans and the seabed are unstable. As a result, the oceans turn out to be young, and their study is limited in time (although it is likely that after some time it will become clear that even in young zones there are traces of a rather ancient past).

9.7 Conclusion: Earth and Other Planets of the Solar System

Geological evolution is, in a sense, a bridge between cosmic evolution and evolution of the Earth. I mean that there are quite a lot of similarities between the evolution of the Earth and other planets in the Solar System in geological terms. This is particularly evident with respect to the early phases of the development of Earth and planets. As already mentioned, there are many similarities in the development of Earth and other planets. However, the closer the era is to us, the greater the differences become (for details see Grinin, 2017, 2020). It makes sense to say a few words about the differences between the evolution of Earth and that of other celestial bodies. But first, it is worth noting that from time to time it is discovered that what was previously thought to be unique to the Earth also occurs on other planets in the Solar System. First it was thought about the presence of water on the planet; but later water was discovered in various celestial bodies. For example, the availability of water, and in particular the hydrosphere, was once thought to be unique to the Earth. However, water has been found in many places, even on the Moon in the form of ice, and the existence of ancient oceans has been discovered on Mars (Ksanfomality, 2012c; Mitrofanov, 2015). Some time ago, they also believed that rifting was a specific terrestrial phenomenon. The search for analogues of modern rift zones of the Earth has led to the identification of a wide distribution of rift and rift-like zones in the geological past, starting with the Late Precambrian, and structures with somewhat similar features starting even from the Early Precambrian (Archaean) period; and thus there has been established a large role of the horizontal extension of the Earth's crust throughout almost all of geological history. Then rifting was once thought to be a specific terrestrial phenomenon. But then rifting structures were found on other planetary bodies, mainly on Mars, but also on Venus, some of Jupiter's moons, etc. It turned out that rifting is not a specifically terrestrial phenomenon, but a process that took place at different stages of the evolution of the planets of the Solar System.

If we talk about the original, or rather formed at a very early stage, features and advantages of the Earth over other planets, these are location, size and mass, and the presence of a lunar satellite. Other features appeared at later stages of Earth's evolution: the hydrosphere, plate tectonics, and changes in the atmosphere.

Similarities Between the Evolution of the Earth and That of Other Planets:

1. Cooling processes took place everywhere.
2. Like other planets, Earth experienced the consequences of bombardment by planetesimals and, in a later period, by meteorites (the so-called Late Heavy Bombardment [see about it in Workshop on Early Solar System... 2012; Gomes et al., 2005; Johnson & Melosh, 2012]).
3. Active volcanic processes took place on all the planets of the Earth group (Head & Solomon, 1981; Katz et al., 1984; Ksanfomality, 1997; Adushkin & Vityazev, 2007; Sirotin, 2009).
4. The differentiation of matter took place everywhere.

Differences:

1. Not all the planets and their satellites were able to differentiate to the same extent as the Earth. The Moon and Mars are much less differentiated (see Zharkov, 2013; about Jupiter's Satellites see Kuskov & Kronrod, 2001).
2. The planets (especially Mercury and the moons, but also to a large extent Mars [Ksanfomality, 1997, 2004, 2012b, 2012c; Sirotin, 2009]), which cooled relatively quickly, compared to the Earth, have retained very numerous traces of collisions with meteorites almost from the moment of their formation. On Earth, due to various processes, the earliest are 2 billion years old. The Earth and the Venus are similar in this respect: on Venus, there are almost no traces of collisions earlier than 500 million years ago, when almost the entire surface of the planet was covered by lava eruptions from Venusian volcanoes (Ksanfomality, 2012a).

3. The first billion years (this period includes the two major eons: the Hadean and part of the Archean) on Earth are the least known. Paradoxically, the opposite is true for other planets. Although we don't know much about them in general, we know more about the early periods of some planetary bodies than we do about the later ones. This is explained by the fact that where the planets cooled early (Mercury, the Moon, and, to a lesser extent, Mars), traces of ancient transformations have been preserved; in particular, rocks 4.4 billion years old have been found on the Moon. On Earth, geological evolution has been more or less reconstructed only for the last billion years; the deeper the period, the less we know about it (Hazen, 2012).
4. Plate tectonics is a unique event in the Solar System, at least for now it is considered unique. Its presence makes the geology of Earth very different from the geology of other planets. And it is precisely this that allows the face of Earth to diversify constantly. It accelerates the development of evolution. At the same time, plate tectonics (and even pre-tectonic evolution) is associated with the constant build-up of plates, which also accelerates evolution. Plate tectonics is known to be responsible not only for active volcanic activity but also for many changes in relief, as well as for ocean floor renewal and continental drift. It contributed to the formation and break-up of supercontinents. Plate tectonics provided a mechanism for crustal rejuvenation (through spreading and subduction), mainly oceanic, due to the circulation of substances. That's why there is no crust at the bottom of the oceans older than 180 million years.

In this chapter, I cannot go into sufficient detail about such similarities and differences, but much can be seen in the attached Table in the *Appendix*.

Appendix: History of the Solar System

Time from present, billion years (approx.)	Sun and Solar System in general	Mercury	Venus	Earth	Moon	Mars	Giant planets (Jupiter, Saturn, Uranus, and Neptune)
4.57	*Approximate time of the beginning of the formation of the solar system*						
4.56–4.45	*The stage of mass accumulation by all planets, the primary differentiation of the terrestrial planets and silicate satellites*						
4.56–4.0				Hadean eon			
4.52–4.47	Formation of the terrestrial planets				Formation of the moon as a result of the collision of the proto-earth with another body (Ryder et al., 2000; Canup & Asphaug, 2001)		
4.51–4.4					The first period of evolution: Formation of the macrostructure of the lunar interior		
About 4.5	The sun's exit to the Main sequence						
4.5–4.1/4.0	*The first phase of the evolution of the planets of the solar system*						

4.5–4.1		Primary processes of differentiation	Primary processes of differentiation	Primary processes of differentiation	Primary processes of differentiation	Primary processes of differentiation	Probably, processes of differentiation
4.5–4.3					Age of primordial lunar rocks		
4.5–3.8						Noachian period, when all the major impact basins emerged; formation of the primary crust	Powerful release of energy
4.5–3.0		Formation of a vast iron core	Heavy meteorite bombardment		The most active geological period		
4.45–4.4		Loss of atmosphere			Loss of atmosphere		
4.45–4.35				Formation of the primary crust			
4.45–4.25				The emergence of the hydrosphere			
No later than 4.4					Crystallization processes in the upper mantle—Source zones of marine basalts		
4.4–4.0					The second period of the development of the moon		
4.35					Completion of the formation of the lunar primary crust		

(continued)

Time from present, billion years (approx.)	Sun and Solar System in general	Mercury	Venus	Earth	Moon	Mars	Giant planets (Jupiter, Saturn, Uranus, and Neptune)
More than 4.3					Primary rock preservation; crustal solidification		
No later than 4.25				Formation of secondary crust, sedimentary cover, hypergene minerals, including clay minerals, necessary for the origin of life			
4.2–4.0		Heavy meteorite bombardment, resulting in the formation of smooth plains of buried craters					
4.1/4.0–3.8	*The second phase of the evolution of the planets of the solar system*						
4.1	The beginning of the “late bombardment” epoch					A planetary structural reorganization that created a dichotomy between the southern and northern hemispheres, divided by “dichotomy boundary”	

4.1–3.2	The epoch of late heavy bombardment						
No later than 4.0				Four billion years old rocks have been found in Western Canada	Completion of crystallization of the upper mantle and the beginning of the formation of circular seas. Completion of the formation of the macrostructure of the continents.		
Starting from 4.0				The beginning of tectonic activity			
About 4.0	Completion of the first stage of geological development of silicate planets; completion of the formation of the primary crust	The presence of a powerful magnetic field, which later disappeared			Formation of the south pole–Aitken basin, the largest impact crater in the solar system	Differentiation into crust, mantle and core	
4.0–3.9					Formation of circular impact basins		
4.0–3.5	Particularly high solar wind levels			Recycling of primary and secondary crust and formation of microcontinents (tertiary crust)			

(continued)

Time from present, billion years (approx.)	Sun and Solar System in general	Mercury	Venus	Earth	Moon	Mars	Giant planets (Jupiter, Saturn, Uranus, and Neptune)
From 4.0 till 2.6				The period of formation of the earth's core			
About 4.0–2.5		High level of heat generation		Archean eon			
About 4.0–2.0						Complete transformation of the primary crust in the northern hemisphere by basaltic areal volcanism	
About 3.97–3.87	The period of gravitational excitation of the solar system caused by the orbital resonance between Jupiter and Saturn						Period of orbital resonance between *Jupiter* and *Saturn* *Uranus* and *Neptune* change orbits
About 3.9		Formation of the Caloris Basin, or plain of heat, as a result of the fall of a large celestial body					

3.9–3.16					The third period of evolution (“marine phase”): Filling of lunar seas with basaltic lava and completion of the formation of inhomogeneities associated with gravity anomalies (mascons)		
About 3.86–3.85					Impact formation of seas (sea of rains, Eastern Sea, sea of serenity, etc.)		
3.85–3.75							*Uranus* and *Neptune* become fixed in modern orbits
3.85–3.0					Formation of the last giant impact basins (Imbrium and Orientale). Reduction of meteorite bombardment intensity to modern levels		

(continued)

Time from present, billion years (approx.)	Sun and Solar System in general	Mercury	Venus	Earth	Moon	Mars	Giant planets (Jupiter, Saturn, Uranus, and Neptune)
Up to 3.8							*Jupiter* and *Saturn* had high heat generation
About 3.8	*Completion of the formation of the solar system in its current form*						
3.8–3/2.8	*The third phase of the evolution of the planets of the solar system*						
About 3.8				The earliest stromatoliths (fossil remains of ancient organisms)			
3.8–3.75				This is the reliably established age of the most ancient rocks			
3.8–3.5						Traces of water flow activity	
About 3.8–3.2	Migration of the asteroid belt from orbits at a distance of 1.7 AU to orbits of 2.1 AU						

3.8–2.8						Hesperian period. Global catastrophic climate changes (from warm humid to cold dry) and the formation of huge lava fields	
About 3.8–2.5				Period of transformation of the continental crust			
More than 3.7						Formation of the early Mars Ocean known as Arabia (today the Arabia Terra)	
About 3.7					The first peak of lunar volcanic activity	Formation of the Tharsis plateau	
About 3.6					Weakening of the magnetic field	Aeolis dorsa a region with numerous traces of liquid water availability	
3.6–3.4			Maximum heat dissipation	Maximum heat dissipation			
More than 3.5		Decline in activity				The existence of a dense atmosphere and water cover	Formation of the giant crater Valhalla on Jupiter's moon *Callisto*
About 3.5						The beginning of the formation Martian polar ice caps	

(continued)

Time from present, billion years (approx.)	Sun and Solar System in general	Mercury	Venus	Earth	Moon	Mars	Giant planets (Jupiter, Saturn, Uranus, and Neptune)
3.5–2.0						Plains filled with basalts were formed	
About 3.2	The end of the era of the late heavy bombardment, a sharp reduction in impact energy, the beginning of active cooling of some planets				Second peak of lunar volcanic activity		
About 3	The current level of solar luminosity has been established					A catastrophic rupture in the Martian crust, forming a grandiose canyon—Valles Marineris	
3/2.8–1	*The fourth phase of the evolution of the planets of the solar system*						
3.16—Present time					"The post-marine period." decline of geological activity		
3.0—No later than 2.5				Formation of the plate tectonics system			

3.0–2.0		Cooling and loss of magnetic field		The emergence of global plate tectonics			
2.8—Present time						The Amazonian period. Formation of surfaces with few craters; lava flows spread over the surface	
About 2.6					Minimum age of mare basalts		
About 2.5				"The great oxidation event"—The beginning of accumulation of free oxygen in the atmosphere (Lyons et al., 2014)			
About 2.5–0.5						Maximum heat emission	
About 2				Late collisions of large asteroids with earth			
2.0–1.0						Formation of giant volcanoes in the equatorial zone	
1—Present time	*The fifth phase of the evolution of the planets of the solar system*						
About 1					Disappearance of magnetic field	The displacement of the Martian crust relative to its polar axis and redistribution of ice deposits on the surface	

(continued)

Time from present, billion years (approx.)	Sun and Solar System in general	Mercury	Venus	Earth	Moon	Mars	Giant planets (Jupiter, Saturn, Uranus, and Neptune)
0.55				The second great oxidation event. The emergence of multicellular organisms			
0.5					Cessation of volcanism	The last powerful manifestation of area volcanism. Formation of four major shield volcanoes (up to 20 km high and up to 700 km in diameter) on the Farsida plateau	
0.5–0.3			Completion of the formation of surface relief; peak and gradual decline of volcanic activity. Grandiose renewal of the surface as a result of massive lava outpourings				

0.4–0.3						Formation of the giant Martian volcanoes Arsia, Pavlina, Askrian, and Olympus	
0.115–0.004/ 0.002						"The epoch of the last eruptions"	
0.11						The last short-term episode of major warming	
0.07–0.04						Cessation of activity of large area volcanoes on the Tharsis plateau	
About 0.05		Formation of the "youngest" scarps (pits, grabens)		Formation of the earth's youngest crater in Arizona			
0.05–0.03			Formation of the "youngest" craters with radio-dark parabolas			The last eruption of Olympus Mons	
0.002							
Present times							
From 100 years ago				The warming process		The warming process	

References

Adushkin, V. V., & Vityazev, A. V., (2007). The origin and evolution of the Earth: A contemporary view. *Herald of the Russian Academy of Sciences, 77*(3), 232–238. https://doi.org/10.1134/S1019331607030045

Alekseyev, F. A., Kukharenko, N. K., Voytov, G. I., & Galkin, Y. A. (1972). Initial results of geophysical and geochemical investigations of very deep boreholes. *International Geology Review, 14*(3), 275–280. https://doi.org/10.1080/00206817209475693

Andrei, M. (2014, June 3). The World's Deepest Hole Lies Beneath this Rusty Metal Cap—The Kola Superdeep Borehole. *ZME Science*. https://www.zmescience.com/feature-post/natural-sciences/geology-and-paleontology/planet-earth/worlds-deepest-hole-lies-beneath-rusty-metal-cap-kola-superdeep-borehole/.

Armand, A. D. (1975). *Information models of natural complexes USSR*. Nauka. (In Russian).

Ault, A. (2015, February 19). Ask Smithsonian: What's the deepest hole ever dug? *Smithsonian Magazine*. https://www.smithsonianmag.com/smithsonian-institution/ask-smithsonian-whats-deepest-hole-ever-dug-180954349/.

Boyko, S. V., & Prokaten, E. V. (2014). *General geology*. Publishing House of the Siberian Federal University. (In Russian).

Bozhko, N. A. (2009). Supercontinental cyclicity in the history of the Earth. *Vestnik Moskovskogo Universiteta. Series 4. Geology, 2*, 13–28. (In Russian).

Canup, R. M., & Asphaug, E. (2001). Origin of the moon in a giant impact near the end of the earth's formation. *Nature, 412*, 708–712.

Gibson, B., & Ibata, R. (2007). Ghosts of dead galaxies. *V Mire Nauki, 6*, 29–35. (In Russian).

Gomes, R., Levison, H. F., Tsiganis, K., & Morbidelli, A. (2005). Origin of the cataclysmic late heavy bombardment period of the terrestrial planets. *Nature, 435*(7041), 466–469. https://doi.org/10.1038/nature03676

Grinin, L., & Korotayev A. (2023). Social evolution: Theoretical aspects and general outlines. *Social Evolution & History, 22*(2), 111–158. https://doi.org/10.30884/seh/2023.02.08

Hazen, R. M. (2012). *The story of earth: The first 4.5 billion years, from stardust to living planet*. Viking.

Grinin, L. E. (2017). *Big History of the world development: History and evolution of the solar system*. Uchitel. (In Russian).

Grinin, L. E. (2020). *Big History of the world development: Planets of the solar system. Their history and evolution. Chemical evolution in space and on earth*. Uchitel. (In Russian).

Grinin, L. E. (2025). Geological evolution. *Journal of Big History, 8*(1), 32–44. https://doi.org/10.22339/jbh.v8i1.8106

Grinin, L., & Grinin, A. (2019). The star-galaxy era in terms of Big History and universal evolution. *Journal of Big History, 3*(4), 69–92.

Gumerova, N. V. (2010). *Historical geology* (2nd ed.). Tomsk Polytechnic University. (In Russian).

Head, J. W., & Solomon, S. C. (1981). *Tectonic evolution of the terrestrial planets. Science, 213*(4503), 62–76. https://doi.org/10.1126/science.213.4503.62

Johnson, B. C., & Melosh, H. J. (2012). Impact spherules as a record of an ancient heavy bombardment of earth. *Nature, 485*(7396), 75–77. https://doi.org/10.1038/nature10982

Katz, Ya. G., Kozlov, V. V., Makarova, N. V. & Sulidi-Kondratyev, E. D. (1984). *Geologists Study Planets*. Nedra. In Russian.

Khain, V. I. (2010). On the basic principles of constructing a truly global model of the earth's dynamics. *Geology and Geophysics, 51*(6), 753–760. In Russian.

Khain, E. V., & Koronovsky, N. V. (2007). *Planet earth. From the core to the ionosphere*. KDU. (In Russian).

Krut, I. V. (1978). *Introduction to the general theory of the earth: Levels of organization of geosystems*. Mysl. (In Russian).

Ksanfomality, L. V. (1997). *Parade of Planets*. Nauka. In Russian.

Ksanfomality, L. V. (2004). Mountain streams and basins of Mars. In V. G. Surdin (Ed.), *Mars: The great opposition* (pp. 199–207). Fizmatlit. (In Russian).

Ksanfomality, L. V. (2012a). Venus. In Surdin, V. G. (Ed.), *Solar System (ch. 5)*. Fizmatlit. (In Russian).

Ksanfomality, L. V. (2012b). Mercury. In V. G. Surdin (Ed.), *Solar system (ch. 4)*. Fizmatlit. (In Russian).

Ksanfomality, L. V. (2012c). Mars. In V. G. Surdin (Ed.), *Solar System (ch. 6)*. Fizmatlit. (In Russian).

Kuskov, O. & Konrod, V. (2001) Core sizes and internal structure of earth's and Jupiter's satellites. *Icarus, 151*(2), 204–227. https://doi.org/10.1006/icar.2001.6611

Lyons, T. W., Reinhard, C. T., & Planavsky, N. J. (2014). The rise of oxygen in earth's early ocean and atmosphere. *Nature, 506*, 307–315.

Mitrofanov, I. G. (2015). Water and life on Mars. Solar system studies: Space milestones. In A. V. Zakharov (Ed.), *Proceedings of the scientific session dedicated to the 80th anniversary of academician M. Ya. Marov* (pp. 155–167). IKI RAS. In Russian.

Puchkov, V. N. (2016). Interrelation of plate tectonic and plume processes. *Geotektonika, 4*, 88–104. (In Russian).

Ryder, G., Koeberl, C. H., & Mojzsis, S. J. (2000). *Origin of the earth and moon*. University of Arizona Press.

Sirotin, V. I. (2009). *Comparative planetology (essays on the results of research of the earth, the solar system and space)*. VSU. (In Russian).

Smythe, D. K., Smithson, S. B., Humphreys, C., Gillen, C., Kristoffersen, Y., Karaev, N. A., Garipov, V. Z., & Pavlenkova, N. I. (1994). Project images crust, collects seismic data in world's largest borehole. *Eos, Transactions American Geophysical Union, 75*(41), 473–476. https://doi.org/10.1029/94EO01089

Sorokhtin, O. G., & Ushakov, S. A. (2002). *Land development*. MSU. (In Russian).

Vinkovetsky, Y. A. (1971). *Geology and the general theory of the evolution of nature*. Nedra. (In Russian).

Workshop on Early Solar System Impact Bombardment III. LPI Contribution No. 1826. (2012). Lunar and Planetary Institute.

Yanvarev, G. S. (1993). *Historical and regional geology*. Nedra. (In Russian).

Zharkov, V. N. (2013). *The Internal structure of the Earth and planets. Elementary introduction to planetary and satellite geophysics*. Nauka i obrazovanie. (In Russian).

Zwart, S. P. (2016). How the sun abducted dwarf planets from an alien solar system. *New Scientist, 8*. https://www.newscientist.com/article/mg23030770-300-how-the-sun-abducted-dwarf-planets-from-an-alien-solar-system/

Leonid Grinin is Senior Research Professor at the Center for Stability and Risk Analysis at the HSE University in Moscow, as well as Director of Eurasian Center for Big History & System Forecasting at the Institute of Oriental Studies, Russian Academy of Sciences. He is co-editor of the international journals *Social Evolution & History* and the *Journal of Globalization Studies*, as well as co-editor of the international almanacs *Evolution, History & Mathematics*, and *Kondratieff Waves*. He is author of more than 700 publications, including 35 monographs in Russian, English, and Chinese. His current research focuses on comparative political studies, theory of revolution, political anthropology, global economy, global history, historical sociology, and futurology. Among his monographs are *The Cybernetic Revolution and the Forthcoming Epoch of Self-Regulating Systems* (Uchitel, 2016, with Anton Grinin), *Big History of Globalization. The Emergence of a Global World System* (Springer 2019, in co-authorship with Julia Zinkina, David Christian, and others), and *Islamism, Arab Spring, and the Future of Democracy. World System and World Values Perspectives* (Springer, 2019; with Andrey Korotayev and Arno Tausch); *Cybernetic Revolution and Global Aging. Humankind on the Way to Cybernetic Society, or the Next Hundred Years* (Springer 2024 with Anton Grinin and Andrey Korotayev). In 2004, he was awarded as a winner of the best publication of *Socis* journal during the year for the article "People of Publicity—a New Social Layer?." In 2012, he was awarded the Gold Kondratieff Medal by the International N. D. Kondratieff Foundation. In 2016, he received the Belyaev award for the book *From Bifaces to Nanorobots* (the book published in Russian in co-authorship with Anton Grinin). 12, Rozhdestvenka, Moscow, 107031, Russian Federation. e-mail: leonid.grinin@gmail.com. ORCID 0000-0003-0278-2619

Part IV

Biological Complexity

Viruses in Big History: Evolutionary Roles and Complex Systems

10

Anton Grinin

Abstract

Viruses play a crucial yet underexplored role in the context of Big History and evolutionary development. Despite their ubiquitous presence, viruses are difficult to study due to their microscopic nature and lack of physical traces in the geological record. This chapter examines current theories on the evolutionary origins of viruses, including regressive, protobiont, and symbiotic hypotheses, and explores the debate over whether viruses can be considered alive. The role of viruses as parasites is also scrutinized, revealing a complex and contradictory nature. Additionally, viruses are discussed from an energetic and informational perspective, highlighting their influence on ecosystems and genetic evolution. This chapter argues that viruses are not merely dangerous pathogens but essential molecular tools that drive evolution and may offer significant insights into biotechnology and genetic modification. Finally, viruses are compared to other complex self-regulating systems, such as computer viruses and cancer, emphasizing the similarities in their behavior and adaptation.

Keywords

Big history · Universal evolution · Evolution · Biological evolution · Evolutionary roles · Viruses · Complex systems · Complex adaptive systems

This chapter is an extensively re-worked version of the text originally published in Journal of Big History (Grinin, 2025).

A. Grinin (✉)
HSE University, Moscow, Russia

Lomonosov Moscow State University, Moscow, Russia

D. J. LePoire et al. (eds.), *Complexity in Universal Evolution*, World-Systems Evolution and Global Futures, https://doi.org/10.1007/978-3-032-20517-9_10

10.1 What Are Viruses?

Even though we hear about viruses daily, it is not easy to define what a virus is.

The classical definition of a virus is that it is a submicroscopic infectious agent that replicates only inside the living cells of an organism. According to the Merriam-Webster dictionary, a virus is "Any of a large group of submicroscopic infectious agents that are usually regarded as non-living extremely complex molecules."[1] The term "agents" is often used in virology, because it is hard to find a common word for viruses. A virus is not an organism, as we used to understand it, nor a chemical compound as it has a very complex structure and behavior. One of the biggest problems regarding the understanding of viruses is that they are very different from everything we know. For the theory of evolution, viruses also present a problem. They are like a piece of a puzzle that does not fit into the whole picture. It is clear, but at the same time it is hard to realize that there are an unimaginably huge number of viruses, probably more than stars in the universe. Astronomers estimate that the observable(?) universe could contain up to one septillion stars which in numbers is 10 to the 24th power.[2] An estimated 10 nonillion (10 to the 30th power) individual viruses exist on our planet, enough to assign one to every star in the universe millions of times over (Wu, 2020). This number is so large that our brain is incapable of understanding it. Undoubtedly, viruses are the most widespread agents on Earth. However, at present only over 6000 virus species have been described in detail.

It may also be surprising that the vast majority of viruses on Earth are marine viruses, the diversity of which is almost astronomic. A teaspoon of seawater typically contains about 50 million viruses. Most of them are bacteriophages which infect and destroy marine bacteria, and control the growth of *phytoplankton.*

Viruses are not only the most widespread but also the most changeable of all known forms of life. The presence of viral diversity and its importance is evidenced by the fact that each new sequenced virome includes new sequences. Like every new wave on the seashore, every virus is different from the other. At the same time viruses prove not only that they are very widespread but also demonstrate a great diversity in size. There are smaller viruses, for example MS2, the size of which is about 27 nanometers. Other viruses, such as the mimiviruses, are so large that they are actually larger than some of the smallest cells we know. For example, pithovirus can be up to 1500 nanometers in size (Colson et al., 2017), which is comparable to the average cell size of prokaryotes (Campbell et al., 2009).

10.1.1 Alive or Not?

Probably the most debatable question about viruses is whether they are alive or not.

---

[1]URL: https://www.merriam-webster.com/dictionary/virus.

[2]URL: https://www.nationalgeographic.co.uk/science-and-technology/2020/04/there-are-more-viruses-than-stars-in-the-universe-why-do-only-some-infect-us.

First, viruses are often considered non-living because they do not have the metabolic activity that is required to maintain cellular structures and various processes. Second, they cannot reproduce without a host organism. The seemingly simple question of whether or not viruses are alive raises a fundamental issue: What exactly defines "life?" "A precise scientific definition of life is an elusive thing, but most observers would agree that life includes certain qualities in addition to an ability to replicate. A living entity is in a state bounded by birth and death" (Villarreal, 2008).

Living organisms are thought to require a degree of biochemical autonomy, carrying on the metabolic activities that produce the molecules and energy needed to sustain the organism' (*Ibid.*). On the other hand, not all living cells have metabolic activity at all times during their existence. Many include potential dormant states that are practically considered alive but are no less inert than viral particles outside their hosts. In these dormant systems, no growth or detectable *metabolism* will take place over very long periods of time. Are dormant seeds or spores alive? Yet they have the potential for life, and they may be destroyed. In this regard, viruses resemble seeds more than they do live cells. They have a certain potential, which can be snuffed out, but they do not attain the more autonomous state of life' (*Ibid.*).

As for reproduction, it is even more complicated because there are different types of reproduction. But it would also be a mistake to argue that every organism is independent in propagation, since sexual reproduction intrinsically needs another organism for genetic exchange. We will see below that there is a reason to believe that the cell nucleus itself is of viral origin. Since the nucleus plays a central role in sexual reproduction in eukaryotes, this suggests that reproduction itself was most likely not invented by eukaryotes, but rather "borrowed" from viruses.

10.1.2 Parasites or Not?

According to the generally accepted definition, a virus is an "obligate intracellular parasite." Although it is not so obvious, viruses maintain many types of relationships with their hosts, from mutualism to pure *parasitism*. A parasite usually uses the host for resources, typically food, and does not need the host to replicate. Viruses do not need any food but they use the host for replication. Viruses are only active when inside a cell where they take control of its processes and energy to replicate itself.

There is another important question: Can an organism be defined as a parasite that does not harm its host[3]? We believe it is more reasonable to consider such relations as cooperation. Another quite surprising fact is that most of known viruses are persistent and *innocuous*, not pathogenic (*Ibid.*). According to some data only one virus of 20 is pathogenic. They take up residence in cells, where they may remain dormant for long periods or take advantage of the cells' replication apparatus to

[3]Usually "parasitism" is defined as "organisms that consume resources from only one host during a stage in their life and also cause harm, but not usually or necessarily the death of its host."

reproduce at a slow and steady rate. These viruses have developed many clever ways to avoid detection by the host immune system—essentially every step in the immune process can be altered or controlled by various genes found in one virus or another (Villarreal, 2008).

Although there is a dualistic question (Are viruses parasites or not?), any parasite, as any other organisms, originates from the emergence of the replicators. With a virus being a possible source of the genetic material for all other organisms, it seems that the relationships between viruses and other organisms are more complex and important to all parties than the usual parasite-host interrelation.

"In fact, viruses can have beneficial effects on their hosts, creating a symbiotic relationship. And this might be far from anecdotic. Many examples of viruses that provide functional benefits to their hosts are known to create a mutualistic relationship" (Solé & Santiago, 2019, p. 117).

Other remarkable examples include a long-term coevolution between a large class of viruses (e.g., the polydnaviruses) and their host wasps. These wasps are parasitoids: they are different from standard predators, since they lay their eggs inside the larvae of their prey species, which develop inside the body of the living **larvae** by eating them from inside. The normal outcome of this should be an immune response capable of encapsulating the injected eggs and inhibiting egg development. However, the endogenous virus carried by the wasp egg suppresses this response. The coevolutionary ties are very strong and some authors have questioned how appropriate it is to consider the polydnavirus as an independent virus in this case, or whether we are dealing with a biochemical mechanism of suppression of the victim's immunity.

10.2 Evolution of Viruses

Viruses have shaped the evolution of cells, organisms, ecosystems, and even the biosphere. From an evolutionary perspective, self-replication became one of the most important landmarks of life.

We know that viruses are non-cellular structures, consisting of a piece of DNA or RNA surrounded by a slim protein coating.

Although all living organisms on the globe have a DNA genome, some viruses use RNA as a genome. This can mean that viruses could have separated from DNA organisms or preceded them. Viruses are the *quintessence* of genetic information, since their main and often only function is to deliver and spread genetic information. Given the fact that viruses outnumber all other organisms, they produce a great flow of genetic information, providing genetic diversity and its incredible variability. Just as bits of digital information are transmitted over the Internet, in the web of life, viruses exchange genetic information within living organisms. From this point of view, there are good reasons to believe that viruses also affect energy flows within living organisms.

Genetic information is key to understanding viruses. Viruses are literally part of the genome of all organisms. For example, there are over 60,000 proviruses in the human genome.[4] From the energetic point of view, viruses are an important part of information and energy flows. They strongly influence energy flows in complex ecosystems. "Viruses live at the edge of disorder, where high instability but also *adaptability* occur" (*Ibid.*).

Viruses are probably one of the main sources of mutations in organisms and thus a locomotive of entire evolution. As is well known, mutation is a crucial component of evolution because genetic variability is the fuel on which natural selection operates to adapt populations to their environment.

Viruses are found infecting all forms of life and have probably been around since the first cells arose, or perhaps even before them. Tracing back the origin of viruses is a titanic, almost impossible, *endeavor* because they do not form fossils.

Viruses are perhaps the first organisms which developed a mechanism of reproduction up to the level which is capable of life. The amazing diversity of reproductive mechanisms among viruses demonstrates specialization of viruses in reproduction.

There are many theories of the origin of viruses. Some of the most well-known include:

1. *The regressive hypothesis.* It suggests that *viruses were once small cells that parasitized larger cells.* Over time, as the parasite became more dependent on the host cell to complete its life cycle, genes not strictly necessary for their acquired *parasitism* were lost.
2. *Vagrancy hypothesis, or the escape hypothesis. The second classic hypothesis for the origin of viruses states that some viruses may have evolved from pieces of DNA or RNA that "escaped" from the genome of cells.* The escaped DNA could have come from plasmids (pieces of naked DNA that can move between cells) or from transposons (molecules of DNA that replicate and move around the cellular genomes to different positions)' (Solé & Santiago, 2019, p. 177).
3. *Protobiont hypothesis.* This is also called the virus-first hypothesis and suggests that viruses may have evolved from complex molecules of protein and *ribonucleic* acids at the same time as cells first appeared on Earth, and would have been dependent on cellular life from the very beginning. In the primitive precellular soup, as in any other replicating system, parasites would have also evolved that grew at the expense of other more complex molecular systems' (Solé & Santiago, 2019, p. 179).

"Some researchers think that the cell nucleus itself is of viral origin. The advent of the nucleus—which differentiates eukaryotes, including humans, from prokaryotes, such as bacteria—cannot be satisfactorily explained solely by the gradual adaptation of prokaryotic cells until they became eukaryotic. Rather the nucleus may have evolved from a persisting large DNA virus that made a permanent home within prokaryotes" (Villarreal, 2008).

[4]URL: https://bigthink.com/health/human-virus-herv-genome/.

As complex systems, viruses are similar to other complex systems.[5]

10.3 Computer Viruses

Computer viruses share some similarities to biological ones. Like flu viruses that cannot reproduce without a host cell, computer viruses cannot reproduce and spread without a file or a document.[6]

The list of tricks adopted by programmers in order to make their viruses more effective or even invisible is very large. In addition, biological viruses have numerous ways to bypass the immune system. At the beginning, antivirus software was mainly based on detecting specific sequences of bytes that somewhat defined the signature of a given virus. As the number of potential viruses increased, this approach became more and more involved and less efficient. New viruses became more difficult to detect and remove. Old tricks had to be replaced by a great invention: polymorphism (Nachenberg, 1997). Polymorphic computer viruses incorporated a powerful weapon mutation. As it occurs with real viruses, constantly escaping from the selection forces imposed by immune responses, detection, and removal strategies is a major driver of genetic variability. As RNA viruses escape from the immune system, polymorphic viruses are able to keep their basic functions intact and yet appear "different" to the infected machine. The new challenge was responded to by novel antivirus software, but the arms race is still unfolding. Antivirus software just as our immunity has to be constantly adapting to new viral tricks. While some computer viruses can be playful in intent and effect, others can have profound and damaging effects. This includes erasing data or causing permanent damage to your hard disk. Worse yet, some viruses are designed with financial gains in mind.[7]

Computer viruses can hide disguised as attachments of socially shareable content such as funny images, greeting cards, or audio and video files and replicate themselves by modifying other computer programs and inserting their own code.

Both biological and computer viruses are capable of mutations.

However, we know that most mutations affecting a virus genome are harmful, impeding or threatening their replication potential. How many mutations in man-made counterparts are lethal? None. Thus, there is a sharp separation between

[5]Usually a complex system is understood as a system composed of many components which may interact with each other and whose behavior is intrinsically difficult due to the dependencies, competitions, relationships, or other types of interactions between their parts or between a given system and its environment. Viruses have a complex system for avoiding the immune response, especially when they are inside the host organism organizing a special production and control center for the creation of new viruses. At this stage, viruses demonstrate the characteristics of a complex adaptive system. For more details on complex systems in Big History, see also LePoire et al., 2025, as well as Chapter "Introduction to Complexity Theory for Big History" (LePoire et al., 2026) in this volume.

[6]URL: https://wireless4u.ca/virus-removal/.

[7]URL: https://wireless4u.ca/virus-removal/.

these two types of viruses' (Solé & Santiago, 2019, p. 195). And Darwinian natural selection does not work within computer viruses as it does in nature.

10.4 Viruses in Big History

Despite the fact that viruses are tiny objects, their role in Big History may be more significant than ~~it~~ might be expected. First, they show that not everything in Big History has similar patterns and fits into one simple narrative.

Second, they demonstrate that even such basic phenomena as life, death, reproduction, parasitism, and others are in fact very conditional. Viruses as a transitional link between all these phenomena show that reality is much more diverse and complex. In the evolutionary phases of big history, we place them between the chemical *abiogenic* and biological phases.

Viruses also demonstrate involvement in the evolution of all types of organisms as well as all basic principles of evolution such as diversity and specialization. The role of viruses in evolution is obviously underestimated. A breakthrough in virology in the coming decades could be no less important than a breakthrough in understanding the nature of bacteria, both pathogenic and beneficial.

Viruses are an important component from the perspective of information evolution, because in some respect viruses can be regarded as pure information particles. Viruses can be considered as basic information agents that transmit genetic information, serve as a resource for mutations, and thus are a major molecular tool of evolution. Moreover, this tool is beginning to be used by humans for the genetic alteration of many organisms, including themselves.

Viruses are self-regulating systems similar to many other complex systems. An integrated approach to the study of self-regulating systems can become key to exploring the interconnections and direction of Big History (Fig. 10.1).

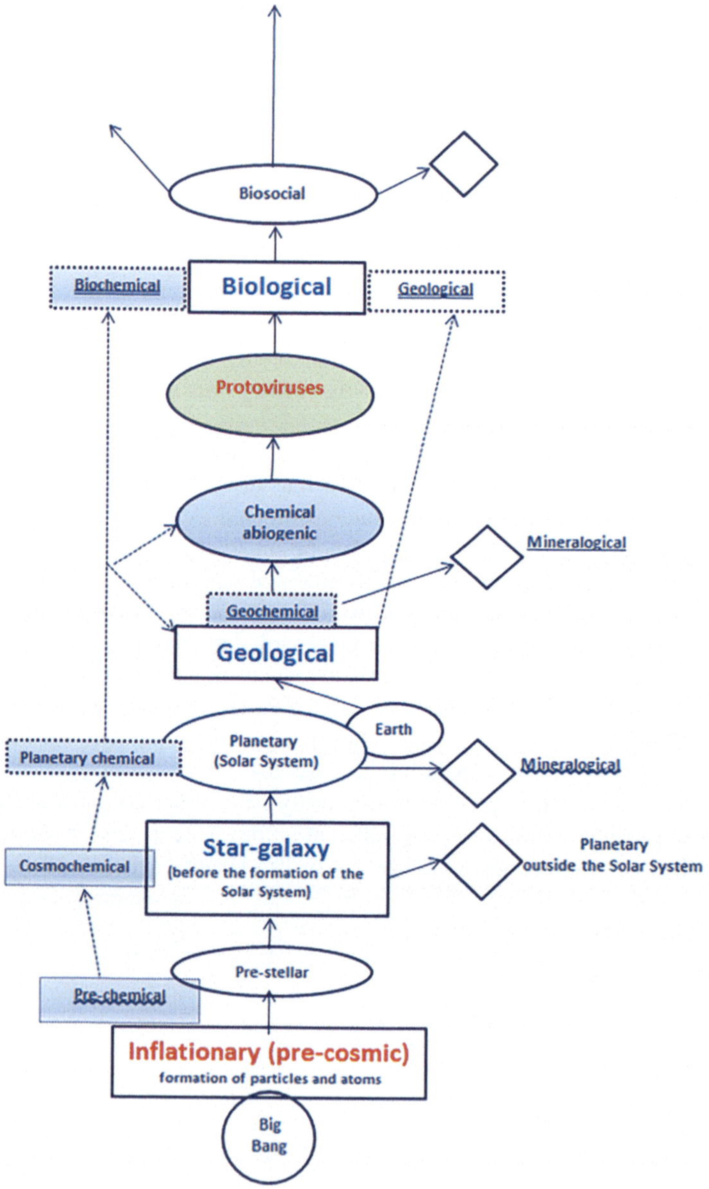

Fig. 10.1 Evolutionary phases of Big History

References

Campbell, N. A., Reece, J. B., & Taylor, M. R. (2009). *Biology concepts and connections*. Pearson Education.

Colson, P., La Scola, B., Levasseur, A., Caetano-Anollés, G., & Raoult, D. (2017). Mimivirus: Leading the way in the discovery of giant viruses of amoebae. *Nature Reviews Microbiology, 15*(4), 243–254. https://doi.org/10.1038/nrmicro.2016.197

Grinin, A. L. (2025). Viruses and evolution: The role of viruses in big history. *Journal of Big History, 8*(1), 51–56.

LePoire, D. J., Grinin, L., & Korotayev, A. (2025). Complexity in big history. An introductory exploration. *Journal of Big History, 8*(3), 98–139. https://doi.org/10.22339/jbh.v8i3.8304

LePoire, D. J., Grinin, L., & Korotayev, A. (2026). Introduction to complexity theory for Big History. In D. J. LePoire, L. Grinin, & A. Korotayev (Eds.), *Complexity in universal evolution – A Big History perspective* (pp. 29–90). Springer Nature (in this volume). https://doi.org/10.1007/978-3-032-20516-2_2

Nachenberg, C. (1997). Computer virus-antivirus coevolution. *Communications of the ACM, 40*(1), 46–51. https://doi.org/10.1145/242857.242869

Solé, R., & Santiago, F. E. (2019). *Viruses as complex adaptive systems*. Princeton University Press.

Villarreal, L. P.(2008). Are Viruses Alive? *Scientific American*. www.scientificamerican.com/article/are-viruses-alive-2004/

Wu, K. J. (2020). There are more viruses than stars in the universe. Why do Only Some Infect Us? *National Geographic*. https://www.nationalgeographic.co.uk/science-and-technology/2020/04/there-are-more-viruses-than-stars-in-the-universe-why-do-only-some-infect-us

Anton Grinin PhD in Biological Sciences, is Research Fellow at HSE University and Moscow State University, as well as Senior Research Fellow at International Center for Education and Social and Humanitarian Studies. His main research interests include biotechnologies, global technological transformations, and forecasts. He is the author/coauthor of more than 150 works, including six monographs in English and Russian such as *From Bifaces to Nanorobots: The World on the Way to the Epoch of Self-Regulating Systems* (2015 in Russian), *The Cybernetic Revolution and the Forthcoming Epoch of Self-Regulating Systems* (2016; with Leonid Grinin); *BRICS development in the context of world dynamics: challenges and perspectives* (2024; with Victor Sadovnichy, Askar Akaev et al.); *Reconsidering the Limits to Growth: A Report to the Russian Association of the Club of Rome* (Springer Nature, 2023; with Andrey Korotayev, Leonid Grinin, and others), and *Cybernetic Revolution and Global Aging: Humankind on the Way to Cybernetic Society, or the Next Hundred Years* (Springer Nature, 2024, with Andrey Korotayev and Leonid Grinin); *The Struggle for a New World Order: History, Present, and Future* (2025).

Biology *for* Life: A Big History Approach to Complex Systems Thinking in Biology Education

11

Paul J. Narguizian

> *"A system is never the sum of its parts; it's the product of their interactions."—Russell Ackoff*

Abstract

This chapter explores the integration of Complex Systems Thinking (CST) into undergraduate biology education, advocating for a more holistic approach to understanding living systems. Traditional reductionist methods, while valuable for dissecting specific biological components, often fail to capture the complexity, interconnectedness, and emergent properties inherent in biological systems. By embracing CST, biology educators can provide students with the tools needed to comprehend and navigate the ever-growing complexity of life. This chapter delves into the significance of CST in biological education, discussing key concepts such as complexity, emergence, and systems thinking, and highlighting the pedagogical implications for both educators and students. It also offers practical strategies for incorporating these interdisciplinary approaches into the curriculum. Ultimately, the goal is to equip future biologists and scientifically literate citizens with a deeper, more integrated understanding of biological systems, preparing them to address the multifaceted challenges of the natural world.

This chapter is an extensively re-worked version of the text originally published in Journal of Big History (Narguizian, 2025). https://deming.org/ackoff-on-systems-thinking-and-management/

P. J. Narguizian (✉)
California State University, Los Angeles, CA, USA
e-mail: pnargui@calstatela.edu

D. J. LePoire et al. (eds.), *Complexity in Universal Evolution*, World-Systems Evolution and Global Futures, https://doi.org/10.1007/978-3-032-20517-9_11

Keywords

Big History · Biology · Complex Systems Thinking

11.1 Introduction

In an era defined by accelerating ecological crises, technological disruption, and global interdependence, biology education must evolve to equip students not only with foundational knowledge but also with the ability to think systemically, historically, and ethically about life. This chapter introduces *Biology for Life*, a systems-based pedagogical model that reframes general biology through the dual lenses of Big History and Complexity Thinking. This approach situates biological phenomena within a broader cosmic and planetary narrative, where life is understood as a dynamic product of energy flows, emergent properties, and threshold transitions across scales, from molecules to ecosystems and from stardust to consciousness.

Drawing on the narrative structure of Big History, *Biology for Life* organizes biological concepts in relation to the major evolutionary thresholds of increasing complexity, such as the origin of life, the emergence of eukaryotes, and the rise of human culture. These transitions are not presented as isolated milestones, but as interconnected nodes within a living, evolving system. At the same time, this chapter leverages principles from Complexity Science, including feedback loops, hierarchy, self-organization, and resilience, to help students analyze living systems as adaptive, multi-layered, and context-dependent.

By weaving these frameworks into a coherent curriculum, this systems-based pedagogy fosters deep conceptual understanding, transdisciplinary thinking, and civic scientific literacy. Students come to see biology not merely as a catalogue of facts but as a way of interpreting the world's complexity—preparing them to navigate uncertainty, evaluate claims, and contribute meaningfully to a sustainable future. This chapter outlines the theoretical foundation, modular structure, and real-world case studies that make *Biology for Life* both an intellectually rigorous and pedagogically transformative approach to teaching general biology.

This chapter is centered around the key question: how do we address teaching about biology, based on complex systems? In the dynamic landscape of biological sciences, the exploration of life's intricacies demands a holistic and interconnected approach. Complexity science and systems thinking offer a transformative framework for understanding the dynamic and interconnected nature of biological systems. As educators in the field of biology, it is imperative to equip undergraduate students with the tools to navigate and comprehend the ever-growing complexity of living systems. Traditional reductionist views, while valuable in understanding specific components of biological phenomena, often fall short in capturing the intricate interconnections and emergent properties that characterize living systems. In the pursuit of understanding these complexities, undergraduate biology education must transcend traditional *reductionistic* silos and embrace *holistic* interdisciplinary approaches such as complexity science and systems thinking. This chapter delves into the integration of *Complex Systems Thinking* (CST) in undergraduate biology courses, highlighting its significance, pedagogical implication, and practical strategies for implementation. Furthermore, it describes the key characteristics, and proposes a few pedagogical concepts associated with complexity, emergence, and systems thinking. The primary goal is to provide biology educators with the tools and enough basic information about the "*features of complexity*," utilized in Big History, so that they can consider why and how such an approach might be applied in the education of both biologists and a scientifically literate citizenry.

Contemporary biology education is often siloed, fragmented into disconnected units, molecular biology, ecology, evolution, physiology, without scaffolding that helps students synthesize how these parts function together. As a result, learners may accumulate facts but lack a coherent systems-level understanding of life. A Big History–informed approach restores narrative cohesion, while (CST) offers the conceptual tools needed to interpret life not as a static catalogue, but as a dynamic, interdependent web of evolving systems. Together, these frameworks guide students toward integrated, durable learning.

The concept of Big History provides a context of time for Complex Systems Thinking (CST). Scholars of Big History, for example, David Christian and Fred Spier to name two of the most prominent ones, have divided Big History into a series of stages of increasing complexity, the beginning of each period being hallmarked by some punctuating event; there are eight such stages, the first four being associated with Complex Physical Systems (CPS) and the latter four with Complex Adaptive Systems (CAS). This aspect of Big History as it applies to the application of Complex Systems Thinking (CST) in biological education will be further described in this chapter. The infusion of (CST) into the twenty-first century undergraduate biological education, all under the umbrella of a Big History context, suggests the following:

1. Systems thinking is adapted to the interdisciplinary nature and hierarchical organization of biology.
2. The five core ideas of biology as presented in *The Vision and Change in Undergraduate Biology Education: Call to Action report* (American Association for the Advancement of Science, 2011) are amenable to a (CST) approach.
3. The overarching theme of Big History provides not only an historic context but also examples of emergent change.

The aim of biology is to understand the natural world. The history of life is a story of epic proportions about the changes in the physical and biological features of our planet over its 4.5-billion-year history. It is the work, or it should be, of biology educators to use the complex evidence of these changes, combined with emergent properties known today, to bring life's history to light.

Biology is an exploration of the intricate web of life, where organisms interact within complex ecosystems shaped by dynamic processes. The interdisciplinary field of *Big History* offers a unique lens through which to examine the unfolding narrative of life on Earth, from its cosmic origins to the emergence of complex organisms and ecosystems. Within this vast landscape, the study of biology occupies a vital role, providing insights into the mechanisms driving evolution, biodiversity, and ecological dynamics. In recent years, there has been a growing recognition of the need to incorporate increasing complexity into undergraduate biology education, mirroring the multifaceted nature of biological systems and preparing students for the challenges of the twenty-first century (AAAS, 2011).

"Complex systems exist at various levels of organization that range from the subatomic realm to individual organisms to whole populations and beyond. They include, for example, molecules, cells, organisms, ecosystems, and human societies" (Mazzocchi, 2008, p. 1).

Within recent history, both science research and science education have been largely reductionist and mechanistic in perspective. While the reductionist approach has resulted in a significant increase in our knowledge of the natural world and in great technological advances, it is not sufficient for addressing global world challenges, such as climate change, sustainability, habitat destruction, and social justice (Orgill et al., 2019).

Living systems—from proteins to ecosystems—are not linear or deterministic. They are shaped by energy flows, thresholds, feedback loops, and emergent properties. Yet biology is often taught in ways that prioritize isolated outcomes over systems–level interactions. Big History contextualizes the emergence of life within broader cosmological and planetary processes, while Complexity Thinking reveals the *how* and *why* behind biological phenomena. This pedagogical shift mirrors the reality that biology is a story of increasing complexity over deep time.

In the twenty-first century, understanding biology requires more than content mastery, it demands systems literacy. Students must learn to think across scales, reason with uncertainty, and analyze the cascading effects of environmental, technological, and genetic change. Teaching biology through Big History and complexity enables learners to:

- Grasp patterns and principles that recur across biological systems (e.g., self-organization, resilience, transformation);
- Understand the deep-time context of contemporary issues (e.g., climate change, extinction, pandemics);
- Cultivate interdisciplinary thinking that connects biology to physics, chemistry, and human systems.

These capacities are essential for nurturing informed citizens, future scientists, and ethical decision-makers in a complex world.

Narratives are powerful cognitive tools. Big History's sweeping, evidence-based story of the universe invites students into biology as part of a larger human quest to understand our origins, evolution, and impact. Rather than viewing biology as a decontextualized subject, students begin to see themselves within the systems they are studying—as biological entities, as stewards of the biosphere, and as participants in an ongoing evolutionary narrative. This fosters a sense of personal relevance, curiosity, and responsibility. Perhaps most importantly, a Big History and complexity-infused biology curriculum cultivates transferable epistemological skills. Students learn to:

- Identify and analyze complex systems;
- Apply threshold-based and dynamic models to novel phenomena;
- Evaluate scientific explanations in historical and societal contexts.

These are not only skills for scientists, they are essential for lifelong learners and active citizens navigating the entangled systems of modern life, from public health to climate policy to biotechnology.

Finally, this approach supports equity and interdisciplinarity. Big History naturally bridges the sciences, humanities, and social sciences, allowing students with diverse strengths and backgrounds to engage with biology through multiple entry points: data, narrative, ethics, and systems analysis. It opens up science for all learners, not just those already inclined toward traditional biology pathways. In sum, this systems-based pedagogy rooted in Big History and Complexity Thinking does not replace disciplinary biology, it deepens, contextualizes, and rehumanizes it. It helps students grasp not just *what* life is, but *how* it emerges, changes, and connects us all.

11.2 Theoretical Foundations: Complexity and Thresholds Across Time and Scale

The pedagogical framework presented in *Biology for Life* is deeply informed by the interdisciplinary insights of Big History theorists and complexity scientists who have worked to explain how complexity arises, organizes, and persists across cosmic, biological, and cultural systems.

At the core of Big History lies the idea of thresholds of increasing complexity, first articulated by David Christian (2011), who describes these as pivotal moments when the universe becomes more complex due to the right combination of conditions, such as the formation of stars, the emergence of life, or the rise of human societies. Each threshold represents a leap in organizational structure, energy flow, and informational capacity. Christian argues that education must reflect this overarching story, helping learners see how life fits into a continuous narrative of systemic transformation over deep time.

Building on this, Spier (2015) developed a more refined framework by introducing the concept of regimes of complexity, shaped by three interdependent variables: energy flows, matter organization, and stability mechanisms. Spier emphasizes that life, especially biological life, can only persist when sufficient energy is harnessed and organized in a way that promotes both stability and innovation. His work reinforces the pedagogical value of teaching biology not in isolation, but as a system nested within other systems, shaped by universal conditions and constraints.

Complementing these historians, Eric Chaisson (2001, 2014) brings a physics-based perspective to complexity by quantifying it through the concept of energy rate density, the flow of energy per unit mass per unit time, as a universal metric for the emergence and maintenance of complex systems. From stars and galaxies to living organisms and human brains, Chaisson shows that higher levels of complexity are consistently associated with higher energy flow densities. His work offers a measurable, scientific basis for understanding life as an energy-dependent system, thereby bridging Big History's narrative insights with the mechanistic underpinnings of biology and physics.

Together, these scholars provide a theoretical foundation for a curriculum that treats life as a dynamic system, embedded within broader energetic and evolutionary frameworks. By drawing from their work, this chapter situates biology education within a scientifically coherent and pedagogically powerful story—one that unites the empirical rigor of biology with the explanatory richness of Big History and complexity science.

11.3 Complex Systems Thinking (CST) Is a Core Idea in Biology Education: Understanding Complex Systems

Living systems have both physiochemical and biological components. We can reconstruct many aspects of life's physical and biological history, including its major evolutionary transition and mass extinction events. We can also trace the evolution of life from an early world primarily inhabited by prokaryotes through the origin of eukaryotes. We cannot understand the history of the Earth and life in isolation, however, because the two systems are closely intertwined: the physical environment has influenced life, and the biological environment, in turn, has influenced the physical environment (Stanley, 2005).

At its core, Complex Systems Thinking (CST) involves viewing biological systems as interconnected networks of components, where emergent properties and behaviors arise from the interactions between these components. CST emphasizes the holistic analysis of biological phenomena, recognizing the interdependence of several factors and the dynamic nature of living systems. It involves understanding biological phenomena not as isolated entities but as interconnected networks of components that exhibit emergent properties and behaviors. CST in biology encompasses the hierarchical organization of living systems, from the molecular and cellular levels to organisms, populations, and ecosystems. At each level of organization, emergent properties arise from the interactions between constituent elements, leading to phenomena that cannot be fully understood by examining individual components in isolation. By incorporating the concept of CST, educators can provide students with a framework for exploring the interconnectedness of biological processes and understanding of the emergent properties that characterize living systems. CST in biology refers to an approach that recognizes and embraces the complexity inherent in biological systems.

Understanding complex systems involves grasping the intricate interconnections, emergent behaviors, and nonlinear dynamics inherent in systems composed of numerous interconnected components. In his book, *Big History and the Future of Humanity* (2015), Fred Spier defines complex systems under two broad categories: (1) *Complex Nonadaptive Systems* (CNAS) and (2) *Complex Adaptive Systems* (CAS). It should be stated here that Fred Spier deliberately distinguishes between a system and a regime, because to him, as a social scientist, the word "system" connotates stability—where "there are no forms of complexity that are completely stable over time." He suggests "regime" as a general term for both **complex nonadaptive systems** (CNAS) (those systems that do not adapt in response to changing conditions, such as stars or galaxies) and **complex adaptive systems** (CAS) (those systems that adapt in response to changing conditions, such as organisms or civilizations).

11.4 What Is a Complex System?

The precise definitions of *complexity*, let alone *complex systems*, are currently being debated by scientists, scholars, and academics. It is important to note here that CAS are a complex class of complex systems that *can* and *do* learn and evolve. Spier's use of the term complex non-adaptive systems (CNAS) can potentially cause confusion, since the alternative to CAS is commonly recognized as complex physical systems (CPS) and has been professionally recognized and used in the field of Big History for some time. Furthermore, distinguishing between *complex physical systems* (CPS) versus *complex adaptive systems* (CAS) provides another layer of, forgive the pun, complexity! For simplicity, this chapter will highlight a few definitions provided by scholars and academics, which will set the foundation for my pedagogical argument.

Complexity: "A catch-all term for all phenomena that have arisen in the Universe over 13.8 billion years, from the cosmos to culture, typified by many diverse and interacting component parts forming a part of a system that is created, sustained, and increased in intricacy by its flows of energy" (Christian & Baker, n.d.).[1]

Complex Physical Systems (CPS): "Shorthand term for systems that are studied primarily in the discipline of physics, e.g., stars and planets. Typified for their lack of "animate" activity, or those behaviors seen in biological organisms or human cultures. Instead of active and adaptive responses to their environments, these systems are non-adaptive and are governed by the natural laws of physics and chemistry. Complex physical systems vary in the degree of their complexity, but both in terms of diversity of component parts and density of energy flows, they are less complex than biological or cultural systems." (Spier, 2015 and Christian & Baker, n.d.)

[1] For more detail on various definitions of complexity, see also Chapter "Introduction to Complexity Theory for Big History" (LePoire et al., 2026) in this volume.—*Editors.*

Complex Adaptive Systems (CAS): "A system where component parts have the capacity to adapt to changes in the environment. Contrast with complex nonadaptive systems. Complex Adaptive Systems are generally taken to refer to systems that arise out of biology and natural selection, but the term also can apply to the next level tier of systems that arise out of culture, which is itself a complex adaptive system. Complex adaptive systems tend to have a greater number of component parts, actively seek out matter and energy to sustain or increase their own complexity, and in the long term their behaviors are much more difficult to predict." (Spier, 2015 and Christian & Baker, n.d.)

In his book, *Origin Story: A Big History of Everything* (2018, pp. 77–78), David Christian provides further context in distinguishing between complex physical systems and complex adaptive systems which can be used in biology courses to help further clarify the concept of *entropy* or the *2nd Law of Thermodynamics* as it relates to complex systems:

> Life, with its never-ending attempts to push back against entropy, represents a new type of complexity. Complexity theorists sometimes describe entities at this level as *complex adaptive systems*. Unlike the complex physical systems. . .the components of which behave in ways that can usually be predicted from the universe's basic operating rules, the components in complex adaptive systems seem to have a will of their own. They appear to follow additional rules that are harder to detect. Indeed, complex adaptive systems, such as bacteria, your dog, or multinational companies, function as if every component is constantly adjusting to the behavior of many other components. And yields extremely complex and unpredictable behaviors.

Regardless of the context in which it is used, there is an implicit understanding that "complex systems" encompass both the entities, or *ingredients* it comprises and the operational rules or *specific arrangements*, which govern how these entities interact (Momsen et al., 2022).

11.5 Pedagogical Rationale

The incorporation and integration of Complexity Systems Thinking (CST) into undergraduate biology education align with constructivist theories of learning (Brooks & Brooks, 1999; Bruner, 1961; Dewey, 1938; Jonassen, 1999; Papert, 1993; Piaget, 1970; Vygotsky, 1978), which emphasize active engagement, inquiry-based exploration, and the construction of knowledge through authentic experiences. By exposing students to the complexities of biological systems, educators foster critical thinking skills, metacognitive awareness, and a deeper appreciation for the nuances of biological phenomena, including *emergent properties*. Moreover, complexity-based learning prepares students for the interdisciplinary nature of modern scientific research and innovation, along with careers in fields such as medicine, biotechnology, conservation biology, and ecological restoration, where an understanding of complexity and systems thinking is essential.

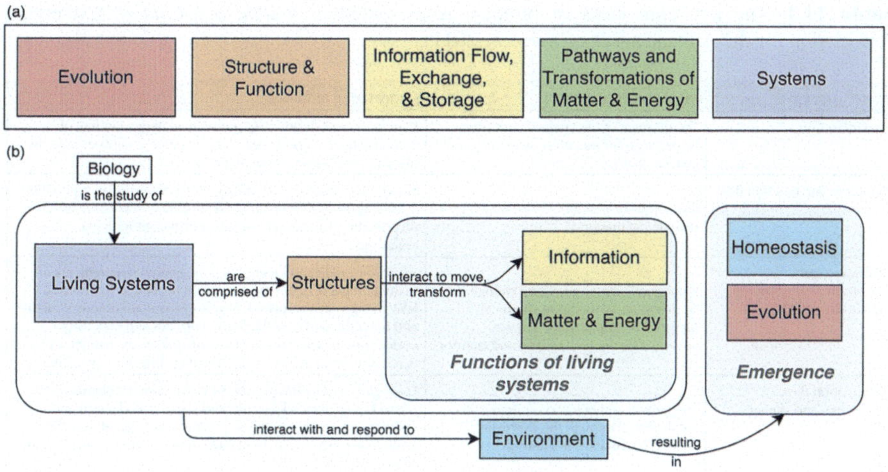

Fig. 11.1 Five core ideas in biology (Momsen et al., 2022)

In 2009, the AAAS (American Association for the Advancement of Science) published a report entitled: *Vision and Change in Undergraduate Biology Education: A Call to Action*, a guiding document which outlined recommendations for transforming undergraduate biology education to better prepare students for the challenges of the twenty-first century. Regarding the expectations for learning about systems, the Vision and Change document emphasizes several key points, including *Systems Biology*, which stated that:

> As defined in A New Biology for the 21st Century (NRC, 2009), systems biology seeks a deep quantitative understanding of complex biological processes through an elucidation of the dynamic interactions among components of a system at multiple functional scales. A systems approach to biological phenomena focuses on emergent properties at all levels of organization, from molecules to ecosystems to social systems. Mathematical and computational tools and theories grounded in the physical sciences enable biologists to discover patterns and construct predictive models that inform our understanding of biological processes. Through these models, researchers seek to relate the dynamic interactions of components at one level of biological organization to the functional properties that emerge at higher organizational levels. Systems biology provides rich opportunities for all students to learn about scientific inquiry and, because of the complex nature of the research involved, to practice in a multidisciplinary context. For example, early applications of systems biology to ecosystem processes resulted in useful simulation models.

Furthermore, *Vision and Change* emphasized the "Five Core Ideas in Biology" including (1) Evolution; (2) Structure and Function; (3) Information Flow, Exchange, and Storage; (4) Pathways and Transformation of Matter and Energy; and (5) Systems (see Fig. 11.1 and Table 11.1).

The five core ideas in biology represent the foundational pedagogical and content approach further outlined in this chapter. In addition, undergraduate biology education forms the foundation of a student's academic journey, laying the foundation for

Table 11.1 The five core ideas of biology as presented in *Vision and Change* (American Association for the Advancement of Science, 2011)

Core Concept	Main Description	Importance in Biology
Evolution (EV)	The diversity of life evolved over time by processes of mutation, selection, and genetic change.	Principles of evolution help promote an understanding of natural selection and genetic drift and their contribution to the diversity and history of life on Earth.
Structure and function (SF)	Basic units of structure define the function of all living things.	Structural complexity, together with the information it provides, is built upon combinations of subunits that drive increasingly diverse and dynamic physiological responses in living organisms.
Pathways and transformation of energy and matter (PTEM)	Biological systems grow and change by processes based upon chemical transformation pathways and are governed by the laws of thermodynamics.	An understanding of kinetics and the energy requirements of maintaining a dynamic steady state is needed to understand how living systems operate, how they maintain orderly structure and function, and how the laws of physics and chemistry underlie such processes as metabolic pathways, membrane dynamics, homeostasis, and nutrient cycling in ecosystems.
Information flow, exchange, and storage (IFES)	The growth and behavior of organisms are activated through the expression of genetic information in context.	From gene expression networks to endocrine mechanisms for physiological regulation, and from signal transduction and cellular homeostasis to biogeochemical cycling, all may be understood in terms of the storage, transmission, and utilization of biological information.
Systems (SS)	Living systems are interconnected and interacting.	Mathematical and computational tools and theories grounded in the physical sciences enable biologists to discover patterns and construct predictive models that inform our understanding of biological processes. Through these models, researchers seek to relate the dynamic interactions of components at one level of biological organization to the functional properties that emerge at higher organizational levels.

future specialization and professional growth. However, the traditional approach to teaching biology often focuses on reductionistic and mechanistic concepts without delving deeply into the complexities that characterize living systems. In today's rapidly evolving data-science landscape, there is a growing recognition of the importance of incorporating and integrating increasing complexity into undergraduate biology education. This change in thinking not only enriches students' learning experiences but also equips them with the skills needed to navigate the complexities of modern biology.

It is to note here that students of biology need to understand that ***the*** core principle of biology is that biological systems evolve. The theory of evolution by means of natural selection should be considered part-and-parcel of systems thinking. Furthermore, biology instructors need to highlight the distinction between learning at the individual level of biological organization and evolution at the population level. The bland use of the term, evolution, implies but does not explain the mechanisms behind evolutionary change, and there will be a need for the elucidation of mechanisms involved in both learning and evolution at some point in the system thinking approach to biological education. At the individual level of biological organization, learning refers to the process by which an organism acquires new knowledge or skills through experience or interaction with its environment. This learning can occur through various mechanisms such as classical conditioning, operant conditioning, observation, or cognitive processes. It is an adaptive process that allows individuals to adjust their behavior based on past experiences, which can enhance their survival and reproductive success.

On the other hand, evolution at the population level involves changes in the genetic makeup of a population over successive generations. These changes occur through mechanisms such as natural selection, genetic drift, mutation, and gene flow. Evolutionary processes result in the accumulation of genetic variations within a population, leading to changes in the frequency of traits or alleles over time. These changes may ultimately lead to the emergence of new species or the adaptation of populations to their environments.

While learning at the individual level can influence behavior and traits within a single organism's lifetime, evolution at the population level operates on longer timescales and involves changes in the genetic composition of populations over generations. Learning can contribute to the overall fitness of individuals within a population, influencing their ability to survive and reproduce, which in turn can shape the evolutionary trajectory of the population. However, learning itself is not a direct mechanism of evolution; rather, it is a process that operates within individuals and can influence their interactions with their environment and other organisms.

In 2018, the U.S. National Academies of Sciences, Engineering, and Medicine published a report entitled: *The Integration of the Humanities and Arts with Sciences, Engineering, and Medicine in Higher Education: Branches from the Same Tree*, in which they stated that: "Given that today's challenges and opportunities are at once technical and human, addressing them calls for the full range of human knowledge and creativity. Future professionals and citizens need to see when specialized approaches are valuable and when they are limiting, find synergies at the intersections between diverse fields, create and communicate novel solutions, and empathize with the experiences of others" (NAS, 2018, p. 8). The importance of actively integrating and involving undergraduate biology students in the various subdisciplines of science as a part of their learning experiences cannot be overstated, as it not only enhances their understanding of complex biological concepts but also cultivates critical thinking skills. Biology is inherently interdisciplinary, drawing from fields such as chemistry, physics, mathematics, computer science, and sociology. Integrating increasing complexity into undergraduate biology education provides opportunities for students to explore the intersections between different disciplines. By breaking down disciplinary boundaries, students develop a more holistic understanding of biology and are better equipped to tackle multifaceted scientific challenges (Fig. 11.2).

Finally, evolutionary change is a dynamic process that occurs against the backdrop of time, shaping the diversity of life on Earth where; change (= rate) = some difference/time. Evolutionary change occurs over time through the accumulation of genetic variations within populations. The rate of evolutionary change, or the pace at which genetic frequencies shift, is influenced by several factors including mutation rates, natural selection, genetic drift, gene flow, and the generation time of the organisms involved.

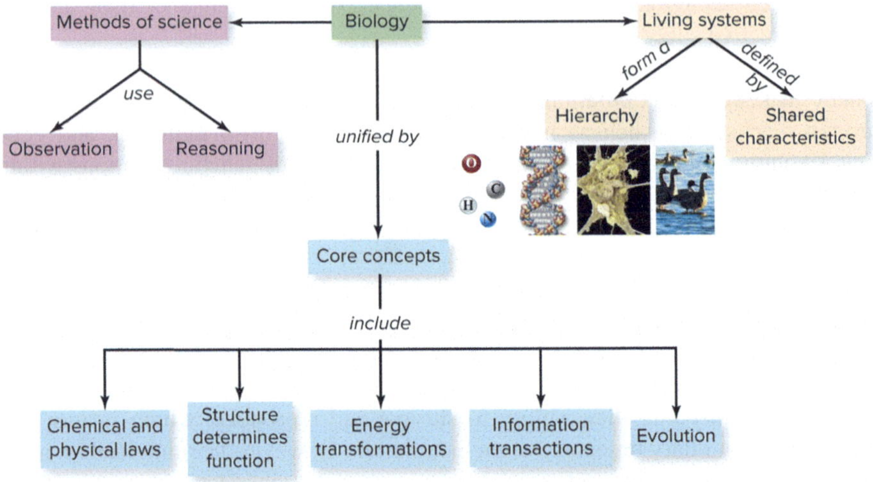

(Nerve cell): SPL/Science Source; (Ducks and geese on pond): George Ostertaga/gefotostock/Alamy Stock Photo

Fig. 11.2 By breaking down disciplinary boundaries, students develop a more holistic understanding of biology

1. **Mutation**: Mutations are the ultimate source of genetic variation. They introduce new alleles into a population's gene pool, leading to the potential emergence of new traits. The rate of mutation varies among species and can be influenced by factors such as environmental conditions and the fidelity of DNA replication.
2. **Natural Selection**: Natural selection acts on the variation present within a population, favoring traits that confer higher fitness in a given environment. Over time, individuals with advantageous traits are more likely to survive and reproduce, passing on their genes to future generations. Natural selection can drive evolutionary change by increasing the frequency of beneficial alleles and reducing the frequency of deleterious ones.
3. **Genetic Drift**: Genetic drift refers to random fluctuations in allele frequencies due to chance events, especially in small populations. It can lead to the loss of genetic diversity over time and can have a significant impact on evolutionary trajectories, particularly in small or isolated populations.
4. **Gene Flow**: Gene flow occurs when individuals or alleles migrate between different populations. It can introduce new genetic variations into a population or homogenize genetic differences between populations, depending on the direction and magnitude of the gene flow. Gene flow can accelerate or slow down the rate of evolutionary change depending on its effects on genetic diversity.
5. **Generation Time**: The generation time of a species influences the rate of evolutionary change by determining how quickly new generations replace older ones. Species with shorter generation times tend to evolve more rapidly because they experience more generations in a given period, allowing for faster accumulation of genetic changes.

Overall, the rate of evolutionary change is determined by the interplay of these factors over time. While some changes may occur relatively quickly, especially in response to strong selective pressures or environmental changes, others may take longer periods to manifest.

The title *Biology for Life* is intentionally layered. At face value, it signals an introductory biology course concerned with the fundamental characteristics and processes that define life. Yet on a deeper, pedagogical level, the phrase "for life" articulates a broader mission: to design and deliver a biology curriculum that cultivates cognitive tools, conceptual habits, and epistemological flexibility that students can apply throughout their lives. These are not merely tools for mastering exams or earning academic credentials but essential frameworks for navigating the increasingly complex, interdependent, and fragile systems that define the twenty-first century.

At the core of this approach is the integration of two powerful frameworks: Big History and Complex Systems Thinking. Big History, as developed by scholars such as David Christian (2011), Spier (2015), and Eric Chaisson (2001, 2014), situates the emergence of life within a cosmological narrative of thresholds—milestones of increasing complexity driven by specific enabling conditions and sustained by flows of energy and information. In this framework, life is not an isolated miracle; it is the logical, though astonishing, outcome of processes that began 13.8 billion years ago and continue to unfold. To study biology, then, is to enter a story already in motion, one that connects atoms to ecosystems, and molecular evolution to human culture.

Biology for Life uses this narrative structure as a pedagogical scaffold to help students locate biology content within a larger story that is both temporally vast and conceptually unified. Each biological topic—whether the structure of a cell, the process of evolution, or the functioning of an ecosystem is treated not in isolation, but in relation to the thresholds that preceded and enabled it. Students trace how energy flows gave rise to life, how information began to self-replicate, and how biological innovation set the stage for cultural and technological evolution. This approach brings coherence to what is often perceived as a fragmented curriculum, helping students to see biology not as a set of disconnected facts but as an emergent, dynamic system.

At the same time, Complex Systems Thinking (CST) provides the analytical toolkit necessary for understanding how living systems function, and malfunction, in the real world. Grounded in concepts such as emergence, self-organization, feedback loops, resilience, and tipping points, CST allows students to model living systems not as static entities, but as adaptive networks of components and relationships. When applied to biology education, Complexity Thinking moves students beyond rote memorization, encouraging them to think in terms of interdependencies, non-linearity, and multiple causal pathways. For example, instead of simply describing the parts of a food web, students learn to anticipate how disruptions in one part of the system (e.g., pollinator collapse or ocean warming) reverberate through others—an insight with immediate ecological, agricultural, and societal implications.

This dual framework, narrative from Big History and analysis from complexity science, lays the groundwork for what we might call a systems literacy for life. In this sense, the "for life" in *Biology for Life* refers not just to the content (i.e., the science of living systems), but to the cultivation of a worldview and a method of reasoning that is urgently needed in today's world. Students who engage deeply with this pedagogy are better prepared to:

- Assess real-world phenomena such as pandemics, climate change, biodiversity loss, and biotechnological innovation not as isolated events, but as outcomes of complex, interacting systems.
- Recognize the structural and historical conditions under which systems evolve, persist, or collapse.
- Make informed decisions, both personally and civically, with a clearer understanding of systems interdependencies, unintended consequences, and ethical trade-offs.

In this sense, *Biology for Life* prepares students not only to understand life, but to live more thoughtfully within it. It empowers them with narrative coherence, systemic insight, and analytical flexibility—three capacities that are not only pedagogically transformative but existentially urgent.

Moreover, this pedagogy aligns with the broader goals of Big History as an educational movement. As David Christian (2018) and Spier (2015) argue, Big History is not just a way to organize knowledge; it is a means of cultivating a planetary consciousness—a sense of connectedness and responsibility that emerges when learners see themselves as part of a 13.8-billion-year story. In a world increasingly shaped by human systems that exceed planetary boundaries, this kind of consciousness is not a luxury; it is a necessity. When combined with Complexity Thinking, Big History becomes more than a narrative—it becomes a toolkit for navigating systemic uncertainty and building more resilient futures.

Thus, for Big History and Complexity scholars, *Biology for Life* offers a compelling extension of both fields into the domain of biology education. It brings their ideas into conversation with foundational science instruction and demonstrates how transdisciplinary thinking can enrich disciplinary content. Most importantly, it models an educational approach that helps students cultivate the intellectual and ethical capacities they will need—not just to pass a course but to thrive in a world of complexity—*for life*.

11.6 Strategies for Pedagogical Implementation

Integrating complex physical systems (CPS) and complex adaptive systems (CAS) perspectives into an undergraduate biology course using the principles of Big History's features of complexity: (1) Diverse Components, (2) Specific Arrangements, (3) Flows of Energy, (4) Goldilocks Conditions, and (5) Emergent Properties requires a strategic pedagogical approach. In their book, *Big History: Between Nothing and Everything* (2014), Christian, Brown, and Benjamin define the *features of complexity* in the following way:

(1) **Diverse Components**: The most basic units, building blocks, or ingredients that comprise the particular form of complexity.
(2) **Specific Arrangements**: The precise structures into which those components, building blocks, or ingredients are arranged or organized, and the connections among them.
(3) **Flows of Energy**: The ways that energy flows through those diverse components in those precise arrangements, including the amounts of energy flowing through them over a set period of time.
(4) **Goldilocks Conditions**: The specific set of circumstances or parameters that are "just right" for the emergence and sustenance of life or other complex phenomena. These conditions include factors such as temperature, chemical composition, energy availability, and environmental stability. The concept of Goldilocks conditions is central to understanding the habitability of planets, the origins of life, and the development of complex systems throughout the history of life and
(5) **Emergent Properties**: The new properties characteristic only of such a form of complexity, with just such flows of energy (Goldilocks levels) through those particular components in those precise arrangements. These emergent properties may lead to new forms (or higher orders) of complexity (Table 11.2).

Table 11.2 The features of complexity

Features of complexity	Description
Diverse components	Biology encompasses a vast array of components, including particles, atoms, molecules, cells, organisms, societies, and galaxies. It explores the interconnectedness and evolution of these components over time.
Specific arrangements	Specific arrangements refer to the unique configurations and interactions of components that give rise to emergent properties and complex systems. Examples include the arrangement of nucleotides in DNA molecules, the organization of cells into tissues and organs, and the formation of ecosystems.
Flows of energy	Energy exchange and transformation play a crucial role in biology, driving the processes of change and evolution across various scales. From the fusion of hydrogen atoms in stars to the metabolic processes within living organisms, energy flow shapes the dynamics of the universe.
Goldilocks conditions	Goldilocks conditions refer to the specific set of circumstances or parameters that are "just right" for the emergence and sustenance of life or other complex phenomena. These conditions include factors such as temperature, chemical composition, energy availability, and environmental stability. The concept of Goldilocks conditions is central to understanding the habitability of planets, the origins of life, and the development of complex systems throughout the history of life.
Emergent properties	Emergent properties are characteristics that arise from the interactions and organization of components within complex systems, often exhibiting behaviors that cannot be predicted from the properties of individual components alone. Examples include consciousness in living organisms, self-organization in ecosystems, and societal structures in human civilizations.

11.7 Pedagogical Tools for Complex Systems Thinking (CST) Implementation

Big History organizes the various major evolutionary transitions of the universe and life on Earth into eight (8) "*Thresholds of Increasing Complexity.*" These thresholds include:

(1) The Big Bang: Origin of the Universe
(2) Stars
(3) Heavier Chemical Elements
(4) Planets
(5) Life
(6) *Homo sapiens*
(7) Agriculture
(8) Modern World/Anthropocene

Organizing the eight (8) thresholds of increasing complexity by first subdividing them into two broad systems including **System 1: Complex Physical Systems (CPS)** and **System 2: Complex Adaptive Systems (CAS)** would assist educators in designing their course. Figure 11.3 represents the organizational model for doing so.

Effective implementation of increasing complexity in undergraduate biology education requires a multifaceted approach. This includes:

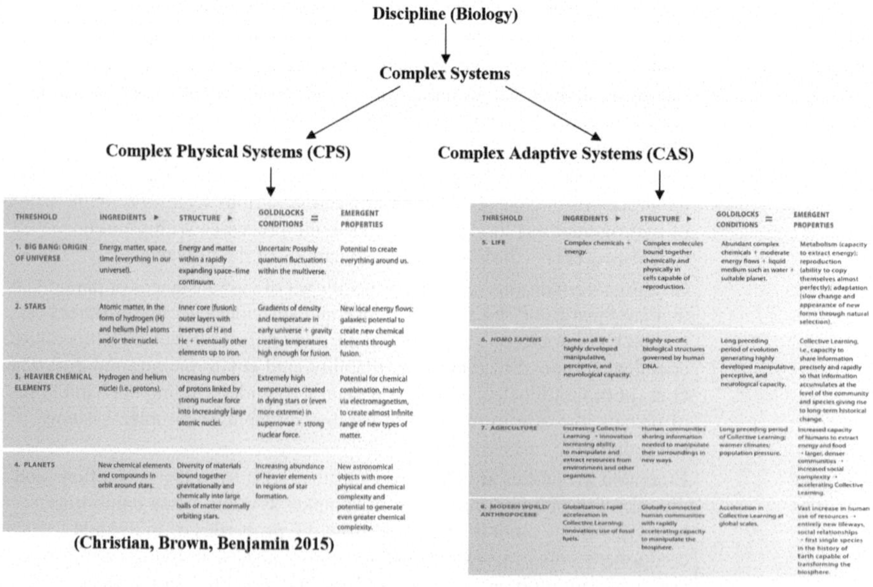

Fig. 11.3 Organizational model for Complex Systems Thinking (CST) Approach

1. **Scaffolded Learning Experiences:** Begin with foundational concepts and gradually introduce complexity through scaffolded learning experiences that build upon students' prior knowledge and skills.
2. **Active Learning Pedagogies:** Engage students in active learning activities, such as problem-based learning, case studies, and collaborative projects, which mirror the complexity of real-world biological phenomena. These active learning experiences empower students to apply theoretical knowledge to practical scenarios, enhancing their understanding and retention.
3. **Interdisciplinary Connections:** Explore interdisciplinary connections between biology and other fields, such as physics, chemistry, mathematics, computer science, and anthropology, to provide students with a holistic understanding of complex systems.
4. **Technology-Enhanced Learning:** Leverage digital tools, such as virtual labs, simulations, modeling software, and online databases, to provide students with immersive learning experiences and facilitate exploration of complex biological concepts.
5. **Collaborative Learning Opportunities**: Foster collaborative learning environments where students collaborate on group projects, participate in peer teaching activities, and engage in interdisciplinary discussions. By working collaboratively, students gain diverse perspectives and develop communication skills essential for navigating complex scientific landscapes.

Integrating complex physical systems (CPS) and complex adaptive systems (CAS) perspectives into an undergraduate biology course, drawing on the features of complexity from Big History, requires a thoughtful pedagogical approach. Here are a few strategies for pedagogical implementation:

1. **Interdisciplinary Approach**: Begin by introducing students to the interdisciplinary nature of biology, highlighting how it intersects with fields such as physics, chemistry, geology, and sociology. Emphasize the interconnectedness of biological systems with other scientific domains, setting the stage for exploring CPS and CAS perspectives.
2. **Contextualize Biological Concepts**: Use examples from Big History to contextualize biological concepts within broader historical and environmental contexts. Show how major events such as the formation of Earth, the evolution of life, and the rise of human civilizations have shaped biological systems and influenced their dynamics.
3. **Case Studies and Real-World Examples**: Incorporate case studies and real-world examples that illustrate both CPS and CAS perspectives in biology. For CPS, explore ecological networks, metabolic pathways, and physiological systems, demonstrating how emergent properties arise from interactions between components. For CAS, examine evolutionary processes, ecosystem dynamics, and population genetics, highlighting adaptability, feedback loops, and self-organization.

4. **Hands-On Activities and Simulations**: Engage students in hands-on activities and simulations that allow them to explore complex biological phenomena firsthand. Use interactive simulations to model ecological dynamics, population genetics, and evolutionary processes, providing students with opportunities to experiment with different variables and observe emergent behaviors.
5. **Collaborative Learning and Group Projects**: Encourage collaborative learning and group projects that promote active engagement and peer-to-peer interaction. Assign students to interdisciplinary teams tasked with exploring complex biological systems from CPS and CAS perspectives, fostering critical thinking, problem-solving, and teamwork skills.
6. **Integration of Technology**: Leverage technology to enhance student learning experiences and facilitate exploration of complex systems in biology. Use data visualization tools, modeling software, and online resources to analyze biological data, simulate dynamic processes, and visualize complex phenomena, allowing students to interact with complex systems in virtual environments.
7. **Reflection and Synthesis**: Provide opportunities for reflection and synthesis, where students can connect their understanding of complex biological systems to broader themes in Big History and complex systems theory. Encourage students to critically evaluate their own learning processes, identify connections between different concepts, and articulate their insights into the complexities of living systems.

By incorporating these strategies, educators can effectively integrate CPS and CAS perspectives into an undergraduate biology course, leveraging the *features of complexity* from Big History to enhance student learning experiences and foster a deeper understanding of the intricacies of biological systems.

1. **Diverse Components**:
   - **Strategy**: Begin by introducing students to the diverse components that make up biological systems, emphasizing the multitude of organisms, cells, molecules, and environmental factors involved.
   - **Pedagogical Approach**: Use multimedia resources, such as videos and interactive presentations, to highlight the diversity of life forms and ecological environments on Earth. Incorporate case studies and examples that highlight the complexity arising from the interactions between diverse biological components.
2. **Specific Arrangements**:
   - **Strategy**: Illustrate how specific arrangements of components within biological systems contribute to their structure, function, and emergent properties.
   - **Pedagogical Approach**: Engage students in hands-on activities, such as dissections or laboratory experiments, to explore the specific arrangements of biological structures and their functional significance. Use visual aids, diagrams, and three-dimensional (3D) models to demonstrate how molecular

structures, cellular organelles, and organismal systems are organized to perform specific functions.

3. **Flows of Energy**:
   - **Strategy**: Highlight the importance of energy flow in driving biological processes and maintaining system dynamics.
   - **Pedagogical Approach**: Integrate discussions on energy flow in biological systems into lectures and activities. Use examples such as photosynthesis, cellular respiration, and food webs to illustrate how energy flows through ecosystems and sustains life processes. Incorporate demonstrations and simulations to visualize energy transfer and conversion within biological systems.
4. **Goldilocks Conditions**:
   - **Strategy**: Discuss the concept of Goldilocks Conditions and how optimal conditions are necessary for the emergence and maintenance of life.
   - **Pedagogical Approach**: Integrate discussions on environmental factors and conditions that are conducive to life, such as temperature, pH, moisture, and nutrient availability. Use examples from astrobiology, ecology, and microbiology to explore extreme environments where life thrives under specific Goldilocks Conditions. Encourage students to consider how changes in environmental factors can impact biological systems.
5. **Emergent Properties**:
   - **Strategy**: Explore how emergent properties arise from the interactions between components within complex biological systems.
   - **Pedagogical Approach**: Provide case studies and examples that demonstrate emergent properties in biological systems, such as self-organization in ecosystems, synchronization in biological rhythms, or collective behaviors in social insect colonies. Encourage students to analyze how interactions at lower levels of organization give rise to emergent phenomena at higher levels.

By incorporating complex systems into an undergraduate biology course, students can develop a deeper appreciation for the intricate dynamics of biological phenomena and gain valuable insights into the interconnectedness of living systems. This approach encourages critical thinking, systems-level analysis, and a holistic understanding of biology.

Table 11.3 incorporates Big History's concepts of Diverse Components, Specific Arrangements, Flows of Energy, Goldilocks Conditions, and Emergent Properties into various biology content: This table provides a framework for organizing different biology topics based on Big History's concepts. It demonstrates how diverse components interact and are arranged in specific ways, how energy flows through biological systems, the emergent properties that arise from these interactions, and the optimal conditions required for biological processes.

Overall, by incorporating these strategies, educators can effectively integrate CPS and CAS perspectives into an undergraduate biology course, leveraging the

Table 11.3 For use with any biology content related which incorporates Big History's (1) Diverse Components, (2) Specific Arrangements, (3) Flows of Energy, (4) Goldilocks Conditions, and (5) Emergent Properties

Biology topic	Diverse components	Specific arrangements	Flows of energy	Goldilocks conditions	Emergent properties
Ecosystem dynamics	Biotic and abiotic components, organisms, interactions	Food webs, trophic levels, nutrient cycling	Solar energy, chemical energy transfer	Optimal temperature, nutrient availability	Stability, resilience, biodiversity
Cellular respiration	Glucose, oxygen, mitochondria	Electron transport chain, ATP synthesis	Chemical energy in bonds, electron carriers	Optimal pH, temperature, oxygen availability	ATP production, waste product elimination
Genetics	DNA, genes, alleles, chromosomes	Gene expression, DNA replication, meiosis	Genetic variation, inheritance	Optimal gene regulation, DNA fidelity	Phenotypic traits, genetic disorders
Photosynthesis	Chloroplasts, pigments, sunlight	Light-dependent and light-independent reactions	Solar energy, chemical energy conversion	Optimal light intensity, carbon dioxide concentration	Oxygen production, glucose synthesis
Evolution	Genetic variation, mutation, selection	Reproduction, adaptation, speciation	Genetic diversity, fitness	Optimal environmental conditions, selective pressures	Species diversity, adaptation

principles of Big History to enhance students' understanding of the complexities of biological systems and their interactions with the environment.

Moving Beyond the Basics. While understanding foundational principles is essential, undergraduate biology education should not stop at the surface level. Embracing increasing complexity means challenging students to explore biological systems at various scales, from molecular interactions to ecosystems. By introducing students to complex topics such as systems biology, evolutionary dynamics, and ecological resilience, educators can provide a more comprehensive understanding of how living organisms function and interact with their environments.

Enhancing Critical Thinking and Problem-Solving Skills. Navigating complex biological systems requires critical thinking and problem-solving skills. By engaging students in hands-on research projects, case studies, and collaborative exercises, educators can cultivate these essential skills. Encountering and grappling with real-world complexities challenge students to think analytically, creatively, and adaptively. Whether it's deciphering intricate genetic networks or devising strategies to mitigate the impacts of climate change, students learn to approach complex problems with confidence and resilience.

Narrative Coherence: Teaching Biology within a Story of Increasing Complexity. The term *Biology for Life* signals more than subject matter—it reflects a deeper pedagogical commitment: to teach biology in a way that connects content to meaningful frameworks students can carry into their lives. At the heart of this approach is the narrative structure provided by Big History, a transdisciplinary synthesis that organizes knowledge through a series of thresholds of increasing complexity (Christian, 2011; Spier, 2015). These thresholds, ranging from the Big Bang to the emergence of life, consciousness, and society, offer a coherent framework through which students can understand life not as an isolated phenomenon, but as part of a dynamic, evolving universe.

This narrative approach serves as a powerful scaffold for teaching biological content. Concepts such as cellular structure, genetic inheritance, photosynthesis, and biodiversity are not introduced as decontextualized units, but as milestones in a much larger evolutionary arc. Big History thus situates biology as one moment in a grander cosmic narrative, and each biological threshold—such as the rise of eukaryotic cells or multicellular life—becomes an opportunity to explore how complexity emerges from simpler components under specific enabling conditions.

For students, this narrative coherence makes learning more memorable, more relevant, and more connected. The Big History framework helps resolve the fragmentation often experienced in introductory biology courses and fosters a broader appreciation for life's place in the universe. Students are not just learning about cells and ecosystems; they are learning how life, in all its complexity, is part of an ongoing story—a story in which they, too, are participants.

Systems Literacy: Cultivating Complexity Thinking for a Complex World. While narrative provides the scaffolding, Complex Systems Thinking offers the tools for analysis. Life, as both subject and experience, is deeply systemic. Biological systems—from organelles to organisms, food webs to biomes—are not linear or deterministic. They are dynamic, interdependent, and shaped by feedback loops, thresholds, and emergent properties. However, many traditional biology curricula emphasize memorization over relational thinking, inadvertently obscuring how living systems function and change over time.

Biology for Life directly addresses this gap by grounding instruction in systems principles derived from complexity science. Concepts such as resilience, energy rate density (Chaisson, 2001), self-organization, and tipping points are not reserved for advanced courses—they are introduced early as foundational ways of seeing and reasoning about biology. Through case-based learning (e.g., coral reef collapse, antibiotic resistance), students practice modeling systems, identifying causal loops, and predicting dynamic outcomes.

This shift enables students to move from what to how and why—from observing biological structures to understanding the behaviors of systems under stress, adaptation, and transformation. The result is a form of systems literacy: the ability to perceive and interpret interdependent relationships, assess systemic vulnerabilities, and envision change across multiple scales. Such literacy is essential not only for scientific competence but also for reasoning within real-world challenges—climate change, pandemics, ecosystem disruption, and more.

Civic Reasoning: Preparing Learners to Engage Ethically and Systemically with the World. The ultimate goal of *Biology for Life* is not simply to produce students who can recall biological facts, but to nurture systems-literate, ethically reflective citizens prepared to act thoughtfully in an increasingly complex world. This is where the phrase "for life" reveals its full significance—not just as subject matter (biology) but as a life skill, a worldview, and a way of thinking that extend beyond the classroom.

By learning biology within a Big History framework and through the lens of complexity, students begin to understand that they themselves are participants in systems that are biological, ecological, and social. They come to see that human decisions affect planetary systems, and that planetary processes in turn shape the conditions for life, including human life. This fosters not only scientific awareness but also civic consciousness—an understanding of agency, responsibility, and ethical complexity.

For example, when students explore CRISPR and genome editing through systems thinking, they are encouraged to ask not only "what can we do?" but "what should we do?" They consider not just immediate effects but long-term systemic consequences, trade-offs, and emergent risks. This form of civic reasoning is central to preparing learners for leadership in a world where biological knowledge intersects with global policy, technological acceleration, and sustainability challenges.

Looking Ahead: Teaching for the Anthropocene. In the Anthropocene, the ability to think systemically, reason ethically, and act collectively may be as essential as any technical skill. Biology education—particularly at the undergraduate and general education levels—has a unique role to play in preparing students for this future. *Biology for Life* demonstrates how a systems-based pedagogy informed by Big History and complexity can rise to that challenge.

This approach does not reject disciplinary rigor. On the contrary, it enhances it—by showing students how biological knowledge fits into a wider network of meaning, method, and consequence. It encourages learners to connect molecular biology to ecosystems, to link evolution with emergence, and to view their own lives as part of a larger web of interdependence that stretches across time and space.

11.8 Conclusion: Overcoming Challenges and Barriers

As we continue to unravel the mysteries of life through the lens of Big History, it is essential to equip undergraduate students with the cognitive tools and analytical skills necessary to navigate the complexities of biological systems. While integrating Complex Systems Thinking (CST) into undergraduate biology education offers immense benefits, educators may encounter challenges such as faculty readiness, curriculum constraints, and resource limitations. Addressing these barriers requires institutional support, faculty development initiatives, and a commitment to innovation in teaching and learning. "All human institutions, professions, programs, and activities must now be judged primarily by the extent to which they inhibit, ignore, or foster a mutually enhancing human-earth relationship."—(Thomas Berry, quoted in O'Sullivan, 1999, p. 43).

Table 11.4 provides a structured summary overview of how CST can be integrated into undergraduate biology education, along with the associated pedagogical rationale, implementation strategies, benefits for students, and potential challenges.

In a world increasingly defined by complexity, uncertainty, and accelerating change, it is no longer sufficient to teach biology as a series of isolated facts or fragmented units. To meet the challenges of the twenty-first century, biology education must evolve—both conceptually and pedagogically. *Biology for Life* represents one such evolution: a systems-based approach grounded in the integrative frameworks of Big History and Complex Systems Thinking, designed not only to teach biology *as life* but also to teach biology *for life*.

This chapter has argued for a pedagogy built upon three mutually reinforcing pillars: narrative coherence, systems literacy, and civic reasoning. Each plays a critical role in reimagining biology education for a generation of learners who must contend with interlocking global crises and opportunities—from climate instability to genetic engineering, from pandemics to planetary ethics.

In closing, *Biology for Life* is more than a curriculum; it is a commitment to teaching biology in a way that is coherent, connected, and consequential. It offers students not only a deeper understanding of life's complexity, but also the tools—and the mindset—to apply that understanding *for life*—for their personal growth, for their communities, and for the continued flourishing of the biosphere we all share.

As our understanding of biology continues to evolve, so too must our approach to undergraduate education in the life sciences. By embracing CST, educators can enrich students' learning experiences, foster systems thinking, promote interdisciplinary learning, enhance critical thinking and problem-solving skills, and prepare

Table 11.4 Overview of challenges and barriers

Aspect	Description
Conceptual framework	Understanding the hierarchical organization of biological systems, from molecules and cells to organisms and ecosystems. Recognizing *emergent properties* that arise from interactions between components.
Pedagogical rationale	Embracing constructivist theories of learning to promote active engagement and inquiry-based exploration. Stimulating critical thinking skills and metacognitive awareness through exposure to progressively complex challenges. Preparing students for careers in scientific research and innovation.
Strategies for implementation	– **Scaffolded Learning Experiences:** Starting with foundational concepts and gradually introducing complexity through scaffolded learning activities. – **Active Learning Pedagogies:** Utilizing problem-based learning, case studies, and collaborative projects to mirror real-world complexity. – **Interdisciplinary Connections:** Exploring connections between biology and related disciplines to provide a holistic understanding. – **Technology-Enhanced Learning:** Leveraging digital tools such as simulations, virtual labs, and online databases to facilitate exploration of complex concepts.
Benefits for students	– **Deep Understanding:** Students gain a deeper understanding of biological systems and their interconnectedness. – **Critical Thinking Skills:** Students develop critical thinking skills by analyzing complex biological phenomena. – **Preparation for Careers:** Students are better prepared for careers in fields such as medicine, biotechnology, and ecology, where an understanding of complex systems is essential.
Challenges and considerations	– **Faculty Readiness:** Faculty may require training and support to effectively implement complexity-based teaching strategies. – **Curriculum Constraints:** Incorporating increasing complexity may require revisions to existing curriculum and instructional materials. – **Resource Limitations:** Access to technology and digital resources may be limited, particularly in under-resourced educational settings.

students for the multifaceted challenges of the future. By equipping the next generation of biologists with the tools and knowledge needed to navigate complex biological systems, we empower them to make meaningful contributions to scientific discovery and innovation. Furthermore, by harnessing CST in the undergraduate biology classroom, we empower students to unravel the mysteries of life, foster a deeper appreciation for the interconnectedness of living organisms, and inspire a lifelong pursuit of scientific inquiry. Together, let us embark on this transformative journey towards enhancing learning in undergraduate biology, one "ingredient" at a time.

Final Reflection Questions:

1. How might the story of life change for students if biology is taught as a series of thresholds of increasing complexity rather than as isolated facts?

2. What elements of Big History's narrative structure could be most effectively integrated into existing biology curricula at the secondary or undergraduate level?
3. How can narrative frameworks like Big History foster greater emotional or existential connection between students and the subject of biology?
4. In what ways does your current biology teaching (or curriculum design) help students think in systems? Where might it fall short?
5. How can feedback loops, emergent properties, or nonlinearity be incorporated into standard biology topics such as genetics, physiology, or ecology?
6. What challenges arise when teaching Complexity Thinking to students unfamiliar with systems concepts—and how might Big History help address them?
7. How can biology education better prepare students to make ethical and systems-informed decisions in the face of real-world challenges like climate change or biotechnology?
8. What does it mean to foster civic reasoning in the context of scientific literacy? How might systems thinking be essential to this process?
9. How can educators avoid reinforcing deterministic or reductionist views of biology and instead emphasize agency, contingency, and responsibility?
10. In what ways could a course like *Biology for Life* transform your own practice—as an educator, scholar, or learner?
11. What would a truly "for life" biology education look like in your context—one that students remember and apply long after the semester ends?
12. If we take seriously the claim that biology should be taught *for life*, what new priorities, partnerships, or pedagogical experiments might that inspire?

References

American Association for the Advancement of Science (AAAS). (2011). *Vision and change in undergraduate biology education*. AAAS Publications.

Brooks, J. G., & Brooks, M. G. (1999). *In search of understanding: The case for constructivist classrooms. Association for Supervision and Curriculum Development.*

Bruner, J. S. (1961). The act of discovery. *Harvard Educational Review, 31*(1), 21–32.

Chaisson, E. (2001). *Cosmic evolution: The rise of complexity in nature*. Harvard University Press.

Chaisson, E. (2014). Practical applications of cosmology to human society. *Natural Science, 6*, 767–796. https://doi.org/10.4236/ns.2014.610077

Christian, D. (2011). *Maps of time: An introduction to big history* (2nd ed.). University of California Press.

Christian, D. & Baker, D. (n.d.). *Analysing Complexity* [MOOC]. Coursera. https://www.coursera.org/learn/analysing-complexity

Christian, D., Brown, C. S., & Benjamin, C. (2014). *Big history: Between nothing and everything*. McGraw Hill.

Dewey, J. (1938). *Experience and education*. Simon and Schuster.

Jonassen, D. H. (1999). Designing constructivist learning environments. In C. M. Reigeluth (Ed.), *Instructional-design theories and models: A new paradigm of instructional theory* (Vol. 2, pp. 215–239). Lawrence Erlbaum Associates.

LePoire, D. J., Grinin, L., & Korotayev, A. (2026). Introduction to complexity theory for Big History. In D. J. LePoire, L. Grinin, & A. Korotayev (Eds.), *Complexity in universal evolution –*

A Big History perspective (pp. 29–90). Springer Nature (in this volume). https://doi.org/10.1007/978-3-032-20516-2_2
Mazzocchi, F. (2008). Complexity in biology. Exceeding the limits of reductionism and determinism using complexity theory. *EMBO Reports, 9*(1), 10–14. https://doi.org/10.1038/sj.embor.7401147
Momsen, J., Bray Speth, E., Wyse, S., & Long, T. (2022). Using systems and systems thinking to unify biology education. *CBE—Life Sciences Education, 21*(2), es3. https://doi.org/10.1187/cbe.21-05-0118
Narguizian, P. J. (2025). A pedagogical proposal for enhancing undergraduate biology education through complex systems thinking (CST). *Journal of Big History, 8*(1), 57–70.
National Research Council. (2018). *The integration of the humanities and arts with sciences, engineering, and medicine in higher education: Branches from the same tree*. National Academies Press.
Orgill, M., York, S., & MacKellar, J. (2019). Introduction to systems thinking for the chemistry education community. *Journal of Chemical Education, 96*(12), 2720–2729. https://doi.org/10.1021/acs.jchemed.9b00169
Papert, S. (1993). *The children's machine: Rethinking school in the age of the computer*. Basic Books.
Piaget, J. (1970). *Science of education and the psychology of the child*. Oxford University Press.
Spier, F. (2015). *Big history and the future of humanity*. Wiley Blackwell.
Stanley, S. M. (2005). *Earth system history* (2nd ed.). Freeman.
Vygotsky, L. S. (1978). *Mind in society: The development of higher psychological processes*. Harvard University Press.

Paul J. Narguizian is a Professor of Biological Sciences and Science Education at California State University, Los Angeles. His research examines how information processing, energy flows, and increasing complexity shape biological systems, with a focus on integrating these concepts into undergraduate biology curricula. He explores both genetic and cultural informational inheritance as drivers of evolutionary and societal change, and develops courses that frame the genetic code as an algorithm of life while drawing parallels to cultural algorithms. His teaching emphasizes building students' mental immunity and digital literacy, enabling them to recognize pseudoscience and navigate disinformation. Inspired by the work of David Christian, Brian Swimme, and Fred Spier. He designs pedagogical strategies that highlight the dynamic interplay of matter, energy, and information across scales, helping students see biology as part of a larger story of emergent complexity.

Biological and Social Phases of Big History and Complexity Growth

12

Leonid Grinin, Alexander Markov, and Andrey Korotayev

Abstract

In the first half of this chapter, Grinin et al. survey general similarities and differences between biological and social macroevolution. They undertake a systematic comparison between biological and social evolution at different levels of analysis and in various aspects, formulating a considerable number of general principles and rules of evolution, and working to develop a common terminology to describe some key processes in biological and social evolution. In the second half, they consider a specific mathematical model capable of describing important features of both biological and social macroevolution. In mathematical models of historical macrodynamics, a hyperbolic pattern of world population growth arises from nonlinear, second-order positive feedback between demographic growth and technological development. Based on diverse paleontological data and an analogy with macrosociological models, Grinin et al. suggest that the hyperbolic character of biodiversity growth can be similarly accounted for by nonlinear, second-order positive feedback between biodiversity growth and the complexity of community structure. They further discuss how such positive feedback mechanisms can be modeled mathematically.

This chapter is an extensively re-worked version of the texts originally published in *Evolution* (Grinin, Markov et al., 2012) and *Journal of Big History* (Grinin et al., 2025).

L. Grinin · A. Korotayev (✉)
HSE University, Moscow, Russia

Eurasian Center for Big History and System Forecasting, Institute of Oriental Studies of the Russian Academy of Sciences, Moscow, Russia

A. Markov
Haifa, Israel

D. J. LePoire et al. (eds.), *Complexity in Universal Evolution*, World-Systems Evolution and Global Futures, https://doi.org/10.1007/978-3-032-20517-9_12

Keywords

Big History · Complexity · Universal evolution · Social evolution · Biological evolution · Mathematical model · Biodiversity · Population growth · Positive feedback · Hyperbolic growth

12.1 Introduction

This chapter represents an attempt to move further in our research on the similarities and differences between social and biological evolution (see Grinin et al., 2009, 2011, 2012, 2013, 2020; Korotayev & Markov, 2014, 2015). We have endeavored to make a systematic comparison between biological and social evolution at different levels of analysis and in various aspects. We have formulated a considerable number of general principles and rules of evolution, and worked to develop a common terminology to describe some key processes in biological and social evolution. In particular, we have introduced the notion of "social aromorphosis" to describe the process of widely diffused social innovation that enhances the complexity, adaptability, integrity, and interconnectedness of a society or social system (Grinin et al., 2009; Grinin & Korotayev, 2023). This work has convinced us that it might be possible to find mathematical models that can describe important features of both biological and social macroevolution. In the first half of this chapter, we survey general similarities and differences between the two types of macroevolution. In the second half, we consider a specific mathematical model that we deem capable of describing important features of both biological and social macroevolution.

In the recent decades, the research of the Universe as a whole has made a significant progress. This is especially clearly manifested in the development of the Big History.[1] Big History provides a unique opportunity to consider the development of the Universe as a single process. Big History approach regards the process of evolution as a continuous and unified process from the origins of the Universe to the present state of humankind and further to the future. Cosmic, geological/biological, and social phases of Universal evolution have a genetic and structural continuity.

The importance of this approach is doubtless. Within Big History researchers distinguish some main evolutionary laws and principles (concerning energy and complexity, for example). However, it is very important to recognize that there are

[1] Christian et al. (2013), Grinin et al. (2014), Spier (2015), Rodrigue et al. (2015, 2016, 2017), Christian (2017, 2019), Henriques et al. (2019), Voros (2019), Panov et al. (2020), Henriques and Volk (2023), Rodrigue (2023a, 2023b), LePoire et al. (2025a, 2025b, 2025c).

many more such integrating principles, laws, mechanisms, and patterns of evolution at all its levels than it is usually supposed.[2]

In our chapter, we will demonstrate that one can find common traits in development, functioning, and interaction of apparently rather different processes and phenomena of Big History if we regard this in connection with the macroevolutionary paradigm (see Grinin, Carneiro et al., 2011).

On the one hand, Big History is very similar in content to a coherent description of the universal macroevolutionary process, because it shows the formation of new levels of complexity in the process of historical development of the Universe (see LePoure et al., 2025a, 2025b, 2025c). At the same time, the general features of the development, operation, and interaction are found in many seemingly dissimilar processes and phenomena, showing a particular aspect of universal evolution as real similarities that occur in a variety of manifestations at all levels and in all lines. Therefore, the combining of the potential of the Big History with evolutionary approaches can open new horizons in this direction.

We plan to analyze the biological and social phases of the Big History and global evolution to give ideas about the driving forces and patterns of transition from one phase to another. We will discuss common, that is, cross phasing, evolutionary and Big History rules, principles, patterns, and laws; at the same time, we will highlight the peculiarities of realization of those rules at each phase.

We believe that it is possible to identify a considerable number of such similarities and to group them into large blocks. In particular, we will consider such features as:

- Ability for self-preservation and self-organization
- Law of life cycle of various objects
- Rule of evolutionary block assemblage
- Unevenness and catastrophes
- Typical and unique objects
- Recombination or circulation of matter in nature

Comparison between different phases of Big History is especially important, because such an analysis suggests new promising possibilities to deepen our understanding of the course, trends, mechanisms, and peculiarities of different phases of Big History.

Each subsequent phase of Big History is accompanied by the emergence of new evolutionary mechanisms. For example, certain prerequisites and respective preadaptations can be normally detected within the previous phase; emergence of new evolutionary mechanisms does not imply the invalidation of those active during the previous phases.

One can observe the emergence of a complex system of interaction between various forces and mechanisms determining the evolution of new forms. For example, biological organisms operate in the framework of certain physical, chemical, and

[2]Grinin et al. (2009, 2011, 2012), Grinin (2014, 2017, 2018, 2020, 2023), Grinin and Grinin (2019a, 2020a, 2023), Korotayev (2018, 2020a, 2020b, 2024), LePoire (2020), Panov et al. (2020), Korotayev and LePoire (2025), LePoire et al. (2025a, 2025b, 2025c).

geological laws; behavior of social systems and people has certain biological limitations and so on. Some models (similar in principle) may emerge not only in the breakthrough points, changing the main line of evolution, but also in some directions that may be regarded as dead-ends (from the Big History leading line perspective). For example, the emergence of social life forms took place in many phyla and classes—from bacteria to insects, birds, and mammals. Additionally, among insects, we can find rather high forms of social organization. See Fig. 12.1 for the place of this lateral phase as a side line in the general scheme of Big History.

In this chapter, we primarily analyze similarities and differences between social and biological evolution. This is moreover important because the comparison of biological and social evolution is a regrettably understudied subject.

We are still at the stage of a vigorous discussion about the applicability of Darwinian evolutionary theory to social/cultural evolution. Regrettably, we all are mostly dealing with a polarization of views. On the one hand, we confront a total rejection of Darwin's theory of social evolution. On the other, we deal with those who stress that cultural evolution demonstrates all the key Darwinian evolutionary characteristics.

We believe that, instead of following the outdated objectivist principle of "either—or," we should concentrate on the search for methods that could allow us to apply achievements of biological evolutionary science to social evolution and *vice versa*. In other words, we should concentrate on the search for productive generalizations and analogies for the analysis of evolutionary mechanisms. The Big History approach aims for inclusion of all megaevolution within a single paradigm.[3] Hence, this approach provides an effective means to address the above-mentioned task.

12.2 On Evolutionary Mechanisms

As is known, not only systems evolve, but mechanisms of evolution evolve too. This concept also appears rather fruitful as regards the development of Big History itself. Let us consider some of the parameters and examples that we might consider.

The emergence of each new phase of Big History is accompanied by the emergence of new evolutionary mechanisms; therefore, certain prerequisites and preadaptations can be detected within the previous phase. So, development of new mechanisms of evolution does not imply invalidation of evolutionary mechanisms that were active during previous phases (cf. LePoire et al., 2025b, 2025c). As a result, one can observe the emergence of a complex system of interaction of forces and mechanisms determining the evolution of new forms. Biological organisms operate in the framework of certain physical, chemical, and geological laws (see Kutter, 2015 on this topic and also on the comparison between physical and biological evolution).

[3] This paradigm is discussed in Grinin, Carneiro et al., 2011.

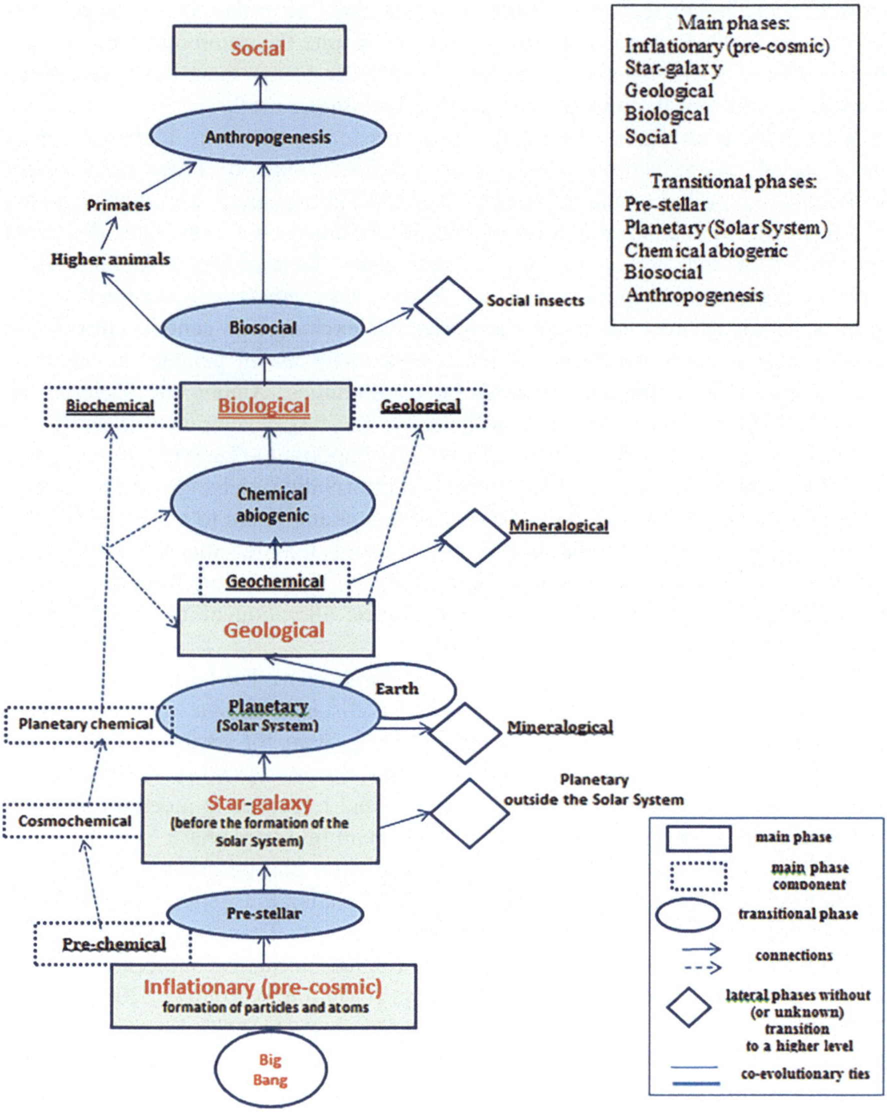

Fig. 12.1 Main and transitional phases of Big History

Likewise, the behaviors of social systems and people have certain biological limitations. New forms of evolution that determine Big History transition into a new phase may result from activities going in different directions. Some evolutionary forms that are similar in principle may emerge not only at a breakthrough point but may also result in a dead-end—from the overall view of Big History. For example, the emergence of social forms of life took place in many phyla and classes—bacteria, insects, birds, and mammals. Additionally, among insects, we can find rather high forms of socialization (see, *e.g.*, Chittka & Rossi, 2022; Reznikova, 2011; Ryabko & Reznikova, 2009; Robson & Traniello, 2002). Despite the common trajectory and interrelation of social behaviors by these various life forms, there has been a large overall difference in the impact that each has had on the Earth.

What is more, as regards information transmission mechanisms, it appears possible to speak about certain "evolutionary freaks." Some of those mechanisms (in particular, the horizontal exchange of genetic information) were spread rather widely in the biological evolution of simple organisms but were later discarded (or transformed into highly specialized mechanisms, for example, sexual reproduction) among more complex organisms. Today, they are mostly confined to the simplest forms of life. We mean the horizontal exchange of genetic information (genes) among microorganisms, which makes many useful genetic "inventions" literally a sort of "commons" of microbe communities. Among the bacteria, the horizontal transmission of genes contributes to the fast development of antibiotic resistance (*e.g.*, Markov & Naymark, 2009). It is important to keep in mind here that diversity, symbiosis, division of functions, and information exchange are the original properties of earthly life (Markov, 2014, p. 32). It is important to keep in mind that viruses can be considered as basic information agents that transmit genetic information and are thus the basic molecular tool of evolution. Moreover, this tool is beginning to be used by humans for the genetic alteration of many organisms, including ourselves (Grinin A. 2025).

For this chapter, the following turns out to be important: The horizontal exchange of genetic information (in its general function) is distantly similar to those forms of information exchange that became extremely important for social evolution—the direct borrowing of innovations and their introduction into social life. Hence, principles and mechanisms that appear of marginal relevance at a certain phase of Big History may turn out to be extremely important in a later phase.[4]

These parallels suggest that analysis of similarities and differences between the mechanisms of evolution may help us to understand the general principles of megaevolution[5] and Big History in a much fuller way. They may also help us to better understand their driving forces and supra-phase mechanisms (mechanisms that operate in two or more phases of Big History) (Grinin & Korotayev, 2008, 2009a, 2009b; Grinin et al., 2009, 2011, 2013, 2022; Grinin et al., 2020; Korotayev, 2018, 2020b, 2024, 2025a; LePoire et al., 2025a, 2025b, 2025c; Grinin & Grinin, 2025a).

Let us return now to a comparison of biological and social evolution. It is important to describe similarities and differences between these two types of macroevolution—at various levels and in various aspects. This is necessary because such

[4]Note that in the biological macroevolution, the "borrowing" is found mostly at lower levels of the biological evolution, whereas it is found much less frequently at higher levels. The opposite situation is observed in social macroevolution—in general, the older the society, the lower its borrowing rate (incidentally, this accounts to a considerable extent for a low rate of change in the majority of ancient societies).

[5]We denote as *mega*evolution all the process of evolution throughout the whole of Big History, whereas we denote as *macro*evolution the process of evolution during one of its particular phases.

comparisons tend to be deformed by conceptual extremes[6] and tend to be incomplete. These limitations are true even in respect of the above-mentioned paper by Mesoudi et al. (2006), as well as a rather thorough monograph by Christopher Hallpike, *Principles of Social Evolution* (1986). There, Hallpike offers a fairly complete analysis of similarities and differences between social and biological organisms, but does not provide a clear and systematic comparison between social and biological evolution.

12.3 Biological and Social Organisms: A Comparison at Various Levels of Evolution

There are a few important and understandable differences between biological and social macroevolution; nonetheless, it is possible to identify a number of fundamental similarities. One may single out at least three basic sets of shared factors.

- First of all, there are similarities that stem from very complex, nonequilibrium, but stable systems whose principles of function and evolution are described by General Systems Theory, as well as by a number of cybernetic principles and laws.
- Second, we are not dealing with isolated systems but with a complex interaction between organisms and their external environment. As a result, the reaction of systems to external challenges can be described in terms of general principles that express themselves within abiological reality and a social reality.
- Third, it is necessary to mention a direct "genetic" link between the two types of macroevolution and their mutual influence.

It is important to emphasize that similarity between the two types of macroevolution does not imply commonality. Rather significant similarities are frequently accompanied by enormous differences. For example, the genomes of chimpanzees and the humans are 98% similar. However, there are enormous intellectual and social differences between chimpanzees and humans that arise from the apparently "insignificant" variations between the two genomes.[7]

[6]This is typical, for example, for a very interesting and controversial article by Mesoudi, Whiten, and Laland "Towards a Unified Science of Cultural Evolution" (2006), where we clearly deal with an attempt to impose the Darwinian methodology on the study of social evolution (see also Mesoudi, 2021). The importance of the above-mentioned differences (including such fundamental differences as the absence in social evolution of a clear distinction between genotype and phenotype) is downplayed by a statement that those differences are either illusory or unimportant (*Ibid.*: 345). Such an approach also reduces the value of a rather interesting methodology that they propose.

[7]It appears appropriate to mention that the genomes of the humans and the chimpanzees differ by ten major genome reorganizations. Some time ago it turned out to be possible to sequence the genome of the rhesus macaque (a special issue of the *Science* was devoted to this subject; see in particular RMGSAC, 2007). This is the third primate genome that was sequenced (after the human and chimpanzee genomes). Up to that moment, when detecting differences between the genomes of

It appears reasonable to continue the comparison between the two types of macroevolution on the basis of the analysis used by Hallpike (1986, p. 33), who singles out the following **similarities between social and biological organisms**:

1. "The institutions of societies are interrelated in a manner analogous to the organs of the body, and preserve their continuity despite changes of individual membership, just as individual cells are renewed in organs."
2. "There is a specialization of organic functions analogous to the social division of labor."
3. "In both cases self-maintenance and feedback processes occur."
4. "There are adaptive responses to the physical environment."
5. "In organisms we find the transmission of matter, energy, and information analogous to trade, communication, *etc.*, in societies."

According to Hallpike (1986, pp. 33–34), **societies are unlike organisms in the following respects**:

1. "They are not physical entities at all, since their individual members are linked by information bonds, not by those of a purely physical nature."
2. "Societies are not clearly bounded, for example, two societies may be distinct politically, but not culturally or religiously."
3. "Societies do not reproduce, so that cultural transmission from generation to generation is indistinguishable from general processes of self-maintenance."[8]
4. "Societies are capable of metamorphosis to a degree only found in organic phylogeny."
5. "The individual members of a society, unlike cells, are capable of acting with purpose and foresight, and of learning from experience."
6. "Structure and function are far less closely related in societies than in organisms."

Hallpike also comes to the sound conclusion that similarities between social and biological organisms are in general determined by similarities in organization and structure (we would say similarities between different types of systems). As a result, Hallpike believes that one can use certain analogies when institutions can be represented as similar to some organs. In this way, cells may be regarded as similar to individuals; central government similar to the brain and so on. Herbert Spencer

---

the humans and the chimpanzees, specialist could not determine which of those differences emerged in the human evolutionary line, and which appeared in the evolutionary line of the chimpanzees. The reading of the rhesus macaque genome substantially facilitated this task. The comparison with the macaque genome allowed detecting that three of those differences happened in the human evolutionary line, whereas the other seven occurred in the evolutionary line of the chimpanzees (see Markov & Naymark, 2009 for more detail).

[8]However, there are cases when societies create new societies rather similar (with basically the same "memotype") to the "maternal" ones, for example, with the establishment of settler colonies. See the next section for more information on the differences in ways of information transmission.

Table 12.1 Similarities and differences between social and biological organisms, as described by Hallpike (1986)

Similarities	Differences
Social institutions are interrelated in a manner analogous to the organs of the body.	Individual societies do not have clear boundaries. For example, two societies may be distinct politically, but not culturally or religiously.
Despite changes in membership, social institutions maintain continuity, as do biological organs when individual cells are replaced.	Unlike organic cells, the individuals within a society have agency and are capable of learning from experience.
The social division of labor is analogous to the specialization of organic functions.	Social structure and function are far less closely related than in organic structure and function.
Self-maintenance and feedback processes characterize both kinds of system.	Societies do not reproduce. Cultural transmission between generations cannot be distinguished from the processes of system maintenance.
Adaptive responses to the physical environment characterize both kinds of system.	Societies are more mutable than organisms, displaying a capacity for metamorphosis only seen in organic phylogeny.
The trade, communication, and other transmission processes that characterize social systems are analogous to the processes that transmit matter, energy, and information in biological organisms.	Societies are not physical entities, rather their individual members are linked by information bonds.

(1898) and Émile Durkheim (1893/1991) are important representatives of this tradition (see also Heylighen, 2011). Hallpike also has sufficient grounds to add Alfred Radcliffe-Brown and Talcott Parsons (Table 12.1).

When comparing biological species and societies, Hallpike (1986, p. 34) singles out the following similarities:

1. "Species, like societies, do not reproduce."
2. "Both have phylogenies and metamorphosis."
3. "Both are composed of competing individuals."

He also singles out the following difference: "*Unlike* species, however, societies are organized systems, whereas species are simply collections of individual organisms" (Hallpike, 1986, p. 34).

Further, Hallpike tries to demonstrate that, because of such differences between biological and social organisms,[9] the very idea of natural selection does not appear to be very productive with respect to social evolution. We believe that his proofs are not

[9]Because the systems of transmission of traits within biological and social systems are rather different, because of the higher degree of complexity of social systems, and so on.

very convincing, although they make some sense in certain respects. In addition, his analysis is concentrated mostly at the level of an individual organism and an individual society. He hardly moves at the supra-organism level (though he, of course, discusses the evolution of species). We believe that with this, Hallpike (notwithstanding his desire to demonstrate the sterility of the application of Darwinian theory to social evolution) involuntarily amplifies the effect of similarity between biological and social evolution, because the analogy between the biological organism and society (as Hallpike admits himself) is rather salient indeed.

On the other hand, Hallpike does not take into account the point in social evolution where a few substantially new supra-socium levels of development emerge. We contend that it is very important to consider not only evolution at the level of a society but also at the level above individual societies, as well as the point at which both levels are interconnected. The supra-organism level is very important, as regards biological evolution (but, perhaps, less so with respect to social evolution). Thus, it might be more productive to compare societies with ecosystems rather than with organisms or species. However, this would demand the development of special methods, as in this case it would be necessary to consider the society not as a social organism, but as a part of a wider system, which includes the natural and social environment.

We identify the following differences between the social and biological evolution.

A. At the Level of an Individual Society and an Individual Biological Organism

1. As Hallpike (1986, p. 33) notes, societies are capable of such rapid evolutionary metamorphoses that they were not observed in the pre-human organic world. However, social systems are not only capable to change and transform but also to borrow innovations and new elements.
2. They may be also transformed consciously and with a certain purpose. Such characteristics are almost absent in natural biological evolution in any form.[10] In the process of social evolution, the same social organism may experience radical transformation more than once.
3. Key information transmission differs significantly in biological and social evolution (we shall consider this point in more detail in the next section).
4. In biological evolution, the acquired characteristics are not inherited; thus, they do not influence the biological evolution that proceeds very slowly. This point will be also considered in more detail in the next section.

[10]Though it cannot be excluded that this is still possible in primates who can borrow some patterns. At least we can talk about borrowing technologies during the period of anthropogenesis. And since some properties do not appear suddenly, but are the result of pre-adaptations, we can assume that to some extent the property of borrowing and transferring to new generations by learning could have existed before the hominids. It has also been suggested that "just as a biological species is characterized by its genetic information (genotype) and the biological traits of the corresponding organism (phenotype), a 'species in a broader sense' is characterized by a certain general information and the traits of the resulting form" (Glötzl, 2025, p. 268).

5. It appears very important to note that, though biological and social organisms are significantly (actually "systemically") similar, they are radically different in their capabilities to evolve. The biological organism does not evolve by itself; evolution may only take place at a higher level (population, species, *etc.*), whereas social evolution can often well be traced at the level of an individual social organism. What is more, it is frequently possible to trace the evolution of particular institutions and subsystems within a social organism.

B. As Regards the Results of Social/Natural Selection

1. Biological evolution is more additive (cumulative) than substitutive; put in another way: "the new is added to the old." In contrast, social evolution (especially during the two recent centuries) is more substitutive than additive: "the new replaces the old" (Grinin et al., 2011, 2022). However, it should be borne in mind that 97% of all species that existed on Earth no longer exist today, although they still had a certain influence on further evolution in the form of their contribution to the genome of currently existing species.
2. Since social evolution is different from biological evolution, in respect of mechanisms of emergence, fixation, and diffusion of evolutionary breakthroughs (aromorphoses), this leads to long-term restructuring in size and complexity of social organisms. It is important to note that, in contrast to biological evolution, where some growth of complexity is also observed, such social reorganization becomes continuous. In recent decades, societies that do not experience a constant and significant evolution look inadequate and risk extinction. In addition, size of societies (and systems of societies) tends to grow constantly through more and more tightly integrative links (this trend has become especially salient in recent millennia), whereas the trend toward increase in the size of biological organisms in nature is rather limited and far from general.
3. Within social evolution, we observe the formation of special suprasocietal systems that also tend to grow in size. This can be regarded as one of the results of social evolution and serves as a method of aromorphosis fixation and diffusion.

C. At Supra-organic (Suprasocietal) Level

As a result of the above-mentioned differences, within the process of social evolution, we observe the formation of two types of special suprasocietal systems: (A) amalgamations of societies with varieties of complexity that have analogies to biological evolution; (B) emergence of elements and systems that do not belong to any society, in particular that lack many analogies to biological evolution.

Naturally, type B needs a special comment. The first type of supra-organic amalgamation is rather typical not only for social but also for biological evolution.[11]

[11] There is, however, a major difference: any large enough society usually consists of a whole hierarchy of social systems (*e.g.*, with respect to a typical agrarian empire these would be: nuclear family—extended family—clan community—village community—primary district—secondary

However, within biological evolution, amalgamations of organisms with more than one level of organization[12] are usually very unstable and are especially unstable among highly organized animals. Of course, analogs do exist within the communities of some social animals (*e.g.*, social insects, primates). Neither should we forget that scale is important: while we might compare a society with an individual biological organism, we must also consider groups of organisms bound by cooperative relationships (see, *e.g.*, Boyd & Richerson, 1996; Reeve & Hölldobler, 2007). Such groups are quite common among bacteria and even among viruses. These caveats aside, it remains the case that within social evolution, one observes.[13] Within social evolution, we observe the emergence of more and more levels: from groups of small sociums to humankind as a whole. Of course, it makes sense to recollect analogies with social animals: social insects, primates, and so on. Neither should we forget to compare society with the individual biological organism nor with groups of organisms bound by cooperative relationships. Such groups are widely present among bacteria and even among viruses (see Grinin A. 2025; see also Solé & Santiago, 2019; Wu, 2020).

It should be noted that modern biologists have developed well respected theories that account for the emergence of intragroup cooperation and altruism, including competition, kin selection, group selection, and so on (see, *e.g.*, Reeve & Hölldobler, 2007; Muzammal et al., 2025). However, it is not clear if societies should be really compared with groups of organisms rather than individual organisms, whether we should not consider societies within the system of numerous intersocietal links?

In any case, it is clear that the level of analysis is very important for the comparison of biological and social evolution. Which systems should be compared? Such analogies are more frequent when society (the social organism) is compared to a biological organism or species. However, in many cases, it may turn out to be more productive to compare societies with other levels of the biota's system organization: with populations, ecosystems, and communities; with particular structural elements or blocks of communities (*e.g.*, with particular fragments of trophic networks or with particular symbiotic complexes); with colonies (with respect to colonial organisms); or finally—and this is the closest analogy—with groups of highly organized animals (cetaceans, primates, and other social mammals or termites, ants, bees, and other social insects).

Thus, here we are confronting a rather complex and hardly studied methodological problem: which levels of biological and social processes are most congruent? What are the levels whose comparison could produce the most interesting results? In

district—province), so that it can hardly be compared with a single biological organism (though both systems can still be compared functionally, as is correctly noted by Hallpike [1986]).

[12] We could mention various flocks and packs of animals as examples of such amalgamations with one level of organization.

[13] More complex superorganic amalgamations may be found in the biological evolution among less complex organisms. This trend seems to be opposite to what is observed in the social evolution, though, say, village communities in more complex societies tend to be less complex than in more simple ones (see, *e.g.*, Korotayev, 1995, 2003, pp. 75–90; 2022; Korotayev et al., 2000, 2011).

general, it seems clear that such an approach should not be a mechanical equation of "social organism = biological organism" at all times and in every situation. The comparisons should be operational and instrumental. That means that we should choose the scale and level of social and biological phenomena, forms, and processes that are adequate for their respective tasks.

We would say again that sometimes it is more appropriate to compare an individual biological organism with a society, whereas in other cases it could well be more appropriate to compare a society with a community (of, say, ants or bees), a colony, a population, or a species. We believe that the issue of the "presence" of the social ontogenesis (and its comparison with the biological ontogenesis) should be studied in this framework (see Grinin et al., 2022: Chap 6 for more details). However, there are some cases when it seems more appropriate to compare social ontogenesis with biological phylogenesis. Hence, different scales and types of scientific problems need special approaches. This subject will be discussed further in the subsequent section of this chapter.

12.4 Similarities and Differences at the Level of Evolutionary Mechanisms

12.4.1 Biological and Social Aromorphoses

In certain respects, it appears reasonable to consider biological and social macroevolution as a single macroevolutionary process. This implies the necessity to comprehend the general laws and regularities that describe this process, though their manifestations may display significant variations, depending on properties of a specific evolving entity (biological or social). We believe that many similarities and differences in laws and driving forces in the biological and social phases of Big History can be comprehended more effectively if we apply the concepts of biological and social aromorphosis.

The application of the notion of biological and social aromorphosis has helped us to detect a number of regularities and rules that are common for biological and social evolution—"payment for the arogenic progress," "special conditions for the aromorphosis emergence," and so on. Such rules and regularities are similar for both biological and social phases of Big History. However, we shall not analyze them in this chapter.

12.4.2 On the Peculiarities of Key Information Transmission at Various Phases of Big History

Replication on the basis of the matrix principle is a fundamental feature of all forms of life (see, *e.g.*, Timofeev-Ressovsky et al., 1969, pp. 15–16). However, the process of such replication cannot be conducted with a 100% accuracy; hence, the replication of a complete genome without any errors is virtually impossible. That is why the emergence of practically any new biological organism is accompanied by random change in genes (*i.e.*, mutations). However, a significant change of the genotype occurs extremely rarely. Yet, the role of mutations in biological evolution is extremely important and very well known, because the mutations are one of the main sources providing "raw materials" for evolution (Timofeev-Ressovsky et al., 1969, p. 72). However, there is also an opinion that the importance of mutations for evolution has been exaggerated, whereas the main source of new genetic material for major morphobiological reorganizations was provided by the gene duplication (see, *e.g.*, Shatalkin, 2005, p. 30). However, we still prefer to keep to the classical point of view on the role of mutation in the process of biological evolution.

However, it is important to emphasize that the number of distortions by which transmission of information is accompanied from generation to generation within social evolution (especially in complex societies) is orders of magnitude higher than that observed within biological evolution. There are grounds to maintain that the role of such "distortions" in social macroevolution tends to increase (in addition to conscious and purposeful alteration of cultural information). In the meantime, it appears that we observe just the opposite within biological macroevolution. For example, among viruses and some bacteria, mutational variability is constantly necessary for their mere survival; on the other hand, in complex biological organisms, it is necessary only up to a very limited extent.

Within social evolution, some unconscious distortions of transmitted cultural information always take place, which may be regarded to some extent as analogous to biological mutations.[14] This, by itself, may lead to certain socioevolutionary shifts (Korotayev, 2003; Grinin & Korotayev, 2007, 2009b; Grinin & Grinin, 2020b, 2023, 2025b; Kradin, 2023). However, the conscious directed alteration of the information by its carriers is significantly more important. Though many are still sure that "history never teaches anything to anybody," already the elites of many complex agrarian societies quite often tried to take into account errors made by their predecessors and to modify the "socio-cultural genotype" accordingly in order to avoid them in future.

One may recollect, for example, the conscious alteration of the social position of the military elite by the founders of the Sung dynasty in China (960–1279 CE), in order to prevent the military coups that jeopardized the political stability of their predecessors (Wright, 2001). Similarly, there was the conscious and purposeful replacement of traditional military systems with the modernized military systems of Western Europe by Peter the Great in Russia and Muhammad Ali in Egypt (see, *e.g.*, Grinin, 2006a; Grinin & Korotayev, 2009c, 2009d), and so on. Modernization is a very broad concept in relation to processes that have found their place in the history of many dozens of countries affecting very significantly their evolution, as well as the global social evolution in general.[15]

---

[14] Close results are arrived at by Dawkins (1993) in his theory of the "evolution of memes."

[15] See, for example, Grinin (2012a, 2012b, 2013b, 2022), Grinin et al. (2024), Grinin and Grinin (2025b), Khokhlova and Korotayev (2023), Korotayev, Bilyuga et al. (2018), Korotayev, Shadrova

Thus, the major part of fixed sociocultural alterations (supported by social selection) emerge not as a result of "random errors of copying" (though, of course, such random errors do exist), but as a result of purposeful alteration of respective memes. Such "mutations" are directional from the very beginning and do not seem to have any analogs in natural biological evolution.

However, in a broad universal evolutionary sense, attempts to represent biological and social evolution in the general series of information development as the most characteristic through line of evolution are not without interest. Evolution is characterized by the fact that new types of information have developed in leaps and bounds, each with new storage technologies, new duplication technologies, and new processing technologies. This unified concept of evolution makes it possible, among other things, to achieve a unified view of biological and cultural evolution (Glötzl, 2025). In this respect, it makes sense to recollect that also viruses can be considered as basic information agents (Grinin A. 2025; see above). Also, we can mention one more idea in megaevolution line that the mindedness of animals is akin to the livingness of organisms and the cumulative cultural processes in human persons (Volk & Henriques, 2025).

12.4.3 On the Inheritance of Acquired Characteristics

The other (and perhaps even more important) difference is that, in the process of biological (but not social) evolution, the acquired characteristics are not inherited.[16] That is why socioevolutionary changes are accumulated much faster than biologically useful changes of phenotype determined by mutation processes.

Thus, because the acquired characteristics do not influence biological evolution, biological evolutionary processes go extremely slowly (in comparison with social evolution). On the other hand, within social evolution, the acquired characteristics can be inherited, and, hence, social evolution goes "according to Lamarck" rather than "according to Darwin." This point has been noted many times by a number of evolutionists (see, *e.g.*, Mesoudi et al., 2006, pp. 345–346). Consequently, social evolution proceeds much faster. In addition, as social evolution tended to go more and more "according to Lamarck," it became more and more Lamarckian rather than Darwinian, which was one of the main factors for the acceleration of social evolution.

Still, it appears necessary to mention that in some rare cases one can observe the inheritance of acquired characteristics in complex biological organisms (Zhivotovsky, 2002a). For example, somatic mutations may well be inherited in

et al. (2024), Korotayev, Slinko et al. (2020), Korotayev, Zinkina et al. (2019), Sawyer et al. (2022), Zinkina et al. (2018).

[16] As one of the differences between social and biological evolution is connected with the absence in the former of clear equivalents of genotype and phenotype (see, *e.g.*, Mesoudi et al., 2006, pp. 344–345), it appears evident that the expressions "sociocultural genotype" and "sociocultural phenotype" should be regarded as metaphors rather than as exact scientific terms.

plants both with vegetative and sexual reproduction. In animals, viruses can insert themselves into the genome of gametes—subsequently the offspring inherit quite an "acquired characteristic," the virus infection. The ability to inherit acquired characteristics is found in many plant-eating insects, in which specialized symbiotic bacteria live. Biochemical and ecological characteristics of such symbiotic complexes are determined up to a very large extent by bacteria (see, *e.g.*, Dunbar et al., 2007).

The possibility of inheritance of acquired characteristics through special particles (pangenes) was proposed by Darwin (1883) himself. Within the genomes of complex biological organisms one can find a very large number of retropseudogenes and even working copies of genes that emerged as a result of the "copying" of genetic information from RNA molecules to the chromosome with special enzymes (such genes are characterized by the absence of introns). Thus, in biological evolution, one may observe the "copying" into the genome of information on the structure of mature matrix RNA. Because the alternative splicing is quite a controlled process, regulated by the cell and subject of intermediate influence of external conditions (see, *e.g.*, Lareau et al., 2007), mature mRNA may actually carry some (albeit rather incomplete and fragmentary) information on "acquired phenotypic characteristics," and this information may be transmitted to the genome of the germ line.

The impossibility of genetic inheritance "according to Lamarck" postulated by the Synthetic Theory of Evolution exists because the mechanism of reverse translation does not appear to have emerged. That is why there is no way for changes that occur in an organism during its lifetime, at the level of proteins, to be recorded back into the genome.[17] On the other hand, at present, we know that the phenotype at the cellular level is determined not only by proteins but also by a great variety of functional RNAs, whereas intravital changes of those molecules may well be written into the genome because here the mechanism of reverse transcription exists and is rather widely spread in biological organisms (including complex organisms). Hence, the point is not that within the biological evolution the "Lamarckian" inheritance is totally impossible; rather the point is that such an inheritance is rather disadvantageous in most cases (see also Steele et al., 2002; Zhivotovsky, 2002b). Consequently, such an inheritance is not usually an important mechanism of evolution (and, especially, of arogenic evolution).

For example, it is evident that the hereditary fixation of adaptive modifications ("modification genocopying") is disadvantageous in many cases. Note that this includes those very consequences of the organ exercise whose inheritance played such an important role in Lamarck's theory. In order for an adaptive modification to appear, we should observe first a genetically determined capability for such a modification (*e.g.*, the muscles' ability to grow as a result of exercise or the lymphocytes' ability to develop immunity against new pathogens). However, if

[17]On the other hand, there is a hypothesis that such a mechanism may have existed at the earliest phases of biological evolution. What is more, scientists have experimentally obtained RNA molecules that can perform certain stages of reverse translation (Nashimoto, 2001).

such a genetically determined ability has appeared, the firm fixation in the genotype (the *genocopying*) of only one of many possible versions of the final state of the trait (*e.g.*, a precise size of a muscle or an immunity toward a specific pathogens) will not be a progressive evolutionary change; it will be a degenerative evolutionary change, accompanied by a decrease in the organic complexity and a loss of one of the ontogenetic regulatory circuits. In biological evolution, such events take place rather frequently, but this is not the arogenic evolutionary pathway.

Within social evolution, there is no significant difference in the inheritance mechanisms between those traits that have been inherited from "ancestral" societies and the ones that have been acquired throughout the history of existence of a given society. There could be some insignificant difference as regards the firmness of the fixation of the respective alterations, the easiness of their acceptance, and so on, but it is impossible to say that acquired social characteristics are transmitted to new generations with significantly more difficulties (especially in complex societies).

A serious obstacle for the operation of the "Lamarckian" mechanism can be seen in traditionalism, which holds negative attitudes toward innovation and glorifies everything inherited from ancestors. This was very typical for simple traditional societies. However, such attitudes have weakened in a significant way in modern complex social systems.[18] This might be connected with the development of the means, methods, and technologies of forecasting, which is the conscious evaluation of innovation. Forecasting makes those characteristics that might be considered dangerous or disadvantageous by traditionalists (in particular, a very low precision of the "memotype" replication and "Lamarckian" inheritance) to become more acceptable in a society.

12.4.4 On the Nature of Hereditary Variation

Hereditary variation is a key issue in the theory of evolution. This is the issue, around which the main discussions between representatives of various schools of evolutionary thought (classical Darwinism, Synthetic Theory of Evolution, Orthogenesis, Nomogenesis, Neo-Lamarckism, and so on) are concentrated. *Variation is the main material basis of evolution; its character, mechanisms, factor, and emergence rates determine to a very high extent the character of the evolutionary process.*

---

[18]On the other hand, we observe another trend in connection with some sorts of regulation mechanisms. One should not think that the only evolutionary mechanism in social evolution is a conscious change of existing objects. There is also an opposite trend that may be denoted as institutionalization. In many cases certain relationships are fixed by customs or laws in order to avoid excessive variation/equivocality that may often be harmful for a social system. For example, one could observe the development of rather rigid marriage institutions, various legal codes and constitutions that can be only altered with significant difficulties (that are usually consciously established by respective norms aimed at the provision of the stability of respective codes and constitutions). In this respect the trend toward the canalization of changes may be also traced in the social macroevolution.

These mechanisms of variation are one of the most fundamental areas of difference between biological and social evolution.[19]

Starting with Darwin, biologists have based their evolutionary theories on the idea that hereditary variation is basically "indeterminate" or undirected, that is, random. However, as we have noted, within biological evolution, one can still detect a trend toward a decrease in randomness, both in mutational and recombinational variation. In some sense, this trend continues into social evolution, where variation is even less random and more directed.[20]

As mentioned above, there are significant differences between biological and social evolution in regard to the accuracy of copying (reproduction of replicators), because in general the precision of copying of genes (and, correspondingly, periods of their existence in a recognizable form) exceeds by orders of magnitude values of analogous indicators for memes. That is why "memetics" (in contrast with genetics) has to deal with a much lower precision of replication and with a much higher speed of mutagenesis, though some replicators (memes) may have rather long periods of life.

For example, according to some recent estimates, roots of some most widely used words may be preserved in a recognizable form for about 10,000 (and even more) years of linguistic evolution (Pagel et al., 2007). Another example can be provided by "long-lived" folklore-mythological motifs that can survive for dozens thousand years (see, e.g., Korotayev & Khaltourina, 2011; Berezkin, 2007; Korotayev, 2006; Korotayev et al., 2006, 2017). The same can be said about a very long life of some technical methods, for example, the production of stone tools. However, it makes sense to distinguish between various types of information transmission, depending on the number of copies in which the information is stored and reproduced (as well as the forms of that reproduction).

There could be situations in which there is just a single carrier of important information. An ancient engineer could take his secrets of construction to the grave

[19] It appears that this is relevant not only for the biological and social phases of Big History but also to all its preceding phases.

[20] When we make such comparisons, we compare genotype with that totality of sociocultural information (it may be denoted as "memotype"), which is transmitted from generation to generation and determines main characteristics of social systems. In social systems, in addition to biological generations, parents and children, we find other types of continuity (that could be sometimes even more important) like institutional and legal continuity whose role increases constantly. That is, we observe the growth of the importance of information transmission in the framework of institutions, corporations, organizations, and so on, that is conducted not between biological generations (from parents to children), but, say, from an experienced worker to an inexperienced one, or from a teacher to a pupil. In addition, the emergence of external information carriers (in form of books, electronic records, and so on) allows conducting a distance transmission of information without any direct contact between respective people, which, incidentally, contributes to the growth of the sociocultural evolution rate. Actually, as a result, in complex social systems the number of information transmission channels grows by orders of magnitude (especially with the emergence of external information carriers). In some sense, this growth already starts with the development of social life among the animals.

so that nobody could continue his techniques any more. There are lots of historical facts known to us from just one source; and if, in the process of transmission, there was distortion of the initial text; this could affect our current knowledge of the past. Those unique ancient books that disappeared in fire did not let us know the important information contained in them, and so on. These are examples of distortion or loss of information by functioning social systems.

It seems appropriate here to recollect the **information irreplaceability principle (Lyapunov principle)**. According to this principle, information that has entirely disappeared cannot be reconstructed in its entirety—what can be replaced are portions of information coming from a common source (see Rautian, 1988a, 1988b). We confront a different case when we deal with information that is used by numerous carriers (as in the case of the use of a mass language). In such cases, changes in a living language should not be always regarded as information distortion; we should rather speak about some drift in the use of linguistic matrices and patterns (similar to gene drift in populations), because language carriers may well know older forms, but prefer new ones. One may even observe the coexistence of persons using differently linguistic forms and lexemes (similarly within one population there could be different phenotypes). However, with time, some forms win the competition and language changes.

When we speak about the accuracy of transmission of biological information, it is necessary to take into account that biological evolution has worked out rather effective molecular mechanisms that allow for sharply reduced precision of DNA replication when necessary (e.g., *SOS-response* among bacteria). For some primitive biological objects, such as viruses, too high a precision of replication can even be lethal; in order to successfully go through their life cycles they need very low precision of replication or, in other words, a very high rate of mutation (mutagenesis), especially when some viruses use RNA as a genome (Solé & Santiago, 2019). For such organisms, evolutionary changes turn out to be necessary components of their everyday life (Vignuzzi et al., 2006; Grinin A., 2025)!

Generally, however, in biological evolution, replication accuracy increases rather than decreases with the growth of the organismal complexity. In this sense, the reduction in precision that is observed in the transition from biological to social phase of the Big History looks as if this was a "step backward." However, this observation is rather superficial, as it does not take into account the nature of those errors that emerge in the process of replication, notably the degree of their randomness/directionality.

Within biological systems, replication errors are basically random. Taking into consideration the decrease in randomness, this may be interpreted in the following way: nature has not developed any biological mechanisms that allow us to forecast results of concrete genetic changes and to plan them. Though a cell (e.g., a lymphocyte) may "know" in advance that, in order to achieve a needed result, it should alter some particular part of the genome; it, however, lacks mechanisms that would allow it to forecast results of a concrete genetic alteration. That is why, in the framework of biological evolution, the acceleration of adaptation-genesis through a radical reduction of the precision of replication is a very expensive and risky strategy that can be afforded only by very primitive forms of life. The situation changes radically if the replication "errors" become not random, but actually purposeful, based on forecast of the possible results of concrete changes introduced into the "memotype" of a social system.

The presence of "directed mutations" (in addition to undirected ones) **radically distinguishes the process of "mutational variation" in the evolution of memes from what is observed within the evolution of genes, where ALL the mutations are basically undirected.**

That is why we believe that the difference between biological and social evolution in respect to randomness/directionality of hereditary variation is more fundamental than the differences in precision of replicator copying or mutation rate. In the process of "sociocultural mutagenesis," the element of randomness is significantly smaller, because people possess the ability (albeit limited) to foresee results of certain concrete "mutations." That is why human creativity (say, in development of new judicial laws or new technologies) may differ qualitatively from the "creativity" of biological evolution—especially, as regards the effectiveness and the speed with which the respective results are achieved.

On the other hand, one should not exaggerate the role of conscious planning in relation to social innovation. Random search, trial, and error remain very important in social evolution (Grinin, 1997, 2006b, 2007, 2011, 2012a, 2013a, 2017, 2020; Korotayev 2003, 2004), although there has been a clear decreasing trend in randomness in recent centuries (see, *e.g.*, Korotayev, 2003, 2004, 2007; Korotayev et al., 2006a, 2006b; Grinin, 1997, 2007, 2009a; Grinin et al., 2024). Thus, it is not sufficient just to have respective challenges in order that serious transformations could take place. Most societies "respond" to new problems in old, habitual, tested and familiar ways, as they choose—not from a set of hypothetical alternatives—but from a set of accessible alternatives (Van Parijs, 1981, p. 51). In other words, they use actually known measures instead of potential ones (Claessen, 1989). In these situations, their behavior is often quite similar to that of social animals. Naturally, not all such "responses" are effective. As a result, many societies perish, disappear, or lose their independence (Grinin, 2011).

For example, after the Roman regiments were withdrawn from Britain in 410 CE, the Britons (Romanized British Celts) sought protection from the raids of their Irish and Scottish neighbors. They invited Saxons to defend them in return for plots of land in Britain. Actually, this was a variation in the very well-known Roman method "to *use barbarians to fight barbarians*." However, the Saxons, after they had seen the Britons military weakness, stopped obeying local authorities and became masters of the country (together with Angles and Jutes). In this way, the Britons, notwithstanding their fierce and long resistance, were partly evicted, partly destroyed, and partly enslaved. As a result, barbarian Anglo-Saxon states were found in place of the state of the Britons (Blair, 1966, pp. 149–168; Chadwick, 1987, p. 71; Philippov, 1990, p. 77).

If we take into account general historical contexts, we see that an extremely small fraction of all responses to various challenges turned out to be capable of becoming sources for system aromorphoses. This implies that most societies turned out to be incapable to move to new qualitative levels: they did not have the necessary potential for change, their construction had certain "defects," the system might have been too rigid to transform easily, or some necessary conditions were lacking, and so on (Grinin, 2011; Grinin & Korotayev, 2009e).

12.4.5 The Ability to Borrow and the Horizontal Exchange of Genetic Information

These facts illustrate a rather strange situation. There are similarities in biological and social evolution, such as the transmission of information, variability, community complexity, *etc.* However, these similarities occur at the lower stages of biological evolution (involving simple biological organisms), whereas they are absent in higher stages of biological evolution (involving complex biological organisms). It is important to take into consideration that far from every organism is independent in propagation, since sexual reproduction intrinsically needs another organism for genetic exchange. We will see below that there is a reason to believe that the cell nucleus itself is of viral origin. Since the nucleus plays a central role in sexual reproduction in eukaryotes, this suggests that reproduction itself was most likely not invented by eukaryotes, but rather "borrowed" from viruses. (Grinin A., 2025, p. 52).[21]

One of the main differences between social and biological evolution is the ability of social systems to not just change and transform, but also to borrow new elements. However, in this respect, social evolution resembles the biological processes that prevailed during the epoch of the "prokaryotic biosphere" (and those processes continue up to the present among prokaryotes and monocellular eukaryotes). Among the prokaryotes, we find the ability to "transform naturally"—to absorb DNA from the environment and to insert it into their genome, which leads to an immediate transformation of the phenotype. There is also, of course, a significant difference between this biological and social analog: in society the borrowings are usually made consciously.

Horizontal gene transfer produces many useful genetic "inventions," a sort of commons for microbe communities. For example, communities of marine planktonic microbes use the genes of proteorhodopsins—proteins that allow them to partly utilize sunlight. In contrast to the proteins that participate in real photosynthesis, proteorhodopsins do not need the help of many other specialized proteins. Thus, in order to acquire a useful function, it is sufficient for a microorganism to borrow a single gene (Frigaard et al., 2006).

[21] We do not have a full explanation of this phenomenon, but one may think about the application to the macro- and even megaevolution of **the law of the negation of the negation**, which in this case may be interpreted in the following way: "*From a free borrowing of information to its rigid isolation and canalization, and then again to its free (but now conscious) borrowing.*" "*From contraposition of biological (genetic) and social mechanisms of evolution (within the process of anthropogenesis and sociogenesis) to genetic evolution controlled by the humans.*"

Complex borrowing of entire gene systems is observed much less frequently, but when they occur, they have more significant consequences. An original and widespread version of such "borrowing" results in the emergence of symbiotic systems, which sometimes actually leads to the formation of a new organism out of several other organisms. The role of such systems is often underestimated, but all functioning of the modern biosphere are based on them.

There are many examples. *Terrestrial plants* would not have been able to achieve evolutionary success without symbiosis with mycorrhizal fungi and nitrogen-fixing bacteria. *Herbivorous animals,* both insects and vertebrates, are unable to digest plant food without symbiosis with specialized microorganisms. Indeed, the principal ecological, biospheric role of animals is precisely to process plant food!

In highly complex biological organisms, in contrast to social organisms and human societies, large-scale "borrowings" in the form of symbiotic relations or alien genetic material **rarely take place, but many of the most important aromorphoses are connected just with them.**

12.4.6 Analogs of "Suprasocietal Institutions" in Biological Evolution

Let us come back to the question: Are there analogs of such structures in the evolution of the biosphere? The answer will depend on the level of the biosphere's system organization. Society is frequently compared with biological organisms, but—in this case—we are comparing suprasocietal amalgamations with supra-organic systems: populations, species, ecosystems, groups of social animals, and so on. However, this is probably not quite an appropriate scale of analysis, so we need to compare suprasocietal institutions of a global scale (like the United Nations) with biological objects of immeasurably smaller scale, for example, with particular "casts" of the ant family.[22]

At any scale, it is difficult to find good analogies to the formation of suprasocietal institutions within biological evolution. This becomes even more evident if we compare societies, not with organisms, but with supra-organic biological systems (*e.g.*, populations or species). Although those biological systems (like societies) can amalgamate into systems of a higher order (ecosystems or the biosphere), these higher-order systems are not centralized but are weakly integrated—nothing like suprasocietal institutions as the World Health Organization, UNESCO, or even a complex tribal confederation with its own supra-tribal regulation organs. For example, one can observe the formation of rather complex links between species in ecosystems; certain key species may produce a decisive influence on other species in the community, but this does not result in the formation of any "supra-species institutions."

[22] On the other hand, a large anthill or termitary may well be compared with a large village community.

On the one hand, it is possible to see in this comparison one of the fundamental differences between social and biological macroevolution. On the other hand, some biological analogs of "suprasocietal institutions" did emerge. In the Holocene (the last 10,000 years, starting with the Agrarian Revolution), human societies developed suprasocietal institutions. In the course of the sociobiological evolution of the resulting "anthroposphere," we observe a parallel growth in the integration of humankind and integration and coordination of evolutionary changes of biological populations, species, and ecosystems. In other words, the development of the global human community (the World System) may be regarded as a factor of integration of biological evolution at its upper level.

Thus, social and biological evolutions are related processes that supplement and maintain each other. Indeed, there is a tendency toward their fusion into a single complex process, one leading to the development of an "anthropo-biosphere." In this respect, it appears to be possible to speak about the co-evolution of biological and social development.

12.4.7 On the Role of Selection in Biological and Social Evolution

The role of selection in social evolution differs significantly from the one in biological evolution. In the biological world, the main source of stable, heritable innovations (mutational and recombinational variation) is characterized by a high degree of randomness and unpredictability (although, of course, it is also necessary to take into consideration all the above-mentioned qualifications about the means of optimization). In this situation, "post factum selection," the selection among the deviations that have already emerged and have found their realization in the phenotype, becomes the only way to give the process a certain directionality (in this case—to secure the additive character of changes). (For our views on the role of selection in evolution, see Grinin, 2020; Grinin & Grinin, 2020a; Markov & Naymark, 2019.)

In the social world, the main sources of heritable innovations are not random errors of copying and reproduction but conscious and purposeful correction and alteration of memes. However, such purposefulness is unable to foresee not only all the consequences of its actions but even the near consequences. That is why intentional actions may appear random. Throughout human history, failures of some societies have been a sort of payment for the success of others (what we denote as "a rule of payment for the arogenic progress"), from which the role of selection in the search for successful aromorphic variants acquires an especially important meaning (Grinin et al., 2011, 2022; Grinin et al., 2020; Grinin, 1997, 2007; Grinin & Korotayev, 2009b; Korotayev, 2004). Societies frequently confront such situations when an old system does not work. Those who do not change or look for more effective means perish.

Selection at the gene/meme level plays a less important role in social evolution than it does in biological evolution. However, selection in social evolution takes place not so much at the level of memes but more **at the level of organizations, institutions and social systems**. At the level of intersocietal competition, until recently, social selection acted in an especially tough way: "the victor gets more or everything; the defeated may lose himself" (Grinin, 2003, 2004, 2009a, 2009b, 2010a, 2011, 2020; Grinin & Grinin, 2023; Grinin & Korotayev, 2009e, 2015). So, this is a selection mechanism that is rather different from the one found in biological evolution.

One more important aspect of social selection that is absent in biological evolution is the struggle for the selection of a certain model (model of reforms, model of unification, ideological model) at the level of individual societies, as well as at the intersocietal level. Everywhere, we can observe the selection of leaders, models, courses, central positions, and so on. The decisive advantage could be rather different in different cases. In some cases, this could be a very capable and talented leader (Grinin, 2010b); in others this could be an advantageous geographic position; in still other situations this could be just a lucky contingency.

Thus, although we are dealing with rather different mechanisms of selection in biological and social evolution, their roles are very important in both cases. Still, within biological evolution, selection process is more important, because there is no alternative, whereas such an alternative exists within social evolution.

12.5 Some Preconditions of the Transition from Biological to Social Phase of the Big History

12.5.1 Social Evolution as a Logical Result of the Development of Adaptiogenesis Mechanisms

In addition to what has been already said about the organic links between biological and social evolution, one should consider another aspect of adaptiogenesis. The process of adaptation that constitutes the principal contents of biological evolution may proceed at different levels: (1) the organism structure; (2) its behavior; (3) structure and behavior of a socium as a superorganic amalgamation.

At all those levels, one may observe the transition from primary, primitive and slow methods of adaptiogenesis based on random mutations, recombination and selection to more progressive, effective and rapid ways of evolutionary change. **Not only organisms, species and societies evolve; mechanisms of** evolution evolve too. The general direction of this evolutionary movement may be characterized as a trend to the reduction of the role of random processes and the growth of systematic controlled processes. The initial and primary evolutionary algorithm is the random search, the trial-and-error method. However, at all levels of adaptiogenesis, one may observe a gradual development of such mechanisms that decrease the role of randomness and, thus, optimize this algorithm; though it appears impossible to exclude entirely an element of randomness either from biological or from social evolution.

(1) The organism structure level. Even at the basic level of biochemistry, physiology, and morphology, many forms of life have developed ways of adaptiogenesis that are faster and more effective than the random search conducted according to the scheme of "random mutations + selection." One of these mechanisms is regulation of the mutagenesis rate, depending on available conditions: under favorable conditions, the mutagenesis rate is low; in unfavorable conditions it increases (Grinin et al., 2022: Chap. 6, §6.8).

It is also appropriate to mention epigenetic changes of hereditary material that are transmitted to a number of generations, in particular parental genomic imprinting that became especially developed in the most complex organisms, such as mammals and flowering plants (Jablonka & Lamb, 1999). Imprinting is actually a sort of purposeful manipulation of hereditary properties of offspring. With the maturation of male and female gametes, certain parts of the genome are marked in a special way, for example, through methylation. The methylation of DNA is not a chaotic process but is regulated by complex molecular systems. What is especially important is that methylation of particular nucleotides increases the probability of their mutating. Thus, through the methylation (or nonmethylation) of particular nucleotides, cells can in principle regulate the probability of their mutation (Vanyushin, 2004).

Another example of the purposeful change of hereditary information is provided by the development of adaptive (acquired) immunity through combining genetic blocks, subsequent somatic hypermutation, and clonal selection. Both the combining of DNA fragments (V-(D)-J recombination) and hypermutation are processes that are only partly random. In other words, the limits of randomness in this case are rather accurately demarcated (Grinin et al., 2022: Chap. 4, §4.2.4). The combination of DNA fragments is conducted from a precisely defined set and the hypermutation takes place at a rather accurately demarcated part of a gene, while the selection of lymphocyte clones makes the whole process unequivocally directional. As a result, the final outcome of such a "sequence of random events" turns out to be quite deterministic.

Such a mechanism may be designated as "**optimized** random search."[23] Note that in the case of the acquired immunity, from a "technical" point of view, the achieved result may well be transmitted to the offspring, for example, via the mechanism of reverse transcription and transmission of the genetic material from lymphocytes to gametes through endogenous retroviruses (Steele et al., 2002). However, this does not happen, because it is more advantageous to transmit not a concrete immunity to a particular pathogen to the offspring but a universal capability to develop immunity against any pathogen.

In general, such mechanisms of purposeful genome alteration do not have a universal presence in biological organisms, and the overwhelming majority of mutations takes place in a quite random way.

[23] In this way, a more flexible reaction to unknown situations develops; this may be compared with multifunctional institutions in human societies that while remaining apparently the same institutions may allow social systems to behave differently in different situations, whereas respective institutions would experience certain changes with the change of situations. Thus, army may be relatively small during the time of peace, and then it would grow sharply in size as a result of mobilization, whereas its functions also substantially change. The same can be said about the flexibility of the family, the village community, and many other social groups and institutions.

Biologists rarely consider that assortative (selective) mating, mediated sometimes through extremely complex mechanisms of mate-choice; this is nothing but an extremely effective mechanism for the management of recombinational variation. However, in the real biological world, absolutely unselective, random mating is hardly ever observed. Indeed, random mating is a scientific abstraction, like an "ideal gas," or an "absolutely dark body." With growth in the level of organization of biological organisms, the complexity and effectiveness of mate-choice also grew, whereas the recombinational variation became less random as a result.

(2) Level of individual behavior. One can trace the transition from predominantly hereditary and genetically determined behavioral patterns to more flexible learning-based ones. As we saw above, in the case of immunity, it was more advantageous to transmit to the offspring a universal capability to "learn" instead of a rigidly determined means of resistance to a particular pathogen. In an analogous way, in the general course of evolution, it has turned out to be more advantageous to transmit the ability to learn rather than to transmit rigidly fixed behavioral stereotypes.[24] No doubt, the emergence of the capability to learn is a major aromorphosis, though it is very stretched over time. Even unicellular organisms have some inchoate abilities to learn (sensitization, habituation), let alone such highly organized animals as ants or bees.

(3) Biological socium level (social adaptiogenesis). A wide variety of living organisms—from bacteria to mammals—lead a social way of life. The socium as a whole has certain system characteristics that can be more or less adaptive (Popov, 2006). Here, we also observe the transition from rigidly genetically determined forms of social relationships to more flexible versions, within which a social system may adequately (adaptively) react to changes in its environment. For example, the size of subsidiary colonies of an anthill may change in a reasonable, that is, adaptive way, depending on resource availability (Zakharov, 1978, p. 49). However, in general, for all the pre-human forms of life, such possibilities are limited. The human development of the ability to evolve socially, which implies the possibility of an almost limitless change in the structure of social systems, appears to be a natural (though qualitatively higher) continuation of this evolutionary trend.

12.5.2 One of the "Preadaptations" That Facilitated the Transition from Biological to Social Evolution

The issue of how biological evolution transformed into social evolution is among the most important questions of Big History and Evolutionary Studies. What

[24] It appears necessary to note that in both cases the ability to learn does not replace entirely the genetically determined concrete adaptations; the former is added to the latter. In the immunity system of higher organisms, the system of innate immunity is preserved in addition to a new system of adaptive (acquired throughout the life) immunity; similarly, in the behavior of higher animals, behavioral patterns developed throughout the life through the learning are combined with innate genetically determined behavioral traits.

"preadaptations" were needed for the transition from biological to social phase of the Big History? This is a very complex subject. And here we shall restrict ourselves to consideration of just one of those preconditions.

Social macroevolution (and, hence, the start of the Big History social phase) became possible due to the emergence of a uniquely human ability denoted as "ultra-sociality" (Boyd & Richerson, 1996). This is only found among humans and designates the ability to change their social organization radically and almost limitlessly in response to internal and external challenges. Only humans are capable of forming collectivities that could be entirely different as regards their structure, their traditions, their norms of behavior, their modes of subsistence, their systems of intragroup relationships, their family types, *etc.*

Whatever the complexity of the collectivities of nonhuman primates, they do not have such flexibility (Butovskaya et al., 2020). Each species usually has only one type of social organization; some cultural differences are observed, but they are incomparable with the ones observed in *Homo sapiens*. Yet, some animals possess a limited ability to adaptively change the structure of their socium. For example, in disadvantageous circumstances, one may observe growth in the rigidity of social hierarchy (the "power vertical"), whereas the relationships become more egalitarian under more favorable conditions. Sometimes the transition to a social way of life occurs during unfavorable conditions, whereas the same animals may return to solitary life with improvement of conditions (Popov, 2006). Those adaptive modifications of social structure in animal communities are still significantly inferior in their scale to what is observed in human societies; in addition, among other animals, they are characterized by a much higher degree of predictability.

The emergence of ultra-sociality was a natural result of the preceding *co-development of intellect and social relations* among our ancestors. The progressive development of the brain and intellectual capacities in primates is inseparably linked with a social way of life—with the necessity to predict actions of other members of their group, to manipulate them, to learn from them, to achieve an optimum combination of altruism and egoism in their behavior. At present, this is the point of view of the majority of primatologists (*e.g.*, Byrne, 2022; Byrne & Whiten, 1988; Byrne & Bates, 2007).

The idea that the primates intellect developed first of all for, say, effective search for fruit (the "ecological intellect hypothesis") does not now have many supporters. It cannot explain why primates need such a large brain, if many other animals, such as squirrels, perfectly manage similar tasks, though their brain remains small. In contrast, the "social intellect hypothesis" is supported by facts. Scientists have detected a significant positive correlation between brain size in primates and the size of their social groups (Dunbar, 2003, 2024). It is necessary to note that primates (in contrast to the majority of other social animals) know all the members of their group "by sight" and have particular relationships with each of them. There are grounds to maintain that individualized pair relationships are the most intellectually "resource-intensive" (Dunbar & Shultz, 2007).

A positive feedback appears to have existed between the development of social intelligence and the growth of complexity in social relationships of hominids.[25]

[25] This social intellect is also called the "Machiavellian intellect" (*e.g.*, Byrne & Whiten, 1988).

Those individuals that managed to achieve a higher status within a social hierarchy, due to a higher intellect or a better ability to foresee actions of others, left more numerous offspring, which in turn led to the general intellectual growth of the socium. As a result, in subsequent generations, in order to move up the social ladder, it was necessary for individuals to possess an even more developed social intellect, and so on (Speechley et al., 2024).

Interesting experimental facts have been obtained. They indicate that intellectual abilities of a "social" character, which allow for resolution of social tasks, developed in our ancestors earlier in comparison with the intellectual capabilities of the other types (*e.g.*, the ones that allow to solve "physical" and instrumental tasks) (Herrmann et al., 2007).

In order to function effectively in a complex, constantly changing sociocultural environment, our pre-human ancestors had to develop intellectual abilities of a rather concrete type: abilities of effective communication, learning and—most importantly—of understanding not only actions, but also thoughts and desires of members of their groups (Vygotsky, 1978; Markov & Markov, 2020). It is quite evident that abilities of this kind should become apparent in early childhood, in the period of active learning and social adaptation. There are two alternative hypotheses about possible mechanisms in the evolutionary development of these social abilities.

The first hypothesis suggests that they emerged as a result of the uniform development of the intellect as a whole (*general intelligence hypothesis*). The second hypothesis suggests that this was the directed development of specific socially-oriented abilities, whereas all the other abilities (such as abilities to think logically, to detect cause-and-effect links in the physical world, and so on) developed later, as something additional and secondary. This is called the *cultural intelligence hypothesis* (Barkow, et al., 1992; Shettleworth, 1998; Herrmann et al., 2007; Henrich, 2015).

At first glance, the *general intelligence hypothesis* looks more plausible, but, it is also possible to provide evidence in support of the *cultural intelligence hypothesis*. For example, it is known that specific intellectual abilities develop locally in many animals, but their overall intellectual level does not grow (or grows insignificantly). One can mention, for example, the birds' unique orientation abilities (Shettleworth, 1998). Special experiments have been conducted in order to test these hypotheses.

The experiments were based on the following reasoning: If the *cultural intelligence hypothesis* is true, then there should be an age in the individual development of humans when we are not different in our "physical" intellect from the apes, even though we are already far above them in our "cultural-social" intellect. Experiments have confirmed the *cultural intellect hypothesis*: it turns out that 2.5 year old children have the same level of development as adult chimpanzees and orangutans in respect to solving physical tasks (spatial, quantitative, detection of cause-and-effect relationships, and so on), but they are significantly superior as regards the effectiveness with which they solve tasks of a social nature, such as those connected with the prediction of others' actions, communication, learning, and so on (Herrmann et al., 2007; Henrich, 2015; Markov & Markov, 2020, 2021).

In general, present-day anthropological data suggests the following:

(1) The development of social relationships and intellectual abilities in the higher primates (in general) and the hominids (in particular) proceeded within a single evolutionary process that was accelerated by the above-mentioned positive feedback.
(2) This process tended to lead to the growth of complexity and flexibility of social relationships. Thus, the development of ultra-sociality and the ability to evolve socially within one of the groups of primates was a natural and logical result of the development of a trend that started among the primates long before the emergence of *Homo sapiens*.

12.6 Transitional Sub-Phases to Biological and Social Phases

The transition to both the biological and social phases did not occur immediately.

In both cases, we are dealing with transitional (sub)phases, in which there were many aromorphoses, although not all of them are known to us. See Fig. 12.1 above for the place of these intermediate phases in the general scheme of Big History. Let us briefly consider these transitional phases.

The **abiogenic chemical phase.** The abiogenic chemical phase began after the emergence of our planet and lasted for several hundred million years, as a result of geological and geochemical processes on Earth. It is very important to emphasize this, because the abiogenic chemical phase was pivotal for the emergence of life. There are various hypotheses about abiogenic chemical evolution and the origin of life including the so-called RNA world. Although some progress has been made in many respects, especially in the last 15 years, none of them seems to be completely satisfactory yet. We believe that the first pre-living systems (the so-called protobionts) should not be considered as direct ancestors of the first living organisms, but as their analogs. These analogues were already comparable to the most primitive living systems in a number of functions. But in general, they were organized differently (it is now extremely difficult to say exactly how). In addition, one should also take into account that the conditions on the young Earth were peculiar. Consequently, such structures could have formed, but modern scientists are unlikely to believe in their existence until concrete facts are available. The evolution of abiogenic organic matter occurred in the following different directions:

(a) Increasing complexity of chemical compounds and structures
(b) Increasing energy output and reaction rate
(c) Selection of elements and compounds according to certain parameters
(d) Concentration of substances
(e) The ability of complex compounds and proto-organisms to expand and grow fast

(f) The selectivity and recognition of some substances by others, according to the important evolutionary pattern for self and nonself-discrimination

For detail, see (Grinin, 2020). For processes developing in the abiotic phase and points of view on these processes, see (Anisimov, 2013; Markov, 2013, 2014, 2015, 2016; Markov et al., 2010; Mitrofanov, 2017).

The **biosocial phase.** After the emergence and development of life, evolution begins to develop in the direction of social processes among living organisms. This is the biosocial phase.

The biosocial phase is of great interest from the point of view of the growing complexity of Big History. In the previous volume of our series, a great attention is paid to many of the most important processes of this phase: the development of the capacity for brain activity, the continued growth of the brain, the increase in psychological and cognitive complexity, and so on (Volk & Henriques, 2025; Hoggard, 2025; LePoire, 2025; see also Aunger, 2007; Gilbert 2014; LePoire, 2016).

The biosocial phase shows a powerful breakthrough in **cephalization** that evolution has made. In essence, at this phase we see evolution gradually searching for new directions of development. The first of these is the activity of the brain, psyche, and consciousness. We see this in higher animals, especially in primates, whose behavioral lability is exceptionally high. Second, this is growth in terms of collectivism, a superstructure over ordinary biology in terms of creating a "super-organism" in the form of a large and organized group of animals. In fact, this is another step in terms of enlarging organisms, no less important than the transition from unicellular to multicellular organisms. This should be called **socialization**. In the biological world, we can distinguish several attempts at socialization. These are social insects. In Fig. 12.1, this line is a dead end. However, another line develops through higher animals and becomes more complex. First, there are the herd animals, then there is a complication in the primates and then the hominids. Thus, in this line there was a connection between two directions: cephalization and socialization, which led to a new transitional phase—anthropogenesis.

Anthropogenesis came as a transition to the social phase of Big History. It was the second transitional phase from biological to social phase.

The transition from the biological to social phase of Big History was a very complex process that we do not quite understand even now. *Within this transition it appears possible to speak about a phase change of a few subtypes of macroevolution: the biological type of macroevolution was first transformed into the biological-social type, then the biological-social type was transformed into the social-biological type; and, finally, the latter was transformed into the social type of macroevolution already in the framework of the unequivocally human society* (see chapter "Human Evolution in the Complexity Growth Perspective. Toward Periodization of the Big History Biosocial Era" [Korotayev, 2026] in this volume; see also Grinin & Korotayev, 2009b: Chap. 1; Grinin & Korotayev, 2025; Korotayev, 2025b).

In the course of anthropogenesis, biological macroevolution was transformed into biosocial evolution. The discoveries of recent decades have moved the dating of the emergence of our species deep in the past to about 200,000–300,000 BP (see, *e.g.*, Callaway, 2017; Culotta, 1999; Gibbons, 1997; Guran et al., 2024; Hublin et al., 2017; Kaufman, 1999; Korotayev, 2025b; Pääbo, 1995; Richter et al., 2017; Stringer, 2016; Stringer & Galway-Witham, 2017; Stringer & Crété, 2022; White et al., 2016).

However, the borderline around 60,000–40,000 BP (the so-called Upper Paleolithic Revolution) still retains an immense importance (Barton & Clark, 2021; Bar-Yosef, 2002, 2007; Bar-Yosef & Kuhn, 2016; Goder-Goldberger et al., 2023; Sánchez-Yustos et al., 2024; Shea, 2013; Zwyns, 2021). This is the point from which we can speak with a complete confidence about the global presence of not only anatomically, but also behaviorally modern humans, "really human" culture (*e.g.*, Conard et al., 2009; Cuenca-Solana et al., 2013; Giacobini, 2007; Korotayev, 2025b; Lbova, 2021; Lewis-Williams, 2002; Marshack, 1991; Oktaviana et al., 2024; Pettitt, 2011; Stiner, 2014; Zilhão, 2005). There is, of course, some hypothesis that human language appeared long before 50,000–40,000 BP. Although this is contested by other scientists, everybody agrees that by 40,000 BP fully developed languages existed wherever humans lived (*e.g.*, Holden, 1998, p. 1455; Rozov, 2023).[26]

We believe that the evolutionary driving forces were still mostly biological when modern humans first emerged, but that the social forces gradually increased in their importance and prevailed over the biological ones at a certain point. Naturally, this was a rather prolonged process, within which the breakthrough point could hardly be identified. We contend that the social component became dominant after 50,000–40,000 BP. However, it did not become absolutely dominant, as biological adaptation and physical anthropological transformation continued in many important ways. The point is that they did not disappear, but their role significantly decreased.[27]

---

[26] Richard Klein, an anthropologist from Stanford University, proposed the following hypothesis to explain the gap between the emergence of anatomically modern *Homo sapiens sapiens* and the emergence of full-fledged languages and cultural artifacts that appear to have taken place much later. According to Klein, the modern brain is a result of rapid genetic changes. He hypothesized that such changes took place around 50,000 BP, pointing out that the affluence of cultural artifacts starts just after that date, as well as the migration of anatomically modern humans out of Africa (see Zimmer, 2003, p. 41ff). Thus, according to him, the emergence of *Homo sapiens* did not automatically result in social macro-evolution proper.

[27] On the one hand, in no way the biological evolution of the humans stopped 50 thousand years ago; in fact, it definitely continues, and in recent millennia (especially, after the Neolithic Revolution) it has even very significantly accelerated (Cochran & Harpending, 2009; Hawks, 2014; Hawks et al., 2007; Kerner et al., 2023; Wilde et al., 2014). However, there is no evidence that this resulted in any increase in biological complexity of humans. After 50,000 BP human evolution mostly led to idioadaptations (sometimes very significant, like emergence of lactase persistence [or the ability to digest milk after weaning, e.g., Check, 2006; Curry, 2013]), sometimes degenerations, but never to aromorphoses that would have increased the biological complexity of humans. There is no evidence that at present human babies are born with more complex brain than Upper Paleolithic babies. In fact, the average brain size of the Upper Paleolithic Cro-Magnons in Europe was a bit larger than the one of present-day Europeans (e.g., Balzeau et al., 2013). This, of course, does not mean that Upper Paleolithic people were more complex biologically than present-day humans, but, naturally, this is

Thus, starting with the Upper Paleolithic Revolution, we may speak about *the transition from socio-biological evolution to social evolution*, a process that was finalized by the Agrarian Revolution.

12.7 The Formation of Social Evolution's Own Mechanisms

There were not so many major aromorphoses in the hunter–gatherer production principle (Grinin, 2006b, 2009a; Grinin & Grinin 2015; Grinin et al., 2023, 2024), which is why the overall rates of socioevolutionary processes were slow and their directionality rather vague. However, they were both in technical sphere and in social one. *Such a type of social macroevolution may be denoted as socio-natural.* As a result of a system of interrelated aromorphoses connected with the agrarian revolution, one could observe the transition to the sociohistorical type of macroevolution. As a result of this, social macroevolution changed its algorithm in a rather significant way, resulting in modification of certain evolutionary laws. We shall consider below how the significance of laws of evolution and the process of social macroevolution changed as a result of the Agrarian Revolution.

Main factors of social change in foraging societies were the result of adaptation to new and various environments—from the deserts of Australia to the pack ice of the Arctic. This was only possible through the modification of sociocultural systems. This made it possible for humans to people most of the world's landmass,[28] to create an enormous variety of tools and crafts, as well as social and other institutions. Effective adaptations let people not only survive, but sometimes also live relatively "comfortable" lives (Sahlins, 1972). The character of human relations with their environment varied significantly, but generally these were ones of human adaptation to the natural world (see, *e.g.*, Grinin, 2006b, pp. 82–83; Grinin & Grinin 2015, 2019a, 2020a; Rodrigue, 2023a, 2023b).

not congruent with any substantial increase of biological complexity of humans after 50,000 BP. Finally, 50,000 years do not look like a sufficient time for the development of a serious biological aromorphosis leading to a serious increase in biological complexity. On the other hand, it is quite clear that the complexity of human planetary system has increased enormously since 50,000 BP; and it has increased exclusively (or almost exclusively) through the increase in its sociocultural, not biological complexity.

[28] It involved crossing the Bab-el-Mandeb straits and movement along the Arabian Peninsula coast to Iran, then to India, South-East Asia, New Guinea, and Melanesia, reaching Australia by around 50,000 BP (see, e.g., Kealy et al., 2018; O'Connell et al., 2018; Korotayev, 2025a). Around 55,000 BP Homo sapiens reached southern parts of China (Hublin, 2021, p. 3), around 45,000 BP modern humans appeared in Europe (Hublin et al., 2020; Prüfer et al., 2021; Teyssandier, 2023) and Siberia (Fu et al., 2014; Li et al., 2020; Rybin et al., 2023), about 40,000 BP they reached Northern China (Bae et al., 2017; Li et al., 2020), around 38,000 BP they arrived in Japan (e.g., Cooke et al., 2021), and so on. In the later Upper Paleolithic Homo sapiens peopled the New World, most likely they did this in a few waves (e.g., Becerra-Valdivia & Higham, 2020; Gruhn, 2023; Korotayev, 2006; Korotayev et al., 2006a, 2017; Stanford & Bradley, 2012).

In the agrarian epoch, the character of those interrelations changed significantly through the transition to much more conscious and effective change of the environments at a rather wide scale (irrigation, clearing of forests, plowing of steppes, soil fertilization, construction of cities, roads, and so on). Natural forces (animal, wind and water energy) started to be used on a much wider scale (earlier humans actively used only fire). Natural raw materials started to be transformed into entirely new products (metals, fabrics, ceramics, glass) (Grinin & Grinin, 2015, 2019b, 2020a).

Thus, within social evolution process a more and more significant role started to be played by peculiarly social factors that (in contrast with natural factors) are connected to conscious goal-setting and goal-achieving. Gradually, with economic–technological progress, the growth of surplus accumulation capacities, as well as general cultural complexity of social systems, their evolution became almost purely social. As a result, the "vector" of evolutionary selection turned out to be directed toward societal capabilities to adapt to social (rather than natural) environments, which implies the capacity to compete with neighboring social systems in economic, military, commercial, cultural, ideological, and other spheres (Bondarenko D., 2025; Bondarenko V., 2024; Grinin & Korotayev, 2023; Komlosy, 2024).

Finally, we would like to mention the following important changes in the "algorithm" of social evolution:

- *The start of the mechanism for resource accumulation.*

In the tens of thousands of years of the human foraging epoch, long-term material resource accumulation was relatively insignificant when compared to subsequent epochs. There was, of course, a certain amount of accumulation, of knowledge, traditions and technologies, albeit at a limited scale. This accumulation took place not in every society, but was observed at the global scale and was due to the overall demographic growth, increase in numbers of social systems, emergence of new tools, products, *etc.* (see, e.g., Grinin et al., 2023). There was practically no special accumulation sector prior to the Agrarian Revolution.[29]

In many cases, people could produce more than they actually needed, and sometimes even so-called "original affluent societies" could emerge (Sahlins, 1972).

For example, with respect to the gatherers of sago in New Guinea, people would spend a minor part of their time securing food for themselves, whereas they would spend the rest of the time at other activities and leisure (Shnirelman, 1983, 1989).

[29] With a possible exception of some highly specialized hunters (usually of large aquatic animals), gatherers, and fishers—for example, some social systems described ethnographically for the North-Western Coast of America (see, *e.g.*, Averkieva, 1978; Shnirelman, 1986; Townsend, 1985). See also about other most complex societies of the Hunter-Gatherer Production Principle: Ammerman & Cavalli-Sforza, 2014; Conte et al., 2018; Lozovskaya et al., 2018; March, 2013; Shnirelman, 1983; Tanno et al., 2013).

The impossibility to accumulate and/or the absence of the desire to accumulate slowed down development, which contributed to the slow speed of social evolution (Grinin, 2006b, 2009a). *In simple social systems of agriculturalists and pastoralists, the emergence of the possibility (and, later, the desire) to accumulate led to numerous transformations in the spheres of functional differentiation, distribution, social stratification, exchange, trade, development of property relationships, increasing political complexity, and so on.*

- *Strengthening of the ability of social systems to change.*

Agrarian societies turned out to be more capable of serious social transformations than hunter–gatherers, while complex agrarian societies turned out to be much more capable of such transformations than simple agriculturalist and pastoralist systems. The growth of social systems' ability to change provides a vivid demonstration of the main difference between social and biological evolution—that humans were capable of consciously transforming their social systems, with preconceived goals.

- *Intersocietal contacts become the leading factor of social evolution.*

The importance of various contacts increased sharply, and this contributed to a more active adaptation of social systems to their environments. The growth of the role of contacts dramatically raised the importance of external social driving forces (Grinin, 1997, p. 23; 2007, p. 177). Note that this had an enormous importance for the development of the World System and for the humankind as a whole. Military and other interactions stimulated improvements in administration, defense, culture, technology, and so on. All this contributed to the development of a single global process involving numerous societies and peoples.

It is also appropriate to note that the growth of societal size is not only due to natural demographic growth, but is more importantly due to the integration and unification of social systems. Thus, external contact factors become most important with respect to societal evolution.

12.8 Mathematical Modeling of Biological and Social Macroevolution

In this part, we consider a specific mathematical model capable of describing important features of both biological and social macroevolution. In mathematical models of historical macrodynamics, a hyperbolic pattern of world population growth arises from nonlinear, second-order positive feedback between demographic growth and technological development. Based on diverse paleontological data and an analogy with macrosociological models, we suggest that the hyperbolic character of biodiversity growth can be similarly accounted for by nonlinear, second-order positive feedback between diversity growth and the complexity of community structure. We further discuss how such positive feedback mechanisms can be modeled mathematically.

The authors of this chapter met for the first time in 2005, in the town of Dubna (near Moscow), at what seems to have been the first ever international conference dedicated specifically to Big History studies (see Korotayev & Markov, 2014, 2015). Without advance knowledge of one another, we found ourselves in a single session. During the course of the session, we presented two different diagrams. One illustrated population dynamics in China between 700 BCE and 1851 CE, and the other illustrated the dynamics of marine Phanerozoic biodiversity over the past 542 million years (Fig. 12.2).

The similarity between the two diagrams was striking. This, despite the fact that they depicted the development of very different systems (human population *vs.* biota) at different timescales (hundreds of years *vs.* millions of years), and had been generated using the methods of different sciences (historical demography *vs.* paleontology) with different sources (demographic estimates *vs.* paleontological data). What could have caused similarity of developmental dynamics in very different systems?

* * *

In 1960, von Foerster *et al.* published a striking discovery in the journal *Science* (von Foerster et al., 1960). They showed that between 1 and 1958 CE, the world's population (N) dynamics could be described in an extremely accurate way with an astonishingly simple equation:[30]

$$N_t = \frac{C}{t_0 - t}, \tag{12.1}$$

where N_t is the world population at time t, and C and t_0 are constants, with t_0 corresponding to an absolute limit ("singularity" point) at which N would become infinite. Parameter t_0 was estimated by von Foerster and his colleagues as 2026.87, which corresponds to November 13, 2026; this made it possible for them to supply their article with a title that was a public-relations masterpiece: "Doomsday: Friday, 13 November, A.D. 2026."

Of course, von Foerster and his colleagues did not imply that the world population on that day could actually become infinite. The real implication was that the world population growth pattern that operated for many centuries prior to 1960 was about to end and be transformed into a radically different pattern. This prediction began to be fulfilled only a few years after the "Doomsday" paper was published as World System growth (and world population growth in particular) began to diverge more and more from the previous blow-up regime. Now no longer hyperbolic, the world population growth pattern is closer to a logistic one (see, *e.g.*, Korotayev, Malkov et al., 2006a; Korotayev, 2009; Korotayev et al., 2023).

[30] To be exact, the equation proposed by von Foerster and his colleagues looked as follows: $N_t = \frac{C}{(t_0 - t)^{0.99}}$. However, as von Hoerner (1975) and Kapitza (1996) showed, it can be simplified as $N_t = \frac{C}{t_0 - t}$.

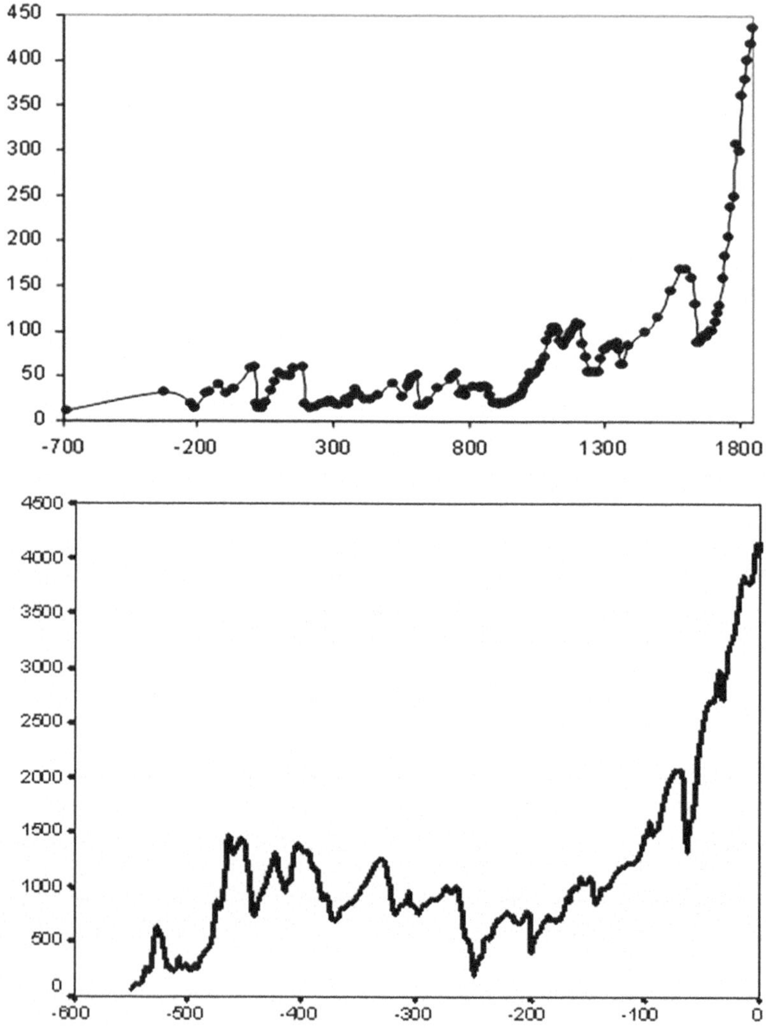

Fig. 12.2 Similarity between the long-term population dynamics of China (top: millions of people, following Korotayev, Malkov, et al. 2006b, pp. 47–88) and the dynamics of Phanerozoic marine biodiversity (bottom: number of genera, *N*. Data source: Markov & Korotayev, 2007)

Figure 12.3 presents the overall correlation between the curve generated by von Foerster *et al.*'s equation and the most detailed series of empirical estimates of world population (McEvedy & Jones, 1978, for the period 1000–1950; U.S. Bureau of the Census, 2025 and UN Population Division, 2025, for 1950–1970). The formal characteristics are: $R = 0.998$; $R^2 = 0.996$; $p = 9.4 \times 10^{-17} \approx 1 \times 10^{-16}$. For readers unfamiliar with mathematical statistics: R^2 can be regarded as a measure of the fit between the dynamics generated by a mathematical model and the empirically

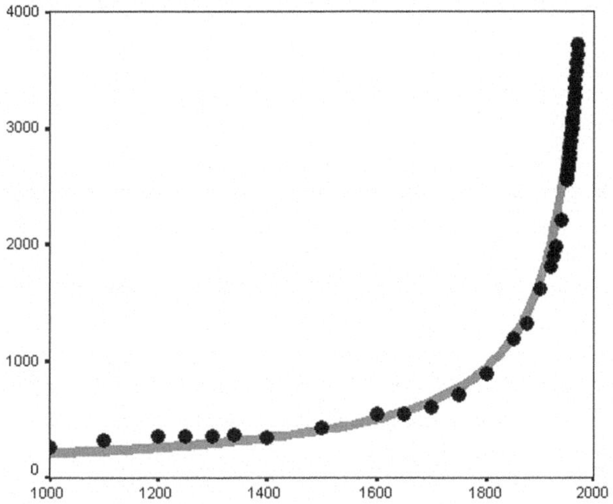

Fig. 12.3 Correlation between empirical estimates of world population (black, in millions of people, 1000–1970) and the curve generated by von Foerster *et al.*'s equation (gray)

observed situation, and can be interpreted as the proportion of the variation accounted for by the respective equation. Note that 0.996 also can be expressed as 99.6%.[31] Thus, von Foerster *et al.*'s equation accounts for an astonishing 99.6% of all the macrovariation in world population, from 1000 CE through 1970, as estimated by McEvedy and Jones (1978) and UN Population Division (2025). Note also that the empirical estimates of world population find themselves aligned in an extremely neat way along the hyperbolic curve, which convincingly justifies the designation of the pre-1970s world population growth pattern as "hyperbolic."

The von Foerster *et al.'s* equation $N_1 = C/(t_0 - t)$ is the solution for the following differential equation (see, *e.g.*, Korotayev, Malkov et al., 2006a, pp. 119–120):

$$\frac{dN}{dt} = \frac{N^2}{C}. \tag{12.2}$$

This equation can be also written as:

[31] The second characteristic (p, standing for "probability") is a measure of the correlation's statistical significance. A bit counter-intuitively, the lower the value of p, the higher the statistical significance of the respective correlation. This is because p indicates the probability that the observed correlation could be accounted solely by chance. Thus, $p = 0.99$ indicates an extremely low statistical significance, as it means that there are 99 chances out of 100 that the observed correlation is the result of a coincidence, and, thus, we can be quite confident that there is no systematic relationship (at least, of the kind that we study) between the two respective variables. On the other hand, $p = 1 \times 10^{-16}$ indicates an extremely high statistical significance for the correlation, as it means that there is only one chance out of 10,000,000,000,000,000 that the observed correlation is the result of pure coincidence (a correlation is usually considered statistically significant once $p < 0.05$).

$$\frac{dN}{dt} = aN^2, \qquad (12.3)$$

Where $a = \frac{1}{C}$.

What is the meaning of this mathematical expression? In our context, *dN/dt* denotes the absolute population growth rate at a certain moment in time. Hence, this equation states that the absolute population growth rate at any moment in time should be proportional to the square of world population at this moment. This significantly demystifies the problem of hyperbolic growth. To explain this hyperbolic growth, one only needs to explain why for many millennia the world population's absolute growth rate tended to be proportional to the square of the population.

The main mathematical models of hyperbolic growth in the world population (Cohen, 1995; Golosovsky, 2010; Grinin, Markov et al., 2013; Korotayev, 2005, 2007, 2008, 2009, 2013, 2018, 2020b, 2024, 2025a; Korotayev, Malkov et al., 2006a, pp. 21–36; Korotayev & Malkov, 2012; Kremer, 1993; Podlazov, 2017; Taagapera, 1976, 1979; Tsirel, 2004) are based on the following two assumptions:

(1) "the Malthusian (Malthus, 1978 [1798]) assumption that population is limited by the available technology, so that the growth rate of population is proportional to the growth rate of technology" (Kremer, 1993, pp. 681–682)[32] and
(2) The idea that "[h]igh population spurs technological change because it increases the number of potential inventors. . . In a larger population there will be proportionally more people lucky or smart enough to come up with new ideas," thus, "the growth rate of technology is proportional to total population" (Kremer, 1993, p. 685).[33]

Here Kremer uses the main assumption of Endogenous Technological Growth theory (see, *e.g.*, Kuznets, 1960; Grossman & Helpman, 1991; Aghion & Howitt, 1998; Simon, 1977, 2000; Komlos & Nefedov, 2002; Jones, 1995, 2005).

The first assumption looks quite convincing. Indeed, throughout most of human history the world population was limited by the technologically determined ceiling of the carrying capacity of land. For example, with foraging subsistence technologies the Earth could not support more than a few million people because the amount of naturally available useful biomass on this planet is limited. The world population could only grow over this limit when people started to apply various means to artificially increase the amount of available biomass, that is with the transition from foraging to food production. Extensive agriculture is also limited in terms of the

[32] In addition to this, the absolute growth rate is proportional to the population itself. With a given relative growth rate, a larger population will increase more in absolute number than a smaller one.

[33] Note that "the growth rate of technology" here means the relative growth rate (*i.e.*, the level to which technology will grow in a given unit of time in proportion to the level observed at the beginning of this period).

number of people that it can support. Thus, further growth of the world population only became possible with the intensification of agriculture and other technological improvements (see, *e.g.*, Turchin, 2003; Korotayev, Malkov et al., 2006a, 2006b; Korotayev & Khaltourina, 2006).

However, as is well known, the technological level is not constant, but variable (see, *e.g.*, Grinin 2006b; Grinin & Korotayev, 2015; Grinin et al., 2024), and in order to describe its dynamics the second basic assumption is employed. In general, we find this assumption rather plausible—in fact, it is quite probable that, other things being equal, within a given period of time, 5 million people will make approximately five times more inventions than 1 million people.

This assumption was expressed mathematically by Kremer in the following way:

$$\frac{dT}{dt} = kNT. \qquad (12.4)$$

This equation simply says that the absolute technological growth rate (dT/dt) at a given moment of time is proportional to the technological level (T) observed at this moment (the wider the technological base, the higher the number of inventions that can be made on its basis). On the other hand, this growth rate is also proportional to the population (N): the larger the population, the larger the number of potential inventors.[34]

When united in one system, Malthusian and Kuznetsian equations account quite well for the hyperbolic growth of the world population observed before the early 1990s (see, *e.g.*, Korotayev 2005, 2007, 2008, 2013, 2018, 2020b, 2024, 2025a; Korotayev, Malkov et al., 2006a). The resultant models provide a rather convincing explanation of *why,* throughout most of human history, the world population followed the hyperbolic pattern with the absolute population growth rate tending to be proportional to N^2. For example, why would the growth of population from, say, 10 million to 100 million, result in the growth of dN/dt 100 times? The above-mentioned models explain this rather convincingly. The point is that the growth of world population from 10 to 100 million implies that human subsistence technologies also grew approximately 10 times (given that it will have proven, after all, to be able to support a population ten times larger). On the other hand, the tenfold population growth also implies a tenfold growth in the number of potential inventors, and, hence, a tenfold increase in the relative technological growth rate. Thus, the absolute technological growth rate would expand $10 \times 10 = 100$ times as, in accordance with Eq. (12.4), an order of magnitude higher number of people having at their disposal an order of magnitude wider technological base would tend to make two orders of magnitude more inventions. As throughout the Malthusian epoch, the world population (N) tended toward the

[34] Kremer did not test this hypothesis empirically in a direct way. Note, however, that our own empirical test of this hypothesis has supported it (see Korotayev, Malkov et al., 2006b, pp. 141–146).

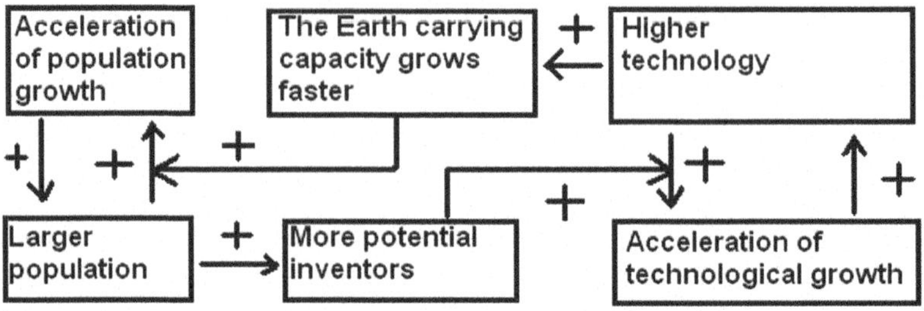

Fig. 12.4 Cognitive scheme of the nonlinear second-order positive feedback between technological development and demographic growth

technologically determined carrying capacity of the Earth, we have good reason to expect that *dN/dt* should also grow just by about 100 times.

In fact, it can be shown (see, *e.g.*, Korotayev, Malkov et al., 2006a, 2006b; Korotayev & Khaltourina, 2006) that the hyperbolic pattern of the world's population growth could be accounted for by a nonlinear second-order positive feedback mechanism that was long ago shown to generate just the hyperbolic growth, also known as the "blow-up regime" (see, *e.g.*, Kurdyumov, 1999). In our case, this nonlinear second-order positive feedback looks as follows: more people—more potential inventors—faster technological growth—faster growth of the Earth's carrying capacity—faster population growth—more people allow for more potential inventors—faster technological growth, and so on (see Fig. 12.4).

Note that the relationship between technological development and demographic growth cannot be analyzed through any simple cause-and-effect model, as we observe a true dynamic relationship between these two processes—each of them is both the cause and the effect of the other.

The feedback system described here can be identified with the process of "collective learning" described, principally, by Christian (2005, pp. 146–148, see also Baker, 2016). The mathematical models of World System development discussed in this chapter can be interpreted as models of the influence that collective learning has on global social evolution (*i.e.*, the evolution of the World System). Thus, the rather peculiar hyperbolic shape of accelerated global development prior to the early 1970s may be regarded as a product of global collective learning. We have also shown (Korotayev, Malkov et al., 2006a, pp. 34–66) that, for the period prior to the 1970s, World System economic and demographic macrodynamics, driven by the above-mentioned positive feedback loops, can simply and accurately be described with the following model:

$$\frac{dN}{dt} = aSN, \tag{12.5}$$

$$\frac{dS}{dt} = bNS. \tag{12.6}$$

The world GDP (G) can be calculated using the following equation:

$$G = mN + SN, \tag{12.7}$$

where G is the world GDP, N is population, and S is the produced surplus per capita, over the subsistence amount (m) that is minimally necessary to reproduce the population with a zero growth rate in a Malthusian system (thus, $S = g - m$, where g denotes per capita GDP); a and b are parameters (e.g., Korotayev & Malkov, 2016).

The mathematical analysis of the basic model (not described here) suggests that up to the 1970s, the amount of S should be proportional, in the long run, to the World System's population: $S = kN$. Our statistical analysis of available empirical data has confirmed this theoretical proportionality (Korotayev, Malkov et al., 2006a, pp. 49–50). Thus, in the right-hand side of Eq. (12.6), S can be replaced with kN, resulting in the following equation:

$$\frac{dN}{dt} = kaN^2.$$

Recall that the solution of this type of differential equations is:

$$N_t = \frac{C}{t_0 - t},$$

which produces a simple hyperbolic curve.

As, according to our model, S can be approximated as kN, its long-term dynamics can be approximated with the following equation:

$$S = \frac{kC}{t_0 - t}. \tag{12.8}$$

Thus, the long-term dynamics of the most dynamic component of the world GDP, SN, the "world surplus product," can be approximated as follows:

$$SN = \frac{kC^2}{(t_0 - t)^2}. \tag{12.9}$$

This suggests that the long-term world GDP dynamics up to the early 1970s must be approximated better by a quadratic hyperbola, rather than by a simple one. As shown in Fig. 12.5, this approximation works very effectively indeed.

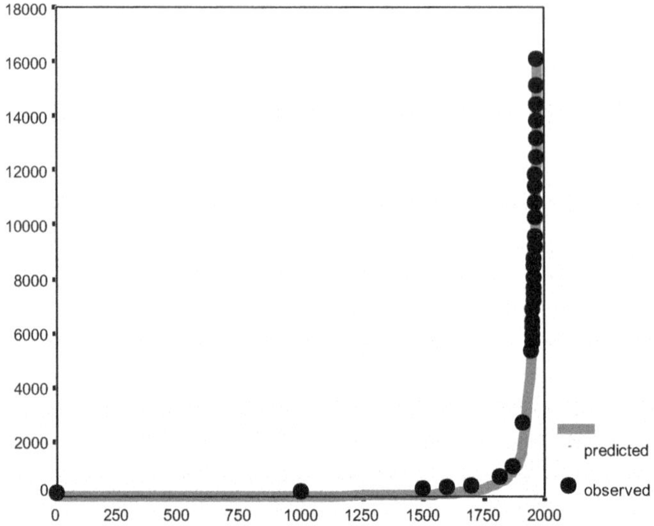

Fig. 12.5 The fit between predictions of a quadratic–hyperbolic model and observed world GDP dynamics, 1–1973 CE (in billions of 1990 international dollars, PPP). *Note:* $R = 0.9993$, $R^2 = 0.9986$, $p << 0.0001$. The black markers correspond to Maddison's (2001) estimates (Maddison's estimates of the world per capita GDP for 1000 CE has been corrected on the basis of [Meliantsev, 2004]). The gray solid line has been generated by the following equation:

$$G = \frac{1.8 \times 10^6}{(2006 - t)^2}.$$

Thus, up to the 1970s the hyperbolic growth of the world population was accompanied by the quadratic–hyperbolic growth of the world GDP, as suggested by our model. Note that the hyperbolic growth of the world population and the quadratic–hyperbolic growth of the world GDP are very tightly connected processes, actually two sides of the same coin, two dimensions of one process propelled by nonlinear second-order positive feedback loops between the technological development and demographic growth (see Fig. 12.6).

We have also demonstrated (Korotayev, Malkov et al., 2006a, pp. 67–80) that the World System population's literacy (l) dynamics are rather accurately described by the following differential equation:

$$\frac{dl}{dt} = aSl(1 - l), \tag{12.10}$$

where l is the proportion of the population that is literate, S is per capita surplus, and a is a constant. In fact, this is a version of the autocatalytic model. Literacy growth is proportional to the fraction of the population that is literate, l (potential teachers), to the fraction of the population that is illiterate, $(1 - l)$ (potential pupils), and to the

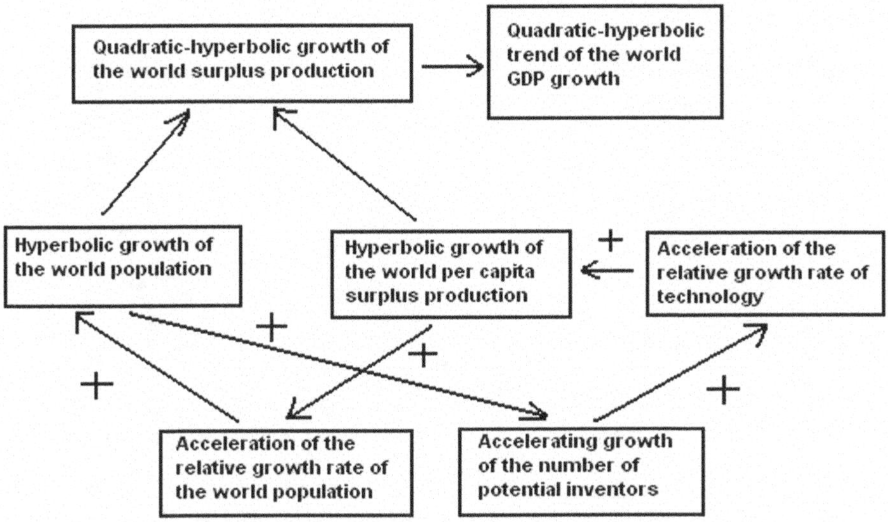

Fig. 12.6 Cognitive scheme of the world economic growth generated by nonlinear second-order positive feedback between technological development and demographic growth

amount of per capita surplus S, since it can be used to support educational programs. (Additionally, S reflects the technological level T that implies, among other things, the level of development of educational technologies.) From a mathematical point of view, Eq. (12.9) can be regarded as logistic where saturation is reached at literacy level $l = 1$. S is responsible for the speed with which this level is being approached.

It is important to stress that with low values of l (which correspond to most of human history, with recent decades being the exception), the rate of increase in world literacy generated by this model (against the background of hyperbolic growth of S) can be approximated rather accurately as hyperbolic (see Fig. 12.7).

The overall number of literate people is proportional both to the literacy level and to the overall population. As both of these variables experienced hyperbolic growth until the 1960s/1970s, one has sufficient grounds to expect that until recently the overall number of literate people in the world (L)[35] was growing not just hyperbolically, but rather in a quadratic-hyperbolic way (as was world GDP). Our empirical test has confirmed this—the quadratic–hyperbolic model describes the

[35] Since literacy appeared, almost all of the Earth's literate population has lived within the World System; hence, the literate population of the Earth and the literate population of the World System have been almost perfectly synonymous.

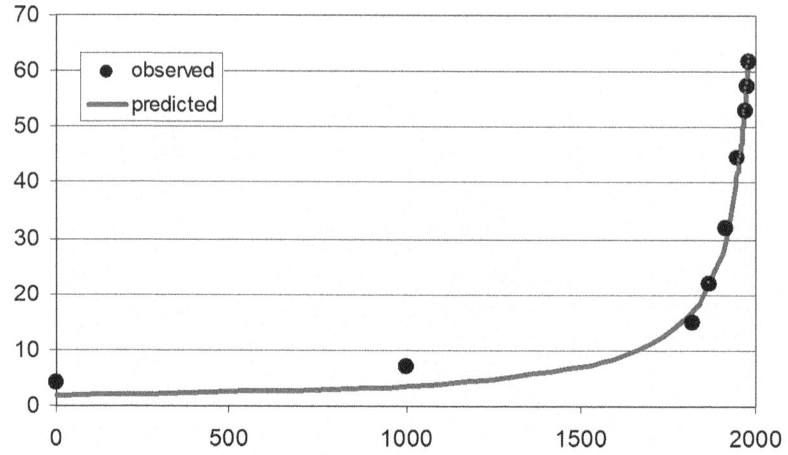

Fig. 12.7 The fit between predictions of the hyperbolic model and observed world literacy dynamics, 1–1980 CE (%%). *Note:* $R = 0.997$, $R^2 = 0.994$, $p << 0.0001$. Black dots correspond to World Bank (2013) estimates for the period since 1970, and to Meliantsev's (2004) estimates for the earlier period. The gray solid line has been generated by the following equation: $l_t = \frac{3.8 \times 10^3}{(2040-t)^2}$. The best-fit values of parameters C (3.8×10^3) and t_0 (2040) have been calculated with the least squares method

growth of the literate population of this planet with an extremely good fit indeed (see Fig. 12.8).

Similar processes are observed with respect to world urbanization, the macrodynamics of which appear to be described by the differential equation:

$$\frac{du}{dt} = bSu(u_{\lim} - u), \qquad (12.11)$$

where u is the proportion of the population that is urban, S is per capita surplus produced with the given level of the World System's technological development, b is a constant, and $u_{\lim}$ is the maximum possible proportion of the population that can be urban. Note that this model implies that during the Malthusian-Kuznetsian era of the blow-up regime, the hyperbolic growth of world urbanization must have been accompanied by a quadratic–hyperbolic growth of the urban population of the world, as supported by our empirical tests (see Figs. 12.9 and 12.10).

Within this context it is hardly surprising to find that the general macrodynamics of largest settlements within the World System are also quadratic–hyperbolic (see Fig. 12.11).

As was long ago demonstrated by cross-cultural anthropologists (see, *e.g.*, Naroll & Divale, 1976; Levinson & Malone, 1980, p. 34), for pre-agrarian, agrarian, and early industrial cultures the size of the largest settlement is a rather effective indicator of the general sociocultural complexity of a social system. This, of course, suggests

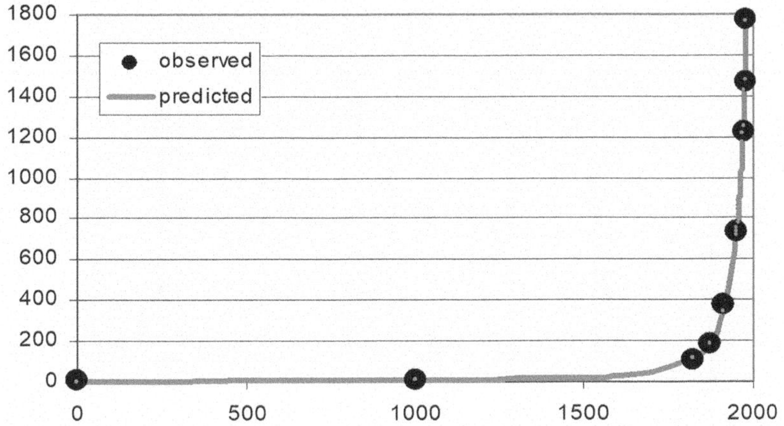

Fig. 12.8 The fit between predictions of the quadratic–hyperbolic model and observed world literate population dynamics, 1–1980 CE (*L*, millions). *Note:* $R = 0.9997$, $R^2 = 0.9994$, $p << 0.0001$. The black dots correspond to UNESCO/World Bank (2025) estimates for the period since 1970, and to Meliantsev's (2004) estimates for the earlier period; we have also taken into account the changes of age structure on the basis of UN Population Division (2025) data. The gray solid line has been generated by the following equation: $L_t = \frac{4.96 \times 10^6}{(2033-t)^2}$. The best-fit values of parameters C (4.96×10^6) and t_0 (2033) have been calculated with the least squares method

that the World System's general sociocultural complexity also grew, in the Malthusian-Kuznetsian era, in a generally quadratic–hyperbolic way.[36]

Turning to a more concrete case study, as suggested at the beginning of this section, the hyperbolic model is particularly effective for describing the long-term population dynamics of China, the country with the best-known long-term demographic history. The Chinese population curve reflects not only a hyperbolic trend but also cyclical and stochastic dynamics. These components of long-term population dynamics in China, as well as in other complex agrarian societies, have been discussed extensively (see, *e.g.*, Alexander, 2016; Chu & Lee, 1994; Goldstone, 1991; Hoyer et al., 2024; Komlos & Nefedov, 2002; Kondor & Turchin, 2024; Korotayev & Khaltourina, 2006; Korotayev, Malkov et al., 2006b; Korotayev & Zinkina, 2015; Korotayev, Zinkina et al., 2016; Korotayev, 2017; Nefedov, 2004, 2012, 2013, 2025; Turchin, 2003, 2005; Turchin & Korotayev, 2006; Turchin & Nefedov, 2009; Usher, 1989; Wittmann & Kuehn, 2024).

As we have observed with respect to world population dynamics, even before the start of its intensive modernization, the population dynamics of China were characterized by a pronounced hyperbolic trend (Figs. 12.12 and 12.13). The

[36]Dynamics of this variable is also considered in chapter "Punctuated Growth and Complexity in Urban Evolution: Insights from Historical City Size Distributions" (Harper, 2026) in this volume.

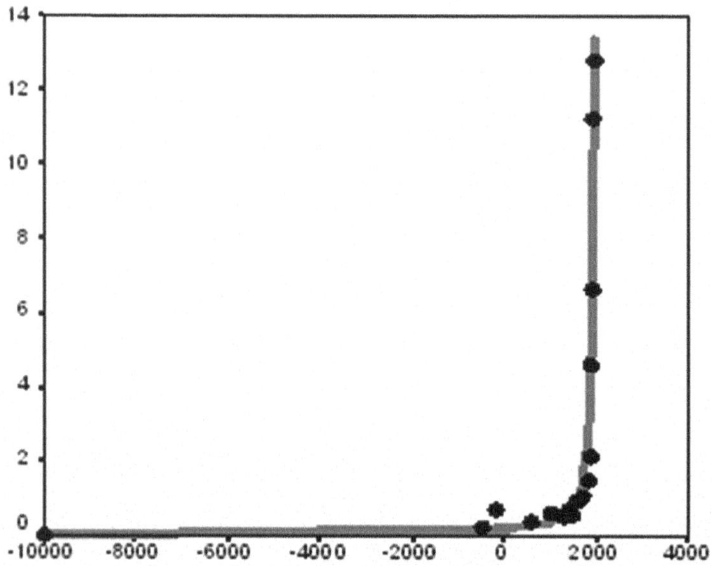

Fig. 12.9 The fit between predictions of the hyperbolic model and empirical estimates of world megaurbanization dynamics (% of the world population living in cities with > 250,000 inhabitants), 10,000 BCE—1960 CE. *Note:* $R = 0.987$, $R^2 = 0.974$, $p << 0.0001$. The black dots correspond to estimates of Chandler (1987), UN Population Division (2025), Modelski (2003), and Gruebler (2006). The gray solid line has been generated by the following equation: $u_t = \frac{403}{1990-t}$. The best-fit values of parameters C (403) and t_0 (1990) have been calculated with the least squares method. For comparison, the best fit (R^2) obtained here for the exponential model is 0.492

hyperbolic model describes traditional Chinese population dynamics *much* more accurately than either linear or exponential models.

The hyperbolic model describes the population dynamics of China in an especially accurate way if we take the late modern period into account (Fig. 12.14).

It is curious that, as we noted above, the dynamics of marine biodiversity are strikingly similar to the population dynamics of China. The similarity probably derives from the fact that both curves are produced by the interference of the same three components (the general hyperbolic trend, as well as cyclical and stochastic dynamics). In fact, there is a lot of evidence that some aspects of biodiversity dynamics are stochastic (Raup et al., 1973; Sepkoski, 1994; Markov, 2001; Cornette & Lieberman, 2004), while others are periodic (Raup & Sepkoski, 1984; Rohde & Muller, 2005; Roberts & Mannion, 2019). In any event, the hyperbolic model describes marine biodiversity (measured by number of genera) through the Phanerozoic much more accurately than an exponential model (Fig. 12.15).

When measured by number of species, the fit between the empirically observed marine biodiversity dynamics and the hyperbolic model becomes even better (Figs. 12.15 and 12.16).

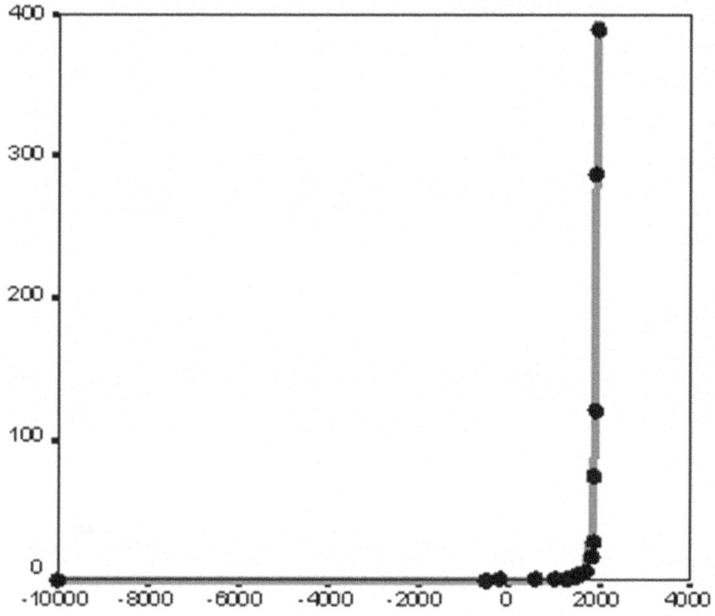

Fig. 12.10 The fit between predictions of the quadratic–hyperbolic model and the observed dynamics of world urban population living in cities with > 250,000 inhabitants (millions), 10,000 BCE–1960 CE. *Note:* $R = 0.998$, $R^2 = 0.996$, $p << 0.0001$. The black markers correspond to estimates of Chandler (1987) and UN Population Division (2025). The gray solid line has been generated by the following equation: $U_t = \frac{9.1 \times 10^5}{(2008-t)^2}$. The best-fit values of parameters C (9.1×10^5) and t_0 (2008) have been calculated with the least squares method. For comparison, the best fit (R^2) obtained here for the exponential model is 0.637

Likewise, the hyperbolic model describes continental biodiversity in an especially accurate way (Fig. 12.17).

However, the best fit between the hyperbolic model and the empirical data is observed when the hyperbolic model is used to describe the dynamics of total (marine and continental) global biodiversity (Fig. 12.18).

The hyperbolic dynamics are most prominent when both marine and continental biotas are considered together. This fact can be interpreted as a proof of the integrated nature of the biosphere. But why, throughout the Phanerozoic, did global biodiversity tend to follow a hyperbolic trend similar to that which we observed for the World System in general and China in particular?

As we have noted above, in sociological models of macrohistorical dynamics, the hyperbolic pattern of world population growth arises from nonlinear second-order positive feedback (more or less identical with the mechanism of collective learning) between demographic growth and technological development. Based on analogy with these sociological models and diverse paleontological data, we suggest that the hyperbolic character of biodiversity growth can be similarly accounted for by

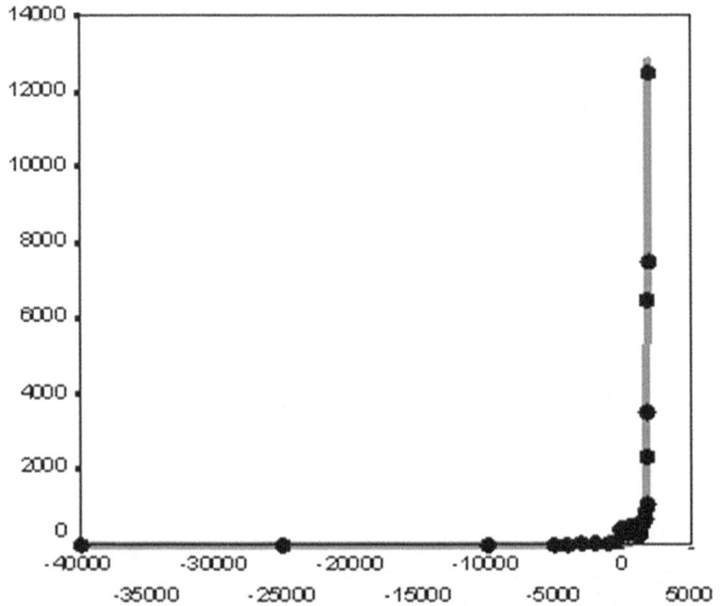

Fig. 12.11 The fit between predictions of the quadratic–hyperbolic model and the observed dynamics of size of the largest settlement of the world (thousands of inhabitants), 10,000 BCE–1950 CE. *Note:* $R = 0.992$, $R^2 = 0.984$, $p << 0.0001$. The black markers correspond to estimates of Modelski (2003) and Chandler (1987). The gray solid line has been generated by the following equation: $U_{\max t} = \frac{1.04 \times 10^8}{(2040-t)^2}$. The best-fit values of parameters C (1.04×10^8) and t_0 (2040) have been calculated with the least squares method. For comparison, the best fit (R^2) obtained here for the exponential model is 0.747

nonlinear second-order positive feedback between biodiversity growth and the complexity of community structure: more genera—higher alpha diversity—enhanced stability and "buffering" of communities—lengthening of average life span of genera, accompanied by a decrease in the extinction rate—faster diversity growth—more genera—higher alpha diversity, and so on. Indeed, this begins to appear as a (rather imperfect) analog of the collective learning mechanism active in social macroevolution.

The growth of genus richness throughout the Phanerozoic was mainly due to an increase in the average longevity of genera and a gradual accumulation of long-lived (stable) genera in the biota. This pattern reveals itself in a decrease in the extinction rate. Interestingly, in both biota and humanity, growth was facilitated by a decrease in mortality rather than by an increase in the birth rate. The longevity of newly arising genera was growing in a stepwise manner. The most short-lived genera appeared during the Cambrian; more long-lived genera appeared in Ordovician to Permian; the next two stages correspond to the Mesozoic and Cenozoic (Markov, 2001, 2002). We suggest that diversity growth can facilitate the increase in genus longevity via progressive stepwise changes in the structure of communities.

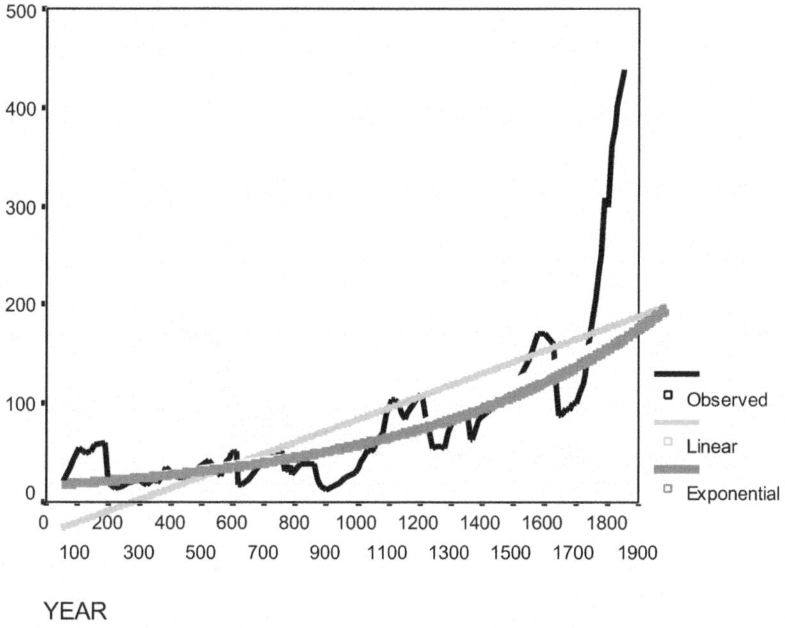

Fig. 12.12 Population dynamics of China (million people, following Korotayev, Malkov et al., 2006b, pp. 47–88), 57–1851 CE. Fit with Linear and Exponential Models. *Note:* Linear model: $R^2 = 0.469$. Exponential model: $R^2 = 0.600$

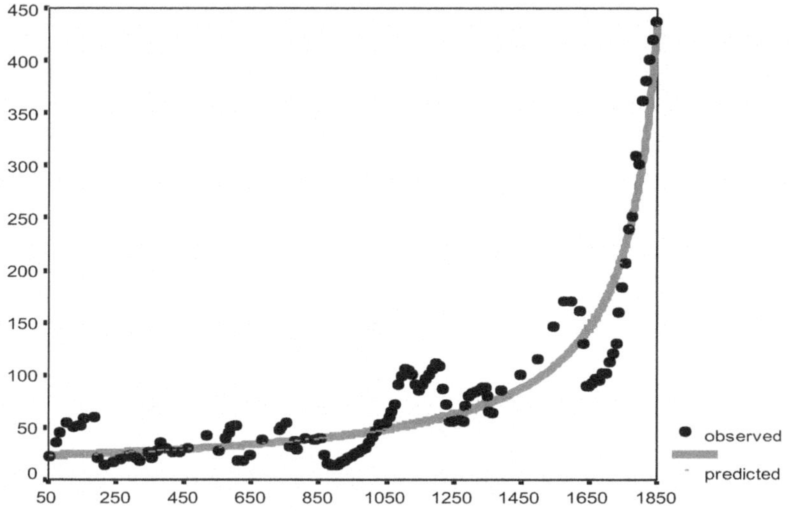

Fig. 12.13 Fit between a hyperbolic model and observed population dynamics of China (million people), 57–1851 CE. *Note:* $R^2 = 0.884$. The gray solid line has been generated by the following equation: $N_t = \frac{3.34 \times 10^4}{1915 - t}$

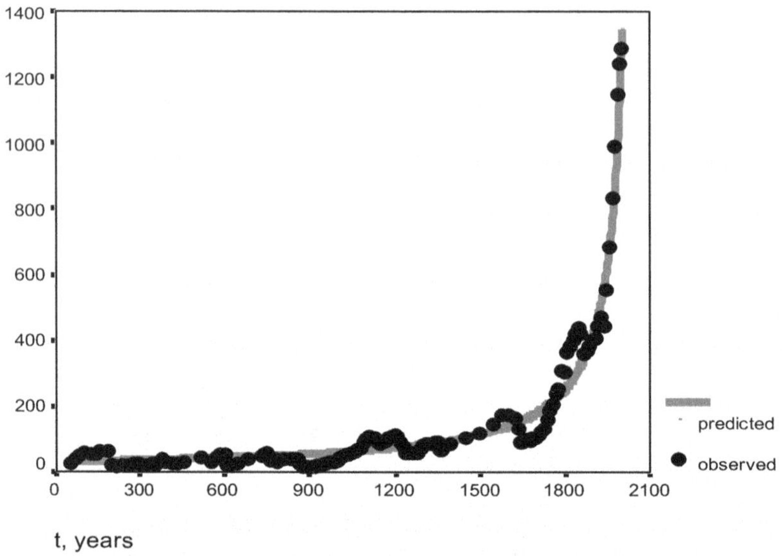

Fig. 12.14 Fit between a hyperbolic model and observed population dynamics of China(million people, following Korotayev, Malkov, et al., 2006b, pp. 47–88), 57–2003 CE. *Note*: R^2 = 0.968. The gray solid line has been generated by the following equation: $N_t = \frac{6.3\times10^4}{2050-t}$

Fig. 12.15 Global change in marine biodiversity (number of **genera**, *N*) through the Phanerozoic (following Korotayev & Markov, 2014). *Note*: Exponential model: R^2 = 0.463. Hyperbolic model: R^2 = 0.854. The hyperbolic line has been generated by the following equation: $N_t = \frac{1.83\times10^5}{37-t}$

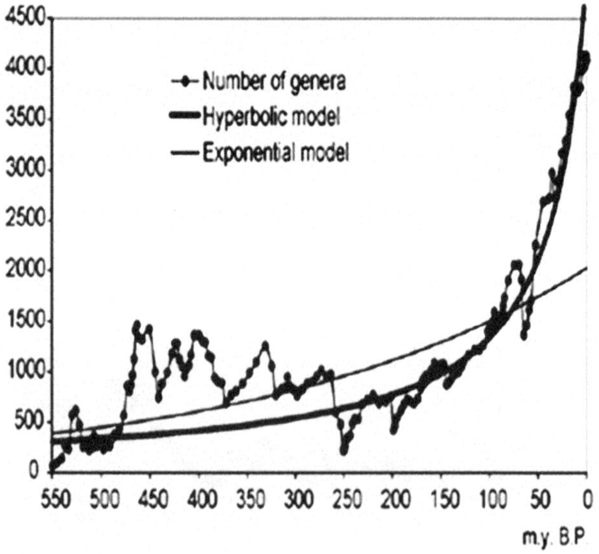

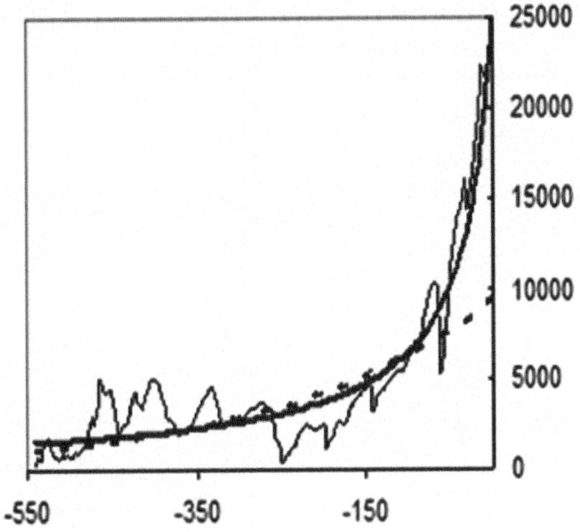

Fig. 12.16 Global change in marine biodiversity (number of **species**, N) through the Phanerozoic (following Korotayev & Markov, 2014). *Note:* Exponential model: $R^2 = 0.51$. Hyperbolic model: $R^2 = 0.91$. The hyperbolic line has been generated by the following equation: $N_t = \frac{8.93 \times 10^5}{35-t}$

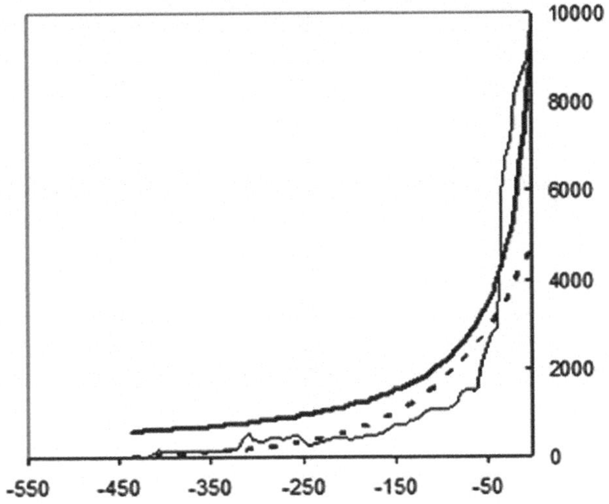

Fig. 12.17 Global change in continental biodiversity (number of genera, N) through the Phanerozoic (following Korotayev & Markov, 2014). *Note:* Exponential model: $R^2 = 0.86$. Hyperbolic model: $R^2 = 0.94$. The hyperbolic line has been generated by the following equation: $N_t = \frac{2.72 \times 10^5}{29-t}$

Most authors agree that three major biotic changes resulted in the fundamental reorganization of community structure during the Phanerozoic: Ordovician radiation (also known as "the Great Ordovician Biodiversification Event" [GOBE]), end-Permian extinction, and end-Cretaceous extinction (Bambach, 1977; Sepkoski et al., 1981; Sepkoski, 1988; Markov, 2001; Bambach et al., 2002; Benton, Wu, 2022; Wu et al., 2024). Generally, after each major crisis, the communities became more complex, diverse, and stable. The stepwise increase of alpha diversity (*i.e.*, the average number of species or genera in a community) through the Phanerozoic was demonstrated by Bambach (1977) and Sepkoski (1988). Although Powell and Kowalewski (2002) have argued that the observed increase in alpha diversity

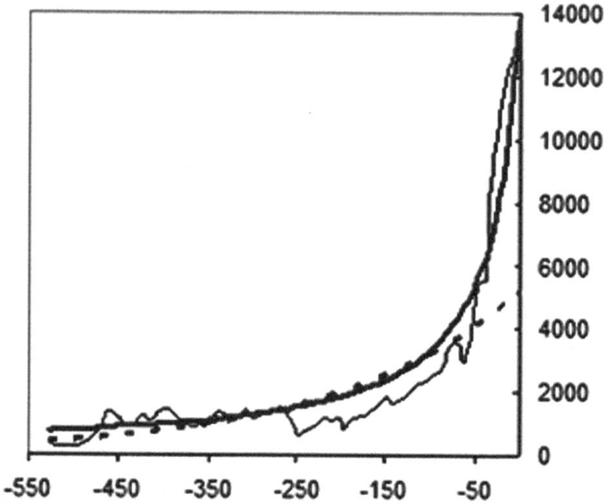

Fig. 12.18 Global change in total biodiversity (number of genera, *N*) through the Phanerozoic (following Korotayev & Markov, 2014). *Note:* Exponential model: $R^2 = 0.67$. Hyperbolic model: $R^2 = 0.95$. The hyperbolic line has been generated by the following equation: $N_t = \frac{4.35 \times 10^5}{30 - t}$

might be an artifact caused by several specific biases that influenced the taxonomic richness of different parts of the fossil record, there is evidence that these biases largely compensated for one another so that the observed increase in alpha diversity was probably underestimated rather than overestimated (Bush & Bambach, 2004; Markov, 2009).

Another important symptom of progressive development of communities is an increase in the evenness of species (or genus) abundance distribution. In primitive, pioneer, or suppressed communities, this distribution is strongly uneven: the community is overwhelmingly dominated by a few very abundant species. In more advanced, climax, or flourishing communities, this distribution is more even (Magurran, 1988). The former type of community is generally more vulnerable. The evenness of species richness distribution in communities increased substantially during the Phanerozoic (Powell & Kowalewski, 2002; Bush & Bambach, 2004). It is most likely there was also an increase in habitat utilization, total biomass, and the rate of trophic flow in biota through the Phanerozoic (Powell & Kowalewski, 2002).

The more complex the community, the more stable it is due to the development of effective interspecies interactions and homeostatic mechanisms based on the negative feedback principle. In a complex community, when the abundance of a species decreases, many factors arise that facilitate its recovery (*e.g.*, food resources rebound while predator populations decline) (Foote, 2023). Even if the species becomes extinct, its vacant niche may "recruit" another species, most probably a related one that may acquire morphological similarity with its predecessor and thus will be assigned to the same genus by taxonomists. So a complex community can facilitate the stability (and longevity) of its components, such as niches, taxa and morphotypes. This effect reveals itself in the phenomenon of "coordinated stasis." The fossil record contains many examples in which particular communities persist for million years while the rates of extinction and taxonomic turnover are minimized (Brett et al., 1996, 2007; Roopnarine, 2025).

Selective extinction leads to the accumulation of "extinction-tolerant" taxa in the biota (Sepkoski, 1991b). Although there is evidence that mass extinctions can be nonselective in some aspects (Jablonski, 2005), they are obviously highly selective with respect to the ability of taxa to endure unpredictable environmental changes (Payne et al., 2023). This can be seen, for instance, in the selectivity of the end-Cretaceous mass extinction with respect to the time of the first occurrence of genera. In younger cohorts, the extinction level was higher than that of the older cohorts (see Markov & Korotayev, 2007: fig. 2). The same pattern can be observed during the periods of "background" extinction as well. This means that genera differ in their ability to survive extinction events, and that extinction-tolerant genera accumulate in each cohort over the course of time. Thus, taxa generally become more stable and long-lived through the course of evolution, apart from the effects of communities. The communities composed of more stable taxa would be, in turn, more stable themselves, thus creating positive feedback.

The stepwise change of dominant taxa plays a major role in biotic evolution. This pattern is maintained not only by the selectivity of extinction (discussed above) but also by the selectivity of the recovery after crises (Bambach et al., 2002). The taxonomic structure of the Phanerozoic biota was changing in a stepwise way, as demonstrated by the concept of three sequential "evolutionary faunas" (Sepkoski, 1992). There were also stepwise changes in the proportion of major groups of animals with different ecological and physiological parameters. There was stepwise growth in the proportion of motile genera to nonmotile, "physiologically buffered" genera to "unbuffered," and predators to prey (Bambach et al., 2002). All these trends should have facilitated the stability of communities. For example, the diversification of predators implies that they became more specialized. A specialized predator regulates its prey's abundance more effectively than a nonspecialized predator.

There is also another possible mechanism of second-order positive feedback between diversity and its growth rate. Some research has demonstrated a shift in typical relative-abundance distributions in paleocommunities after the Paleozoic (Wagner et al., 2006). One possible interpretation of this shift is that community structure and the interactions between species in the communities became more complex. In post-Paleozoic communities, new species probably increased ecospace more efficiently, either by facilitating opportunities for additional species or by niche construction (Wagner et al., 2006; Solé et al., 2002; Laland et al., 1999; Antell & Saupe, 2021). This possibility makes the mechanisms underlying the hyperbolic growth of biodiversity and human population even more similar, because the total ecospace of the biota is analogous to the "carrying capacity of the Earth" in demography. As far as new species can increase ecospace and facilitate opportunities for additional species entering the community, they are analogous to the "inventors" of the demographic-economic models whose inventions increase the carrying capacity of the Earth.

Exponential and logistic models of biodiversity imply several possible ways in which the rates of origination and extinction may change through time (Sepkoski, 1991a; Foote, 2023). For instance, exponential growth can be derived from constant per-taxon extinction and origination rates, the latter being higher than the former. However, actual paleontological data suggest that origination and extinction rates did not follow any distinct trend through the Phanerozoic, and their changes through time look very much like chaotic fluctuations (Cornette & Lieberman, 2004). Therefore, it is more difficult to find a simple mathematical approximation for the origination and extinction rates than for the total diversity. In fact, the only critical requirement of the exponential model is that the difference between the origination and extinction through time should be proportional to the current diversity level:

$$(N_o - N_e)/\Delta t \approx kN, \tag{12.12}$$

where N_o and N_e are the numbers of genera with, respectively, first and last occurrences within the time interval Δt, and N is the mean diversity level during the interval. The same is true for the hyperbolic model. It does not predict the exact way in which origination and extinction should change, but it does predict that their difference should be roughly proportional to the square of the current diversity level:

$$(N_o - N_e)/\Delta t \approx kN^2. \tag{12.13}$$

In the demographic models discussed above, the hyperbolic growth of the world population was not decomposed into separate trends of birth and death rates. The main driving force of this growth was presumably an increase in the carrying capacity of the Earth. The way in which this capacity was realized—either by decreasing death rate or by increasing birth rate, or both—depended upon many factors and may varied from time to time.

The same is probably true for biodiversity. The overall shape of the diversity curve depends mostly on the differences in the mean rates of diversity growth in the Paleozoic (low), Mesozoic (moderate), and Cenozoic (high). The Mesozoic increase was mainly due to a lower extinction rate (compared to the Paleozoic), while the Cenozoic increase was largely due to a higher origination rate (compared to the Mesozoic) (see Markov & Korotayev, 2007, p. 316, Figs. 3a and b). This probably means that the acceleration of diversity growth during the last two eras was driven by different mechanisms of positive feedback between diversity and its growth rate. Generally, the increment rate ($(N_o - N_e)/\Delta t$) was changing in a more regular way than the origination rate $N_o/\Delta t$ and extinction rate $N_e/\Delta t$. The large-scale changes in the increment rate correlate better with N^2 than with N (see Markov & Korotayev, 2007, p. 316, Figs. 3c and d; see also Markov & Korotayev, 2008), thus supporting the hyperbolic rather than the exponential model.

12.9 Conclusion

In mathematical models of historical macrodynamics, a hyperbolic pattern of world population growth arises from nonlinear second-order positive feedback between the demographic growth and technological development. Based on the analogy with macrosociological models and diverse paleontological data, we suggest that the hyperbolic character of biodiversity growth can be similarly accounted for by nonlinear second-order positive feedback between the diversity growth and the complexity of community structure. This hints at the presence, within the biosphere, of a certain analog to the collective learning mechanism. The feedback can work via two parallel mechanisms: (1) a decreasing extinction rate (more surviving taxa—higher alpha diversity—communities become more complex and stable—extinction rate decreases—more taxa, and so on), and (2) an increasing origination rate (new taxa—niche construction—newly formed niches occupied by the next "generation" of taxa—new taxa, and so on). The latter possibility makes the mechanisms underlying the hyperbolic growth of biodiversity and human population even more similar, because the total ecospace of the biota is analogous to the "carrying capacity of the Earth" in economic demography. As far as new species can increase ecospace and facilitate opportunities for additional species entering the community, they are analogous to the "inventors" of the economic–demographic models whose inventions increase the carrying capacity of the Earth.

The hyperbolic growth of Phanerozoic biodiversity suggests that "cooperative" interactions between taxa can play an important role in evolution, along with generally accepted competitive interactions. Due to this "cooperation" (which may be roughly analogous to "collective learning"), the evolution of biodiversity acquires some features of a self-accelerating process. The same is naturally true of cooperation/collective learning in global social evolution. This analysis suggests that we can trace rather similar macropatterns within both the biological and social phases of Big History. These macropatterns can be represented by relatively similar curves and described accurately with very simple mathematical models (see Korotayev, 2018, 2020b, 2024, 2025a for more detail).

References

Aghion, P., & Howitt, P. (1998). *Endogenous growth theory*. MIT Press.

Alexander, M. A. (2016). Application of mathematical models to English secular cycles. *Cliodynamics, 7*(1), 76–108. https://doi.org/10.21237/C7clio7128325.

Ammerman, A. J., & Cavalli-Sforza, L. L. (2014). *The neolithic transition and the genetics of populations in Europe*. Princeton University Press.

Anisimov, V. A. (2013). Gipoteza emnogoabiogeneza v svete dannykh paleontologii, molekulyarnoy biologii i analiza khimicheskogo sostava molekul RNK. *Evolyutsiya, 5*, 14–31.

Antell, G. T., & Saupe, E. E. (2021). Bottom-up controls, ecological revolutions and diversification in the oceans through time. *Current Biology, 31*(19), 1237–1251.

Aunger, R. (2007). A rigorous periodization of 'big' history. *Technological Forecasting and Social Change, 74*(8), 1164–1178. https://doi.org/10.1016/j.techfore.2007.01.007

Averkieva, Y. (1978). The Indians of the Northwest Coast of the North America (the Tlingit). In Y. Averkieva (Ed.), *The North American Indians* (pp. 318–360). Progress.

Bae, C. J., Douka, K., & Petraglia, M. D. (2017). On the origin of modern humans: Asian perspectives. *Science, 358*(6368), eaai9067. https://doi.org/10.1126/science.aai9067

Baker, D. (2016). Collective learning: A potential unifying theme of human history. *Journal of World History, 26*(1), 77–104. https://doi.org/10.1353/jwh.2016.0006

Balzeau, A., Grimaud-Hervé, D., Détroit, F., Holloway, R. L., Combès, B., & Prima, S. (2013). First description of the Cro-Magnon 1 endocast and study of brain variation and evolution in anatomically modern Homo sapiens. *Bulletins et Mémoires de la Société d'anthropologie de Paris, 25*(1), 1–18. https://doi.org/10.1007/s13219-012-0069-z

Bambach, R. K. (1977). Species Richness in Marine Benthic Habitats through the Phanerozoic. *Paleobiology, 3*, 152–167.

Bambach, R. K., Knoll, A. H., & Sepkoski, J. J., Jr. (2002). Anatomical and ecological constraints on Phanerozoic animal diversity in the marine realm. *Proceedings of the National Academy of Sciences, 99*(10), 6854–6859. https://doi.org/10.1073/pnas.092150999

Barkow, J., Cosmides, L., & Tooby, J. (Eds.). (1992). *The adapted mind: Evolutionary psychology and the generation of culture*. Oxford University Press.

Barton, C. M., & Clark, G. A. (2021). From artifacts to cultures: Technology, society, and knowledge in the upper paleolithic. *Journal of Paleolithic Archaeology, 4*, article 6. https://doi.org/10.1007/s41982-021-00091-8

Bar-Yosef, O. (2002). The Upper Paleolithic revolution. *Annual review of anthropology, 31*(1), 363–393. https://doi.org/10.1146/annurev.anthro.31.040402.085416

Bar-Yosef, O. (2007). The archaeological framework of the Upper Paleolithic revolution. *Diogenes, 54*(2), 3–18. https://doi.org/10.1177/0392192107076869

Bar-Yosef, O., & Kuhn, S. L. (2016). The big deal about blades: Laminar technologies and human evolution. In *Human Evolution Source Book* (pp. 597–610). Routledge.

Becerra-Valdivia, L., & Higham, T. (2020). The timing and effect of the earliest human arrivals in North America. *Nature, 584*(7819), 93–97. https://doi.org/10.1038/s41586-020-2491-6

Benton, M.J., & Wu, F. (2022). Triassic revolution. *Frontiers in Earth Science, 10*, article 899541.

Berezkin, Y. (2007). *Myths Populate America. Areal allocation of folk motives and early migration to the new world.* OGI. In Russian.

Blair, P. (1966). *Roman Britain and Early England 55 B.C. – A.D. 871*. W. W. Norton.

Bondarenko, V. M. (2024). The present and future of technological development and the construction of the information society. *Journal of Globalization Studies, 15*(2), 53–64. https://doi.org/10.30884/jogs/2024.02.03

Bondarenko, D. M. (2025). Evolution and forms of socio-political organization: Historical and anthropological approaches (introductory remarks). *Social Evolution & History, 24*(2), 118–126. https://doi.org/10.30884/seh/2025.02.06

Boyd, R., & Richerson, P. J. (1996). Why culture is common, but cultural evolution is rare. *Proceedings of the British Academy, 88*, 77–93.

Brett, C. E., Hendy, A. J., Bartholomew, A. J., Bonelli, J. R., Jr., & McLaughlin, P. I. (2007). Response of shallow marine biotas to sea-level fluctuations: A review of faunal replacement and the process of habitat tracking. *Palaios, 22*(3), 228–244. https://doi.org/10.2110/palo.2005.p05-028r

Brett, C. E., Ivany, L. C., & Schopf, K. M. (1996). Coordinated stasis: An overview. *Palaeogeography, Palaeoclimatology, Palaeoecology, 127*(1–4), 1–20. https://doi.org/10.1016/S0031-0182(96)00085-5

Bush, A. M., & Bambach, R. K. (2004). Did alpha diversity increase during the Phanerozoic? Lifting the veils of taphonomic, latitudinal, and environmental biases. *The Journal of Geology, 112*(6), 625–642. https://doi.org/10.1086/424576

Butovskaya, M. L., Korotayev, A., & Kazankov, A. A. (2020). Dominance styles and variability of social relationships in non-human and human primates. *Evolution, 7*, 69–163.

Byrne, R., & Bates, L. (2007). Sociality, evolution, and cognition. *Current Biology, 17*(16), R714–R723.

Byrne, R., & Whiten, A. (Eds.). (1988). *Machiavellian intelligence*. Oxford University Press.

Byrne, R. W. (2022). Machiavellian intelligence. In J. Vonk & T. K. Shackelford (Eds.), *Encyclopedia of animal cognition and behavior* (pp. 4033–4038). Springer Nature. https://doi.org/10.1007/978-3-319-55065-7

Callaway, E. (2017). Oldest Homo sapiens fossil claim rewrites our species' history. *Nature, 546*, 289–293. https://doi.org/10.1038/nature.2017.22114

Chadwick, N. (1987). *The celts*. Penguin.

Chandler, T. (1987). *Four thousand years of urban growth: An historical census*. The Edwin Mellen Press.

Check, E. (2006). How Africa learned to love the cow: The development of lactose tolerance in sub-Saharan Africa, a fascinating tale of genetic convergence. *Nature, 444*(7122), 994–996. https://doi.org/10.1038/444994a993

Chittka, L., & Rossi, N. (2022). Social cognition in insects. *Trends in Cognitive Sciences, 26*(7), 578–592. https://doi.org/10.1016/j.tics.2022.04.001

Christian, D. (2005). *Maps of time: An introduction to Big History*. University of California Press.

Christian, D. (2017). What is Big History? *Journal of Big History, 1*(1), 4–19.

Christian, D. (2019). *Origin story: A Big History of everything*. Little, Brown Spark.

Christian, D., Brown, C., & Benjamin, C. (2013). *Big History: Between nothing and everything*. McGraw Hill Education.

Chu, C. Y. C., & Lee, R. D. (1994). Famine, revolt, and the dynastic cycle: Population dynamics in historic China. *Journal of Population Economics, 7*, 351–378.

Claessen, H. (1989). Evolutionism in development. *Vienne Contributions to Ethnology and Anthropology, 5*, 231–247.

Cochran, G., & Harpending, H. (2009). *The 10,000 year explosion: How civilization accelerated human evolution*. Basic Books.

Cohen, J. E. (1995). Population growth and earth's carrying capacity. *Science, 269*(5222), 341–346.

Conard, N. J., Malina, M., & Münzel, S. C. (2009). New flutes document the earliest musical tradition in southwestern Germany. *Nature, 460*(7256), 737–740. https://doi.org/10.1038/nature08169

Conte, C., Ibáñez Estévez, J. J., Gibaja Bao, J. F., Mazzucco, N., Terradas, X., MozotaHolgueras, M., & Borrell, F. (2018). Cereal use-wear traces and harvesting methods. In: O. Lozovskaya, A. Vybornov, & E. Dolbunova (Eds.), *Subsistence strategies in the stone age, direct evidence of fishing and gathering. Materials of the international conference dedicated to the 50th anniversary of Vladimir Mikhailovich Lozovski 15–18 May 2018* (pp. 192–193). Institute for the History of Material Culture and the State Hermitage Museum.

Cooke, N. P., Mattiangeli, V., Cassidy, L. M., Okazaki, K., Stokes, C. A., Onbe, S., et al. (2021). Ancient genomics reveals tripartite origins of Japanese populations. *Science Advances, 7*(38), eabh2419. https://doi.org/10.1126/sciadv.abh2419

Cornette, J. L., & Lieberman, B. S. (2004). Random walks in the history of life. *Proceedings of the National Academy of Sciences, 101*(1), 187–191. https://doi.org/10.1073/pnas.2637009100

Cuenca-Solana, D., Gutiérrez-Zugasti, F. I., González-Morales, M. R., Setién-Marquinez, J., Ruiz-Martinez, E., García-Moreno, A., & Clemente-Conte, I. (2013). Shell technology, rock art, and the role of marine resources during the Upper Paleolithic. *Current Anthropology, 54*(3), 370–380. https://doi.org/10.1086/670325

Culotta, E. (1999). A new human ancestor? *Science, 284*, 572–573.

Curry, A. (2013). Archaeology: The milk revolution. *Nature, 500*(7460), 20–22. https://doi.org/10.1038/500020a

Darwin, C. (1883). *The variation of animals and plants under domestication.* 2nd ed., rev. D. Appleton.

Dawkins, R. (1993). *The selfish gene.* Mir. In Russian.

Dunbar, H. E., Wilson, A. C. C., Ferguson, N. R., & Moran, N. A. (2007). Aphid thermal tolerance is governed by a point mutation in bacterial symbionts. *PLoS Biology, 5*(5), e96. https://doi.org/10.1371/journal.pbio.0050096

Dunbar, R. (2003). The social brain: Mind, language, and society in evolutionary perspective. *Annual Review of Anthropology, 32*(1), 163–181.

Dunbar, R. (2024). The social brain hypothesis – thirty years on. *Annals of Human Biology, 51*(1), article 2359920.

Dunbar, R., & Shultz, S. (2007). Evolution in the social brain. *Science, 317*, 1344–1347.

Durkheim, E. (1991) [1893]. *The division of labor in society. Sociological method.* Nauka. In Russian.

Foote, M. (2023). Diversity-dependent diversification in the history of marine animals. *American Naturalist, 201*(5), 680–693.

Frigaard, N. U., Martinez, A., Mincer, T. J., & DeLong, E. F. (2006). Proteorhodopsin lateral gene transfer between marine planktonic Bacteria and Archaea. *Nature, 439*(7078), 847–850. https://doi.org/10.1038/nature04435

Fu, Q., Li, H., Moorjani, P., Jay, F., Slepchenko, S. M., Bondarev, A. A., et al. (2014). Genome sequence of a 45,000-year-old modern human from western Siberia. *Nature, 514*(7523), 445–449. https://doi.org/10.1038/nature13810

Giacobini, G. (2007). Richness and diversity of burial rituals in the Upper Paleolithic. *Diogenes, 54*(2), 19–39. https://doi.org/10.1177/0392192107077649

Gibbons, A. (1997). Chromosome shows that Adam was an African. *Science, 278*, 804–805.

Gilbert, K. (2014). The universal breakthroughs of Big History: Developing a unified theory. In L. Grinin, D. Baker, E. Quaedackers, & A. Korotayev (Eds.), *Teaching and researching Big History: Exploring a new scholarly field* (pp. 122–146). Uchitel.

Glötzl, E. (2025). Unification of biological and cultural evolution through natural periodization. In D. J. LePoire, L. Grinin, & A. Korotayev (Eds.), *Navigating complexity in Big History—Exploring periodization across cosmic and biosocial dimensions* (pp. 265–318). Springer Nature. https://doi.org/10.1007/978-3-031-85410-1_9

Goder-Goldberger, M., Barzilai, O., & Boaretto, E. (2023). Innovative technological practices and their role in the emergence of initial Upper Paleolithic technologies: A view from BokerTachtit. *Journal of Paleolithic Archaeology, 6*(1), article 11. https://doi.org/10.1007/s41982-023-00137-z

Goldstone, J. A. (1991). *Revolution and rebellion in the early modern world.* University of California Press.

Golosovsky, M. (2010). Hyperbolic growth of the human population of the earth: Analysis of existing models. *History & Mathematics, 3*, 188–204.

Grinin, A. (2025). Viruses and evolution: The role of viruses in Big History. *Journal of Big History, 8*(1), 51–56. https://doi.org/10.22339/jbh.v8i1.8108

Grinin, A., & Grinin, L. (2025a). Modernization in African countries and destabilization processes. *Journal of Globalization Studies, 16*(2), 3–28. https://doi.org/10.30884/jogs/2025.02.01

Grinin, L. (1997). Formations and civilizations. Ch. 2. *Filosofia i obschestvo, 2*, 5–89. In Russian.
Grinin, L. (2003). The early state and its analogues. *Social Evolution & History, 2*(1), 131–176.
Grinin, L. (2004). Early state and democracy. In L. Grinin et al. (Eds.), *The early state, its alternatives and analogues* (pp. 419–463). Uchitel.
Grinin, L. (2006a). The transformation of the egypt state system in the 19th – Early 20th centuries: From developed state to the mature one. In A. Logunov (Ed.), *Egypt, near east and global world* (pp. 123–132). Krankes. In Russian.
Grinin, L. (2006b). *Productive forces and historical process* (3rd ed.). KomKniga. In Russian.
Grinin, L. (2007). Problems in the analysis of the driving forces of historical development: Social progress and social evolution. In Y. Semenov et al. (Eds.), *Philosophy of history: Problems and prospects* (pp. 148–247). KomKniga/URSS. In Russian.
Grinin, L. (2009a). *State and historical process. Political cut of historical process* (2nd ed.). LIBROCOM. In Russian.
Grinin, L. (2009b). The pathways of politogenesis and models of the early state formation. *Social Evolution & History, 8*(1), 92–132.
Grinin, L. (2010a). *State and historical process. The evolution of the statehood: From an early to a mature state* (2nd ed.). LIBROCOM.
Grinin, L. (2010b). The role of the individual in history: A reconsideration. *Social Evolution & History, 9*(2), 95–136.
Grinin, L. (2011). *The evolution of statehood: From early states to global society*. Lambert Academic Publishing.
Grinin, L. (2012a). *Macrohistory and globalization*. Uchitel.
Grinin, L. (2012b). State and socio-political crises in the process of modernization. *Cliodynamics, 3*(1), 124–157.
Grinin, L. (2013a). *Bolshaya istoriya razvitiya mira: kosmicheskaya evolyutsiya*. Uchitel.
Grinin, L. (2013b). State and socio-political crises in the process of modernization. *Social Evolution & History, 12*(2), 35–76.
Grinin, L. (2014). The star-galaxy era of Big History in the light of universal evolutionary principles. In L. Grinin, D. Baker, E. Quaedackers, & A. Korotayev (Eds.), *Teaching and researching Big History: Exploring a new scholarly field* (pp. 163–187). Uchitel.
Grinin, L. (2017). *Bolshaya istoriya razvitiya mira: istoriya i evolyutsiya Solnechnoy sistemy*. Moskovskaya redaktsiya izdatelstva Uchitel.
Grinin, L. (2018). Evolution of the early solar system in terms of Big History and universal evolution. *Journal of Big History, 2*(1), 15–26.
Grinin, L. (2020). *Bolshaya istoriya razvitiya mira: planet Solnechnoy sistemy. Ikh istoriya i evolyutsiya. Khimicheskaya evolyutsiya v kosmose i na Zemle*. Moskovskaya redaktsiya izdatelstva Uchitel.
Grinin, L. (2022). Revolution and modernization traps. In J. A. Goldstone, L. Grinin, & A. Korotayev (Eds.), *Handbook of revolutions in the 21st century: The new waves of revolutions, and the causes and effects of disruptive political change* (pp. 219–238). Springer Nature. https://doi.org/10.1007/978-3-030-86468-2_8
Grinin, L. (2023). Chemical evolution in Big History. *Journal of Big History, 7*(2), 97–105. https://doi.org/10.22339/jbh.v7i2.7208
Grinin, L., Baker, D., Quaedackers E., & Korotayev A. (2014). (Eds.). *Teaching & Researching Big History: Exploring a New Scholarly Field*. Uchitel.
Grinin, L., Carneiro, R. L., Korotayev A., & Spier F. (Eds.) (2011). *Evolution: Cosmic, biological, and social*. Uchitel.
Grinin, L., & Grinin, A. (2015). *Ot rubil do nanorobotov. Mir na puti k epokhe samoupravlyayemykh sistem (istoriya tekhnologiy i opisaniye ikh budushchego)*. Uchitel.
Grinin, L., & Grinin, A. (2019a). Technological dimension of Big History and the cybernetic revolution. *Evolution, 6*, 185–213.

Grinin, L., & Grinin, A. (2019b). The Star-galaxy era in terms of Big History and universal evolution. *Journal of Big History, 3*(4), 69–92.
Grinin, L., & Grinin, A. (2020a). Big History and the future of technology. In: C. Benjamin, E. Quaedackers, & D. Baker. (Eds.), *The Routledge companion to Big History*. Routledge.
Grinin, L., & Grinin, A. (2020b). Social evolution as an integral part of universal evolution. *Social Evolution & History, 19*(2), 19–45.
Grinin, L., & Grinin, A. (2023). Evolutionary phases in Big History and complexity. *Journal of Big History, 6*(3), 5–17. https://doi.org/10.22339/jbh.v6i3.6302
Grinin, L., & Grinin, A. (2025b). Big History periodization. Complexity, directions and phase characteristics. In D. J. LePoire, L. Grinin, & A. Korotayev (Eds.), *Navigating complexity in Big History—Exploring periodization across cosmic and biosocial dimensions* (pp. 319–342). Springer Nature. https://doi.org/10.1007/978-3-031-85410-1_10
Grinin, L., Grinin, A., & Korotayev, A. (2023). Demographic transformations in the light of technological development: Types of demographic reproduction in the past and in the future. *Social Evolution & History, 22*(2), 203–248. https://doi.org/10.30884/seh/2023.02.09
Grinin, L., Grinin, A., & Korotayev, A. (2024). *Cybernetic revolution and global aging. Humankind on the way to cybernetic society, or the next hundred years*. Springer Nature. https://doi.org/10.1007/978-3-031-56764-3
Grinin, L., & Korotayev, A. (2007). Social macroevolution and historical process. *Filosofiya i obschestvo, 2*, 19–66. In Russian.
Grinin, L., & Korotayev, A. (2008). Macroevolution and the world-system: New facets of conceptualization. *Istoriya i sovremennost, 1*, 3–31. In Russian.
Grinin, L., & Korotayev, A. (2009a). Social macroevolution: Growth of the world system integrity and a system of phase transitions. *World Futures, 65*(7), 477–506. https://doi.org/10.1080/02604020902733348
Grinin, L., & Korotayev, A. (2009b). *Social macroevolution: Genesis and transformations of the world system*. LIBROKOM. In Russian.
Grinin, L., & Korotayev, A. (2009c). On some peculiarities of social-political development of Ottoman Egypt (the 16th – 18th centuries). *Vostok (Oriens), 1*, 46–62. In Russian.
Grinin, L., & Korotayev, A. (2009d). On the typological characteristics of statehood of Ottoman Egypt (the 16th – 19th centuries). *Vostok (Oriens), 3*, 35–51.
Grinin, L., & Korotayev, A. (2009e). The Epoch of the initial politogenesis. *Social Evolution & History, 8*(1), 52–91.
Grinin, L., & Korotayev, A. (2015). *Great divergence and great convergence. A global perspective*. Springer Nature. https://doi.org/10.1007/978-3-319-17780-9
Grinin, L., & Korotayev, A. (2023). Social evolution: Theoretical aspects and general outlines. *Social Evolution & History, 22*(2), 111–158. https://doi.org/10.30884/seh/2023.02.07
Grinin, L., & Korotayev, A. (2025). Possible methodological ideas for periodizing big history. In D. J. LePoire, L. Grinin, & A. Korotayev (Eds.), *Navigating complexity in Big History. Exploring periodization across cosmic and biosocial dimensions* (pp. 49–71). Springer Nature. https://doi.org/10.1007/978-3-031-85410-1_3
Grinin, L., Korotayev, A., & Markov, A. (2020). Big History and macro-evolution. Evolutionary principles and mechanisms at biological and social phases of the Big History. In: C. Benjamin, E. Quaedackers, & D. Baker (Eds.), *The Routledge companion to Big History* (pp. 72–105). Routledge.
Grinin, L., Malkov, S., & Korotayev, A. (2023). High income and low income countries. Toward a common goal at different speeds. In V. Sadovnichy et al. (Eds.), *Reconsidering the limits to growth. A report to the Russian Association of the Club of Rome* (pp. 207–224). Springer. https://doi.org/10.1007/978-3-031-34999-7_12

Grinin, L., Markov, A., & Korotayev, A. (2009). Aromorphoses in biological and social evolution: Some general rules for biological and social forms of macroevolution. *Social Evolution & History, 8*(2), 6–50.

Grinin, L., Markov, A., & Korotayev, A. (2011). Biological and social aromorphoses: A comparison between two forms of macroevolution. In L. Grinin, R. L. Carneiro, A. Korotayev, & F. Spier (Eds.), *Evolution: Cosmic, biological, and social* (pp. 162–211). Uchitel.

Grinin, L., Markov, A., & Korotayev, A. (2012). Biological and social phases of Big History: Similarities and differences of evolutionary principles and mechanisms. *Evolution, 2*, 158–198.

Grinin, L., Markov, A., & Korotayev, A. (2013). On similarities between biological and social evolutionary mechanisms: Mathematical modeling. *Cliodynamics, 4*(2), 185–228. https://doi.org/10.21237/C7clio4221334

Grinin, L., Markov, A., & Korotayev, A. (2022). *Macroevolution in biological and social systems*. URSS.

Grinin, L., Markov, A., & Korotayev, A. (2025). Biological and social phases of Big History and universal evolution: Similarities, differences, and the growth of complexity. *Journal of Big History, 8*(3), 1–51. https://doi.org/10.22339/jbh.v8i3.8301

Grossman, G., & Helpman, E. (1991). *Innovation and growth in the global economy*. MIT Press.

Gruebler, A. (2006). *Urbanization as core process of global change: The last 1000 years and next 100.* In Paper presented at the International Seminar 'Globalization as Evolutionary Process: Modeling, Simulating, and Forecasting Global Change', International Institute for Applied Systems Analysis (IIASA), Laxenburg, Austria, April 6–8.

Gruhn, R. (2023). An anthropological conception of the initial peopling of the Americas. *PaleoAmerica, 9*(3), 167–173. https://doi.org/10.1080/20555563.2023.2278948

Guran, S. H., Yousefi, M., Kafash, A., & Ghasidian, E. (2024). Reconstructing contact and a potential interbreeding geographical zone between Neanderthals and anatomically modern humans. *Scientific Reports, 14*(1), article 20475. https://doi.org/10.1038/s41598-024-70206-y

Hallpike, C. R. (1986). *Principles of social evolution*. Clarendon.

Harper, A. (2026). Punctuated growth and complexity in urban evolution: Insights from historical city size distributions. In: D. J. LePoire, L. Grinin, & A. Korotayev (Eds.), *Complexity in universal evolution – A Big History perspective* (pp. 519–536). Springer Nature (in this volume). https://doi.org/10.1007/978-3-032-20516-2_18

Hawks, J. (2014). Still evolving (after all these years). *Scientific American, 311*(3), 86–91.

Hawks, J., Wang, E. T., Cochran, G. M., Harpending, H. C., & Moyzis, R. K. (2007). Recent acceleration of human adaptive evolution. *Proceedings of the National Academy of Sciences, 104*(52), 20753–20758. https://doi.org/10.1073/pnas.0707650104

Henrich, J. (2015). *The secret of our success: How culture is driving human evolution, domesticating our species, and making us smarter*. Princeton University Press.

Henriques, G. R., Michalski, J., Quackenbush, S., & Schmidt, W. (2019). The tree of knowledge system: A new map for Big History. *Journal of Big History, 3*(4), 1–17. https://doi.org/10.22339/jbh.v3i4.3410

Henriques, G. R., & Volk, T. (2023). Toward a Big History 2.0: A brief position paper. *Journal of Big History, 6*(3), 1–4. https://doi.org/10.22339/jbh.v6i3.6301

Herrmann, E., Call, J., Hernàndez-Lloreda, M., Hare, B., & Tomasello, M. (2007). Humans have evolved specialized skills of social cognition: The cultural intelligence hypothesis. *Science, 317*, 360–366.

Heylighen, F. (2011). Conceptions of a global brain: An historical review. In L. Grinin, R. L. Carneiro, A. Korotayev, & F. Spier (Eds.), *Evolution: Cosmic, biological, and social* (pp. 274–289). Uchitel.

Hoggard, N. (2025). A new theory of evolution is supported by the teaching patterns of human ancestors. In D. J. LePoire, L. Grinin, & A. Korotayev (Eds.), *Navigating complexity in Big History— Exploring periodization across cosmic and biosocial dimensions* (pp. 217–263). Springer Nature. https://doi.org/10.1007/978-3-031-85410-1_8

Holden, C. (1998). No last word on language origins. *Science, 282*, 1455–1458.

Hoyer, D., Bennett, J. S., Whitehouse, H., Francois, P., Reddish, J., Davis, D., et al. (2024). Crises averted. How a few past societies found adaptive reforms in the face of structural-demographic crises. *Cliodynamics, 15*(2), 1–50. https://doi.org/10.21237/C7clio.38365

Hublin, J. J. (2021). How old are the oldest Homo sapiens in Far East Asia? *Proceedings of the National Academy of Sciences, 118*(10), e2101173118. https://doi.org/10.1073/pnas.2101173118

Hublin, J. J., Ben-Ncer, A., Bailey, S. E., Freidline, S. E., Neubauer, S., Skinner, M. M., et al. (2017). New fossils from Jebel Irhoud, Morocco and the pan-African origin of Homo sapiens. *Nature, 546*(7657), 289–292. https://doi.org/10.1038/nature22336

Hublin, J. J., Sirakov, N., Aldeias, V., Bailey, S., Bard, E., Delvigne, V., et al. (2020). Initial upper palaeolithic homo sapiens from BachoKiro Cave, Bulgaria. *Nature, 581*(7808), 299–302. https://doi.org/10.1038/s41586-020-2259-z

Jablonka, E., & Lamb, M. (1999). *Epigenetic inheritance and evolution: Lamarckian dimension.* Oxford University Press.

Jablonski, D. (2005). Mass extinctions and macroevolution. *Paleobiology, 31*(2), 192–210.

Jones, C. I. (1995). R&D-based models of economic growth. *The Journal of Political Economy, 103*, 759–784.

Jones, C. I. (2005). The shape of production functions and the direction of technical change. *The Quarterly Journal of Economics, 120*, 517–549.

Kapitza, S. (1996). The phenomenological theory of world population growth. *Physics-Uspekhi, 39*(1), 57–71. https://doi.org/10.1070/PU1996v039n01ABEH000127

Kaufman, D. (1999). *Archeological perspectives on the origins of modern humans. A view from Levant.* Bergin & Garvey.

Kealy, S., Louys, J., & O'Connor, S. (2018). Least-cost pathway models indicate northern human dispersal from Sunda to Sahul. *Journal of Human Evolution, 125*, 59–70. https://doi.org/10.1016/j.jhevol.2018.10.003

Kerner, G., Neehus, A. L., Philippot, Q., Bohlen, J., Rinchai, D., Kerrouche, N., et al. (2023). Genetic adaptation to pathogens and increased risk of inflammatory disorders in post-Neolithic Europe. *Cell Genomics, 3*(2).

Khokhlova, A., & Korotayev, A. (2023). The dimensions of sociopolitical destabilization in MENA region in comparative global perspective. *Journal of Globalization Studies, 14*(1), 67–94. https://doi.org/10.30884/jogs/2023.01.05

Komlos, J., & Nefedov, S. (2002). A compact macromodel of pre-industrial population growth. *Historical Methods, 35*, 92–94.

Komlosy, A. (2024). Re-examining the concept of secular transformation. How can we capture the turn of an era? Theoretical approaches, methods and implications. *Journal of Globalization Studies, 15*(2), 65–80. https://doi.org/10.30884/jogs/2024.02.04

Kondor, D., & Turchin, P. (2024). Soil fertility depletion is not a credible mechanism for population boom/bust cycles in early agricultural societies. *Human Ecology, 52*(4), 731–741. https://doi.org/10.1007/s10745-024-00541-1

Korotayev, A. (1995). Mountains and democracy: An introduction. In N. Kradin & V. Lynsha (Eds.), *Alternative pathways to early state* (pp. 60–74). Dalnauka – Vladivostok.

Korotayev, A. (2003). *Social evolution: Factors, regularities, trends.* Vostochnaya literatura. In Russian. https://elibrary.ru/item.asp?id=22456735

Korotayev, A. (2004). *World religions and social evolution of the old world Oikumene civilizations: A cross-cultural perspective*. The Edwin Mellen Press.

Korotayev, A. (2005). A compact macromodel of world system evolution. *Journal of World-Systems Research, 11*(1), 79–93. https://doi.org/10.5195/jwsr.2005.401

Korotayev, A. (2006). 'Midwest-Amazonian' Folklore-Mythological Parallels? *Acta Americana, 14*(1), 5–24. https://www.academia.edu/3552083/Midwest_Amazonian_Folklore_Mythological_Parallels

Korotayev, A. (2007). Compact mathematical models of world system development, and how they can help us to clarify our understanding of globalization processes. In G. Modelski, T. Devezas, & W. R. Thompson (Eds.), *Globalization as evolutionary process: Modeling global* (pp. 133–160). Routledge.

Korotayev, A. (2008). Globalization and mathematical modeling of global development. In L. Grinin, D. Beliaev, & A. Korotayev (Eds.), *Hierarchy and power in the history of civilizations: Political aspects of modernity* (pp. 225–240). LIBROCOM/URSS. https://elibrary.ru/item.asp?id=29412931

Korotayev, A. (2009). Compact mathematical models of the world system development and their applicability to the development of local solutions in third world countries. In J. Sheffield (Ed.), *Systemic development: Local solutions in a global environment* (pp. 103–116). ISCE Publishing.

Korotayev, A. (2013). Globalization and mathematical modeling of global evolution. *Evolution, 3*, 69–83. https://elibrary.ru/item.asp?id=20433939

Korotayev, A. (2017). Malthusian cycle. In F. M. Moghaddam (Ed.), *The SAGE Encyclopedia of political behavior* (pp. 457–462). SAGE. https://doi.org/10.4135/9781483391144.n208

Korotayev, A. (2018). The 21st century singularity and its Big History implications: A re-analysis. *Journal of Big History, 2*(3), 71–118. https://doi.org/10.22339/jbh.v2i3.2320

Korotayev, A. (2020a). How singular is the 21st century singularity? In A. Korotayev & D. LePoire (Eds.), *The 21st century singularity and global futures. A big history perspective* (pp. 571–595). Springer. https://doi.org/10.1007/978-3-030-33730-8_26

Korotayev, A. (2020b). The 21st century Singularity in the Big History perspective. A re-analysis. In A. Korotayev & D. LePoire (Eds.), *The 21st century singularity and global futures. A Big History perspective* (pp. 19–75). Springer. https://doi.org/10.1007/978-3-030-33730-8_2

Korotayev, A. (2022). Political centralization and communal complexity in cross-cultural and historical perspective. *History & Mathematics, 9*, 181–200. https://www.elibrary.ru/item.asp?id=49052766

Korotayev, A. (2024). Complexity growth patterns in the Big History. Preliminary results of a quantitative analysis. *Journal of Big History, 7*(2), 9–27. https://doi.org/10.22339/jbh.v7i2.7203

Korotayev, A. (2025a). Comparing and contrasting Big History singularity trends of the Big Bang and terrestrial evolution. In D. J. LePoire, L. Grinin, & A. Korotayev (Eds.), *Navigating complexity in Big History — Exploring periodization across cosmic and biosocial dimensions* (pp. 101–157). Springer Nature. https://doi.org/10.1007/978-3-031-85410-1_5

Korotayev, A. (2025b). Toward periodization of the biosocial era of the Big History. A reanalysis. *Journal of Big History, 8*(2), 1–37. https://doi.org/10.22339/jbh.v8i2.8201

Korotayev, A. (2026). Human evolution in the complexity growth perspective. Toward periodization of the Big History biosocial era. In: D. J. LePoire, L. Grinin, & A. Korotayev (Eds.), *Complexity in universal evolution – A Big History perspective* (359–410). Springer Nature (in this volume). https://doi.org/10.1007/978-3-032-20516-2_13

Korotayev, A., Malkov, S., & Musieva, J. (2023). Demography. Toward optimization of demographic processes. In: V. Sadovnichy et al. (Eds.), *Reconsidering the limits to growth. A report to the Russian Association of the Club of Rome* (pp. 97–116). Springer Nature. https://doi.org/10.1007/978-3-031-34999-7_6

Korotayev, A., Berezkin, Y., Borinskaya, S., Davletshin, A., & Khaltourina, D. (2017). Genes and myths: Which genes and myths did the different waves of the peopling of Americas bring to the new world? *History & Mathematics, 6*, 9–77. https://www.researchgate.net/publication/351364727_Genes_and_Myths_Which_Genes_and_Myths_did_the_Different_Waves_of_the_Peopling_of_Americas_Bring_to_the_New_World?

Korotayev, A., Berezkin, Y., Kozmin, A., & Arkhipova, A. (2006). Return of the White Raven: Postdiluvial Reconnaissance Motif A2234.1.1 reconsidered. *Journal of American Folklore, 119*(472), 203–235. https://doi.org/10.2307/4137924

Korotayev, A., Bilyuga, S., & Shishkina, A. (2018). GDP Per capita and protest activity: A quantitative reanalysis. *Cross-Cultural Research, 52*(4), 406–440. https://doi.org/10.1177/1069397117732328

Korotayev, A., & Khaltourina, D. (2006). *Introduction to social macrodynamics: Secular cycles and millennial trends in Africa*. KomKniga/URSS.

Korotayev, A., & Khaltourina, D. (2011). *Myths and genes: A deep historical reconstruction*. Librokom/URSS. https://www.academia.edu/32351421/Мифы_и_гены_Глубокая_историческая_реконструкция_М_Либроком_2011_

Korotayev, A., Kradin, N., Munck, V., & Lynsha, V. (2000). Alternatives of social evolution: Introductory notes. In N. Kradin et al. (Eds.), *Alternatives of Social Evolution* (pp. 12–51). FEB RAS.

Korotayev, A., Kradin, N., Munck, V., & Lynsha, V. (2011). Alternatives of social evolution: Introductory notes. In N. Kradin et al. (Eds.), *Alternatives of social evolution* (pp. 12–51). Lambert Academic Publishing.

Korotayev, A., & LePoire, D. J. (2025). Complexity growth patterns in the Stelliferous Era of Big History. Preliminary observations. *Journal of Big History, 8*(1), 1–12. https://doi.org/10.22339/jbh.v8i1.8102

Korotayev, A., & Malkov, A. (2016). A compact mathematical model of the World System economic and demographic growth, 1 CE – 1973 CE. *International Journal of Mathematical Models and Methods in Applied Sciences, 10*, 200–209. https://elibrary.ru/item.asp?id=27160462

Korotayev, A., Malkov, A., & Khaltourina, D. (2006a). *Introduction to social macrodynamics: Compact macromodels of the world system growth*. KomKniga/URSS.

Korotayev, A., Malkov, A., & Khaltourina, D. (2006b). *Introduction to social macrodynamics: Secular cycles and millennial trends*. KomKniga/URSS.

Korotayev, A., & Malkov, S. (2012). Mathematical models of the world-system development. In S. Babones & C. Chase-Dunn (Eds.), *Routledge handbook of world-systems analysis* (pp. 158–161). Routledge.

Korotayev, A., & Markov, A. (2014). Mathematical modeling of biological and social phases of Big History. In L. Grinin et al. (Eds.), *Teaching & researching Big History: Exploring a new scholarly field* (pp. 188–219). Uchitel.

Korotayev, A., & Markov, A. (2015). Mathematical modeling of biological and social phases of big history. *Globalistics and Globalization Studies, 4*, 319–343.

Korotayev, A., Slinko, E., Meshcherina, K., & Zinkina, J. (2020). Variation of human values and modernization: Preliminary results. *Cross-Cultural Research, 54*(2–3), 238–272. https://doi.org/10.1177/1069397119874781

Korotayev, A., & Zinkina, J. (2015). East Africa in the Malthusian Trap? *Journal of Developing Societies, 31*(3), 385–420. https://doi.org/10.1177/0169796X15590322

Korotayev, A., Zinkina, J., & Andreev, A. (2016). Secular cycles and millennial trends. *Cliodynamics, 7*(2), 204–216. https://doi.org/10.21237/C7CLIO7230484

Korotayev, A., Zinkina, J., Slinko, E., & Meshcherina, K. (2019). Human values and modernization: A global analysis. *Journal of Globalization Studies, 10*(1), 44–71. https://doi.org/10.30884/jogs/2019.01.04

Korotayev, A., Shadrova, A., Sokovnina, E., Kaygorodova, M., Musieva, J., & Vasiliev, A. (2024). Echo of the arab spring in asia: A quantitative analysis. *Journal of Globalization Studies, 15*(1), 110–138. https://doi.org/10.30884/jogs/2024.01.07

Kradin, N. N. (2023). Origins of the early Russian state: Anthropological perspectives. *Social Evolution & History, 22*(2), 67–93. https://doi.org/10.30884/seh/2023.02.05

Kremer, M. (1993). Population growth and technological change: One million B.C. to 1990. *The Quarterly Journal of Economics, 108*(3), 681–716. https://doi.org/10.2307/2118405

Kurdyumov, S. P. (Ed.). (1999). *Blow-up Regimes*. Nauka. In Russian.

Kutter, G. S. (2015). A brief account of the science of Big History. In B. Rodrigue, L. Grinin, & A. Korotayev (Eds.), *From big bang to galactic civilizations. Vol.1: Our place in the universe. An introduction to Big History* (pp. 24–41). RatnaSagar.

Kuznets, S. (1960). Population change and aggregate output. In *Demographic and economic change in developed countries* (pp. 324–340). Princeton University Press.

Laland, K. N., Odling-Smee, F. J., & Feldman, M. W. (1999). Evolutionary consequences of niche construction and their implications for ecology. *Proceedings of the National Academy of Sciences, 96*(18), 10242–10247. https://doi.org/10.1073/pnas.96.18.10242

Lareau, L. F., Inada, M., Green, R. E., Wengrod, J. C., & Brenner, S. E. (2007). Unproductive splicing of SR genes associated with highly conserved and ultraconserved DNA elements. *Nature, 446*(7138), 926–929. https://doi.org/10.1038/nature05676

Lbova, L. (2021). Personal ornaments as markers of social behavior, technological development and cultural phenomena in the Siberian early upper Paleolithic. *Quaternary International, 573*, 4–13. https://doi.org/10.1016/j.quaint.2020.05.035

LePoire, D. J. (2016). Exploring temporal patterns in big history dynamics. *KronoScope, 16*(2), 229–249. https://doi.org/10.1163/15685241-12341358

LePoire, D. J. (2020). *Time patterns in Big History. Cycles, fractals, waves, transitions, and singularities*. David LePoire Consulting, LLC.

LePoire, D. J. (2025). A simple, compatible, and extensible Big History framework and periodization based on previous findings. In D. J. LePoire, L. Grinin, & A. Korotayev (Eds.), *Navigating complexity in Big History—Exploring periodization across cosmic and biosocial dimensions* (pp. 159–193). Springer Nature. https://doi.org/10.1007/978-3-031-85410-1_6

LePoire, D. J., Grinin, L., & Korotayev, A. (2025a). Navigating complexity in Big History—Exploring periodization across cosmic and biosocial dimensions: An introduction. In D. J. LePoire, L. Grinin, & A. Korotayev (Eds.), *Navigating complexity in Big History. Exploring periodization across cosmic and biosocial dimensions* (pp. 1–27). Springer Nature. https://doi.org/10.1007/978-3-031-85410-1_1

LePoire, D. J., Grinin, L., & Korotayev, A. (2025b). Discussion: Cross-cutting topics and a variety of periodizations. In D. J. LePoire, L. Grinin, & A. Korotayev (Eds.), *Navigating complexity in Big History. Exploring periodization across cosmic and biosocial dimensions* (pp. 471–494). Springer Nature. https://doi.org/10.1007/978-3-031-85410-1_15

LePoire, D. J., Grinin, L., & Korotayev, A. (2025c). Conclusions: Exploring Big History periodizations across cosmic and biosocial dimensions with ideas for further research. In D. J. LePoire, L. Grinin, & A. Korotayev (Eds.), *Navigating complexity in Big History. Exploring periodization across cosmic and biosocial dimensions* (pp. 495–525). Springer Nature. https://doi.org/10.1007/978-3-031-85410-1_16

Levinson, D., & Malone, M. (1980). *Toward explaining human culture*. HRAF Press.

Lewis-Williams, D. (2002). *The mind in the cave: Consciousness and the origins of art*. Thames and Hudson.

Li, F., Petraglia, M., Roberts, P., & Gao, X. (2020). The northern dispersal of early modern humans in eastern Eurasia. *Science Bulletin, 65*(20), 1699–1701. https://doi.org/10.1016/j.scib.2020.06.026

Lozovskaya, O. V., Vybornov, A. A., & Dolbunova E. V. (Eds.) (2018). *Materialy mezhdunarodnoy konferentsii, posvyashchennoy 50-letiyu V. M. Lozovskogo, 15–18 maya*. Institut Istorii Material'noy Kul'tury RAN; Gosudarstvennyy Ermitazh.

Maddison, A. (2001). *Monitoring the world economy: A millennial perspective*. OECD.

Magurran, A. E. (1988). *Ecological diversity and its measurement*. Princeton University Press.

Malthus, T. (1978 [1798]). *Population: The first essay*. University of Michigan Press.

March, R. J. (2013). Searching for the functions of fire structures in Eynan (Malahha) and their formation processes; A geochemical approach. In O. Bar-Yosef & F. R. Valla (Eds.), *Natufian Foragers in Levant: Terminal Pleistocene Social Changes in Western Asia. International Monographs in Prehistory* (pp. 227–283). Archaeological Series 19.

Markov, A. (2001). Dynamics of the Marine Faunal Diversity in the Phanerozoic: A New Approach. *Paleontological Journal, 35*, 1–9.

Markov, A. (2002). Mechanisms responsible for the increase in the taxonomic diversity in the Phanerozoic marine biota. *Paleontological Journal, 36*, 121–130.

Markov, A. (2009). Alpha diversity of Phanerozoic marine communities positively correlates with longevity of genera. *Paleobiology, 35*(2), 231–250.

Markov, A. (2013). Teoriya RNK mira. *Youtube*. https://www.youtube.com/watch?v=T4RfQX9p-ng

Markov, A. (2014). *Rozhdeniye slozhnosti*. Corpus (AST). In Russian.

Markov, A. (2015). Problemy proiskhozhdeniya zhizni. Abiogenez. *Youtube*. https://www.youtube.com/watch?v=3YTkFsWw4_4

Markov, A. (2016). Dobiologicheskaya evolyutsiya. *Youtube*. https://www.youtube.com/watch?v=2s6KB5MLq1I

Markov, A., Anisimov, V., & Korotayev, A. (2010). Relationship between genome size and organismal complexity in the lineage leading from prokaryotes to mammals. *Paleontological Journal, 44*(4), 363–373. https://doi.org/10.1134/S0031030110040015

Markov, A., & Korotayev, A. (2007). Phanerozoic marine biodiversity follows a hyperbolic trend. *Palaeoworld, 16*(4), 311–318. https://doi.org/10.1016/j.palwor.2007.01.002

Markov, A., & Korotayev, A. (2008). Hyperbolic growth of marine and continental biodiversity through the Phanerozoic and community evolution. *Zhurnal Obshchei Biologii, 69*(3), 175–194.

Markov, A., & Markov, M. (2020). Runaway brain-culture coevolution as a reason for larger brains: Exploring the "cultural drive" hypothesis by computer modeling. *Ecology and Evolution, 10*(12), 6059–6077.

Markov, A., & Markov, M. (2021). Coevolution of brain, culture, and lifespan: Insights from computer simulations. *Biochemistry, 86*, 1503–1525.

Markov, A., & Naymark, E. (2009). On some of the newest achievements in evolutionary biology. *Evolution, 1*, 306–363. In Russian.

Markov, A., & Naymark, E. (2019).*Prospects for selection. From green warblers and senseless complication to naked mole rats*. Corpus. In Russian.

Marshack, A. (1991). *The roots of civilization: The cognitive beginnings of man's first art, symbol and notation*. Moyer Bell.

McEvedy, C., & Jones, R. (1978). *Atlas of world population history*. Facts on File.

Meliantsev, V. A. (2004). Russia's comparative economic development in the long run. *Social Evolution & History, 3*(1), 106–136.

Mesoudi, A. (2021). Cultural selection and biased transformation: Two dynamics of cultural evolution. *Philosophical Transactions of the Royal Society B, 376*(1828), 20200053. https://doi.org/10.1098/rstb.2020.0053

Mesoudi, A., Whiten, A., & Laland, K. (2006). Towards a unified science of cultural evolution. *Behavioral and Brain Sciences, 29*, 329–383.

Mitrofanov, I. G. (2017). Poiski vnezemnoy zhizni v Solnechnoy sisteme: status i perspektivy. *Astronomicheskiy zhurnal, 94*(4), 315–322.

Modelski, G. (2003). *World Cities: –3000 to 2000*. Faros2000.

Muzammal, M., Bibi, A., Shafiq, M., Malik, H., Istiaq, S., Asad, S., Khan, H., Saba, H., Raoon, U., Ismail, M., Burki, N. F., Hashmi, H. B., Gul, H., & Ahmad Khan, M. (2025). The genetic basis of altruism and cooperative behaviour in human societies. *Social Evolution & History, 24*(2), 103–117. https://doi.org/10.30884/seh/2025.02.05

Naroll, R., & Divale, W. T. (1976). Natural selection in cultural evolution: Warfare versus peaceful diffusion. *American Ethnologist, 3*(1), 97–129.
Nashimoto, M. (2001). The RNA/Protein symmetry hypothesis: Experimental support for reverse translation of primitive proteins. *Journal of Theoretical Biology, 209*, 181–187.
Nefedov, S. (2025). Grain yields and the causes of the russian revolution. *Cliodynamics, 16*(1), 73–82. https://doi.org/10.21237/C7clio.48378
Nefedov, S. A. (2004). A model of demographic cycles in traditional societies: The case of ancient China. *Social Evolution & History, 3*(1), 69–80.
Nefedov, S. A. (2012). Neo-Malthusianism in the modern methodology of history. *Herald of the Russian Academy of Sciences, 82*(6), 453–460. https://doi.org/10.1134/S1019331612060081
Nefedov, S. A. (2013). Modeling Malthusian dynamics in pre-industrial societies. *Cliodynamics, 4*(2), 229–240. https://doi.org/10.21237/C7clio4221335
O'Connell, J. F., Allen, J., Williams, M. A., Williams, A. N., Turney, C. S., Spooner, N. A., et al. (2018). When did Homo sapiens first reach Southeast Asia and Sahul? *Proceedings of the National Academy of Sciences, 115*(34), 8482–8490. https://doi.org/10.1073/pnas.1808385115
Oktaviana, A. A., Joannes-Boyau, R., Hakim, B., Burhan, B., Sardi, R., Adhityatama, S., et al. (2024). Narrative cave art in Indonesia by 51,200 years ago. *Nature, 631*(8022), 814–818. https://doi.org/10.1038/s41586-024-07541-7
Pääbo, S. (1995). The Y-Chromosome and the origin of all of us (men). *Science, 268*, 1141–1142.
Pagel, M., Atkinson, Q. D., & Meade, A. (2007). Frequency of word-use predicts rates of lexical evolution throughout Indo-European history. *Nature, 449*(7163), 717–720. https://doi.org/10.1038/nature06176
Panov, A., LePoire, D., & Korotayev, A. (2020). The 21st century singularity in the Big History perspective. An overview. In A. Korotayev & D. LePoire (Eds.), *The 21st century Singularity and global futures. A Big History perspective* (pp. 1–18). Springer Nature. https://doi.org/10.1007/978-3-030-33730-8_1
Payne, J. L., Al Aswad, J. A., Deutsch, C., Monarrez, P. M., Penn, J. L., & Singh, P. (2023). Selectivity of mass extinctions: Patterns, processes, and future directions. *Cambridge Prisms: Extinction, 1*, article e12.
Pettitt, P. (2011). *The paleolithic origins of human burial*. Routledge.
Philippov, I. (1990). The emergence of feudalism in Western Europe. In Z. Udaltsova & S. Karpov (Eds.), *History of the middle ages* (pp. 42–84). Vysshaya Shkola.
Podlazov, A. (2017). A theory of the global demographic process. *Herald of the Russian Academy of Sciences, 87*(3), 256–266. https://doi.org/10.1134/S1019331617030054
Popov, S. (2006). The problem of adaptation in social structure studies. *Jurnalobscheybiologii, 67*(5), 335–343.
Powell, M. G., & Kowalewski, M. (2002). Increase in evenness and sampled alpha diversity through the Phanerozoic: comparison of early Paleozoic and Cenozoic marine fossil assemblages. *Geology, 30*(4), 331–334. https://doi.org/10.1130/0091-7613(2002)030<0331:IIEASA>2.0.CO;2
Prüfer, K., Posth, C., Yu, H., Stoessel, A., Spyrou, M. A., Deviese, T., et al. (2021). A genome sequence from a modern human skull over 45,000 years old from Zlatý kůň in Czechia. *Nature Ecology & Evolution, 5*(6), 820–825. https://doi.org/10.1038/s41559-021-01443-x
Raup, D. M., Gould, S. J., Schopf, T. J., & Simberloff, D. S. (1973). Stochastic models of phylogeny and the evolution of diversity. *The Journal of Geology, 81*(5), 525–542. https://doi.org/10.1086/627905
Raup, D. M., & Sepkoski, J. J., Jr. (1984). Periodicity of extinctions in the geologic past. *Proceedings of the National Academy of Sciences, 81*(3), 801–805. https://doi.org/10.1073/pnas.81.3.801

Rautian, A. (1988a). Paleontology as a source of the data on evolutionary laws and factors. *Modern Paleontology, 2*, 76–118. In Russian.
Rautian, A. (1988b). Dictionary of terms and subjects. *Modern Paleontology, 2*, 356–372. In Russian.
Reeve, H. K., & Hölldobler, B. (2007). The emergence of a superorganism through intergroup competition. *Proceedings of the National Academy of Sciences, 104*(23), 9736–9740.
Reznikova, Z. (2011). Evolutionary and behavioural aspects of altruism in animal communities: Is there room for intelligence? In L. Grinin, R. L. Carneiro, A. Korotayev, & F. Spier (Eds.), *Evolution: Cosmic, biological, and social* (pp. 162–211). Uchitel.
Richter, D., Grün, R., Joannes-Boyau, R., Steele, T. E., Amani, F., Rué, M., et al. (2017). The age of the hominin fossils from Jebel Irhoud, Morocco, and the origins of the Middle Stone Age. *Nature, 546*(7657), 293–296. https://doi.org/10.1038/nature22335
RMGSAC [= Rhesus Macaque Genome Sequencing and Analysis Consortium]. (2007). Evolutionary and biomedical insights from the Rhesus Macaque Genome. *Science, 316*, 222–234.
Roberts, G. G., & Mannion, P. D. (2019). Timing and periodicity of Phanerozoic marine biodiversity and environmental change. *Scientific Reports, 9*, article 6116.
Robson, S. K., & Traniello, J. F. A. (2002). Transient division of labour and behavioral specialization in the Ant Formica Schaufussi. *Naturwissenschaften, 89*, 128–131.
Rodrigue, B. H. (2023a). The bedrock of survival: humanity, history, and adaptation. *Journal of Globalization Studies, 14*(1), 137–169. https://doi.org/10.30884/jogs/2023.01.09
Rodrigue, B. H. (2023b). The study of all existence: Towards global symbiosis. *Journal of Globalization Studies, 14*(2), 171–185. https://doi.org/10.30884/jogs/2023.02.12
Rodrigue, B. H., Grinin, L., & Korotayev, A. (2015). (Eds.). *From Big Bang to galactic civilizations*. Vol. 1: *Our place in the universe. An introduction to Big History*. RatnaSagar.
Rodrigue, B. H., Grinin, L., & Korotayev, A. (2016). (Eds.). *From Big Bang to galactic civilizations*. Vol. 2: *Education and understanding Big History around the world*. RatnaSagar.
Rodrigue, B. H., Grinin, L., & Korotayev, A. (2017). (Eds.). *From Big Bang to galactic civilizations*. Vol. 3: *The ways that Big History works: Cosmos, life, society and our future*. RatnaSagar.
Rohde, R. A., & Muller, R. A. (2005). Cycles in fossil diversity. *Nature, 434*(7030), 208–210.
Roopnarine, P. D. (2025). Selection, evolution and persistence of paleoecological systems. *Frontiers in Earth Science, 13*, article 1528448.
Rozov, N. S. (2023). *The origin of language and consciousness. How social orders and communicative concerns gave rise to speech and cognitive abilities*. Springer Nature. https://doi.org/10.1007/978-3-031-30630-3
Ryabko, B., & Reznikova, Z. (2009). The use of ideas of information theory for studying 'Language' and intelligence in ants. *Entropy, 11*(4), 836–853.
Rybin, E. P., Mescherin, M. N., Klementiev, A. M., & Khatsenovich, A. M. (2023). Beyond 55 North latitude: Expanding the northern boundary of initial upper Paleolithic dispersal. *Archaeological Research in Asia, 36*, article 100481. https://doi.org/10.1016/j.ara.2023.100481
Sahlins, M. (1972). *Stone age economics*. Aldine de Gruyter.
Sánchez-Yustos, P., Marín-Arroyo, A. B., Arnold, L. J., Luque, L., Kehl, M., López-Sáez, J. A., et al. (2024). Initial upper palaeolithic lithic industry at Cueva Millán in the hinterlands of Iberia. *Scientific Reports, 14*(1), 21705. https://doi.org/10.1038/s41598-024-69913-3
Sawyer, P., Romanov, D., Slav, M., & Korotayev, A. (2022). Urbanization, the youth, and protest: A cross-national analysis. *Cross-Cultural Research, 56*(2–3), 125–149. 2022. https://doi.org/10.1177/10693971211059762
Sepkoski, J. J. (1994). Limits to randomness in paleobiologic models: The case of Phanerozoic species diversity. *Acta PalaeontologicaPolonica, 38*(3–4), 174–198.

Sepkoski, J. J., Jr. (1988). Alpha, beta, or gamma: where does all the diversity go? *Paleobiology, 14*(3), 221–234. https://doi.org/10.1017/S0094837300011969

Sepkoski, J. J., Jr. (1991a). Population biology models in macroevolution. In N. L. Gilinsky & P. W. Signor (Eds.), *Analytical paleobiology. Short courses in paleontology* (Vol. 4, pp. 136–156). The Paleontological Society.

Sepkoski, J. J., Jr. (1991b). Diversity in the phanerozoic oceans: A Partisan review. In E. C. Dudley (Ed.), *The unity of evolutionary biology* (pp. 210–326). Dioscorides Press.

Sepkoski, J. J., Jr. (1992). Phylogenetic and ecologic patterns in the Phanerozoic history of marine biodiversity. In N. Eldredge (Ed.), *Systematics, ecology and the biodiversity crisis* (pp. 77–100). Columbia University Press.

Sepkoski, J. J., Jr., Bambach, R. K., Raup, D. M., & Valentine, J. W. (1981). Phanerozoic marine diversity and the fossil record. *Nature, 293*(5832), 435–437. https://doi.org/10.1038/293435a0

Shatalkin, A. (2005). Molecular phylogenies– revolutionary breakthrough in systematics. In: E. Vorobyeva & B. Striganova (Eds.), *Evolutionary factors of forming animal world* (pp. 30–42). KMK.

Shea, J. J. (2013). *Stone tools in the paleolithic and Neolithic near east: A guide*. Cambridge University Press.

Shettleworth, S. (1998). *Cognition, evolution, and behavior*. Oxford University Press.

Shnirelman, V. (1983). Gatherers of Sago. *Voprosy istorii, 11*, 182–187. In Russian.

Shnirelman, V. (1986). Late primitive community of agriculturalists, herders, and advanced hunters, fishers and gatherers. In Y. Bromley (Ed.), *The history of primitive society. The epoch of the primitive clan community* (pp. 236–426). Nauka. In Russian.

Shnirelman, V. (1989). *Emergence of food production*. Nauka.

Simon, J. (1977). *The economics of population growth*. Princeton University Press.

Simon, J. (2000). *The great breakthrough and its cause*. University of Michigan Press.

Solé, R., & Santiago, F. E. (2019). *Viruses as complex adaptive systems*. Princeton University Press.

Solé, R. V., Montoya, J. M., & Erwin, D. H. (2002). Recovery after mass extinction: Evolutionary assembly in large–scale biosphere dynamics. *Philosophical Transactions of the Royal Society of London. Series B: Biological Sciences, 357*(1421), 697–707. https://doi.org/10.1098/rstb.2001.0987

Speechley, E. M., Ashton, B. J., Foo, Y. Z., Simmons, L. W., & Ridley, A. R. (2024). Meta-analyses reveal support for the social intelligence hypothesis. *Biological Reviews, 99*(5), 1889–1908.

Spencer, H. (1898). *Principles of sociology. Complete work.* Vol. 1. St. Petersburg: T-vo I. D. Sytina, Otd. N. A. Rubakina. In Russian.

Spier, F. (2015). *Big History and the future of humanity*. John Wiley & Sons.

Stanford, D. J., & Bradley, B. (2012). *Pre-Clovis First Americans: The origin of America's Clovis culture*. University of California Press.

Steele, E., Lindley, R., & Blanden, R. (2002). *What if Lamark is Right? Immunogenetics and evolution.* Mir. In Russian.

Stiner, M. C. (2014). Finding a common bandwidth: Causes of convergence and diversity in Paleolithic beads. *Biological Theory, 9*, 51–64. https://doi.org/10.1007/s13752-013-0157-4

Stringer, C. (2016). The origin and evolution of Homo sapiens. *Philosophical Transactions of the Royal Society B: Biological Sciences, 371*(1698), article 20150237. https://doi.org/10.1098/rstb.2015.0237

Stringer, C., & Crété, L. (2022). Mapping interactions of H. neanderthalensis and Homo sapiens from the Fossil and Genetic Records. *PaleoAnthropology, 2022*(2), 401–412. https://doi.org/10.48738/2022.iss2.130

Stringer, C., & Galway-Witham, J. (2017). On the origin of our species. *Nature, 546*(7657), 212–214. https://doi.org/10.1038/546212a

Taagepera, R. (1976). Crisis around 2005 AD? A technology-population interaction model. *General Systems, 21*, 137–138.

Taagepera, R. (1979). People, skills, and resources: An interaction model for world population growth. *Technological Forecasting and Social Change, 13*, 13–30. https://doi.org/10.1016/0040-1625(79)90003-9

Tanno, K., Willcox, G., Muhesen, S., Nishiaki, Y., Kanjo, Y., Akazawa, T. (2013). Preliminary results from analysis of Charred Plant remains from a Burnt Natufian Building at Dederiyeh Cave in Northwest Syria. In: O. Bar-Yosef & F. R. Valla (Eds.), *Natufian foragers in Levant: Terminal Pleistocene social changes in Western Asia. International Monographs in Prehistory* (pp. 83–87). Archaeological Series 19.

Teyssandier, N. (2023). The beginnings of the Aurignacian in a European framework: where do we stand? Abridged version. *Gallia Préhistoire, 63*, article 4154. https://doi.org/10.4000/galliap.4154

Timofeev-Ressovsky, N., Vorontsov, N., & Yablokov, A. (1969). *Brief essay on evolution theory*. Nauka. In Russian.

Townsend, J. B. (1985). The autonomous village and the development of chiefdoms. In H. J. M. Claessen, P. van de Velde, & E. M. Smith (Eds.), *Development and decline. The evolution of sociopolitical organization* (pp. 141–155). Bergin & Garvey.

Tsirel, S. (2004). On the possible reasons for the hyperexponential growth of the earth population. In M. Dmitriev & A. Petrov (Eds.), *Mathematical modeling of social and economic dynamics* (pp. 367–369). Russian State Social University.

Turchin, P. (2003). *Historical dynamics: Why states rise and fall*. Princeton University Press.

Turchin, P. (2005). *War and peace and war: Life cycles of imperial nations*. Pi Press.

Turchin, P., & Korotayev, A. (2006). Population density and warfare: A reconsideration. *Social Evolution & History, 5*(2), 121–158.

Turchin, P., & Nefedov, S. (2009). *Secular cycles*. Princeton University Press.

UN Population Division. (2025). *United Nations Population Division database*. http://www.un.org/esa/population

US Bureau of the Census. (2025). *World Population Information*. http://www.census.gov/ipc/www/world.html.

Usher, D. (1989). The dynastic cycle and the stationary state. *The American Economic Review, 79*, 1031–1044.

Van Parijs, P. (1981). *Evolutionary explanation in the social sciences: An emerging paradigm*. Rowman and Littlefield.

Vanyushin, B. (2004). Materialization of epigenetics, or little changes and their big consequences. *Himiya i zhizn, 2*, 32–37. In Russian.

Vignuzzi, M., Stone, J. K., Arnold, J. J., Cameron, C. E., & Andino, R. (2006). Quasispecies diversity determines pathogenesis through cooperative interactions in a viral population. *Nature, 439*(7074), 344–348. https://doi.org/10.1038/nature04388

Volk, T., & Henriques, G. (2025). Integratingthetreeofknowledge and combogenesisapproachesto Big History. In D. J. LePoire, L. Grinin, & A. Korotayev (Eds.), *Navigating complexity in Big History—Exploring periodization across cosmic and biosocial dimensions* (pp. 195–215). Springer Nature. https://doi.org/10.1007/978-3-031-85410-1_7

von Foerster, H., Mora, P. M., & Amiot, L. W. (1960). Doomsday: Friday, 13 November, AD 2026. *Science, 132*(3436), 1291–1295. https://doi.org/10.1126/science.132.3436.1291

von Hoerner, S. J. (1975). Population explosion and interstellar expansion. *Journal of the British Interplanetary Society, 28*, 691–712.

Voros, J. (2019). Big History in its cosmic context. *Journal of Big History, 3*(3), 58–80. https://doi.org/10.22339/jbh.v3i3.3340
Vygotsky, L. (1978). *Mind in society: The development of higher psychological processes*. Harvard University Press.
Wagner, P. J., Kosnik, M. A., & Lidgard, S. (2006). Abundance distributions imply elevated complexity of post-Paleozoic marine ecosystems. *Science, 314*(5803), 1289–1292. https://doi.org/10.1126/science.1133795
White, T., Asfaw, B., DeGusta, D., Gilbert, H., Richards, G. D., Suwa, G., & Howell, F. C. (2016). Pleistocene homo sapiens from Middle Awash, Ethiopia. In *Human evolution source book* (pp. 557–565). Routledge.
Wilde, S., Timpson, A., Kirsanow, K., Kaiser, E., Kayser, M., Unterländer, M., et al. (2014). Direct evidence for positive selection of skin, hair, and eye pigmentation in Europeans during the last 5,000 y. *Proceedings of the National Academy of Sciences, 111*(13), 4832–4837. https://doi.org/10.1073/pnas.1316513111
Wittmann, L., & Kuehn, C. (2024). The demographic-wealth model for cliodynamics. *Plos One, 19*(4), e0298318. https://doi.org/10.1371/journal.pone.0298318
World Bank. (2025). *World Development Indicators Online*. http://data.worldbank.org/indicator
Wright, D. (2001). *The history of China*. Greenwood.
Wu, K. J. (2020). There are more viruses than stars in the universe. Why do only some infect us? *National Geographic*. https://www.nationalgeographic.co.uk/science-and-technology/2020/04/there-are-more-viruses-than-stars-in-the-universe-why-do-only-some-infect-us.
Wu, Q., Zhang, H., Ramezani, J., Zhang, F. F., Erwin, D. H., Feng, Z., et al. (2024). The terrestrial end-Permian mass extinction in the paleotropics postdates the marine extinction. *Science Advances, 10*(5), eadi7284. https://doi.org/10.1126/sciadv.adi7284
Zakharov, A. (1978). *Ants, Family, Colony*. Nauka.
Zhivotovsky, L. (2002a). On inheritance of acquired characteristics. In *Materials of Academic Genetic Conference. February 26–27, 2002* (pp. 110–119). Timiryazev Agricultural Academy.
Zhivotovsky, L. (2002b). A model of the early evolution of Soma-to-Germline feedback. *Journal of Theoretical Biology, 216*, 51–57.
Zilhão, J. (2005). Burial evidence for social differentiation of age classes in the Early Upper Paleolithic. *Etud Res Archéol Univ Liège, 111*, 231–241.
Zimmer, C. (2003). Great mysteries of human evolution. *Discover, 24*(9), 34–44.
Zinkina, J., Slinko, E., Bykanova, D., & Korotayev, A. (2018). Dynamics of values and modernization: a quantitative analysis. *Zhurnal sotsiologii i sotsialnoy antropologii = The Journal of Sociology and Social Anthropology, 21*(1), 44–72. https://doi.org/10.31119/jssa.2018.21.1.3
Zwyns, N. (2021). The Initial Upper Paleolithic in Central and East Asia: blade technology, cultural transmission, and implications for human dispersals. *Journal of Paleolithic Archaeology, 4*(3), article number 19. https://doi.org/10.1007/s41982-021-00085-6

Leonid Grinin is Senior Research Professor at the Center for Stability and Risk Analysis at the HSE University in Moscow, as well as Director of Eurasian Center for Big History & System Forecasting at the Institute of Oriental Studies, Russian Academy of Sciences. He is co-editor of the international journals *Social Evolution & History* and the *Journal of Globalization Studies*, as well as co-editor of the international almanacs *Evolution, History & Mathematics*, and *Kondratieff Waves*. He is author of more than 700 publications in Russian, English, German, Spanish, and Chinese, including 35 monographs. His current research focuses on comparative political studies, theory of revolution, political anthropology, global economy, global history, historical sociology, and futurology. Among his monographs are *The Cybernetic Revolution and the Forthcoming Epoch of Self-Regulating Systems* (Uchitel, 2016, with Anton Grinin), *Big History of Globalization: The Emergence of a Global World System* (Springer 2019, in co-authorship with Julia Zinkina, David

Christian and others); *Islamism, Arab Spring, and the Future of Democracy: World System and World Values Perspectives* (Springer, 2019; with Andrey Korotayev and Arno Tausch); and *Cybernetic Revolution and Global Aging: Humankind on the Way to Cybernetic Society, or the Next Hundred Years* (Springer, 2024 with Andrey Korotayev and Anton Grinin). In 2004, he was awarded as a winner of the best publication of *Socis* journal during the year for the article "People of Publicity—a New Social Layer?." In 2012, he was awarded the Gold Kondratieff Medal by the International N. D. Kondratieff Foundation. In 2016, he received the Belyaev award for the book *From Bifaces to Nanorobots* (the book published in Russian in co-authorship with Anton Grinin). 12, Rozhdestvenka, Moscow, 107031, Russian Federation. e-mail: leonid.grinin@gmail.com. ORCID 0000-0003-0278-2619

Alexander Markov is a Russian biologist, paleontologist, and popularizer of science. He is prize winner (2011) of the main Russian prize for popular science ("Prosvetitel"). Markov graduated from the Moscow State University (Faculty of Biology) in 1987. He was working in the Paleontological Institute of the Russian Academy of Sciences since 1987. He has received Doctor of Biological Sciences. He is a professor of the Russian Academy of Sciences. ORCID 0000-0002-7113-7181

Andrey Korotayev heads the Center for Stability and Risk Analysis at the HSE University in Moscow, Russia. He is also Senior Research Professor at Eurasian Center for Big History & System Forecasting at Institute of Oriental Studies, Russian Academy of Sciences and Institute for African Studies, Russian Academy of Sciences, as well as at the Faculty of Global Studies of the Lomonosov Moscow State University. Beginning as an historian of Arabia, he made focused studies, such as seen in "Two Social Ecological Crises and Genesis of Tribal Organization in the Yemeni North-East" (1996). Korotayev researches topics in quantitative cross-cultural anthropology and seeks to understand biological and social macroevolution, as in "A Compact Macromodel of World System Evolution" (2005) and "Mathematical Modeling of Biological and Social Phases of Big History" (2014). He has authored or co-authored over 650 scholarly publications, including such monographs as *Ancient Yemen* (Oxford University Press, 1995), *World Religions and Social Evolution of the Old World Oikumene Civilizations: A Cross-Cultural Perspective* (The Edwin Mellen Press, 2004), *Introduction to Social Macrodynamics: Compact Macromodels of the World System Growth* (URSS, 2006; with Daria Khaltorina and Artemy Malkov), *Introduction to Social Macrodynamics: Secular Cycles and Millennial Trends* (URSS, 2006; with Daria Khaltorina and Artemy Malkov), *Great Divergence and Great Convergence. A Global Perspective* (Springer, 2015; with Leonid Grinin), *Economic Cycles, Crises, and the Global Periphery* (Springer, 2016; with Leonid Grinin), and *Islamism, Arab Spring, and the Future of Democracy: World System and World Values Perspectives* (Springer, 2019; with Leonid Grinin and Arno Tausch). He is a founding member of the Eurasian Center for Megahistory & System Forecasting and founder of Evolution, an almanac dedicated to the study of Universal Evolution, and thus is closely linked with Big History. In 2020, he produced, with David LePoire, a collective monograph, *The 21st Century Singularity and Global Futures: A Big History Perspective* (Springer Nature, 2020). Recently, he has produced, with David LePoire and Leonid Grinin, another Big History collective monograph, *Navigating Complexity in Big History: Exploring Periodization Across Cosmic and Biosocial Dimensions*

(Springer Nature, 2025). He is a laureate of a Russian Science Support Foundation in "The Best Economists of the Russian Academy of Sciences" Nomination (2006); in 2012, he was awarded the Gold Kondratieff Medal by the International N. D. Kondratieff Foundation. He serves as a board member of the International Big History Association. 20, Myasnitskaya, Moscow 101000, Russian Federation. e-mail: akorotayev@gmail.com. ORCID 0000-0003-3014-2037

Part V

Human and Social Complexity

Human Evolution in the Complexity Growth Perspective: Toward Periodization of the Big History Biosocial Era

13

Andrey Korotayev

Abstract

We have undertaken an attempt to propose a periodization of the Big History Biosocial (Anthropogenesis) Era on the basis of the most recent scientific data. This periodization is complexity-based, that is, the boundaries of the identified epochs are marked with complexity jumps, that is, in our case, such phase transitions that result in significant increases in both biological and sociocultural complexity of the evolving planetary biosocial systems. This has allowed us to identify the following epochs: (1) *Australopithecine Epoch*—from *c*. 4.2 million before present (BP) up to *c*. 2.9–2.4 million BP—emergence of the australopithecines with subsequent (rather vague) indications of increases in sociocultural complexity. (2) Oldowan-Habilis Epoch—from *c*. 2.9–2.4 million BP up to *c*. 2 million BP—emergence of the Oldowan stone tool industry and the first Homo species. (3) Early Acheulean-Erectus Epoch—from *c*. 2 million BP to *c*. 900–500 thousand BP—emergence of *Homo erectus* followed by the invention of Early Acheulean stone tool technology. (4) Late Acheulean-Heidelbergensis Epoch—from *c*. 900–500 thousand BP to *c*. 300,000 BP—emergence of *Homo heidelbergensis* followed by the invention of Late Acheulean stone tool technology, including production of first composite tools, and other major advances in sociocultural complexity—systematic use of fire, construction of dwellings, etc. (5) Middle Paleolithic Epoch—epoch of Neanderthals, Denisovans, and early *Homo sapiens*—from *c*. 300,000 BP to *c*. 50,000 BP—emergence of the three

This chapter is an extensively re-worked version of the text originally published in *Journal of Big History* (Korotayev, 2025b).

A. Korotayev (✉)
HSE University, Moscow, Russia

Eurasian Center for Big History and System Forecasting, Institute of Oriental Studies of the Russian Academy of Sciences, Moscow, Russia

D. J. LePoire et al. (eds.), *Complexity in Universal Evolution*, World-Systems Evolution and Global Futures, https://doi.org/10.1007/978-3-032-20517-9_13

most advanced and complex Homo species (Neanderthals, Denisovans, and early *Homo sapiens*), in comparison with all the previous species of our genus; major advances in sociocultural complexity (Middle Paleolithic Levallois-based stone technologies, intentional burials with ritualistic elements, personal ornaments and pigment use, engraved art and mark-making, structured use of space, and so on). Note that, within the proposed Big History periodization, the Upper Paleolithic Revolution (*c.* 60–40 thousand years ago) marks the end of both the Middle Paleolithic Epoch—epoch of Neanderthals, Denisovans, and early *Homo sapiens*—and the end of the BH Biosocial Era. It is also shown that most extent complexity-based periodizations of the Big History (including the ones proposed by Theodore Modis and Alexander Popov) contradict significantly the presently available scientific data on human evolution and need serious amendments as regards the periodization of the BH Biosocial Era. On the other hand, it has been shown that this does not invalidate our earlier calculations of the planetary complexity growth patterns based on the mathematical analysis of the Modis and Panov series. Indeed, it has been demonstrated that after the replacement of Modis' and Panov's dates of the Big History Biosocial Era complexity jumps with more accurate dates, the calculation continues to detect essentially the same patterns of the hyperbolic acceleration of the planetary biosocial complexity growth.

Keywords

Big History · Complexity · Human evolution · Biological evolution · Social evolution · Periodization · Hyperbolic acceleration

Abbreviations

BH Big History
BP [years] before present
cc cubic centimeter

13.1 Introduction

The idea of this chapter came to my mind when we, together with David LePoire and Leonid Grinin, were working on the previous Springer Nature collective monograph in our series—*Navigating Complexity in Big History: Exploring Periodization across Cosmic and Biosocial Dimensions* (LePoire et al., 2025a, 2025b, 2025c). Within this project, I suggested my own Big History periodization founded on the analysis of the patterns, phases, and results of the complexity growth within the Big History on the basis of ideas, proposed by Aunger (2007, 2025), LePoire (2020a, 2020b, 2025), Panov (2005a, 2005b, 2020), and the Grinins (2025).

I have found it appropriate to follow the Big History period typology suggested by Robert Aunger on the basis of the typology used by geologists who "have divided earth history into eons (e.g., the Precambrian), eras (e.g., Paleozoic), periods (e.g., Cambrian) and epochs (e.g., Pleistocene), each a subcategory of the previous, based on changes in Earth's own composition and largescale changes in the biota inhabiting the planet's surface" (Aunger, 2007, p. 1165). Following this logic, he divides the Big History into two eons, with eons subdivided into eras and eras subdivided into periods (Aunger, 2007; see also Aunger, 2025).

Within the periodization that I have proposed, the whole course of the Big History is subdivided into three eons identified on the basis of the complexity growth pattern that is characteristic for the respective eon; each eon is subdivided into eras identified on the basis of the complexity growth driver that was typical for the respective era. Note that following conventions of geologists, Aunger subdivide eras into periods. However, I have preferred to denote the subdivisions of the Early Universe Era as "epochs" rather than as "periods," because the denomination of most those subdivisions as "epochs" (Planck Epoch, Grand Unification Epoch, Electroweak Epoch, Quark Epoch, Hadron Epoch, Lepton Epoch, and Photon Epoch) appears to be rather well established by the moment (see, e.g., Brandenburg et al., 2017; Gustafson, 2020; Holmes, 2023; Meyer-Ortmanns & Thurner, 2011; Muratov et al., 2013; Perrone et al., 2021; Yang & Rafelski, 2022). Then, for the sake of uniformity, I have subdivided all the Big History eras in epochs, rather than periods.

Thus, in my Big History periodization, each era is subdivided into epochs identified on the basis of the highest level of complexity achieved within the respective epoch (thus, the borders between epochs correspond to complexity jumps such as the Big Bang nucleosynthesis, recombination, emergence of the first stars, "Neoproterozoic Revolution," Cambrian explosion, Upper Paleolithic Revolution, transition from foraging to food production, Axial Age, and so on) (see Korotayev, 2025a for detail, see also Table 13.1 below that reproduces a scheme of the Big History eons and eras suggested in Korotayev, 2025a with some modifications).

Within the proposed approach, the second Big History eon is denoted as the "Eon of the Hyperbolic Acceleration of the Global Complexity Growth." It corresponds almost entirely to the "Terrestrial [Accelerating] Eon" identified by Aunger (2007). Robert Aunger subdivides this eon into three eras: (1) Biological, (2) Cultural, and (3) Technological. This more or less corresponds to the subdivision by David LePoire of his "second cone"/eon into (1) the Era of Life Evolution, (2) the Era of Mind Evolution, and (3) the Era of Civilization Evolution (2015, 2016, 2020a, 2020b), to Gregg Henriques' "Cones" of (1) Life, (2) Mind, and (3) Culture (Henriques, 2003, 2011; Henriques et al., 2019; Henriques & Volk, 2023; see also Volk & Henriques, 2025), as well as to eras of Biological, Cultural, and Information Technology Evolution identified by Nicholas Hoggard (2024, 2025).

On the other hand, Tyler Volk and the Grinins find here only two main "eras." Tyler Volk calls them (1) "Realm of Biological Evolution" and (2) "Realm of Cultural Evolution" (Volk, 2017, 2020; see also Henriques & Volk, 2023; Volk & Henriques, 2025). The Grinins denote them as (1) "Biological Main Evolutionary

Table 13.1 Overall Big History periodization on the basis of the complexity growth patterns and phases

Eon	Date of the start	Era	Date of the start
Eon of Hyperbolically Decelerating Growth of the Universal Complexity ("Universal Deceleration Eon") = Cosmological "Decelerating" Eon according to Aunger (2007)[a]	0	Early Universe Era	0
		Stelliferous Era[b]	150 million years after the Big Bang Singularity
Eon of Hyperbolically Accelerating Growth of the Global Complexity ("Global Acceleration Eon")	*c.* 4 billion [years] *before present* (BP throughout) (emergence of life on the Earth)	Era of Hyperbolically Accelerating Growth of Biological Complexity ("Biological Era")	c. 4 billion BP (emergence of life on the Earth)
		"Biosocial Era"	*c.* 6 million BP (?)
		Era of Hyperbolically Accelerating Growth of Sociocultural Complexity ("Sociocultural Era")	*c.* 50,000 BP (Upper Paleolithic Revolution)
The Third Eon[c] = The Third Cone of David LePoire (2016, 2020a, 2020b)	*c.* the early 1970s–the late 1990s (?)	Era 1 (???)	Early 1970s–1990s (?)
		Era 2 (???)	???

[a]It also corresponds to "the First Cone"/ "Cosmic Phase" of David LePoire (2016, 2020a, 2020b, 2025), to Erich Jantsch (1980, pp. 77–96) Cosmic Prelude, to the "Matter Cone" of Gregg Henriques (2003, 2011, Henrique et al., 2019) and to the "Realm of Physical Law" of Tyler Volk (2017, 2020); see also Henriques & Volk, 2023; Volk & Henriques, 2025), as well as to "Physical Evolution" of Nicholas Hoggard (2024, 2025)

[b]The two eras of the first eon appear to be clearly distinguished by the principal drivers of the universal complexity growth. Within the Early Universe Era, the growth of complexity was very tightly connected with its cooling. It was this cooling that allowed the formation in the Universe of more and more complex entities—quarks, then hadrons, then atomic nuclei, then atoms, then molecules. At the very beginning, the cooling of the Universe proceeded very fast, and the complexity in the Universe grew extremely fast (with a few phase transitions just within the first second after the Big Bang Singularity). Then the cooling of the Universe slowed down, which caused the slowing down of the growth of complexity in the Universe (Korotayev, 2024, 2025a, 2025b). However, around 150 million years after the Big Bang, with the end of the cosmological Dark Ages and the start of the Stelliferous Era, the further cooling of the Universe stopped playing any significant role in the further growth of the universal complexity. Against this background, a new driver of the universal complexity growth emerged. There are some grounds to maintain that this was the nucleosynthesis in the emerging stars that produced elements more complex than hydrogen and helium that appeared in the Epoch of Big Bang Nucleosynthesis (between 10 and 1000 seconds after the BBS) and that increased systematically the physical–chemical complexity of the Universe for billions of years afterwards, see Korotayev and LePoire (2025) for more detail

[c]Eon of Hyperbolically (?) Decelerating Growth Rate (?) of the Global Complexity ("Global Deceleration Eon"). Note that the recent deceleration of the global complexity growth appears to be connected among other things with the global aging (e.g., Blešić et al., 2024; Grinin, 2025; Grinin et al., 2023, 2024; Grinin & Grinin, 2023a; Korotayev et al., 2023; Ustyuzhanin et al., 2025)

Phase of Big History" and (2) "Social Main Evolutionary Phase of Big History" (Grinin & Grinin, 2023b, 2025).

Note that within our Big History periodization on the basis of the complexity growth patterns, the Big History second eon, the Eon of Hyperbolically Accelerating Growth of the Global Complexity (= the Global Acceleration Eon), can be reasonably subdivided both in two and three eras. On the one hand, its subdivision into (1) the Era of Hyperbolically Accelerating Growth of Biological Complexity and (2) the Era of Hyperbolically Accelerating Growth of Sociocultural Complexity appears rather logical, as the growth of global complexity was driven during the first era mostly by the mechanisms of biological evolution; during the second era, this was mostly done by the mechanisms of sociocultural evolution, whereas it has been shown that though mechanisms of both types of evolution possess significant similarities, the differences between them are very substantial indeed (Grinin et al., 2009, 2011, 2012, 2013, 2020; Grinin & Korotayev, 2023; Korotayev & Markov, 2014, 2015; Markov & Korotayev, 2007, 2008; Markov et al., 2010), which by itself justifies the subdivision of the second eon into the two respective eras.

On the other hand, at the conjunction of the Era of Hyperbolically Accelerating Growth of Biological Complexity and the Era of Hyperbolically Accelerating Growth of Sociocultural Complexity (for the period stretching from the onset of human evolution/"anthropogenesis" a few million years before present (BP) and up to the Upper Paleolithic Revolution about 50 thousand years BP), we can identify an intermediate era when the growth of complexity was achieved both through biological evolution (without which, for example, the radical increase of the brain complexity among the hominins / hominini would be impossible) and through sociocultural evolution (without which we would not see, e.g., an impressive development of stone industries, or domestication of fire). During this era, biological evolution and sociocultural evolution appear to have contributed more or less equally to the growth of complexity, being tightly interwoven and constituting a single process of gene–culture coevolution (though, with the passage of time, the comparative contribution of biological evolution to the complexity growth tended to decrease, and the one of sociocultural evolution tended to increase). This part of the second Big History eon can be denoted as an intermediate *Era of Hyperbolically Accelerating Growth of Biosocial Complexity*. It more or less corresponds to the second era of the second Big History eon identified by Aunger, LePoire, and Henriques. But especially well does it correspond to the Grinins' "Biosocial Intermediate Phase of Big History" (Grinin & Grinin, 2023a, 2023b, 2025).

It was with this Big History era, with which I encountered the most serious problems when trying to subdivide it into epochs. With respect to the subdivision of the Big History eras into epochs I mostly relied on the identification of the Big History epochs within respective eras proposed by the other big historians; and mostly this worked. However, this has not worked with respect to the second ("biosocial") era of the second Big History eon (= the Global Acceleration Eon = "Terrestrial [Accelerating] Eon").

Indeed, by the moment, big historians have proposed the following periodizations of the era in question.

13.2 Extant Periodizations of the BH Biosocial Era

13.2.1 Theodore Modis

Theodore Modis (2002, 2003) proposed the following periodization of the stretch of time corresponding to what we denote as the "Biosocial Era" of the Big History:[1]

1. Chimpanzees and humans diverge, earliest hominid bipedalism—*5.1 million years ago.*
2. First stone tools, first humans, *Homo erectus—2.2 million years ago.*
3. Emergence of *Homo sapiens—555,000 years ago.*
4. Domestication of fire/ *Homo heidelbergensis—325,000 years ago.*
5. Differentiation of human DNA types—*200,000 years ago.*
6. Emergence of "modern humans"/earliest burial of the dead—*105,700 years ago.*
7. Rock art/protowriting—*35,800 years ago.*

On the one hand, at the first glance, this periodization might look very relevant for us, because it is explicitly complexity based, as Modis (2002, 2003) quite explicitly identifies his Big History milestones as "complexity jumps." However, to my mind, the periodization of the Biosocial Era proposed by Modis (as well as almost all of the "complexity jumps" identified by him for the respective stretch of time) mostly does not make much sense.

Of course, this is mostly explained by the point that the study of the human evolution has been one of the most fast-developing fields of knowledge, and many hypotheses in this field that looked true in 2002 have been shown to be wrong by 2025. However, many points in Modis periodization (like the emergence of *Homo sapiens* 230,000 years before *Homo heidelbergensis*) could have been very easily identified as plainly wrong already in 2002. It is also very remarkable that the periodization of the Big History era in question suggested by Erich Jantsch (1980, p. 154) looks *much* more adequate than the one proposed by Modis 22 years later.

To start with, according to recent research, the earliest hominid bipedalism has been attested much earlier than 5.1 million years ago, not later than 6 million (e.g., Almécija et al., 2013), or rather even around 7 million BP as regards *Sahelanthropus tchadensis* (Daver et al., 2022; Lebatard et al., 2008).

As regards chimpanzee-human divergence, though indeed, by 2002, there were some estimates of its divergence happening 5–7 million years ago (e.g., Chen & Li, 2001), now it is generally estimated to have happened around 6–8 million BP if not

[1]Note that this periodization was occasionally used by Raymond Kurzweil (2005); so, sometimes we denote it as Modis–Kurzweil series. Note also that we reproduce below the respective part of time series as it was published in Modis' essay in *the Futurist* (2003, p. 29), as it is this version of Modis' series that is reproduced by Kurzweil and that has been analyzed mathematically (Korotayev, 2018, 2020b, 2024, 2025a, 2025b); however, we sometimes use fuller versions of the description of some Modis "complexity jumps" / "milestones" from his 2002 article in the *Technological Forecasting & Social Change* (2002, pp. 401–402).

earlier (Besenbacher et al., 2019; Buffalo & Kern, 2024; Langergraber et al., 2012; Moorjani et al., 2016; Staes et al., 2019; Yousaf et al., 2021). On the other hand, later dates for chimpanzee–human divergence are also still proposed (see Osozawa, 2023 and, especially, Poszewiecka et al., 2022), so, with respect to this divergence (though not to the earliest hominid bipedalism), the date suggested by Modis cannot be totally excluded.

The main problem is rather with identification by Modis of both chimpanzee-human divergence and transition to bipedalism as a "complexity jump." The point is that there does not seem to be any evidence that either chimpanzee-humans divergence or emergence of the earliest hominid bipedalism was per se accompanied by any significant growth of complexity. No doubt, that both the chimpanzee–humans divergence and the transition of our ancestors[2] to bipedalism formed important prerequisites for a very significant increase in biosocial complexity in future, but they could hardly be regarded as complexity jumps per se.

Consider now the next complexity jump identified by Modis ("First stone tools, first humans, *Homo erectus—2.2 million years ago*").

On the one hand, there is no doubt that the emergence of both the genus *Homo* and the first stone tool industries constituted a very significant increase in both biological and sociocultural complexity. On the other hand, by the moment, the oldest archeologically identified stone tools appear to be as old as 3.3 million BP, more than 1 million years earlier than was suggested by Modis (see Harmand et al., 2015; Harmand & Lewis, 2023). Of course, the respective artefacts can well be regarded as products of experimentations with tool making by late australopiths that did not produce any lasting influence on the further global evolution. However, around 2.9 million BP a stone tool technology emerged that produced quite a lasting influence on the further global evolution (Plummer et al., 2023). This was the Oldowan stone tool archaeological industry that continued to be used for more than 1 million years by at least two australopith species, by some forms that were transitional between Australopithecines and Homo, as well as by two Homo species—*Homo habilis* and *Homo erectus* (Heinzelin et al., 1999; Mussi et al., 2023; Plummer et al., 2023; Rightmire, 2025; Villmoare et al., 2015), and had a truly global diffusion (see Fig. 13.1):

The emergence of the first humans (that is the first Homo species) is now dated earlier than was suggested by Modis (Kimbel, 2009, p. 35; Rightmire, 2025), and, of course, Modis cannot be blamed for this; however, the point that the first Homo species was *Homo habilis* rather than *Homo erectus* (cp. Spoor et al., 2015) was known well before 2002 (e.g., Tobias, 1983). On the other hand, *Homo erectus* emerged significantly later than indicated by Modis (Herries et al., 2020), and this was also well known before 2002.

At this point, we would stop a detailed analysis of the Modis series. Otherwise, it would take too many more pages. At this point, we will limit ourselves to saying that, as could be seen in the subsequent sections of this chapter, Modis' identification

[2]Who, incidentally, do not appear to have been the first bipedal hominids.

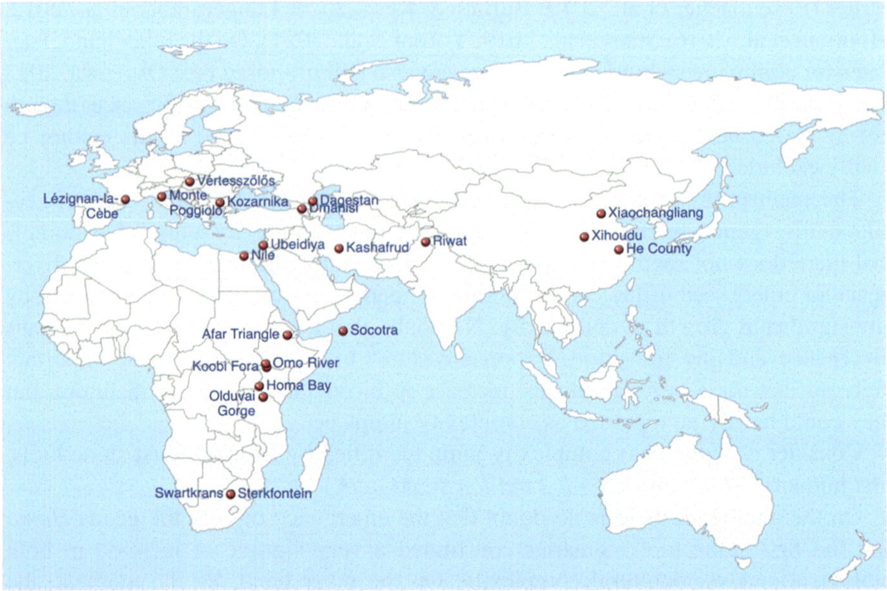

Fig. 13.1 Map of Afro-Eurasia showing important sites of the Oldowan industry. *Source*: https://en.wikipedia.org/wiki/Oldowan (accessed on May 6, 2025)

of "complexity jumps" for the Biosocial Era of the Big History is clearly wrong—with a partial exception for the last milestone in this series ("Rock art/protowriting – *35,800 years ago*"), which appears to correspond to quite a real big-history complexity jump generally known as "Upper Paleolithic Revolution" that, however, is mostly considered to have happened much earlier, around 50,000 BP (this will be discussed in detail below in the subsequent sections of this chapter).

13.2.2 Alexander Panov

Alexander Panov (2005a, pp. 124–127; 2005b, p. 221; 2020, p. 442) proposed the following periodization of what we call the "Biosocial Era" of the Big History:

6. The beginning of Quaternary period (Anthropogene)—*4.4 · 10^6 years ago.*[3] The first primitive Homo genus (hominidae) separated from hominoidae.
7. Paleolithic Revolution—*2.0–1.6 · 10^6 years ago.*[4] *Homo habilis*, the first stone implements.
8. The beginning of Chelles period—0.7–0.6 · 10^6 years ago.[5] Fire, *Homo erectus*.

[3] 4–5 · 10^6 years ago according to Panov (2005b).

[4] 2–1.5 · 10^6 years ago according to Panov (2005b).

[5] 0.7 · 10^6 years ago according to Panov (2005b).

9. The beginning of Acheulean period—*0.4* · 10^6 *years ago*. Standardized symmetric stone implements.
10. The culture revolution of neanderthal (Mustier culture)—*150–100* · 10^3 *years ago*. *Homo sapiens neanderthalensis*. Fine stone implements, burial of dead men (a sign of primitive religions).
11. The Upper Paleolithic Revolution—*40* · 10^3 *years ago*. *Homo sapiens sapiens* became the leader of cultural evolution. Development of advanced hunter instruments—spears, snares. Imitative art is widespread.

There are lots of problems with this periodization too. To start with, the Quaternary period began 2.58 million BP (e.g., International Commission on Stratigraphy, 2024), not 4.4 million years ago. *Homo habilis* formed 2.3–2.4 million years ago (Kimbel, 2009, p. 35; Rightmire, 2025) not 2.0 million BP. The first stone implements appeared not 2.0 but 3.3 million years ago if we start with the Lomekwian (Harmand et al., 2015; Harmand & Lewis, 2023), or 2.9 million BP if we start with the Oldowan (Plummer et al., 2023). *Homo erectus* emerged not later than ~2.04 million BP (Herries et al., 2020), not 700,000–600,000 years ago (and there is no mention of *Homo heidelbergensis* at all!). The Acheulean archaeological industry of stone tool manufacture emerged not around 400,000 BP, but not later than 1.95 million years ago (Mussi et al., 2023). As will be seen below there are lots of problems with "the cultural revolution of neanderthal" that according to Panov must have happened at 150,000–100,000 BP. As with the Modis series, the only exception is constituted by the last "phase transition"/"biospheric revolution" in the Panov series—"the Upper Paleolithic Revolution"—that is dated by Panov to 40,000 BP, which, as we will see below, falls within the range of presently accepted dates.

13.2.3 David LePoire

David LePoire (2016, pp. 236–237) proposed the following periodization of what we call the "Biosocial Era" of the Big History:

1. 5 Million [BP]—Humans—With the capability for walking to expand the range of natural resources, humans adapt to a changing climate by leaving the forest for the savannah.
2. 1.5 Million [BP]—Speech—Further social organization leads to an expanded food sources, including scavenging.
3. 500,000 [BP]—Fire improves the energy availability from food.
4. 150,000 [BP]—Ecoadaptation—Humans move out into other ecosystems, expanding the range of energy resources.
5. 50,000 [BP]—Modern humans—The benefits of specialization and social organization are realized during the Ice Age.

As can be seen from the evidence presented both above and below, there are lots of problems with this periodization too. And again, the least problematic is the last phase transition, corresponding to the Upper Paleolithic Revolution.

In addition to the rather detailed periodizations of the Biosocial Big History Era discussed above, I would mention a couple of less detailed periodizations aimed at the identification of the most significant complexity jumps (omitting some less significant phase transitions).

13.2.4 Robert Aunger

His periodization (Aunger, 2007, p. 1168) of what we call the "Biosocial Era" of the Big History looks as follows:

1. 2.5 million years ago—[onset of] "Fire" period—[characterized in the field of] energy innovation by fire use—organization in bands—normative social systems (culture)
2. 500,000 years ago—[onset of] "Multi-tool" period—[characterized in the field of] energy innovation by tool kits—organization in tribes—grammatical language
3. 50,000 years ago—[onset of] "Complex tool" period—[characterized in the field of] energy innovation by compound tools—organization in "Big Man" societies—iconic representation (cave art)

It becomes clear at the first glance that this periodization too has numerous problems. There is no evidence at all for the intentional use of fire by australopiths, whereas at about 2.5 million BP all the most advanced hominid genera were australopithecines, and, as we will see below, the first Homo species was still in the process of formation. In fact, the first firm evidence for controlled use of fire only comes from about 1 million BP (for *Homo erectus*) (see, e.g., Berna et al., 2012; Gowlett, 2016), and only from about 400,000 BP for the systematic use of fire (for *Homo heidelbergensis*) (Roebroeks & Villa, 2011; Stahlschmidt et al., 2020; Trájer, 2024). There is no evidence at all that about 500,000 years ago, *Homo heidelbergensis* (and this was the most advanced *Homo* species that time) communities were organized as tribes. Most specialists would say that the idea that at 500,000 BP *Homo heidelbergensis* groups communicated in "grammatical" language is clearly overoptimistic (e.g., Rozov, 2023); compound tools are attested since about 400,000 BP (Doronichev, 2016; Wilkins et al., 2012), and so on. Still again, the least problematic is the last phase transition, corresponding to the Upper Paleolithic Revolution.

13.2.5 Erhard Glötzl

Erhard Glötzl (2025) proposes the following periodization of the stretch of time corresponding to what we denote as the "Biosocial Era" of the Big History:

4.1. 6,000,000 years ago—[onset of Age of] "Neural network/storage of causal relations" = Age of Hominins
4.2. 900,000 years ago—[onset of Age of] "Simple language/duplication of experience" = Age of Homo
4.3. 60,000 years ago—[onset of epoch of] "Cognitive revolution/logical reasoning" = Age of *Homo sapiens*

It is easy to see that there are lots of problems with this periodization too. To start with, there is no evidence at all that around 6 million BP there was any dramatic increase in the effectiveness of neural networks or storage of causal relations that could justify the identification of 6,000,000 BP as the time of the start of a new age. In addition, Glötzl (2025, p. 274) considers this time point as the time of the start of not only the "Age of Hominins" but also a more general age (I would rather say "Era")—Era/"Age of Cerebrum" (it includes "Aga of Hominins," "Age of Homo," and "Age of Homo sapiens" is its constituent parts). This also does not make much sense, as the first cerebra emerged hundreds of millions of years before 6 million BP. I would also add that the Hominini emerged millions of years before 6,000,000 BP (let's not forget that the taxonomic tribe Hominini/hominins includes, in addition to genus Homo and australopith genera, also genus Pan, consisting at present of chimpanzees and bonobos).

Note also that the first Homo species emerged not 900,000 years ago, but around 2.3—2.4 million BP (Kimbel, 2009, p. 35; Rightmire, 2025), and the period between 2 and 1 million BP was clearly dominated by another Homo species, *Homo erectus* (see below).

As with all the periodizations considered above, the only exception is constituted by the last phase transition in the series, around 60 thousand BP, denoted Glötzl as "Cognitive revolution," which clearly corresponds to what the others mostly call "Upper Paleolithic Revolution." I do not have much complaints about Glötzl calling the epoch that followed this phase transition "Age of Homo sapiens." Indeed, though *Homo sapiens* emerged not later than about 300 thousand BP (e.g., Callaway, 2017), before 60,000 BP our species was present almost exclusively in Africa, and it only became dominant in Afro-Eurasia after this date in direct connection with the Upper Paleolithic Revolution (see below for detail). I have no much complain either about this date, though the Upper Paleolithic Revolution is usually dated to 50–40,000 BP, because, as we will see below, there are some grounds to suppose that it may have started around 60,000 BP.

13.3 On a Possible New Periodization of the Biosocial Era of the Big History

As we could see it above, by now, there is no acceptable periodization of the Biosocial Era of the Big History, which makes it necessary to develop a scientifically grounded periodization.

Note, that due to the reasons spelled out in the first part of this chapter, the proposed periodization of the Biosocial Era is complexity-based, that is, the boundaries of the identified epochs are marked with what Theodore Modis calls "complexity jumps," that is, in our case, such phase transitions that result in significant increases in both biological and sociocultural complexity of the evolving planetary biosocial systems.

13.3.1 Australopithecine Epoch (~4.2 Million BP–2.9–2.4 Million BP)

I believe it makes sense to start the Big History Biosocial Era with the Australopithecine Epoch, whose onset around 4.2 million BP is suggested by the earliest findings of australopithecines[6] (Guil-Guerrero & Manzano-Agugliaro, 2024; White et al., 2006). In fact, this is a sort of transitional epoch, for which we have some evidence for a substantial increase in the biological complexity of our ancestors, but we have very thin evidence for an increase in their sociocultural complexity. The former is suggested by quite a substantial increase in the australopith brain size:

> "The last advances on australopiths brain evolution demonstrate that this group of species is characterized by brain sizes in average 30% bigger than the means measured among common chimpanzee" (Hublin & Changeux, 2022, p. 58; Kimbel & Villmoare, 2016), with average chimpanzee brains volume being c. 360 cc versus c. 466 cc among the australopiths.

The first substantive evidence of the increase in the sociocultural complexity in the australopiths only appears for the late australopithecines at about 3.3 million BP when we have evidence for the emergence of the first (though very short-lived), *Lomekwian*, stone tool technology possibly created by Kenyanthropus[7] (Harmand et al., 2015; Harmand & Lewis, 2023). So, in some respects, one may argue that the Australopithecine Epoch was the last epoch of the BH Biological Era, rather than the first epoch of the BH Sociocultural Era.

[6] Australopithecines = australopiths, that is Australopithecus and their close relatives, any species in the related genera of Australopithecus, Paranthropus, Kenyanthropus, Ardipithecus, and Praeanthropus (see, e.g., Guil-Guerrero & Manzano-Agugliaro, 2024).

[7] A late australopithecine genus closely related to Australopithecus.

13.3.2 Oldowan-Habilis Epoch (2.9–2.4 Million BP[8]–2 Million BP)

The onset of this epoch evidenced significant increases in both sociocultural and biological complexity. A bit paradoxically, the documented increase in sociocultural complexity in this case preceded the biological complexity jump. The start of the transition to the Oldowan-Habilis Epoch appears to be connected with the invention of the *Oldowan stone tool industry* that emerged around 2.9 million BP. Its earliest finds are dated to ~2.9 million BP, and are associated with *Paranthropus*[9] fossils (Plummer et al., 2023). Thus, between 2.9 and 2.4 the Oldowan stone tool industry appears to have been practiced by late Australopithecines (including possibly *Paranthropus* and *Australopithecus garhi* [Heinzelin et al., 1999]), as well as by some forms that were transitional between Australopithecines and Homo (Villmoare et al., 2015). However, *Homo habilis* (who emerged around 2.4 million BP, as suggested by the earliest findings [Kimbel, 2009, p. 35; Rightmire, 2025]) substantially developed this technology (e.g., Rightmire, 2025).

The emergence of *Homo habilis* constituted a significant increase in biological complexity, as brain size experienced a very substantial increase with the emergence of *Homo habilis*. Their average brain size is estimated to be around 640 cc, which is 37% larger than the means measured among the australopiths (466 cc) (Hublin & Changeux, 2022, p. 58; Tuttle, 2025; Wakita, 2020); the complexity of brain also increased substantially (e.g., Bruner & Beaudet, 2023), which suggests a breakthrough in cognitive abilities.

Note that the point that the first long-lasting stone tool archaeological industry emerged at the onset of transition to the Oldowan-Habilis Epoch appears to have created evolutionary pressures toward the emergence of individuals with larger and more complex brains, as the formation of *Homo habilis* appears to have taken place among the australopithecine populations that practiced systematic stone tool production, and better cognitive abilities (associated with a larger and more complex brain) could have yielded substantive selection advantages within such a context.

13.3.3 Early Acheulean-Erectus Epoch (~2 Million BP–900–500 Thousand BP)

The transition from the Oldowan-Habilis to Early Acheulean-Erectus Epoch was associated with both biological and sociocultural complexity jumps. The onset of this BH epoch was precipitated by the emergence of a new Homo species, *Homo erectus*, that appears to have taken place around 2 million BP (Rightmire & Tobias,

[8] Note that for calculation purposes, we pinpoint the date of the transition from Australopithecine to Oldowan-Habilis Epoch to the middle point of this range (2.65 million BP).

[9] This might not be a coincidence in view of the point that among the *Paranthropus*, the average brain size (519 cc) was more than 10% larger than among the australopiths in general (466 cc) (Hublin & Changeux, 2022; Kimbel & Villmoare, 2016; Tuttle, 2025), and, thus, almost one and a half times larger than among the chimpanzee.

2025) as suggested by the earliest findings of *Homo erectus* remains that are dated to about 2.04 million years ago (Herries et al., 2020).

On average, *Homo erectus* had significantly larger and more complex brains than *Homo habilis*. Thus, among Asian *Homo erectus*, brain size was on average 1000 cc (Antón et al., 2016), 56% larger than the means measured among *Homo habilis* (640 cc), and already much closer to the average brain size of the present-day humans (1350 cc) than to the chimpanzees (360 cc) (Tuttle, 2025). In addition, *Homo erectus* brain was much more complex than the one of *Homo habilis*, with expanded frontal lobes (associated with planning, decision-making) and increased brain asymmetry (similar to modern humans), suggesting specialization in different brain regions (Bisconti & Tartarelli, 2025; Holloway et al., 2004; Ponce de León et al., 2021; Rightmire, 2004). In particular, *Homo erectus* brain was characterized by "a pronounced Broca's cap, as is characteristic for modern humans" (Luef, 2018, p. 43; Wu et al., 2011), which suggests rather developed communication capacities, some speech-like communication, using a sort of "proto-language," indicating a very significant increase in not only biological but also sociocultural complexity.

A marked documented increase in sociocultural complexity in this epoch is clearly evidenced by the emergence (in about 1.95 million BP, that is, not long after the formation of *Homo erectus*) of the *Acheulean stone industry* (Key et al., 2021; Mussi et al., 2023), much more advanced than the previous Oldowan one. In particular, it included bifacial flaking (shaping both sides of the tool), requiring advanced planning. The Acheulean stone industry was characterized by more standardization, tools followed a consistent, intentional design (e.g., symmetry in hand axes); greater skill required for hierarchical reduction (multiple steps to shape the core); planning ahead: selection of high-quality raw materials (e.g., basalt, quartzite) and transport over long distances. Acheulean toolmaking suggests advanced spatial reasoning, motor control, and working memory (linked to *Homo erectus*'s larger brain) (Ember et al., 2015; Gowlett, 2011; Larsen, 2017; Stout, 2024; Stout et al., 2008, 2015). A very important sociocultural innovation of *Homo erectus* was occasional intentional use of fire (e.g., Berna et al., 2012; Gowlett, 2016). Rather advanced culture of *Homo erectus* allowed them to undertake a rather impressive dispersion out of Africa, reaching Transcaucasia by about 1.85 million BP (Ferring et al., 2011), Indonesia about 1.8 million BP, China and Southern Europe around 1.7 million years ago (Arzarello et al., 2007; Boaz & Ciochon, 2004; Rightmire, 2001; Zaim et al., 2011) (see Fig. 13.2):

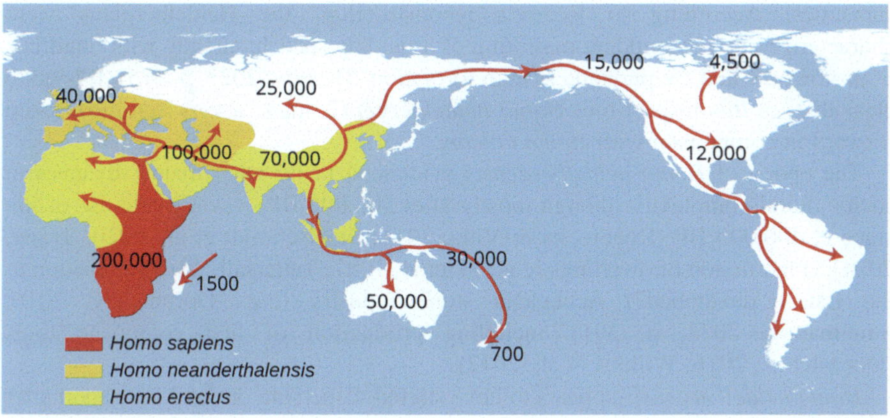

Fig. 13.2 Diffusion of *Homo erectus* (yellow zone), *Homo neanderthalensis*, and *Homo sapiens*. *Source*: https://en.wikipedia.org/wiki/Homo_erectus#/media/File:Spreading_homo_sapiens_la.svg (accessed on May 14, 2025). *Note*: Note that *Homo erectus* appears to have peopled the southern part of the zone that later was peopled by the Neanderthals (the brown zone above). Red zone in the map above was also peopled by *Homo erectus* before *Homo sapiens*

13.3.4 Late Acheulean-Heidelbergensis Epoch (900–500 Thousand BP[10]–300 Thousand BP)

The transition from the Early Acheulean-Erectus to Late Acheulean-Heidelbergensis Epoch was associated first of all with the emergence of a new Homo species—*Homo heidelbergensis*.

Its earliest findings are dated to ~850,000 BP (?) (Profico et al., 2016)/900,000 BP (?) (Suleimanov, 2023). *Homo heidelbergensis* may be considered as the common ancestor for Neanderthals, Denisovans, and modern humans (e.g., Di Vincenzo & Manzi, 2023; McHenry, 2025; Meyer et al., 2014; Wilkins et al., 2012), though this point is sometimes disputed (see, e.g., Stringer & Crété, 2022).

The brain size of *Homo heidelbergensis* is estimated to have been on average ~ 1200 cc (Rightmire, 2004; Wakita, 2020), 20% larger than the means measured among Asian *Homo erectus* (1000 cc), and just 11% smaller than the average for *Homo sapiens* (1350 cc) (Tuttle, 2025). "Endocast prints show that their brains were left-occipital-right-frontal asymmetric and showed a prominent bulge in the left hemisphere over Broca's area" (Luef, 2018, p. 43; Holloway et al., 2004), which suggests rather advanced speech-like communication capacities. "With Heidelbergians... researchers associate the sapient development of the larynx, the hyoid nerve canal, and the respiratory muscles" (Rozov, 2023, p. 175; see also Wood & Bauernfeind, 2012), which are also very strongly associated with advanced speech

[10] Note that for calculation purposes we pinpoint the date of the transition from Early Acheulean-Erectus to Late Acheulean-Heidelbergensis Epoch to the middle point of this range (700 thousand BP).

capacities. According to Rozov's reconstruction, the Heidelbergians were characterized by a "confident mastering of articulate speech: already with hundreds of protowords and the phonetic system formation" (Rozov, 2023, p. 175). It is quite clear that, in this respect too, *Homo heidelbergensis* made a very significant step forward in comparison with *Homo erectus*.

The epoch of *Homo heidelbergensis* evidenced many other major advances in sociocultural complexity (though mostly after 500,000 BP)—systematic use of fire since c. 400,000 BP (Roebroeks & Villa, 2011; Stahlschmidt et al., 2020; Trájer, 2024), construction of dwellings (e.g., Sklenář, 1987; Tattersall, 2010), transition to the Late ("developed") Acheulean stone industry (e.g., Doronichev, 2016; Suleimanov, 2023, p. 391), including production of first composite tools (Doronichev, 2016; Wilkins et al., 2012).

Homo heidelbergensis appears to have started dispersing out of Africa soon after their emergence from ~800,000 years ago (e.g., Carrión et al., 2011; Suleimanov, 2023); they are very well evidenced in Europe (e.g., Carrión et al., 2011); in Asia there is often a problem to distinguish them from the Denisovans. Much larger expansion zone than for the previous two waves (that could not penetrate beyond the Tropical belt) is clearly connected with much more advanced culture and cognitive abilities of *Homo heidelbergensis* that allowed them to penetrate into Sub-Tropical and Temperate belts.

13.3.5 The Middle Paleolithic Epoch—Epoch of Neanderthals, Denisovans, and Early *Homo Sapiens*

The onset of the Middle Paleolithic is usually dated to about 300,000 BP (e.g., Ember et al., 2015; Larsen, 2017; Strauss, 1989). This period is characterized by the prevalence of the three most advanced and complex Homo species (in comparison with all the previous species of our genus). And this is relevant not only for early *Homo sapiens* but also for the Neanderthals and Denisovans (for the phylogenetic relationships between these three Homo species see Guran et al., 2024; Ongaro & Huerta-Sanchez, 2024; Stringer, 2016; Stringer & Barnes, 2015; Stringer & Crété, 2022; Viola, 2022; see also Fig. 13.3).

"Neanderthals had a brain (approximately 1500 cubic centimeters)... larger than those of modern humans[11]" (Luef, 2018, p. 43; see also Holloway, 2015; Macciardi & Martini, 2022, p. 91 etc.). "Studies beginning in the late 1980s of the hypoglossal canal (one of two small openings in the lower part of the skull) and a hyoid (the bone located between the base of the tongue and the larynx) from the paleoanthropological site at Kebara, Israel, suggested that the Neanderthal vocal tract could have been similar to that found in modern humans. Moreover, genetic studies in the early 2000s involving the Neanderthal FOXP2 gene (a gene thought to allow for the capacity for speech and language) indicated that Neanderthals probably used language in the

[11] C. 1350 cc (e.g., Tuttle, 2025).

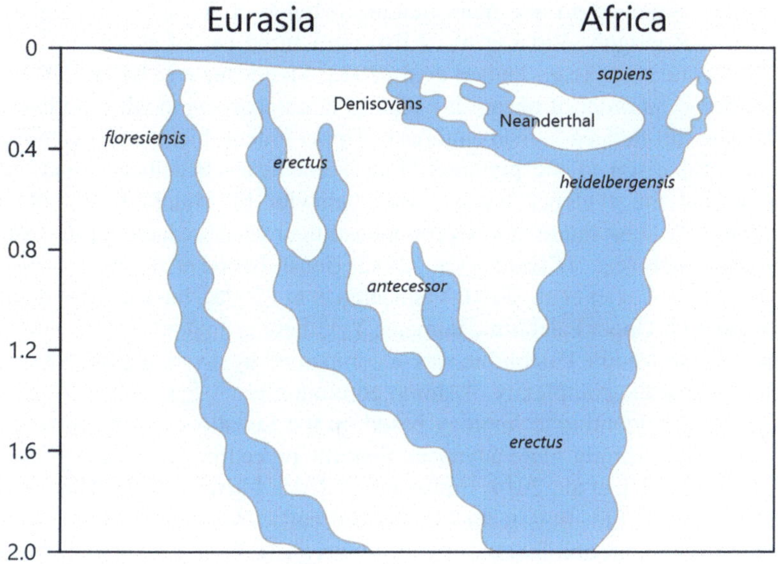

Fig. 13.3 A model of the origins of *Homo sapiens*. The Y-axis represents time in millions of years ago. *Source*: https://en.wikipedia.org/wiki/Human_evolution#/media/File:Homo-Stammbaum,_Version_Stringer-en.svg (accessed on May 29, 2025)

same way that modern humans have" (Williams & Tuttle, 2025); of course, their languages were less developed than the ones of the behaviorally modern humans (e.g., Claessen, 2020, p. 13; Rozov, 2023), but the same can well be said about early anatomically modern humans (before the Upper Paleolithic Revolution [e.g., Rozov, 2023]), not to mention Denisovans.[12]

The transition to the Middle Paleolithic marked a very significant advance in the development of cultural symbolic complexity—symbolic behaviors, protoartistic expression, and so on, suggesting complex cognitive abilities. And this was not only restricted to early *Homo sapience* who emerged just around the start of the Middle Paleolithic (Callaway, 2017; Hublin et al., 2017; Richter et al., 2017; Stringer & Galway-Witham, 2017), but can also be said about the Neanderthals (and very likely supposed about Denisovans, though we know so little about them by now [e.g., Viola, 2022], but there are still some hints even at their symbolic behavior [e.g., Baldwin, 2015; Douka et al., 2019; Shunkov et al., 2020]). Indeed, with respect to the Neanderthals we have evidence on intentional burials with possible ritualistic elements (e.g., Jaubert et al., 2022; Pomeroy et al., 2020; Rendu et al., 2014), personal ornaments and pigment use (Heyes et al., 2016; Hoffmann, Angelucci, et al., 2018; Hoffmann, Standish, et al., 2018; Radovčić et al., 2015; Roebroeks

[12]The presence of FOXP2 gene in the Denisovan genome suggests substantial language abilities (Luef, 2018; Meyer et al., 2012), though, of course, there are no grounds to maintain that Denisovan languages were more developed than the ones of the early anatomically modern humans.

et al., 2012), engraved art and mark-making (Majkić et al., 2018; Meneganzin & Killin, 2024; Rodríguez-Vidal et al., 2014), structured use of space (e.g., Jaubert et al., 2016), and so on (e.g., Jaubert et al., 2022; Meneganzin & Killin, 2024). Yes, this evidence is an order of magnitude scarcer in comparison to what we have with respect to the full-fledged human cultures of Upper Paleolithic, but it is significantly richer in comparison to the previous "Late Acheulean-Heidelbergensian" phase. Note also that the evidence on symbolic behavior of Middle Paleolithic early *Homo sapiens* is substantially more pronounced than what we have for the synchronous Neanderthals (e.g., D'Errico et al., 2012; Henshilwood et al., 2011; McBrearty & Brooks, 2000; Texier et al., 2010; Vanhaeren et al., 2006), but even it still pales in comparison with Upper Paleolithic humans. Yet, there is no doubt that the transition from the Low to Middle Paleolithic was accompanied by a very significant increase in cultural–symbolic complexity of Homo communities.

The Middle Paleolithic technology based on the Levallois / prepared-core technique[13] was significantly more advanced than the preceding Acheulean stone technology (see, e.g., Hu et al., 2019; Ember et al., 2015; Larsen, 2017). The use of the Middle Paleolithic Levallois technique by the early *Homo sapiens* is attested in North-West Africa around 300,000 BP (Callaway, 2017); this technique, however, appears to have been invented by the Neanderthals (White & Ashton, 2003)—hence, we might be dealing with the borrowing by the early humans of a very important technology from the Neanderthals. Levallois technique was also used by Denisovans (Jacobs et al., 2019).

In fact, the zone of diffusion of the Middle Paleolithic Levallois-based stone technologies can well be regarded as the area of the Middle Paleolithic World System (see Fig. 13.4). Actually, Fig. 13.4 grossly underrepresents the spread of Levallois-based stone technologies in Africa, as they are also actually attested in:

- Eritrea (site: Asfet, Gulf of Zula [Beyin, 2023, pp. 236, 238]),
- Ethiopia (sites: The Gademotta Formation [Sahle, 2023, p. 333]; Goda Buticha [Pleurdeau et al., 2023, p. 345]; Gona [Rogers et al., 2023, p. 367]; Herto [Beyene et al., 2023, p. 391]; Kesem-Kebena-Dulecha Study Area [Gilbert et al., 2023, p. 417]; Mochena Borago Rockshelter [Brandt et al., 2023, pp. 469, 472]),
- Kenya (sites: Cartwright's Site [Waweru, 2023, p. 532]; Kapthurin Formation [Tryon, 2023, p. 556]; Panga ya Saidi [Shipton et al., 2023, p. 669]),
- Namibia (site: Pockenbank [Schmidt et al., 2023, p. 906]),
- Senegal (site: Ndiayene Pendao and the Lower Senegal Valley [Blinkhorn et al., 2023, p. 936]),
- Tanzania (site: Kilwa Basin [Beyin & Ryano, 2023, pp. 1062, 1063]),
- Zambia (site: Kalambo Falls [Barham & Kleindienst, 2023, p. 1174]),

[13] Unlike earlier techniques (e.g., Oldowan flakes or Acheulean bifaces), Levallois involved careful preparation of a stone core before striking off a predetermined flake.

Fig. 13.4 Selected Middle Paleolithic sites from the ROAD database (CC BY-SA 4.0 ROCEEH). *Source*: https://en.wikipedia.org/wiki/Middle_Paleolithic#/map/0 (accessed on May 28, 2025)

- South Africa (sites: Border Cave [Backwell et al., 2023, pp. 1304, 1305]; Die Kelders [Grine, 2023, p. 1375]; Erfkroon [Bousman et al., 2023, p. 1445]; Olieboomspoort [Val et al., 2023, p. 1608]).

Note that the Middle Paleolithic globalization (and, hence, the Middle Paleolithic World System) appears to have been substantially different from the earlier globalization waves:

- The first wave, in the late third million BP, associated with the diffusion of the Oldowan stone industry out of Africa in connection with migrations of (most likely) *Homo habilis*.
- The second wave, in the second million BP, associated with the diffusion of the early Acheulean stone industry out of Africa in connection with migrations of *Homo erectus*.
- The third wave, in the first half of the first million BP, associated with the diffusion of the late Acheulean stone industry out of Africa in connection with migrations of *Homo heidelbergensis*.

The Middle Paleolithic globalization (unlike all the other Paleolithic globalizations, including the Upper Paleolithic one, which we will describe below) does not appear to have been connected with migrations of a certain particular Homo species. It is rather connected with the diffusion of a particular Middle Paleolithic Levallois-based stone technology that appears to have been invented by the Neanderthals in north-western Europe (White & Ashton, 2003), but later diffused throughout the Middle Paleolithic World System among the representatives of three Homo species, that is not only Neanderthals of Eurasia but also early *Homo sapiens* in Africa and Denisovans in Asia (see above). In this respect, the Middle Paleolithic globalization appears to resemble more the post-Paleolithic globalization waves (that were connected with the diffusion of technologies more than with human migrations[14] [Zinkina et al., 2019]) rather than the other Paleolithic migration waves.

13.3.6 Onset of Upper Paleolithic (the "Upper Paleolithic Revolution")

As we could see above, there is an unusual consensus among the big historians as regards the dating and high importance of the Upper Paleolithic Revolution. On the other hand, there are still some doubts about this (e.g., Christian, 2011; McBrearty & Brooks, 2000). Thus, as in my periodization, the Upper Paleolithic Revolution occupies a place of very important milestone marking not only a border between two epochs but also the end of the BH Biosocial Era and the onset of Era of Hyperbolically Accelerating Growth of Sociocultural Complexity ("Sociocultural Era"); it appears appropriate to provide systematic evidence in support of such a classification.

The beginning of the Upper Paleolithic (the "Upper Paleolithic Revolution") is usually dated to 40,000–50,000 BP (e.g., Ember et al., 2015, p. 164; Rozov, 2023, p. 241; Grinin et al., 2024, pp. 3, 73; Larsen, 2017, p. 416).

The transition from Middle to Upper Paleolithic marked the finalization of the formation of modern *Homo sapiens* species (e.g., Neubauer et al., 2018); however, this transition resulted mostly in the dramatic increase of sociocultural rather than biological complexity of humans. In this respect, it was very different from the above described phase transitions in the complexity growth within the sociobiological era of the Big History (on this Big History era see, e.g., Grinin & Korotayev, 2025; Korotayev, 2025a), as within them the increase in biological complexity was as important as the one in sociocultural complexity. For example, the substantial increase in sociocultural complexity achieved by the *Homo erectus* communities (e.g., the invention of the Acheulean stone industry that was far superior and more complex than the preceding Oldowan one) would not have been possible without the increase in the size and complexity of the *Homo erectus* brain.

In this respect, the "Upper Paleolithic Revolution" marked not only the transition from the Middle to Upper Paleolithic but also the transition from the Sociobiological to Sociocultural Era of the Big History (Grinin & Korotayev, 2025; Korotayev, 2025b).

The transition from the Middle to Upper Paleolithic was accompanied by a very significant increase in the technological complexity. This includes, for example, immense increase in the variety of tools, tools becoming more uniform and specialized (e.g., burins, end-scrapers, backed blades), development of pressure flaking and soft hammer techniques (finer retouching for sharper, more refined

[14] On the other hand, of course, migrations appear to have played an important role in all the globalization waves without exception.

tools), increased use of organic materials for needles, harpoons, and figurines, spear-throwers (securing increased range and force), and barbed harpoons and developed fishing tools (allowing improved exploitation of aquatic resources). The speed of technological development increased dramatically: before the Upper Paleolithic, rare inventions alternated with dozens thousand years of virtual stagnation, whereas after the Upper Paleolithic Revolution, we deal with almost constant technological development, with certain technological innovations observed almost every millennium (Barton & Clark, 2021; Bar-Yosef, 2002, 2007; Bar-Yosef & Kuhn, 2016; Goder-Goldberger et al., 2023; Sánchez-Yustos et al., 2024; Shea, 2013; Zwyns, 2021).

Upper Paleolithic Revolution also manifested itself through a dramatic increase in the cultural–symbolic–informational complexity of human communities. Indeed, for Upper Paleolithic, we have immense evidence for the symbolic behavior of humans—numerous manifestations of cave art and rock engravings, portable art and figurines, personal adornment and jewelry, systematic intentional burial with developed ritualistic elements, musical instruments, and even abstract symbols and notation (e.g., Conard et al., 2009; Cuenca-Solana et al., 2013; Giacobini, 2007; Lbova, 2021; Lewis-Williams, 2002; Marshack, 1991; Oktaviana et al., 2024; Pettitt, 2011; Stiner, 2014; Zilhão, 2005); there is also evidence on the increased social complexity and even certain social stratification in some Upper Paleolithic communities (e.g., Hayden, 2021; Trinkaus & Buzhilova, 2018; Vanhaeren & d'Errico, 2005; Wengrow & Graeber, 2021). On the one hand, most of the abovementioned was attested in some form in the Middle Paleolithic, especially among the early *Homo sapiens*, but also among the Neanderthals and even Denisovans (see above). However, in the Middle Paleolithic, manifestations of symbolic behavior were occasional and comparatively primitive, whereas in the Upper Paleolithic they become systematic and comparatively developed. Finally, the Upper Paleolithic Revolution also appears to have manifested itself in the finalization of the formation of fully developed syntactic human languages (e.g., Rozov, 2023). The dramatic increase in cultural, symbolic, informational, and social complexity of human communities that took place with the transition from the Middle to Upper Paleolithic appears to be evidenced rather well.

The new globalization wave (and much more pronounced and significant than the previous ones) was, in fact, an important component of the Upper Paleolithic Revolution. In fact, the out-of-Africa migrations of *Homo sapiens* that took place during the transition to the Upper Paleolithic and later within the Upper Paleolithic were not the first "attempts" of the representatives of our species to migrate out of the "Black Continent." Actually, such attempts seem to have been attested as early as 270,000 BP (Posth et al., 2017), and later around 210,000 BP (Harvati et al., 2019), 150,000–130,000 BP (Armitage et al., 2011), and 120,000–80,000 BP (Grün et al., 2005; Langdon, 2023; Stringer et al., 1989). However, during all those "attempts" the early modern humans did not manage to move far from Africa, their existence in some areas of Western Asia and South West Europe did not last very long and eventually came to an end (through extinction, assimilation by the Neanderthals, or back-migration to Africa), and they (unlike the Early Upper Paleolithic human migrations) did not produce any significant impact on the global evolution.

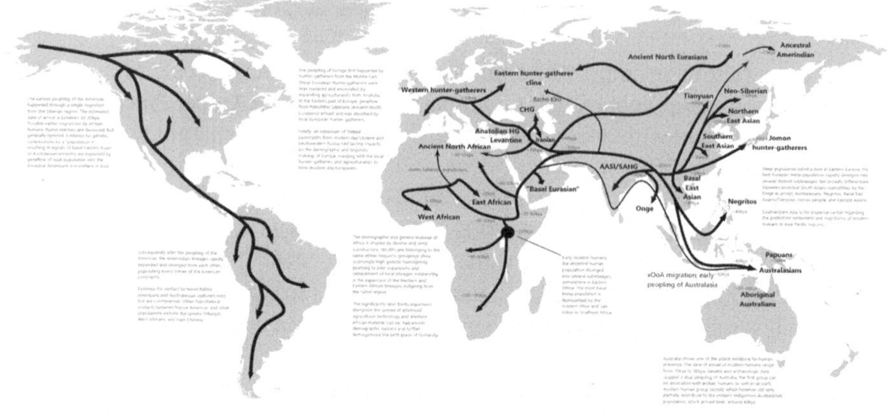

Fig. 13.5 Migration routes of modern humans, showing the northern route populating Western Eurasia, and the southern/coastal route populating Eastern Eurasia. https://en.wikipedia.org/wiki/Recent_African_origin_of_modern_humans#/media/File:Migration_routes_of_modern_humans_(2023).png

The first really long-distance out-of-Africa migration of the modern humans with really long-lasting global impact appears to have taken place around 60,000–50,000 BP along the so-called "coastal route." It involved crossing the Bab-el-Mandeb Strait and movement along the Arabian Peninsula coast to Iran, then to India, South-East Asia, New Guinea, and Melanesia, reaching Australia by around 50,000 BP[15] (Kealy et al., 2018; Metspalu et al., 2004; O'Connell et al., 2018; Petraglia & Allchin, 2007; Wells, 2003; Zhivotovsky et al., 2003). Around 55,000 BP *Homo sapiens* reached southern parts of China (Hublin, 2021, p. 3), around 45,000 BP modern humans appeared in Europe[16] (Hublin et al., 2020; Prüfer et al., 2021; Teyssandier, 2023) and Siberia (Fu et al., 2014; Li et al., 2020; Rybin et al., 2023), about 40,000 BP they reached Northern China (Bae et al., 2017; Li et al., 2020), around 38,000 BP they arrived in Japan (e.g., Cooke et al., 2021), and so on (see Fig. 13.4 for detail). In the later Upper Paleolithic Homo sapiens peopled the New World, most likely in a few waves (e.g., Becerra-Valdivia & Higham, 2020; Gruhn, 2023; Korotayev, 2006; Korotayev, Berezkin, et al., 2006; Korotayev, Malkov, & Khaltourina, 2006; Korotayev et al., 2017; Meltzer, 2009; Stanford & Bradley, 2012; Jablonski, 2002) (see Fig. 13.5 for detail).

It is quite clear that this globalization wave was very tightly connected with the Upper Paleolithic cultural and technological revolution, as it supplied the humans with a very flexible technological culture which allowed them during the Upper Paleolithic expansion to people almost all of the habitable landmass of our planet

[15] Though there are some indications that this might have happened earlier (e.g., Clarkson et al., 2015, 2017); yet, there are significant doubts about this (O'Connell et al., 2018).

[16] Though there might have been earlier failed incursions of modern humans into Europe (Slimak et al., 2022).

(with an exception of a few islands that were peopled a bit later), as this implied the necessity to adapt rather quickly (by the geological scale) to very different environments.

The Upper Paleolithic globalization wave produced much more serious impact on the global development than all the previous waves, as it laid the foundation for all the subsequent globalization waves by spreading the human culture all around the globe, establishing the nodes, between which more and more intensive connections could be established during new globalization waves in the future.

Note that we believe that the Upper Paleolithic Revolution marked not only the end of the Middle Paleolithic Epoch (= epoch of Neanderthals, Denisovans, and early *Homo sapiens*) but also the end of the BH Biosocial Era and the onset of the Era of Hyperbolically Accelerating Growth of Sociocultural Complexity (BH "Sociocultural Era"). On the one hand, in no way the biological evolution of the humans stopped 50,000 years ago; in fact, it definitely continues, and in recent millennia (especially, after the Neolithic Revolution) it has even very significantly accelerated (Cochran & Harpending, 2009; Hawks, 2014; Hawks et al., 2007; Kerner et al., 2023; Wilde et al., 2014). However, there is no evidence that this resulted in any increase in biological complexity of humans. After 50,000 BP human evolution mostly led to idioadaptations (sometimes very significant, like emergence of lactase persistence [or the ability to digest milk after weaning, e.g., Check, 2006; Curry, 2013]), sometimes degenerations, but never to aromorphoses that would have increased the biological complexity of humans. There is no evidence that at present human babies are born with more complex brain than Upper Paleolithic babies. In fact, the average brain size of the Upper Paleolithic Cro-Magnons in Europe was a bit larger than the one of present-day Europeans (e.g., Balzeau et al., 2013). This, of course, does not mean that Upper Paleolithic people were more complex biologically than present-day humans, but, naturally, this is not congruent with any substantial increase in biological complexity of humans after 50,000 BP. Finally, 50,000 years do not look like a sufficient time for the development of a serious biological aromorphosis leading to a serious increase in biological complexity.

On the other hand, it is quite clear that the complexity of human planetary system has increased enormously since 50,000 BP; and it has increased exclusively (or almost exclusively) through the increase in its sociocultural, not biological complexity (see, e.g., LePoire, 2023). This allows us to maintain with a high level of confidence that the Upper Paleolithic Revolution marked not only the end of the Middle Paleolithic Epoch but also the end of the BH Biosocial Era and the onset of the Era of Hyperbolically Accelerating Growth of Sociocultural Complexity (BH "Sociocultural Era").

13.4 New Periodization of the Biosocial Era of the Big History and Some of Its Implications

Thus, we arrive at the following periodization of the BH Biosocial Era (see Table 13.2, where it is described together with the periodization of Era of Hyperbolically Accelerating Growth of Sociocultural Complexity ["Sociocultural Era"]):

Table 13.2 BH periodization of "Biosocial" and "Sociocultural" eras on the basis of the complexity growth phases

Era	Date of the onset	Epoch	Date of the onset
"Biosocial era"	*c.* 4.2 million BP (emergence of the australopithecines)	Australopithecine epoch	*c. 4.2 million BP* (emergence of the australopithecines with subsequent [rather vague] indications of increases in sociocultural complexity)
		Oldowan-Habilis epoch	*c. 2.9–2.4 [2.65]* million BP (emergence of the Oldowan stone tool industry and the first homo species)
		Early Acheulean-erectus epoch	*c. 2 million BP* (emergence of *Homo erectus* followed by the invention of early Acheulean stone tool technology)
		Late Acheulean-Heidelbergensis epoch	*c. 900–500 [700] thousand BP* (emergence of *Homo heidelbergensis* followed by the invention of late Acheulean stone tool technology, including production of first composite tools, and other major advances in sociocultural complexity—systematic use of fire, construction of dwellings, and so on)
		Middle Paleolithic epoch—Epoch of Neanderthals, Denisovans, and early *Homo sapiens*	*c. 300,000 BP* (emergence of the three most advanced and complex homo species [in comparison with all the previous species of our genus]—Neanderthals, Denisovans, and early *Homo sapiens*; major advances in sociocultural complexity—Middle Paleolithic Levallois-based stone technologies, intentional burials with ritualistic elements, personal ornaments and pigment use, engraved art and mark-making, structured use of space, and so on)

(continued)

Table 13.2 (continued)

Era	Date of the onset	Epoch	Date of the onset
Era of hyperbolically accelerating growth of sociocultural complexity ("sociocultural era")	*c*. 50,000 BP (upper Paleolithic revolution)	Upper Paleolithic	*c. 50,000 BP* (upper Paleolithic revolution)
		Early (pre-civilization) agrarian epoch	*c*. 12,000–9000 BP (Neolithic revolution)
		Epoch of early agrarian civilizations	*c*. 4000–3000 BCE (urban revolution)
		Epoch of advanced agrarian civilizations	*c*. 800–500 BCE (axial age revolutions)
		Early modern epoch	*c*. 1450–1550 CE (Panov's "first industrial revolution")
		Industrial epoch	*c*. 1830–1840 (Panov's "second industrial revolution")
		Postindustrial epoch	*c*. 1950 (Panov's "information revolution")

Note: The periodization of the BH Sociocultural Era above generally follows Panov (2005a, 2005b, 2020), with one important exception. The point is that Panov (2005a, 2005b), LePoire (2016), as well as Modis (2002, 2003) and Kurzweil (2005) find a phase transition/"complexity jump" in the mid-first century CE. This does not appear to be supported by any empirical evidence that rather suggests a certain decline in global complexity between 150 and 600 CE (e.g., Zinkina et al., 2019). Hence, I had to omit the respective phase transition and corresponding phase/epoch from my Big History periodization. If one can find anything like a "complexity jump" between 150 and 1450 CE, it would be rather the eleventh century when one could see a simultaneous significant increase in complexity in the most important World System centers of that time (e.g., Zinkina et al., 2019); the last point, however, can hardly be regarded as widely recognized yet

Note that earlier we used periodizations proposed by Modis and Panov to calculate equations describing acceleration of the global complexity growth (Korotayev, 2018, 2020a, 2020b, 2024, 2025a, 2025b). Let us see now how the rectification of dates of BH Biosocial Era complexity jumps in both series will affect the results of such calculations.

13.4.1 Modis–Kurzweil Series

For the Modis–Kurzweil series, this analysis yielded the following results (see Fig. 13.6; for a detailed description of this calculation algorithm, see Korotayev, 2018, 2020a, 2020b):

As we see, our power-law regression of this time series identifies the following best fit equation describing this time series in an almost ideally accurate ($R^2 = 0.999$) way:

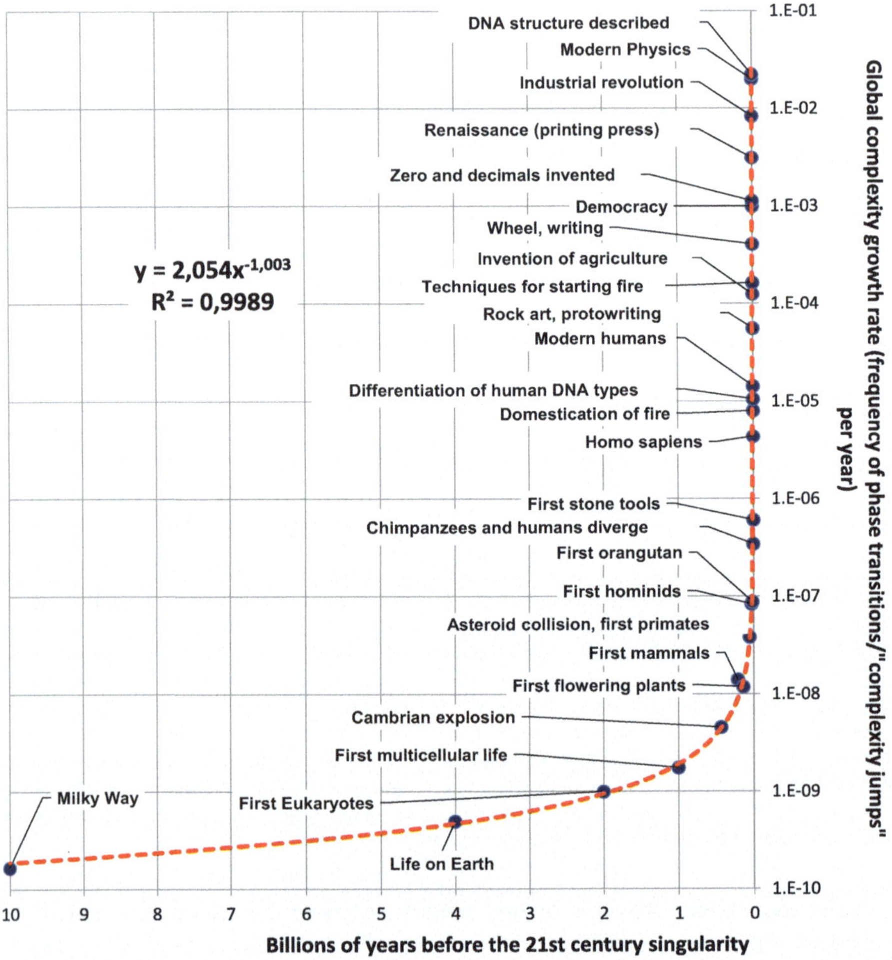

Fig. 13.6 Scatterplot of the "complexity jumps" from the Modis–Kurzweil list with the fitted power-law regression line (with a logarithmic scale for the Y-axis)—for the Singularity date identified as 2029 CE with the least squares method

$$y = \frac{2.054}{x^{1.003}}, \tag{13.1}$$

where y is the global complexity growth rate, x is the time remaining till the Singularity, and 2.054 and 1.003 are constants.

Note that the denominator's exponent (1.003) turns out to be only negligibly different from 1 (well within the error margins); hence, there are all grounds to use this equation in the following simplified form:

$$y = \frac{2.054}{x}, \tag{13.2}$$

where y is the global macrodevelopment rate, x is the time remaining before the Singularity, and 2.054 is a constant.

Of course, x (the time remaining till the Singularity) at the moment of time t equals t^*—t, where t^* is the time of Singularity. Thus,

$$x = t^* - t.$$

Hence, Eq. (13.2) can be re-written in the following way:

$$y_t = \frac{2.054}{t^* - t}, \tag{13.3}$$

where y_t is the global macrodevelopment rate at time t, t^* is the time of Singularity, and 2.054 is a constant.

Finally, let us recollect that our power-law analysis of the transformed Modis–Kurzweil series has identified the Singularity date as 2029 CE. Thus, Eq. (13.3) can be further re-written in the following way:

$$y_t = \frac{2.054}{2029 - t}. \tag{13.4}$$

Of course, in a more general form it should be written as follows:

$$y_t = \frac{C}{t^* - t}, \tag{13.5}$$

where C and t^* are constants.

Now let us repeat this analysis replacing Modis' dates for "complexity jumps" of the BH Biosocial Era with the dates identified by our analysis (see Figs. 13.7 and 13.8):

As we see the replacement of Modis' dates of BH Biosocial Era complexity jumps with more accurate dates does not produce any dramatic changes (we will discuss causes of this a bit later). The series continues to be described extremely well with an equation of the following type:

$$y_t = \frac{C}{(t^* - t)^{\beta}}, \tag{13.6}$$

where y_t is the global macrodevelopment rate at time t, t^* is the time of singularity, and C and β are constants. The singularity date (t^*) changes by just 4 years (from 2029 to 2025 CE), which is entirely negligible in the billion-year perspective. Constant C changes from 2.05 to 1.94, and β alters from 1.003 to 0.999:

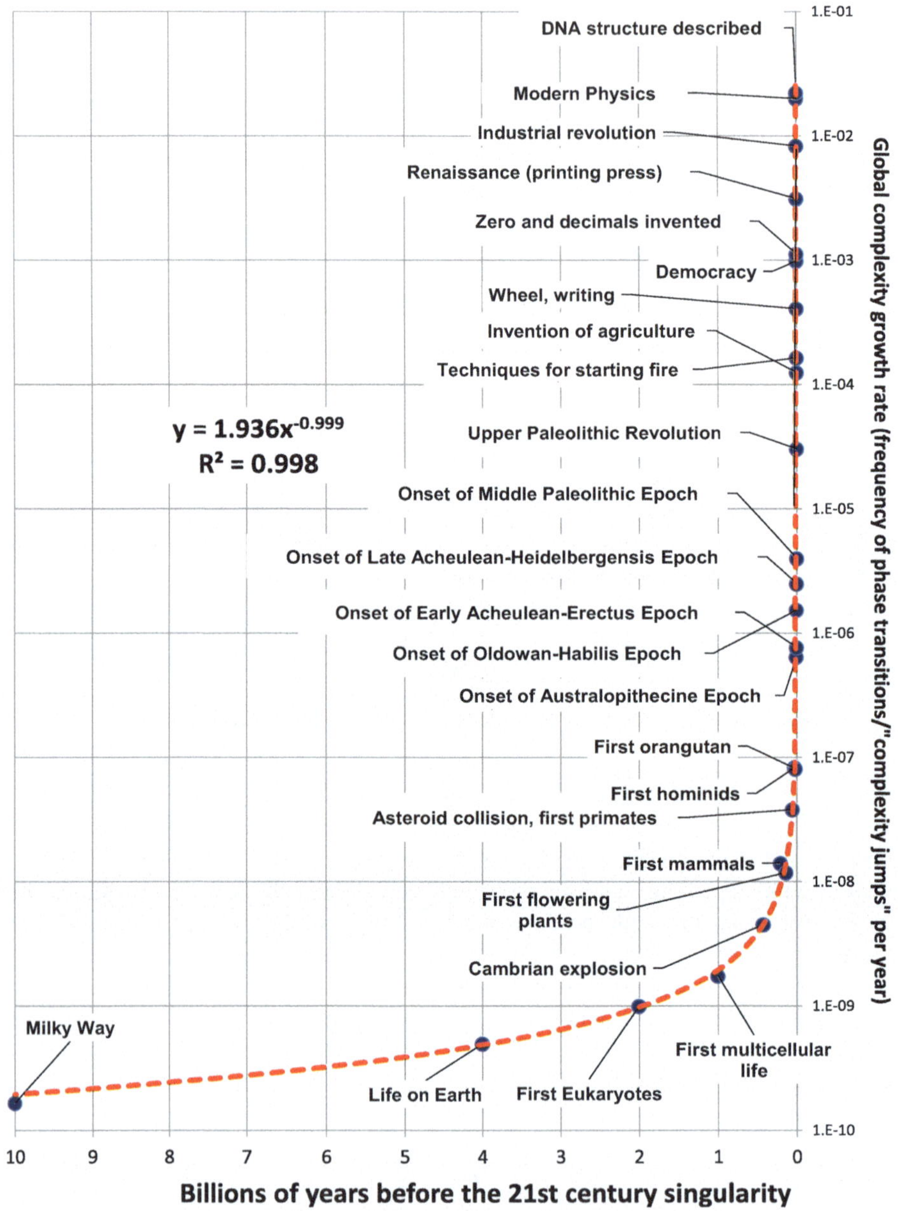

Fig. 13.7 Scatterplot of the "complexity jumps" from the *amended* Modis–Kurzweil list with the fitted power-law regression line (*with a logarithmic scale for the Y-axis*)—for the Singularity date identified as *2025 CE* with the least squares method

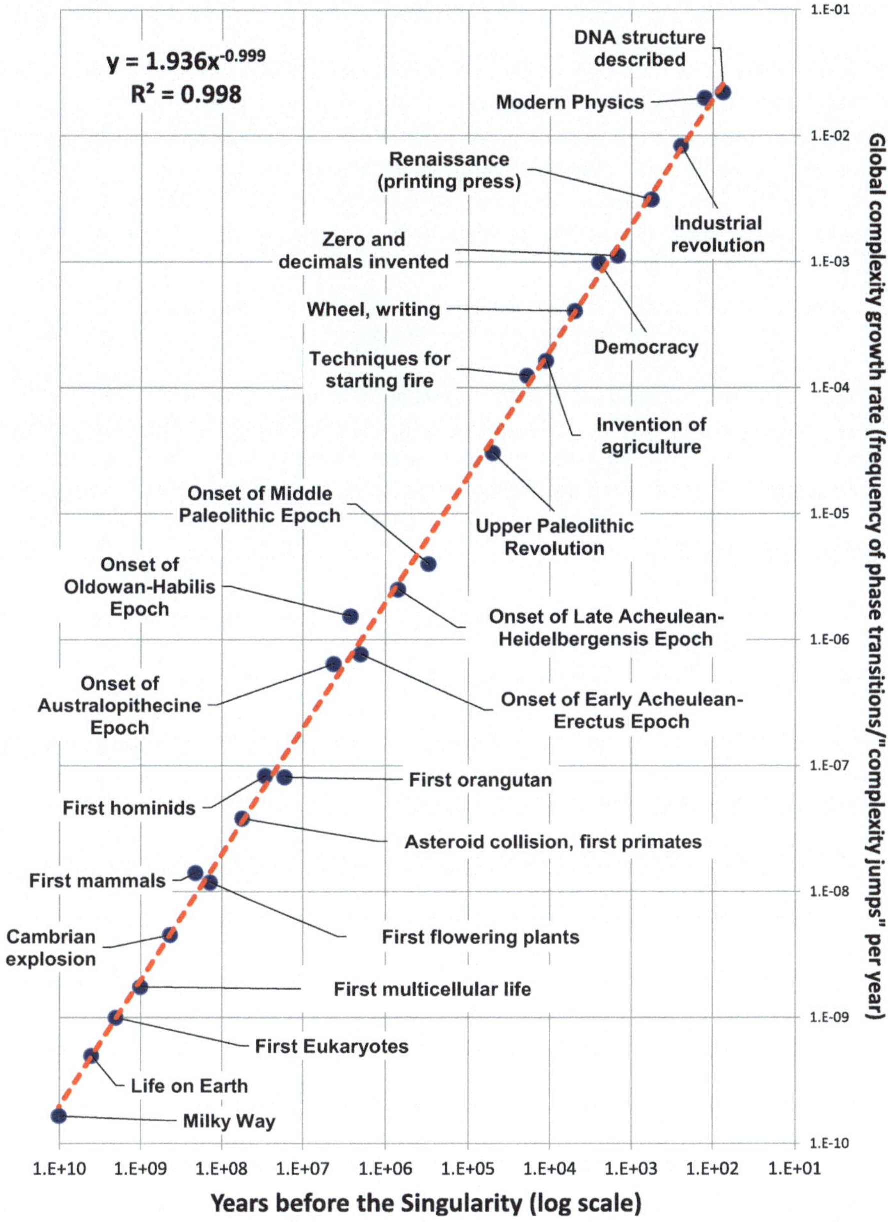

Fig. 13.8 Scatterplot of the "complexity jumps" from the *amended* Modis–Kurzweil list with the fitted power-law regression line (*with a logarithmic scale for the BOTH axes*)—for the Singularity date identified as *2025 CE* with the least squares method

$$y_t = \frac{1.936}{(2025 - t)^{0.999}}. \tag{13.7}$$

The resultant very slightly altered hyperbolic equation describes with practically the same extremely high accuracy ($R^2 = 0{,}998$) the acceleration of the global complexity growth in the amended Modis–Kurzweil series.[17]

In Eq. (13.7), the value of exponent β of Eq. (13.6)–0.999—turns out to be even closer to 1 than in Eq. (13.1); hence, there are even more grounds to write it down in the following simplified form:

$$y_t = \frac{1.936}{2025 - t} \approx \frac{1.9}{2025 - t}. \tag{13.7a}$$

Thus, after the rectification of dates of BH Biosocial Era complexity jumps in the Modis series it continues to be described extremely accurately with a very simple hyperbolic equation of the same type as was identified earlier, which supports our identification of the hyperbolic pattern of acceleration of the global complexity growth described in our earlier publications (Korotayev, 2018, 2020a, 2020b, 2024, 2025a, 2025b).

13.4.2 Panov Series

Next, let us test how the replacement of Panov dates of BH Biosocial Era complexity jumps with more accurate dates will affect the calculations of the general pattern of acceleration of the global complexity growth. For the unamended Panov series our calculations yielded the following results (see Fig. 13.9; for a detailed description of the respective calculation algorithm, see Korotayev, 2018, 2020a, 2020b):

As we see, the power-law equation best describing the acceleration pattern in the unamended Panov series looks as follows:

$$y = \frac{1.886}{(2027 - t)^{1.01}}. \tag{13.8}$$

In general form, the respective equation looks as follows:

$$y = \frac{C}{(t^* - t)^{\beta}}. \tag{13.6a}$$

This equation has three parameters—C, t^*, and β. Note that all the three parameters turn out to be extremely close for both Modis–Kurzweil and Panov.

[17] Note that the best-fit equation does not change substantially if we delete from the series a rather dubious mid-first millennium CE complexity jump, or especially if we move it (as suggested above) to the mid-eleventh century CE. In the last case, the singularity year, parameter β, and accuracy of fit do not change at all, whereas constant C experiences a negligible change from 1.936 to 1.942.

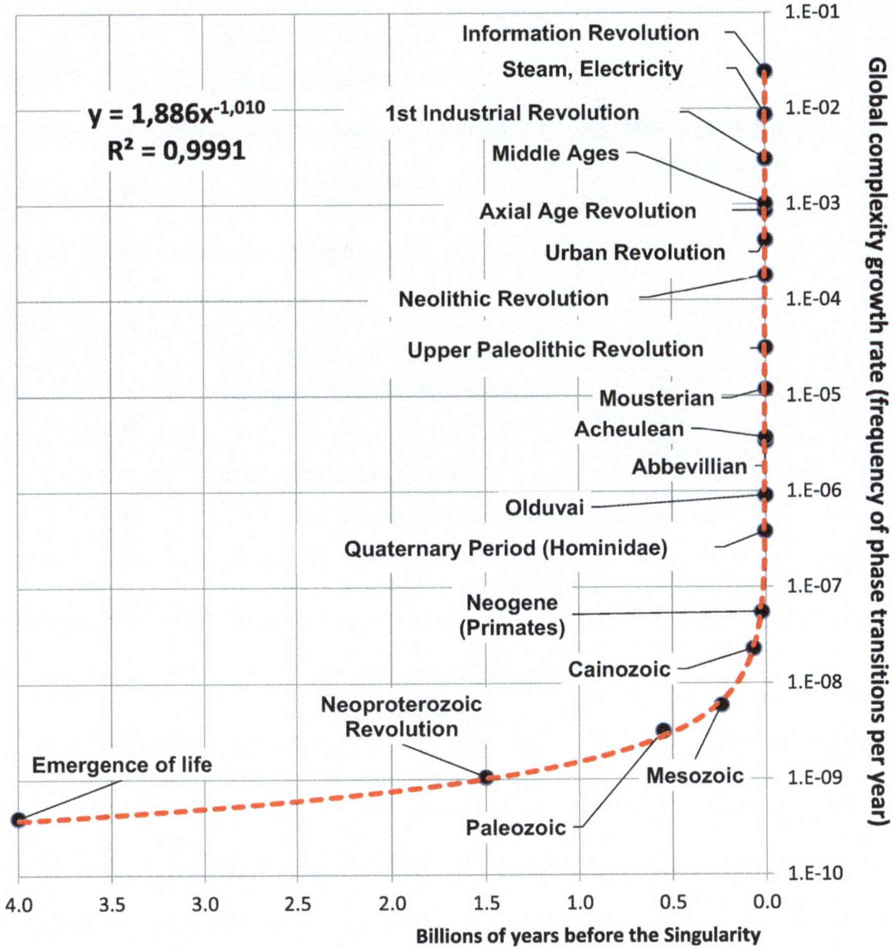

Fig. 13.9 Scatterplot of the "biospheric revolutions" from the *original* Panov list with the fitted power-law regression line (with a logarithmic scale for the Y-axis)—for the Singularity date identified as *2027* CE with the least squares method

As we have already mentioned, there are sufficient grounds to simplify Eq. (13.8)

$$y = \frac{1.886}{(2027 - t)^{1.01}} \tag{13.8a}$$

to the simple hyperbolic version (9):

$$y = \frac{1.9}{2027 - t}. \tag{13.9}$$

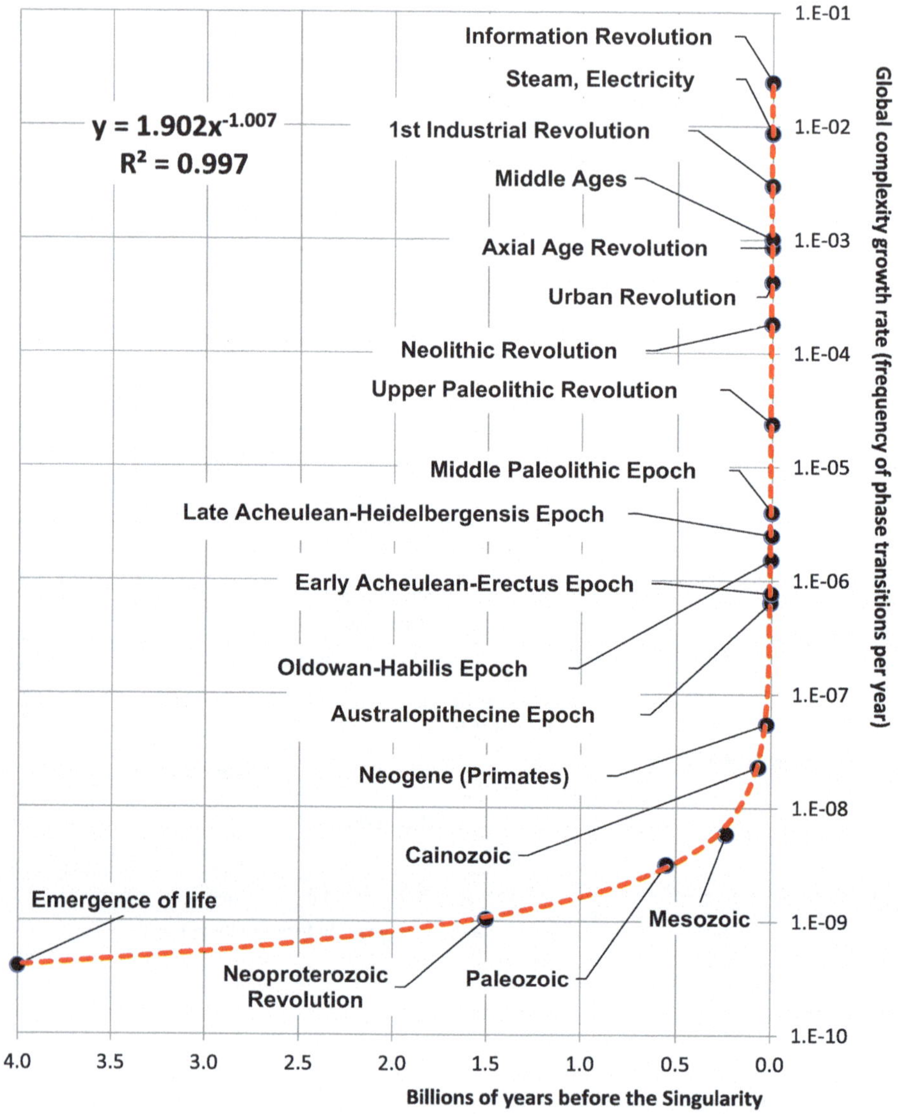

Fig. 13.10 Scatterplot of the "biospheric revolutions" from the *amended* Panov list with the fitted power-law regression line (*with a logarithmic scale for the Y-axis*)—for the Singularity date identified as *2030* CE with the least squares method

Now let us repeat this analysis replacing Panov's dates for "biospheric revolutions" of the BH Biosocial Era with the dates identified by our analysis (see Figs. 13.10 and 13.11):

As we see, as we have already observed this with the Modis series, the replacement of Panov's dates of BH Biosocial Era "biospheric revolutions/phase

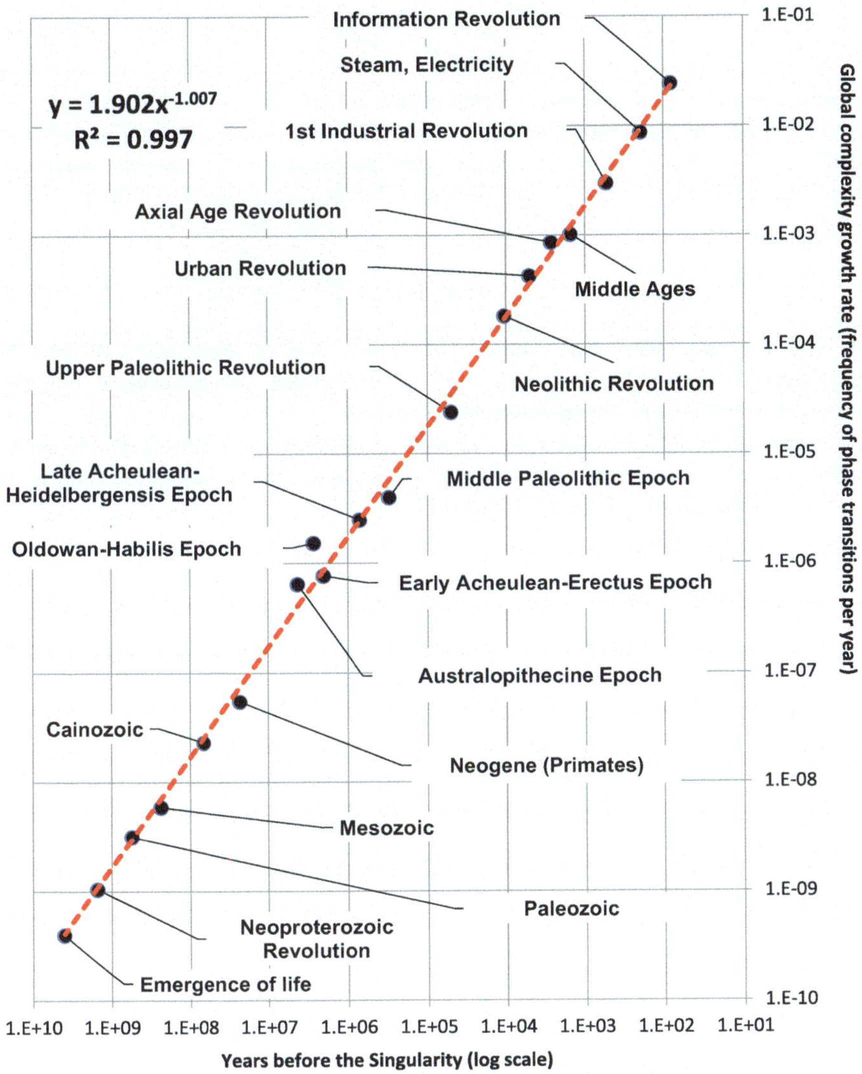

Fig. 13.11 Scatterplot of the "biospheric revolutions" from the *amended* Panov list with the fitted power-law regression line (*with a logarithmic scale for the BOTH axes*)—for the Singularity date identified as *2030* CE with the least squares' method

transitions" with more accurate dates does not produce any dramatic changes. The series continues to be described extremely well with an equation of the following type:

$$y_t = \frac{C}{(t^* - t)^{\beta}}, \tag{13.6b}$$

where y_t is the global macrodevelopment rate at time t, t^* is the time of Singularity, and C and β are constants. The Singularity date (t^*) changes by just 3 years (from 2027 to 2030 CE), which is entirely negligible in the billion-year perspective. Constant C changes from 1.87 to 1.90, and β alters from 1.010 to 1.007:

$$y_t = \frac{1.902}{(2030 - t)^{1.007}}. \tag{13.10}$$

The resultant very slightly altered hyperbolic equation describes with practically the same extremely high accuracy ($R^2 = 0{,}997$) the acceleration of the global complexity growth in the amended Panov series.[18]

Again, in Eq. (13.10), the value of exponent β of Eq. (13.10)—1.007—turns out to be even closer to 1 than in Eq. (13.8); hence, there are even more grounds to write it down in the following simplified form:

$$y = \frac{1.9}{2030 - t}. \tag{13.11}$$

Thus, similarly to what we have observed above with the Modis series, after the rectification of dates of BH Biosocial Era complexity jumps in the Panov series it continues to be described extremely accurately with a very simple hyperbolic equation of the same type as was identified earlier, which provides additional support for our identification of the hyperbolic pattern of acceleration of the global complexity growth described in our earlier publications (Korotayev, 2018, 2020a, 2020b, 2024, 2025a, 2025b).

Quite predictably, after the rectification of dates, the values of constant C in both equations become even more similar to each other and can be rounded to 1.9. Thus, we have

$$y \approx \frac{1.9}{2030 - t}. \tag{13.7b}$$

for the rectified Modis–Kurzweil series and

[18] Note that the best-fit equation does not change substantially if we delete from the series a rather dubious mid-first millennium CE "biospheric revolution"/phase transition/complexity jump, or especially if we move it (as suggested above) to the mid-eleventh century CE. In the last case, the singularity year (2028) moves quite closely back to 2027, parameter β (= 1.006) moves even closer to 1, accuracy of fit does not change at all, whereas constant C experiences a negligible change from 1.902 to 1.884 (still $\approx$ 1.9).

$$y \approx \frac{1.9}{2025 - t}. \tag{13.11a}$$

for the Panov one. Note that the mean value of constant t^* (the Singularity date) in the two equations is 2027.5, which is extremely close to the value of this constant (2027) in the equation $y_t = C/(2027—t)$ that was discovered by Heinz von Foerster and his colleagues as early as in 1960 to describe almost perfectly the world population dynamics between 1 and 1958 CE (see von Foerster et al., 1960; on the implications of this striking similarity see Korotayev, 2018, 2020a, 2020b, 2024, 2025a, 2025b).

As we have already noted on a number of occasions (Korotayev, Berezkin, et al., 2006, pp. 118–120; Korotayev, 2018, 2020b), such algebraic equations as (11) and (12) of the type $y_t = C/(t^*—t)$ can be regarded as a solution of the following differential equation:

$$\frac{dy}{dt} = \frac{y^2}{C} = \frac{y^2}{1.9} \approx 0.5y^2. \tag{13.12}$$

Thus, the overall pattern of acceleration of planetary macroevolution that describes so accurately the rectified Modis and Panov series of complexity jumps turns out to be identical with the one that was detected originally (Korotayev, 2018, 2020a, 2020b, 2024, 2025a, 2025b): the increase in the planetary biosocial complexity growth rate a times is accompanied by a^2 increase in the acceleration speed of this development rate; thus, a twofold increase in planetary biosocial complexity growth rate tends to be accompanied by a fourfold increase in the acceleration speed of this development rate; an increase in planetary biosocial complexity growth rate 10 times tended to accompanied by 100 times increase in the acceleration speed of this development rate; and so on...

To my mind, all these indicate the existence of sufficiently rigorous global macroevolutionary regularities (describing the evolution of complexity on our planet for a few billions of years), which can be surprisingly accurately described by extremely simple mathematical functions.

13.5 Conclusion

To conclude, we have undertaken an attempt to propose a periodization of the BH Biosocial Era on the basis of the most recent scientific data. This periodization is complexity-based, that is, the boundaries of the identified epochs are marked with complexity jumps, that is, in our case, such phase transitions that result in significant increases in both biological and sociocultural complexity of the evolving planetary biosocial systems. This has allowed us to identify the following epochs:

1. Australopithecine Epoch—from *c*. 4.2 million BP up to *c*. 2.9–2.4 million BP—emergence of the australopithecines with subsequent (rather vague) indications of increases in sociocultural complexity.
2. Oldowan-Habilis Epoch—from *c*. 2.9–2.4 million BP up to *c*. 2 million BP—emergence of the Oldowan stone tool industry and the first Homo species.
3. Early Acheulean-Erectus Epoch—from *c*. 2 million BP to *c*. 900–500 thousand BP—emergence of *Homo erectus* followed by the invention of Early Acheulean stone tool technology.
4. Late Acheulean-Heidelbergensis Epoch—from *c*. 900–500 thousand BP to *c*. 300,000 BP—emergence of *Homo heidelbergensis* followed by the invention of Late Acheulean stone tool technology, including production of first composite tools, and other major advances in sociocultural complexity—systematic use of fire, construction of dwellings, etc.
5. Middle Paleolithic Epoch—epoch of Neanderthals, Denisovans, and early *Homo sapiens*—from *c*. 300,000 BP to *c*. 50,000 BP—emergence of the three most advanced and complex Homo species (Neanderthals, Denisovans, and early *Homo sapiens*) in comparison with all the previous species of our genus; major advances in sociocultural complexity (Middle Paleolithic Levallois-based stone technologies, intentional burials with ritualistic elements, personal ornaments and pigment use, engraved art and mark-making, structured use of space, and so on).

Note that, within the proposed BH periodization, the Upper Paleolithic Revolution (*c*. 60–40 thousand years ago) marks the end of both the Middle Paleolithic Epoch—epoch of Neanderthals, Denisovans, and early *Homo sapiens*—and the end of the BH Biosocial Era.

It is also shown that most extent complexity-based periodizations of the Big History (including the ones proposed by Theodore Modis and Alexander Popov) contradict significantly the presently available scientific data on human evolution and need serious amendments as regards the periodization of the BH Biosocial Era.

On the other hand, it has been shown that this does not invalidate our earlier calculations of the planetary complexity growth patterns based on the mathematical analysis of the Modis and Panov series (Korotayev, 2018, 2020a, 2020b, 2024, 2025a, 2025b). Indeed, it has been demonstrated that after the replacement of Modis' and Panov's dates of BH Biosocial Era complexity jumps with more accurate dates, the calculation continues to detect essentially the same patterns of the hyperbolic acceleration of the planetary complexity growth.[19]

[19] How can this be explained? The point is that, on the one hand, I would definitely not advise to use the BH Biosocial Era periodizations of both Modis and Panov for pedagogical purposes. But, on the other hand, though very many points of human evolution have been very substantially clarified after 2005, the basic facts of human evolution had been established very well by that time. And it should be noted that both periodizations reflected more or less adequately the general rhythm of the BH Biosocial Era—a noticeable increase in the frequency of complexity jumps during the first millions of years of this era in comparison with the last dozens million years of the previous BH era, as well as the further substantial increase in the frequency of biosocial complexity jumps in the final million

References

Almécija, S., Tallman, M., Alba, D. M., Pina, M., Moyà-Solà, S., & Jungers, W. L. (2013). The femur of Orrorin tugenensis exhibits morphometric affinities with both Miocene apes and later hominins. *Nature Communications, 4*(1), 2888. https://doi.org/10.1038/ncomms3888

Antón, S. C., Taboada, H. G., Middleton, E. R., Rainwater, C. W., Taylor, A. B., Turner, T. R., et al. (2016). Morphological variation in *Homo erectus* and the origins of developmental plasticity. *Philosophical Transactions of the Royal Society. B: Biological Sciences, 371*(1698), 20150236. https://doi.org/10.1098/rstb.2015.0236

Armitage, S. J., Jasim, S. A., Marks, A. E., Parker, A. G., Usik, V. I., & Uerpmann, H. P. (2011). The southern route "out of Africa": Evidence for an early expansion of modern humans into Arabia. *Science, 331*(6016), 453–456. https://doi.org/10.1126/science.1199113

Arzarello, M., Marcolini, F., Pavia, G., Pavia, M., Petronio, C., Petrucci, M., et al. (2007). Evidence of earliest human occurrence in Europe: The site of Pirro Nord (Southern Italy). *Naturwissenschaften, 94*, 107–112. https://doi.org/10.1007/s00114-006-0173-3

Aunger, R. (2007). A rigorous periodization of 'big' history. *Technological Forecasting and Social Change, 74*(8), 1164–1178. https://doi.org/10.1016/j.techfore.2007.01.007

Aunger, R. (2025). Theorizing in big history: What we are learning. In D. LePoire, L. Grinin, & A. Korotayev (Eds.), *Navigating complexity in big history—Exploring periodization across cosmic and biosocial dimensions* (pp. 73–98). Springer Nature. https://doi.org/10.1007/978-3-031-85410-1_4

Backwell, L., d'Errico, F., de la Peña, P., & Wadley, L. (2023). Border cave, South Africa. In A. Beyin, D. K. Wright, J. Wilkins, & D. I. Olszewski (Eds.), *Handbook of Pleistocene archaeology of Africa* (pp. 1297–1310). Springer Nature. https://doi.org/10.1007/978-3-031-20290-2_84

Bae, C. J., Douka, K., & Petraglia, M. D. (2017). On the origin of modern humans: Asian perspectives. *Science, 358*(6368), eaai9067. https://doi.org/10.1126/science.aai9067

Baldwin, T. (2015). Denisovan bracelet: Advanced technological skills in early human groups is still resisted. *Pleistocene Coalition News, 7*(3), 4–5.

Balzeau, A., Grimaud-Hervé, D., Détroit, F., Holloway, R. L., Combès, B., & Prima, S. (2013). First description of the Cro-Magnon 1 endocast and study of brain variation and evolution in anatomically modern Homo sapiens. *Bulletins et Mémoires de la Société d'anthropologie de Paris, 25*(1), 1–18. https://doi.org/10.1007/s13219-012-0069-z

Barham, L., & Kleindienst, M. R. (2023). Kalambo Falls, Zambia. In A. Beyin, D. K. Wright, J. Wilkins, & D. I. Olszewski (Eds.), *Handbook of Pleistocene archaeology of Africa* (pp. 1067–1080). Springer. https://doi.org/10.1007/978-3-031-20290-2_76

---

of years of this era. Finally, both Modis and Panov identified a number of BH Biosocial Era complexity jumps and their dates more or less adequately, but grossly misnamed them. For example, Panov was right identifying a major complexity jump around 2.0–1.6 million BP, but he was plainly wrong connecting it with *Homo habilis*, whereas as we could see this above, it was rather connected with *Homo erectus* and the emergence of Acheulean stone industry (while, according to the wrong identification of Panov, both *Homo erectus* and the Acheulean stone industry only appeared more than a million years letter—a mistake, that could hardly be called negligible). Both Modis and Panov correctly identified a significant complexity jump in the first half of the first million BP, but Modis wrongly connected it with *Homo sapiens*, and Panov wrongly connected it with *Homo erectus*, whereas in reality, as we could see this above, it was clearly connected with *Homo heidelbergensis*. On the other hand, Modis connects the complexity jump around 300,000 BP with *Homo heidelbergensis*, whereas in reality it was connected with the emergence of early *Homo sapiens, Homo neanderthalensis*, *Homo denisova*, and so on.

Barton, C. M., & Clark, G. A. (2021). From artifacts to cultures: Technology, society, and knowledge in the upper paleolithic. *Journal of Paleolithic Archaeology, 4*, 6. https://doi.org/10.1007/s41982-021-00091-8

Bar-Yosef, O. (2002). The upper paleolithic revolution. *Annual Review of Anthropology, 31*(1), 363–393. https://doi.org/10.1146/annurev.anthro.31.040402.085416

Bar-Yosef, O. (2007). The archaeological framework of the Upper Paleolithic revolution. *Diogenes, 54*(2), 3–18. https://doi.org/10.1177/0392192107076869

Bar-Yosef, O., & Kuhn, S. L. (2016). The big deal about blades: Laminar technologies and human evolution. In *Human evolution source book* (pp. 597–610). Routledge.

Becerra-Valdivia, L., & Higham, T. (2020). The timing and effect of the earliest human arrivals in North America. *Nature, 584*(7819), 93–97. https://doi.org/10.1038/s41586-020-2491-6

Berna, F., Goldberg, P., Horwitz, L. K., Brink, J., Holt, S., Bamford, M., & Chazan, M. (2012). Microstratigraphic evidence of in situ fire in the Acheulean strata of Wonderwerk Cave, Northern Cape province, South Africa. *Proceedings of the National Academy of Sciences, 109*(20), E1215–E1220. https://doi.org/10.1073/pnas.1117620109

Besenbacher, S., Hvilsom, C., Marques-Bonet, T., Mailund, T., & Schierup, M. H. (2019). Direct estimation of mutations in great apes reconciles phylogenetic dating. *Nature Ecology & Evolution, 3*(2), 286–292. https://doi.org/10.1038/s41559-018-0778-x

Beyene, Y., Defleur, A., Asfaw, B., WoldeGabriel, G., & Sahle, Y. (2023). Herto, Ethiopia. In A. Beyin, D. K. Wright, J. Wilkins, & D. I. Olszewski (Eds.), *Handbook of Pleistocene archaeology of Africa* (pp. 387–398). Springer Nature. https://doi.org/10.1007/978-3-031-20290-2_23

Beyin, A. (2023). Asfet (Gulf of Zula), Eritrea. In A. Beyin, D. K. Wright, J. Wilkins, & D. I. Olszewski (Eds.), *Handbook of Pleistocene archaeology of Africa* (pp. 235–242). Springer Nature. https://doi.org/10.1007/978-3-031-20290-2_13

Beyin, A., & Ryano, K. P. (2023). Kilwa Basin, Tanzania. In A. Beyin, D. K. Wright, J. Wilkins, & D. I. Olszewski (Eds.), *Handbook of Pleistocene archaeology of Africa* (pp. 1055–1067). Springer Nature. https://doi.org/10.1007/978-3-031-20290-2_69

Bisconti, M., & Tartarelli, G. (2025). Morphological evolution of the hominid brain. *Academia Biology, 3*(2), 1–19. https://doi.org/10.20935/AcadBiol7710

Blešić, I., Tretiakova, T. N., & Syromiatnikova, J. A. (2024). Age discrimination in the hospitality industry: How do hospitality students view their elderly colleagues? *Journal of Globalization Studies, 15*(1), 199–206. https://doi.org/10.30884/jogs/2024.01.11

Blinkhorn, J., Niang, K., Groucutt, H. S., & Scerri, E. M. L. (2023). Ndiayène Pendao and the Lower Senegal Valley, Senegal. In A. Beyin, D. K. Wright, J. Wilkins, & D. I. Olszewski (Eds.), *Handbook of Pleistocene archaeology of Africa* (pp. 929–939). Springer Nature. https://doi.org/10.1007/978-3-031-20290-2_59

Boaz, N. T., & Ciochon, R. (2004). *Dragon Bone Hill: An ice-age saga of Homo erectus*. Oxford University Press.

Bousman, B., et al. (2023). Erfkroon, South Africa. In A. Beyin, D. K. Wright, J. Wilkins, & D. I. Olszewski (Eds.), *Handbook of Pleistocene archaeology of Africa* (pp. 1431–1450). Springer Nature. https://doi.org/10.1007/978-3-031-20290-2_92

Brandenburg, A., Kahniashvili, T., Mandal, S., Roper Pol, A., Tevzadze, A. G., & Vachaspati, T. (2017). Evolution of hydromagnetic turbulence from the electroweak phase transition. *Physical Review D, 96*(12), 123528. https://doi.org/10.1103/PhysRevD.96.123528

Brandt, S. A., et al. (2023). Mochena Borago Rockshelter, Ethiopia. In A. Beyin, D. K. Wright, J. Wilkins, & D. I. Olszewski (Eds.), *Handbook of Pleistocene archaeology of Africa* (pp. 461–482). Springer Nature. https://doi.org/10.1007/978-3-031-20290-2_28

Bruner, E., & Beaudet, A. (2023). The brain of Homo habilis: Three decades of paleoneurology. *Journal of Human Evolution, 174*, 103281. https://doi.org/10.1016/j.jhevol.2022.103281

Buffalo, V., & Kern, A. D. (2024). A quantitative genetic model of background selection in humans. *PLoS Genetics, 20*(3), e1011144. https://doi.org/10.1371/journal.pgen.1011144

Callaway, E. (2017). Oldest Homo sapiens fossil claim rewrites our species' history. *Nature, 546*, 289–293. https://doi.org/10.1038/nature.2017.22114

Carrión, J. S., Rose, J., & Stringer, C. (2011). Early human evolution in the western Palaearctic: Ecological scenarios. *Quaternary Science Reviews, 30*, 1281–1295. https://doi.org/10.1016/j.quascirev.2011.04.003

Check, E. (2006). How Africa learned to love the cow: The development of lactose tolerance in sub-Saharan Africa, a fascinating tale of genetic convergence. *Nature, 444*(7122), 994–996. https://doi.org/10.1038/444994a993

Chen, F. C., & Li, W. H. (2001). Genomic divergences between humans and other hominoids and the effective population size of the common ancestor of humans and chimpanzees. *The American Journal of Human Genetics, 68*(2), 444–456.

Christian, D. (2011). *Maps of time: An introduction to big history*. University of California Press.

Claessen, H. J. (2020). Neandertals–Life, work, and vanishing. *Social Evolution & History, 19*(1), 3–20.

Clarkson, C., Jacobs, Z., Marwick, B., Fullagar, R., Wallis, L., Smith, M., et al. (2017). Human occupation of northern Australia by 65,000 years ago. *Nature, 547*(7663), 306–310. https://doi.org/10.1038/nature22968

Clarkson, C., Smith, M., Marwick, B., Fullagar, R., Wallis, L. A., Faulkner, P., et al. (2015). The archaeology, chronology and stratigraphy of Madjedbebe (Malakunanja II): A site in northern Australia with early occupation. *Journal of Human Evolution, 83*, 46–64. https://doi.org/10.1016/j.jhevol.2015.03.014

Cochran, G., & Harpending, H. (2009). *The 10,000 year explosion: How civilization accelerated human evolution*. Basic Books.

Conard, N. J., Malina, M., & Münzel, S. C. (2009). New flutes document the earliest musical tradition in southwestern Germany. *Nature, 460*(7256), 737–740. https://doi.org/10.1038/nature08169

Cooke, N. P., Mattiangeli, V., Cassidy, L. M., Okazaki, K., Stokes, C. A., Onbe, S., et al. (2021). Ancient genomics reveals tripartite origins of Japanese populations. *Science Advances, 7*(38), eabh2419. https://doi.org/10.1126/sciadv.abh2419

Cuenca-Solana, D., Gutiérrez-Zugasti, F. I., González-Morales, M. R., Setién-Marquinez, J., Ruiz-Martinez, E., García-Moreno, A., & Clemente-Conte, I. (2013). Shell technology, rock art, and the role of marine resources during the Upper Paleolithic. *Current Anthropology, 54*(3), 370–380. https://doi.org/10.1086/670325

Curry, A. (2013). Archaeology: The milk revolution. *Nature, 500*(7460), 20–22. https://doi.org/10.1038/500020a

D'Errico, F., Backwell, L., Villa, P., Degano, I., Lucejko, J. J., Bamford, M. K., et al. (2012). Early evidence of San material culture represented by organic artifacts from Border Cave, South Africa. *Proceedings of the National Academy of Sciences, 109*(33), 13214–13219. https://doi.org/10.1073/pnas.1204213109

Daver, G., Guy, F., Mackaye, H. T., Likius, A., Boisserie, J. R., Moussa, A., et al. (2022). Postcranial evidence of late Miocene hominin bipedalism in Chad. *Nature, 609*(7925), 94–100. https://doi.org/10.1038/s41586-022-04901-z

Di Vincenzo, F., & Manzi, G. (2023). Homo heidelbergensis as the Middle Pleistocene common ancestor of Denisovans, Neanderthals and modern humans. *Journal of Mediterranean Earth Sciences, 15*, 303–315. https://doi.org/10.13133/2280-6148/18074

Doronichev, V. (2016). The pre-Mousterian industrial complex in Europe between 400 and 300 ka: Interpreting its origin and spatiotemporal variability. *Quaternary International, 409*, 222–240. https://doi.org/10.1016/j.quaint.2015.05.063

Douka, K., Slon, V., Jacobs, Z., Ramsey, C. B., Shunkov, M. V., Derevianko, A. P., et al. (2019). Age estimates for hominin fossils and the onset of the Upper Palaeolithic at Denisova Cave. *Nature, 565*(7741), 640–644. https://doi.org/10.1038/s41586-018-0870-z

Ember, C. R., Ember, M., & Peregrine, P. N. (2015). *Anthropology* (14th ed.). Pearson.

Ferring, R., Oms, O., Agustí, J., Berna, F., Nioradze, M., Shelia, T., et al. (2011). Earliest human occupations at Dmanisi (Georgian Caucasus) dated to 1.85–1.78 Ma. *Proceedings of the National Academy of Sciences, 108*(26), 10432–10436. https://doi.org/10.1073/pnas.1106638108

Foerster, H. v., Mora, P. M., & Amiot, L. W. (1960). Doomsday: Friday, 13 November, AD 2026. *Science, 132*(3436), 1291–1295. https://doi.org/10.1126/science.132.3436.1291

Fu, Q., Li, H., Moorjani, P., Jay, F., Slepchenko, S. M., Bondarev, A. A., et al. (2014). Genome sequence of a 45,000-year-old modern human from western Siberia. *Nature, 514*(7523), 445–449. https://doi.org/10.1038/nature13810

Giacobini, G. (2007). Richness and diversity of burial rituals in the Upper Paleolithic. *Diogenes, 54*(2), 19–39. https://doi.org/10.1177/0392192107077649

Gilbert, W. H., et al. (2023). Kesem-Kebena-Dulecha Study Area, Ethiopia. In A. Beyin, D. K. Wright, J. Wilkins, & D. I. Olszewski (Eds.), *Handbook of Pleistocene archaeology of Africa* (pp. 399–419). Springer Nature. https://doi.org/10.1007/978-3-031-20290-2_24

Glötzl, E. (2025). Unification of biological and cultural evolution through natural periodization. In D. J. LePoire, L. Grinin, & A. Korotayev (Eds.), *Navigating complexity in Big History—Exploring periodization across cosmic and biosocial dimensions* (pp. 265–318). Springer Nature. https://doi.org/10.1007/978-3-031-85410-1_9

Goder-Goldberger, M., Barzilai, O., & Boaretto, E. (2023). Innovative technological practices and their role in the emergence of initial Upper Paleolithic technologies: A view from Boker Tachtit. *Journal of Paleolithic Archaeology, 6*(1), article 11. https://doi.org/10.1007/s41982-023-00137-z

Gowlett, J. A. (2011). Innovation and the evolution of human behavior. The vital sense of proportion: Transformation, golden section, and 1: 2 preference in Acheulean Bifaces. *Paleo-Anthropology, 2011*, 174–187. https://doi.org/10.4207/PA.2011.ART51

Gowlett, J. A. (2016). The discovery of fire by humans: A long and convoluted process. *Philosophical Transactions of the Royal Society B: Biological Sciences, 371*(1696), 20150164. https://doi.org/10.1098/rstb.2015.0164

Grine, F. E. (2023). Die Kelders, South Africa. In A. Beyin, D. K. Wright, J. Wilkins, & D. I. Olszewski (Eds.), *Handbook of Pleistocene archaeology of Africa* (pp. 1367–1382). Springer. https://doi.org/10.1007/978-3-031-20290-2_88

Grinin, A. (2025). Global aging in the process of the world order transformations. *Journal of Globalization Studies, 16*(1), 181–192. https://doi.org/10.30884/jogs/2025.01.13

Grinin, L., & Grinin, A. (2023a). Demographic transformations in the historical process in the light of technological development: Theoretical approach. *Social Evolution & History, 22*(2), 159–202. https://doi.org/10.30884/seh/2023.02.08

Grinin, L., & Grinin, A. (2023b). Evolutionary phases in Big History and complexity. *Journal of Big History, 6*(3), 5–17. https://doi.org/10.22339/jbh.v6i3.6302

Grinin, L., & Grinin, A. (2025). Big History periodization. Complexity, directions and phase characteristics. In D. J. LePoire, L. Grinin, & A. Korotayev (Eds.), *Navigating complexity in Big History—Exploring periodization across cosmic and biosocial dimensions* (pp. 319–342). Springer Nature. https://doi.org/10.1007/978-3-031-85410-1_10

Grinin, L., & Korotayev, A. (2023). Social evolution: Theoretical aspects and general outlines. *Social Evolution & History, 22*(2), 111–158. https://doi.org/10.30884/seh/2023.02.07

Grinin, L., & Korotayev, A. (2025). Possible methodological ideas for periodizing Big History. In D. J. LePoire, L. Grinin, & A. Korotayev (Eds.), *Navigating complexity in Big History. Exploring periodization across cosmic and biosocial dimensions* (pp. 49–71). Springer. https://doi.org/10.1007/978-3-031-85410-1_3

Grinin, L., Markov, A., & Korotayev, A. (2009). Aromorphoses in biological and social evolution: Some general rules for biological and social forms of macroevolution. *Social Evolution & History, 8*(2), 6–50.

Grinin, L., Markov, A., & Korotayev, A. (2011). Biological and social aromorphoses: A comparison between two forms of macroevolution. *Evolution, 1*, 162–211.

Grinin, L., Markov, A, & Korotayev, A. (2012). Biological and social phases of big history: similarities and differences of evolutionary principles and mechanisms. *Evolution, 2*, 158–198.

Grinin, L., Markov, A., & Korotayev, A. (2013). On similarities between biological and social evolutionary mechanisms: mathematical modeling. *Cliodynamics, 4*(2), 185–228. https://doi.org/10.21237/C7clio4221334

Grinin, L. E., Korotayev, A., & Markov, A. (2020). Big History and macro-evolution. Evolutionary principles and mechanisms at biological and social phases of the Big History. In C. Benjamin, E. Quaedackers, & D. Baker (Eds.), *The Routledge companion to big history* (pp. 72–105). Routledge.

Grinin, L., Grinin, A., & Korotayev, A. (2023). Demographic transformations in the light of technological development: Types of demographic reproduction in the past and in the future. *Social Evolution & History, 22*(2), 203–248. https://doi.org/10.30884/seh/2023.02.09

Grinin, L., Grinin, A., & Korotayev, A. (2024). *Cybernetic revolution and global aging. Humankind on the way to cybernetic society, or the next hundred years.* Springer Nature. https://doi.org/10.1007/978-3-031-56764-3

Gruhn, R. (2023). An anthropological conception of the initial peopling of the Americas. *PaleoAmerica, 9*(3), 167–173. https://doi.org/10.1080/20555563.2023.2278948

Grün, R., Stringer, C., McDermott, F., Nathan, R., Porat, N., Robertson, S., et al. (2005). U-series and ESR analyses of bones and teeth relating to the human burials from Skhul. *Journal of Human Evolution, 49*(3), 316–334. https://doi.org/10.1016/j.jhevol.2005.04.006

Guil-Guerrero, J. L., & Manzano-Agugliaro, F. (2024). 2024 worldwide research on Australopiths. *African Archaeological Review, 41*, 477–503. https://doi.org/10.1007/s10437-024-09580-x

Guran, S. H., Yousefi, M., Kafash, A., & Ghasidian, E. (2024). Reconstructing contact and a potential interbreeding geographical zone between Neanderthals and anatomically modern humans. *Scientific Reports, 14*(1), 20475. https://doi.org/10.1038/s41598-024-70206-y

Gustafson, L. S. (2020). Foreword. In A. Korotayev & D. LePoire (Eds.), *The 21st century singularity and global futures. A Big History perspective* (pp. v–vii). Springer Nature.

Harmand, S., & Lewis, J. (2023). Lomekwi 3, West Turkana, Kenya. In A. Beyin, D. K. Wright, J. Wilkins, & D. I. Olszewski (Eds.), *Handbook of Pleistocene archaeology of Africa* (pp. 635–649). https://doi.org/10.1007/978-3-031-20290-2_40

Harmand, S., Lewis, J. E., Feibel, C. S., Lepre, C. J., Prat, S., Lenoble, A., et al. (2015). 3.3-million-year-old stone tools from Lomekwi 3, West Turkana, Kenya. *Nature, 521*(7552), 310–315. https://doi.org/10.1038/nature14464

Harvati, K., Röding, C., Bosman, A. M., Karakostis, F. A., Grün, R., Stringer, C., et al. (2019). Apidima Cave fossils provide earliest evidence of Homo sapiens in Eurasia. *Nature, 571*(7766), 500–504. https://doi.org/10.1038/s41586-019-1376-z

Hawks, J. (2014). Still evolving (after all these years). *Scientific American, 311*(3), 86–91.

Hawks, J., Wang, E. T., Cochran, G. M., Harpending, H. C., & Moyzis, R. K. (2007). Recent acceleration of human adaptive evolution. *Proceedings of the National Academy of Sciences, 104*(52), 20753–20758. https://doi.org/10.1073/pnas.0707650104

Hayden, B. (2021). Foragers or «feasters?» inequalities in the Upper Palaeolithic. *PALEO. Revue d'archéologie préhistorique, 32*, 36–49. https://doi.org/10.4000/paleo.6607

Heinzelin, J. D., Clark, J. D., White, T., Hart, W., Renne, P., WoldeGabriel, G., et al. (1999). Environment and behavior of 2.5-million-year-old Bouri hominids. *Science, 284*(5414), 625–629. https://doi.org/10.1126/science.284.5414.625

Henriques, G. R. (2003). The tree of knowledge system and the theoretical unification of psychology. *Review of General Psychology, 7*, 150–182.

Henriques, G. R. (2011). *A new unified theory of psychology*. Springer.

Henriques, G., Michalski, J., Quackenbush, S., & Schmidt, W. (2019). The tree of knowledge system: A new map for big history. *Journal of Big History, 4*(4), 1–17. https://doi.org/10.22339/jbh.v3i4.3410

Henriques, G., & Volk, T. (2023). Toward a big history 2.0: A brief position paper. *Journal of Big History, 6*(3), 1–4. https://doi.org/10.22339/jbh.v6i3.6301

Henshilwood, C. S., d'Errico, F., Van Niekerk, K. L., Coquinot, Y., Jacobs, Z., Lauritzen, S. E., et al. (2011). A 100,000-year-old ochre-processing workshop at Blombos Cave, South Africa. *Science, 334*(6053), 219–222. https://doi.org/10.1126/science.1211535

Herries, A. I., Martin, J. M., Leece, A. B., Adams, J. W., Boschian, G., Joannes-Boyau, R., et al. (2020). Contemporaneity of australopithecus, paranthropus, and early homo erectus in South Africa. *Science, 368*(6486), eaaw7293. https://doi.org/10.1126/science.aaw7293

Heyes, P. J., Anastasakis, K., De Jong, W., Van Hoesel, A., Roebroeks, W., & Soressi, M. (2016). Selection and use of manganese dioxide by Neanderthals. *Scientific Reports, 6*(1), 22159. https://doi.org/10.1038/srep22159

Hoffmann, D. L., Angelucci, D. E., Villaverde, V., Zapata, J., & Zilhão, J. (2018). Symbolic use of marine shells and mineral pigments by Iberian Neandertals 115,000 years ago. *Science Advances, 4*(2), eaar5255. https://doi.org/10.1126/sciadv.aar5255

Hoffmann, D. L., Standish, C. D., García-Diez, M., Pettitt, P. B., Milton, J. A., Zilhão, J., et al. (2018). U-Th dating of carbonate crusts reveals Neandertal origin of Iberian cave art. *Science, 359*(6378), 912–915. https://doi.org/10.1126/science.aap7778

Hoggard, N. (2024). How chaos theory brings order to the evolution of intelligence. *Journal of Big History, 7*(2), 43–65. https://doi.org/10.22339/jbh.v7i2.7205

Hoggard, N. (2025). A new theory of evolution is supported by the teaching patterns of human ancestors. In D. J. LePoire, L. Grinin, & A. Korotayev (Eds.), *Navigating complexity in Big History—Exploring periodization across cosmic and biosocial dimensions* (pp. 217–263). Springer Nature. (in this volume). https://doi.org/10.1007/978-3-031-85410-1_8

Holloway, R. L. (2015). Brain evolution. In M. P. Muehlenbein (Ed.), *Basics in human evolution* (pp. 235–250). Elsevier.

Holloway, R. L., Broadfield, D. G., & Yuan, M. S. (2004). *The human fossil record. Brain endocasts: The paleoneurological evidence*. Wiley & Sons.

Holmes, R. B. (2023). Matter-antimatter asymmetry from preon condensation prior to the hadron epoch. *Journal of Modern Physics, 14*(11), 1437–1451. https://doi.org/10.4236/jmp.2023.1411083

Hu, Y., Marwick, B., Zhang, J. F., Rui, X., Hou, Y. M., Yue, J. P., et al. (2019). Late Middle Pleistocene Levallois stone-tool technology in Southwest China. *Nature, 565*(7737), 82–85. https://doi.org/10.1038/s41586-018-0710-1

Hublin, J. J. (2021). How old are the oldest Homo sapiens in Far East Asia? *Proceedings of the National Academy of Sciences, 118*(10), e2101173118. https://doi.org/10.1073/pnas.2101173118

Hublin, J. J., Ben-Ncer, A., Bailey, S. E., Freidline, S. E., Neubauer, S., Skinner, M. M., et al. (2017). New fossils from Jebel Irhoud, Morocco and the pan-African origin of Homo sapiens. *Nature, 546*(7657), 289–292. https://doi.org/10.1038/nature22336

Hublin, J. J., & Changeux, J. P. (2022). Paleoanthropology of cognition: An overview on Hominins brain evolution. *Comptes Rendus. Biologies, 345*(2), 57–75. https://doi.org/10.5802/crbiol.92

Hublin, J. J., Sirakov, N., Aldeias, V., Bailey, S., Bard, E., Delvigne, V., et al. (2020). Initial upper palaeolithic homo sapiens from Bacho Kiro Cave, Bulgaria. *Nature, 581*(7808), 299–302. https://doi.org/10.1038/s41586-020-2259-z

International Commission on Stratigraphy. (2024). International Chronostratigraphic Chart. V2024/12. https://stratigraphy.org/ICSchart/ChronostratChart2024-12.pdf

Jablonski, N. G. (2002). *The first Americans: The pleistocene colonization of the new world*. California Academy of Sciences.

Jacobs, Z., Li, B., Shunkov, M. V., Kozlikin, M. B., Bolikhovskaya, N. S., Agadjanian, A. K., et al. (2019). Timing of archaic hominin occupation of Denisova Cave in southern Siberia. *Nature, 565*(7741), 594–599. https://doi.org/10.1038/s41586-018-0843-2

Jantsch, E. (1980). *The self-organizing universe: Scientific and human implications of the emerging paradigm of evolution*. Pergamon.

Jaubert, J., Maureille, B., & Peresani, M. (2022). Spiritual and symbolic activities of Neanderthals. In F. Romagnoli, F. Rivals, & S. Benazzi (Eds.), *Updating Neanderthals. Understanding behavioural complexity in the Late Middle Palaeolithic* (pp. 261–274). Academic Press.

Jaubert, J., Verheyden, S., Genty, D., Soulier, M., Cheng, H., Blamart, D., et al. (2016). Early Neanderthal constructions deep in Bruniquel Cave in southwestern France. *Nature, 534*(7605), 111–114. https://doi.org/10.1038/nature18291

Kealy, S., Louys, J., & O'Connor, S. (2018). Least-cost pathway models indicate northern human dispersal from Sunda to Sahul. *Journal of Human Evolution, 125*, 59–70. https://doi.org/10.1016/j.jhevol.2018.10.003

Kerner, G., Neehus, A. L., Philippot, Q., Bohlen, J., Rinchai, D., Kerrouche, N., et al. (2023). Genetic adaptation to pathogens and increased risk of inflammatory disorders in post-Neolithic Europe. *Cell Genomics, 3*(2).

Key, A. J., Jarić, I., & Roberts, D. L. (2021). Modelling the end of the Acheulean at global and continental levels suggests widespread persistence into the Middle Palaeolithic. *Humanities and Social Sciences Communications, 8*(1), 1–12. https://doi.org/10.1057/s41599-021-00735-8

Kimbel, W. H. (2009). The origin of homo. In F. E. Grine, J. G. Fleagle, & R. E. Leakey (Eds.), *The first humans—Origin and early evolution of the genus homo* (Vertebrate paleobiology and paleoanthropology) (pp. 31–37). Springer. https://doi.org/10.1007/978-1-4020-9980-9_4

Kimbel, W. H., & Villmoare, B. (2016). From Australopithecus to Homo: The transition that wasn't. *Philosophical Transactions of the Royal Society B: Biological Sciences, 371*(1698), 20150248.

Korotayev, A. (2006). "Midwest-Amazonian" folklore-mythological parallels? *Acta Americana, 14*(1), 5–24. https://www.academia.edu/3552083/Midwest_Amazonian_Folklore_Mythological_Parallels

Korotayev, A. (2018). The 21st century singularity and its big history implications: A re-analysis. *Journal of Big History, 2*(3), 71–118. https://doi.org/10.22339/jbh.v2i3.2320

Korotayev, A. (2020a). How singular is the 21st century singularity? In A. V. Korotayev & D. LePoire (Eds.), *The 21st century singularity and global futures. A Big History perspective* (pp. 571–595). Springer Nature. https://doi.org/10.1007/978-3-030-33730-8_26

Korotayev, A. (2020b). The 21st century singularity in the Big History perspective. A re-analysis. In A. V. Korotayev & D. LePoire (Eds.), *The 21st century singularity and global futures. A Big History perspective* (pp. 19–75). Springer Nature. https://doi.org/10.1007/978-3-030-33730-8_2

Korotayev, A. (2024). Complexity growth patterns in the Big History. Preliminary results of a quantitative analysis. *Journal of Big History, 7*(2), 9–27. https://doi.org/10.22339/jbh.v7i2.7203

Korotayev, A. (2025a). Comparing and contrasting Big History singularity trends of the big bang and terrestrial evolution. In D. J. LePoire, L. Grinin, & A. Korotayev (Eds.), *Navigating complexity in Big History. Exploring periodization across cosmic and biosocial dimensions* (pp. 101–157). Springer Nature. https://doi.org/10.1007/978-3-031-85410-1_5

Korotayev, A. (2025b). Toward periodization of the biosocial era of the Big History: A reanalysis. *Journal of Big History, 8*(2), 1–37. https://doi.org/10.22339/jbh.v8i2.8201

Korotayev, A., & Markov, A. (2014). Mathematical modeling of biological and social phases of big history. In: L. Grinin et al. (Eds.), *Teaching & Researching Big History: Exploring a New Scholarly Field* (pp. 188–219). Uchitel.

Korotayev, A., & Markov, A. (2015). Mathematical modeling of biological and social phases of big history. *Globalistics and Globalization Studies, 4*, 319–343.

Korotayev, A., Berezkin, Y., Arkhipova, A., & Kozmin, A. (2006). Return of the white raven: Postdiluvial reconnaissance motif A2234.1.1 reconsidered. *Journal of American Folklore, 119*(472), 203–235. https://doi.org/10.2307/4137924

Korotayev, A., Berezkin, Y., Borinskaya, S., Davletshin, A., & Khaltourina, D. (2017). Genes and myths: Which genes and myths did the different waves of the peopling of Americas bring to the new world? *History of Mathematics, 6*, 9–77. https://www.sociostudies.org/upload/socionauki.ru/book/files/iim_6_en/009-077.pdf

Korotayev, A., Butovskaya, M., Shulgin, S., & Zinkina, J. (2023). How can the global ageing affect the global value system? An evolutionary perspective. *Social Evolution & History, 22*(1), 57–76. https://doi.org/10.30884/seh/2023.01.03

Korotayev, A., & LePoire, D. J. (2025). Complexity growth patterns in the stelliferous era of Big History. Preliminary observations. *Journal of Big History, 8*(1), 1–12. https://doi.org/10.22339/jbh.v8i1.8102

Korotayev, A., Malkov, A., & Khaltourina, D. (2006). *Introduction to social macrodynamics: Compact macromodels of the world system growth.* KomKniga/URSS. https://www.academia.edu/32757085/Introduction_to_Social_Macrodynamics_Models_of_the_World_System_Development_Moscow_KomKniga_URSS_2006

Kurzweil, R. (2005). *The singularity is near: When humans transcend biology.* Viking Penguin.

Langdon, J. H. (2023). *Human evolution: Bones, cultures, and genes.* Springer Nature. https://doi.org/10.1007/978-3-031-14157-7

Langergraber, K. E., Prüfer, K., Rowney, C., Boesch, C., Crockford, C., Fawcett, K., et al. (2012). Generation times in wild chimpanzees and gorillas suggest earlier divergence times in great ape and human evolution. *Proceedings of the National Academy of Sciences, 109*(39), 15716–15721. https://doi.org/10.1073/pnas.1211740109

Larsen, C. S. (2017). *Our origins. Discovering physical anthropology* (4th ed.). W. W. Norton.

Lbova, L. (2021). Personal ornaments as markers of social behavior, technological development and cultural phenomena in the Siberian early upper Paleolithic. *Quaternary International, 573*, 4–13. https://doi.org/10.1016/j.quaint.2020.05.035

Lebatard, A. E., Bourlès, D. L., Duringer, P., Jolivet, M., Braucher, R., Carcaillet, J., et al. (2008). Cosmogenic nuclide dating of Sahelanthropus tchadensis and Australopithecus bahrelghazali: Mio-Pliocene hominids from Chad. *Proceedings of the National Academy of Sciences, 105*(9), 3226–3231. https://doi.org/10.1073/pnas.0708015105

LePoire, D. J. (2015). Potential nested accelerating returns logistic growth in big history. *Evolution, 4*, 46–60.

LePoire, D. J. (2016). Exploring temporal patterns in big history dynamics. *KronoScope, 16*(2), 229–249. https://doi.org/10.1163/15685241-12341358

LePoire, D. J. (2020a). Exploring the singularity concept within Big History. In A. V. Korotayev & D. LePoire (Eds.), *The 21st century singularity and global futures. A Big History perspective* (pp. 77–97). Springer. https://doi.org/10.1007/978-3-030-33730-8_3

LePoire, D. J. (2020b). *Time patterns in Big History. Cycles, fractals, waves, transitions, and singularities.* David LePoire Consulting, LLC.

LePoire, D. J. (2023). Exploring phenomenological models for societal and technological transitions of the neolithic revolution and early civilization formation. *Cliodynamics, 14*(1), 5–26. https://doi.org/10.21237/C7clio14162728

LePoire, D. J. (2025). A simple, compatible, and extensible Big History framework and periodization based on previous findings. In D. J. LePoire, L. Grinin, & A. Korotayev (Eds.), *Navigating complexity in Big History—Exploring periodization across cosmic and biosocial dimensions* (pp. 159–193). Springer Nature. https://doi.org/10.1007/978-3-031-85410-1_6

LePoire, D. J., Grinin, L., & Korotayev, A. (2025a). Navigating complexity in Big History—Exploring periodization across cosmic and biosocial dimensions: An introduction. In D. J. LePoire, L. Grinin, & A. Korotayev (Eds.), *Navigating complexity in Big History. Exploring periodization across cosmic and biosocial dimensions* (pp. 1–27). Springer Nature. https://doi.org/10.1007/978-3-031-85410-1_1

LePoire, D. J., Grinin, L., & Korotayev, A. (2025b). Discussion: Cross-cutting topics and a variety of periodizations. In D. J. LePoire, L. Grinin, & A. Korotayev (Eds.), *Navigating complexity in Big History. Exploring periodization across cosmic and biosocial dimensions* (pp. 471–494). Springer Nature. https://doi.org/10.1007/978-3-031-85410-1_15

LePoire, D. J., Grinin, L., & Korotayev, A. (2025c). Conclusions: Exploring Big History periodizations across cosmic and biosocial dimensions with ideas for further research. In D. J. LePoire, L. Grinin, & A. Korotayev (Eds.), *Navigating complexity in Big History. Exploring periodization across cosmic and biosocial dimensions* (pp. 495–525). Springer. https://doi.org/10.1007/978-3-031-85410-1_16

Lewis-Williams, D. (2002). *The mind in the cave: Consciousness and the origins of art*. Thames and Hudson.

Li, F., Petraglia, M., Roberts, P., & Gao, X. (2020). The northern dispersal of early modern humans in eastern Eurasia. *Science Bulletin, 65*(20), 1699–1701. https://doi.org/10.1016/j.scib.2020.06.026

Luef, E. M. (2018). Tracing the human brain's classical language areas in extant and extinct hominids. In E. M. Luef & M. M. Marin (Eds.), *The talking species. Perspectives on the evolutionary, neuronal and cultural foundations of language* (pp. 29–56). Uni-Press Graz Verlag GmbH.

Macciardi, F., & Martini, F. (2022). The Neanderthal brain: Biological and cognitive evolution. In F. Romagnoli, F. Rivals, & S. Benazzi (Eds.), *Updating Neanderthals. Understanding behavioural complexity in the late middle palaeolithic* (pp. 89–108). Academic Press.

Majkić, A., d'Errico, F., Milošević, S., Mihailović, D., & Dimitrijević, V. (2018). Sequential incisions on a cave bear bone from the Middle Paleolithic of Pešturina Cave, Serbia. *Journal of Archaeological Method and Theory, 25*, 69–116. https://doi.org/10.1007/s10816-017-9331-5

Markov, A., & Korotayev, A. (2007). Phanerozoic marine biodiversity follows a hyperbolic trend. *Palaeoworld, 16*(4), 311–318. https://doi.org/10.1016/j.palwor.2007.01.002

Markov, A., & Korotayev, A. (2008). Hyperbolic growth of marine and continental biodiversity through the Phanerozoic and community evolution. *Zhurnal Obshchei Biologii, 69*(3), 175–194.

Markov, A., Anisimov, V., & Korotayev, A. (2010). Relationship between Genome Size and Organismal Complexity in the Lineage Leading from Prokaryotes to Mammals. *Paleontological Journal, 44*(4), 363–373. https://doi.org/10.1134/S0031030110040015

Marshack, A. (1991). *The roots of civilization: The cognitive beginnings of man's first art, symbol and notation*. Moyer Bell.

McBrearty, S., & Brooks, A. S. (2000). The revolution that wasn't: A new interpretation of the origin of modern human behavior. *Journal of Human Evolution, 39*(5), 453–563. https://doi.org/10.1006/jhev.2000.0435

McHenry, H. (2025). *Homo heidelbergensis*. Encyclopedia Britannica. https://www.britannica.com/topic/Homo-heidelbergensis

Meltzer, D. J. (2009). *First peoples in a new world: Colonizing ice age America*. University of California Press.

Meneganzin, A., & Killin, A. (2024). Beyond reasonable doubt: Reconsidering Neanderthal aesthetic capacity. *Phenomenology and the Cognitive Sciences*. (online first). https://doi.org/10.1007/s11097-024-10003-0

Metspalu, M., Kivisild, T., Metspalu, E., Parik, J., Hudjashov, G., Kaldma, K., et al. (2004). Most of the extant mtDNA boundaries in south and Southwest Asia were likely shaped during the initial settlement of Eurasia by anatomically modern humans. *BMC Genetics, 5*, 25. https://doi.org/10.1186/1471-2156-5-26

Meyer, M., Fu, Q., Aximu-Petri, A., Glocke, I., Nickel, B., Arsuaga, J. L., et al. (2014). A mitochondrial genome sequence of a hominin from Sima de los Huesos. *Nature, 505*(7483), 403–406. https://doi.org/10.1038/nature12788

Meyer, M., Kircher, M., Gansauge, M. T., Li, H., Racimo, F., Mallick, S., et al. (2012). A high-coverage genome sequence from an archaic Denisovan individual. *Science, 338*, 222–226. https://doi.org/10.1126/science.1224344

Meyer-Ortmanns, H., & Thurner, S. (Eds.). (2011). *Principles of evolution: From the Planck epoch to complex multicellular life*. Springer.

Modis, T. (2002). Forecasting the growth of complexity and change. *Technological Forecasting and Social Change, 69*(4), 377–404. https://doi.org/10.1016/S0040-1625(01)00172-X

Modis, T. (2003). The limits of complexity and change. *The Futurist, 37*(3), 26–32.

Moorjani, P., Gao, Z., & Przeworski, M. (2016). Human germline mutation and the erratic evolutionary clock. *PLoS Biology, 14*(10), e2000744. https://doi.org/10.1371/journal.pbio.2000744

Muratov, A. L., Gnedin, O. Y., Gnedin, N. Y., & Zemp, M. (2013). Revisiting the first galaxies: The epoch of population III stars. *The Astrophysical Journal, 773*(1), 19. https://doi.org/10.1088/0004-637X/773/1/19

Mussi, M., Skinner, M. M., Melis, R. T., Panera, J., Rubio-Jara, S., Davies, T. W., et al. (2023). Early Homo erectus lived at high altitudes and produced both Oldowan and Acheulean tools. *Science, 382*(6671), 713–718. https://doi.org/10.1126/science.add9115

Neubauer, S., Hublin, J. J., & Gunz, P. (2018). The evolution of modern human brain shape. *Science Advances, 4*(1), eaao5961. https://doi.org/10.1126/sciadv.aao5961

O'Connell, J. F., Allen, J., Williams, M. A., Williams, A. N., Turney, C. S., Spooner, N. A., et al. (2018). When did Homo sapiens first reach Southeast Asia and Sahul? *Proceedings of the National Academy of Sciences, 115*(34), 8482–8490. https://doi.org/10.1073/pnas.1808385115

Oktaviana, A. A., Joannes-Boyau, R., Hakim, B., Burhan, B., Sardi, R., Adhityatama, S., et al. (2024). Narrative cave art in Indonesia by 51,200 years ago. *Nature, 631*(8022), 814–818. https://doi.org/10.1038/s41586-024-07541-7

Ongaro, L., & Huerta-Sanchez, E. (2024). A history of multiple Denisovan introgression events in modern humans. *Nature Genetics, 56*, 2612–2622. https://doi.org/10.1038/s41588-024-01960-y

Osozawa, S. (2023). Geologically calibrated mammalian tree and its correlation with global events, including the emergence of humans. *Ecology and Evolution, 13*(12), e10827. https://doi.org/10.1002/ece3.10827

Panov, A. (2005a). Scaling law of the biological evolution and the hypothesis of the self-consistent Galaxy origin of life. *Advances in Space Research, 36*(2), 220–225. https://doi.org/10.1016/j.asr.2005.03.001

Panov, A. (2005b). Singulyarnaya tochka istorii. *Obshchestvennyye nauki i sovremennost, 1*, 122–137.

Panov, A. (2020). Singularity of evolution and post-singular development in the Big History perspective. In A. V. Korotayev & D. LePoire (Eds.), *The 21st century singularity and global futures. A Big History perspective* (pp. 439–465). Springer. https://doi.org/10.1007/978-3-030-33730-8_20

Perrone, L. M., Gregori, G., Reville, B., Silva, L. O., & Bingham, R. (2021). Neutrino-electron magnetohydrodynamics in an expanding universe. *Physical Review D, 104*(12), 123013. https://doi.org/10.1103/PhysRevD.104.123013

Petraglia, M. D., & Allchin, B. (Eds.). (2007). *The evolution and history of human populations in South Asia: Inter-disciplinary studies in archaeology, biological anthropology, linguistics and genetics*. Springer.

Pettitt, P. (2011). *The paleolithic origins of human burial*. Routledge.

Pleurdeau, D., et al. (2023). Goda Buticha, Ethiopia. In A. Beyin, D. K. Wright, J. Wilkins, & D. I. Olszewski (Eds.), *Handbook of pleistocene archaeology of Africa* (pp. 337–352). Springer Nature. https://doi.org/10.1007/978-3-031-20290-2_20

Plummer, T. W., Oliver, J. S., Finestone, E. M., Ditchfield, P. W., Bishop, L. C., Blumenthal, S. A., et al. (2023). Expanded geographic distribution and dietary strategies of the earliest Oldowan hominins and Paranthropus. *Science, 379*(6632), 561–566. https://doi.org/10.1126/science.abo7452

Pomeroy, E., Bennett, P., Hunt, C. O., Reynolds, T., Farr, L., Frouin, M., et al. (2020). New Neanderthal remains associated with the 'flower burial' at Shanidar Cave. *Antiquity, 94*(373), 11–26. https://doi.org/10.15184/aqy.2019.207

Ponce de León, M. S., Bienvenu, T., Marom, A., Engel, S., Tafforeau, P., Alatorre Warren, J. L., et al. (2021). The primitive brain of early Homo. *Science, 372*(6538), 165–171. https://doi.org/10.1126/science.aaz0032

Posth, C., Wißing, C., Kitagawa, K., Pagani, L., Van Holstein, L., Racimo, F., et al. (2017). Deeply divergent archaic mitochondrial genome provides lower time boundary for African gene flow into Neanderthals. *Nature Communications, 8*(1), 16046. https://doi.org/10.1038/ncomms16046

Poszewiecka, B., Gogolewski, K., Stankiewicz, P., & Gambin, A. (2022). Revised time estimation of the ancestral human chromosome 2 fusion. *BMC Genomics, 23*(Suppl 6), 616.

Profico, A., Di Vincenzo, F., Lorenza, G., Marcello, P., & Manzi, G. (2016). Filling the gap. Human cranial remains from Gombore II (Melka Kunture, Ethiopia; ca. 850 ka) and the origin of Homo heidelbergensis. *Journal of Anthropological Sciences, 94*, 41–63. https://doi.org/10.4436/JASS.94019

Prüfer, K., Posth, C., Yu, H., Stoessel, A., Spyrou, M. A., Deviese, T., et al. (2021). A genome sequence from a modern human skull over 45,000 years old from Zlatý kůň in Czechia. *Nature Ecology & Evolution, 5*(6), 820–825. https://doi.org/10.1038/s41559-021-01443-x

Radovčić, D., Sršen, A. O., Radovčić, J., & Frayer, D. W. (2015). Evidence for Neandertal jewelry: Modified white-tailed eagle claws at Krapina. *PLoS One, 10*(3), e0119802. https://doi.org/10.1371/journal.pone.0119802

Rendu, W., Beauval, C., Crevecoeur, I., Bayle, P., Balzeau, A., Bismuth, T., et al. (2014). Evidence supporting an intentional Neandertal burial at La Chapelle-aux-saints. *Proceedings of the National Academy of Sciences, 111*(1), 81–86. https://doi.org/10.1073/pnas.1316780110

Richter, D., Grün, R., Joannes-Boyau, R., Steele, T. E., Amani, F., Rué, M., et al. (2017). The age of the hominin fossils from Jebel Irhoud, Morocco, and the origins of the Middle Stone Age. *Nature, 546*(7657), 293–296. https://doi.org/10.1038/nature22335

Rightmire, G. P. (2001). Patterns of hominid evolution and dispersal in the Middle Pleistocene. *Quaternary International, 75*(1), 77–84. https://doi.org/10.1016/S1040-6182(00)00079-3

Rightmire, G. P. (2004). Brain size and encephalization in early to Mid-Pleistocene Homo. *American Journal of Physical Anthropology: The Official Publication of the American Association of Physical Anthropologists, 124*(2), 109–123. https://doi.org/10.1002/ajpa.10346

Rightmire, G. P. (2025). Homo habilis. *Encyclopedia Britannica*. https://www.britannica.com/topic/Homo-habilis

Rightmire, G. P., & Tobias, P. V. (2025). Homo erectus. *Encyclopedia Britannica*. https://www.britannica.com/topic/Homo-erectus

Rodríguez-Vidal, J., d'Errico, F., Pacheco, F. G., Blasco, R., Rosell, J., Jennings, R. P., et al. (2014). A rock engraving made by Neanderthals in Gibraltar. *Proceedings of the National Academy of Sciences, 111*(37), 13301–13306. https://doi.org/10.1073/pnas.1411529111

Roebroeks, W., Sier, M. J., Nielsen, T. K., De Loecker, D., Parés, J. M., Arps, C. E., & Mücher, H. J. (2012). Use of red ochre by early Neandertals. *Proceedings of the National Academy of Sciences, 109*(6), 1889–1894. https://doi.org/10.1073/pnas.1112261109

Roebroeks, W., & Villa, P. (2011). On the earliest evidence for habitual use of fire in Europe. *Proceedings of the National Academy of Sciences, 108*(13), 5209–5214. https://doi.org/10.1134/S1019331623010057

Rogers, M. J., Semaw, S., Stinchcomb, G. E., Quade, J., Levin, N. E., & Cauche, D. (2023). Gona, Ethiopia. In A. Beyin, D. K. Wright, J. Wilkins, & D. I. Olszewski (Eds.), *Handbook of pleistocene archaeology of Africa* (pp. 353–372). Springer Nature. https://doi.org/10.1007/978-3-031-20290-2_21

Rozov, N. S. (2023). *The origin of language and consciousness. How social orders and communicative concerns gave rise to speech and cognitive abilities*. Springer Nature. https://doi.org/10.1007/978-3-031-30630-3

Rybin, E. P., Mescherin, M. N., Klementiev, A. M., & Khatsenovich, A. M. (2023). Beyond 55 north latitude: Expanding the northern boundary of initial upper Paleolithic dispersal. *Archaeological Research in Asia, 36*, 100481. https://doi.org/10.1016/j.ara.2023.100481

Sahle, Y. (2023). The Gademotta formation, Ethiopia. In A. Beyin, D. K. Wright, J. Wilkins, & D. I. Olszewski (Eds.), *Handbook of Pleistocene archaeology of Africa* (pp. 327–336). Springer Nature. https://doi.org/10.1007/978-3-031-20290-2_19

Sánchez-Yustos, P., Marín-Arroyo, A. B., Arnold, L. J., Luque, L., Kehl, M., López-Sáez, J. A., et al. (2024). Initial Upper Palaeolithic lithic industry at Cueva Millán in the hinterlands of Iberia. *Scientific Reports, 14*(1), 21705. https://doi.org/10.1038/s41598-024-69913-3

Schmidt, I., et al. (2023). Pockenbank, Namibia. In A. Beyin, D. K. Wright, J. Wilkins, & D. I. Olszewski (Eds.), *Handbook of Pleistocene archaeology of Africa* (pp. 897–911). Springer Nature. https://doi.org/10.1007/978-3-031-20290-2_57

Shea, J. J. (2013). *Stone tools in the Paleolithic and Neolithic Near East: A guide*. Cambridge University Press.

Shipton, C., Boivin, N., & Petraglia, M. (2023). Panga ya Saidi, Kenya. In A. Beyin, D. K. Wright, J. Wilkins, & D. I. Olszewski (Eds.), *Handbook of Pleistocene archaeology of Africa* (pp. 665–672). Springer Nature. https://doi.org/10.1007/978-3-031-20290-2_42

Shunkov, M. V., Kozlikin, M. B., & Derevianko, A. P. (2020). Dynamics of the Altai Paleolithic industries in the archaeological record of Denisova Cave. *Quaternary International, 559*, 34–46. https://doi.org/10.1016/j.quaint.2020.02.017

Sklenář, K. (1987). The lower Paleolithic dwelling structure at Přezletice and its significance. *L'Anthropologie, 25*(2), 101–103.

Slimak, L., Zanolli, C., Higham, T., Frouin, M., Schwenninger, J. L., Arnold, L. J., et al. (2022). Modern human incursion into Neanderthal territories 54,000 years ago at Mandrin, France. *Science Advances, 8*(6), eabj9496. https://doi.org/10.1126/sciadv.abj9496

Spoor, F., Gunz, P., Neubauer, S., Stelzer, S., Scott, N., Kwekason, A., & Dean, M. C. (2015). Reconstructed Homo habilis type OH 7 suggests deep-rooted species diversity in early Homo. *Nature, 519*(7541), 83–86. https://doi.org/10.1038/nature14224

Staes, N., Smaers, J. B., Kunkle, A. E., Hopkins, W. D., Bradley, B. J., & Sherwood, C. C. (2019). Evolutionary divergence of neuroanatomical organization and related genes in chimpanzees and bonobos. *Cortex, 118*, 154–164. https://doi.org/10.1016/j.cortex.2018.09.016

Stahlschmidt, M. C., Mallol, C., & Miller, C. E. (2020). Fire as an artifact—Advances in paleolithic combustion structure studies: Introduction to the special issue. *Journal of Paleolithic Archaeology, 3*, 503–508. https://doi.org/10.1007/s41982-020-00074-1

Stanford, D. J., & Bradley, B. (2012). *Pre-Clovis first Americans: The origin of America's Clovis culture*. University of California Press.

Stiner, M. C. (2014). Finding a common bandwidth: Causes of convergence and diversity in Paleolithic beads. *Biological Theory, 9*, 51–64. https://doi.org/10.1007/s13752-013-0157-4

Stout, D. (2024). Stability and change in paleolithic toolmaking. In M. Charbonneau (Ed.), *The evolution of techniques: Rigidity and flexibility in use, transmission, and innovation* (pp. 139–158). MIT Press.

Stout, D., Hecht, E., Khreisheh, N., Bradley, B., & Chaminade, T. (2015). Cognitive demands of Lower Paleolithic toolmaking. *PLoS One, 10*(4), e0121804. https://doi.org/10.1371/journal.pone.0121804

Stout, D., Toth, N., Schick, K., & Chaminade, T. (2008). Neural correlates of Early Stone Age toolmaking: Technology, language and cognition in human evolution. *Philosophical Transactions of the Royal Society B: Biological Sciences, 363*(1499), 1939–1949. https://doi.org/10.1098/rstb.2008.0001

Strauss, L. G. (1989). On early hominid use of fire. *Current Anthropology, 30*, 488–491.

Stringer, C. (2016). The origin and evolution of Homo sapiens. *Philosophical Transactions of the Royal Society B: Biological Sciences, 371*(1698), 20150237. https://doi.org/10.1098/rstb.2015.0237

Stringer, C. B., & Barnes, I. (2015). Deciphering the denisovans. *Proceedings of the National Academy of Sciences, 112*(51), 15542–15543. https://doi.org/10.1073/pnas.1522477112

Stringer, C., & Crété, L. (2022). Mapping interactions of H. neanderthalensis and Homo sapiens from the fossil and genetic records. *Paleo Anthropology, 2022*(2), 401–412. https://doi.org/10.48738/2022.iss2.130

Stringer, C., & Galway-Witham, J. (2017). On the origin of our species. *Nature, 546*(7657), 212–214. https://doi.org/10.1038/546212a

Stringer, C. B., Grün, R., Schwarcz, H. P., & Goldberg, P. (1989). ESR dates for the hominid burial site of Es Skhul in Israel. *Nature, 338*(6218), 756–758. https://doi.org/10.1038/338756a0
Suleimanov, R. K. (2023). Culture and industry of the Acheulean Age. *Herald of the Russian Academy of Sciences, 93*(6), 390–398. https://doi.org/10.1134/S1019331623010057
Tattersall, I. (2010). A putative role for language in the origin of human consciousness. In R. K. Larson, V. Déprez, & H. Yamakido (Eds.), *The evolution of human language: Biolinguistic perspectives* (pp. 193–198). Cambridge University Press.
Texier, P. J., Porraz, G., Parkington, J., Rigaud, J. P., Poggenpoel, C., Miller, C., et al. (2010). A Howiesons Poort tradition of engraving ostrich eggshell containers dated to 60,000 years ago at Diepkloof Rock Shelter, South Africa. *Proceedings of the National Academy of Sciences, 107*(14), 6180–6185. https://doi.org/10.1073/pnas.09130471
Teyssandier, N. (2023). The beginnings of the Aurignacian in a European framework: Where do we stand? Abridged version. *Gallia Préhistoire, 63*, 4154. https://doi.org/10.4000/galliap.4154
Tobias, P. V. (1983). Hominid evolution in Africa. *Canadian Journal of Anthropology, 3*(2), 163–183.
Trájer, A. J. (2024). The habitat utilization and environmental resilience of Homo heidelbergensis in Europe. *Archaeological and Anthropological Sciences, 16*(5), 81. https://doi.org/10.1007/s12520-024-01986-7
Trinkaus, E., & Buzhilova, A. P. (2018). Diversity and differential disposal of the dead at Sunghir. *Antiquity, 92*(361), 7–21. https://doi.org/10.15184/aqy.2017.223
Tryon, C. A. (2023). Kapthurin formation, Kenya. In A. Beyin, D. K. Wright, J. Wilkins, & D. I. Olszewski (Eds.), *Handbook of Pleistocene archaeology of Africa* (pp. 549–561). Springer Nature. https://doi.org/10.1007/978-3-031-20290-2_34
Tuttle, R. H. (2025). Human evolution. *Encyclopedia Britannica.* https://www.britannica.com/science/human-evolution
Ustyuzhanin, V., Zinkina, J., & Korotayev, A. (2025). Global aging and demographic policies: A reconsideration. *Journal of Globalization Studies, 16*(1), 47–60. https://doi.org/10.30884/jogs/2025.01.04
Val, A., de la Peña, P., & Stratford, D. (2023). Olieboomspoort, South Africa. In A. Beyin, D. K. Wright, J. Wilkins, & D. I. Olszewski (Eds.), *Handbook of pleistocene archaeology of Africa* (pp. 1603–1613). Springer Nature. https://doi.org/10.1007/978-3-031-20290-2_104
Vanhaeren, M., & d'Errico, F. (2005). Grave goods from the Saint-Germain-la-Rivière burial: Evidence for social inequality in the Upper Palaeolithic. *Journal of Anthropological Archaeology, 24*(2), 117–134. https://doi.org/10.1016/j.jaa.2005.01.001
Vanhaeren, M., d'Errico, F., Stringer, C., James, S. L., Todd, J. A., & Mienis, H. K. (2006). Middle Paleolithic shell beads in Israel and Algeria. *Science, 312*(5781), 1785–1788. https://doi.org/10.1126/science.1128139
Villmoare, B., Kimbel, W. H., Seyoum, C., Campisano, C. J., DiMaggio, E. N., Rowan, J., et al. (2015). Early Homo at 2.8 Ma from Ledi-Geraru, Afar, Ethiopia. *Science, 347*(6228), 1352–1355. https://doi.org/10.1126/science.aaa1343
Viola, B. (2022). Different species on the horizon: The Denisova hominins. In F. Romagnoli, F. Rivals, & S. Benazzi (Eds.), *Updating Neanderthals. Understanding behavioural complexity in the Late Middle Palaeolithic* (pp. 55–69). Academic Press.
Volk, T. (2017). *Quarks to culture: How we came to be.* Columbia University Press.
Volk, T. (2020). The metapattern of general evolutionary dynamics and the three dynamical realms of big history. *Journal of Big History, 4*(3), 31–53.
Volk, T., & Henriques, G. (2025). Integrating the tree of knowledge and combogenesis approaches to Big History. In D. J. LePoire, L. Grinin, & A. Korotayev (Eds.), *Navigating complexity in Big History—Exploring periodization across cosmic and biosocial dimensions* (pp. 195–215). Springer Nature. https://doi.org/10.1007/978-3-031-85410-1_7
Wakita, M. (2020). Language evolution from a perspective of Broca's area. In N. Masataka (Ed.), *The origins of language revisited.* Springer. https://doi.org/10.1007/978-981-15-4250-3_5

Waweru, V. (2023). Cartwright's site, Kenya. In A. Beyin, D. K. Wright, J. Wilkins, & D. I. Olszewski (Eds.), *Handbook of Pleistocene archaeology of Africa* (pp. 521–535). Springer Nature. https://doi.org/10.1007/978-3-031-20290-2_32

Wells, S. (2003). *The journey of man: A genetic odyssey*. Random House.

Wengrow, D., & Graeber, D. (2021). *The dawn of everything. A new history of humanity*. Farrar, Strauss & Giroux.

White, M., & Ashton, N. (2003). Lower Palaeolithic core technology and the origins of the Levallois method in North-Western Europe. *Current Anthropology, 44*(4), 598–609. https://doi.org/10.1086/377653

White, T. D., WoldeGabriel, G., Asfaw, B., Ambrose, S., Beyene, Y., Bernor, R. L., et al. (2006). Asa Issie, Aramis and the origin of Australopithecus. *Nature, 440*(7086), 883–889.

Wilde, S., Timpson, A., Kirsanow, K., Kaiser, E., Kayser, M., Unterländer, M., et al. (2014). Direct evidence for positive selection of skin, hair, and eye pigmentation in Europeans during the last 5,000 y. *Proceedings of the National Academy of Sciences, 111*(13), 4832–4837. https://doi.org/10.1073/pnas.1316513111

Wilkins, J., Schoville, B. J., Brown, K. S., & Chazan, M. (2012). Evidence for early hafted hunting technology. *Science, 338*(6109), 942–946. https://doi.org/10.1126/science.1227608

Williams, F. L., & Tuttle, R. H. (2025). Neanderthal. *Encyclopedia Britannica*. https://www.britannica.com/topic/Neanderthal

Wood, B., & Bauernfeind, A. (2012). The fossil record: Evidence for speech in early hominins. In M. Tallerman & K. Gibson (Eds.), *Language evolution* (pp. 258–272). Oxford University Press.

Wu, X., Holloway, R. L., Schepartz, L. A., & Xing, S. (2011). A new brain endocast of Homo erectus from Hulu Cave, Nanjing, China. *American Journal of Physical Anthropology, 145*, 452–460. https://doi.org/10.1002/ajpa.21527

Yang, C. T., & Rafelski, J. (2022). Cosmological strangeness abundance. *Physics Letters B, 827*, 136944. https://doi.org/10.1016/j.physletb.2022.136944

Yousaf, A., Liu, J., Ye, S., & Chen, H. (2021). Current progress in evolutionary comparative genomics of great apes. *Frontiers in Genetics, 12*, 657468. https://doi.org/10.3389/fgene.2021.657468

Zaim, Y., Ciochon, R. L., Polanski, J. M., Grine, F. E., Bettis, E. A., III, Rizal, Y., et al. (2011). New 1.5 million-year-old Homo erectus maxilla from Sangiran (Central Java, Indonesia). *Journal of Human Evolution, 61*(4), 363–376. https://doi.org/10.1016/j.jhevol.2011.04.009

Zhivotovsky, L. A., Rosenberg, N. A., & Feldman, M. W. (2003). Features of evolution and expansion of modern humans, inferred from genomewide microsatellite markers. *The American Journal of Human Genetics, 72*(5), 1171–1186. https://doi.org/10.1086/375120

Zilhão, J. (2005). Burial evidence for social differentiation of age classes in the Early Upper Paleolithic. *Etud Res Archéol Univ Liège, 111*, 231–241.

Zinkina, J., Christian, D., Grinin, L., Ilyin, I., Andreev, A., Aleshkovski, I., Shulgin, S., & Korotayev, A. (2019). *A Big History of globalization. The emergence of a global world system.* Springer Nature. https://doi.org/10.1007/978-3-030-05707-7

Zwyns, N. (2021). The Initial Upper Paleolithic in Central and East Asia: Blade technology, cultural transmission, and implications for human dispersals. *Journal of Paleolithic. Archaeology, 4*(3), 19. https://doi.org/10.1007/s41982-021-00085-6

Andrey Korotayev heads the Center for Stability and Risk Analysis at the HSE University in Moscow, Russia. He is also Senior Research Professor at Eurasian Center for Big History & System Forecasting at the Institute of Oriental Studies, Russian Academy of Sciences and Institute for African Studies, Russian Academy of Sciences, as well as at the Faculty of Global Studies of the Lomonosov Moscow State University. Beginning as an historian of Arabia, he made focused studies, such as seen in "Two Social Ecological Crises and Genesis of Tribal Organization in the Yemeni North-East" (1996). Korotayev researches topics in quantitative cross-cultural anthropology and seeks to understand biological and social macroevolution, as in "A Compact Macromodel of World System Evolution" (2005) and "Mathematical Modeling of Biological and Social Phases of Big History" (2014). He has authored or co-authored over 650 scholarly publications, including such monographs as *Ancient Yemen* (Oxford University Press, 1995), *World Religions and Social Evolution of the Old World Oikumene Civilizations: A Cross-Cultural Perspective* (The Edwin Mellen Press, 2004), *Introduction to Social Macrodynamics: Compact Macromodels of the World System Growth* (URSS, 2006; with Daria Khaltorina and Artemy Malkov), *Introduction to Social Macrodynamics: Secular Cycles and Millennial Trends* (URSS, 2006; with Daria Khaltorina and Artemy Malkov), *Great Divergence and Great Convergence: A Global Perspective* (Springer Nature, 2015; with Leonid Grinin), *Economic Cycles, Crises, and the Global Periphery* (Springer Nature, 2016; with Leonid Grinin), *Islamism, Arab Spring, and the Future of Democracy: World System and World Values Perspectives* (Springer Nature, 2019; with Leonid Grinin and Arno Tausch). He is a founding member of the Eurasian Center for Big History & System Forecasting and founder of Evolution, an almanac dedicated to the study of Universal Evolution, and thus is closely linked with Big History. Recently, he has produced, with David LePoire and Leonid Grinin, a collective monograph, *Navigating Complexity in Big History: Exploring Periodization Across Cosmic and Biosocial Dimensions* (Springer Nature, 2025). He is a laureate of a Russian Science Support Foundation in "The Best Economists of the Russian Academy of Sciences" Nomination (2006); in 2012, he was awarded the Gold Kondratieff Medal by the International N. D. Kondratieff Foundation. He serves as a board member of the International Big History Association. 20, Myasnitskaya, Moscow 101,000, Russian Federation. e-mail: akorotayev@gmail.com. ORCID 0000-0003-3014-2037

The Role of Artificial Intelligence in Big History: Understanding Human Evolution and Future Scenarios

14

Ekaterina Sazhienko and David J. LePoire

Abstract

Artificial intelligence (AI) is rapidly becoming integral to our daily lives, offering transformative insights into both human evolution and future trajectories. This chapter explores how AI enhances Big History by shedding light on the complexity of human evolution and offering new perspectives on the evolving relationship between humans and AI. The discussion includes the potential of AI tools to advance scientific discovery and provide broader access to information through improved search capabilities and communication systems. By comparing AI development with brain evolution, this chapter delves into how AI mirrors aspects of brain architecture, leading to possible future integrations of AI and humans. It examines various scenarios, such as the potential for a parent–child dynamic between humans and AI, and explores pathways where AI will be embedded in virtual realities, human minds, and organizational processes, shaping the future of this intricate relationship.

Keywords

Artificial intelligence · Future · Evolution · Scenarios · Complexity · Big History · Universal evolution

This chapter is an extensively re-worked version of the text originally published in Journal of Big History (LePoire & Sazhienko, 2025).

E. Sazhienko
International University of Humanity and Nature, Dubna, Russia

D. J. LePoire (✉)
Argonne National Laboratory, Argonne, IL, USA
e-mail: dlepoire@anl.gov

D. J. LePoire et al. (eds.), *Complexity in Universal Evolution*, World-Systems Evolution and Global Futures, https://doi.org/10.1007/978-3-032-20517-9_14

14.1 Introduction

Today's change (evolution) is often associated with information technology. A particular form of this technology that attracts a lot of attention is artificial intelligence (AI). Even though artificial intelligence is often associated with reproducing the human mind, the majority of its applications aim to enhance human abilities. Because what humans find easy is usually difficult for computers, it is a good match. In contrast to humans, computers have a difficult time walking, grasping, understanding, and making sense of common sense situations. Computers, however, can perform massive computations for simulations of a variety of physical systems.

Although we have only begun measuring consciousness, we really don't know what it is and have no idea how to measure it in computers. How do AI and robotics evolve in contrast to natural evolution, in which all complexity arises from the bottom-up? AI, on the other hand, evolves through a combination of top-down design and bottom-up exploration of information technology (IT) applied to various tasks. Although AI has made great strides, planning for an uncertain future remains an intractable problem for both humans and advanced AI (Norvig & Russell, 2021).

AI can provide Big History with insight into the evolution of the human brain, which rapidly increased in size during human evolution and led to the emergence of self-awareness (Azarian, 2022). This rapid evolution occurred along with greater social interaction and tool use. In general, insights into AI will expand our understanding of how information was used in evolution, and how long-term stability was sacrificed in order to maintain greater speed and capacity.

The general study of AI and the human brain can contribute greatly to understanding complexity and how it evolves. It is often asserted that the brain is the most complex system in the universe. Or in the words of Carl Sagan (1977), "The cosmos is within us. We are made of star-stuff. We are a way for the universe to know itself." While some project that artificial general intelligence (AGI) is near, many argue that we have only begun to understand the intricacies of the brain (Hameroff, 2022). Some alternative ways to approach artificial intelligence, other than artificial neural networks, might provide key insights into the way that brain integrates memory, processing, and consciousness (Friston, 2010; Kanerva, 2009; LeCun & Friston, 2024).

Research in interdisciplinary fields, such as big history, can also benefit from AI applications as a new tool. Among its applications is literature search, which is particularly relevant to the field. Search engines have been very helpful, but they are continuing to improve through language models such as ChatGPT (Ruby, 2023), which can already address some of the basic understandings in Big History.

It is currently being applied in the variety of fields contributing to Big History (Helland, 2019) including physics, where AI is being used to select experimental "events" to investigate. One of the most difficult problems in computational biology, that is, protein folding, has already been solved by AI. By scanning animal brains, AI is used to build a complete neuronal connectome to study human evolution. For studying civilizations, AI is used for speeding up intensive models like climate models. For future energy technology, artificial intelligence is being explored to control plasma chaos in fusion energy machines.

The third aspect of AI and Big History is trying to understand the various scenarios for the future evolution. What will be the role of AI? How will humans and AI further collaborate and evolve? Will AI reach consciousness? A major question being discussed is the time that many of these questions will need to be answered. One perspective is that technology has been improving at a rate faster than an exponential growth and will soon surpass the complexity of the human brain and therefore show emergent properties real soon. This is the "technological singularity" as viewed by people like Ray Kurzweil (2005, 2012), where the rate of improvement is so fast that humans can no longer make sense of it. Another perspective, taken in this chapter, is that while the technology is rapidly progressing, there are many technological, scientific, and social questions to be explored, which will take a longer time as trust is developed between societies and AI systems.

A general model of an evolutionary system with the ability to sense, process information, act, and learn has been proposed by (Friston, 2010). This is possibly a general model for use in various phases of big history evolution. The Friston model includes many general aspects of collecting information, predicting with a model based on previous experiences, what might happen, mix it with the goals of the system to arrive at plans for action to achieve those goals, take the action and see how the system responds and learn from the differences. This is very similar to the scientific method of having a theory, predicting what might happen if there is a controlled manipulation of the environment (experiment), and then learning from the results by modifying the parameters of the model or reconstructing the model altogether. Other aspects include attention, intermediate objectives, competing models, communications, planning, common sense, uncertainty, maintain different projections into the future, learn to anticipate actions of others, transfer learning through analogies, and emotions (quick assessments of situation and goals).

14.2 Recent Advances

We can marvel at the progress of artificial intelligence. Depending on the definition of artificial intelligence, it could be considered to have been with us for quite some time, helping many with decisions, designs, and planning. The post office used an early neural net system to read handwriting on letters. One fascinating application under development is driverless vehicles, where the computer uses a range of sensors to identify the situation and act accordingly in the midst of human-driven traffic. However, it was only less than a decade ago that the big news in AI was that it could distinguish between cats and dogs in images on the Internet.

While great progress has been made in recent years, it still seems like the comparison with human-level intelligence is still a long way off. However, for most AI researchers, the goal is not to duplicate the human-level intelligence but instead to identify areas where the computer can leverage its advantages when combined with human abilities. So, the objectives are (1) to extend capabilities to sense, process, decide, and act in conjunction with humans; (2) make the AI more general and robust; and (3) explore improved ways to connect to humans. This is another phase of tool development which has been happening since human ancestors began with wood and stone tools. This set of technology adds new elements to compute.

This compares with evolution up to human evolution in that evolution has to form complete systems that were able to sense, process, act, extract energy, and maintain organization while growing, reproducing, and evolving. The technology evolution has quite different constraints and objectives. This was a bottom-up development from simple cells (prokaryotes) and complex cells with organelles (eukaryotes) which comprise about 90% of age of the Earth. A complexity hurdle was how to grow and reproduce beyond one cell. Once this capability evolved, a variety of multicellular organisms could explore new environments including land (amphibians/reptiles/plants) and night (warm-blooded animals such as mammals). The capabilities of these complex adaptive systems had to evolve in parallel because any one component (energy extraction, reproduction, sensing) could lead to extinction.

Predictions of human-level intelligence in the next few decades (e.g., the technological singularity) can be misleading in that (1) that is not the goal of most AI research; (2) the number of computing elements in the brain is not really known, especially if subcellular processing is done in neurons (Craddock et al., 2012; Hameroff, 2022); and (3) if by human-level intelligence means consciousness, then we do not even know what it consists of, how to measure it, or why it is helpful.

One often mentioned test of human-level intelligence is the ability to maintain a conversation where a person cannot tell if they are communicating with a person or computer. Again, great progress has been made with this in recent years. The various AI assistants (Siri, ALexa, and Cortana) can parse words and understand commands. The debut of IBM's Watson (Ferrucci et al., 2010) on the game show Jeopardy in 2011 was quite remarkable as it was able to outcompete the stars of the game in answering general questions before the other players could respond. In the IBM Debate Project, a computer was able to compete (although not consistently win) against a high-level formal debater. Still a typical conversation carries over much context from beginning to end with many principles (common sense) assumed to be understood. So far the communication has been done by collecting and interpreting words to and from sentences by comparing with previous human writing, such as Internet pages, and really does not form an "understanding."

Yet, while AI is being applied in wider circumstances, Maravec's Paradox, that what is easy for humans is difficult for computers and vice versa, is still true (Minsky, 1988). For example, computers have a difficult time walking, grasping, understanding, using common sense, and discriminating between cats and dogs. However, they do great at performing massive sets of computations for such things as physical system simulation. Usually the more focused the subject, the better a computer could perform such as with expert systems in medical diagnosis. However, these very narrow applications are becoming more generalized. For example, the narrow fields of game playing have been greatly expanded from the late 1950s ability to play checkers. Just recently, the Alpha Go AI with supervised training was able to beat the best human player at this extremely complex game. Then the AI was quickly extended to Alpha Go Zero, where the AI learned in a much more general reinforcement learning by playing a copy of itself. This version was a large step toward generalization in game-playing.

14.3 Complex Adaptive System Components

A viable system needs to be able to (1) interact with the environment through sensing and actions to extract resources such as energy and materials, (2) process the information to guide actions based on the information, and (3) thrive by maintaining order (removing waste and repairing), growing, reproducing, and evolving. With large complex systems, interaction with the environment includes communication between units to coordinate actions whether it is between cells in an animal or humans within a community. Sensing includes chemical detection (smell), interpretation of light and sound, and determining dangers such as hot or cold temperatures.

Processing can be quite simple but often include aspects of memory, pattern recognition, model formation, learning, decision-making, planning (including projecting future consequences of different actions), monitoring to compare the predictions of an action with what really happened. However, sensing, processing, and acting need to be towards achieving some goal even if it is to survive through eating. AI is usually provided with some narrow objective by humans, often with supervision. In this way, AI is much like raising children, where knowledge is gained and interpreted with various supervised lessons and monitoring to ensure that the objective has been communicated correctly and constraints specified. This kind of supervised learning in narrow aspects would subvert any attempt for an AI system to enact the "stapler" catastrophe, that is, if an AI is given the goal to make as many staplers as possible, it could interpret that objective by commandeering all plastic and metal resources to produce staplers at the expense of other important products. This communication of objective is often difficult even between humans since the language is vague and much common-sense knowledge is assumed. The children's humorous fiction book of Amelia Bedelia illustrates this by Amelia's actions to carry out actions as she interprets them, not as the requester means (Wolchock, 1990).

14.3.1 Bottom-Up Evolution

Nature has built up the brain from the bottom-up (Ogas & Gaddam, 2022; Ekstig, 1994; Solis & LePoire, 2020). This includes the beginning of life and the information needed to self-organize, the chemical signals that can go in simple cells (amoebas can learn), the microtubules to help larger cells that most multicellular animals are made from, neurons which are specialized cells from transmission of signals, groups of neuron which make up some function such as a reflex, groups that make up memories, groups that make up neocortex cells, and the combination of neocortex that makes consciousness (Jantsch, 1980; LePoire, 2020).

One theory about the origin of life is that it began with autocatalytic chemical cycles, for example, with a three chemical cycle, A catalyzes B, B catalyzes C, and C catalyzes A. Somewhere in the cycle there must be some energy input to keep the cycle going along with some materials being fed in. This is then a dissipative system since energy is used (dissipated) to keep the cycle going. Simple real systems like these include the BZ reaction which is known to organize into dynamic patterns.

This might have happened in an alkaline hydrothermal vent in the early earth . The vents provide thermal gradients for energy flow, water rich in minerals, pore spaces for membranes to form on surfaces that could be later released, ionic potential within the FeS chemical environment similar to cells today, protection from the ultraviolet light of the more active early sun. The membranes (dual layered soap bubbles) might have been able to contain certain combinations of the autocatalytic cycles. Those that grew faster and split the membrane were able to outcompete slower versions. It was determined by Manfred Eigen, Nobel Prize Chemist, that under evolutionary conditions, the growth is not just exponential since the process improves over time it becomes hyperbolic leading to a time with very rapid change. He suggested that this may have been the reason that life occurred so fast after the Earth formed and cooled. He also showed that such a multicomponent autocatalysis mechanism is robust against the rapid mutations in the exchanging chemicals, avoiding the Eigen Paradox of formation of long enough genetic strings.

In an autocatalytic cycle, each of the chemical catalysts can be replaced. One such replacement might have been a general way of generating and testing new enzymes through an RNA structure which is formed when nucleotides are dried (removal of water) or “digested” when water reacts. The RNA is mostly single stranded and is much less stable than the later DNA. However, the single stranded RNA can form many shapes since the nucleotides interact with others along the one strand. RNA snippets have been observed to form catalysts. The RNA can then also act as an information template and later evolve into the more complex and controlled DNA information storage for genetics.

Eukaryotic cells are much larger than the simpler and earlier Prokaryotic cells. It is believed that these eukaryotic cells are formed by combining specialized eukaryotic cells. The genetics had to be integrated into one system. However, there was one component, the mitochondria involved in energy metabolism that was able to keep its own DNA. In humans, the mitochondrial DNA is used to trace matrilineal lines since the mother’s mitochondrial DNA is passed to all offspring.

So, internal compartments were needed along with necessary ways to share information and control the movements of resources inside a eukaryotic cell. This was a major infrastructure project using microtubules as the tracks. The microtubules are quite strange geometric chemicals formed from pairs of proteins in cylinders with about 13 protein pairs around the circumference. The microtubules are also known to help organize the dance of the chromosome separation during cell division. It is also known that there is a higher concentration of microtubules in neuron cells. Their role is not quite determined although some indications are that they participate in memory systems using their ordered surface with makers of Ca ions on spider like molecules. These neuron cells can collect, process, and then distribute results to other neurons, allowing very nice nonlinear processing which is needed to make decisions, learn, and form models.

While Neurons take information from sensors and process the information and eventually connect to actuators like muscles. While some basic processes could be handled locally (reflexes) other information needed to be coordinated. However, it could be better accomplished in a centralized location (the brain). While the brain at first handled coordination of senses and movement, it later developed more complex layers to handle decisions that needed to be further integrated into experience and at an organism level (emotions). Only later was the neocortex developed in mammals to address needs of caring for young, working in a group, and maintaining social coherence.

During the evolution of modern humans, the neocortex greatly expanded. This was facilitated in part by the ability to cook food which allowed energy diversion from digestion to support the brain and develop perspiration for cooling. In modern humans, 20% of the energy is used for the brain. The expansion enabled more social connections to be made as documented by Dunbar who extrapolated the correlation between brain size and social group size to predict the natural human group size was at most about 150 people. This included the "theory of mind" feature in humans to be able to view the perspective of other people. These fundamental aspects enabled developments including the ability to recursively manipulate symbols (e.g., for language) and the ability to time travel with the mind to the past and create future scenarios, which is instrumental for controlling more of nature such as tool formation and fire. But most of the social aspects led to working in teams, specialization of labor, and creating a common culture through music and storytelling.

14.3.2 Width and Depth

Our efforts to reconstruct information with technology are not one of being from the top-down, that is, start with the problems and solutions first and then work out the details. However, it can be better described as developing the width and depth to expand human capabilities. That is, AI technology allows human expertise to be expanded with technology, just as scientific instruments allow us to extend our senses to higher (X-ray) and lower (radio) frequencies of the electromagnetic spectrum.

The new technologies are at first very much like a small child is not knowing the basic rules or consequences of their actions. Definitely bad things can happen in both cases. For IT, the integration is slowing due to adverse effects such as virus, blackmail, misinformation, bias, weaponization (Stuxnet), deep fakes, infrastructure vulnerability, privacy loss (China), unexplored model assumptions (Black-Scholes: Asia), job loss, obsolescence, material and energy use, connection replacement, and dependence–resilience.

However, children around the age of 2 very quickly start to learn natural laws and social norms. This key aspect of learning led the DARPA to call for the advanced exploration of trying to simulate 2-year-old human (Knight, 2019). This can be seen as the ability to learn how to learn as in contrast with medical doctors where the rules and uncertainties are more constrained and can follow more rule-like behavior as exemplified in early and improving expert decision systems. Other very difficult aspects include the ability to form common sense (i.e., a very general sense of how things behave), the ability to integrate signals with appropriate attention and set goals, and the ability to integrate the many perspectives of others in cases where the problems are too uncertain and complicated, that is, wicked problems. It is not clear when the emergent property of consciousness arises. It is clearly not just the combination of enough processing size but the capabilities and arrangement of those capabilities to learn, set goals, test, under uncertainty.

Some of the aspects being explored are from the complex adaptive systems (CAS) model of being able to sense (Lidar accelerometers), model (common sense, physics, economics, games) process, and act (move, play games, communicate). These can be done at a small scale or integrated at larger scales such as self-driving cars or even large scales of smart cities where the vehicles, roads, and infrastructure communicate with each other.

At any scale, a large component of the processing is the ability to handle uncertainty, integrate common sense, and be able to explain and communicate its decision to work with others (Besold et al., 2022). More basic maintenance tasks include controlling the environment (heat and energy use) through necessary diagnostics and learning, forming appropriate data structures where connectionists or symbology (Bajada, 2019) and to use memory as in relational databases (RDB) or object oriented associated databases.

14.4 Current Explorations

Much of the recent success in AI has come through the application of neural nets, a connectionist computing strategy inspired by the brain's neurons. (However, it was determined that a real biological neuron would require thousands of artificial neuron to simulate.) Each neuron has multiple inputs similar to the biological dendrites and one output similar to the biological axons. The connections of other neuron's axons to the dendrites of a neuron cause their output signals coming along the axon to be weighted and summed. This weight sum is passed through a nonlinear function (i.e., the output is not just proportional to the weighted sum).

The neurons are often organized in layers, with the preceding layer feeding their outputs to be processed by a layer and then passed on to the next layer. The final layer has some form of output. However, there are many nuances to this though. In a training mode, the final output is compared against what the real answer is. This information is then fed backwards along the neuron's chain of layers to adjust the dendrite–axon connections, so that next time the result would be better. This is done with many test cases to tune the parameters. When images are used, the layer might consist of an array similar to the image's pixels. The number of adjustable parameters can become huge (billions) with a few parameters per connection throughout the chain.

There are also different ways to represent the data. For an image, it could be the strength of the red, green, and blue components from 0 to 255, for example. Or it could be specified as the B/W intensity along with two color fractions. The choice of representation is often more of an art than a science. In fact much of the architecture—the number of layers, the types of layers, and the number of neurons in the layers are often just a guess. There are tools to test how important different design features are, and there are also tools that can perform a rough guess at what architecture might work best (hyperparameterization tools).

Besides the use of faster computers and graphic cards used for video games, the major breakthrough came with the substitution of the nonlinear function in the neuron with one based on a simple transition to one that continues to grow in one direction (e.g., the output is zero if the input is less than some threshold, and otherwise is proportional to the difference of the input and threshold). This allowed error signals to propagate further back towards the raw data, enhancing the learning capability.

So, we see many differences between the artificial and natural neural networks: the nonlinear function applied to the weighted inputs, the need for many more artificial neurons to simulate one biological neurons, the great number of training cases needed in artificial networks compared to the few in biological systems, and the energy required to operate them with the artificial networks needing their own electrical generator and water cooling stations compared to the 20 W of energy used to run the brain.

However, at another level there are similarities in how information is processed. In recent years, a number of different artificial neural network architectures have been identified that are similar to how the brain might work. It is believed that the brain does not interpret the raw sensory input but instead processes the raw input to some extent and then compares that to what it might expect.

For image processing, that is, seeing, one of the simplest architectures is the convolutional neural network (CNN) where each layer works on finding more complex features building from the previous layers. The first layer might look for simple features like edges to isolate an object. The next might look for the combinations of edges in certain shapes. The main demonstration of the breakthrough came in an CNN that was able to identify cats from dogs. In the human brain, the visual cortex at the back of the brain is divided into six layers (V1–V6). The raw visual information from the eyes is processed starting with V1.

After breaking the raw signals into segments and objects, the brain needs to reconstruct the full image by first generating interpretations of what is going on and then testing them against the processed raw signals. This might be done with expected probabilities, that is, when there are people mingling in a room, it is not expected that a gorilla will walk through. Any differences between what is expected and sensed is then fed back to refine the expectations. This refinement might be either consideration of different known scenarios or the learning of a new situation.

Neural net architecture does a similar process (Repetto, 2018). It consists of two parts, one that generates some output similar to a training set (e.g., a new face) and the other that tries to determine if that output is real or artificial, that is, it acts as an adversary of the first part by catching it in a lie. This is the generative adversarial network (GAN) that generates deep fakes or be a bit creative in generating paintings or music similar to some artists' style.

Recursion is where an output is fed back as an input, as done in recurrent neural networks (RNN). One reason for doing this is to form a type of memory especially in a sequential set of information such as a sentence. Often the words need to be interpreted as a whole and not just in sequence, so in the feedback, a word in a sentence can be held and used later for making sense in the context of the full sentence. This memory architecture is the long, short-term memory neural network architecture (LTSM).

Memory in general is a difficult task. Often we know just a few characteristics of an object we are trying to recall, for example, starting with the characteristics of brown hair, tall, accountant within some context we'd like to recall the name or other characteristics. This type of memory is associative memory, quite different from the memory in a computer which is directly addressable by its location in the chip. A different recurrent neural network that is fully connected, that is, each artificial neuron connected to all the others, is called a Hopfield network. Its behavior is similar to a physics problem, where some measure of closeness (energy) is minimized. Besides associate memory, this architecture also works on some optimization problems, such as the classic traveling salesman route determination.

The difference in energy usage is being explored in many ways. Two approaches are to mimic biological neurons in either processing or in communication. For example, the current artificial neural networks are based on the transistor either the mains computer card (CPU) or graphics card (GPU) of a typical computer. Clearly, biological neurons do not have this structure, so efforts are made, called neuromorphic computing, to simplify the computational process to improve the match with biological systems. Another leftover from traditional computers that is quite different from the biological ones is that the signals of the biological neurons use short bursts (spikes) signals. If the signal is large, then there might be a long string of such spikes. In contrast, the traditional computing signals are on or off for a computing timestep. Great progress has been made in exploring the use of such artificial neural networks that not only require less energy but also seem to learn quicker and are more robust to noisy environments.

14.5 Analogy of Raising Children to Maturity

Instead of rapid takeover by AI, the situation may be more like raising children to a maturity level to be independent but still function within a society. The Big History of the future might be that humans, with all their limitations might develop the next higher complexity level through the interaction of humans and AI. But just as in raising children, they must experience the world in a guarded fashion to develop such difficult things for computers as common sense, values, and trust all under the context of the highly uncertain world. Eventually, maybe, enough experience and trust have been gained in specific areas to increase the autonomy of the children AI along with some approval (check and balance) procedures any adult experiences when proposing doing major actions such as taking out a loan or starting a new business.

The connections are already being explored in virtual environments, closer integration of computer enhancing human thinking and actions, better at communication with natural languages, and being autonomous when they need to be. The virtual environments like video games, augmented reality, and virtual reality allow the AI to function in a world where things can be tested without much risk. The closer integration from standalone desktops to laptops, Internet enabled laptops, wireless smartphones, monitoring watches facilitates the trend to becoming smaller, more energy efficient, but yet better connected. Elon Musk (2019) is working on a brain implant. A part of any of these human–computer interfaces is communication, which has progressed greatly with the large language apps such as ChatGPT.

And there are occasions where AI systems need to work autonomously. Such situations include space vehicles where the signal to communicate would take too long for decisions that need to be more immediate. Another situation might be in any operating machinery where again the decision might need to be quicker than humans can analyze the situation, compose plans, select a decision, and implement it. This includes things like anti-lock brakes in cars but is also a possible breakthrough in controlling chaotic systems like the plasma needed for fusion energy.

Even at the "mature" level, there are decisions which require a level of ethical consideration. These decisions are usually difficult since all the options have a mix of advantage and disadvantages in a probabilistic way. Humans are beginning this exploration of AI ethics, which are often similar to the human ethical issues, in the implicit biases in the data being fed to them to learn and their ability to be used in situations where the assumptions do not match. For example, the financial system has in the past become unstable to the point of collapse by applying a model where all the correlations were not accounted for. Some governments use AI to monitor the activities of citizens not only through monitoring access to the Internet but also while they are walking around the streets by using surveillance cameras to perform face recognition. Then there are issues similar to traditional ones, for example, plagiarism, that have a slightly different spin. For example, should ChatGPT be banned from education since it can write essays and answer questions that can pass law and medical exams? Or should its use be encouraged as a new tool, similar to how the calculator was used to focus on the methods (after the basics had been learned).

14.6 Interaction Between AI and Human Intelligence

The active introduction of technologies in various spheres of life is increasingly immersing a person in an artificial environment, a part of which is artificial intelligence. This environment is increasingly alienating man from biological nature.

Today, artificial intelligence is understood as "the result of human activity, which is a complex set of communication and technological relationships that has the ability to think logically, manage their actions and correct their decisions in case of changing external conditions" (Yastrebov, 2018, p. 41).

These technologies combine the ability to think logically, solve creative problems, understand oral and written speech, conduct a meaningful dialogue, learn, self-learn, and make decisions based on constantly updated knowledge about the situation and the world as a whole. Of course, today there are technologies that can write a musical composition, beat grandmasters in a game of chess or a game of Go, understand and reproduce spoken language, and so on. There are also more complex projects that can perform multiple tasks, but no artificial general intelligence or consciousness has been developed.

In general, there are three main approaches to the future interaction between human and technology (Barrat, 2019):

1. The fusion of technology and the biological species *Homo sapiens*.
2. Technology continues to make our lives easier, but a person remains a person.
3. Humanity will be replaced by machines completely.

The technologies of AI are based on the principle of self-learning and they may become a particular danger to humans. If restrictions aren't introduced into the functioning algorithms, then the self-development of artificial intelligence can lead to the fact that it will quickly surpass human intelligence, which, of course, can also hardly be called completely natural. The formation of neural connections occurs under the influence of the artificial environment that surrounds a person from the moment of his birth. Nevertheless, these technologies can be used for "cognitive enhancement" of a person. Modern medicine is already actively using the latest developments to improve the quality of sick people lives. Artificial intelligence or its individual elements could be used to improve the lives of people with cognitive disabilities eventually. For example, today there are special programs that help people with autism spectrum disorders. Some of them have difficulty to recognize the emotions of other people; there is already a way to help them, which undoubtedly simplifies social interaction (Yonck, 2019).

Artificial intelligence is being applied in many disparate fields. For example, many environmental sciences have a need to model complex systems, and "by providing assistance and automation and by ensuring proper use of models, AI has immense potential to make modeling more efficient by orders of magnitude" (Gil et al., 2021, p. 45). It can be argued that modeling is an important component of the social sciences too.

Yet, most scientists, inventors, and public figures have very different opinions about the AI future. For example, Elon Musk, an apologist for scientific research in various fields, believes that it is an existential threat to all humanity. A self-learning system can quickly outperform a human if the process of self-learning is not restrained by anything. One of the ways to ensure safety can be associated "with the desire to simulate safe behavior and human thinking, that is, the physiological foundations of the brain, which are transformed into a digital model or artificial intelligence" (Sokolova, 2016). The issue of responsibility and legal regulation of relations because of the threat from developing technology, arising from its use, also becomes acute. That fact, who will be responsible for the actions of systems using artificial intelligence, deserves the greatest attention.

All changes related to AI are the extrapolation of the technological power growth as megatrends of social evolution. A.P. Nazaretyan wrote that "during this century, the planetary civilization will face unprecedented turn of 'removal from nature' (which in all critical epochs ensured the overcoming of the anthropogenic crisis), or an equally unprecedented collapse in scale" because of singularity (Nazaretyan, 2001, p. 191). Both scenarios of the civilization collapse and the beginning new technologically enhanced evolution look real due to the rapid development of technology, as well as progressive development scenarios associated with the transition of the system to a new level of complexity (improvement of robotics, development of artificial intelligence, space expansion, even the creation of a complete brain-computer interface). There is the other hypothesis of the AI development: "the line of singularity does not lead to superintelligence and exponential growth of technology, but to a 'broken intelligence' without anybody who owns it" (Plebe & Perconti, 2020, p. 132).

It is also impossible to exclude development options where a person remains a biological being, which, nevertheless, can change significantly because of genome improvement, replacement of internal organs, or bioprosthetics. Perhaps, this is the most realistic approach to the future, which preserves the "hope" for the preservation of human. Some see human degenerating as a biological species due to natural selection being almost completely blocked by humanistic culture, because of this, subsequent generations gradually become genetically weaker than the previous ones (Bochkov, 1978). The survival of our species requires the use of more and more new technologies for medical purposes, which, in turn, can lead to longer life and the improvement in life quality. Although some studies in this area are prohibited for ethical reasons, because endowing a person with properties that are completely alien to his nature remains taboo even in those scientific fields "where the 'demarcation of the boundaries' of the morally permissible actions has not yet been completed" (Milo, 2011, p. 66). Only the ethical norms remain an important limitation of research in some areas (e.g., experiments on human cloning are prohibited).

14.7 Ethics and Regulation

Some moral norms, generally accepted in modern science, are the logical result of tragic events. Nowadays, human trials of medications are carried out only after careful preliminary testing and with the consent of the experiment participants themselves. Apparently, the greatest influence on the emergence of this rule was exerted by numerous experiments carried out by German doctors on concentration camp prisoners and by Japanese doctors on the civilian population of Southeast Asian countries. A significant part of these experiments generally looks senseless from a scientific point of view. They resemble the "entertainment" of inborn sadists. The possibility of using artificial intelligence technologies for medical purposes poses serious sociopolitical, ethical, and legal questions to humanity.

Today, some scientific fields are controlled to a very high degree by ethics committees, which limit the scope and direction of research. The influence of moral norms is so strong that we can talk about the presence of signs of its institutionalization (Razin, 2017).

In our opinion, the existing ethical principles in the field of medical experiments are quite sufficient for the conducted research to be considered humanistically oriented and not violating human rights. The same principles may form the basis of the ethics in the sphere of the medical use of AI. The ethics and openness of studies were recognized as important principles for useful artificial intelligence during the Asilomar Conference in 2017. Despite the fact that research in the field of artificial intelligence has been ongoing for several decades, there is no clear answer to the question: "Are such research safe for humans and humanity as a whole?"

All discussions about the possible AI harm may be categorize into four groups "related to complexity:

1. Technical vulnerabilities and errors in AI systems
2. Issues or harms that emerge through, or are created by, the use of AI
3. Issues or harms that are magnified considerably by the use of AI
4. Mismatches between AI and social expectations or empirical realities" (Weissinger, 2021, p. 5).

AI is often developed in commercial organizations, it means that we need not only moral but also legal norms. Regulation should not be limited exclusively to the medical field, where any untested technology poses a threat to human life and health. The solution of many social problems caused by the AI technology development may be associated with the development of the international regulation system (Weissinger, 2021). Perhaps a new level of the social system complexity will be associated with the maximum inclusion of AI in it.

Modern society (the global social system) is going through a period of instability, which is associated with many difficulties, including in the system of global regulation. During a period of instability of the organization, of course, the management system collapses or makes a transition to a higher level of complexity as a result of successfully overcoming the bifurcation point. At the same time, some researchers note that "it is at the bifurcation point that control turns out to be most effective: small impacts can change the trajectory of the controlled system" (Malinetskiy, 2015, p. 157).

Artificial intelligence does not create new problems as such; it only reveals new facets of existing ones or makes them more obvious and noticeable. At the same time, artificial intelligence can be successfully used in decision-making, especially when it comes to natural systems. Of course, technologies that in the future can surpass humans cannot be left uncontrolled, but they can be partially used, including to solve global problems that humanity faces.

The issue of artificial intelligence and regulation is therefore twofold. On the one hand, it can be argued that international instruments for regulating the interaction between humans and artificial intelligence are needed. Over time, the need for such management mechanisms will increase. On the other hand, artificial intelligence can help in decision-making, including in the sphere of regulating processes of various scales, both local or regional and global.

14.8 AI and Immortality Perspectives

There are different views about humanity's future. Some argue that we should move beyond the physical body as the basis of human identity, but it seems more likely that humans will continue in a close partnership with technology (Nazaretyan, 2001). A few years ago, I thought about the purpose of such aspirations, because it seemed to me that it might be possible to transfer only the rational part of the personality to electronic media in the future, but emotional programming technologies have been actively developing in recent years. However, they are aimed at teaching gadgets to recognize human emotions. Perhaps, artificial intelligence will learn to imitate them over time. Yet the imitation and the real experience of emotions are very different from each other. This may help a person to begin to perceive artificial intelligence as part of his social environment, but it will not make artificial intelligence a full-fledged personality.

In addition, our individuality is formed under the influence of other people, some minor circumstances, and traumatic childhood memories. Those who will be engaged in transferring the human intelligence to an electronic medium may not know all these things. As a result, we get a completely different personality.

Yet, the ideologists of transhumanism believe that artificial mechanisms will appear that can become the forefront of the mind evolution and lead people. In addition, they think that it will happen in the near future. This is no longer a simple creation of robots; these are attempts by a certain part of inventors to create a superman capable of replacing Homo sapiens. The intelligence of this new species may surpass human intelligence, but it is difficult enough to talk about the presence of personality in avatars seriously. Yet such an ideology doesn't have a serious impact on our contemporaries, although it originated more than a century ago. Many people don't aspire to immortality, but to prolong the period of a healthy life.

It is likely that the next technological breakthroughs will be associated with the development of medicine and robotics. Interest in medicine and research in this area will probably be the most relevant today, since the pandemic coronavirus disease 2019 (COVID-19) has once again clearly demonstrated the fragility of the human body and its insecurity to the natural environment.

14.9 Recent Developments

The rapid advancement of AI technology, particularly through large language models (LLMs), has reshaped public engagement with AI while raising critical questions about employment, ethics, and the path to artificial general intelligence (AGI):

- *Large Language Models and White-Collar Job Disruption:* LLMs like GPT-4 have democratized AI access but intensified concerns about automation in knowledge-based roles. A 2023 study found that 80% of US workers could see at least 10% of their tasks impacted by LLMs, with 19% facing ≥50% task exposure, spanning all wage levels (Eloundou et al., 2023). White-collar roles involving repetitive text processing (e.g., legal documentation, financial reporting) are particularly vulnerable.
- *The AGI Debate: Complexity and Timelines:* While some researchers argue AGI could emerge by the late 2020s through scaling existing architectures, others emphasize fundamental gaps (Raman et al., 2025; Trigyn Insights, 2025). The pro-AGI advocates cite improved reasoning in models like OpenAI's o1, which solve multistep problems in coding and biomedicine. The skeptics highlight the human brain's neurobiological complexity, arguing that neural networks lack intrinsic reasoning, adaptability, and consciousness. Hybrid approaches (e.g., neurosymbolic AI) are proposed as alternatives to pure deep learning.
- *Ethical Challenges and Governance Initiatives:* AGI's potential risks have spurred global calls for oversight (Glenn, 2023) including the existential risk of uncontrolled AGI self-improvement which might lead to systems operating beyond human comprehension.
- *Personal AI Companions and Assistants:* LLMs with long-term memory (LTM) capabilities are enabling personalized AI applications (Lee, 2024): These systems prioritize adaptability but face ethical scrutiny over data privacy and dependency risks, such as companions and advisors.
- *Testing AGI in Simulated Environments:* To mitigate real-world risks, organizations are developing controlled testing frameworks. For example, simulated physical world model environments allow AI agents to refine decision-making in scenarios like disaster response or logistics without real-world consequences (Bastian, 2025).

14.10 Conclusions

Artificial Intelligence has quite a bit to offer Big History. It is another window into both the past and the future. Just as radioactive decay offered insights into critical aging determinations, AI can allow us to explore biological information processing and decision-making, the evolution of consciousness, the limitations of humans, and the future possibilities of interactions with AI in this continuing evolutionary experiment on Earth.

However, it is not likely that full integration or dominance will be as soon as others have predicted. While the singularity trend in Big History has been developed, the trend is expected to break (or have done so already) just like the singularity trend in the global population (Korotayev, 2020; LePoire, 2020; Modis, 2002). Kurweil, John Smart, and Diex Faixart have seen the technology trends as culminating soon in untold possibilities of very smart and very small engineered systems that improve the way they are made, accelerating even further the progress. However, there are fundamental uncertainties that cannot follow this trend. For example, the continuation to smaller and smaller physics down to the Planck scale will require tremendously large and expensive experimental equipment if it is even possible. This will not happen in the coming decades (if ever). While it is OK for AI to take control in virtual video games or suggest what products might interest someone, it is different for it to develop the nuances and understanding for driving consensus to solutions of wicked problems, since, by definition, wicked problems require some form of trial and error (and interpretation and negotiation) to resolve. Furthermore, while this singularity trend might seem mysterious to some resulting in some great moment of 13.7 billion years of evolution, the mathematical model used to help understand it is one of utmost simplicity. Just as many growth patterns follow exponential patterns, evolving systems can follow singularity trends as noted by the Nobel Prize winning evolutionary scientist Manfred Eigen in the 1970s.

Instead of being on the brink of human-level thought, the rapid progress in machine learning is more of a highly advanced correlation learning mechanism. However, when applied to communication, it can just guess based on what it has learned from former speech and does not understand the relationships of consequences of its responses. In this way AI is more like reptiles in relying on instantaneous memory, learning, but it is alone, dependent on energy and information from outside, and does not move with grace.

Then the next stage might be more mammalian with the ability not only to learn but to teach and communicate with some social interactions and then the human level with symbolic reasoning, under uncertainty, with common sense, empathy, and emotion, with values, goals, wonder, and creativity. The AI still has a long way to go to get there and be effective and efficient in economic and environmental terms in the ability to use less energy and resolve the issue of what to do with replaced workers. But in the meantime AI has and continues to give great insights and help in filling in where humans are not very good with special topics such as protein folding, optimization, and classification. In the big picture besides these concrete applications, AI has provided us with insight into who we are and how we came to be. And it seems like we are just beginning to learn and be surprised by what we find.

There are three options left:

The first option is the cessation of research in this area and a ban on the improvement of such technologies, similar to the ban on human cloning. However, many difficulties can arise in this case, because such developments are difficult to control. The group of enthusiasts is enough to continue the study. In addition, millennial trends can't be changed by a simple managerial decision: "all attempts to delay the development of technologies for various social reasons (mainly to prevent unemployment) ended in failure" (Akimov, 2017, p. 181).

Second, the development of AI elements, that will improve the quality of human life, revolutionize, for example, the labor market, makes living conditions easier and helps people with serious mental disorders become a full-fledged part of society, etc.

Final option is the development of artificial intelligence in full.

The AI evolutionary prospects will also depend on the choice of one of these options. However, it is not a direct threat. Any technology can be used for both evil and good.

The situation with AI is more complicated than with the previous technologies, because with its maximum development and ability to self-learn, it can surpass the human one. Then a person can lose the place of the "engine" of evolution; it can be occupied by artificial intelligence as a new evolutionary leader. At the same time, all the developments in this area can be used for the development of the person himself, if you are careful and do not succumb to either panic or euphoria.

References

Akimov, A. (2017). Robotics and labor-saving technologies: Prospects for the impact on socio-economic development. *Historical Psychology and the Sociology of History, 1*, 173–192.

Azarian, S. B. (2022). *The romance of reality*. BenBella Books.

Bajada, J. (2019, April 8). Symbolic vs Connectionist A.I. *Towards Data Science*.

Barrat, J. (2019). *Our final invention. Artificial intelligence and the end of the human era*. Alpina.

Bastian, M. (2025). *Google Deepmind CEO Demis Hassabi says world models are making progress toward AGI*. The Decoder.

Besold, T. R., et al. (2022). Neural-symbolic learning and reasoning: A survey and interpretation. In P. Hitzler & M. K. Sarker (Eds.), *Neuro-symbolic artificial intelligence: The state of the art*. IOS Press.

Bochkov, N. (1978). *Human genetics. Heredity and pathology*. Medicine.

Craddock, T. J. A., Tuszynski, J. A., & Hameroff, S. (2012). Cytoskeletal signaling: Is memory encoded in microtubule lattices by CaMKII phosphorylation? *PLoS Computational Biology, 8*(3), e1002421.

Ekstig, B. (1994). Condensation of developmental stages and evolution. *Bioscience, 44*(3).

Eloundou, T., Manning, S., Mishkin, P., & Rock, D. (2023). GPTs are GPTs: An early look at the labor market impact potential of large language models. *arXiv. arXiv:2303.10130*.

Ferrucci, D., et al. (2010). Building Watson: An overview of the DeepQA project. *AI Magazine, 31*(3).

Friston, K. (2010). The free-energy principle: A unified brain theory? *Nature Reviews. Neuroscience, 11*, 127–138. https://doi.org/10.1038/nrn2787

Gil, Y., Garijo, D., Khider, D., Knoblock, C. A., Ratnakar, V., et al. (2021). Artificial intelligence for modeling complex systems: Taming the complexity of expert models to improve decision making. *ACM Transactions on Interactive Intelligent Systems, 11*(2), 49. https://doi.org/10.1145/3453172

Glenn, J. (2023). *International governance issues of the transition from artificial narrow intelligence to artificial general intelligence*. The Millennium Project.

Hameroff, S. (2022). Consciousness, cognition and the neuronal cytoskeleton—A new paradigm needed in neuroscience. *Frontiers in Molecular Neuroscience, 15*, 869935.

Helland, B. (2019). AI for science. *US. Department of Energy Biological and Environmental Research Advisory Committee (BERAC)*, October 24–25. https://science.osti.gov/-/media/ber/berac/pdf/201910/Helland_BERAC_Oct2019.pdf

Jantsch, E. (1980). *The self-organizing universe: Scientific and human implications of the emerging paradigm of evolution*. Pergamon.

Kanerva, P. (2009). Hyperdimensional computing: An introduction to computing in distributed representation with high-dimensional random vectors. *Cognitive Computation, 1*, 139–159. https://doi.org/10.1007/s12559-009-9009-8

Knight, W. (2019, April 8). Two rival AI approaches combine to let machines learn about the world like a child. *MIT Technology Review*.

Korotayev, A. (2020). The 21st century singularity in the Big History perspective. A re-analysis. In A. V. Korotayev & D. LePoire (Eds.), *The 21st century singularity and global futures. A Big History perspective* (pp. 19–75). Springer Nature. https://doi.org/10.1007/978-3-030-33730-8_2

Kurzweil, R. (2005). *The singularity is near*. Viking.

Kurzweil, R. (2012). *How to create a mind: The secret of human thought revealed*. Viking.

LeCun, Y. & Friston, K. (2024, January 19). Beyond the hype cycle what AI is today, and what it can become. Casper Labs Interview. *World Economic Forum's Annual Meeting*, Davos Switzerland. https://www.youtube.com/watch?v=SYQ8Siwy8Ic

Lee, E. (2024). Towards ethical personal AI applications: Practical considerations for AI assistants with long-term memory. *arXiv*. https://arxiv.org/abs/2409.11192

LePoire, D. J. (2020). Exploring the singularity concept within Big History. In A. V. Korotayev & D. LePoire (Eds.), *The 21st century singularity and global futures. A Big History perspective* (pp. 77–97). Springer Nature. https://doi.org/10.1007/978-3-030-33730-8_3

LePoire, D. J., & Sazhienko, E. (2025). AI and Big History: Unveiling complexity insights into human evolution and future scenarios. *Journal of Big History, 8*(1), 71–82.

Malinetskiy, G. G. (2015). Interdisciplinary ideas in sociology and challenges of future. *Sociological Studies, 4*, 152–161.

Milo, P. (2011). *Our flying car awaits*. Lomonosov.

Minsky, M. (1988). *The society of mind*. Simon & Schuster.

Modis, T. (2002). Forecasting the growth of complexity and change. *Technological Forecasting and Social Change, 69*, 377–404. https://doi.org/10.1016/S0040-1625(01)00172-X

Musk, E. (2019). An integrated brain-machine interface platform with thousands of channels. *Journal of Medical Internet Research, 21*(10), e16194.

Nazaretyan, A. (2001). *Civilization crises within the context of universal history. Self-organization, psychology and futurology*. PerSe.

Norvig, P., & Russell, S. (2021). *Artificial intelligence: A modern approach* (4th ed.). Pearson.

Ogas, O., & Gaddam, S. (2022). *Journey of the mind: How thinking emerged from chaos*. W. W. Norton & Company.

Plebe, A., & Perconti, P. (2020). Plurality: The end of singularity? In A. Korotayev & D. LePoire (Eds.), *The 21st century singularity and global futures. A Big History perspective*. Springer Nature.

Raman, R., Kowalski, R., Achuthan, K., et al. (2025). Navigating artificial general intelligence development: Societal, technological, ethical, and brain-inspired pathways. *Scientific Reports, 15*, 8443.

Razin, A. V. (2017). Historical forms of morality and modern ethics. *Philosophy and Society, 1*, 61–91.

Repetto, A. (2018, August 25). Neural networks: Analogies. *Towards Data Science*.

Ruby, M (2023, January 30). How ChatGPT works: The model behind the bot. *Medium*.

Sagan, C. (1977). *The dragons of Eden: Speculations on the evolution of human intelligence*. Random House.

Sokolova, S. (2016). Artificial intelligence and the social safety. *Bulletin of the Polissya State University. Social Sciences and Humanities Series, 1*, 63–68.

Solis, K., & LePoire, D. J. (2020). Big History trends in information processes. In A. V. Korotayev & D. LePoire (Eds.), *The 21st century singularity and global futures. A Big History perspective* (pp. 19–75). Springer Nature.
Trigyn Insights. (2025). Artificial Intelligence (AI) in 2025. https://www.trigyn.com/insights/artificial-intelligence-ai-2025
Weissinger, L. (2021). AI, complexity, and regulation. In *OUP handbook on AI governance*. https://ssrn.com/abstract=3943968
Wolchock, C. (1990). In the classroom: Interpreting idioms. *The Reading Teacher, 43*(8), 614.
Yastrebov, O. (2018). The legal capacity of electronic persons: Theoretical and methodological approaches. *Proceedings of the Institute of State and Law of the RAS, 13*(2), 36–52.
Yonck, R. (2019). *Heart of the machine: Our future in a world of artificial emotional intelligence*. Bombora.

Ekaterina Sazhienko was born in Gubakha, a small industrial city in the Ural Mountains, in the center of Russia. She teaches at the State University Dubna, which is located in one of the Russian "science cities." The subject of her PhD thesis was the evolutionary model for forecasting and regulation of global social processes.

David J. LePoire researches, develops, and applies science principles in environmental issues, Big History evolutionary trends, and particle scattering. He has a BS in Physics from CalTech, a PhD in Computer Science from DePaul University, and over 30 years of experience at the Argonne National Laboratory in the development of scientific analyses, software, training, and modeling. His research includes Big History synergistic trends among energy, environment, organization, and information. Currently, he is the President of the International Big History Association (IBHA).

Humanity's Place in the Universe: Peer Complexity, SETI, and the Fermi Paradox

15

J. N. Nielsen

Abstract

What is the Copernican response to the problem of humanity's place in the universe? What is the place of human history within the vastly larger scale of cosmological history? As radical as our Copernican demotions from the center of space and time have been, they have largely left humanity untouched in regard to our constitution in contradistinction to our location. A Copernican conception of humanity as a form of emergent complexity places humanity, and the carbon complexity regime of the terrestrial biosphere, in the context of a universe of emergent complexities of various regimes. We define a conception of peer emergent complexity and place the emergent complexities familiar to human beings—life, consciousness, intelligence, technology, civilization—within this context of peer emergent complexity specific to the Stelliferous Era. Placing the Earth-originating carbon complexity regime within a cosmologically conceived class of peer emergent complexities converges on a less anthropocentric conception of what we can expect to find in the pursuit of biosignatures and technosignatures, which categories may prove to be too restrictive. The consequences of these considerations for the search for extraterrestrial intelligence (SETI) will then be explored, with stronger and weaker formulations of the Fermi paradox considered.

Keywords

Big History · Emergent complexity · Universal evolution · Fermi paradox · Anthropocentrism · SETI · Copernicanism · Stelliferous Era · Carbon regime

J. N. Nielsen (✉)
Portland, OR, USA

D. J. LePoire et al. (eds.), *Complexity in Universal Evolution*, World-Systems Evolution and Global Futures, https://doi.org/10.1007/978-3-032-20517-9_15

*This chapter is an extensively re-worked version of the text originally published in Journal of Big History (*Nielsen, 2024*).*

> The question of questions for mankind—the problem which underlies all others, and is more deeply interesting than any other—is the ascertainment of the place which Man occupies in nature and of his relations to the universe of things.
>
> Thomas Henry Huxley, *Evidence as to Man's Place in Nature*, 1863

15.1 Ever Since Copernicus

What is humanity's place in the universe? How are we to locate ourselves within the reaches of space and time that have been opened to us by natural science? Two principles adjacent to natural science have a schematic answer to this question: the Copernican[1] and the cosmological[2] principles. The Copernican principle asserts that there are no privileged observers; the cosmological principle asserts that the universe is isotropic and homogeneous. The two principles complement each other; any observer anywhere in the universe will observe essentially the same universe because the universe is isotropic and homogeneous, and a universe that is isotropic and homogeneous has no place for a privileged observer. Thus, Copernicanism implicitly introduces the idea of peer observers. In a Copernican universe without any privileged observers, any observer of the cosmos other than ourselves is our peer in observation. What a peer observer sees of the universe is essentially the same as what humanity sees of the universe, and vice versa. The transformation of humanity from being the sole observer of the cosmos to being an observer among observers is part of the Copernican Revolution. Humanity's place in the universe has suffered with each iteration of the Copernican Revolution, with human beings exiled to a position ever more peripheral to the wider world we now understand to be the proper object of scientific knowledge. Man, as it turns out, and as against Protagoras, is not the measure of all things, and he isn't even the proper study of man. The proper study of man, or, rather, of *Homo scientificus*, is the more fundamental world that makes the human world possible.

Our Copernican demotions to date have been primarily decenterings in space and time. Copernicus began this process by decentering us in space. Subsequent cosmology bore this out and expanded upon it. The appearance of scientific geology—only a century younger than the Copernican Revolution if we trace its origins to the work of Nicolas Steno (1638–1686)—was part of the formulation of Copernican

[1] "(The strong Copernican principle). There are no privileged observers in the universe" (Ringström, 2013: 6).

[2] "The *cosmological principle* is the hypothesis that, on sufficiently large scales, the Universe can be considered spatially homogeneous and isotropic" (Mo et al., 2010: 102).

natural history, by which we are decentered in time. The Darwinian Revolution was a Copernican Revolution that decentered the kind of beings we are, but emergent complexity holds the dubious promise of going much further than this, decentering us as biological beings from the center of the story of the emergent complexity in the universe, in which biology cannot be presumed to be the rule rather than the exception. This decentering of the complexity distinctive to human beings may be called emergent complexity pluralism.

15.2 Periodizations of Big History

Human beings seek to understand our place in the universe in part through reconstructing the history by which we came to be. As astronomers after Copernicus pushed the boundaries of observation ever outward, demonstrating that we are not in the center of the cosmos, historians push the boundaries of time ever outward, and in so doing demonstrate that we are not at the center of time. Some of the pre-Copernican perspective has been recovered in the discovery of the expansion of the universe, which places each observer at the center of a Hubble sphere, but even this revised recentering of the individual is accompanied by the understanding that there are as many centers and as many Hubble spheres as there are observers, and, as there are no privileged observers, there are no privileged Hubble spheres.

Understanding our place within cosmological history means placing the human period of history in relation to all other periods of time to the extent that we can reconstruct them. For this, we require a periodization that not only embraces human history but also transcends it (LePoire et al., 2025). All historiography employs periodization, but it is a distinctive feature of big history to employ emergent complexity thresholds as the basis for periodization, thus embodying the metaphysical ambition of "carving nature at its joints" (Campbell et al., 2011). How emergent complexity is carved up can vary accordingly as one prefers a coarse or fine parsing of complexity; an ideal periodization would both carve the nature of history at the joints while also serving the pedagogical needs of teachers in the formulation of curricula and the needs of students in keeping materials intuitively accessible.

Perhaps as much for pedagogical as for theoretical reasons, David Christian's periodization of big history into eight thresholds of emergent complexity[3] has come to be adopted by big historians as the basis for narratives in the discipline. A

[3]In *Origins Story: A Big History of Everything*, the eight thresholds of emergent complexity are identified as (1) the big bang, (2) the first stars, (3) nucleosynthesis (new elements), (4) formation of our solar system, (5) life, (6) Homo sapiens, (7) farming, and (8) the fossil fuels revolution, or "The Anthropocene." In other works, these thresholds sometimes are differently identified; for example, in *Maps of Time: An Introduction to Big History*, the eighth threshold is identified as "the Modern Revolution," which Christian notes is "deliberately vague." (p. 335) The extent to which the Anthropocene and the Modern Revolution coincide or overlap is not discussed. Christian does, however, explicitly discuss periodization in an appendix to *This Fleeting World: A Short History of Humanity* (pp. 97–105). Another view of big history periodization can be found in Aunger (2007).

periodization of cosmological history due to Adams and Laughlin (1999) is not unlike big history's periodization based on emergent complexity thresholds, though it might be more accurate to characterize the boundaries between Adams and Laughlin's five ages of the universe as phase transitions in the history of the universe rather than as emergent complexity thresholds (though the two are related, as we will see).

Adams and Laughlin divide the history of the universe into five periods: the Primordial Era, a radiation-dominated era from the big bang through recombination; the Stelliferous Era, dominated by stars and galaxies; the Degenerate Era, dominated by degenerate stellar remains (brown dwarfs, white dwarfs, neutron stars, and black holes), and ending with proton decay; the Black Hole Era, dominated by black holes; and the Dark Era, after the last black holes evaporate, when the universe converges on thermodynamic equilibrium and heat death. The transition from the Primordial Era to the Stelliferous Era corresponds to the second of Christian's emergent complexity thresholds, while the remainder of Christian's eight thresholds occur within the first dozen or so billion years of the Stelliferous Era, which latter Adams and Laughlin project to endure for 10^{14} years.[4]

At this cosmological scale of measuring time, the scale of time employed in big history is but a small fragment of the whole, just as recorded human history is but a small fragment of the whole of big history. Thus, the place of human history within cosmological history is a fragment of a fragment, but it is a fragment that includes the observers who reconstruct cosmological time. Without the observer reconstructing the whole of cosmological history, we lapse into a pre-Copernican conception of history in which the observer's perspective is privileged because other perspectives, potentially occupied by other observers, cannot be formulated without the larger historical context of human observers within big history, and big history within cosmological history. This is the reconstruction of the more fundamental world that makes the human world and our experiences (including observations) possible, and which is therefore "the question of questions for mankind," as Thomas Henry Huxley proclaimed.

Big history's periodization in terms of emergent complexity thresholds is part of a deeper revision to our understanding of our place in nature, since it is part of an as-yet-inchoate and therefore incomplete ontological Copernicanism, such that there is no privileged form of being in the universe. The ontology of a periodization that carves nature at the joints points to an ontology that transcends human observers as privileged observers; if there are no privileged observers, then the *kind* of observer is indifferent, and emergent complexity pluralism points to the kinds of observers there can be, namely peer emergent complexity. The pre-Copernican ontological uniformity of the world is (partially) recovered in the fact that all emergent complexities in

[4]Another periodization due to Eric Chaisson (1987) divides the entirety of cosmological history into three periods: the Energy Era, the Matter Era, and the Life Era. Thus, the first threshold in Chaisson's periodization, between energy and matter, does not appear in the big history periodization of Christian, while the second of Chaisson's thresholds, between matter and life, corresponds to Christian's fifth threshold.

our universe are assembled from the same material components—chemical elements and compounds observable throughout the universe, which account for what kind of observers can appear in a universe like ours.

Christian's eight thresholds of emergent complexity essentially correspond to the Hubble time, with a ninth threshold in the offing at an unspecified date, but with the implication that the date is relevant to human beings or to our successors. As astronomers are said to take galaxies as the basic building blocks of the universe[5] (so that the generalizations of cosmology, such as the expansion of the universe, apply to galaxies, not to stars or to planets, etc.), cosmologists might be said to take the Hubble time as the basic building block of cosmological history, extrapolating the universe many orders of magnitude beyond the age of the universe to date. Generalizations about time in cosmology thus occur at a scale beyond that of big history. But while big history can be nested within a cosmological periodization, the principles of big history, namely periodization by emergent complexity thresholds, can inform a cosmological periodization.

We understand each of Adams and Laughlin's cosmological eras to exhibit a more-or-less coherent and unified character between phase transitions, so that whether or not the phase transitions of this cosmological periodization correspond to emergent complexity thresholds, the material unity of each cosmological period is likely to exhibit its own unique suite of emergent complexities that correspond to the physical processes occurring in each period, and which we will call a *complexity regime*. Certainly the transition from the Primordial Era to the Stelliferous Era represents a transition to a later period in which a greater variety of emergent complexity reigns as compared to the earlier period, but we cannot say whether the transition from the Stelliferous Era to the Degenerate Era will represent a transition to another stage of emergent complexity, or whether the transition will be a simplification in which overall complexity is lost.[6]

It may prove to be the case that the exotic matter that constitutes the degenerate remains of stars will result in emergent complexities during the Degenerate Era we cannot now predict, but we will not consider this question at present. Rather, in the present inquiry we will confine ourselves to the complexity regimes that are likely to be found during the Stelliferous Era of the cosmos, which is characterized by processes involving stars and galaxies and the cycling of matter and energy through stars and galaxies.

[5]This usage is common in cosmology. For example: "...we have also come to regard galaxies as the basic building blocks of the universe, the basic units whereby the largescale structure of the universe is apprehended and quantified." (Faber, ed. 1987: ix)

[6]The last of Adams and Laughlin's periods, if not the last two, correspond to a loss of complexity; if the universe converges on thermodynamic equilibrium and heat death, no emergent complexity will be possible past a given point.

15.3 The Stelliferous Era

When we look up from the surface of Earth into the night sky, we see stars in every direction. When we look farther with telescopes, many orders of magnitude farther than we can see with the naked eye, we see galaxies of stars wherever we look. This distribution of stars and galaxies seems to be homogeneous and isotropic. The homogeneity and isotropy of the cosmos is called the Cosmological Principle. These cosmological homogeneity and isotropy are exhibited to us in the nearly uniform distribution of stars and galaxies in the night sky. In other words, we live in a cosmological age of stars, or the Stelliferous Era. The Stelliferous Era in cosmological history is marked by the formation and evolution of stars and galaxies, which is a function of galactic ecology, which consists, in turn, of the processes by which nucleosynthesis and supernovae transform lighter elements into heavier elements and recycle these heavier elements into later generations of stars and their planets. One of the cosmological arrows of time, in addition to cosmological expansion, is the increasing metallicity of the universe due to the galactic ecology of the Stelliferous Era. This increasing metallicity bears upon the emergent complexities that arise during the Stelliferous Era. (Note that while Adams and Laughlin define the Stelliferous Era about one million—1 Ma.—to ten quadrillion—10,000 Ga.—years after the big bang, the first stars did not form until about 100 million years after the big bang. The 1 Ma starting time is to differentiate it from the earlier Primordial Era from the big bang to recombination.)

The galactic ecology of the Stelliferous Era enriches the universe with heavier elements than those produced by the big bang. Cosmologists distinguish three generations of stars since the big bang, counting backward from population III stars, formed after the appearance of matter (hydrogen and helium); population II stars, formed from the remnants of population III stars; and the current generation of population I stars, formed from the remnants of population II stars. Each successive generation of stars is formed from heavier elements made available from the previous generation of stars' nucleosynthesis and supernovae, though hydrogen still predominates.

According to contemporary theories of stellar evolution, stars and planets form from the same protoplanetary disk, so that later generations of stars incorporating heavier elements also mean later generations of planets incorporating heavier elements. Planets that incorporate heavier elements become more complex and begin to exhibit processes and properties of their own, and these properties and processes of complex planets such as Earth have given rise to the complexity regime familiar to us from Earth.

The human condition in its most general features is defined by the parameters of the Stelliferous Era. We evolved on the surface of a planet orbiting a star, which in turn orbits the Milky Way galaxy. The chemistry of our bodies reflects the chemistry of the clouds of dust and gas that condensed to form our sun and the solar system. Energy from the sun drove the emergence of processes of increasing complexity on Earth, of which we are one result.

15.4 Emergent Complexities of the Stelliferous Era

From our vantage point on Earth, we are able to observe galaxies up to the edge of the visible universe, and, in the case of especially bright supernovae (the kind employed in the attempt to measure the Hubble constant), some stars can be observed at distances in the billions of light years. It is only in the past hundred years that we have possessed the technological capability to distinguish ordinary stars within other galaxies, though improving technologies and techniques of observation push this distance further out every year. Our astronomical observation technologies and techniques have been refined to the point that in the past few decades it has been possible to observe exoplanets in orbit around other stars, thus confirming the non-uniqueness of the planets of our solar system, thus also pointing to the possibility of complex properties and processes on exoplanets, though we are not yet able to observe complexities supervening upon exoplanets.

We, thus, have reason to believe that stars, their planets, and the complex chemical, mineralogical, and geological processes that characterize planetary surfaces during the Stelliferous Era are represented throughout the observable universe. If we reason by analogy, as we reasoned about exoplanets prior to the technological ability to actually detect exoplanets (which is historically extremely recent), we can observe that other nearby galaxies contain stars, and these galaxies are sufficiently similar to the Milky Way galaxy that we would expect the individual stars we can resolve in other galaxies to have planets at about the same frequency with which stars in the Milky Way have planets. Similarly, distant galaxies known to us as a single point of light we assume to be galaxies like the galaxies close to our galaxy, and therefore to be constituted by stars much like the galaxies we know to be constituted by stars, and for these stars to have planets at about the same frequency as those we observe in the Milky Way. There is empirical evidence for none of this, and if someone maintains the contrary, we cannot prove such an individual wrong with empirical evidence. In the absence of empirical evidence, we appeal to the cosmological principle, though the precise extent of validity of the cosmological principle is beyond the scope of our present inquiry.

Further improvements to astronomical technology and observation technique are expected to allow for exoplanet atmospheric spectroscopy in the coming decades (Seager et al., 2012). If we are able to differentiate between false positives of life and unique biosignatures (Schwieterman et al., 2016), we will then be in a position to extend the regime of biochemical emergent complexity throughout the observable universe. We may be able to extend the range of biochemistry within our own solar system if sufficient funding is found to build space probes now on the drawing boards that promise to investigate Mars and the moons and Jupiter and Saturn, all of which present promising opportunities in the search for life beyond Earth.

From what we have seen of the surface of Mars, Io, Titan, and other bodies in the solar system, we know that geological complexity is not exclusive to Earth, and we have no reason to believe this planetary complexity is confined to our solar system exclusively, now that we know there are planets, and many kinds of planets and structures of planetary systems, elsewhere in the universe. Mars once had significant oceans, Io is the most volcanically active body in our solar system, and Titan has liquid lakes on its surface, all of which point to the geomorphological complexity of these bodies, though none of these bodies seems to have the geomorphological complexity of Earth, with its active plate tectonics.

Astronomical observations have also revealed the existence of chemistry, including water and simple organic molecules, in deep space, the study of which is known as astrochemistry.[7] The observations of astrochemistry confirm that organic molecules are present throughout the visible universe, so that the degree of emergent complexity that can be observed as being pervasively present in the universe rises to the level of organic molecules. For the time being, then, our knowledge of the distribution of emergent complexity in the universe stops at astrochemistry, and all further emergent complexities—life, consciousness, intelligence, language, technology, civilization, and so on—are known to us only by what we observe on Earth's surface. Astrochemistry and its emergent complexity precursors obey the cosmological principle, being present homogenously and isotropically throughout the visible universe; all emergent complexities known to us that transcend astrochemistry, that are, or would be, built upon the material foundation of astrochemistry, have the non-Copernican distinction of being observable only geocentrically, on the surface of Earth.[8]

15.5 From Astrochemistry to Planetary Endemism

Earth is an exceptionally complex environment, even without (i.e., before) life. The presence of liquid water on Earth's surface entails the hydrological cycle and many erosion processes that result in sediments and minerals formed from sediments. The presence of magma in the mantle and plate tectonics entails that sedimentary rocks are subjected to temperatures and pressures that transform them into metamorphic rocks, and volcanism transforms the metamorphic rocks into igneous rocks, thus the rock cycle joins the water cycle. The presence of gravity results in gravitational wasting that eventually wears away mountains pushed up by plate tectonic activity. Earth's axial tilt, axial precession, orbital eccentricity, and apsidal precession result in seasonal variations in climate as well as longer variations in the form of Milankovitch cycles. Earth's rotation entails day/night cycles, and seasonal variations and the hydrological cycle entail wet/dry cycles. These cycling processes may be regarded as forcing mechanisms for complexity, both at microscopic and macroscopic levels of selection. (Becker et al., 2019; Campbell et al., 2019).

[7]The greatest mass of water in the observable universe is at a quasar 12 billion light years away in the Lynx constellation (Bradford et al., 2011). The presence of Polyatomic organic molecules including polycyclic aromatic hydrocarbons (PAH) has also been demonstrated at cosmological distances (Kwok & Zhang, 2011).

[8]For some more detail on complexity growth patterns in the Stelliferous Era, see also Chapter "The Evolution of Complexity in the Stelliferous Era: Observations and Patterns" (Korotayev & LePoire, 2026) in this volume.—*Editors.*

The complexity of Earth's surface points to planetary surfaces as especially favorable for the appearance of emergent complexity. Planets themselves are forms of emergent complexity that appear when the boundary conditions of complex planets allow, so that if it is the case that later emergent complexities always supervene upon earlier emergent complexities, we would expect planets to be the locus of further emergent complexity. The surface of Earth in particular is especially complex as a consequence of a geologically active geosphere coupled to liquid water oceans (the hydrosphere) and an atmosphere. Each of these planetary spheres—geosphere, hydrosphere, atmosphere, cryosphere, etc.—because contiguous with the other spheres, possesses an interface where complex chemical interactions take place.

There are a greater number of minerals on Earth than on other bodies in our solar system because of this complexity. Even before there was a terrestrial biosphere, Earth had the rock cycle and the hydrological cycle, which produced distinctive minerals not present elsewhere in the solar system. As the biosphere emerged and evolved, there was also a carbon cycle and a nitrogen cycle, and so on, driving chemical changes and transforming these previously sterile cycles into biogeochemical cycles, with Earth systems bound up in coevolutionary relationships with the biosphere. Planetary complexity of this kind produced minerals like clay,[9] which may have provided a scaffold for early macro-molecules.[10]

Whereas astrochemistry in deep space is detectable by current technologies and techniques of observation, and therefore can be observed at cosmological scales, the more complex chemistries we see on the surface of Earth are not yet accessible to contemporary technologies and techniques at a cosmological scale, thus planetary endemism cannot now be empirically confirmed. Stellar insolation may also drive energy flows on the surfaces of proto-planetary bodies, comets, and asteroids. Organic molecules have been observed in the coma of a comet (Altwegg et al., 2016), and laboratory simulations have shown that polycyclic aromatic hydrocarbons can be synthesized by galactic cosmic ray exposure of low-temperature acetylene ices (Abplanalp et al., 2019). It has also been shown

---

[9] Today clay is primarily the result of biotic processes, but in the Precambrian clay was primarily the result of abiotic processes (Siever, 1992). The action of the biosphere transformed the early abiotic silica cycle from a geochemical cycle into a biogeochemical cycle.

[10] This likelihood of complexity emergent on planetary surfaces can be expressed (Nielsen, 2024) as the Planetary Endemism Thesis (PET), according to which emergent complexities arise on planetary surfaces as a consequence of energy flows in chemical compounds driven by stellar insolation, and the *complexity coupling hypothesis* (CCH), according to which emergent complexities are likely to arise in complex environments even though these environments do not directly enter into the emergent complexity in question. In the example of clay, mentioned above, the mineralogical complexity of clay may provide a catalyst for the complexity of life: "The transition from non-living to living matter may have resulted from the self-organizing properties of organic molecules and their interactions with a chemically rich inorganic environment" (Hanczyc et al., 2006) Clay, as a chemically rich inorganic environment, does not itself come to life, but it may provide an environment in which life can emerge, thus the complexity of life may be coupled with the complexity of clay.

that peptide nucleic acid precursors can form in a wide variety of environments (Rodriguez et al., 2019). Moreover, energy flows might also result from sources other than stellar flux, wherever there is an energy gradient to be exploited, such as gravitational tidal forces, hydrothermal vents, naturally arising fission reactions, and so on. To correct for the many possible qualifications that may need to be made to the presupposition of planetary endemism, we can generalize from planets to any astronomical body of sufficient size for chemical reactions, and we can generalize away from the exclusive reliance on stellar sources of energy.

The idea of planetary endemism captures a widely held if rarely expressed assumption about emergent complexities during the Stelliferous Era.[11] Even if we allow for counter-examples like the possibility of life on rogue planets, deep under a layer of ice, or if life originated at hydrothermal vents without the benefit of energy flows driven by insolation, the planetary-scale biosphere of Earth is unambiguously a result of energy from the sun and matter from Earth. While there may be local exceptions to planetary endemism, in the big picture, stars largely organize the distribution of matter in the universe, and stars themselves are largely organized into galaxies. Most rogue planets will have formed in planetary systems, and most rogue stars outside galaxies will have been thrown clear of an originating galaxy by gravitational forces. Generalizing beyond planetary endemism allows us to exapt our biological intuitions about the origins of life to other conditions that may obtain in our solar system or in other planetary systems. This will expand the scope of possible alternative emergent complexities based on the variety of conditions under which emergent complexities may arise elsewhere. Not all worlds will enjoy energy and matter in abundance, and the conditions that bring these factors into a sustained relationship, as on Earth.

15.6 The Carbon Complexity Regime

The complexity of the terrestrial biosphere, coupled as it is to the terrestrial atmosphere and geosphere, has converged on a stable biochemical regime. The lineage of complexity of Earth is not a form of complexity necessarily *distinctive* to our planet or our universe, but rather one form of complexity among other forms of complexity. We will call this terrestrial lineage the *carbon complexity regime*, in view of the centrality of carbon in terrestrial biochemistry (which is organic chemistry in virtue

[11] As against planetary endemism we can cite Freeman Dyson: "…planets compare unfavourably with other places as habitats. Planets have many disadvantages. For any form of life adapted to living in an atmosphere, they are very difficult to escape from. For any form of life adapted to living in vacuum they are death-traps, like open wells full of water for a human child. And they have a more fundamental defect: their mass is almost entirely inaccessible to creatures living on their surface" (Dyson, 2003). Dyson's reasons for favoring life independent of planets does not alter the fact that distinctive chemistry occurs on planets that does not occur elsewhere because other environments do not have large-scale geomorphological processes; however, Dyson's observations do point to the selective value of life that can adapt to habitats without planets.

of its carbon component). The carbon complexity regime is here defined as the laws and boundary conditions under which carbon-based life appears and flourishes, including the interconnected domains of life, the ecological structures of life, and the biogeochemical structures (themselves biogeochemical regimes) that underlie the carbon regime. The carbon regime has dominated Earth's biosphere since the origins of life based on organic biochemistry. The major divisions of life on Earth—Paleozoic, Mesozoic, and Cenozoic—are particular ecological regimes nested within the larger carbon regime that has prevailed on Earth since the origins of life. None of these regime shifts (Rocha et al., 2014) in the history of life rose to the level of the displacement of the reigning carbon regime, although we cannot rule out this possibility a priori, that is, we cannot exclude the possibility of regime shifts as profound and transformative as the origins of life itself.

The carbon complexity regime is, again, probably not *the* carbon complexity regime in our universe, that is, not the one and the only carbon regime, but one carbon regime among other such regimes. The stable carbon regime on Earth constitutes an attractor in the parameter space of life that excludes other attractors and therefore other regimes. We can identify more specific regimes nested within this larger carbon regime (such as the biogeochemical regimes noted above), and regime shifts within the larger carbon regime, both of which admit of degrees. For example, the marsupial mammalian regime of Australia differs from the placental mammalian regime of the other inhabited continents, but it differs to a lesser degree than the mammalian regime differs from the dinosaurian regime. Some regime shifts seem to be an expression of the internal feedback of gradually increasing complexity that becomes manifest at an inflection point (as with the Cambrian Explosion), while other regime shifts conform to the model of a disturbance due to an external shock—literally so with the Chicxulub impactor and the following K–Pg mass extinction, which ended the dinosaurian regime and heralded the mammalian regime.[12] If the mass extinctions of the Permo-Triassic, which divided the Paleozoic regime from the Mesozoic regime, and the K-Pg, which divided the Mesozoic regime from the Cenozoic regime, are both due to external shocks that resulted in regime shift, this means that the major divisions of life on Earth are contingent and not likely to be duplicated on another planet with a biosphere.

15.7 Peer Complexity Beyond the Carbon Regime

Given an ontological Copernicanism in which there are no privileged forms of being, degrees of complexity can still be recognized as an objective feature of the world (Nielsen, 2025), and the concept of peer complexity then becomes important, since

[12] If, as hypothesized, the End Permian mass extinction was due to super volcano eruptions, whether we regard this as internal feedback from the extant biogeochemical regime, or an external shock that disturbs this biogeochemical regime but lying outside that biogeochemical regime proper is a function of how we define the scope of a regime in a given context.

any peer to human complexity could be, like us, an observer who defines a perspective on the universe (a peer observer in the Copernican sense discussed previously).

Two instances of complexity are here identified as peers when they belong to distinct complexity regimes yet possess approximately equivalent capacities. This informal definition relies heavily on the unexplicated notion of equivalent capacities.[13] Intuitively, these capacities might include familiar features of life such as locomotion, perception, niche construction, and intelligence, but in order to avoid an explicit reference to life and life-specific properties, it would be preferable to formulate equivalent capacities more abstractly. For example, equivalency could be expressed in terms of position within a trophic pyramid (still a biological concept, but less specific than, say, perception), effective range of the emergent complexity (related to locomotion, but not the same), or having a comparable number of emergent thresholds between the origin of the universe and the instance of complexity in question (i.e., the *generation* of emergent complexity in question). These distinct conceptions of equivalency could yield different definitions of peer complexity.

The distinctness of one complexity regime from another is a matter of degree, with less comprehensive regimes nested within more comprehensive regimes. Thus the dinosaurian regime prior to the K-Pg mass extinction is distinct from the mammalian regime that followed the K-Pg mass extinction, but both are stable states within the carbon regime of the terrestrial biosphere, or, even more narrowly, the DNA regime of Earth (as there may be carbon regimes that are not DNA regimes). In these two distinct regimes, the dinosaurian and the mammalian, we can identify peer species as those occupying the same position in the trophic period or, what is related, inhabiting the same or similar ecological niche or niches. There have been many speculative and some practical attempts to extend and expand the known carbon regime (cf. note 16 below), and even within the carbon regime there are multiple distinct definition of life (Nielsen, 2023) that could be taken to identify peer forms of life insofar as distinct definitions do not have precisely the same extension.

There has been speculation on whether a dinosaur species could have followed an evolutionary pathway of encephalization similar to hominids among mammals (Lumsden & Wilson, 1983: 154). Such a species, had it evolved, would have constituted a peer in the sense defined above as having a comparable number of emergent thresholds between the origin of the universe and the instance of complexity in question. These stages might need to be defined both quantitatively (as in the

[13]The hand of a human being might be considered a rough functional equivalent to the chelae (claws or pincers) of a crustacean insofar as both a hand and a pincer can manipulate an organism's environment. This is a qualitative comparison. The relative utility of hand and pincer could be characterized quantitatively by a number metrics, such as fineness of motor skills, adaptability to a variety of tasks, grip strength (here the coconut crab, *Birgus latro*, has the upper hand, so to speak), or number of moving parts (whether parts of an endoskeleton or an exoskeleton), which latter is a measure of mechanical complexity. The pincer is a gripping appendage in the exoskeletal regime; the hand is a gripping appendage in the endoskeletal regime.

case of encephalization) and qualitatively (as in the case of equivalent capacities), but lacking a general theory of emergence, we don't have the analytical framework necessary to describe this with precision, so we are forced to employ qualitative conceptions until quantitative concepts become available to us. (Field building in emergent complexity theory is a necessary requisite to progress here.)

15.8 Peer Emergent Complexities

While the search for biosignatures is effectively the search for the carbon complexity regime wherever it may occur in the universe, regardless of whether any further emergent complexities arise from the terrestrial lineage, the search for technosignatures is the search for human *peers* on the cosmological scale. The search for biosignatures regresses us backward in the sequence of emergent complexities to whatever constitutes the cosmological scale distribution of post-astrochemical emergent complexities.[14] The search for technosignatures begins with technological peers (sometimes defined by the ability to build radio telescopes) and progresses from that point of development further toward whatever fate may hold for human civilization and the technologies we produce. Biosignatures and technosignatures, then, are understood to represent particular nodes on a sequence of emergent complexities that are linked in time through descent with modification.

Implicit in the project of SETI, the search for extraterrestrial intelligence, is that the intelligence for which we search is a *peer* intelligence, not entirely incommensurable with human intelligence, in possession of a peer technology and a peer civilization, built by a peer species having evolved in a peer biosphere of a peer planet. The idea of a peer planet is often expressed as an Earth twin, mirror Earth, or Earth 2.0; the ideas of peer intellects, peer technologies, peer civilizations, peer species, and peer biosphere are less familiar than that of a peer planet. Part of the idea of a peer biosphere is captured in the idea of biosignatures, while part of the idea of a peer civilization (or, at very least, a peer technology) is captured in the idea of technosignatures.

SETI, as the search for technosignatures of extraterrestrial intelligence, if successful, would allow us to skip the sequence of intervening stages of emergent complexity between astrochemistry and radio telescopes (the skipped stages being the less familiar peers of our terrestrial sequence of emergent complexities) and thus arrive suddenly at an observational confirmation that all of the precursor emergent complexities prior to radio telescopes are distributed as widely in the universe as stars, galaxies, planets, and organic molecules. The absence of obvious technosignatures that would follow from terrestrial emergent complexities being

[14]"Post-astrochemical complexity" is a generational conception of complexity as complexity having at minimum surpassed the observable astrochemical background complexity, that is, having attained a post-astrochemical emergent complexity threshold and thus having defined a new complexity regime.

distributed according to the cosmological principle is the occasion of the Fermi paradox. This we will set aside for the moment, but will return to later.

If we were to discover technosignatures that indicated emergent complexities such as those known from Earth are also inferred to exist elsewhere in the universe, we would recognize the source of these technosignatures as peers of some sort—peer emergent complexity. But how widely (or narrowly) should the idea of peer emergent complexity be construed? Who, or what, constitute our peers during the Stelliferous Era of the universe? Ought these peers to be limited to some carbon regime? Ought these peers be limited to what human beings can recognize as living? What peer emergent complexities ought to be expected to populate the Stelliferous Era?

In so far as astrochemistry is known to be distributed according to the cosmological principle, astrochemistry should constitute the building blocks for further emergent complexities that supervene upon astrochemistry. Organic molecules exist throughout the visible universe, so what kinds of complexity can be built up from organic molecules, assuming also the availability of other forms of matter, known to exist in the form of planets or even interstellar objects such as "Oumuamua, and the availability of energy flows, known to exist primarily in the form of radiation from stars?"

15.9 The Entangled Bank of Emergent Complexity

The most common alternative conception to that of the wide distribution assumption of peer emergent complexities in formulations of the SETI scenario is that of life being widely represented in the universe, but none of this life being sufficiently complex to communicate (i.e., none of it sufficiently complex to build radio telescopes). The failure to observe unambiguous technosignatures suggests that other technological civilizations are rare or non-existent. Knowing that the astrochemical precursors of biology are distributed pervasively throughout the universe, and assuming that the developmental sequence that links astrochemistry to radio telescopes is tightly constrained by the possible emergent complexities at each threshold, implies that the universe may be populated with many living worlds, but few technological worlds. An intuitive picture of this state of affairs may be obtained by the image of a universe populated with stromatolites, in which Earth is the only planet that continued beyond this rudimentary stage of biological development. However, this scenario fails to consider the possibility of quasi-biological[15] or

[15]"Quasi-biological" is here used to indicate emergent complexities that possess most of the qualitative properties of life (cf. the discussion of definitions of life in sect. 15.13, below) but with a biochemistry distinct from Earth's DNA regime. Thus, quasi-biological entities could span a spectrum from life at the outer edges of the carbon regime (the potential "surprises" of Martian life in Salisbury, 1962), through the further divergencies from terrestrial life anticipated by Barricelli (Barricelli, 1961) and Ward's "life as we do not know it" (Ward, 2005), to recognizable forms of life not part of the carbon complexity regime.

non-biological post-astrochemical peers branching off from the same point at which biology (i.e., biochemistry) emerges from astrochemistry.

The assumption here is that, if the astrochemical precursors to life do not cross the emergent complexity threshold of life, then these precursors remain simple and no other form of emergent complexity arises in place of life. In the scenario in which emergent complexity is the most tightly constrained in its developmental sequence, astrochemistry *exclusively* serves as the precursor to life, and if life does not emerge from astrochemistry, then no further complexity occurs; life serves *exclusively* as the precursor to sensation (the evolution of a central nervous system—CNS), and if sensation does not evolve, then no further complexity occurs; sensation serves *exclusively* as the precursor to consciousness (subjective self-awareness supervening on the CNS of a biological being), and if consciousness does not evolve, then no further complexity occurs; consciousness serves exclusively as the precursor to language... and so on, up through the developmental sequence of emergent complexity familiar to us from Earth and the construction of radio telescopes. Projecting indistinguishable peers at a cosmological scale implies re-tracing this developmental sequence of emergent complexity without branching off into alternative peers.

Implicit in tightly-constrained developmental sequences of emergent complexity is the idea of a ladder of progress rather than a branching bush. Writing specifically about biology, Stephen Jay Gould[16] observed,"...evolution is a copiously *branching bush* with innumerable present outcomes, not a highway or a ladder with one summit." (Gould, 2011, p. 21) And, "...with Darwinism, you don't even have the chain anymore; you have this accidentally constructed luxuriant, tangled bank—the *branching bush.*" (Gould & Hamrum, 1983, p. 107) Gould in this passage is implicitly invoking the "entangled bank" passage in the final paragraph of the *Origin of Species*:

> It is interesting to contemplate an entangled bank, clothed with many plants of many kinds, with birds singing on the bushes, with various insects flitting about, and with worms crawling through the damp earth, and to reflect that these elaborately constructed forms, so different from each other, and dependent on each other in so complex a manner, have all been produced by laws acting around us.[17]

Gould's reasoning (and Darwin's reasoning) can also be applied to emergent complexity, which should be understood as a branching bush rather than as a ladder

---

[16] Gould formulated a "big picture" conception of history like that of big history, but he represents, in a sense, the antithesis of big history, as in Gould's big picture conception of natural history emergent complexity is relegated to a long tail that remains mostly irrelevant to the bulk of life, which is microbial. (We find a similar conception in E. O. Wilson, who once said of ants, "we need them to survive, but they don't need us at all," emphasizing the asymmetry of the relationship between life in the small and more complex life.) Gould de-emphasizes emergent complexity, making it an inevitable emergence due to the "drunkard's walk." Big history, by contrast, focuses on emergent complexity as the central interest we take in history, as well as providing a scaffolding for big history periodization. Cf. section 15.2 above.

[17] Darwin 1859; this is the language of the first edition, which was modified in subsequent editions.

of progress. In contrast to the tightly-constrained developmental sequence sketched above, in which each precursor threshold exclusively (and perhaps also inevitably) defines and facilities one and only one successor threshold, at each threshold in the process of emergence we must postulate the possibility of alternative emergent complexities branching away from the emergent complexities of the carbon regime that are familiar to us. There may be cases of thresholds that *do* tightly constrain successor thresholds, but also cases of thresholds that do *not* tightly constrain their successors, so that multiple distinct forms of complexity may emerge from one and the same precursor.[18]

As in evolution distinct species may have a common ancestor, so too distinct emergent complexities may share a common emergent complexity precursor, and given our state of knowledge of the distribution of astrochemistry on cosmological scales, and the likely role of organic molecules in the origins of life, it would be natural for us to look to astrochemistry as the common precursor to multiple forms of emergent complexity that may populate the Stelliferous Era. The entangled bank of emergent complexity entails quasi-biological and non-biological post-astrochemical emergent complexities as peers of terrestrial biology, insofar as they share the common ancestor of astrochemistry and emerged by different pathways from it.

15.10 Wide Peer Classes: Peer Mammalian Intelligence

From our experience of emergent complexity on Earth's surface, we have become accustomed to making fine distinctions among classes of existents. Our system of biological taxonomy is almost Byzantine in its subtlety, a reflection of the biodiversity of the terrestrial biosphere. Most human beings look only to other human beings as being our peers, though some who have studied natural history would be willing to perhaps include other primates and cetaceans in our terrestrial peer group as intelligent mammals.

Human beings are much more intelligent than our nearest biological relatives, the chimpanzees, and we have built far more sophisticated technology than any other species on the planet, but measured in terms of encephalization (for lack of a better measure), we do not enjoy an order of magnitude advantage over chimpanzees or dolphins. If the peer groups of intelligent mammals were to be divided according to order of magnitude differences in encephalization, we would be classed with a number of primates and cetaceans as our intelligent peers, and, from the perspective

[18]We predict that emergent complexities on a cosmological scale will exemplify a normal distribution, so that both exclusively constrained thresholds and completely unconstrained thresholds will be rare, whereas the bulge in the bell curve of emergent complexity will be constituted by loosely constrained thresholds that may give rise to a handful of emergent complexities, and this we term the *Threshold Constraint Hypothesis* (TCH). We further predict that exploration of potential biospheres within our solar system will provide limited evidence for TCH, but that a full understanding of threshold constraint will not be obtained until the universe entire has been surveyed by scientific methods.

of an intellect removed from ours by an order of magnitude, or by several orders of magnitude, this peer group would seem to have ample justification.

We can cast our net even wider by conceptualizing mammalian intelligence on the whole, all supervening on a common mammalian brain structure inherited from a common ancestor, as *peer mammalian intelligence*, with whatever metric we employ to measure the intelligence of the many species of mammal probably describing a bell curve, with intelligence among mammals representing a normal probability distribution, and our previous peer group—human beings, primates, and cetaceans—occupying the high intelligence end of the curve, all of the latter perhaps the same number of standard deviations from the mean, while the mammals with the lowest encephalization occupy the low intelligence end of the curve, and the rest of the mammal species distributed throughout the bulge in the bell curve. While high-end mammalian intelligence is noticeably greater than low-end mammalian intelligence, the average mammal is more intelligent that the average reptile, and the average reptile is more intelligent than the average arthropod. In other words, mammalian intelligence constitutes a recognizable and coherent peer class.

Here we have employed encephalization as a proxy for intelligence. Even if encephalization is a poor or misleading proxy for intelligence, whatever basis we may employ for the measurement and quantification of intelligence, we could employ this measure narrowly or broadly. Some other proxy for intelligence—say, complexity of tools and machines manufactured by that intelligence—could be similarly employed to define a narrow peer class of intelligence or a wide peer class of intelligence. For example, taking tool complexity as a measure, we know that tool use is pervasive in the biosphere; other primates, several bird species, and even sea otters routinely make use of tools. However, these tools are simple. Sea otters use unmodified rocks to crack open shells, but even this simple tool use leaves archaeological traces (Haslam et al., 2019). Chimpanzees represent greater complexity in tool use, as they will strip a branch of leaves before using this tool to extract termites from a termite mound.[19]

There is an ongoing discussion among archaeologists and anthropologists about the relative complexity of Neanderthal tool-making, especially in comparison to the known complexity of tool-making among *Homo sapiens*. Two recent papers appeared one after another each arguing a different side of the question (Schmidt et al., 2019; Niekus et al., 2019). The Schmidt et al. paper argues that, "...in archaeological science in general, arguing for abstract concepts like modernity or complex cognition in past populations should not rely solely on highly interpretative models of the production pathways of specific material finds. It should rather rely on the interpretation of the actually performed steps, as proven by direct archaeological data." The Niekus et al. paper holds, on the contrary, "Neandertals invested

[19]Chimpanzee tool use is not restricted to stripped branches (cf. Sugïyama et al., 1988, Ant-Catching Wands of Wild Chimpanzees at Bossou, Guinea. Folia Primatologica; Boesch & Boesch, 1990, Tool Use and Tool Making in Wild Chimpanzees. Folia Primatologica). For a wider consideration of chimpanzee tool use in the context of chimpanzee culture, cf. Whiten et al., 1999, Cultures in chimpanzees, Nature.

considerable time and resources in domestic tools and activities. The regular performance of logistically complex, cognitively demanding production processes provides important evidence on the evolution and transmission of complex technology." What the two papers share is a conception of the increasing complexity of technology serving as a proxy for cognitive complexity.

Other measures instead of, or in addition to, encephalization and technology could include social group interaction (presumably inferred from material culture), complexity of social customs (again inferred from material culture), the presence of particular brain structures (e.g., layers of the neocortex or the internal structure of these layers), complexity of signaling (which, like tool use, is pervasive in the biosphere), and so on. Whatever metric is employed, a narrow peer concept is entailed by taking human examples as normative, while a wide peer concept is possible though less anthropocentric definitions of some quantifiable measure.

15.11 Peer Emergent Complexity as a Wide Peer Class

The utility of wide peer classes will become evident when we consider the observation of emergent complexities at cosmological scales. Over cosmological scales of observation, detail is lost and we are left with a low resolution profile of that which we observe. Wide peer classes naturally apply to a variety of phenomena that are identical at low resolution, understanding that further distinctions can be made within a wide peer class if and when our technology and techniques of observation improve or when our ability to travel to some phenomena (or *closer* to some phenomena) for the purpose of direct observation improves. When we apply the concept of a wide peer class reflexively to ourselves, which is implicit in the remit of SETI, we obtain a wide peer class of emergent complexities like ourselves, with "like" being construed as widely as possible, and these are our peer emergent complexities of the Stelliferous Era (peer observers in the Copernican sense).

Our first approximation of peer emergent complexities of the Stelliferous Era—other biological beings that have evolved intelligence and technology sufficient to build radio telescopes and therefore to engage in electromagnetic signaling over cosmological distances—constitutes a narrow peer class that assumes a developmental sequence of emergent complexity that paralleled the development of emergent complexity of Earth. This narrow peer class we will call *indistinguishable peers*, as when observed over interstellar distances these peers would be indistinguishable from ourselves.

We observe here without elaborating that the narrow peer emergent complexity concept implicit in SETI points to civilization as the missing piece of the SETI puzzle. Any artifact or technosignature detectable on interstellar distances would have to be product of industry at the scale of civilization—if not the product of civilization as we know it, the product of something that is the indistinguishable peer of civilization. SETI presupposes civilization and often mentions civilization, but has formulated no theory of civilization to explain exactly what it is we are looking for. The remarks on civilization found in the SETI literature are non-falsifiable due to their lack of detail, and yet they implicitly underlie the entire enterprise. Expansion upon this observation must be left for another occasion.

Narrowly conceived peers are indistinguishable at cosmological scales of magnification, and the search for biosignatures and technosignatures takes place at just such scales. For an intuitive picture of indistinguishable peers, we can invoke the *squint test*: step back from a given phenomenon and squint; features previously distinguishable become indistinguishable while squinting. Our distance from potential SETI sources, and the imperfection of our observation, means that we observe at a low resolution over interstellar scales. There is a scale of magnification at which indistinguishable peers are distinguishable, but this is not the scale here under consideration—the scale of cosmological visibility and interstellar interaction.

Indistinguishable peers have a special fascination for us, as we have already imagined the possibility of communication with indistinguishable peers taking the form of sharing science, philosophy, literatures, and so forth.[20] However, when observing for emergent complexity distributed on cosmological scales, we should not confine ourselves to a search for indistinguishable peers, but rather we should construct a wide peer concept for emergent complexities of the Stelliferous Era in order to cast a wider net than a search for indistinguishable peers.

What would a wide peer class of emergent complexity, that is, emergent complexities sufficiently like us that they would fall within the remit of SETI, broadly construed, look like? The wide peer emergent complexity concept might include:

- Branching away from an astrochemical point of origin that constitutes our common ancestor (being, like us, post-astrochemical emergent complexity)
- Serving as a point of origin for other forms of emergent complexity (being, like us, in turn, a threshold for further emergent complexities)
- Exhibiting planetary-scale behavior (emergent complexities that exemplify planetary endemism) or planetary system-scale behavior (emergent complexities that exemplify complexity coupled to an underlying environment, independent of planetary endemism)
- Producing detectable signatures over interstellar distances

This list of peer emergent complexity properties is to be understood as provisional, not exhaustive or definitive.

At each threshold of emergent complexity, there will be peers in the sense of belonging to the same wide peer class but which would be identical at a low level of resolution, that is, over interstellar distances. Our peers in this wide sense would span a variety of biological (the carbon regime), quasi-biological (life beyond the carbon regime), and non-biological emergent complexities (non-living alternative complexities with shared astrochemical ancestors), as well as the further emergent

[20] Freeman Dyson calls this the "Philosophical Discourse Dogma," in which, "...the universe is filled with societies engaged in long-range philosophical discourse" (Dyson, 1979: 207).

complexities that arise from these alternative possibilities. This gives us not only different kinds of entities that we could count as peers on a cosmological scale but also different sequences of development from astrochemistry up through a galaxy-spanning presence.

The wide peer class of Stelliferous Era emergent complexities conceived as a normal distribution would span a range from the rarity of some emergent complexity being endemic to a single planet and not being observable beyond that planet, to a common case of being visible on a cosmological scale (given adequate technological means and technical skills), and the again rare instances of emergent complexity not only visible on a cosmological scale but also distributed on a cosmological scale. Given this normal distribution, the emergent complexities of Earth likely represent the common case of cosmological visibility to some degree, and therefore exemplify the principle of mediocrity.[21]

15.12 Alternative Emergent Complexities Populating the Wide Peer Class

There have been many intimations of alternative emergent complexities, in the imaginations of scientists and science fiction writers alike, but it is only with emergent complexity pluralism, given exposition in this chapter, that we possess a theoretical framework within which divergent forms of emergent complexity can be studied in one unified context and integrated into the history of the universe.

The history of the universe is significant in this context as distinct emergent complexity developmental sequences appear to correspond with distinct stages in the history of the universe, as implied earlier in the discussion of the distinct complexity regimes of the Stelliferous Era. Abraham Loeb (2014) has postulated the possibility of life in the earliest history of the universe, 10–17 million years following the big bang, bypassing some of our prejudices about the Stelliferous Era, in which we find ourselves, and in which we expect life to appear only after other developments have occurred before life appears.

While Loeb's conception of life early in the history of the universe does not explore alternative forms of life—Loeb specifies only, "the chemistry of 'life as we know it'"—it would be expected that the lower metallicity regime of the early universe, with its different abundances of elements, would skew toward forms of life, perhaps even quasi-biological or non-biological emergent complexities, distinct from those of the later universe, as we also would expect that even later in the history of the universe, perhaps during the Degenerate Era, forms of emergent complexity arising at this time would differ from those emergent complexities of the Stelliferous Era. Loeb's formulations are concerned with the Stelliferous Era, as we are here, but

[21] The principle of mediocrity can be understood as another principle adjacent to natural science, like the cosmological and Copernican principles, which raises the question of whether some complementary holds between mediocrity and the Copernican or cosmological principles, or both.

at a period of time so early in the Stelliferous Era that different conditions would have obtained than later in the Stelliferous Era.

First proposed by Sagan and Salpeter, Atmospheric Habitable Zones (AHZ) in gas giant atmospheres were conceptualized as biology as we know it, but biology conceived in a radically distinct context, without being predicated upon a planetary surface or the chemical interactions of a planetary surface (a conception of life that contravenes planetary endemism). The idea of an AHZ has been further elaborated in Yates et al. (2017) and Lingam and Loeb (2019b). Lingam and Loeb make this scenario for life even more exotic by taking it out of the planetary context first proposed by Sagan and Salpeter and postulating the possibility of an AHZ in the atmosphere of a brown dwarf substellar body.

Stellivores—"a civilization that feeds actively on stars"—have been proposed by Vidal (2016) as a form of emergent complexity that maximizes energy rate density.[22] An account of Stellivores could be given in terms of further development from an indistinguishable peer of humanity (Vidal's "developmental path," that is, an account consistent with the exclusively biological origins of further emergent complexity), but Vidal also allows for the possibility of a non-biological developmental sequence culminating in stellivores: "...a non-organic form of life may thus be advanced or not compared to us, it would just be different, and happened to have self-organized at a stellar scale."

Because we are biological beings with a biological bias, we have formulated even alternative forms of emergent complexity in terms of life, even when these conceptions have little to do with biology. Thus, it was with "crystalline life" as conceived by Desiderius Papp almost a hundred years ago, who, drawing on the research of Otto Lehmann on liquid crystals, postulated the possibility of crystals that had made the transition to a further level of complexity, beyond the growth, regeneration, and reproduction that they exhibit and that they share with biological systems (Brandstetter, 2012).

The point of these thought experiments in life early in the universe, in AHZs, at increasing levels of energy density, and crystalline life, is to expand the conceptual framework we employ in trying to grasp the permutations that emergent complexity may take in our universe. The expansion of our conceptual framework as a result of critical reflection upon problematic cases of quasi-biological and non-biological emergent complexities embodies Einstein's hopeful observation that, "There could be no fairer destiny for any physical theory than that it should point the way to a more comprehensive theory, in which it lives on as a limiting case" (Einstein, 2006). Our current distinction between the living and the non-living, between biochemistry and inorganic chemistry, may someday be understood as a limiting case within a more comprehensive theory. Life would then be a node in a network of complex

[22] There is some resemblance between Vidal's stellivores and John Smart's Transcension Hypothesis (Smart, 2012), which latter also emphasizes increasing energy rate density as the likely developmental trajectory of emergent complexity.

chemistries with interesting properties, rather than the one, true higher emergent complexity arising from anterior astrochemical complexity.

If we were able to study such ambiguous phenomena up close, eventually specialist sciences would be formulated that would focus on each distinctive form of emergent complexity; each form of emergent complexity will necessitate a science of its own. Just as there are Egyptologists who specialize in the study of ancient Egyptian civilization, Byzantinists who specialize in the study of Byzantine civilization, and Sinologists who study Chinese civilization, so there would be crystallists who would specialize in crystalline life and stellivorists who would specialize in the study of stellivores. It is not necessary to try to reduce these possible forms of emergent complexity to a generic conception of life (namely, the carbon regime), which may be peculiar to Earth and closely similar worlds (perhaps limited to other worlds within our own solar system, connected to each other by lithopanspermia).

15.13 A Counterintuitive Scenario for Peer Emergent Complexity

In the spirit of formulating thought experiments that expand our conceptual framework for emergent complexities in the universe, we will pursue the same course in formulating a thought experiment intended to expand our conceptual framework for peer emergent complexity during the Stelliferous Era. To this end, we formulate a speculative scenario for peer emergent complexity without life, without consciousness, without intelligence, and without technology. The intention here is to radically diverge from our expectations for peer emergent complexity as embodied in past SETI practice. Let us set the stage for our thought experiment by considering, in turn, non-life that resembles life (quasi-biological and non-biological emergent complexity), non-consciousness that resembles consciousness, non-intelligence that resembles intelligence, and non-technology that resembles technology.

In order to understand the possibility of quasi-biological or non-biological emergent complexity, as in the case of crystalline life discussed above in Sect. 15.12, we note that life has been defined in a number of ways, none of which have proved to be entirely satisfactory in a theoretical sense (Nielsen, 2023). In the actual practice of identifying life we refer to a litany of properties usually possessed by living systems, and, if a sufficient number of properties are present, we identify something as living, effectively employing a recursive definition in which we proceed something like, "x is a living system if and only if x has F and x has G and x has H. . ." and so on for some set of properties {F, G, H,. . .}that characterize living systems. If we add to this recursive definition of life an extremal clause—"and nothing else is a living system"—we have then excluded the possibility of quasi-biological and non-biological systems being identified as life.

Excluding non-conforming properties (or the non-conforming absence of properties) of quasi-biological or non-biological post-astrochemical emergent complexities simplifies the task of constraining our concepts within conventional bounds, but it does not resolve the problem of how we classify emergent complexities that do not fit into a conventional conceptual framework. There are always problematic cases, as, for example, with viruses. Whether or not viruses are living organisms has been long debated. Most biologists do not consider viruses to be living organisms because they can reproduce only within the cell of a host; in other words, viruses lack the biological property of autonomous replication. However, virology is understood to be a specialization of microbiology, while microbiology is a specialization of biology, meaning that viruses are studied as a part of biology. Moreover, a minority of biologists identify viruses as "non-cellular life."

The taxonomic problem posed by viruses may be multiplied manifold by artificial biology, which may produce unclassifiable variations on biology here on Earth before we achieve the deprovincialization of biology that will be afforded by our first glimpse of life on another world.[23] We may, by our own industry, produce materials that are lifelike but that are not life as we know it. An example of this already has been created in the laboratory:

> We have created a dynamic material powered by artificial metabolism using simultaneous processes of biochemical synthesis and dissipative assembly... We succeeded in constructing machines from this novel dynamic biomaterial with emergent regeneration, locomotion, and racing behaviors... (Hamada et al., 2019)

The ability to manufacture quasi-biological artificial DNA (Doi et al., 2008) and non-biological materials such as DNA-based Assembly and Synthesis of Hierarchical materials (DASH) (Hamada et al., 2019) and inorganic chemical cells (iCHELLs) (Cooper et al., 2011) that overlap but do not coincide with biology, because they do not have all the same properties of biological systems, suggests the possibility of these or similar materials occurring naturally under conditions that overlap but do not coincide with the conditions under which terrestrial biology involved.[24] This, then, we will understand to be non-life that is nevertheless *like* life.

As for non-consciousness that is like consciousness, and non-intelligence that is like intelligence, the distinctive character of human intelligence can be defined by an overlap of biological individuality, consciousness, emotion, technological adeptness, language, shared mammalian instincts and inclinations, as well as shared kinesthetic sensations and sensory experiences with other mammals. Take away any of these properties, or several of them, or add to these properties, and the resulting intelligence would depart from peculiarly human intelligence to the degree that we have modified the distinctive baseline by which our intelligence is defined. As with life, the human mind is implicitly understood in terms of a recursive

[23] "The study of a single instance of extraterrestrial life, no matter how humble, will deprovincialize biology." (Sagan, 1980, p. 41) Sagan used both "deprovincialization" and "deparochialization" to express what is more commonly known as the principle of mediocrity (on which cf. note 22).

[24] "The conditions for developing organisms with many of the properties considered characteristic of living beings, by evolutionary processes, do not have to be similar to those prevailing on Earth" (Barricelli, 1961).

definition, though we are much less clear about the properties that characterize mind than we are about the properties that characterize life.

Given that human consciousness and human intelligence are emergents from human biology, any change in the underlying biology would imply a change in the supervening consciousness or intelligence. Thus, in the case of non-life that resembles life, should any such instance become sufficiently complex that consciousness or intelligence supervenes upon it, one could expect that these forms of consciousness and intelligence would depart from human consciousness and intelligence to some degree proportionally as non-life departs from life.

Examples of non-consciousness that resemble consciousness and non-intelligence that resemble intelligence are also widely represented in the literature of artificial intelligence (AI). This issue has been treated so extensively in other contexts that we will only briefly touch upon the matter here, and anyone interested in more detailed discussions of consciousness and intelligence vis-à-vis AI will find an ample bibliography in any current paper on the topic. The Turing test, in particular, has been a point of fascination. Some are willing to cut the Gordian Knot that asserts that any machine that passes the Turing test ought to be identified as conscious, regardless of whether or not it conforms to the human (or, more widely, biological) norms of consciousness.

Finally, Raup has identified the possibility of non-conscious intelligence—noting that, "The kind of problem-solving used by animals may be called *nonconscious intelligence* in contrast to the *conscious intelligence* of humans" (Raup, 1992)—which would constitute a kind of intelligence distinct from human intelligence, which is implemented by a conscious mind. There are, then, many ways of thinking about consciousness and intelligence that suggest alternatives to consciousness and intelligence as they are possessed and utilized by human beings.

In regard to non-technology that resembles technology, many of the ideas from Raup (1992) apply here as well. Raup states, "Many animals on Earth exhibit feats of engineering which are functionally indistinguishable from the technology produced by human intelligence." This would hold true not only for animal life not originating on Earth but also for quasi-biological and non-biological emergent complexities, which could produce functional equivalents of technology. An indirect suggestion of what this might look like in practice can be found in Bialy and Loeb (2018), who sought to explain the apparent acceleration of the interstellar object "Oumuamua":

> If radiation pressure is the accelerating force, then 'Oumuamua represents a new class of thin interstellar material, either produced naturally, through a yet unknown process in the ISM or in protoplanetary disks, or is of an artificial origin (Bialy & Loeb, 2018).

Bialy and Loeb focused on the possibility of 'Oumuamua being of artificial origin, and suggested that the acceleration might be due to, "...a lightsail, floating in interstellar space as debris from advanced technological equipment," but the alternative that they have offered here of a natural origin for a material sufficiently thin that it could serve as a solar sail points to a kind of emergent complexity that could distribute itself on a cosmological scale. If a material such as that described in the Hamada et al. paper was to emerge and take the form of a membrane that could act as a light sail, such a material would distribute itself first within its home planetary system. Subject to selection pressures for those entities that arrived at the farthest reaches of its home system, some eventually would have the capacity to travel to the nearest stars. Again, under selection pressures, interstellar journeys would select for those entities that could make longer journeys, eventually populating the home galaxy, and possibly spreading to other galaxies.

If such a material evolved from the astrochemistry present in the halo of a comet, in a planetary system well stocked with unusually large comets, so that these comets passed material among themselves as they passed in and out of the inner planetary system where stellar insolation could drive the processes of emergent complexity in the material substrate of the comets. The periodicity of comets could, in addition, serve as the kind of forcing mechanism discussed above in Sect. 15.5, so that any emergent complexity in this environment would be repeatedly stimulated to respond to the changes in its environment.

A form of emergent complexity that distributes itself from its point of origin throughout its home system, and then its home galaxy, and then the universe, would benefit from the coordination of its activities. We know that radio signals can be propagated over interstellar and even inter-galactic distances, as natural radio sources are detectable over cosmological scales. The most distant radio galaxy yet discovered was found near the observation limit of the observable universe, about 12 billion light years away (Saxena et al., 2018). Given this capacity for signaling over cosmological distances, radio would be an ideal coordination mechanism, perhaps produced by a process similar to bioelectricity, with the light sail surface itself having a dual purpose as the equivalent of an antenna array.[25] Thus, there would be a selective effect in terms of the medium of communication employed in the distribution of peer emergent complexity over interstellar and intergalactic distances. Even an emergent complexity without consciousness, intelligence, or civilization as we know it that nevertheless had to fulfill some coordinating function could evolve radio signals to do so on cosmological scales.

Raup, again, anticipated this, explicitly discussing the phenomenon of bioelectricity, stating, "...the ingredients for effective radio communication are clearly available in the animal world and it may only be a matter of time until true radio communication is discovered in our own biosphere." Radio as a coordinating mechanism, first for planetary system-wide distribution of some emergent complexity, and then later for interstellar and intergalactic distribution of that emergent complexity, could serve a function not unlike the function of pheromones in a termite mound. (Ocko et al., 2019) One could imagine a colony of quasi-biological emergent complexity adapted to life in space, with something like the queen of the

[25]Radio signaling is not the only possibility for a coordinating function among peer emergent complexities; other portions of the electromagnetic spectrum could serve a similar purpose. Infrared (IR) radiation can penetrate through the gas and dust of the interstellar medium and thus has certain advantages as a signaling mechanism.

colony emitting radio signals instead of pheromones. Each time the "queen" molts the process of ecdysis allows it to grow in size, the radio signals become more powerful and so could direct the coordination of the colony on system-wide and then interstellar scales. Only the most successful colonies would grow to a scale that these radio signals would be detectable over interstellar distances, but the planetary system of origin of such a peer would select for this kind of success, or for extinction.

As a biological analogy of close coordination of complex entities, the superorganism represented by the hives of eusocial honey bees, which coordinate foraging by means the "waggle dance," is not understood to be a language in the form of grammatically-structured communication, but nevertheless serves a coordinating function for a hive and constitutes an effective signaling mechanism in the absence of more complex signaling as in human language and its analogues.[26] Something like the waggle dance translated into electromagnetic signals, directing the workers of the colony to energy and material resources, would be the kind of signal transmitted by the expanding colony such as we have described in our present thought experiment. A quasi-biological superorganism of this kind, with mechanisms for growth, cosmological scale distribution, and coordination, would be an emergent complexity radically different from terrestrial emergent complexity, but still our peer in terms of a wide peer class of emergent complexities, and would be moreover positioned for aggressive expansion in the cosmos.

It should be emphasized that this thought experiment must not be considered to be any more plausible than any number of alternative emergent complexity scenarios. This is merely a thought experiment to suggest that kind of wide peer class emergent complexities that might be possible. The human imagination is limited in its ability to conceive emergent complexities that diverge significantly from terrestrial emergent complexities, as human minds have been shaped by the selection pressures of these terrestrial emergent complexities; we think in terms of terrestrial emergent complexities. Only a sustained and purposeful effort to transcend our imaginative limitations can deliver us from the prison of our own minds.

15.14 Mediocrity and Emergent Complexity

Our discussion of this scenario for peer emergent complexity during the Stelliferous Era without biology, consciousness, intelligence, technology, or civilization is not to be understood as definitive or distinctive. By definition, scenarios exemplifying peer emergent complexity fulfill the principle of mediocrity, and so our scenario here should be understood as a mediocre exemplification of peer emergent complexity. Our ability to formulate peer emergent complexity scenarios is limited by our biological and anthropocentric imagination, but this is an epistemic limit and not

[26]The bee "waggle dance" may be more sophisticated that was previously believed to be the case. A recent paper (Howard et al., 2019) demonstrates symbolic representation of numerosity in *Apis mellifera*.

an ontological limit. There are more things in the heavens than are dreamt of in our astrobiology.

The point is not to focus on any one scenario, but to multiply peer emergent complexity scenarios through variations on every possible parameter of our thought experiments. The Fermi paradox will not be resolved by the formulation of a single knock-down argument that will banish all other contenders from the field; the Fermi paradox will be resolved by populating the conceptual space of peer emergent complexity as densely as possible, allowing us to explore and eventually to exhaust this conceptual space, which must be transformed from *terra incognita* into a region that is exhaustively mapped and intimately known.

When we have densely populated the conceptual space of peer emergent complexity with the totality of its possibilities, we will then be able to distinguish within these possibilities classifications among them, some of which will be more common than others, some less common. Emergent complexities of the Stelliferous Era will likely eventually come to describe a normal distribution, with the most common forms clustered near the center, and the least common forms occupying the increasingly sparsely populated slopes at each side of the center. With emergent complexities located several standard deviations from the center, we would expect there to be few exact peers, but many low-resolution peers.

At this point, we may be justified in recovering the sense of human exceptionalism if it can be shown that human beings and the complexity regime associated with Earth and its carbon complexity regime constitute an outlier among the forms of emergent complexity. It is entirely possible that humanity and our civilization may be a peer to other post-astrochemical emergent complexities, while at the same time being an outlier in possessing, simultaneously, conscious biological intelligence of a technological bent that augments its capabilities by using tools. As an outlier emergent complexity, we would expect that our cosmological situation would give the appearance of isolation according to a narrow conception of peers, though a mature conception of emergent complexity pluralism reveals us to be part of a wide peer class of Stelliferous Era complexity regimes.

15.15 A Stronger Formulation of the Fermi Paradox

Given the possibility of peer emergent complexity that diverges considerably from the carbon complexity regime familiar on Earth—so divergent that these emergent complexities may not recognizably constitute life or consciousness or intelligence or technology or civilization—it is necessary to re-formulate the Fermi paradox in the light of these possible alternative emergent complexities. While alternative emergent complexity may be unfamiliar and unprecedented from a terrestrial perspective, it may be more widely represented in the universe than indistinguishable peers.

Placing ourselves in the context of a universe rich in peer emergent complexities, even if impoverished in respect to indistinguishable peers, the Fermi paradox can be formulated in such a way as to include peer emergent complexities that are not indistinguishable peers. Thus, not only have we detected no signs of indistinguishable emergent complexity, nor found any traces of indistinguishable emergent complexities within our planetary system, we also have not detected any signs of wide peer class emergent complexity, nor any traces of wide peer class emergent complexity within our planetary system. That is to say, not only have we not received a classic SETI transmission nor have we discovered anything like a Bracewell probe in our planetary system; we also have not received a transmission (or signal) from an alternative or peer emergent complexity, nor have we found any traces on Earth of any previous presence of a galaxy-spanning emergent complexity regime different from ourselves and the carbon regime that characterizes the terrestrial biosphere. To the extent that we have surveyed our solar system with spacecraft, we also have observed nothing from these probes that would betray the presence, now or in the past, of some emergent complexity peer having passed through our solar system.

In this strong formulation of the Fermi paradox, which adds to the absence of indistinguishable peers such as SETI has been seeking, also the absence of entities that would represent the wide peer class of the Stelliferous Era complexity regime, we are not responsible merely for constraining the emergent of indistinguishable peers so as to explain our cosmological isolation in a plausible way, but we must also explain all of the inexplicable emergent complexities that have not found us, nor have we found them. Here we not only need to account for the lack of other technological civilizations that build radio telescopes but also have to account for the lack of incomprehensible and radically divergent emergent complexities such as formulated in our thought experiment of Sect. 15.13.

15.16 A Weaker Formulation of the Fermi Paradox

While a stronger formulation of the Fermi paradox in terms of peer emergent complexity would seem to underline human isolation, in so far as we have no evidence of peer emergent complexity, much less any evidence of indistinguishable emergent complexity, there is another sense of SETI in terms of peer emergent complexity that suggests we cannot assume human isolation in the cosmos. Most SETI efforts focus on technosignatures that would be, by definition, indistinguishable peers of human civilization. Technosignatures as conventionally understood would be a product of a peer civilization employing a peer technology constructed by peer intelligence.

As we have seen in our thought experiment (Sect. 15.13), any peer emergent complexity present on a cosmological scale (or potentially present on a cosmological scale), which would benefit from coordination (which need not be centralized coordination), and this coordination would be made possible over cosmological distances by signaling mechanisms, could produce signatures detectable over interstellar distances. A search for such signatures would not be SETI as it is understood today, but rather the search for alternative or wide peer class emergent complexity. The detection of such a signal by means intended to detect technosignatures would pose significant interpretive problems; the mechanism of radio transmission would be understood, but the content of the signal would not look anything like that which an indistinguishable peer intelligence would transmit either for its own benefit or as an attempt to communicate across cosmological distances.

We do not even know how to begin such a search, or, if some kind of search found some kind of ambiguous signal, how we would go about interpreting the meaning of that signal. Thus, the other side of the coin of a stronger formulation of the Fermi paradox is a weaker formulation of the Fermi paradox in which the signatures of peer emergent complexity may be so different from that which we are expecting that phenomena of which we are already aware and which we attribute to natural, mechanistic causes may be signatures of peer emergent complexity for which we have no cognitive categories in which to assign them a place and thus to integrate them into our conceptual framework. It is possible, after all, that we live in a universe dense with peer emergent complexity, and that we are oblivious to the fact.

Lingam and Loeb (2019a), citing Raup (1992), note that, "...the possibility that the boundaries between biosignatures and technosignatures could become blurred must be borne in mind," and, "...certain biosignatures could end up being conflated with technosignatures." The bulk of the paper, however, focuses on the proportionality of spending on biosignature versus technosignature detection and does not fundamentally question the distinction between biosignatures and technosignatures. The authors are more effective in noting how different technosignatures may be conflated ("...we observe that the proposed distinction between intentional electromagnetic signals, artifacts at the level of human technology, and megastructures ought not be regarded as being clear-cut") and how artificial technosignatures may be mistaken for natural EM signals ("It is quite conceivable that artifacts, especially megastructures, are already present in our catalogs but have not been identified yet").

Referring to the above quoted paper as LL18, Ćirković et al. (2019) affirm a fundamental distinction between biosignatures and technosignatures:

> ...we have fully accepted the tacit premise of LL18 (universally accepted in the current astrobiology/SETI discourse) that biosignatures are clearly and unambiguously distinguishable from technosignatures. One interesting speculative possibility which perhaps deserves more attention—and is anyway necessary for the logical closure of discussions of detectability—is that the distinction between biological and technological might eventually be erased.

Ćirković et al. focus on the possibility of the biosignature/technosignature distinction being effaced by future (future for human beings) post-biological developments, which could already have been realized by a civilization older than ours. However, the paper does not discuss the possibility of the biosignature/technological distinction being called into question by alternative forms of emergent complexity that may not be easily classifiable as either biology or technology. We argue that the biosignature/technosignature distinction must be fundamentally questioned; the very idea of a biosignature/technosignature distinction reifies the categories of emergent complexity derived from our experience in the terrestrial biosphere, which may or may not be representative of the Stelliferous Era complexity regime (further, its representativeness is predicated upon how widely the peer class is construed).

Technosignatures could be mistaken for naturally occurring EM radiation, and naturally occurring EM radiation could be mistaken for a technosignature, and, in fact, several times have been so mistaken, or, at least, an artificial source been given as at least one explanation among others for a given EM signal, later determined to be natural sources.[27] This is obvious for technosignatures, but not as obvious for biosignatures. However, artificial biosignatures could be similarly mistaken for naturally occurring life, and it is possible that naturally occurring life could be mistaken for something artificial in origin, or something artificially (i.e., intelligently) cultivated at the location or locations where it is observed.

If we should someday find other life in the universe, we may well find instances of life that occur in what seem to us to be highly unlikely places, as indeed we find life on Earth in what seem to us to be unlikely places. There will be a tendency to attribute apparently unlikely distribution of life to directed panspermia, in the same way that the design argument has appealed to the apparent unlikelihood of life on Earth as proof of divine intervention, which latter is simply directed panspermia given a supernatural formulation.

The question of biosignature gases in exoplanet atmospheres can be extended to other gases—say, gases we attribute to industrial pollution—that might be generated abiotically, biotically, or artificially, so that the search for biosignatures overlaps with the search for planetary-scale technosignatures. Similar concerns hold for other planetary-scale biosignatures. While biosignature detection of vegetation red edge (VRE) has been suggested as a possibility relatively recently (Seager et al., 2005; Takizawa et al., 2017; O'Malley-James & Kaltenegger, 2018), it is plausible that the refinement of VRE biosignature technology and observation techniques could distinguish between a wild biosphere and a cultivated biosphere, that is, a planet with the greater part of its arable land given over to agricultural production, which would constitute a planetary-scale technosignature for an agricultural civilization. VRE detection on multiple planets of a single planetary system could be an indication of natural mechanisms of distributing life within closely-spaced planets within a circumstellar habitable zone[28] or it could be a technosignature of multiple planets having been intentionally bioengineered by a civilization capable of planetary-scale engineering, as well as being capable of interplanetary travel.

---

[27] The discovery of pulsar PSR B1919 + 21 prompted speculation that its signals were artificial in origin (Hewish et al., 1968; Bell Burnell, 1977). Similarly, Kardashev wrote of radio sources CTA-21 and CTA-102, subsequently understood to be quasars, that they "...exhibit a spectrum highly similar to the anticipated artificial spectrum" (Kardashev, 1964). When fast radio bursts and the unusual light curve of Boyajian's star were initially identified, there was speculation of artificial sources for these observations as well.

[28] As has been suggested for systems such as TRAPPIST-1 where there are seven planets located in the CHZ; on fast or enhanced lithopanspermia in such planetary systems cf. Krijt et al., 2017; Lingam & Loeb, 2017.

Emergent complexity signatures will be inherently ambiguous and therefore open to interpretation, making it difficult to maintain any sharp distinction between biosignatures and technosignatures. The interpretation of empirical evidence obtained by scientific methods will occur within a larger scientific, social, cultural, and event biological context. As we have noted above, human beings are biological beings, and we think in biological terms. Our human biological bias will be a factor in our interpretation of evidence about the universe that does not conform to our expectations, that is, evidence that is ambiguous because it does not immediately or obviously exemplify our conceptual framework, or any conceptual framework of an indistinguishable peer.

15.17 Emergent Complexity Pluralism

Our investigation into that emergent complexity likely to characterize the Stelliferous Era complexity regime points to the importance of recognizing *emergent complexity pluralism* as the framework for understanding humanity's place in the universe, and, with greater development, it has not escaped our notice that the framework of emergent complexity pluralism may provide insights into the Fermi paradox that are not to be had from more conventional approaches to the problem working within an established framework (largely defined by the carbon complexity regime). Given the alternative framework of emergent complexity pluralism, an emergent complexity threshold need not uniquely determine the successor form of complexity, so that any one form of emergent complexity constitutes a particular pathway from the simplest elements present after the big bang to later generations of emergent complexity. We prioritize investigations into human precursor emergent complexity because this is the pathway that led to our existence, but it is not the only pathway that nature might take. As noted in Sect. 15.6 above, we cannot exclude the possibility of a planetary-scale regime shift that results in a non-linear complexity sequence.

For all the rigor we attempt to bring to scientific understanding, in the end we are human beings, we are interested in ourselves as human beings, and our science reflects this. However, if our science is to be more than omphaloskepsis, we must find a way to transcend our anthropocentric limitations. What is most needed at the present time is to populate the landscape of wide peer class emergent complexity, and this means developing detailed thought experiments that vary as many parameters of the known as is consistent with emergent complexity in the Stelliferous Era, then iterating this procedure by working through variations of our variations, until the later thought experiments diverge indefinitely from the earlier thought experiments, eventually finding ways to transform these thought experiments into actual empirical experiments to the extent that this possible. It is only when, at long last, we arrive at actual empirical experiments that our hypotheses can be put to the test.

The implied research program into emergent complexity pluralism discussed in the this chapter would not only involve populating the conceptual space of alternative emergent complexities but would also constitute a struggle against the pervasive anthropocentrism reflected in our science in understanding not only our place in the universe but the universe itself. This research program is in part scientific, and may be pursued exclusively in the context of natural science, but it is also in part philosophical insofar as we would be engaged in defining an ontological Copernicanism in which there is no privileged form of being. All sciences in their foundations are unavoidably philosophical, and the research program here proposed is effectively a new science that requires a philosophical foundation. Moreover, the question of questions for mankind, according to T. H. Huxley, leads us to an even more profound question: *What kind of universe do we inhabit?* Placing our universe within a taxonomy of possible universes is both a more comprehensive and a more fundamental parallel to placing human beings in the context of peer emergent complexities. Moreover, we cannot fully place human beings in the context of emergent complexity pluralism until we have a full knowledge of the emergent complexity structure of our universe, which, if we understood it, would define our universe in relation to the parameter space of possible universes. For example, the kind of universe that produces only the carbon complexity regime, or which produces only a single instance of the carbon complexity regime, would be distinct from the kind of universe in which a robust emergent complexity pluralism prevails.

We recall the Quito doubloon nailed to the mast of the Pequod in Melville's *Moby Dick*, "...this round gold is but the image of the rounder globe, which, like a magician's glass, to each and every man in turn but mirrors back his own mysterious self" (Chap. 99). In looking outward at the universe, and seeking something like ourselves with which we might communicate and experience some degree of cosmic fellowship, more often than not we look into ourselves and see mirrored back to us our own mysterious selves. We can retain this sense of mystery while overcoming the anthropocentrism of omphaloskepsis through a rigorous research program into emergent complexity pluralism.

References

Abplanalp, M. J., Frigge, R., & Kaiser, R. I. (2019). Low-temperature synthesis of polycyclic aromatic hydrocarbons in Titan's surface ices and on airless bodies. *Science Advances, 5*(10), eaaw5841. https://doi.org/10.1126/sciadv.aaw5841

Adams, F., & Laughlin, G. (1999). *The five ages of the universe: Inside the physics of eternity*. The Free Press.

Altwegg, K., Balsiger, H., Bar-Nun, A., Berthelier, J.-J., Bieler, A., Bochsler, P., et al. (2016). Prebiotic chemicals—Amino acid and phosphorus—In the coma of comet 67P/Churyumov-Gerasimenko. *Science Advances, 2*(5), e1600285–e1600285.

Aunger, R. (2007). A rigorous periodization of "big" history. *Technological Forecasting and Social Change, 74*(8), 1164–1178. https://doi.org/10.1016/j.techfore.2007.01.007

Barricelli, N. A. (1961). Prospects and physical conditions for life on Venus and Mars. *Scientia, 96*.

Becker, S., Feldmann, J., Wiedemann, S., Okamura, H., Schneider, C., Iwan, K., et al. (2019). Unified prebiotically plausible synthesis of pyrimidine and purine RNA ribonucleotides. *Science, 366*(6461), 76–82. https://doi.org/10.1126/science.aax2747

Bell Burnell, S. J. (1977). Petit Four. *Annals of the New York Academy of Sciences, 302*(1), 685–689. https://doi.org/10.1111/j.1749-6632.1977.tb37085.x

Bialy, S., & Loeb, A. (2018). Could solar radiation pressure explain "Oumuamua"s peculiar acceleration? *The Astrophysical Journal, 868*(1), L1. https://doi.org/10.3847/2041-8213/aaeda8

Boesch, C., & Boesch, H. (1990). Tool use and tool making in wild chimpanzees. *Folia Primatologica, 54* (1–2), 86–99. https://doi.org/10.1159/000156428

Bradford, C. M., Bolatto, A. D., Maloney, P. R., Aguirre, J. E., Bock, J. J., Glenn, J., et al. (2011). The water vapor Spectrum of APM 08279+5255: X-ray heating and infrared pumping over hundreds of parsecs. *The Astrophysical Journal, 741*(2), L37. https://doi.org/10.1088/2041-8205/741/2/l37

Brandstetter, T. (2012). Life beyond the limits of knowledge: Crystalline life in the popular science of Desiderius Papp (1895–1993). *Astrobiology, 12*(10), 951–957. https://doi.org/10.1089/ast.2011.0779

Campbell, J. K., O'Rourke, M., & Slater, M. H. (2011). *Carving nature at its joints: Natural kinds in metaphysics and science*. The MIT Press.

Campbell, T., Febrian, R., McCarthy, J., Kleinschmidt, H., Forsythe, J., & Bracher, P. (2019). Prebiotic condensation through wet-dry cycling regulated by deliquescence. *Nature Communications, 10*, 4508. https://doi.org/10.1038/s41467-019-11834-1

Chaisson, E. (1987). *The life era: Cosmic selection and conscious evolution*. The Atlantic Monthly Press.

Ćirković, M. M., Vukotić, B., & Stojanović, M. (2019). Persistence of Technosignatures: A comment on Lingam and Loeb. *Astrobiology, 19*(10). https://doi.org/10.1089/ast.2019.2052

Cooper, G. J., Kitson, P. J., Winter, R., Zagnoni, M., Long, D. L., & Cronin, L. (2011). Modular redox-active inorganic chemical cells: iCHELLs. *Angewandte Chemie, 123*(44), 10557–10560. https://doi.org/10.1002/anie.201105068

Doi, Y., Chiba, J., Morikawa, T., & Inouye, M. (2008). Artificial DNA made exclusively of nonnatural C-nucleosides with four types of nonnatural bases. *Journal of the American Chemical Society, 130*(27), 8762–8768. https://doi.org/10.1021/ja801058h

Dyson, F. (1979). *Disturbing the universe*. Harper & Row.

Dyson, F. J. (2003). Looking for life in unlikely places: Reasons why planets may not be the best places to look for life. *International Journal of Astrobiology, 2*(2), 103–110.

Einstein, A. (2006). *Relativity: The special and general theories*. Plume.

Faber, S. M. (Ed.). (1987). *Nearly Normal galaxies from the Planck time to the present*. Springer.

Gould, S. J. (2011). *Full House: The spread of excellence from Plato to Darwin*. Harvard University Press.

Gould, S. J., & Hamrum, C. L. (1983). *Darwin's legacy*. Harper & Row.

Hamada, S., Yancey, K. G., Pardo, Y., Gan, M., Vanatta, M., An, D., et al. (2019). Dynamic DNA material with emergent locomotion behavior powered by artificial metabolism. *Science Robotics, 4*(29).

Hanczyc, M. M., Mansy, S. S., & Szostak, J. W. (2006). Mineral surface directed membrane assembly. *Origins of Life and Evolution of Biospheres, 37*(1), 67–82. https://doi.org/10.1007/s11084-006-9018-5

Haslam, M., Fujii, J., Espinosa, S., Mayer, K., Ralls, K., Tinker, M. T., & Uomini, N. (2019). Wild Sea otter mussel pounding leaves archaeological traces. *Scientific Reports, 9*(1). https://doi.org/10.1038/s41598-019-39902-y

Hewish, A., Bell, S. J., Pilkington, J. D. H., Scott, P. F., & Collins, R. A. (1968). Observation of a rapidly pulsating radio source. *Nature, 217*(5130), 709–713. https://doi.org/10.1038/217709a0

Howard, S. R., Avarguès-Weber, A., Garcia, J. E., Greentree, A. D., & Dyer, A. G. (2019). Symbolic representation of numerosity by honeybees (*Apis mellifera*): Matching characters to small quantities. *Proceedings of the Royal Society B, 286*, 20190238. https://doi.org/10.1098/rspb.2019.0238

Kardashev, N. (1964). Transmission of information by extraterrestrial civilizations. *Soviet Astronomy, 8*(2), 217–221.

Korotayev, A., & LePoire, D. J. (2026). The evolution of complexity in the Stelliferous era: Observations and patterns. In D. J. LePoire, L. Grinin, & A. Korotayev (Eds.), *Complexity in universal evolution—A Big History perspective* (pp. 151–172). Springer Nature (in this volume). https://doi.org/10.1007/978-3-032-20516-2_5

Krijt, S., Bowling, T. J., Lyons, R. J., & Ciesla, F. J. (2017). Fast Litho-panspermia in the habitable zone of the TRAPPIST-1 system. *The Astrophysical Journal, 839*(2), L21. https://doi.org/10.3847/2041-8213/aa6b9f

Kwok, S., & Zhang, Y. (2011). Mixed aromatic–aliphatic organic nanoparticles as carriers of unidentified infrared emission features. *Nature, 479*(7371), 80–83. https://doi.org/10.1038/nature10542

LePoire, D. J., Grinin, L., & Korotayev, A. (2025). Navigating complexity in big history—Exploring periodization across cosmic and biosocial dimensions: An introduction. In D. J. LePoire, L. Grinin, & A. Korotayev (Eds.), *Navigating complexity in big history. Exploring periodization across cosmic and biosocial dimensions* (pp. 1–27). Springer Nature. https://doi.org/10.1007/978-3-031-85410-1_1

Lingam, M., & Loeb, A. (2017). Enhanced interplanetary panspermia in the TRAPPIST-1 system. *Proceedings of the National Academy of Sciences, 114*(26), 6689–6693. https://doi.org/10.1073/pnas.1703517114

Lingam, M., & Loeb, A. (2019a). Relative likelihood of success in the search for primitive versus intelligent extraterrestrial life. *Astrobiology, 19*(1), 28–39. https://doi.org/10.1089/ast.2018.1936

Lingam, M., & Loeb, A. (2019b). Brown dwarf atmospheres as the potentially Most detectable and abundant sites for life. *The Astrophysical Journal, 883*(2), 143. https://doi.org/10.3847/1538-4357/ab3f35

Loeb, A. (2014). The habitable epoch of the early universe. *International Journal of Astrobiology, 13*(04), 337–339.

Lumsden, C. J., & Wilson, E. O. (1983). *Promethean fire: Reflections on the origin of mind.* Harvard University Press.

Mo, H., van den Bosch, F., & White, S. (2010). *Galaxy formation and evolution.* Cambridge University Press.

Niekus, M., Kozowyk, P., Langejans, G., Ngan-Tillard, D., van Keulen, H., van der Plicht, J., Cohen, K., van Wingerden, W., van Os, B., Smit, B., Amkreutz, L., Johansen, L., Verbaas, A., & Dusseldorp, G. (2019). Middle Paleolithic complex technology and a Neandertal tar-backed tool from the Dutch North Sea. *Proceedings of the National Academy of Sciences, 116*(44), 22081–22087. https://doi.org/10.1073/pnas.1907828116

Nielsen, J. N. (2023). The life and death of habitable worlds. In C. Tandy (Ed.), *Death and anti-death, volume 21: One year after James lovelock (1919–2022).* Ria University Press.

Nielsen, J. N. (2024). Peer complexity in big history. *Journal of Big History, 8*(1), 83–98. https://doi.org/10.22339/jbh.v8i1.8111

Nielsen, J. N. (2025). An approach to constructing a big history complexity ladder. In D. J. LePoire, L. Grinin, & A. Korotayev (Eds.), *Navigating complexity in Big History. Exploring periodization across cosmic and biosocial dimensions.* Springer Nature. https://doi.org/10.1007/978-3-031-85410-1_12

Ocko, S. A., Heyde, A., & Mahadevan, L. (2019). Morphogenesis of termite mounds. *Proceedings of the National Academy of Sciences, 116*(9), 3379–3384. https://doi.org/10.1073/pnas.1818759116

O'Malley-James, J. T., & Kaltenegger, L. (2018). The vegetation red edge biosignature through time on earth and exoplanets. *Astrobiology, 18*(9), 1123–1136. https://doi.org/10.1089/ast.2017.1798

Raup, D. M. (1992). Nonconscious intelligence in the universe. *Acta Astronautica, 26*, 257–261.

Ringström, H. (2013). *On the topology and future stability of the universe.* Oxford University Press.

Rocha, J. C., Biggs, R., & Peterson, G. (2014). Regime shifts: What are they and why do they matter. *Regime Shifts Database, 20*(1), 681–697.

Rodriguez, L. E., House, C. H., Smith, K. E., Roberts, M. R., & Callahan, M. P. (2019). Nitrogen heterocycles form peptide nucleic acid precursors in complex prebiotic mixtures. *Scientific Reports, 9*(1). https://doi.org/10.1038/s41598-019-45310-z

Sagan, C. (1980). *Cosmos*. Random House.

Salisbury, F. B. (1962). Martian biology: Accumulating evidence favors the theory of life on Mars, but we can expect surprises. *Science, 136*(3510), 17–26. https://doi.org/10.1126/science.136.3510.17

Saxena, A., Marinello, M., Overzier, R. A., Best, P. N., Röttgering, H. J. A., Duncan, K. J., et al. (2018). Discovery of a radio galaxy at $z = 5.72$. *Monthly Notices of the Royal Astronomical Society, 480*(2), 2733–2742. https://doi.org/10.1093/mnras/sty1996

Schmidt, P., Blessing, M., Rageot, M., Iovita, R., Pfleging, J., Nickel, K. G., et al. (2019). Birch tar production does not prove Neanderthal behavioral complexity. *Proceedings of the National Academy of Sciences, 116*(36), 17707–17711. https://doi.org/10.1073/pnas.1911137116

Seager, S., Schrenk, M., & Bains, W. (2012). An astrophysical view of earth-based metabolic biosignature gases. *Astrobiology, 12*(1), 61–82. https://doi.org/10.1089/ast.2010.0489

Seager, S., Turner, E. L., Schafer, J., & Ford, E. B. (2005). Vegetation's red edge: A possible spectroscopic biosignature of extraterrestrial plants. *Astrobiology, 5*(3), 372–390. https://doi.org/10.1089/ast.2005.5.372

Siever, R. (1992). The silica cycle in the Precambrian. *Geochimica et Cosmochimica Acta, 56*(8), 3265–3272. https://doi.org/10.1016/0016-7037(92)90303-z

Sugïyama, Y., Koman, J., & Sow, M. B. (1988). Ant-catching wands of wild chimpanzees at Bossou, Guinea. *Folia Primatologica, 51*(1), 56–60. https://doi.org/10.1159/000156357

Schwieterman, E. W., Meadows, V. S., Domagal-Goldman, S. D., Deming, D., Arney, G. N., Luger, R., et al. (2016). Identifying planetary biosignature impostors: Spectral features of CO and O4 resulting from abiotic O2/O3 production. *The Astrophysical Journal, 819*(1), L13. https://doi.org/10.3847/2041-8205/819/1/l13

Smart, J. M. (2012). The transcension hypothesis: Sufficiently advanced civilizations invariably leave our universe, and implications for METI and SETI. *Acta Astronautica, 78*, 55–68. https://doi.org/10.1016/j.actaastro.2011.11.006

Takizawa, K., Minagawa, J., Tamura, M., Kusakabe, N., & Narita, N. (2017). Red-edge position of habitable exoplanets around M-dwarfs. *Scientific Reports, 7*(1). https://doi.org/10.1038/s41598-017-07948-5

Vidal, C. (2016). Stellivore extraterrestrials? Binary stars as living systems. *Acta Astronautica, 128*, 251–256. https://doi.org/10.1016/j.actaastro.2016.06.038

Ward, P. (2005). *Life as we do not know it: The NASA search for (and synthesis of) alien life*. Penguin.

Whiten, A., Goodall, J., McGrew, W. C., Nishida, T., Reynolds, V., Sugiyama, Y., et al. (1999). Cultures in chimpanzees. *Nature, 399*(6737), 682–685. https://doi.org/10.1038/21415

Yates, J. S., Palmer, P. I., Biller, B., & Cockell, C. S. (2017). Atmospheric habitable zones in Y dwarf atmospheres. *The Astrophysical Journal, 836*(2), 184. https://doi.org/10.3847/1538-4357/836/2/184

J. N. "Nick" Nielsen is an independent scholar from Portland, Oregon, with interests in big history, space philosophy, philosophy of history, technology, astrobiology, civilization, and the future. He has spoken at many events, including the Hundred Year Starship Symposium, Icarus Interstellar Starship Congress, International Big History Association, Network of Researchers on the Chemical Emergence of Life, The Society for Social and Conceptual Issues in Astrobiology, Forming and Exploring Habitable Worlds, and the International Society for the Comparative Study of Civilization. He has published in the Journal of Space Philosophy, Journal of Big History, Heroism Science, and the Journal of Virtual Reality.

Part VI

Civilization Complexity

Myth, Complexity, and Big History: Rethinking Our Shared Map for the Future

16

Ken Baskin

Abstract

David Christian envisioned Big History as a modern creation myth—a shared story that helps humanity navigate an ever-changing universe. As E.O. Wilson described it, this "evolutionary epic" can offer a unifying perspective on reality, much like traditional myths have done for past societies. In today's world, where existential crises such as climate change demand new ways of thinking, a shared mythic map may be more crucial than ever. This chapter argues that integrating complexity science with Big History can deepen our understanding of this modern myth. Complexity science explores the dynamic patterns that shape evolving systems, while Big History tells the story of those evolving systems over time. By applying complexity science, we can see that societies throughout history—including during the Axial Age and Modernity—have faced transformational challenges and successfully adapted by rewriting their foundational myths. To make this case, this chapter examines four key issues: (1) Christian's concept of a shared mythic map and its neurobiological foundations, (2) Complexity science's insights into the structure of the universe, (3) The dynamic processes that drive the evolution of complex systems, and (4) The potential impact of consciously integrating complexity science into Big History. By rethinking our shared map through this lens, we may gain valuable insights into how societies can navigate today's global challenges and shape a more sustainable future.

This chapter is an extensively re-worked version of the text originally published in Journal of Big History (Baskin, 2024).

K. Baskin (✉)
Philadelphia, PA, USA

D. J. LePoire et al. (eds.), *Complexity in Universal Evolution*, World-Systems Evolution and Global Futures, https://doi.org/10.1007/978-3-032-20517-9_16

Keywords

Myth · Big History · Universal evolution · Complexity · Axial Age

16.1 Introduction

At the heart of Big History is David Christian's vision that the story of our universe from the Big Bang to the present, which E.O. Wilson had described as our "evolutionary epic" (1978),[1] could become a "modern creation myth" (2004, p. 1). In other words, Christian hoped that this "origin story" could function like the myth of more traditional societies, providing the "shared map" that shows people in any society how to live in a "rich, beautiful, and sometimes terrifying universe" (2018, p. 8).

It's hard to imagine a time when such a shared map has been needed more. My morning news feed often seems like a list of existential challenges that, as Einstein explained, can't be solved with the logic that created them. For example, we've known for more than three decades that global warming could make much of the world unlivable for our 8 billion planetary co-inhabitants. Yet, in spite of international conferences and treaties, conditions have only continued to deteriorate. What we need, I want to argue, is a new shared map—with a new logic—that might actually help us make the changes we need in the face of these challenges.

More than that, I want to argue that integrating the principles of complexity science with Big History can give us a deeper understanding of this modern myth. These two disciplines are, after all, complementary. To a large extent, complexity science is the study of the *patterns that emerge* as "complex systems" evolve. And Big History explores *the story that emerges* as those complex systems evolve. Above all, this story is extremely dynamic. Complexity science describes the specifics of these dynamics in ways that can enhance Big History's creation myth. In applying complexity science to our myth, we can see, first, that ours is not the only society to face such existential challenges and, second, that societies that have overcome those challenges have partly done so by reinventing their shared maps!

I reached this conclusion in writing *The Axial Ages of World History* (2014) with Dmitri Bondarenko. In that book, we explored the similarities between the Axial Age (c. 800–200 BCE) and Modernity (1500 CE-the present). Both periods, we agreed were times when increases in population, advances in technology and communication, and rising available wealth combined to overwhelm the dominant social structures. Both periods would witness vast social experimentation, horrifying warfare, and the emergence of new ways of thinking about the world. The most surprising discovery we made was that the transformations we studied in both

[1] For an extensive examination of how a variety of thinkers treat this evolutionary epic, see Genet et al. (2009).

periods were largely driven by the elites in the societies and involved *rewriting their mythologies*.

We also agreed that the twenty-first century seemed similar to the end of the Axial Age. At that time, the societies that successfully transformed themselves—China, India, Israel, and Greece—learned the lessons of their transformative experiences and defined the ways of thinking about the world that would bring them into the age of empires that followed the Axial Age.[2] Moreover, we speculated that, just as their Axial Age transformations made it possible for these societies to thrive in an age of empires, the modern transformation could enable societies around the globe to enter an age of globalization. These conclusions came as a direct result of our application of what we'd learned from complexity science.

All of which suggests a fascinating application of complexity science to Big History. In this chapter, I argue that applying complexity science can deepen our understanding of our shared map. To make my case, I examine four central issues:

- Christian's idea of a mythic shared map of reality and its neurobiological underpinnings
- The implications of complexity science's understanding of the structure of the universe
- The underlying dynamics complexity science finds in the evolution of complex systems
- What happens when we self-consciously integrate complexity science with Big History

I'll conclude with some thoughts on how this approach may enable us to contribute to efforts to address the existential crisis societies around the globe now face.

We begin with a look at Christian's idea of mythic shared maps and why they are so important to us today.

[2] In periodizing human history, we focused on what we thought of as the three stable socio-politico-economic forms that dominated Eurasian politics—hunter-gatherer bands (c. 9000 years ago), mostly agricultural kingdoms (c. 3000–800 BCE), and vast empires (c. 200 BCE–1500 CE), interrupted by transformational periods—the Neolithic Revolution, the Axial Age, and Modernity. Our intent wasn't to suggest that other periodization are mistaken. For example, Leonid Grinin's (2012) periodization—hunter-gatherer, craft-agrarian, industrial, and information-scientific—or Tyler Volk's (2017)—animal social groups, tribal metagroups, transplantable agrovillages, and geopolitical states—both seem accurate *from their point of views*. Bondarenko and I, however, focused on what we saw as the overall patterns of social organization. Consider the differences between kingdoms and empires, for instance. Kingdoms, from Greece to China, generally ruled entities that were limited in geographical size and culturally homogeneous; were governed by semi-divine kings, supported by the loyalty of their inner circles; practiced polytheistic religions; depended on bronze technology; and were still learning how to use writing to manage their cultures. Empires, on the other hand, ruled vast territories with multi-cultural populations; were governed by emperors, supported by bureaucracies; generally practiced religions that were universalistic and moralistic; depended on iron technology; and had become expert at managing their cultures with writing.

16.2 Mythic Maps and the Symbolic Order

To understand the idea of a shared mythic map, we need to begin with a discussion of how neurobiology suggests that we humans organize experience around a *symbolic order*. This ordering is part of the process by which each individual's unconscious mind transforms what their senses report about events into a coherent story that enables them to make decisions (Gazzaniga, 2011 or Ramachandran, 2011). Several thinkers have commented on how symbolic ordering helps make us human. Terrence Deacon, for example, calls *Homo sapiens* "the symbolic species," in his book by that name. For Deacon, this process results in the ability to "inhabit a world full of abstractions, impossibilities and paradoxes," the "defining attribute of human beings" (1997, pp. 21–22). The result, he adds, is that we live in a self-created "virtual reality," much of which members of any society share, due to their common symbolic order. Merlin Donald goes even further. Symbolic ordering he notes made it possible for humans to create the mythic culture that characterizes our species' history, because "symbols 'define' the world (rather than vice versa)" (1991, p. 219). Brian Fagan adds that this ability was "the real edge" that *Homo sapiens* had over the Neanderthals who preceded us, making it possible "to plan ahead and to think of their surroundings as a living, vibrant world ... that changed constantly over the generations" (2010, p. 14).

The perceptual world each of us constructs is unique. However, shared symbolic ordering makes it possible for people in the same society to construct perceptions similar enough that they can cooperate. Discussing this shared symbolism as part of religious rituals—he calls them "Ultimate Sacred Postulates"—Roy Rappaport (1999) notes how they help societies to link the cosmological order they observe to social life. This, I'm convinced, is another way of talking about what Christian's creation myth enables us to do.

The symbolic order of any society seems to respond to critical challenges it faces. For instance, Ancient Egypt relied on the annual flooding of the Nile and the abundant harvests it made possible. As a result, its symbolic order is grounded in a cycle of birth, death, and rebirth. This symbolic order is at the heart of the Egyptian myth that explains the cycle of day and night, as the Sun god Re rides his solar barque across the sky during the day and through the underworld during the night, being reborn into the sky again in the morning. In this way, the Egyptian near-obsession with death, as in the practice of mummification, is actually a reflection of the cycle of birth, death, and rebirth, *as complementary parts of one process* (Assmann, 2002). Or consider the myth of Isis, Osiris, Seth, and Horus. Osiris (the abundance of the Nile) is murdered and dismembered by his brother Seth (chaos), and Osiris' wife Isis (fertility) gathers the pieces of his body and brings him back to life long enough to impregnate her with Horus (protector of the abundance of the Nile), who grows up to engage in an ongoing battle with Seth. In this way, the Pharaoh, represented by Horus, is responsible for fighting off the powers of chaos (Seth) that beset an agricultural state dependent on a thin strip of fertile land on the Nile River, surrounded by desert chaos.

While the symbolic order most of us grew up with is very different, it reflects the key challenges of early Western modernity just as Egypt's symbolic order reflected its challenges. The Western symbolic order—the Clockwork Universe[3]—emerged with the beginnings of scientific astronomy. At a time when mathematics was being perfected, astronomers such as Nicholas Copernicus, Johannes Kepler, Galileo Galilei, and Isaac Newton were astounded at the precision with which the heavenly bodies moved and began to think of the universe as a machine. The key to the map for this symbolic order appeared in Newton's *Principia* (1687), but, like all mythic symbolic orders, would be debated, disputed, and filtered (Donald, 1991) in its society over an extended period. As Stephen Gaukroger notes (2006), this mechanical symbolic order wouldn't be generally accepted until the mid-nineteenth century.

This worldview teaches us to understand reality as a collection of solid, independent "things" that respond to other things in causal chains, just as one gear in Big Ben drives the next. These "things" are passive: Like atoms of gas in a bottle, they can move only when the invariable laws of nature drive them to move. As a result, events unfold deterministically, as one thing responds predictably to another (Berman, 1981). This is the logic that underlies most mainstream thought today. For instance, until recently biological evolution was understood to occur mechanically, with random mutations leading to changes in an organism's phenotype, and then tested by natural selection in the environment. According to this model, it is impossible for organisms to evolve as acts of intentional adaptation. Change is driven purely by chance and the forces created by the laws of nature (Gould, 2002; Jablonka & Lamb, 2014). Grounded in such mechanical thought, our social institutions tend to operate mechanically. In contemporary public education, for instance, this model teaches people to think of students as little machines being programmed with important information. That has resulted in an emphasis on measuring what students learn and teachers having to "teach to the test." And because the virtual reality we share demands that we navigate both such thought and the institutions that grew from it, we learn this mechanical virtual reality automatically, in our homes, schools, and workplaces.

In many ways, this mechanical symbolic order has been wildly successful. It has allowed the West to create scientific medicine, to produce a wide range of consumer products available to the vast majority, to create mass literacy, and to generate an enormous, rapidly growing body of knowledge. With it, people throughout the world can travel widely and be connected to a worldwide network of Internet users.

At the same time, it has driven the evolution of existential challenges, which seem insoluble. For instance, leaders ranging from scientists to oil company CEOs have known, for decades, that burning fossil fuels is driving a climate change that could destroy the ability of our world to support complex societies. Another problem partly created by treating the Earth as a machine that should be used for human benefit is the growing possibility of a mass extinction (Kolbert, 2014). Yet, in spite of

[3]As Kepler noted, "My aim is to show that the machine of the universe is not similar to a divine animated being, but similar to a clock" (as quoted in Dolnick, 2011, p. 182).

international conferences and widely signed treaties, both situations have only gotten worse.

What we need is a different way of thinking about the world—a new symbolic order, with a new shared map, that will create a shared virtual social reality where it *is* possible to address these challenges. Fortunately, scientific advances, in fields ranging from physics to paleoanthropology, have also revolutionized our understanding of the world. That shift is grounded in Einstein's theory of relativity, where $E = mc^2$ and its suggestion that matter is a form of energy. The world, we have learned, is not composed of solid passive "things," as the mechanical worldview insists, but of dynamic energy systems, continually responding to the changes around them, and it is this dynamic world that complexity science examines to explore the patterns which emerge as it evolves.

In the rest of this chapter, I want to examine two of these patterns and suggest how applying them to Big History can enhance our ability to contribute to developing Christian's shared map for our time of existential challenge.

16.3 A Universe of Nested Networks

Once Einstein's theory of relativity was proven, it was clear that matter is a form of energy, structured to store that energy. In the words of Mae-Wan Ho, matter can best be understood as "domain[s] of *coherent* energy storage" and the average time the energy remains in these domains is "a measure of the *organized* complexity of the system" (2008, p. 81; author's italics).

As opposed to the mechanical model's world of distinct, passive things, complexity science portrays the world as a deeply interconnected nested network of domains of energy storage that are continually adapting to each other. At the simplest scale we're aware of, quantum particles, such as quarks, combine to form a variety of larger particles, including neutrons and protons. These entities can then network to form atoms, and atoms, sometimes with less complex particles, can network to form molecules. And so it goes—with networks all the way up and down—into material networks, such as minerals, planets, solar systems, and galaxies; living networks, such as macromolecules, cells, organs, organisms, and ecosystems; and cultural networks, such as hunter-gatherer bands, tribes, chiefdoms, kingdoms, and empires.[4] At each more extensive scale, these systems develop new capabilities, often difficult or impossible to predict by understanding only the smaller networks that make them up.

As an example, consider an oversimplified picture of the human body. The human body is a nested network of organ systems, such as the respiratory system,

[4]This description of our world of nested networks is intentionally oversimplified. For a fuller, more precise discussion of this network formation, see Tyler Volk's *Quarks to Culture* (2017). Also, for a deeper dive into the three varieties of nested networks—material, living, and cultural—see both Volk's book and Gregg Henriques' *A New Synthesis for Solving the Problem of Psychology* (2022).

which, in turn, is a network of organs like the lungs. Those lungs, in turn, are a network of groups of cells, and each cell group is a network of cells. Similarly, each cell is a network of smaller structures, organelles, which are further networked from macromolecules, like DNA, molecules, atoms, and quantum particles. *To be healthy, the body needs the cooperation of structures at all these scales.* Consider the way that the body needs the iron in red blood cells to pick up oxygen in the lungs, travel through the circulatory system, and drop the oxygen off at a cell. The fullness (and messiness) of such complex systems becomes even clearer when we put it in the context of the many other processes in the body—from cognitive to immune systems, from those that control motor activity to waste removal.

But the activities of these sub-systems of the body isn't everything. The body as a whole develops a series of capabilities as these sub-systems interact with each other, giving rise to one common definition of complex systems: The whole is greater than the sum of the parts. In addition, each human body can be part of a variety of more extensive social and ecological systems—from families and organizations to the living ecosystem of New York City. When I think about an energy storage domain like New York and all the scales working down to macromolecules, I can begin to understand what a "complex system," as complexity theorists call it, is, as well as why the term has been so difficult to define.[5]

Perhaps the reality of complex systems is too, well, complex to capture in a short definition. Still, even the example of the body makes a variety of qualities evident. Complex systems are composed of many different component systems, often on many scales, that have access to the information they need to adapt to and innovate in the changing world they're part of, without the assistance of an outside intelligence. At each scale, we can examine these complex systems in three ways—as a functioning whole, a component in a functioning whole, or the environment for other systems. At the scale of each functioning whole, the behavior of the components creates the behavior of the whole; the nature of the whole limits the behavior of its parts; and the conditions of the environment shapes the nature of the whole. As a result, to study the behavior of a family, we must also study the behavior of each member of the family, their interactions, and the environment of communities in which they are grounded.

One interesting controversy concerns at what scale complex systems become full-fledged, decision making agents. The problem here is that most physicists, such as Nobel Laureate Murray Gell-Mann (1994), agree that quantum particles, atoms, and molecules respond mechanically, relying on their structure to determine their behavior. Others insist that it's decision-making agents all the way down. However, starting with macromolecules—that is, at the point that around which they begin

[5]Readers who want to know more about the problems of defining complex systems can consult Landyman et al. (2013). For a quicker insight into some of the ins and outs of defining the term, compare two first-class definitions—in Cilliers (1998, pp. 3–5) and Mitchell (2009, p. 13).

to be living systems—they become what are often called *complex adaptive systems* (CASs), which can learn and adapt, sometimes in unexpected ways.[6] For instance, sub-systems in the body's immune system are able to identify and attack pathogens that their organism has never previously experienced. Similarly, some neural networks are able to filter out select objects that the senses perceive, so they don't appear in conscious perception. These omissions may occur because the unconscious mind judges them as creating contradictions, which may compromise the ability to make decisions (Gazzaniga, 2011; Ramachandran, 2011).

In a world composed of nested networks, many of which can become decision-making agents, events unfold as agent/networks at many scales adapt to the changes in other networks, sometimes cascading to produce surprising combinations. Consider the recent coronavirus disease 2019 (COVID-19) pandemic. It began with changes in the genetic macromolecules of a virus—whether they resulted from random mutations or intentional manipulation in a laboratory—and quickly spread across the globe. What made the pandemic so damaging was the combination of how easy it was to communicate the virus, advances in travel technology, the effort of Chinese officials trying to hide it, and the refusal of some governments and portions of their populations to treat the pandemic as a threat. As a result, the pandemic created extensive social, economic, and political damage, as all these CASs interacted, resulting in cascades of adaptation.

This is a far cry from the mechanical clockwork universe of the late nineteenth century. Consider a few of the most important differences:

- A world of nested networks of energy storage systems is deeply interconnected, as opposed to the collection of distinct things in a mechanical universe. After all, on an atomic scale, there is no "you" and "me," only a single, immensely complex network of atoms.
- Such a world of nested networks is also far more difficult to predict than a mechanical world. Think of all the systems in human bodies—systems for breathing and metabolizing the nutrition we take in, immune systems that keep us healthy, systems of neural networks for processing information from the outside (and inside) world and for responding to what we learn. Add the interactions between humans and their environments, as they adapt to each other, and the determinism of the mechanical model seems vastly oversimplified.
- Largely as a result, the causal chains of the mechanical worldview no longer seem accurate. In a world populated largely by CASs, a wide range of events have many causes. So, it's no longer enough to say that an apple tree grows simply because Johnny Appleseed planted one. Soil, weather, and temperature conditions all have to be in the right range. And if a passing bird digs into the soil and eats the seed, no tree will grow. What is needed is not a chain of causes

[6]For more detail on complex adaptive systems and various definitions of complexity see also Chapter "Introduction to Complexity Theory for Big History" (LePoire et al., 2026) in this volume.—*Editors.*

and effects, but the *emergence* of the apple tree, as a result of a wide variety of interactions under the just-right "Goldilocks conditions" that we in Big History talk so much about. And when we get to a phenomenon like the evolution of human life, the relative simplicity of the emergence of an apple tree is transformed to a level of improbability that approaches the mysterious (Theroux, 2023).

The concept of emergence is so critical to complexity science that mathematical physicist Henrik Jensen titled his textbook *Complexity Science: The Study of Emergence* (2023).[7] As opposed to the linear causality of the mechanical symbolic order, an order grounded in complexity science suggests that such events emerge, as the component systems of any CAS interact with each other and all the complex systems around them. In systems as extensive as an ecosystem, a language, or a large city, so many CASs are continually interacting with other systems that it becomes near-impossible to predict what will happen, as with the COVID-19 pandemic of the 2020s. Ironically, the United States had quite literally written the book, a manual on coping with a pandemic, on how to minimize the effects of such a pandemic. But political concerns—and perhaps some human frailties—made it impossible for authorities to implement that plan. This is emergence with a vengeance.

So, while a clock is an appropriate metaphor for a mechanical symbolic order, an order grounded in emergence might better be represented by a *carnival*. The choice of a carnival as a metaphor may seem frivolous at first. But what is a carnival, other than a platform on which all sorts of agents—each with their particular purposes—come together to play games and enjoy rides? As in life, what happens (emerges) at a carnival is often predictable. But sometimes the interaction of human agents, other animal agents, and mechanisms produces results that no one had imagined. In addition, carnivals are notorious for the con games that get played out at them. And nature is full of con games. For instance, one species of acacia trees grows hollow thorns and provides a sugary mix for the ants that live in the thorns and protect the trees from damage by small mammals. It seems like a simple exchange, but it's really a con game. The sugary mix that the trees supply is actually mixed with a chemical that makes it impossible for the ants to metabolize any other type of sugar![8]

If we develop complexity science's understanding of the world, we arrive in just such a carnival world. I've come to believe that it is also the world depicted in Big History.

[7]For my personal take on emergence, see Baskin, 2022.

[8]For a fuller discussion of the intricacies of such co-evolutionary cooperation in ecosystems, see Margulis (1999).

16.4 Stable States and Phase Transitions

But it isn't just the structure of the universe that differs in a carnival world. The process of evolution, itself, becomes more dynamic, as these energy systems oscillate between periods when they are stable, acting from long-held habits, and those when they are transforming themselves through experimenting with alternatives. Or, as physicist Stuart Kauffman (2000), one of the pioneers of complexity science, puts it, evolution occurs in life's exploration of such alternatives, which he calls the "adjacent possible," in these periods of order and chaos.

With the mechanical symbolic order, evolution had been presented as linear. For example, in school, I'd learned that evolution occurred gradually, one random mutation at a time in chains of cause and effect. For example, *Homo sapiens* evolved in a chain of precursors, one leading inexorably to the next—australopiths, *Homo habilis, Homo erectus,* Neanderthals, and then us. With this symbolic order, human evolution was viewed as "a long, gradual slog from primitiveness to perfection," as Ian Tattersall and Jeffery Schwartz put it. But, in recent decades, scientists realized that human evolution wasn't so linear—that evolution seemed more like life struggling to find its way, in a continually changing world. In this way, *Homo sapiens* emerged in "a history of experimentation, of constant exploration of the very many ways there are to be hominid" (2001, pp. 46, 52). Moreover, current research indicates that the origin of this development is not merely random mutations, but the interaction of a wide variety of environmental shifts during a period of environmental warming in East Africa that drove our ancestors to adapt as they left the rainforests for the savannahs (see Rappaport & Corbally, 2020 and Turner et al., 2018).

Complexity science suggests that with this dynamic, carnival model of the world, evolution follows a pattern alternating stable (orderly) and dynamic (chaotic) periods. Here's the back-of-the-cocktail-napkin diagram I drew when I was thinking about this conception of evolution (see Fig. 16.1):

Some readers will immediately recognize that this figure illustrates punctuated equilibrium, the theory Niles Eldredge and Stephen Jay Gould developed to explain the evolutionary jumps that follow mass extinctions.[9] I call this figure the life cycle of an attractor. "Attractor" is a term from higher math that complexity scientists use to indicate those behaviors, among all possible behaviors, that a complex system tends to evolve toward in a particular set of conditions. That is, an attractor defines a system's habitual behaviors—its default responses for surviving—in its environment.

For instance, when you throw a chunk of ice into a pot and heat it on a stove, it will cycle through three "phases"—solid, liquid, and gas. At first, it will remain ice, until it approaches the melting point, becomes turbulent, and turns into liquid. It then

[9]For an in-depth discussion of punctuated equilibrium, its history, and the controversies it generated, see Stephen Jay Gould's *The Structure of Evolutionary Theory* (2002).

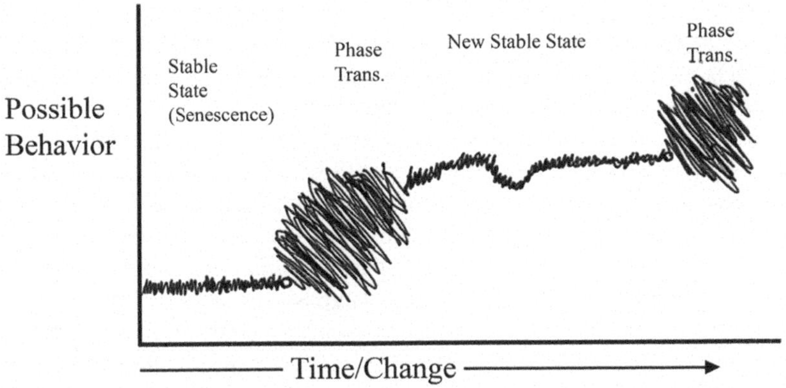

Fig. 16.1 Life cycle of an attractor (Baskin, 2008)

remains liquid until it approaches the boiling point, again becomes turbulent, and becomes gas.

Or consider what happened when Earth was struck by a comet about 65 million years ago. Starting at the far left of the figure, we see that the end of the period during which Earth's ecosystems were dominated by dinosaurs. The comet destroyed those ecosystems, driving them into a 10 million year phase transition, leading to the dominance of mammals. Mammal dominance then proceeded until our own period in the senescence of mammal domination, followed by the likelihood of another phase transition (far right).

Similar dynamics appear in market cycles of boom and bust, or in human development, where a stable childhood personality is interrupted by adolescence, leading to a stable adult personality, often ended by a mid-life crisis. The figure can also represent the period that Bondarenko and I wrote about, starting, from the left with a couple of centuries before the events that undermined the kingdoms, like Ancient Egypt or Zhou China, that preceded the Axial Age. We then see the Axial Age as about 600 years of social experiments, ending around 200 BCE, as vast empires, such as the Roman Empire or the Han Empire became dominant powers. Finally, as the line approaches the far right, we can see another experimental period in modernity, after about 1500 CE. In this way, order begets chaos, which, in turn, begets order, in a positive feedback loop, and this tension between order and chaos "is at the heart of all creativity" (McGilchhrist, 2021, p. 818).

Once we get beyond the simplest examples, such as heating ice, the dynamics of this pattern show how evolution works in a world of nested networks: Toward the end of the stable state, the component systems have a long successful history of following a limited number of all possible behaviors. Their survival had depended on interacting successfully with other component systems, whose behaviors may have had equally long histories. As conditions shift over time new challenges are likely to emerge. Yet the old, habitual behaviors, which allowed them to survive, are likely to continue, in a period biologist Stan Salthe (1993) calls "senescence," until some kind of catastrophe reinvents the environment and breaks the hold of the old "attractor," and a phase transition begins. In animals with sophisticated brains, such as mammals including humans, these habitual behaviors are available in the often-used neural networks that drive much of their behavior.[10]

In the phase transition, these components—free of the constraints of the old attractor—experiment with new behaviors with which they may be able to succeed in their new conditions. Those experimental behaviors may be entirely new—the downward strokes moving forward in time—or the reintroduction of old behaviors—the upward strokes that appear to be going *backward* in time. Which, of courses, is impossible. But I wanted to suggest that component systems don't merely innovate; they also try out old experiments that may not have been successful. Finally, when the component systems find the range of behaviors that do work in current conditions, their interactions will develop a new attractor for the system as a whole. So, in human personality development, people in adolescent phase transition experience new powerful feelings and, most often, are expected to behave in a far less dependent way. To meet these shifts, they often begin experimenting with new behaviors that will lead to new neural networks and an adult personality.

Let me finish up this section with a thought I offer provisionally. The way complex systems evolve by moving through stable and transformational theory is a result of the way existing complex systems combine to form more complex systems with new capabilities. Unstructured energy is chaotic. Order appeared in the universe with the Big Bang, which structured energy as quantum particles, and that order increased as the universe cooled following the Big Bang. Over time, those particles formed matter at scales that became more and more complex, until, today we have galaxies, which average 100 million stars; four-square-mile patches of rainforests that can contain more than a thousand flowering plants, several hundred species each of trees and birds, and more than 100 species of butterflies; and cities with populations as large as 30 million humans, as well as ways of accessing all the goods and services that they need.

However, as Marc Widdowson noted (2023) at a recent International Big History Association (IBHA) conference, the more complex—and highly ordered—these energy storage systems become, the more likely it is that they will not be capable of addressing the challenges emerging *in a continually changing world*. To put it another way, order makes it possible for domains of energy storage systems to remain stable, but it also restricts their ability to adapt in innovative ways. In order for their sub-systems to survive, they may need to be released from the attractors that hold them together. It is only when they are plunged into the resulting chaotic phase transition that they are able to fully explore the environment for new, more appropriate behaviors.

From this perspective, then, order and chaos are not diametrical opposites, but "contraries [that] *fulfil* one another" (McGilchhrist, 2021, p. 816). This, then, is the central significance of the ongoing battle between Horus (order) and Seth (chaos) in Ancient Egyptian myth, as it is in the concept of yin and yang in Chinese philosophy. This understanding, however, is denied in the modern Western symbolic order. On the other hand, the symbolic order embodied in complexity science indicates that such

[10]For an in-depth discussion of how these neural networks form in humans, see Gazzaniga, 2011 or Ramachandran, 2011.

opposites *are complementary*. As one of my mentors in complexity science, Jack Cohen, used to love to point out, being intentionally provocative, in complex systems, opposites are frequently both true. In presentations, he would often show a slide of a photograph of a "lost and found" sign.[11]

This last set of comments is still in a rough form. They require more research and exploration. However, I'm convinced they point toward ways of examining Big History's mythic story that might allow us to address our current existential challenges. Part of those ways is the application of complexity science to Big History, and it is to those ways that we now turn.

16.5 Applying These Complexity Patterns

For me, one of the first advantages of incorporating complexity science into my approach to Big History is its ability to explain today's most disturbing events. I grew up in the 1950s and 1960s, and never could have imagined the violent polarization that dominates world politics. From the perspective of complexity science, societies across the globe have entered the chaos of a phase transition. Especially among political leaders, behaviors demanded by old cultural attractors no longer hold. In just about every institution in every society, people seem to recognize, if only unconsciously, that those old ways don't work, but that no one knows what the new ways will be. Most of us are understandably terrified, and the result is a sort of global psychosis, reflecting a disturbingly wide range of experiments. As my back-of-the-napkin figure predicts, these experiments both look forward and back in efforts to find what will work.

More importantly, these patterns will allow us to refine Big History's origin story. It will, for example, help us explain our current existential challenges as an example of a phenomenon that other societies have faced and overcome. Bondarenko and I agreed that the Axial Age was the best example of a period in human history when societies faced the same sort existential challenges we do today (Baskin & Bondarenko, 2014). But we also speculated that the period leading up to the first agricultural kingdoms emerging about 3000 BCE in Mesopotamia and Egypt was also shaped by an existential challenge. That made sense to us because of the challenges of increasing population, the city Ur reached a population of about 40,000 at this time; the rise of both writing and bronze technology; and the emergence of semi-divine kings across much of Eurasia. Meeting these challenges

[11] Evolutionary biologist Jack Cohen wrote about complexity science with mathematician Ian Stewart, including their primer of the subject, *The Collapse of Chaos* (1994). He presented frequently, emphasizing this point, at workshops sponsored by the Institute for the Study of Coherence and Emergence, between 1998 and 2006.

required exactly the kind of massive social experiments that dominated the Axial Age and modernity.[12]

What we found was that in all these cultural phase transitions, the societies that were able to transform themselves to meet their challenge did so largely by reinventing their shared symbolic orders, with what I've called a "mythic twist."[13] We focused on the transformations of the Axial Age, because its societies were literate, some of the first societies where social elites began using writing to manage their societies (Assmann, 2011). For example, in Axial Age Greece, the fifth century BCE was a century of chaos. First, the Persians invaded Greece in 490 and 480, when the Persians burned Athens, but were eventually defeated by the alliance of Greek city-states. After their victory, those city-states competed for political dominance, with alliances led by Athens and Sparta, resulting in the Peloponnesian Wars (431–404 BCE). Those wars would also cause a plague, further reducing the power of Athens and undermining confidence in the myths of the gods of Olympus. We can see this shift in the criticism of the gods in Greek tragedy, where basically good people, such as Oedipus and Orestes, suffer intensely because of those gods' whims. Combined with Greece's experimentation with science, a new, more rational symbolic order emerged, taking its mythic form in the philosophies of Plato and Aristotle, whose mythic elements became especially important in Medieval Christian theology.

A similar mythic twist helped drive the emergence of the modern scientific symbolic order of the West, a rewriting of the myth of Late Medieval Christianity.[14] The key to this twist came in the works of Thomas Aquinas in the thirteenth Century, which explained that God and his creation were rational (Gillespie, 2009). This position would be largely absorbed by the Late Medieval Catholic Church. The mythic twist that would shape Western science came from Francis Bacon, in works like his *Novum Organum* (1620), during what Peter Turchin (2023) calls a "disintegrative phase" of society that culminated in a century and a half of religious wars. To create a society that would be far more safe and comfortable, he explained, the source of human knowledge should come, not from the speculative methods of Medieval churchmen, but from scientific research. The scientist's job would be to "torture" nature into giving up its secrets, thereby enabling life-enhancing innovations, such as the printing press. In this mythic twist, the attention to nature that Aquinas had praised now moved from the act of knowing God to the Scientist's attempt to improve human life.

[12] I am relying, in this chapter, only on examples from the Axial Age, largely because there are no written records. As Bondarenko and I note in our paper on the subject (2019), there is some solid evidence that a similar dynamic was at play. However, because the Axial Age offers such a rich record, documented in writing, I felt it was best to avoid unnecessary speculation about the Neolithic Revolution, which led to the rise of kingdoms such as Mesopotamia and Egypt, and stick with those Axial Age examples.

[13] For a deeper dive into this dynamic, see Baskin (2003).

[14] For a full discussion of how deeply Western science was grounded in Late Medieval Christianity, see especially Gillespie (2009) and Freely (2012).

This emerging symbolic order would, like all myth, be debated, disputed, and filtered (Donald, 1991), until the mid-nineteenth century, when it finally triumphed (Gaukroger, 2006). We can see this process unfolding in the philosophy of Descartes, Leibnitz, and Hobbes, in the research of scientists such as Newton, Robert Lyle, and Charles Darwin, and social movements, including Christian fundamentalism, the Enlightenment, and Romanticism.

The new mechanical symbolic order that emerged has been almost as successful as Bacon had hoped. However, the ordered stable state it produced has also created the chaos of its unintended consequences, in our current existential challenges. One task that Big History's creation myth can fulfill, by incorporating complexity science, is to help people understand how those challenges arose and how other societies, facing similar challenges, were able to overcome them.

Complexity science suggests that, in a continually shifting world, such consequences are to be expected. From this point of view, we can think about any society's symbolic order, communicated in myth, as its attractor, drawing its members' behavior toward the set of time-honored behaviors (and ways of thought) that have allowed those people to thrive. When the conditions in which that society exists shift so much that the behaviors of the old attractor no longer work, that attractor should collapse, leading to Turchin's disintegrative phase—a chaotic phase transition, whose experimentation can generate new myth, different types of behavior, and, eventually, a new attractor. By focusing on the experience of societies that did reinvent themselves, our origin story can create the map for how we can similarly transform the way we think about the world.

Integrating these insights from a variety of disciplines can also enable Big History's creation myth to highlight an ironic truth: That the religious myths that had been ridiculed in the Enlightenment as superstition are also capable of coming to conclusions that are now being re-examined by contemporary science. The most obvious is the way the battle between order and chaos—so central to myth in Egypt, Mesopotamia, and China, as well as that of both Judaism and Christianity, where it is transformed to the battle between good and evil—has been rediscovered in modern disciplines such as complexity science and cliodynamics, for which Turgin is a chief spokesperson. As Widdowson pointed out (2023) sometimes evolution demands that we fall back before we can go forward.

16.6 Conclusion

Perhaps more importantly, focusing on the universe as a deeply interconnected nested network of matter can help us shape that new shared map. Remember, our reliance on symbolic orders results in our experiencing the world as a virtual reality constrained by the order we accept. So, the shared maps we make, mythic or scientific, are critical in our ability, as individuals and societies, to address all our challenges. The old mechanical symbolic order taught people to think about society as a collection of independent living things and machines that operate through cause-and-effect. In the resulting virtual reality, mental illness would come to be thought of as a misstep in the step-by-step process that drove a person to respond to life. If their behavior became disruptive enough, they could be isolated in asylums where most of us need not look at them. Similarly, the effects of industrial pollution could be overlooked because they were not proper parts of a cause-and-effect manufacturing processes and could be isolated in "dumps," where, once again, no one need pay much attention. This way of thinking about problems in society, I'm convinced, is at the heart of several of the existential challenges we face today.

What Big History's creation myth can teach is that other shared maps for the virtual reality we live in are possible. Complexity theory presents the alternative dynamic that current science has uncovered in one field after another. Some psychologists today suggest that much mental illness emerges, as a result of family dynamics and social pressures such as poverty, as well as the internal mental processes of the individual. And it's become unmistakable that industrial pollution is responsible to a large extent for global warming and a host of health problems. What we need is a different shared map of society, grounded in a more appropriate symbolic order, that can help us build a different social virtual reality. That is, we need the sort of mythic twist that allowed Axial Age Greece move from myth about the gods of Olympus to a more philosophical myth that emphasizes the value of working with the world, rather than trying to control it.[15]

Incorporating complexity science into our work with Big History offers an excellent opportunity to contribute to the process of making such a twist.

References

Assmann, J. (2002). *The mind of Egypt: History and meaning in the time of the pharaohs*. (A. Jenkins, Trans.). Harvard University Press.

Assmann, J. (2011). *Culture, memory, and early civilization: Writing, remembrance, and political imagination*. Cambridge University Press. https://doi.org/10.1017/CBO9780511976041

Baskin, K. (2003). The mything link: Why sacred storytelling is a key human survival strategy. *Comparative Civilizations Review Journal, 89*, 93–120. https://doi.org/10.4206/ccr.2003.1118

Baskin, K. (2008). Storied spaces: The human equivalent of complex adaptive systems. *Emergence: Coherence and Organization, 10*(2), 1–12. https://doi.org/10.3172/emer.10.2.1

Baskin, K. (2022). Big history and the principle of emergence. *Journal of Big History, 5*(1), 77–91. https://doi.org/10.22339/jbh.v5i1.6215

Baskin, K. (2024). Complexity science and myth in big history. *Journal of Big History, 7*(2), 125–136. https://doi.org/10.22339/jbh.v7i2.7210

Baskin, K., & Bondarenko, D. M. (2014). *The axial ages of world history: Lessons for the 21st century*. Emergent Publications.

Baskin, K., & Bondarenko, D. M. (2019). Is modernity a third axial age? *New Trends in Social and Liberal Sciences, 4*(2), 1–23. https://doi.org/10.24887/ntsls.v4i2.213

Berman, M. (1981). *The reenchantment of the world*. Cornell University Press. https://doi.org/10.7591/9780801458920

Christian, D. (2004). *Maps of time: An introduction to big history*. University of California Press. https://doi.org/10.1525/9780520936835

[15]For some further speculations on what that new symbolic order might be like, see Baskin (2003).

Christian, D. (2018). *Origin story: A big history of everything*. Little, Brown and Co.
Cilliers, P. (1998). *Complexity and postmodernism: Understanding complex systems*. Routledge. https://doi.org/10.4324/9780203010372
Cohen, J., & Steward, I. (1994). *The collapse of chaos: Discovering simplicity in a complex world*. Viking Penguin.
Deacon, T. W. (1997). *The symbolic species: The co-evolution of language and the brain*. W.W. Norton & Co.. https://doi.org/10.1111/j.1467-873X.1998.tb00371.x
Dolnick, E. (2011). *The clockwork universe: Isaac Newton, the Royal Society & the birth of the modern world*. HarperCollins.
Donald, M. (1991). *Origins of the modern mind: Three stages in the evolution of culture and cognition*. Harvard University Press. https://doi.org/10.2307/j.ctvjhzr6k
Fagan, B. (2010). *Cro-Magnon: How the Ice Age gave birth to the first modern humans*. Bloomsbury Press. https://doi.org/10.5860/choice.48-1002
Freely, J. (2012). *Before Galileo: The birth of modern science in medieval Europe*. Overlook Duckworth.
Gaukroger, S. (2006). *Civilization and the culture of science: Science and the shaping of modernity 1795-1935*. Oxford University Press.
Gazzaniga, M. S. (2011). *Who's in charge? Free will and the science of the brain*. HarperCollins.
Gell-Mann, M. (1994). *The quark and the jaguar: Adventures in the simple and the complex*. W.H. Freeman. https://doi.org/10.1002/andp.19955070220
Genet, C., Genet, R., Swimme, B., Palmer, L., & Gibler, L. (Eds.). (2009). *The evolutionary epic: Science's story and humanity's response*. Collins Foundation Press.
Gillespie, M. A. (2009). *The theological origins of modernity*. University of Chicago Press. https://doi.org/10.7208/9780226293774
Gould, S. J. (2002). *The structure of evolutionary theory*. The Belknap Press. https://doi.org/10.1002/andp.20035150114
Grinin, L. (2012). *Macrohistory and globalization*. "Uchitel" Publishing House.
Henriques, G. (2022). *A new synthesis of solving the problem of psychology: Addressing the enlightenment gap*. Palgrave Macmillan. https://doi.org/10.1007/978-3-030-91118-2
Ho, M.-W. (2008). *The rainbow and the worm: The physics of organisms* (3rd ed.). World Scientific Publishing. https://doi.org/10.1142/9781860947230
Jablonka, E., & Lamb, M. J. (2014). *Evolution in four dimensions: Genetic, epigenetic, behavioral, and symbolic variation in the history of life*. (Revised ed.). The MIT Press. https://doi.org/10.7551/mitpress/9780262525891.001.0001
Jensen, H. J. (2023). *Complexity science: The study of emergence*. Cambridge University Press. https://doi.org/10.1017/9781108891912
Kauffman, S. (2000). *Investigations*. Oxford University Press. https://doi.org/10.1093/acprof:oso/9780195121054.001.0001
Kolbert, E. (2014). *The sixth extinction: An unnatural history*. Henry Holt. https://doi.org/10.1002/9781119777926.ch18
Landyman, J., Lambert, J., & Wiesman, K. (2013). What is a complex system? *European Journal for Philosophy of Science, 3*(1), 33–67. https://doi.org/10.1007/s13194-012-0051-x
LePoire, D. J., Grinin, L., & Korotayev, A. (2026). Introduction to complexity theory for Big History. In D. J. LePoire, L. Grinin, & A. Korotayev (Eds.), *Complexity in universal evolution—A Big History perspective* (pp. 29–90). Springer Nature (in this volume). https://doi.org/10.1007/978-3-032-20516-2_2
Margulis, L. (1999). *Symbiotic planet: A new look at evolution*. Basic Books. https://doi.org/10.1007/978-1-4614-4113-5_14
McGilchhrist, I. (2021). *The matter with things: Our brains, our delusions, and the unmaking of the world* (Vol. II). Perspectiva Press.
Mitchell, M. (2009). *Complexity: A guided tour*. Oxford University Press. https://doi.org/10.1093/acprof:oso/9780195191729.001.0001

Ramachandran, V. S. (2011). *The tell-tale brain: A neuroscientist's quest for what makes us human*. W.W. Norton & Co.

Rappaport, R. (1999). *Ritual and religion in the making of humanity*. Cambridge University Press. https://doi.org/10.1017/CBO9780511815807

Rappaport, M. B., & Corbally, C. J. (2020). *The emergence of religion in human evolution*. Routledge. https://doi.org/10.4324/9780429317589

Salthe, S. (1993). *Development and evolution: Complexity and change in biology*. The MIT Press.

Tattersall, I., & Schwartz, J. (2001). *Extinct humans*. Westview Press.

Theroux, S. J. (2023). *A most improbable story: The evolution of the universe, life, and humankind*. Self-published.

Turchin, P. (2023). *End times: Elites, counter-elites, and the path of political disintegration*. Penguin Press.

Turner, J. H., Maryanski, A., Petersen, A. K., & Geertz, A. W. (2018). *The emergence and evolution of religion: By means of natural selection*. Routledge. https://doi.org/10.4324/9781315181755

Volk, T. (2017). *Quarks to culture: How we came to be*. Columbia University Press.

Widdowson, M. (2023, July 8). *The last stop on the cosmic journey: An estimated time of arrival* [Conference presentation]. IBHA conference, humanity on a finite planet: A Big History perspective.

Wilson, E. O. (1978). *On human nature*. Harvard University Press. https://doi.org/10.4159/9780674044391

Ken Baskin is an independent researcher whose work integrates insights from complexity science, neuro-anthropology, and Big History. After earning a PhD in English Literature in 1977, he spent 15 years writing public relations material for major firms. His books include *Corporate DNA* (1998), an examination of how to think about organizations as living things rather than just mechanisms, and *The Axial Ages of World History* (2014), an exploration of the similarities between the Axial Age and Modernity that he co-wrote with Moscow anthropologist Dmitri Bondarenko. Ken is currently reinterpreting religion as a way that human groups can know and adapt to the powerful forces that surround us. He lives in Philadelphia, Pennsylvania.

Emergence, Complexity, and Thermodynamics in Resource Bubbles and Historical Dynasties

17

Mark Ciotola

Abstract

This chapter explores how emergence and complexity can help analyze the rise and fall of physical Hewett-Hubbert resource bubbles and historical dynasties. Key concepts such as energy, thermodynamic laws, efficiency, and power are introduced, along with proposed extensions of the Second Law of Thermodynamics. Various definitions of complexity are examined, with a focus on the energy rate density approach and its comparison to entropy-based methods. The relationship between energy rate density and complexity artifacts is also explored. Using principles of extended thermodynamics, we analyze how medium-term dissipative structures—such as resource bubbles and dynasties—emerge and evolve. Finally, we identify patterns in complexity changes over time and outline a characteristic complexity profile for these phenomena.

Keywords

Resource bubbles · Thermodynamics · Efficiency · Entropy · Dissipative structures · Dynasties · Cycles · Complexity · Evolution

---

This chapter is an extensively re-worked version of the text originally published in *Journal of Big History* (Ciotola, 2025).

M. Ciotola (✉)
San Francisco State University, San Francisco, CA, USA
e-mail: ciotola@sfsu.edu

D. J. LePoire et al. (eds.), *Complexity in Universal Evolution*, World-Systems Evolution and Global Futures, https://doi.org/10.1007/978-3-032-20517-9_17

17.1 Introduction

Big History includes a substantially physical approach to the emergence of humanity and civilization (Spier, 2015, pp. 54–60). Although non-physical approaches can also be expressed within the context of Big History, this chapter focuses on approaches on physical approaches. Big Historians might call the work product of these physical approaches "little big histories," a term coined by Esther Quaedackers (Spier, 2015). These physical approaches also provide a methodology for analyzing reoccurring historical patterns within the larger framework of Big History.

Big History sets the stage for itself with cosmology. Modern theory proposes that the universe began at a singularity called the "The Big Bang" at which time and space began. The Singularity contained all of the energy of the universe. Then, about 13.8 billion years ago, the universe began a great expansion into time and space (Hubble, 1929; NASA, 2007; also see Gupta, 2023).

Since then, there have been two trends of interest that continue until the present time. First, as the universe has expanded, its energy has become more diluted, thus overall, it has been cooling down (Ryden, 2003, 22–23). This is fortunate for humans, as the universe has, overall, become sufficiently cool to support life as we know it. Second, as time passed, the universe has become more heterogeneous. Within local regions, gravity causes matter to condense, heat up, and form emergent, complex structures such as galaxies, stars, and planets. (There is also dark matter, which does not seem to become part of these structures or radiate energy, but too little is known to comment further.) In summary, there has developed a contrast between hot, clumpy matter and the cold voids of space. This contrast allows us to begin our story.

17.2 A Review of Relevant Thermodynamics and Terminology

It is possible to understand much of the evolution of the cosmos and the evolution of life and the progression of historical processes by knowing just a few principles of thermodynamics. However, Big History and even thermodynamics are interdisciplinary. Since different fields use identical terms in different ways, this chapter will introduce several principles of thermodynamics and useful terms, with a bias toward definitions from physics, to avoid misunderstandings.

- A *tendency* is something with a greater than random likelihood of happening. A tendency is a broader term than *force*. Forces are tendencies, but not all tendencies are forces.
- *Equilibrium* represents a state of a system in which tendencies can no longer cause any change. When a system has reached equilibrium, it is literally dead.
- D*ynamic equilibrium* is when a potential is replenished over time.
- *Potential* represents the ability of a system to achieve something. A potential is typically the result of a system being pulled away from equilibrium with respect to a tendency. For example, when a spring is stretched, it tends to return to its

original length. That spring has stored potential energy. If the spring is released, it will move back toward its equilibrium position.

- *Thermal potential:* When regions in a system differ in temperature, a thermodynamic potential exists. If those regions are bridged by a thermal conductor, then thermal energy will flow from hot to cold regions (Schroeder, 2000). For example, thermal energy will flow from a tank of hot water through a copper conductor to a tank of ice water. This tendency is not a force in the same manner as is gravity, but is a statistical phenomenon with a positive probability of occurring. As thermal energy flows, potential entropy is achieved. Thermal energy will continue to flow until both regions possess the same temperature. At this point, those regions are said to be in thermal equilibrium and the thermal potential has been exhausted.
- A *flow* occurs while systems achieve their potential over time. In some cases, a flow comprises the transport of a physical element such as water or energy. For example, imagine that snow falls on a mountain in winter. The mass of that snow has been raised well above sea level. The material of the snow has gravitational potential. When summer arrives and the snow slowly melts, a steady flow of water will trickle down from the mountains, and travel through rivers back to the sea. As the melted snow water slowly moves downhill, its gravitational potential is consumed. That water has achieved its "goal" of returning to equilibrium with respect to gravity.
- *Thermodynamics* is a branch of physics that concerns the flow of thermal energy (heat) and the ability to convert energy into work. The term statistical mechanics is an applicable, broader term that includes thermodynamics. When this chapter uses the term thermodynamics, statistical mechanics is also implied.
- *Energy* is meant in its physics sense. Physicists would say that the following possess energy. Moving objects possess *kinetic energy*. If the objects are randomly vibrating atoms and molecules in a solid, liquid, or gas, they possess *thermal energy*. A stretched rubber band would be said to possess *potential energy* that can be converted into either kinetic or thermal energy by un-stretching the rubber band.
- The *First Law of Thermodynamics* states that energy can change forms but never be destroyed. The total amount of energy in an isolated system (energy + potential energy) can never change in quantity, although it can change in form. For example, imagine a pendulum swinging back and forth. As the pendulum falls, it goes faster and possesses more kinetic energy and less potential energy. As it rises, it gains potential energy and loses kinetic energy.
- *Exergy* is what most people really mean when they use the term energy. Exergy refers to the quantity of available, useful energy. For example, substances such as coal and petroleum can be consumed to produce useful work. Although the energy involved in burning coal is still governed by the First Law of Thermodynamics, where chemical energy is transformed into thermal energy, exergy itself is not conserved.
- *Emergy* refers to embodied energy. It is a measure of consumed exergy, although the exact relationship is still being debated (Sciubba & Ulgiati, 2005).

- *Energy rate density*, in this chapter, refers to the rate at which exergy is consumed per unit volume. For example, when thermal energy flows from a hot to a cold region, exergy is consumed, and energy rate density becomes positive. Eric Chaisson describes it more precisely as "a mass-normalized (free) energy flow" (Chaisson, 2010).
- *Entropy* increases as all real systems progress though time. Entropy has a precise definition in physics: It is the log of a quantity called *multiplicity*, multiplied by a constant number (Boltzmann's constant). Entropy is produced when thermodynamic potentials are consumed. Operationally, entropy is a measure of how evened-out things are, such as when thermal energy flows to minimize temperature differences. A real-life system can remain active (alive) so long as it can produce entropy. When it can no longer produce entropy, it has achieved equilibrium. It cannot change in a meaningful way. It is essentially dead.
- The *Second Law of Thermodynamics* states that the total entropy of a system can never decrease.
- A *heat engine* is a device that can transform some of the flow of thermal energy from a hot reservoir to a cold reservoir into useful work.
- *Potential entropy* is the quantity of entropy that remains to be achieved. Just as energy + potential energy = constant, entropy + potential entropy = constant. This equation has not been proven and potential entropy is not a generally accepted term. It is chiefly used by this author for purposes of discussion and modeling. In reality, no known system has ever yet achieved its full potential entropy. Although energy can change back into potential energy, entropy cannot typically change back into potential entropy. For example, it is easy to burn coal to create heat and electric power, but it is difficult to recreate coal from ashes.
- *Bubble* will refer to a rise-then-fall progression due to the emergence of reproducing, dissipative structures operating upon a limited, nonrenewable resource. This is similar to a Hewett-Hubbert (HH) resource bubble (Hewett, 1929; Hubbert, 1949, 1956), as further discussed in Sects. 18.3.7 and 18.3.8.

17.3 Extensions of the Second Law of Thermodynamics, Complexity, and a Thermodynamic Perspective on the Emergence of Life

Is the Second Law of Thermodynamics inherently static or dynamic? Or does the Second Law of Thermodynamics place a prohibition of the total entropy of a system from decreasing, but it says nothing about whether the entropy of a system need to increase. Nevertheless, we typically do see entropy increasing in real-life systems. There is a tendency in systems for thermal energy to evenly distribute itself in a manner that temperature becomes equal for all regions in a system. Sometimes this is called an entropic force. Chemical and pressure differences also involve thermodynamic tendencies that can result in flows and the consumption of potential. Humans are biological organisms involving chemical reactions.

For example, let us consider a thermal difference (a thermal potential) bridged by a heat engine rather than a thermal conductor. In addition to performing work, heat engines typically increase entropy. If the work produced by a heat engine can produce additional engines, and those engines can produce yet more engines, then it is possible to have exponential growth in heat engines. Likewise, this can result in an exponential increase in entropy.

We have obviously moved beyond a strict reading of the Second Law of Thermodynamics, from where entropy cannot decrease to an assertion that entropy is likely to increase. There have been numerous proposed laws to describe the nature of this increase, and this discussion has been substantially the subject of entire conferences such as Thermodynamics 2.0 conference in 2022 (IAISE, 2022). There is not yet a general consensus on what these laws are, but there is a large and significant body of evidence for some of the general themes advocated by many of these extended laws. They shall be characterized below.

17.3.1 Maximum Power Principles

It has been proposed that systems tend to configure themselves to maximize their rate of power production. Alfred Lotka proposed the Maximum Power Principle (MPP) (Odum, 1996, p. 19). Ecologist Howard Odom extended MPP to the Maximum Emergy Principle (MEP) (Odum, 1996, p. 20). Cosmologist Eric Chaisson has studied how Energy Rate Density enables emergence of complexity (Chaisson, 2010, 2014, 2015). Power is typically defined as the change in energy versus the change in time. Most of these principles involve the exergy form of energy.

17.3.2 Maximum Entropy Achievement Rate

Maurice Biot pioneered a variational approach to thermal energy transport and entropy production (Biot, 1955). This approach anticipates that entropy production may be subject to the Principle of Least Time. The most classic case of the Principle of Least Time involved the refraction (bending) of a path of light traveling through differing media to minimize the time taken by the light to travel through those media (Fermat, 1894). In a sense, a bent path of light represents the emergence of a slightly complex structure.

This effect of refraction in this classic case has been shown to apply to the travel of thermal energy. When such paths bend or refract, they are in essence and emergent structure of increased complexity. In mechanics, this principle is a special case of the Principle of Least Action, but entropy is more directly applicable to thermodynamics and dissipative systems.

17.3.3 Emergence of Dissipative Structures

In chemistry, Ilya Prigogine observed that dissipative structures form that increase entropy production (Prigogine, 1967). This principle has been observed by many folks in many fields. For example, when modeling atmospheres, astrophysicists will tend to choose the form of energy transfer that maximizes heat flow, such as convection versus conduction or radiation (Carroll & Ostlie, 2007). Moving beyond implicit statements, it has been explicitly proposed that entropy production is subject to the Principle of Least Time and that that systems will tend to emerge in a manner which maximizes their rate of entropy increase with respect to time (independently by psychologist Rod Swenson (Swenson, 1989, Swenson, 2011, Annila & Salthe, 2010, Spier, 2009, Ciotola, 2010).

17.3.4 Applications of the Extended Laws to Emergence

We have discussed above the case in which the work produced by heat engines can be used to produce additional heat engines. This effect could be generalized to apply to any engine that consumes entropic potential. This effect would be consistent with the application of the Principle of Least Time to entropy production. In fact, we frequently see this phenomenon in nature. For example, fire consumes entropic potential and can easily reproduce, so can bacteria and other forms of biological life.

17.3.5 Emergence and Measures of Complexity

The term emergence is often applied to the spontaneous formation of complex structures. It has been suggested that emergence results in regions of low entropy as compared to their surroundings (Schrödinger, 1956). Examples of emergent structures include convection cells, stars, galaxies, machines, and living organisms. Relatively few persons would deny that such structures are more complex than their surroundings.

Although there is no universal consensus, several definitions of complexity have been proposed. *Kolmogorov complexity* is chiefly used for information, and it is a measure of how random information is (Tice, 2003). The more random information is, the more parameters are required to describe it in a lossless manner, so here, the quantity of required parameters would be a measure of complexity. Kolmogorov complexity is similar to the physical definition of entropy. However, these randomness definitions don't really capture the spirit of the term complexity as commonly used and in fact is literally the opposite of emergence.

Systems theory provides yet another perspective, defining complex systems as "systems with interdependent parts." In this view, the intuitive complexity of a task "is the number of wrong choices for every right choice" (Bar-Yam, 2003). Another information-oriented measure is "time complexity," which is how long an algorithm takes to run. It is typically estimated by how many steps counting the number of elementary operations performed by the algorithm. Time complexity is commonly expressed using big O notation (MIT, 2023; Wikipedia, 2023). Yet another measure is the connectedness in networks. For example, a food chain with "species number and connectance" could be said to be more complex than a simpler chain (Williams & Martinez, 2000).

Another implied measure of complexity is the differential entropy of a region compared to a surrounding region. If a region has less entropy than the surrounding region, it is assumed to have greater complexity. This phenomena is often described under the term "negentropy." For example, a fish would have less entropy than the surrounding water of a lake or ocean, so the fish could be considered to be a complex structure. The term "negentropy" is a shortened version of the term "negative entropy," which was first coined by Edwin Schrödinger (Schrödinger, 1956). That these discussions actually concern entropy density should be noted in anticipation of the next measure.

Eric Chaisson has proposed that energy rate density is an appropriate measure of complexity (Chaisson, 2010, 2015). The energy rate density approach has several advantages. It is an unambiguous physical quantity. It is routinely determined for physical systems and in some cases can even be directly measured. It corresponds closely to what many people would call complexity. It can be used to compare a wide range of systems. Finally, whether one agrees that energy rate density as a true measure of complexity, it is at minimum a reasonable baseline proxy for complexity.[1]

17.3.6 Immediate Thermodynamic Environment of Life and Humanity

Likewise, the development of the space environment since the Big Bang has allowed for more rapid increase of overall entropy. Stars allow for sustained nuclear fusion to release the tremendous amount of potential energy locked into nuclear structures. Likewise, disks of gas and dust formed around stars due to gravitational attraction and dissipated gravitational potential energy. Also, in such disks, planets formed. Although many such planets are giant, hot balls of gas similar to Jupiter, other planets are moderate in size and temperature, such as the Earth. Instead of merely reflecting light back into space, such planets are capable of hosting sophisticated chemical reactions can consume potential entropy from starlight. In the case of the Earth, biological life formed. Eventually, most life on Earth became powered by Sunlight.

When it is said that most life on Earth is powered by Sunlight, really, life on Earth is typically powered by the entropic potentials present on the Earth's environment in

[1]For more detail on various definitions and operationalizations of complexity see also chapter "Introduction to Complexity Theory for Big History" (LePoire et al., 2026) in this volume.—*Editors.*

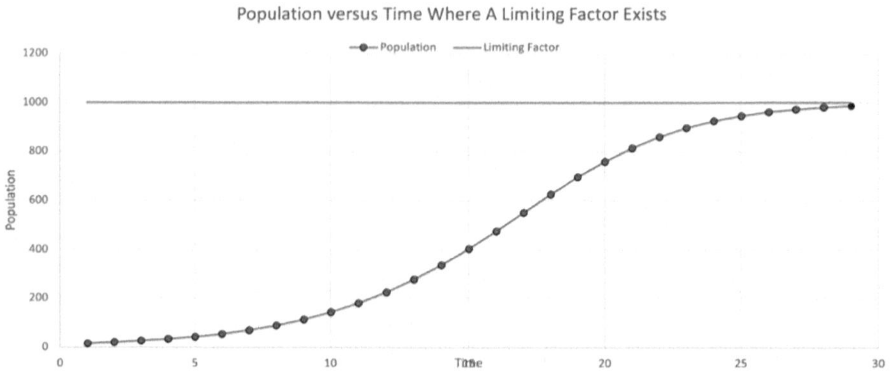

Fig. 17.1 Example of logistics curve

space. A thermodynamic potential exists between photons emitted by the Sun versus the cold, relative darkness of space. Humans make their living off of this Sun-Earth-Space potential, chiefly from Sunlight-driven plant production and related animal production, but also from Sunlight-driven wind and hydro-power. To survive, humans must "go with the flow," literally with the flow of energy from the Sun to the Earth to space. The sum of social interactions must be consistent with harvesting some of this flow for a human society to endure.

17.3.7 Hewett-Hubbert Bubbles and Their Application to Physical Resource Bubbles

There will always be factors that limit the growth in a system. A regime that experiences exponential growth will eventually begin to experience such limiting factors. Work on system dynamics such as the Club of Rome's *Limits to Growth* (Meadows et al., 1972) involves attempts to better understand these limiting factors. Such factors restrain growth and sometimes stop it altogether. Limiting factors usually exist due to a shortage of some essential resource or an excess of some "negative" resource. Turning to biotechnology, an examination of reproducing cells shows that the chief limiting factors are typically a nutrient limitation or an accumulation of a toxic metabolite (Butler, 1996).

We have viewed how exponential expansion is favored by thermodynamics. However, exponential production cannot increase forever. There are limits to available potential. For example, in the Sun-Earth-Space scenario, the Earth receives a limited amount of sunlight per year. Once an area is populated with plants and dependent species of animals and other life, populations cannot generally expand. Then life is first introduced, populations will expand exponentially initially, but then level off to approach an equilibrium. This limited growth is called logistical growth, and a plot of such growth is called a logistics curve or S curve (see Fig. 17.1).

Even in an environment that is rich in resources, there are limits. Scaling issues result in the decrease in surface area to volume ratio of the organism colony. Lack of oxygen can be a limiting factor for large cell cultures. The organisms often cannot get access to abundant resources because they are crowded out by their neighboring organisms. Multi-cell organisms attempt to overcome the surface area limitation with structures such as veins and folding. Yet this approach itself has limits. An elephant still faces many challenges as compared with an ant, such as expelling sufficient body heat. Human civilization meets a similar surface area challenge with similar structures. The great freeways and road networks in cities and even across the countryside resemble the blood circulation system in our own bodies.

If there is a nonrenewable, built-up potential, such as oil, coal or gold, then generally, the intrinsic efficiency at which each additional unit of consumption is transformed into production decreases. This is expected: we go for the low hanging fruit first, then the slightly higher fruit, and only go for the hard-to-reach fruit at the top of the tree last. There are two chief types of efficiency: intrinsic efficiency and overall efficiency. Intrinsic efficiency is limited by the Second Law of Thermodynamics prohibition on entropy decreasing in an isolated system. Overall efficiency cannot exceed intrinsic efficiency, but it can be much lower. Technical improvements and economies of scale can help improve overall efficiency. Overall efficiency functions can be in several forms, depending on the circumstances involved. For example, for a nonrenewable resource, the efficiency function may decay linearly or exponentially. There are ways of determining the overall efficiency function from actual data or known constraints.

17.3.8 The Rise and Fall of a Bubble

We have seen how the emergence of reproducing dissipative structures can lead to exponential growth. We have discussed how intrinsic efficiency decreases as a nonrenewable resource is consumed. Hence, we now have the conceptual means to understand the life cycle of a bubble. To be *clear, from this point further*, a ***bubble*** will refer to *a rise-then-fall progression due to the emergence of reproducing, dissipative structures operating upon a limited, nonrenewable resource*. The term "bubble" here is an analogy of a convective bubble. Such are also known as convection cells or Bénard cells (Swenson, 1989).

1. A thermodynamic potential accumulates. [Complexity: this is a concentration of minor "raw" complexity].
2. A reproducing, dissipative mechanism has emerged to consume the potential. [a single instance of high complexity].
3. The mechanism reproduces exponentially. Consumption increases exponentially. [The instances of high complexity increase, AND the types of high complexity increase; there are more modes of complexity].
4. Growth continues, but intrinsic efficiency decreases. [Complexity is maintained].

5. Eventually either all the potential gets consumed, or the efficiency of exploiting it falls below the ability of the mechanism to maintain itself. [ERD decreases, but complexity artifacts may endure for a while longer].
6. The progression of the bubble ends.

Since exponential growth is involved, but the transformation of consumption into production must be discounted by efficiency, this approach to bubble modeling is called *efficiency-discounted exponential growth (EDEG)*.

A bubble can involve physical or social resources. However, it is easier to begin analyzing physical resource bubbles, since there is often more available data and the examples are fairly straightforward.

It is instructive to apply this bubble paradigm to a few cases involving natural resources such as precious metals or fossil fuels. Mining is both a physical and social activity. Deposits of a commercially-demanded substance represent an economic potential (which is ultimately a thermodynamic potential). Mined substances are typically nonrenewable. If you extract a ton of that material, another ton does not form in the ground. Such substances took millions or even billions of years to form and accumulate. Therefore, there is only so much to extract: the mining bubble must eventually end. In other words:

$$\text{past consumption} + \text{future consumption} = \text{a constant}$$

at any point of time. Further, since the resource is nonrenewable, the intrinsic efficiency of extracting the potential will tend to decrease over time (according the hypothesis above).

It is preferable to examine a case where there is ample data. To apply bubble analysis of a mining region, it is best if the region considered is sufficiently large to initially support many actors, such as mining enterprises, at least at the beginning, so that individual differences average out to clarify the overall pattern. Multiple actors with significant variation provide a system with a flexible "stochastic" capability to allow the system to adapt and progress. In other words, complexity should exhibit some granularity at the beginning.

Then in early stages, but after things get started, no individual mine can determine the fate of the entire region. For example, poor management at one mine is offset by effective management at another. The region should be sufficiently isolated so as not to be too influenced by external factors, but there must nevertheless be an internal or external demand for the substance. Arguably, commercially-viable deposits of minerals represent a concentration of relatively low entropy. Arguably this concentration could likewise be a complex structure in the physical, macroscopic sense.

San Juan Mountain Area Mining Region: The San Juan mountain area (known as the "San Juans") in southwest Colorado is such a region, and is a suitable physical and social example of a bubble, including the mining society that developed in the region. The San Juans region of Colorado produced gold and silver from hundreds of mines, around which towns and communities eventually developed.

The San Juans were ruled by Spain then Mexico until 1848, and then by the United States. Spanish gold mining of placer deposits (pieces of nearly pure gold found on the surface) took place between about 1765–1776. Some US mining took place in 1860, but it was interrupted by US Civil War. At this point, "only the smaller deposits of high-grade ore could be mined profitably." Mining slowly started again in 1869. There were 200 miners by 1870. An Indian Treaty was negotiated in 1873, which removed a major obstacle to an increase of mining (Smith, 1982; Twitty, 2010). As the number of miners and the sophistication of mining operations increase, macroscopic complexity is likewise increasing. Mining facilities, equipment, and infrastructure are emerging and becoming more complex, evolving from tents, mules, and pick-axes to large buildings, deep elevators, and long cable systems. There are more relationship connections and interdependencies.

In 1881, a railroad service was established, resulting in a "decline in ore shipping rates." The regions heydays were between about 1889 and 1900. As the region matured, there was a major consolidation of mining operations as well as significant infrastructure improvements such as rail lines. By 1889, English investors had come to control the major mines. There were also labor troubles. Note that for mining, production refers to the processed metal rather than mere ore. Complexity has increased, perhaps to its peak.

By 1900, the region began to take on more of the characteristics of a settled community. By 1909, "the gilt had eroded," dilapidation set in, and the population decreased. World War I caused production to greatly fall, due to decreased demand from Europe, and the region lost workers. Silver and gold mining all but ceased by about 1921. Today, there is again mining in miscellaneous minerals, but not much in gold, which was the primary economic driver for the "great days" (Smith, 1982; Twitty, 2010).

Hewett-Hubbert (HH) curves are functions that represent the extraction of minerals or petroleum over time for a large region or country. Such curves apparently originated with Donnel Foster Hewett for European metal mining (Hewett, 1929). Such curves were then applied by petroleum geologist M. King Hubbert to US domestic petroleum extraction (Hubbert 1947, 1956). Hewett-Hubbert (HH) bubbles (also called curves or functions) represent statistical distributions regardless of how they are generated. HH curves generally represent situations where there are initially a considerable number of actors, such as in the case of regional or national oil or mining production.

Efficiency-discounted exponential growth (EDEG) is an approach to generating an HH function (Ciotola, 2010). Applying the Principle of Least Time to entropy increase (as we have seen above, is a proxy for the emergence of complexity and maximum energy rate density), it is reasonable to expect exponential growth, which in fact is often observed. The exponential growth can be in terms of exergy consumption, metals production, population, or physical economic activity. In the case of a renewable resource, we observe a logistics curve. However, for the case of HH bubbles, a critical resource is nonrenewable, the curve will rise then slope downwards, and for practical purposes, end in a finite about of time. We shall see several examples below.

Fig. 17.2 Colorado San Juan gold production versus model (data: Henderson, 1926)

Before going further, let's explore some several approaches to analyzing Hewett-Hubbert bubbles in terms of complexity. One way would be to quantity the number of actors and structures, and interconnections between actors. This is akin to network-oriented "connection science" approach pioneered by Sandy Pentland (Pentland, 2023a, 2023b). Another way would be to use a Kolmogorov complexity approach and quantify amount of information to describe the progression of the bubble with respect to time. Yet another way would be to use a method akin to Chaisson's approach to complexity to use power (exergy/time) as a proxy for complexity. This is the primary approach used in this chapter, albeit with handling density somewhat differently.

An EDEG model was generated for the San Juans region and compared with actual mining data (see Fig. 17.2) (Ciotola, 2016). Exponential growth was initially facilitated by an increase in the quantity of mines, then later by the growing size of mines. Intrinsic efficiency decline in mining tends to comprise decreasing quality ore. For example, the early-mined placer deposits might comprise 100% gold, while late-stage ore might comprise less than 1% gold. The model is somewhat higher than the peak, but parameters were adjusted to provide a better fit with the overall data. Deviations shown in the curve occurred due to both random events, social, economic and logistic "turbulence," business cycles, and major external events.

This plot clearly indicates the emergence of a process. Does this plot also represent changes in complexity? Only partially, if macroscopic complexity differs from mere production. There were giant ore processing plants and advanced civic organizations operating well past the peak of the regional production rate. However, macroscopic complexity during the peak of production was clearly much higher than in the early stages of the mining era.

US Petroleum Production: Petroleum (oil) is a mixture of hydrocarbon molecules. The formation of oil is complicated and involves a series of steps and conditions that typically require hundreds of millions of years to complete. First, organic matter from dead plants, animals, and plankton falls to the bottom of shallow sea basins. If organic material is deposited at a high rate in large quantities, it will become preserved. Eventually, the deposited matter sinks deeper due to geologic forces as well as the weight of the accumulating matter above it. As the matter sinks thousands of feet under the ground, it becomes hotter and the organic material is

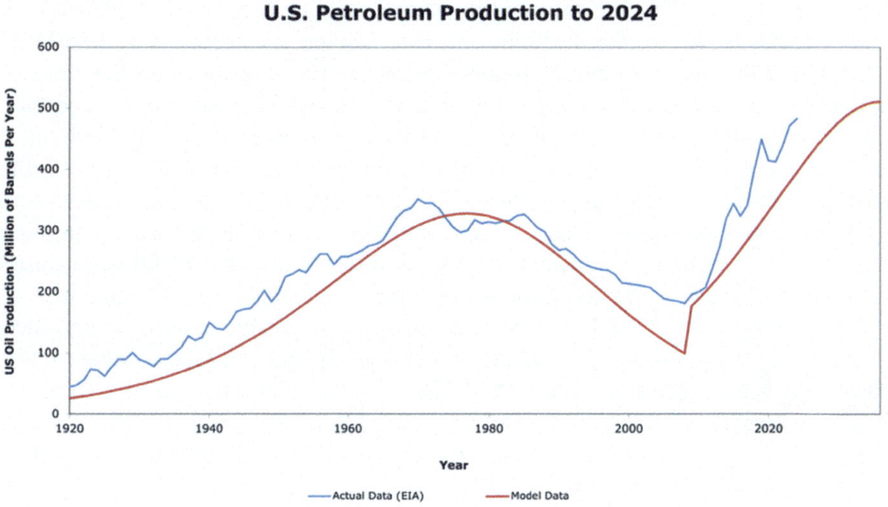

Fig. 17.3 EDEG model for US petroleum production up to 2024 (data: EIA, 2025)

literally "cooked." Heat breaks down the organic matter into long hydrocarbon molecules. Eventually oil is formed (Chapman, 1983; Deffeyes, 2001).

An EDEG model was generated for US domestic petroleum extraction (see Fig. 17.3) (Ciotola, 2010), and updated to reflect "tight oil" (via fracking) that became newly extractable since about 2009 due to breakthrough technology, hence resulting in a double peak. Actual data exceeds model prior and after peak. Parameters were set to match peak, but could have been adjusted for less error elsewhere at the expense of greater peak error.

As oil became more available, the uses and demand for it multiplied, driving and financing an exponential increase in production. That oil wells have had to be drilled increasingly deeper and in unfavorable areas such as at sea is evidence of decreasing intrinsic efficiency. Some of that decrease has been mitigated by improved exploration and drilling technologies as well as industry consolidation.

Petroleum extraction represents an especially interesting case, because energy is literally being extracted. For a dynasty (or modern government) substantially dependent upon petroleum for its energy and economy, the rate of net petroleum production versus time *is* its power progression (in physics, power is the rate of energy expended per unit of time). Net production means the total energy extracted less the energy required for production operations.

A relevant perspective on oil production is energy return on investment (EROI), a concept developed by Cleveland, Hall et al. EROI is the "ratio of gross fuel extracted to economic energy required directly and indirectly to deliver the fuel to society in a useful form" (Cleveland et al., 1984).

This ratio can change over time. For example, petroleum discoveries in the 1940s had an EROI of over 100.0, while petroleum discoveries in the 1970s had an EROI of 8.0. This change, in part, reflects the increased costs of discovery and extraction due to the more challenging nature of the later remaining deposits (Cleveland et al., 1984). EROI is akin to the declining intrinsic efficiency considered in the EDEG approach. EROI is a somewhat broader measure. EROI focuses on the ratio of benefit to cost to produce energy. The efficiency in EDEG focuses on discounting consumption of resources to obtain production. A desirable EROI is a number that is as high as possible (e.g., 100.0), and certainly would be greater than 1. A desirable EDEG would be a number approaching 1 (e.g., 0.8), but never equal to or more than 1. Both concepts capture the fact that a physical quantity of energy (e.g., petroleum in the ground) cannot be obtained for free. A difference is that EROI can capture economies of scale, while the intrinsic efficiency component of EDEG cannot.

Here, complexity viewed as normalized energy rate density is simple to consider. Since the regions considered remain nearly geographically constant, volume (in this case being area) remains constant. Petroleum *is* substantially free energy, so petroleum production versus time represents energy rate density over time. Hence, Fig. 17.3 is a plot of the energy rate density measure of complexity versus time due to domestic oil production.

17.4 Application of Hewett-Hubbert Bubbles to Historical Dynasties

17.4.1 An Intermediate Level Bubble

In between physical resource bubbles and major dynastic bubbles, in terms of time, and possibly scope, are Kondratieff Waves (K-waves). These are characteristics waves for industrial and postindustrial systems with a period of 40–60 years. Examples of these waves with respect to physical commodity production include coal and pig iron, as well as more abstract indicators such as prices, interest rates, and trade (Korotayev & Grinin, 2012). These cycles have upswing and downswing phases, and should in principle be capable of being analyzed via EDEG. Although this chapter does not do so, K-waves involving world gross domestic product appear most promising for EDEG analysis and an area for future research.

17.4.2 Emergence and Progression of a Single Dynasty

We will now express historical dynasties as emergent dissipative structures and generate power progression models of dynasties from fundamental principles. In this discussion, the term *dynasty* is used broadly to refer to a continuous ruling group; it could be a related family but not necessarily so. In contrast, the term *society* will refer to a large group of related people, typically of a single or similar group of ethnicities, such as the Han people in China or the Frankish people in France. Dynasties exist within a society, but they can conquer other societies as well. For example, the society of Russian people produced a series of dynasties, and those dynasties sometimes conquered other societies.

17.4.3 Dynasties as Bubbles

Both physical and social built-up potential can drive the formation of dynasties. Within the context of a civilization progressing over centuries, it is often possible to degrade built-up potential even more quickly with high-level, governing social structures for a society. Hence, dynasties form to accelerate consumption of such potential. Hypothetically, dynasties should result in more rapid degradation of energy than does a more static society. Hence, dynasties are good subjects for complexity analysis.

Each dynasty has a life cycle. A dynasty is born, matures, endures awhile, then ends. A new dynasty will not necessarily follow an old one, nor might it immediately appear. Yet generally, dynasties continue to form, one after another, so long as there exists built-up potential that cannot be more quickly consumed by other means. The dynastic life cycle can be described as a march toward equilibrium in terms of nonrenewable resources, and toward dynamic equilibrium in terms of ongoing flows such as sunlight and rainfall.

We can consider large, independent, robust dynasties to be bubbles. A new dynasty within a society encounters a built-up potential of physical and social resources (e.g., goodwill), albeit of limited magnitude. The society governed by the dynasty fills the role of a collection of heat engines, producing both work and entropy. Prosperity expands exponentially, increasing the consumption of potential exponentially. Eventually, it becomes increasingly difficult for the dynasty to rely upon its store of physical and social resources, decreasing its efficiency. As efficiency decreases, the dynasty will experience social crises and will eventually stop functioning.

17.4.4 Considerations for Modeling Dynasties

It is simpler to model a sufficiently large, robust, independent dynasty than one that existed merely at the whim of its neighbors, for there are less significant dependencies, and thus it can be approximated as a substantially isolated system. We will examine Russia's Romanov dynasty as an example. Widely accepted start and end dates are 1613 and 1917 (Mazour & Peoples, 1975).

Peter the Great and Catherine the Great were the two important rulers of the Romanov dynasty, and the Russian Empire gained much of its most valuable territory by the end of Catherine's reign in 1796. The Romanov dynasty was big, robust and essentially independent. It fought wars, but generally was not under serious threat of extinction. Even Napoleon could not conquer Russia. This Romanov dynasty was reasonably long-lived, rather than just a quick, "flash-in-the-pan" empire.

By developing a fundamental approach to modeling the rise and fall of dynasties, it is possible to accept or reject models (within a range of uncertainty) based upon both qualitative historical evidence and quantitative historical data. We shall discuss generating models of the rise and fall of power of dynasties versus time and how a single dynasty can be modeled using the efficiency-discounted exponential growth (EDEG) approach (Ciotola, 2014). Such an EDEG model of a dynasty can be called a power progression. The models shown should be considered mere first approximations rather than definitive assertions of fact. They are a beginning point of further explorations.

Historical dynasties are consumers of energy and producers of power, so models in terms of such quantities are inherently fundamental in that they can be derived directly from the laws of physics and expressed in physical quantities. Such models are not theories of everything, but rather describe certain aspects of broad macro-historical phenomena rather than the intricate workings of the interactions of individual people.

The term energy is meant in the physical sense here. There are several possible proxies of the physical energy of a dynasty, such as population governed or grain production. Each of these is translatable into physical units of energy. For example, the quantity of people multiplied by the mean Calorie diet per person will result in units of energy. These figures can be estimated for most dynasties over their lifespans, albeit with differing degrees of uncertainty. The proportion of that energy that rulers of a dynasty effectively have at their disposal is beyond the scope of this chapter, but should be considered for improved accuracy.

Power is a physical term. It refers to energy expended per unit of time. Yet it also has meaning within social and political contexts, and it will be discussed in both senses. Absolute power would generally be presented in physical units of power such as Watts. However, it is possible to express any type of power in terms of proportions, such as the ratio of power at a dynasty's peak to its start date. Such a ratio can apply to physical, political, or even military power. Possibly, the EDEG approach can be utilized to model other types of power, such as political power. In fact, the EDEG approach provides a framework to explore the question of how political and physical power are related.

17.4.5 Exponential Growth of Dynasties

A new dynasty will tend to experience exponential growth. A chief characteristic of exponential growth is that growth feeds even more growth, resulting in an increasing rate of growth. Increases in population and power can become explosive. Nevertheless, the growth rate in early stages tends to be relatively flat, while the growth rate later tends to be relatively steep. The transition between "flat" and "steep" can be surprisingly sudden and disruptive (Meadows et al.). It will be assumed that dynasties will strive to grow exponentially. (This chapter does not attempt to prove this assertion, but rather it is a rebuttable presumption.) If so, this certainly explains the rise of a dynasty. Sources of growth can include increased agricultural productivity, geographic expansion, and trade expansion.

The Romanov dynasty is shown with various growth rate models (Fig. 17.4). The plot shapes appear similar, except that a greater rate produces a "sharper" corner. Also, notice the range of power values: a greater growth rate produces a

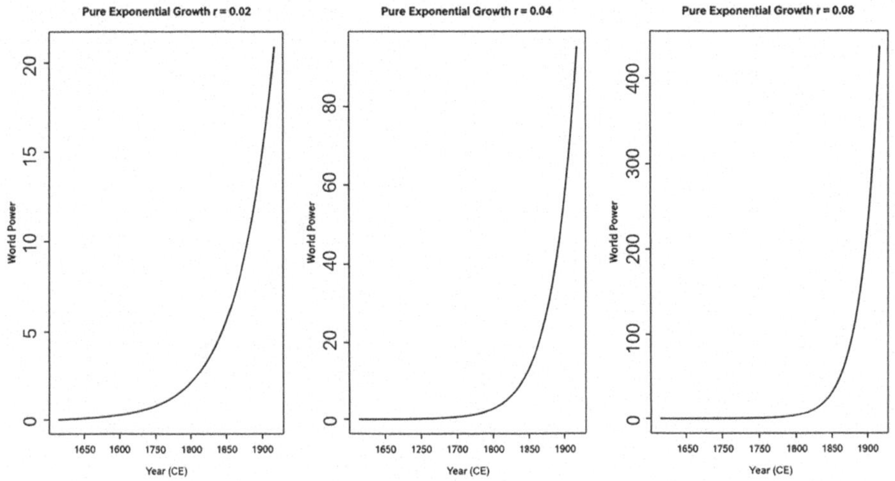

Fig. 17.4 Different rates of exponential growth

disproportionately greater power value at later points of time. The growth rate function can be constrained by the data and an understanding of the growth mechanisms involved.

17.4.6 Limiting Factors and Decreasing Efficiency

Another source of limiting factors is the increasing cost-per-unit to extract nonrenewable resources such as minerals. Societies attempt to use large-scale social and technical structures to shore up efficiency (e.g., San Juans mining case study), but these structures create additional challenges. There are other examples. In the United States, the "closing" of the western frontier marked a limit of growth to homesteading. In petroleum production, the increasing cost of drilling for oil is a limit to growth. Classical economist Thomas Malthus pointed out limiting factors in the growth of agricultural production (Heilbroner, 1980).

A dynasty will typically consume both conserved *and* renewable resources. Yet it is the consumption of one or more critical nonrenewable resources that determine the growth and decline in characteristics of the regime. Production in a dynasty is ultimately dependent upon nonrenewable physical and social resources (otherwise dynasties would not typically end). Dynasties inevitably do end, which is typically preceded by a decline in power.

As the dynasty progresses, nonrenewable resources will be consumed, and efficiency will decrease. There will still be production until the end, but there will be a lower return on investment, so to speak. Physical causes of decay can include overuse of agricultural land leading to nutrient depletion, the build-up of toxins in the environment, and the depletion of old growth forests. Social causes can include

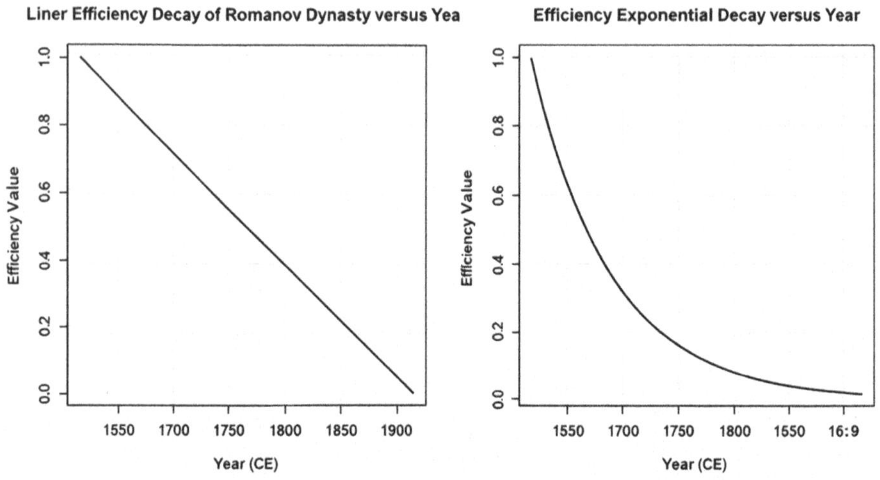

Fig. 17.5 Linear efficiency decay of Romanov dynasty versus year and efficiency exponential decay versus year

running low on social goodwill, the increased dependency on expensive, monopolistic centralized institutions and structures, and the resulting decreased accountability of aristocratic "deadwood." All such may have nonrenewable aspects.

The key impact of limiting factors, whether insufficient positive resources or excessive negative resources, is a decrease in the efficiency of whatever is acting as "heat engines" to do work. There are two types of decay, linear and exponential. Examples are compared. Note that efficiency here is shown as a proportion (multiply by 100 to get a percentage) (Fig. 17.5).

Exponential decline in an EDEG situation can happen more quickly than exponential growth. However, there are two disadvantages of exponential decay within the context of modeling dynasties. First, it is more difficult to set up. For example, exponential decay has an infinitely long tail. While this allows for mathematical immortality, most of the tail is superfluous in the context of a dynasty of limited lifetime. Second, it may not provide the most consistent models with observations, since there can be multiple sources of efficiency decline with differing functions.

A linear approach is simpler to set up. Importantly, it also provides some reflection of overall efficiencies achieved through centralization and economies of scale as the dynasty progresses. Centralization can produce economies of scale that can boost net efficiency, but when a centralized system eventually goes bad, it can go bad fast! Failed central institutions can bring a dynasty crashing down quickly. This is an example of an irreversible process.

A linear approach has unambiguous beginning and end points. Efficiency cannot be greater than 100%, and it is typically not lower than zero. Therefore, as a first approximation, one can set the overall efficiency to 100% at the start date of the dynasty and 0% at the end year (except that the math is simpler if the value 1 is used for 100%). Although physical efficiency is typically lower than 100% for real-life

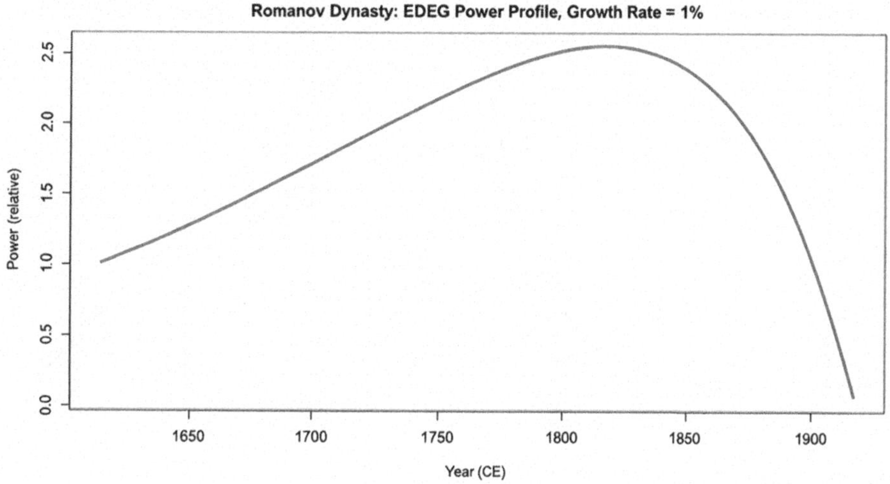

Fig. 17.6 Romanov dynasty: EDEG power profile, growth rate = 1%

heat engines, value 1 provides an easy starting point that also produces the correct shape of curve. (It can be adjusted based on the nature of the power source). Using a value of 0 for ending efficiency ensures that the dynasty ends by its historical end date. It is possible to use a value other than zero for the ending efficiency, but then some other factor must be used to end the dynasty. The following is an example of linear decay function:

$$Efficiency = 1 - \frac{(\text{year} - \text{start year})}{(\text{end year} - \text{start year})}.$$

As the year increases, efficiency will decrease. Using a lower initial efficiency reduces the magnitude of production increase for the dynasty compared to its initial production. It also flattens out the curve (see Fig. 17.6).

17.4.7 Generating a Dynastic Power Progression

We now bring exponential growth and declining efficiency together.

Growth will not only slow down but often will start to reverse. Such growth and decline can be represented by an EDEG function, where the area under the curve represents either the total production or consumption of a conserved resource over time. Note that the critical resource becomes more expensive as each successive unit of it is utilized. In the case of petroleum or a precious ore, the least expensive deposits are extracted first. Then the next least expensive deposits are extracted and so on. The following is an example of an EDEG equation:

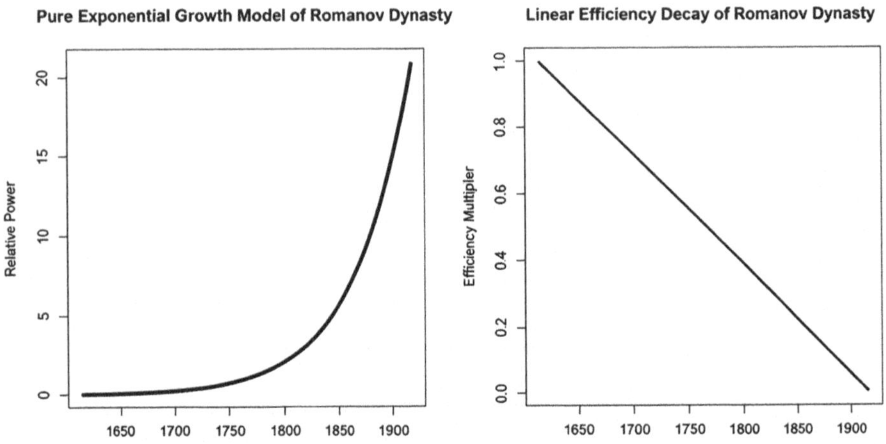

Fig. 17.7 Exponential growth and linear decay

$$y = \text{efficiency function}^{*}\ \text{exponential growth function.}$$

Here is a simple way to generate a quantitative model for a dynasty. It is simplistic, but it generally produces qualitatively correct results. Assume exponential growth:

$$P_{\mathrm{t}} = P_0\, e^{\mathrm{kt}},$$

where P is power, P_0 is initial power, t is time, and k is a growth factor.

Assume that a nonrenewable resource is being consumed that cannot be replaced within the lifetime of the dynasty. Then assume the efficiency of each subsequent unit of resource consumed produces power as a decreasing efficiency. Using the simple linear efficiency decay function from above:

$$\epsilon = 1 - \frac{(\text{year} - \text{start year of dynasty})}{(\text{end year of dynasty} - \text{start year of dynasty})},$$

where ε is efficiency (see Fig. 17.7).

Then the efficiency-discounted power is:

$$P = \epsilon\, P_0\, e^{\mathrm{kt}}.$$

Substituting in our functions (utilizing linear decay):

$$Y = \left(1 - \frac{(\text{year} - \text{start year})}{(\text{end year} - \text{start year})}\right) P_0\, e^{\mathrm{k}\ (\text{year}-\text{start year})}.$$

This produces a steady rise, a level period, and a slightly faster decay. By discounting exponential growth by decreasing efficiency, we then have a rise and fall pattern that is consistent with the rise and fall of a dynasty.

Let us assume a conservative 1% growth rate for the Romanov dynasty. Let us further assume linear decay from 100 to 0% efficiency. A simulation has been written in the Ruby programming language. This language is mathematically robust, yet it involves code that is relatively easy to read and understand. The dynasty is run through the Ruby simulator, using the above parameters. The R program was utilized to generate a plot of the results.

Here the peak is close to 1820. Napoleon had been conquered, and the dynasty had achieved much of its geographic expansion by then. Yet by this time, social unrest began to shake the Romanov dynasty. Also, note how the dynasty power begins at a level of 1 and ends at a level of 0. This is conceptually appropriate, since the dynasty had to begin from something, but typically ends in nothing. (Actual power quantities can be used, but their explanation is beyond the scope of this chapter.) For example, the ancestors of the Romanovs existed before 1613, but the entire immediate family was killed during the Russian Revolution. The peak occurs at a relative power value of height of 2.6, which indicates that the dynasty was over twice as powerful at its peak as it its beginning. Remember, this model is merely a hypothesis that is either valid or not for a particular level of uncertainty.

Note that utilizing a higher growth rate results in a later peak. Also, the total peak to initial power ratio skyrockets as the growth rate is increased. Additional factors can be imposed as adjustment functions. One-time events (such as a rare but large natural disaster) can be superimposed as an event "mask."

It may be of further interest to tie the rise and fall to patterns concerning the production and consumption of resources, to determine what correspondence, if any, there is between physical and political power. This can be explored by utilizing actual physical energy data to produce a model of physical power, and then comparing that model with evidence of political power over time. With the wealth of historical data being gathered in anthropological data warehouses, and other "big data" facilities, this may be accomplished with increasing validity.

17.4.8 Narrative Description of Generic Dynastic Progression

We can use the EDEG approach to describe anecdotally the life cycle of a major dynasty (e.g., China, France, West Africa). It would typically begin with a daring, competent, often unpolished leader, but with effective power loosely distributed. Chaos becomes order and economic production rises. Yet, future generations of rulers will become increasingly desirous of luxurious living and will also demand expensive "trophies" such as palaces, major public works, or optional conquests. This will stress the resources of that society, and the dynasty will experience financial difficulty. Initially, centralization will be used to increase overall efficiency, but at the expense of decreasing individual initiative and decreased ruling class accountability. Eventually, taxes will need to be increased. Bureaucracy will need to be greatly expanded to collect increased taxes. Internal dissatisfaction will increase, so greater internal military effort will be required to suppress rebellions. Dynasty rulers will become increasingly dependent upon their military to maintain internal order and to enforce tax collection.

Meanwhile, the rulers will tend to become increasingly occupied with court etiquette and pursuit of "civilized" activities as art and scholarship; but they will become less competent at governance and further removed from the realities of the population they govern. A large, hungry population and bureaucracy has formed that cannot be downsized without considerable disruption (overshoot). Eventually competing figures from within the society will challenge the rulers. These initial challenges will be put down often brutally, further increasing discontent, and destroying much of the social structure and institutions required for the effective maintenance and defense of the society and its economy.

Due to the chaos and decreasing magnitude of economic activity, the population and its strain on natural resources will decline, allowing for some recovery of productive capabilities and once again the build-up of potential. Nevertheless, it is too late for the dynasty. Further challenges from either within or without the society will replace the dynasty, and a new dynasty will form. Must a dynasty have a life cycle? Could it not last forever, or at least indefinitely? Societies, religions, and some other institutions can last much longer than dynasties. It is conceivable that a dynasty could be managed in a sustainable manner, but this is not what we typically observe in history.

17.4.9 Comparison of Other Approaches to Model Dynasties and EDEG Curves

There have been other quantitative efforts to model dynasties. Joseph Tainter considered economic factors for the rise and fall of societies, including dynasties. Tainter associated collapse with a loss of complexity. While Tainter's approach is more narrative than quantitative, sufficient data are provided to illustrate the principles asserted. Tainter discusses how costs and benefits drove the rise and fall of the Roman Empire. "Costs rise and benefits decline. For a one-time infusion of wealth from each conquered province, Rome had to undertake administrative and military responsibilities that lasted centuries" (Tainter, 1988, 150). Further, as costs increased, "the net value of local autonomy exceeded that of membership in the Empire" (Tainter, 151). By some point in time approaching the collapse of a society, ironically, "a drop in complexity brings with it a corresponding rise in the marginal return on social investment" (Tainter, 1988, 152). Tainter's approach is conceptually akin to Cleveland's EROI approach. However, this author makes a distinction concerning with assertions by Tainter that more complex societies are more expensive to maintain in that, yes, they are more expensive when the entire relevant system is considered, but can be less proportionately expensive when only the society itself is considered, hence a distinction between internal and external costs. Further, the term "expense" takes of different meanings when differentiating between energy versus entropy from different perspectives. From a cosmological perspective, "unspent" entropy is expensive.

Ugo Bardi uses whale oil production as a case study with similar characteristics to the EDEG approach (Bardi, 2020, 61). Whale oil is a resource that cannot be quickly-renewed. Bardi uses this example to critique Hubbert's empirical use of bell-shaped models (Bardi, 2020). Over-hunting whales produces a time-reversed logistics function (a backwards "S") for the whale population, thus resulting in increased per unit costs and an asymmetric consumption function. This author is in general agreement with that reasoning. Bardi's description of Hubbert as "an empiricist who looked at the data and used common sense to interpret them" (Bardi, 2020, 63) is arguably an exaggeration. As noted by this author, Hubbert's models were based upon earlier mathematical mineralogical works rather than mere common sense, but those models were nevertheless empirical. In Hubbert's defense, computers are not available during much of Hubbert's formative years. However, this author argues that Bardi's use of system dynamics is merely one step removed from empirical. While is does capture interdependencies and the dynamic behavior they cause in systems, it does not truly capture the overarching forces.

Peter Turchin and Sergey Nefedov also provide several observations, in part inspired by Tainter, concerning the collapse of societies that those authors argue could provide the basis for the principles of a science of history. Turchin and Nefedov reject the assertions of Karl Popper that "that historical processes were too complex and history was too different from natural sciences to have general laws such as those discovered in physics or biology" (Turchin & Nefedov, 2009, 312). They propose the following general principles. The first is that "neo-Malthusian principle: during periods of sustained population growth, if the output of the agrarian economy does not keep pace with the population, a number of relative price trends will be observed." The second principle concerns how the elite profits from overpopulation. Elites then grow in numbers until their expense exceeds the carrying population of the non-elites; this results in costs that exceed the benefits of being part of that society. (This argument is akin to Tainter's analysis.)

A third principle concerns the causes of sociopolitical instability. "The demographic-structural theory proposes three principal causes of the onset of a disintegrative trend (that is, a lengthy period of heightened instability): overpopulation, elite overproduction, and a fiscal crisis of the state." Whereas "the dominant role in internal warfare appears to be played by elite overproduction leading to intraelite competition, fragmentation, and conflict, and the rise of counterelites who mobilize popular masses in their struggle against the existing order" (Turchin & Nefedov, 2009, 312–314). This author observes that these phenomena are commonplace in the progression of societies, but argues that they are the consequence of higher-level, more fundamental thermodynamic principles related to the cause of emergence.

17.4.10 Series of Dynasties

A society can be modeled as a series of dynasties or EDEG bubbles. Each bubble would typically represent a dynasty for a traditional historic monarchy. Robust, traditional, monarchical, and agricultural-based regimes have historically tended to endure for roughly 300 years. This is an empirical observation and not based upon theory. Not all regimes last for about 300 years. Yet the 300-year pattern has appeared frequently in history from France to China to West Africa. We will focus on dynasties that endure for about that time.

A common error would be to assume that the series of EDEG curves represents a periodic function. It's not. Dynasties might not follow immediately one after another. Not all dynasties last the same amount of time. Or there could be some overlap between older and newer regimes. Each dynasty is its own emergent structure.

Dynasties in major historical civilizations are typically easy to identify. In a sense, dynasties are what fill the pages of historical textbooks. A series of power progression models of Russian dynasties is plotted (Fig. 17.8). Although the plots are each set to a maximum power of one, actual power would vary among dynasties.

The duration of major Chinese dynasties versus their midpoint dates has been plotted (see Fig. 17.9 and Table 17.1). There might have been an exponential decay trend in the duration of major Chinese dynasties over time.

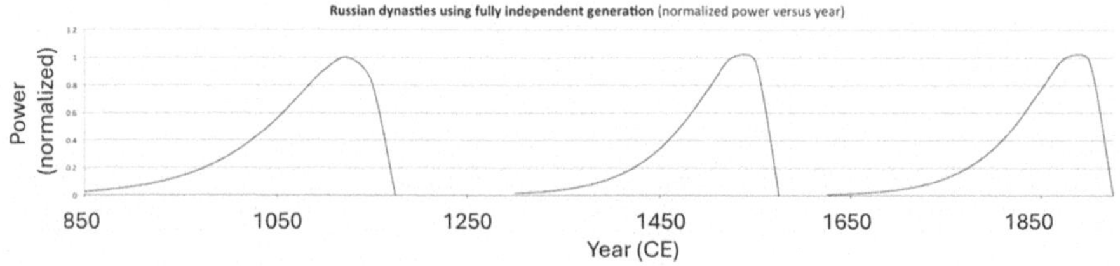

Fig. 17.8 Russian dynasties using fully independent generation (normalized power versus year)

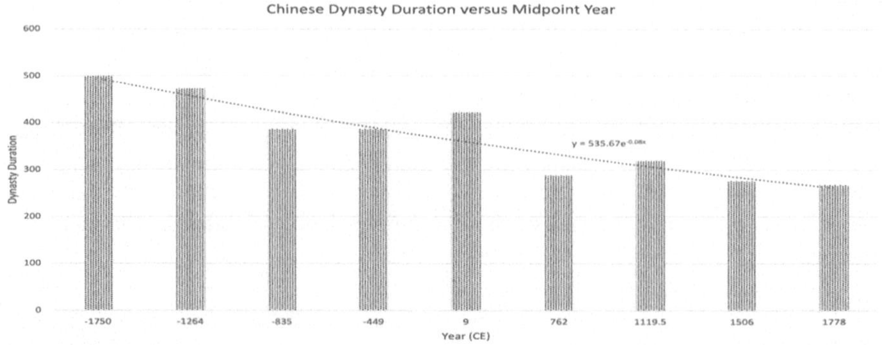

Fig. 17.9 Chinese dynasty duration versus midpoint year

Table 17.1 Major traditional regime series for China

Dates (CE)	Regime	Duration
2000–1500 BCE	Hsia	500 years
1500–1028 BCE	Shang	472 years
1028–642* BCE	Chou 1	386 years
642*–256 BCE	Chou 2	386 years
202 BCE–220 CE	Han	422 years
618–906 CE	Tang	288 years
960–1279 CE	Sung	319 years
1368–1644 CE	Ming	276 years
1644–1912 CE	Manchu	268 years

[a]*Chou dynasty became essentially symbolic by about 700 BC, and China was chiefly ruled by small states during this symbolic "second" Chou dynasty*

Disclaimer: This example only applies to history preceding 1911. Regimes since 1911 may have fundamental characteristics that are different from those of traditional regimes. This same disclaimer could apply to most other contemporary societies as well.

17.4.11 EDEG Approach and Complexity

This chapter has considered thermodynamic approaches to understand and analyze complexity. When structures emerge in systems that are out of thermodynamic equilibrium, this chapter has reviewed the two common scenarios of exponential and logistic growth. Growth beyond linear was modeled as enabled by reproducing engines (physical, biological or other), using heat engines as a specific example. Then a third, less-commonly considered scenario of an exhaustible thermodynamic potential was demonstrated, particularly in the case where intrinsic thermodynamic efficiency drops as an exhaustible resource is consumed. This was shown to result in an efficiency-discounted exponential growth (EDEG) profile, that can them be generalized to a wide range of phenomena such as the rise and fall of physical–economic resource bubbles and major historical dynasties.

This discussion of the EDEG approach is more of a beginning than a complete end. It raises more questions than it answers, but it enables a broad framework to answer these questions. This framework acts as a unifying skeleton to link the humanistic elements of history with the quantitative constraints of the physical universe. This framework also allows for the consideration of various types of complexity. The power of such a framework should not be underestimated. It is possible to gather quantitative data (or quantify qualitative evidence), perform statistical analysis, and accept or reject hypotheses. Yet individually such results, while often important, are merely empirical. They are often hard to use to constrain or illustrate each other. In a unified framework, all results act to constrain all other results. When we learn about one thing, we necessarily learn something about

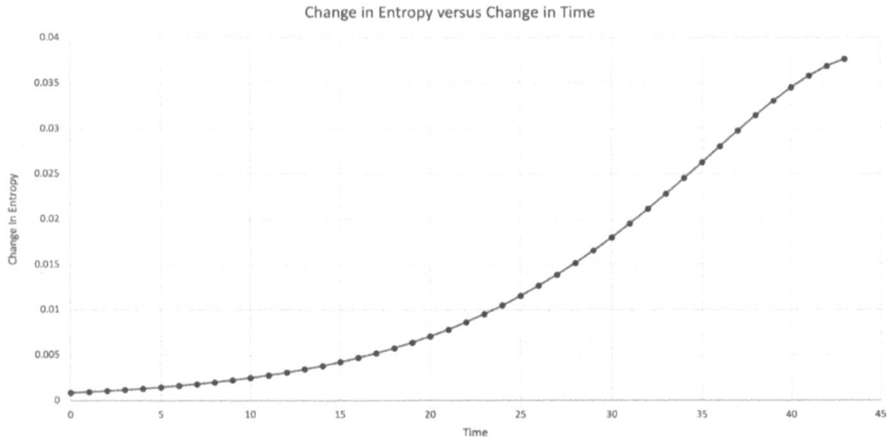

Fig. 17.10 Change in entropy versus change in time

everything else. This is where the physical sciences have derived much of their strength.

Also, what would an HH complexity plot look like? While acknowledging the relevance of Kolmogorov complexity and its similarity to thermodynamic entropy on the microscopic level, this chapter utilizes the more common systems theory approach to complexity of the emergence of macroscopic structures. For example, Fig. 17.10 shows the rate of entropy produced versus time for an HH curve, considering entropy as a proxy for Kolmogorov complexity.

Per the Second Law of Thermodynamics, the rate of entropy increase is monotonic. (Hence, an HH bubble plot using Kolmogorov complexity as a measure of complexity versus time would monotonically increase.) The plot as shown suggests a logistics curve, except that unlike the logistics curve that extends to infinity, an HH curve can end abruptly. While the shape of this curve is conceptually of interest, this form of complexity does not capture the essence of the rise and fall of an HH curve. However, an HH plot of macroscopic complexity (e.g., population or quantity of heat engines) versus time could rise gradually (as shown earlier), then exponentially, level off then fall. The overall plot would be concave downwards. Which approach is intrinsically more meaningful?

There are many immediate apparent enhancements to improve the value of the EDEG approach. Not all stages of an emergent process are equal! So one improvement would be to better understand how efficiency changes as a function of emergent processes. The decrease in intrinsic thermodynamic efficiency (e.g., a heat engine) is relatively easy to model. However, the efficiency changes due to other factors such as economies of scale, overshoot, and evolving characteristics of emergent structures are important, but good models for such do not necessarily exist (except for economies of scale).

Another improvement would be to start using more actual data of physical energy for social phenomena, to the extent such data is available, and to be able to adjust it properly for what is needed. Petroleum production is a more direct proxy. Entropy is produced over time for HH bubble for productive energy, while metal production is less direct and requires a translation into thermodynamic quantities. Physical proxies for energy could be total agricultural production, although apples versus wheat production (literally an energy food) might need to be translated differently.

Another improvement will be to separate the power level of the underlying society from that of the dynasty. For example, Russia did not disappear upon the death of the Romanov dynasty. On the contrary, it is still one of the most powerful societies on Earth. The Chinese people has endured through many dynasties. One approach might be to analyze the human population (energy consumers) represented in each social class, such as farmers who are direct energy producers and soldiers who are indirect energy producers. Arguably, the stratification of societies into classes represents a form of emergence and complexity which can be analyzed thermodynamically.

This brings up the need to be able to model the emergence of a series of dynasties in a way that connects and constrains each dynasty, such as concerning relative strength and timing of emergence. Further, there needs to be a way to compare co-existing dynasties and model their interaction within this framework—literally a matrix over time of interacting emergent structures. While the EDEG approach suggests possible means, the devil will be in the details. More specific future explorations into complexity and the thermodynamic "heat engine" approach would be to explore the types of complexities involved.

Although activity and power increase, activity and control become centralized in fewer actors as industry and power consolidation occurs—in other words, the system could be described more concisely. There is an increasing bureaucracy that perhaps could increase complexity, but due to the need for economies of scale, it apparently would not exceed the complexity decrease caused by fewer actors.

Yet the nature of the fall would depend on the type of structures that had emerged. If structures had become highly centralized, such as to maintain economies of scale, then the decline might be quick and dramatic. Such was the case in the San Juan mining region, where production had become centralized in just a few large facilities. Where centralization was either limited or reversible, the fall might take considerable time and occur relatively gradually. Indeed, the curve might appear relatively symmetric (or reverse asymmetric) as compared to the mining case mentioned. Such was the case of the Roman Empire, which took literally over a thousand years to fall.

17.4.12 Super Engines

Another future direction exploring is how complexity evolves during the process of an HH bubble. This matter has been considered, more generally by many folks: the evolution of life on Earth from single-celled organisms to giant whales and dinosaurs, from individual to productive groupings of hills of ants, hives of bees, hunting pacts of lions, and dynasties of humans. These are all familiar to us. Economies of scale suggest that the nature of production, or characteristics of the engines, changes. Simple replication of identical engines would not necessarily involve economies of scale. True, the first engines would not need to be reinvented. Yet after the third round of reproduction, arguably economies of scale do not increase. However, if engines became bigger, that could invoke economies of scale. Likewise, if engines became more sophisticated (i.e., more complex), then economies of scale could result, manifested by overall increases in efficiency (thermodynamic or otherwise?)

Larger or more sophisticated heat engines might cost less to build per unit of power. Society might require increased social overhead, such as a regulation compliance officer. If each engine (or firm) needs one such officer, then a larger engine (firm) would be more efficiency from just an overhead perspective. Such evolved heat engines could allow for "super" economies of scale. Could they also enable additional modes of complexity? (There is an analogy for gasses when higher energy levels enable additional types of vibrations). In a typical logistics curve model, engine evolution is often not considered, such as for a population of fish in a lake. The fish aren't going to evolve fast enough for sustainable fisheries to have to worry about that in several years or even decades. Will populations of humans physically evolve so quickly in a dynasty that we need to be concerned with taking this into account? (Like more weak "nerds" as a population gentrifies?) That's an open question, but more pressing is the evolution of productive structures, such as farming, roads and groupings of workers, the motivation of workers, and information flows from workers to top decision makers.

Many dynasties did not have especially huge economies of scales, but others did. For example if bigger structures emerge, such as giant factories of the modern area or large aqueducts of ancient times, then these large structures could be considered super engines. Their evolution would involve a changing cost of construction, maintenance, and possibly reproduction, as well as changing efficiencies, productive capacities, and growing minimum total costs. Conversely, many did build better roads and markets (more linear increase).

17.5 Future Work

The current approach by this author has been to treat all engines in a system as being identical in terms of their thermodynamic characteristics. This approach is perhaps suitable for describing reproducing bacteria or a society with uniform, unchanging means of production. However, future work will need to study how the characteristics of engines evolve during historical processes. This author has conducted some work with economies of scale and centralization, but not at a sufficiently fundamental or detailed level to account for the emergence of productive structures with dynamic characteristics. Better models of complex structures and their processes of emergence, especially as applied to human society, could shed tremendous light upon the workings of humanity, and have important implications for social relations, equity, and governance.

Further comparisons with other approaches are merited as well for future work. Energy return on investment is akin to investigations by this author for analyzing economics and business concepts in terms of thermal dynamics. EROI captures a related, but different valuable perspective. Further analysis of the ideas of Tainter may provide for additional case studies. Turchin's and Bardi's works provide an interesting intermediary approach between the primarily ecological approach of emergy researchers such as Odum versus the socioeconomic approach by Tainter. The ideas of Gunderson and Hollings concerning panarchy may be illuminating concerning series of dynasties as well (Gunderson & Holling, 2002), comparing and contrasting it to the eras approach by LePoire (2020). In a similar light, conducting a thermodynamic analysis of the World Systems approach explored by figures such as Immanuel Wallerstein and Chris Chase-Dunn can help capture more macro levels and higher degrees of complexity.

References

Annila, A., & Salthe, S. (2010). Physical foundations of evolutionary theory. *Journal of Non-Equilibrium Thermodynamics, 35*(4), 301–321. https://doi.org/10.1515/jnetdy.2010.021

Bardi, U. (2020). *Before the collapse: A guide to the other side of growth*. Springer. https://doi.org/10.1007/978-3-030-29202-7

Bar-Yam, Y. (2003). Unifying principles in complex systems. In M. C. Roco & W. S. Bainbridge (Eds.), *Converging technologies for improving human performance: Nanotechnology, biotechnology, information technology and cognitive science* (pp. 380–409). Springer. https://doi.org/10.1007/978-94-017-0628-1_24

Biot, M. (1955). Variational principles in lrreversible thermodynamics with application to viscoelasticity. *Physical Review, 97*(6), 1463–1469.

Butler, M. (1996). *Animal cell culture and technology*. IRL Press at Oxford University Press. https://doi.org/10.1093/acprof:oso/9780199634163.001.0001

Carroll, B., & Ostlie, D. (2007). *An introduction to modern astrophysics* (2nd ed.). Pearson Addison-Wesley.

Chapman, R. (1983). *Oil geology*. Elsevier.

Chaisson, E. J. (2010). Energy rate density as a complexity metric and evolutionary driver. *Complexity, 16*(3), 27–40. https://doi.org/10.1002/cplx.20323

Chaisson, E. (2014). The natural science underlying big history. *The Scientific World Journal, 2014*, 384912. https://doi.org/10.1155/2014/384912

Chaisson, E. (2015). Energy flows in low entropy complex systems. *Entropy, 17*(12), 8007–8018. https://doi.org/10.3390/e17127997

Ciotola, M. (2010). *Modeling US petroleum production using standard and discounted exponential growth approaches*. ResearchGate. https://www.researchgate.net/publication/261252578_Modeling_US_Petroleum_Production_Using_Standard_and_Discounted_Exponential_Growth_Approaches

Ciotola, M. (2014). *Efficiency discounted exponential growth (EDEG) approach to modeling the power progression of a historical dynasty* [Unpublished manuscript]. ArXiv.

Ciotola, M. (2016, April 28–30). *Effects upon the progression of a society and its business structures due to dependency upon a nonrenewable resource: The San Juan mining region* [Conference presentation]. Richard Robinson Business History Workshop 2016-Business and Environment in History: Dealing in Nature and Ecology, Pasts and Futures, Portland State University.

Ciotola, M. (2025). Thermodynamic approach to modeling complexity of regional Hewett-Hubbert resource bubbles and large dynasties. *Journal of Big History, VIII*(1), 124–143.
Cleveland, C., Costanza, R., Hall, C., & Kaufmann, R. (1984). Energy and the U.S. economy: A biophysical perspective. *Science, 225*(4665), 890–897. https://doi.org/10.1126/science.225.4665.890
Deffeyes, K. (2001). *Hubbert's peak*. Princeton University Press. https://doi.org/10.1515/9781400824707
Fermat, P. (1894). *Oeuvres* (Vol. 2). Gauthier-Villors and Sons.
Gunderson, L. H., & Holling, C. S. (2002). *Panarchy: Understanding transformations in human and natural systems*. The University of Chicago Press.
Gupta, R. (2023). JWST early Universe observations and ΛCDM cosmology. *Monthly Notices of the Royal Astronomical Society, 524*(3), 3385–3395. https://doi.org/10.1093/mnras/stad2032
Heilbroner, R. L. (1980). *The worldly philosophers* (5th ed.). Touchstone, Simon and Schuster.
Henderson, C. W. (1926). *Mining in Colorado: A history of discovery, development and production* (Professional Paper 138). U.S. Geological Survey, U.S. Department of the Interior.
Hewett, D. F. (1929). *Cycles in metal production* (Technical Publication 183). The American Institute of Mining and Metallurgical Engineers.
Hubbert, M. K. (1949). Energy from fossil fuels. *Science, 109*(2849), 103–109.
Hubbert, M. K. (1956). *Nuclear energy and the fossil fuels (Publication 95)*. Shell Development Company.
Hubble, E. (1929). A relation between distance and radial velocity among extra-galactic nebulae. *Proceedings of the National Academy of Sciences of the United States of America, 15*(3), 168–173. https://doi.org/10.1073/pnas.15.3.168
International Association for the Integration of Science and Engineering. (2022). *Thermodynamics 2.0 conference*. https://iaisae.org
Korotayev, A., & Grinin, L. (2012). Kondratieff waves in the world history perspective. In L. Grinin, T. Devezas, & A. Korotayev (Eds.), *Kondratieff waves: Dimensions and prospects at the dawn of the 21st century*. Uchitel Publishing House.
LePoire, D. (2020). *Time patterns in big history: Cycles, fractals, waves, transitions, and singularities*. David LePoire Consulting, LLC.
LePoire, D. J., Grinin, L., & Korotayev, A. (2026). Introduction to Complexity Theory for Big History. In D. J. LePoire, L. Grinin, & A. Korotayev (Eds.), *Complexity in Universal Evolution—A Big History Perspective* (pp. 29–90). Springer Nature (in this volume). https://doi.org/10.1007/978-3-032-20516-2_2
Massachusetts Institute of Technology. (2023). *Big O notation*.
Mazour, A. G., & Peoples, J. M. (1975). *Men and nations, a world history* (3rd ed.). Harcourt, Brace, Jovanovich.
Meadows, D. H., Meadows, D. L., Randers, J., & Behrens, W. W. (1972). *The limits to growth*. Universe Books.
National Aeronautics and Space Administration. (2007). *WMAP's portrait of the early universe*. https://svs.gsfc.nasa.gov/10123/
Odum, H. T. (1996). *Environmental accounting: Emergy and environmental decision making*. John Wiley & Sons.
Pentland, S. (2023a). *MIT connection science*. https://connection.mit.edu
Pentland, S. (2023b). *Sandy Pentland*. MIT Connection Science. https://connection.mit.edu/alex-sandy-pentland
Prigogine, I. (1967). *Introduction to thermodynamics of irreversible processes*. Wiley.
Ryden, B. (2003). *Introduction to cosmology*. Addison-Wesley.
Schroeder, D. V. (2000). *Introduction to thermal physics*. Addison Wesley Longman.
Schrödinger, E. (1956). *What is life & other scientific essays*. Doubleday Anchor.
Smith, D. A. (1982). *Song of the drill and hammer: The Colorado San Juans, 1860–1914*. Colorado School of Mines Press.

Spier, F. (2009). Big history: The emergence of an interdisciplinary science? *World History Connected, 6*(3).
Spier, F. (2015). *Big history and the future of humanity* (2nd ed.). Wiley-Blackwell.
Swenson, R. (1989). Emergent attractors and the law of maximum entropy production: Foundations to a theory of general evolution. *Systems Research, 6*(3), 187–197. https://doi.org/10.1002/bsr.3190060305
Swenson, R. (2011). Problems with Annila and Salthe's "Physical Foundations of evolutionary theory": Confusing the 2nd and 4th laws. *Journal of Non-Equilibrium Thermodynamics, 36*(4), 311–318.
Tainter, J. A. (1988). *The collapse of complex societies*. Cambridge University Press. https://doi.org/10.1017/CBO9781139462551
Tice, B. (2003). *Aspects of Kolmogorov complexity: The physics of information*. Cuvillier Verlag.
Turchin, P., & Nefedov, S. A. (2009). *Secular cycles*. Princeton University Press. https://doi.org/10.1515/9781400828668
Twitty, E. (2010). *Mining resources of San Juan County*. National Parks Service, U.S. Department of the Interior.
United States Energy Information Administration (EIA)(2025). *Petroleum and other liquids*. https://www.eia.gov/petroleum/data.php
Wikipedia. (2023, November 9). *Big O notation*. https://en.wikipedia.org/w/index.php?title=Big_O_notation&oldid=1184201630
Williams, R. J., & Martinez, N. D. (2000). Simple rules yield complex food webs. *Nature, 404*(6774), 180–185. https://doi.org/10.1038/35004576

Mark Ciotola develops resource-based case studies and models of historical societies. He is a lecturer at San Francisco State University. Recent works include: "Characterizing Historical Macro-Material Utilizing Satellite Imagery," "Using thermodynamic potentials as the prime driver of historical processes," "Emergy and Efficiency Analysis of Historical Bubbles," "Workshop: Feeding Empires—Quantitative, Digital & Graphic Simulations for History Using Food Cases," "Irreversible Processes in History," "Modeling Historical Dynasties as Emergent, Dissipative Mechanisms," "Panel: Unity of Sciences," and "A Comparison of Emergy, Exergy, Entropy and Carbon Accounting: Histories, Definitions, Methodologies, Significance and Synergies." He has obtained a B.A. in Economics from the University of Wisconsin-Madison; B.A. Physics & MBA from San Francisco State University; Graduate Certificate in Space Sciences from the University of South Australia; and a JD from the University of New Hampshire.

Punctuated Growth and Complexity in Urban Evolution: Insights from Historical City Size Distributions

18

Antony Harper

Abstract

Throughout history, the largest urban areas have exhibited hyperbolic growth, yet a logarithmic perspective reveals a more nuanced pattern—periods of stability (stasis) interrupted by rapid expansion. This study examines how this pattern affects the size–frequency distribution and complexity of urban systems relative to the largest city of each era. Two key assumptions guide this research: (1) urban area sizes follow a power-law (Zipf-Pareto) distribution, and (2) the maximum city size in each period reflects historical energy constraints, from agrarian solar-based economies to fossil-fuel-driven industrialization. By fitting historical city size distributions to power laws for each century, a pattern of punctuated equilibria emerges. Here, "complexity" is defined as the diversity of city sizes within an interconnected system. Analyzing this measure over time reveals alternating phases of positive and negative complexity growth. Notably, major increases in complexity coincide with systemic reorganizations, such as the Dark Ages described by Sing C. Chew (*The Recurring Dark Ages: Ecological Stress, Climate Changes, and System Transformation*. Altamira Press, 2006). Furthermore, variations in power-law steepness correspond to key societal transitions, including shifts from agrarian societies to Bronze Age empires and later to industrial and post-industrial urban networks.

Keywords

Punctuated · Urban · Power law · Complexity · Social evolution

A. Harper (✉)
Lake Zurich, IL, USA

D. J. LePoire et al. (eds.), *Complexity in Universal Evolution*, World-Systems Evolution and Global Futures, https://doi.org/10.1007/978-3-032-20517-9_18

This chapter is an extensively re-worked version of the text originally published in Journal of Big History (Harper, 2025).

18.1 Introduction

The pattern of urbanization over the last 5000 years has come into scrutiny over the last several decades and before. Certainly, the data-replete books of both Chandler (1987) and of Modelski (2003) have provided a wealth of empirical evidence geographically at the world-system level and below. More recently, the research of Morris (2013) and of Harper (2010, etc.) has concentrated on maximum urban area size, a term first used by Modelski and meant to imply the area over which a maximum urban population is supported, that is, this urban area is represented by a specific population number; for example, in the case of Rome in the second century of the Common Era, the maximum urban area size is 1.2 million people. [Note: Urban area size is a term created by the late George Modelski and is meant to refer to population size.]

The questions then to answer are: What is the pattern of maximum urban area sizes over time? In what way is maximum urban area part of a network of urban areas at any given point in human history? How does this network change over time? How does the population of maximum urban area size change over time, and is there pattern to this change? Please note that it will be a fundamental assumption of the research reported here that urban areas do not exist in isolation but are by default of the process of human existence and that this implies and requires that interconnectedness with others is part of any urbanized network.

This urbanized network exists in both space and time, so that at a given point in time it may be assumed that there will be a network of urban areas that exist, for example, in trade, communication, or services. Further, that network will include cities with a frequency size (population) distribution of the form: $F = \alpha C^{-\gamma}$, where $F =$ frequency of cities with larger population than C, and α and γ are fitted constants (see Fig. 18.1 for a graphical example of this relationship). This in turn means that there are mechanisms by which, given certain data, the urban area distribution can be constructed; that construction will depend on empirical data on maximum urban area sizes taken from the aforementioned works of Modelski and Chandler. With respect to this specific datum, maximum urban area size, from here on noted as C_{max}, size changes over time will be a focus. As mentioned previously, is there a pattern of change at the top, that is, the size of C_{max}, and how is that pattern related to and reflected in total population and the previously mentioned variables, α and γ?

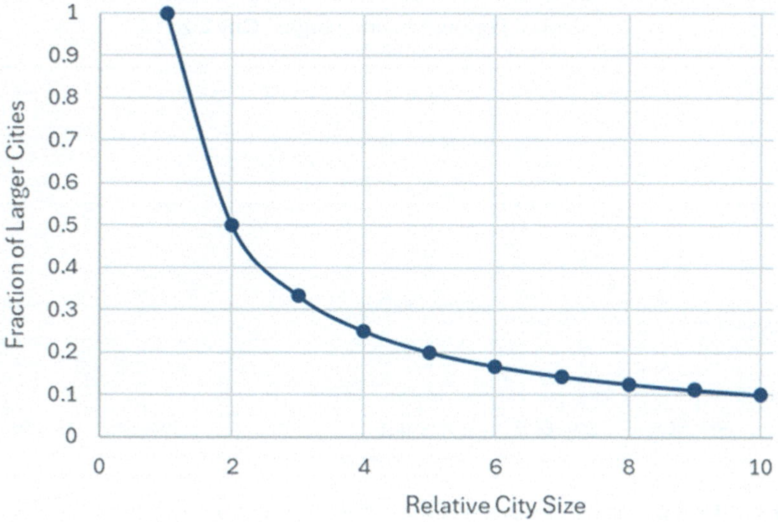

Fig. 18.1 A relationship in which the fraction of cities greater than some population is inversely related to that population ($\gamma = 1$)

18.2 The Pattern of World-System Population Growth over Historic Time

It has been known for some time that the pattern of world-system population growth has not been one of the standard exponential increases. Which world-system is being considered here? Global? Eurasian? Are any areas excluded? This was noted initially by von Forester, et al. (1960) that the rate of human population growth was explosive and, whimsically, these authors suggested a date of November 13, 2026 as the point in time at which population growth would become infinite. Of course, nothing of the sort will happen, but the pattern of this growth is characteristic of a species with such enhanced communication skills and cooperative abilities such as ours. This pattern of population growth, now labeled as hyperbolic (Korotayev et al., 2006; Korotayev, 2018, 2020), has been more elegantly explained and expanded on by these authors. Specifically, the elaborate mathematics of von Forester et al. (1960) can be replaced by a relatively simple differential equation: $dN/dt = rN^2$, which is closely related to the standard differential equation of exponential growth, $dN/dt = rN$, with the solution $N_t = N_o e^{rt}$. Now consider the far simpler solution of the Korotayev et al. to $dN/dt = rN^2$: $N_t = C/[t_o—t]$. [Note: In the previous equations, r = rate of growth, N_t = the population size at time t, e is the base of the natural logarithms, spelling t_o = the upper limit time, a time when growth rate reaches a singularity and becomes infinite, and C is a fitted constant.}

Both global population and largest city size seem to follow a hyperbolic growth pattern. When these are viewed with a logarithmic scale, the curves of world-system

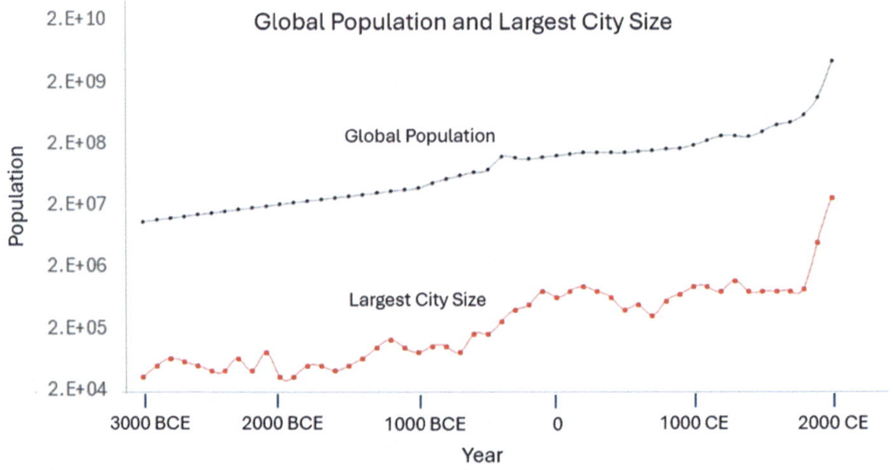

Fig. 18.2 Global population and largest city size. Note that while an overall hyperbolic shape is shared by both the world-system population curve and that of the maximum urban area curve, the latter shows a stepwise or punctuated form, while, with the single exception of a jump around 700 BCE this curve is essentially monotonically increasing

population growth, and that of ln C_{max}, the structure of each curve is quite different (Fig. 18.2). Specifically, C_{max} exhibits a stepwise pattern; phases of stasis alternated with phases of rapid growth, while the ln-transformed world-system population growth data plot as a continuously upward growing curve. The stepwise pattern was given the name, punctuated equilibrium, by the evolutionary biologists, Eldridge and Gould (1972), and that term will be used here. The implications of this pattern applied to temporal changes in C_{max} are that there are contexts in which stasis is favored and others in which the rapid growth of C_{max} is favored. This in turn suggests that this pattern of punctuated equilibrium with respect to C_{max} is both a consequence of and a generator of other trends.

So, while considering a macro-view, the view generated by raw data, both the graph of world-system population and the graph of C_{max} appear hyperbolic in structure; the fine scale structure revealed by ln-transformation of both data sets exhibits quite different patterns at this level of analysis. What is the history of the recognition of this disjunct pattern of C_{max} magnitude over time? While those untransformed graphs have similar, almost identical hyperbolic form, the above two graphs are quite dissimilar. It is the bottom graph that exhibits the form of punctuated equilibrium discussed in the text.

In 2010, Harper noted the potential for the presence of this pattern. A note of scale, graphically, was required in order to compare ancient, classical, and modern C_{max} sizes. When all three of these distributions were placed on the same continuous ln-transformed Y-axis, the emergence of a stepwise pattern emerged. There appeared to be then C_{max} limits to size associated with various periods of time, that is, periods of stasis; in the Ancient World C_{max} approximated 1E5, in the Classical World, C_{max}

Table 18.1 Long-term changes in slope of r value

Time statistic	Slope (/century)	Value at period beginning
3000–700 BCE	0.001	1.24
700–100 BCE	−0.016	1.27
100 BCE–1800 CE	0.002	1.17
1800–2000 CE	−0.048	1.21

is around 1E6 in size with considerable oscillation at times, and currently the maximum is below 4E7. [Note: These time periods are global and are given in centuries, a value-neutral designation, only to avoid the usage of terms such as Ancient or Classical.] This shift in urban area sizes was also noted by Morris (2013). A more expanded treatment of this punctuated distribution of C_{max} was published in 2017 (Harper, 2017). There it was noted that this pattern of change represented, as previously mentioned, a form of punctuated equilibrium not unlike the model of speciation identified by Eldridge and Gould (1972). In that pattern Eldridge and Gould defined two specific sub-patterns, stasis, in which species morphology fluctuated about some mean, and phases of punctuated growth in which rapid change occurred. As mentioned, these terms have been adopted to describe similar trends in C_{max} growth. Over the last 5000 years of historic time, there are three periods of stasis and three periods of punctuated growth, and these time periods are listed in Table 18.1. The implications of this pattern will be given shortly.

It is worthwhile here to indulge in a brief aside regarding mechanism and modeling of both the overall hyperbolic pattern of growth and the stepwise pattern of punctuated equilibrium, both of which are exhibited by the historic time series of C_{max}. First, a potential mechanism will be described to generate hyperbolic growth. It should be noted that this mechanism is *a* mechanism, not *the* mechanism, for generating this pattern of growth per se. It can be quite simply shown that hyperbolic growth can be produced by the actions of two interacting human populations which exhibit specialization in the production of a product beneficial to both populations. Very briefly, the mechanisms involve a shared effect by two populations, in this (fictitious) instance, two farming populations. One population produces farm machinery that can be used by the second population to actually farm, producing enough food for both. A simple math model involving two differential equations was constructed in which the activities of each population enhanced the reproductive efforts of the other community; the total population output was hyperbolic. Is this model by the author or an external source? A cited source would be useful here. Implicit in this mechanism are the human abilities to communicate and cooperate at a high level of effectiveness ultimately resulting in hyperbolic growth for the summed population.

Turning now to a model for the pattern of punctuated equilibrium; this proposed model consists of three differential equations, one each for change in population size, change in carrying capacity, and change in technology. When this model is run for a period of arbitrary time units, say, 60 time units, it exhibits stepwise plots for both population size and changes in technology but exhibits a continuous pattern for change with respect to carrying capacity. With respect to the application of this model, its function as a research tool is in its infancy, but it is mentioned here as a potential model for the generation of a pattern of punctuated equilibrium; it involves broadly three variables and very effectively shows how their interaction produces a system-wide punctuated behavior. There are any number of avenues of further research, for example, frequency and rate of punctuation with respect to initial conditions, nonlinear effects, why change in carrying capacity is not punctuated, and so on. But these are not germane focal points for this particular research. Further, these models are presented here simply as a note of their potential for enhancing (future) understanding of the problems associated with the punctuated pattern of C_{max} change over human history. The implications of this pattern will be considered next.

An assumption without direct proof being made in this research is that C_{max} size during periods of stasis represents some limit determined by either the nature of available energy, the mode of access to that energy, or both. So given this assumption, then during periods of rapid C_{max} growth either new forms of energy become available or new means of energy access develop or some combination of both occur in order to promote rapid C_{max} growth. In other words, when C_{max} stasis occurs, and the longest period of historical stasis to occur extends from 100 BCE to 1800 CE, during which C_{max} begins with a value of 1 million and ends with a value of 1.1 million, it does so due to some sort of energy cap associated with the type of surplus generated by agrarian systems, that is, pre-industrial systems. That energy level was elevated at the end of this 1900 year long period by the occurrence of the Industrial Revolution (IR); other periods of stasis ended with far less explicit events or processes of energy capture elevation, for example, the stasis period from 3000 to 1600 BCE, a period of time approximated by the Bronze Age, and from 1200 to 700 BCE, approximating the Iron Age, have a number of historic occurrences associated with their termination, but none of these occurrences can be directly ascribed to better energy sources or new modes of energy access. However, also of interest are any potential changes in urban area size below C_{max}.

A further assumption of this research, previously mentioned (Fig. 18.1), is that urban area population size distribution can be represented by the size-frequency distribution represented by the following equation: $F = \alpha C^{-\gamma}$, and there is yet a further assumption, a convention actually for simplicity, that changes in the number of urban area population below the level of C_{max} will double per level, as measured by urban area magnitude, that is, for each urban area at level, m, there will be two such urban areas at the level beneath, $m + 1$, that is, for each C_m there are two C_{m+1}'s. However, $C_m > C_{m+1}$, and $2\,C_{m+1} > C_m$. It is easy to show that this regularity can be represented by: $r = 2^{(1-1/\gamma)}$, the derivation of which will be given in the Appendix. The urban area population multiplier, r, is computed for the variable, over the last 5000 years per century and plotted over time. The value of γ for each century is based on the ratio of the largest city size to the global population (see Mathematical Appendix).

There are two sets of points at different r levels that are linked in this time series (Fig. 18.3): The older set of points, extending from 3000 to 700 BCE, and a second

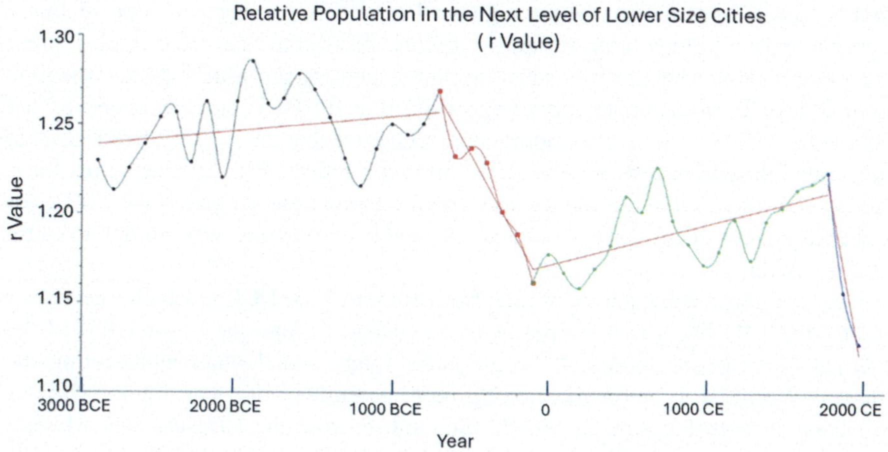

Fig. 18.3 Four linear regressions of the specific data sub-sets mentioned in the text. It can be seen that there are two sub-sets with positive slopes, each connected to the other by a transitional phase, and that the most recent sub-set is a transitional phase to the future. [Please note that all data used to construct this graph are accepted at face value based on the research of both Modelski (2003) and Chandler (1987)]

set extending from 100 BCE to 1800 CE. The linkage between these two sets represents an almost linear transitional bridge from 700 to 100 BCE. That there are two distinct sets of points can be emphasized by applying appropriate linear regression to this entire data set. In Fig. 18.3, these sub-set regressions are displayed and very clearly depict the four distinct temporal regions described above. These regression separations represent the Ancient World, the Classical and Early Modern World, and two transitional phases, the early one from the Ancient World to the Classical and Early Modern Worlds, enclosing what Jaspers called the Axial Age, that is, from 700 to 200 BCE, and the current transitional phase beginning around 1800 CE and extending through the present and constituting both the Industrial World and the ongoing Post-Industrial Technological World. Both of these transitional phases share in common the almost continuously decreasing values of r and increasing values of C_{max}. This implies that both periods of transition are times of significant urbanization at the C_{max} level. For a listing of all pertinent statistics, see Table 18.1.

The periods of time including the Classical and Early Modern World and also the Industrial and Post-Industrial World have two associated regressions, one with a slightly positive slope, the Classical and Early Modern World, and the succeeding one, the phase the world-system is now in, with a distinctly negative slope. The same pattern holds for the Ancient World and, for want of a better term, the Axial Age. If one considers the slopes of these sets of two succeeding phases, then both the variability and slope change about a single temporal point that of 1800 CE for the more recent set and that of 700 BCE for the older set. These conditions about a single point satisfy the context for a tipping point as defined by Lamberson and Page

(2012), and it is, therefore, being suggested that these changes in slope of linear regressions of r values indicate tipping points. Also note that these tipping point changes constitute changes from stasis to punctuated change with respect to the time point, 700 BCE, from punctuated change to stasis at 100 BCE, and the reverse of that change for 1800 CE. It is also important to recognize that a C_{max} shift from stasis to punctuated growth occurs at 1600 BCE but is not reflected in the time series for r. This would suggest that the energy level required to maintain a given size of C_{max} or increasing sizes of C_{max} did not influence the value of r significantly enough to cause a tipping point.

Also of interest is the period of time from 1600 to 1200 BCE in which r decreases in an essentially linear manner but does not extend beyond the lower limits of the data set identified to represent the graphical range and domain representing the Ancient World. It is worth mentioning here, parenthetically, that the oscillations occurring in both the Ancient World data sub-set and the Classical and Modern World data sub-set are within the bounds of the definition of the condition of stasis originally given by Eldridge and Gould (1972).

Higher r values imply that the total population of a given time period is concentrated at lower levels. This is a significant observation as it pertains to both the potential for (heightened) inequality and inequality's equally dark relative, elitism. Since urban area size is inversely related to its frequency, if a particular distribution of urban area sizes has a large gamma, then there will be very few large urban areas, where power is concentrated. This change in relative size of urban areas can be understood quantitatively as follows: If $\gamma = 1.2$, then $r = 1.12$, implying that an individual urban area population is 56% of the size of an individual urban area size of the previous level. However, if $\gamma = 1.6$, then $r = 1.30$, and the size of an individual urban area population is almost 65% of the size of an urban area population of the previous level. This implies that during periods in which r deceases with time that urban area population sizes more quickly decrease with subsequent levels.

However, due to the hyperbolic growth of population, there is greater urbanization currently, even though r values are at their lowest in history. This aspect has significance when system complexity is considered in the next section. This also implies, as mentioned previously, that the bulk of the world-system population will be associated with lower levels of urbanization; however, these levels, when expressed in terms of a center of gravity of population distribution, will actually be higher than those associated with higher r values. This is a product of the interaction between a larger world-system population supporting a larger maximum urban area, and while γ decreases under these circumstances, its inverse increases resulting in a lower value for r, that is, if $r = 2^{(1-1/\gamma)}$, then decreasing γ produces a decrease in the value of the exponent, $1—(1/\gamma)$.

To recap briefly: Increasing r values imply increasing urban area population sizes, although actual populations of these urban areas may be less due to a smaller C_{max}. The total populations of each level also reflect this relationship. Further, a given C_{max} size will affect the absolute sizes of all levels of urbanization below the C_{max} level.

18.3 Complexity

Entering the word, complexity, into the Advanced Search bar in the Amazon Books homepage yields 75 pages of densely packed books. If the search is narrowed to "complex systems," 50 pages of the same sort of information are offered. If the search is further narrowed to complex adaptive systems, there are 40 pages for the browsing reader to muse through. The point here is of course that either in undefined breadth or in a more restricted focus, the topic of complexity has generated a great deal of attention, research, and reporting; the topic of complexity is itself complex.

There are also complexity-related topics spawned by the efforts of diverse scholarship which are, by default of their importance, all part of the bigger picture of complexity. Some of these complexity-related topics are: emergence, chaos (in the mathematical sense), historical contingency, and system learning, aka system evolution, all of which have generated a significant quantity of research in their own right. In following suit from the previous chapter, Amazon Books Advanced Search generates 75 pages of reader-necessary book information.

The concept of complexity has had an interesting definitional history. Most definitions couple complexity with diversity and distinguish the two by diversity representing simply the number of different entities in, for instance, a complex system, while complexity takes into consideration the connections or interactions between entities in a given complex system. This latter is best represented by the mathematics of networks. For a thorough discussion of both diversity and complexity and their relationship, see Page (2011). However, there is one conceptual understanding of complexity (McShea and Brandon, 2010) that simply uses the definition of diversity to represent complexity. While this may seem to be counter-productive and at odds with common understanding, it is worth considering here as a background source of complexity, that is, as a default condition of all complex systems including complex adaptive systems.

If there is a nongenetic source of diversity and complexity for all biological systems, systems that are complex adaptive systems (CAS), then it is certainly worth mentioning here from the perspective of distinguishing between diversity and complexity generated by what McShea and Brandon call ZFEL, a zero force evolutionary law, as a source of complexity versus those sources of complexity that are generated deterministically. The authors note a number of implications of ZFEL-generated diversity and complexity. First, stasis, as part of a punctuated equilibrium pattern, will be difficult to maintain if there is a continuous source of change in complexity; punctuated phases, as phases of accumulation of complexity, will occur regardless of ZFEL. This is particularly pertinent for urban area evolution, as maximum urban area sizes exhibit a pattern of punctuation over the last 5000 years of human history; it is, therefore, important for complexity changes associated with urban area evolution. Second, given ZFEL, then evolutionary change does not (always) imply evolutionary forces such as selection is at work; ZFEL from a general perspective is analogous to the effects of genetic drift on (small) gene pools. This might well be pertinent for understanding the evolution of complexity in urban area systems. Third, McShea and Brandon make a special case of evolutionary equilibria, and note three types of equilibria: stable, unstable, and neutral. They then suggest that these conditions represent contradictions to the notion of evolution (= change) in that no change is occurring. As a rejoinder here it should be noted that biological systems are steady-state systems with continuous input and output, whereas equilibria are characteristic of closed systems. As such, continuous selection is required to maintain the steady state, and it could then be argued that potential change is always in the offing. Further, as a specific example of such a condition, balanced polymorphism automatically implies maintenance of variability, not reduction in variability to some selected for mean; the variability as-a-whole is the result of selection. Interestingly as will be described later in this chapter, urban area complexity exhibits an overall increasing trend during the extent of human history.

My own interest in complexity stems from my research in urban area patterns over historic time (Harper, 2010, 2017). When considering the population sizes of maximum urban areas, the pattern of change over time is one of increases but not of steady increase; the pattern is one of relatively long periods of stasis punctuated by periods of rapid increase (see Fig. 18.2). It should be noted here that I view maximum urban area size as an energy limited threshold, so periods of stasis imply periods of time during which energy availability either in kind or accessibility remains constant, while periods of punctuation are times when either the nature of the energy source fueling the world-system changes, for example, the Industrial Revolution, or the accessibility of energy availability changes, both associated with changes, that is, improvements in technology or in trading efficiency. This latter may be the underlying motive for the period of punctuation from 700 to 100 BCE. Even so, how are these phases associated with change in complexity?

It is a fundamental assumption of the research reported in this chapter and that of research in this area in general that urban area sizes at any point in time during which data for such sizes is available, that urban areas exhibit a Zipf-Pareto size frequency distribution, that is, a distribution that can be represented by the general power function, $F = \alpha C^{-\Upsilon}$, where F = frequency, $\alpha = C_{max}{}^{\Upsilon}$, C = urban area size, C_{max} = maximum urban area size, and Υ = a fitted constant. (Note that the previous formula may also be expressed as: $F = [C_{max}/C]^{\Upsilon}$.) For a thorough discussion of the relationship between the Zipf and Pareto functions; see Newman (2004). Since urban area sizes are not actually continuously distributed, that is, that they do not exhibit a pattern in which C varies continuously from some maximum value down to some limiting size, although they can be represented by the previous equation, it will be assumed that these urban areas from the maximum urban area on down to a minimum of 100, here established as a convenient marker and not a functional one, can be represented by a sequence in which the total population of the system, T, can be represented by: $T = C_{max}[1 + F_1{}^{-1/\Upsilon} + F_2{}^{-1/\Upsilon} + \ldots]$, the solution of which can be represented by: $T = C_{max}(1—r^{m})/(1—r)$, where $r = 2^{(1-1/\Upsilon)}$, and m = the total number of hierarchical levels. [Note that this last equation is a standard equation of the mathematics of series and will not be derived here.] This latter in turn can be solved for by: $m = \ln[T/C_{max}(r—1) + 1]/\ln r$. It is this value that will be used to characterize complexity as represented by urban area data over time. [The derivation of the previous equation is in the Appendix.]

The remainder of this chapter will first consist of an analysis of the variable, *m*, over historic time. Two time periods will be considered for this analysis, the first, from 3000 BCE to the present, that is, 2000 CE, will be considered briefly, and the second from 1600 BCE also to the present; this shortened latter period because the first 1400 years of human history exhibit what I consider to be anomalous behavior with respect to the remainder of human history as reflected by changes in maximum urban area population, hereafter represented as C_{max}. These analyses will be followed by a discussion of entropy changes over the same periods of time. Finally, there will an analysis of the interactions and relationships between complexity and entropy. [Note: All data for maximum urban area sizes were taken from Modelski (2003) and Chandler (1987).]

18.3.1 Historically Derived Complexity

This analysis of historical trends in complexity, complexity as represented by the metric, *m*, per century, will begin with a consideration of the big picture of change in complexity over both the last 5000 years and the last 3600 years.

If we now reconsider Fig. 18.4, it can be seen that the overall trend is one of increasing complexity, but one with oscillations. The regression equation exhibits a slope of approximately 10% per century, suggesting that the mechanisms causing this increase in complexity, at least at the level of changes in C_{max}, are ongoing.

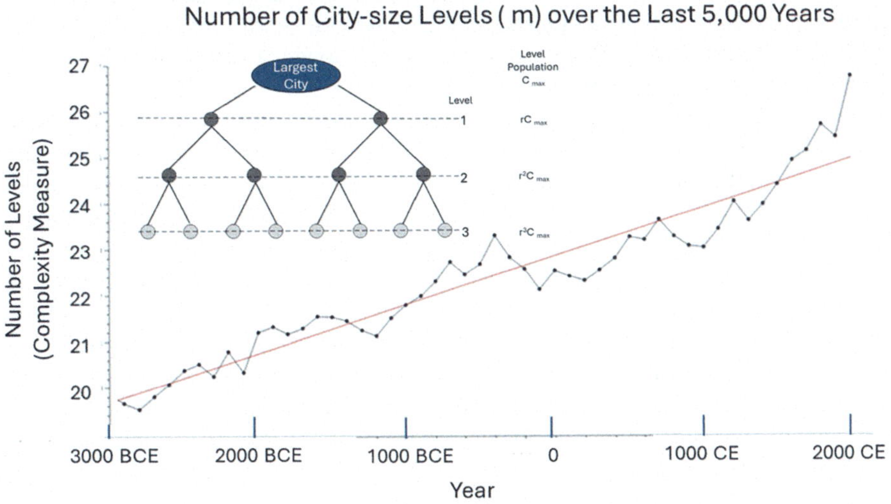

Fig. 18.4 The linear regression is represented by: $m = 0.0989\ t + 19.3965$, where m = complexity as represented by the total number of urban area levels. The metric, *m*, represents the total number of hierarchical levels in an urban area distribution represented by $F = \alpha C^{-\Upsilon}$

Table 18.2 Time periods, directions of slopes of change, and duration of change

Time period	+/–	Duration (years)
1600–1200 BCE	–	400
1200–400 BCE	+	800
400–100 BCE	–	300
100–700 CE	+	800
700–1000 CE	–	300
1000–1200 CE	+	200
1200–1300 CE	–	100
1300–2000 CE	+	700

However, there are a number of peaks and valleys defining periods of complexity increase and decrease which bear closer scrutiny.

These periods of increasing and decreasing complexity are delimited by maximum and minimum values of *m*, and these values are listed in Table 18.2; they span only the last 3600 years. Note also that each time period of Table 18.2 is assigned a +/– assignation depending on the slope of the specific sub-trend occurring during this time, that is, whether or not complexity increases (+) or decreases (–) over the represented period of time.

Each boundary century is also associated with a significant historical occurrence; this is only to give historic context and not, at least initially, to suggest any cause-and-effect relationship between the accruing or decaying complexity and the given historic occurrence. Table 18.2 also shows the duration of each complexity period, and it is interesting to note that for the first 2600 years represented there is an almost periodic set of oscillations occurring between shorter, decaying periods of complexity with periods of longer, increasing complexity, that is, shorter periods of either 400 years or 300 years in length alternating with longer periods of 800 years in length. This pattern continues until 1000 CE where a 200 years period of increasing complexity followed by a period of decreasing complexity of 100 years in length; these two periods are then succeeded by a period of 700 years of increasing complexity to the present. It is important to recognize that $C_{max} \sim 1.5E6$ (Chandler 1987) at 1300 CE, which is then followed by the "calamitous 14th Century" (Tuchman, 1987) from which a continuous if not constant increase in complexity occurs. This 700 years period of increase subsumes three different phases of complexity slope, from 1300 to 1600 CE, from 1600 to 1900 CE, and from 1900 to 2000 CE and beyond, the slopes for which are 0.0041, 0.0016, and 0.0122, respectively. It is important to recognize that this period of time, when compared to changes in C_{max}, includes three periods of change which are not in sync with the two periods of C_{max} change, one a lengthy period of stasis actually beginning at 100 BCE and the second, the most rapid punctuated phase of change in history, one which we are still experiencing.

The first period of complexity change, a decrease a decrease in complexity, matches in duration the phase of punctuation leading to a then historic maximum population of $C_{max} = 1.6E5$ (see Table 18.3).

Table 18.3 Data on the magnitude of C_{max} with respect to boundary centuries for trends in changes of complexity over historic time

Boundary centuries	C_{max}
1600 BCE	80,000
1200 BCE	160,000
400 BCE	500,000
100 BCE	(Rome) 1,000,000
700 CE	400,000
1000 CE	1,200,000
1200 CE	1,000,000
1300 CE	1,500,000
2000 CE	(Tokyo) 35,000,000

At first glance, it would appear to be counterintuitive that an increasing trend in C_{max} would be associated with decreasing complexity; aren't larger urban areas more complex than smaller ones? Yes, but the measure of complexity being used here, the magnitude of m is a system-wide measure of urban areas, not just the largest urban area. So, on face value, a decrease in system-wide complexity in parallel with an increase in C_{max} size might be adaptive, as, a la Tainter (1987) and Bardi (2019), that decrease in complexity reduces the burden of increasing complexity.

Any synchroneity between complexity trends and those of C_{max} change exhibited in the early stages of the last 3600 years of our history is lost for the remainder of this time. To continue temporally, an increase in complexity occurs from 1200 to 400 BCE, while during this time C_{max} magnitude first decreases until 700 BCE and then increases beyond 400 BCE initially to 100 BCE where $C_{max} = 1E6$, and then with a slight hiccup to 8E5 at 1 CE, increases to 1.2E6 at 200 CE. This means that whatever energy cap is being established at this time, as characterized by the magnitude of a given C_{max}, the stasis value being approximately 1E6, complexity changes per se are not also synchronously limited. It should also be recognized that this period of increasing complexity includes 700 BCE, a time point in which a C_{max} minimum occurs, a least with respect to the last 500 years, and also a time encompassing the early years of Jasper's Axial Age (Jaspers 1953). The remainder of this C_{max} growth period is shared by a short period of decreasing complexity; these periods, although not initiating at the same time, are coterminous.

As mentioned previously, the next 2100 years of human history exhibit minimal synchrony between changes in complexity and C_{max}. Complexity increases from 100 BCE to 700 CE, while the magnitude of C_{max} first increases to 1.2E6 at 200 CE and then decreases to 4E5 at 700 CE, a reduction of approximately 67%. This C_{max} minimum does not, however, represent a departure from the 1900 year period of stasis it occurs in that period of stasis having an average C_{max} of approximately 1E6; it is synchronous with a complexity maximum.

The next 600 years can be divided into three periods of complexity change, decreasing complexity for the first 300 years, followed by 200 years of increase beginning, and terminated by 100 years of decrease; this last one is associated with a C_{max} maximum of 1.5E6; this last preliminary to the eructation of the fourteenth century (Tuchman, 1987). There after until the beginning of the nineteenth century,

C_{max} remained approximately at 1E6. But from 1800 forward to the present, the most rapid phase of C_{max} increase in history occurred and is actually ongoing. Over this same period of time in which C_{max} first remained steady and then catapulted to 35E6, complexity increased through three subphases of first increasing rate, then decreasing rate, and then increasing rate (see Fig. 18.4).

If C_{max} can be considered a cap to energy access for civilization in general, then significant changes in complexity can occur under this constraint; changes that involve both rate and duration of change. Further, there are antiparallel changes that do occur between complexity and C_{max} over given periods of time; that of increasing C_{max} and decreasing complexity and vice versa. Previously it was mentioned that this antiparallelism is counterintuitive but only if the entire system is not considered.

As a final comment, it is interesting that these complexity changes associated with the last 700 years occur at the fifteenth and nineteenth centuries, but these changes in rate of complexity change are not synchronized with the Industrial Revolution (IR). They may in fact have been prerequisite to or a harbinger of the IR. Note also that the slope of the complexity curve during this time, the time of initiation of the IR, is at its lowest over this 700-year period of complexity increase.

In summary, a great deal of ink has been spilled over research on complexity, what it is and what its consequences are for innumerable research topics from abstract system analysis to the more concrete reality of assessing the role of complexity in the collapse of a real system, say, Old Winchelsea (Dennis 2023) or an ecological analysis of banking (May et al., 2008). Further, there are at least two broad classes of complexity, complexity by design or by default; this latter class, dubbed ZFEL by McShea and Brandon (2010) has been brought to the attention of readers primarily for completeness. Getting direct data on the effects of or measurement of complexity is sometimes difficult, so proxy data are invoked. With regard to this study, diversity, specifically the diversity of the number, an extent of levels of urbanization below that of C_{max}, is used as a measure of system complexity. A means of computing the number of these levels, m, has been presented, and it was shown that m differs between times of stasis for C_{max} and times of punctuated growth. It was also noted that there were anomalous period s in which m and the pattern of punctuated equilibrium were out of sync with previously established expectations.

It would appear then that both the patterns of r and m, when analyzed over time, suggest a mode of change, alternating change, with respect to the pattern of urbanization. The following is a proposed model of this mode of change over time.

18.3.2 The Urbanization Ratchet

The antiparallel temporal patterns of r and m suggest an alternating pattern of urban growth, urban growth at the top, that is, in the magnitude of C_{max}, alternating with urbanization infill below the magnitude of C_{max}. This pattern of urbanization is being given the name, urbanization pump, as it suggests a leading movement forward followed by a trailing growth of lesser urban sizes. The model follows:

1. During periods of punctuated growth in C_{max}, there is a relative decrease in urbanization sizes at levels below the maximum.
2. Conversely, during periods of stasis there is a relative increase in sizes of urban area population below the maximum.
3. The historic sequence of urbanization is then one of growth at the top as a consequence of a phase of punctuated growth which is followed by relative stasis of C_{max} during which lower levels of urbanization infill, that is, growth at the top creates space underneath that maximum urban area population size which is then filled in during periods of stasis.
4. The transition from stasis to punctuated growth involves crossing a tipping point; a phenomenon which will be now briefly described.

The test for the existence of tipping points used in this research is described in Lamberson and Page (2012). The test data are a change in variability across the tipping point and a change in slope of the process itself. So, if R^2 is significantly greater during stasis than it is during punctuated growth and if the slope changes significantly across the same point, then that particular point can be identified as a tipping point. This is the case for the transitions occurring across 1600 CE, 700 BCE, and 1800 CE.

18.4 Summary and Conclusion

The focus of this research has been twofold: to investigate the relationship between the punctuated equilibrium-like pattern of maximum urban area population over time with respect to the change in urban area sizes below the maximum; this represented by the metric, r, and also to investigate the relationship between this punctuated equilibrium pattern and urban area system diversity in terms of the number of layers of urbanization size below the maximum. The metric used in this investigation is dependent on total world-system population, maximum urban area population, and the variable, γ. Several assumptions were made and include the notion that the distribution of urban areas is Zipf-Pareto-like, that the total population per level is both greater than the previous level but has individual urban area populations which are less than the previous individual urban area population. It is assumed also that the maximum urban area population at any given time is a consequence of the limitation imposed on the whole system by both the nature of and access to energy. It was noted that the total world-system population did not exhibit a punctuated equilibrium pattern even though the maximum urban area populations, on log-transformation, do.

The macro-view of changes in r over historic time exhibits a form reminiscent of hyperbolic growth; however, on closer inspection, a finer scale of change is revealed, one in which there is no pattern of punctuated equilibrium but rather oscillation between positive and negative slopes with the apices of this pattern coinciding with both the beginnings and ends of periods of stasis. Further, there are two distinct segments of oscillation of r, one representing the Ancient World and the second representing the Classical and Early Modern Worlds. There are also two periods of r-transition represented; the first from the Ancient to the Classical and Early Modern Worlds and the second one in which the world-system is currently experiencing and going to the next period of oscillation, clearly as yet to be identified.

Complexity was investigated using the metric, m, a measure of diversity, which is used as a source of proxy data for complexity in that the Zipf-Pareto distribution is hierarchical and that urban areas do not exist in isolation and exist in an interconnected network. It was noted that complexity may be classified as either by default or by design, the former as a consequence of ZFEL, zero force evolutionary law, a law that suggests that over time systems simply by dint of their existence become more complex. Complexity by design is a consequence of the activity of complex adaptive systems, and humans are such systems, although human design may create and does create complexity in unexpected and unpredictable ways. The micropattern of complexity over historic time is one of overall increase with oscillations that are over much of this time span mirrored by the changes in r. The some of the peaks and valleys of changes in m are coincident with the beginnings and ends of periods of stasis. Within the bounds of periods of stasis in particular, there are also oscillations which are in sync or antisync with changes in maximum urban area populations.

The pattern of changes exhibited by both r and m suggests that there are distinctly different consequences for the changes exhibited by the punctuated equilibrium pattern of maximum urban area populations. During periods of punctuated growth, urban areas below the maximum are relatively smaller as a percentage of a given maximum urban area population than those associated with the stasis phase of world-system change. This led to a proposal that there was an urbanization ratchet in which maximum urban area population growth created space at all lower levels of urbanization which were then "filled in" during periods of stasis. Early research with respect to the relationship between changes in complexity with changes in system entropy suggests a potential fruitful area of research. This last was reported in a brief Addendum.

Mathematical Glossary and Appendix

$T =$ Total world-system population.

$C =$ Urban area size as measured by population.

α and g are fitted constants.

If $2 = \alpha C^{-\gamma}$, then $C = 2^{-1/\gamma}$, and since there are two smaller urban areas per single urban area in the higher than $r = 2C = 2^{(1-1/\gamma)}$.

Given: $T = C_{max}[(1—r^{m})/(1—r)]$, a standard formula within the mathematics of sequences and series, then $T/ C_{max} = (1—r^{m})/(1—r)$, and $(1—r^{m}) = T/ C_{max}(1—r)$. Then $r^{m}—1 = T/ C_{max}(r—1)$, followed by $r^{m} = T/ C_{max}(r—1) + 1$. Taking the natural logs of both sides of this equation gives: $m \ln(r) = \ln[T/ C_{max}(r—1) + 1]$, and finally $m = \ln[T/ C_{max}(r—1) + 1]/\ln(r)$.

References

Bardi, U. (2019). *Before the collapse: A guide to the other side of growth*. Springer. https://doi.org/10.1007/978-3-030-29202-7

Chandler, T. (1987). *Four Thousand Years of Urban Growth: An Historical Census*. Edwin Mellen Press.

Chew, S. (2006). *The recurring dark ages: Ecological stress, climate changes, and system transformation*. Altamira Press.

Eldridge, N., & Gould, S. (1972). Punctuated equilibria: An alternative to phyletic gradualism. In T. Schopf (Ed.), *Models in paleobiology* (pp. 82–115). Freeman Cooper.

Harper, A. (2010). The macropattern of urbanization over the course of the last 5000 years of world-system history. *Social Evolution and History, 9*(1), 115–133. https://doi.org/10.30829/1990-2527-2010-9-1-115-133

Harper, A. (2017). The punctuated equilibrium macropattern of world system urbanization and the factors that give rise to that macropattern. *Social Evolution and History, 16*(1), 86–127. https://doi.org/10.30829/2309-6490-2017-1-86-127

Harper, A. (2025). Trends in urbanization and urban size distribution over the last 5000 years. *Journal of Big History, 8*(1), 144–153.

Korotayev, A. (2018). The 21st century singularity and its Big History implications: A re-analysis. *Journal of Big History, 2*(3), 71–118. https://doi.org/10.22339/jbh.v2i3.2320

Korotayev, A. (2020). The 21st century singularity in the Big History perspective. A re-analysis. In A. V. Korotayev & D. LePoire (Eds.), *The 21st century singularity and global futures. A Big History perspective* (pp. 19–75). Springer Nature. https://doi.org/10.1007/978-3-030-33730-8_2

Korotayev, A., Malkov, A., & Khaltourina, D. (2006). *Introduction to social macrodynamics: Compact macromodels of the world system growth*. KomKniga/URSS.

Lamberson, P., & Page, S. (2012). *Tipping points*. Santa Fe Institute Working Paper. https://doi.org/10.2139/ssrn.2069671

May, R., Levin, S., & Sugihara, G. (2008). Ecology for bankers. *Nature, 451*(7181), 893–895. https://doi.org/10.1038/451893a

McShea, D., & Brandon, R. (2010). *Biology's first law: The tendency for diversity & complexity to increase in evolutionary systems*. The University of Chicago Press.

Modelski, G. (2003). *World cities: -3000 to 2000*. Washington DC: Faros 2000.

Morris, I. (2013). *The measure of civilization: How social development decides the fate of nations*, Princeton University Press

Newman, M. (2004). Power laws, Pareto distributions and Zipf's law. *Contemporary Physics, 46*(5), 323–351. https://doi.org/10.1080/00107510500052444

Tainter, J. (1987). *The collapse of complex societies*. Cambridge University Press. https://doi.org/10.1017/CBO9781139462551

Tuchman, B. (1987). *A distant mirror: The calamitous 14th century*. Random House Publishing Group. https://doi.org/10.1007/978-3-642-83563-7

von Forester, H., Mora, P., & Amiot, L. (1960). Doomsday: Friday, November 13, A.D. 2026. *Science, 132*(3436), 1291–1295. https://doi.org/10.1126/science.132.3436.1291

Antony Harper 's initial training was in ecology and evolutionary biology. He taught high school sciences at New Trier High School for over thirty years, during that time developed a relationship with a number of Russian scholars, for example, Andrey Korotayev and Leonid Grinin, who shared his interest in the application of math models to the patterns of human history, and over the past almost twenty years have had the privilege of conducting and publishing research in this general area. Currently, he is an independent scholar with an interest in the pattern of urbanization over historic time, and by default, an interest in the demise of urbanization patterns. Since urbanization is intimately tied to the advent of agriculture and the production of surplus, and since the production of surplus is in turn related to the level of technology, that triad of urbanization, surplus, and technology has led him to investigate a number of topics, one of them being the collapse of human social structures at scale. At the present time, this in turn has directed his research focus to the pattern of survivorship of empires; ongoing, he is also considering survivorship patterns at lower levels of societal structure, for example, city-states.

19 War and the Rise of Social Complexity: A Big History Perspective

Stephen T. Satkiewicz

Abstract

The relationship between war and social complexity is deeply intertwined, forming a reflexive cycle where each influences the other as both cause and effect. This connection extends from the origins of organized violence in the Neolithic period to the present, playing a crucial role in the formation of early states, empires, and socio-political systems. As Charles Tilly (1929–2008) famously stated, "*War made the state and the state made war.*" This study builds on interdisciplinary work within the civilizational field to examine the thresholds of emergent social complexity and the broader processes of social macroevolution. By integrating these insights, we aim to situate Big History more firmly within the human and social sciences. War and organized violence do not occur in isolation; rather, they reflect the complex historical, economic, and political contexts in which they arise. Understanding these dynamics provides a deeper perspective on the evolution of societies and the forces that shape them over time.

Keywords

Civilizations · Social complexity · War · Strategy · Comparative history · Social evolution

This chapter is an extensively re-worked version of the text originally published in Journal of Big History (Satkiewicz, 2025).

S. T. Satkiewicz (✉)
Rochester Hills, MI, USA

D. J. LePoire et al. (eds.), *Complexity in Universal Evolution*, World-Systems Evolution and Global Futures, https://doi.org/10.1007/978-3-032-20517-9_19

19.1 Introduction and Theoretical Outline: Big History, Civilizational Scholarship, and Social Complexity

> *"War is the father of all and king of all,*
> *who manifested some as gods and some as men,*
> *who made some slaves and some freemen."*
> Heraclitus
> (Heraclitus, n.d.)

Since the time of Heraclitus in the fifth–sixth century B.C.E. (Before the Common Era), war remains a part of the world scene, and it remains the responsibility of scholars to properly study its nature and understand this important, if unfortunate, aspect of reality. Already the field of Big History has ventured into this area of study with the pioneering work of Daniel Barreiros utilizing the field of ethnology and the study of apes to better understand human behavior related to war (Barreiros, 2018; Barreiros & Ribera Vainfas, 2019). Anzar Gat has also established a monumental macro-historical model of the study of war, even noting its importance as a field of study:

> *With war being connected to everything else and everything else being connected to war, explaining war and tracing its development in relation to human development in general almost amount to a theory and history of everything. As so much is relevant to the subject, one is required to read pretty much "everything" and become sufficiently expert in many fields.* (Gat, 2008, p. ix)

The scholarly intention of this study is the continuation of previous work related to the connection between social complexity and war (Satkiewicz, 2020a, 2020b, 2023, 2025) It will draw upon interdisciplinary research in the Civilizational field with subsequent contributions from International Relations, Geopolitics, Military History, Grand strategy, Complexity Science, Historical Macrosociology, etc. The Civilizational field has already established an interdisciplinary framework for the macro-scale study of socio-cultural human complexity, which may be of great assistance to Big History, which tends to encounter struggles and obstacles when engaging with the insights of the human and social sciences. It also operates within a wider engagement between Big History and the Civilizational field that has been underway, wishing to further expand this mutual understanding and engagement between the two macro-scale fields—as symbolized by the formal partnership between the International Big History Association (IBHA) and the International Society for the Comparative Study of Civilizations (ISCSC) officially proclaimed on September 14, 2023, at the War Studies University in Warsaw, Poland.

A comprehensive list of examples of the topics involved in the study of war and social complexity could include the importance of war to the rise and fall of states; warfare as a basis for the formation of different forms of social solidarity; war as a basis for cross-cultural exchanges; military revolutions; the place of war within social evolutionary processes; the cross-cultural study of different socio-military systems; warfare as a stimulus for technological and scientific discoveries; and even possibly the effects of war upon the natural environment. The purpose of such would be to accurately assess the place of war within the larger contexts of social complexity. Hopefully, this may serve as a comprehensive introduction.

Since such a broad topic can be addressed in numerous directions, it might be necessary to establish a few preliminary points first. The first important point will be regarding the nature of the Civilizational field and its relevance to a Big History framework. An exhaustive study of the nature of Civilizational scholarship is beyond the scope of this paper, but it is important for Big History to distinguish between two major but distinct intellectual traditions within the field: Civilizational Analysis and Civilizational Science(s). They are not mutually exclusive, but these two traditions do have differing trajectories in regard to the study of civilizations.

Civilizational Analysis is primarily concerned with the macro-scale and comparative study of human socio-cultural configurations across long-term time scales (Arnason & Hann, 2018, p. xiv). An outgrowth of Historical Sociology, it follows the trajectory of the work of Max Weber and Emile Durkheim with subsequent contributions from Benjamin N. Nelson, Vytautas Kavolis, S.N. Eisenstadt, and Johann P. Arnason.

Civilizational Science is the interdisciplinary macro-scale study of the socio-cultural universe with its causal systems and congeries (Alaykin-Izvekov, 2011; Sorokin, 1943). It follows the trajectory of the work of Pitirim Sorokin in establishing an Integral Social Science that represents a more holistic synthesis of the major social, cultural, and human sciences together (Mangone, 2023, p. 19). Among the major representatives of this tradition include Arnold J. Toynbee, Feliks Koneczny, Christopher Dawson, Fernand Braudel, Carroll Quigley, and Samuel P. Huntington (Alaykin-Izvekov, 2011).

One of the major characteristics of the comparative civilizational approach as explained by Johann P. Arnason is the recognition that civilizations "have their distinctive historical contours and rhythms, differing from case to case. Contrasts between ancient Egypt and ancient Greece, or Western and Byzantine Christendom, can serve to illustrate varying patterns of historical experience and historical consciousness" (Arnason & Hann, 2018, p. xiv). As Stephen Kalberg further explains, this means that a more contextual "*mode of analysis*" is necessary in order to properly study "the full complexity, uniqueness, and multi-dimensionality of each civilization..." (Kalberg, 2021, p. 3).

Too often Big History scholarship relies upon a "grand theory of everything" of the type borrowed from physics and other natural sciences. However, this kind of approach is ill-suited for the complexities studied in the human and social sciences. Even the Social Evolution school within Big History relies on mathematical modeling to achieve its results. Therefore, the long-standing dispute between the natural and the human-social sciences persists, with Big History implicitly seeking to subordinate the latter to the former. Whether or not a true "theory of everything" is even feasible in physics has now been seriously questioned. There is no consensus among physicists as exactly what quantum theory and mechanics can discern about the nature of reality, so by default a certain theoretical pluralism of paradigms is the norm in the field (Gibney, 2025). The framework of "effective field theories" offers more contextual forms of analysis, which have proven more fruitful in research (Hartmann, 2025). Jordana Cepelewicz further argues that mathematical models and proofs are grounded in social constructive principles (Cepelewicz, Why Mathematical Proof Is a Social Compact, 2023). This does not mean that mathematical models and proofs should be reduced to their social constructive elements, but rather that they must be critically addressed. These examples demonstrate that subordinating the social and human sciences to the natural sciences is unjustifiable, and that neither has inherent priority over the other. It is common to speak of "physics envy" within the social and human sciences. The natural sciences often fall into a kind of "theology envy," aiming to become the next "queen of the sciences." E.O Wilson's model of *Consilience* built around biology being the basis of all significant knowledge is perhaps the most infamous example of such attempts that has sparked considerable criticism from various fields (Wilson E. O., 1998).

Pointing this out is not meant as any kind of hostility toward Big History, but rather pointing out a potential weak spot to help further its growth as an academic field. Civilizational scholarship can also benefit from greater awareness and engagement with Big History, especially its Social Evolution school (Satkiewicz, 2021a, 2022). The gap between the two forms of science (natural and human) may still continue to exist, but that does not mean that meaningful bridges cannot be built between them. Big History has set about the task of doing so, which is a very noble endeavor. Honest recognition of these differences might help foster more respectful convergences in scholarship—which are already in progress and should continue into the future.

Having established a suitable foundational understanding regarding Civilizational scholarship, it is now necessary to address the nature of social complexity next. Although complexity science and social complexity are well-established academic fields, there is no consensus on an exact definition for them. Although precise definitions may be elusive, certain themes of complexity can be discerned within the field. The Santa Fe Institute, perhaps one of the leading complexity science centers, defines its research as involving

> *[s]ystems with many interacting components, or agents, from which surprising and unpredictable system-wide behaviors can emerge is the hallmark of complexity research. This includes examining the dynamics of these "complex adaptive systems" and how they evolve as agents interact and react to one another their environments.* (SFI in Depth | Santa Fe Institute, n.d.)

In a similar manner, the New England Complex Systems Institute defines its research as:

> *[going] beyond the traditional, reductionist approach of focusing on the parts of a system, to integrating the network of relationships within and between systems. These relationships produce the 'emergent' behaviors we see in all physical, biological, social, economic and technological systems.* (Overview, n.d.)

Robert Axelrod, one of the founders of Game Theory, provides one other definition that is helpful:

> *Complexity theory involves the study of many actors and their interactions. The actors may be atoms, fish, people, organizations, or nations. Their interactions may consist of attraction, combat, mating, communication, trade, partnership, or rivalry.* (Axelrod, 1997, p. 3)

Building upon these foundations, a working definition of *complexity* could be established as *the non-reductionist study of the multifaceted relationships within and between systems and entities that involve (semi)predictable and non-predictable patterns of evolution. Social complexity* could then be expanded to focus on how these apply to social realities, including both humans and non-human animals. This provides a simple enough definition that is understandable but is also applicable to varying contexts as well. Further scholarship into the debates on the precise definitions of *social complexity* can be found elsewhere (Kappeler et al., 2019).

19.2 Further Exploration into the Nature of Human Social and Civilizational Complexity

The important factor here in this study is the focus on *social* complexity, or more specifically still *human* social complexity. Noted social theorist Margaret Archer (1943–2023) explained that the human component adds a peculiar kind of complexity that is contrary to the kinds found in the natural world:

> *Social reality is unlike any other because of its human constitution. It is different from natural reality whose defining feature is self-subsistence: for its existence does not depend upon us, a fact which is not compromised by our human ability to intervene in the world of nature and change it. Society is more different still from transcendental reality, where divinity is both self-subsistent and unalterable at our behest; qualities which are not contravened by responsiveness to human intercession.* (Archer, 1995, p. 1)

A major issue concerning the study of social complexity across large time scales is fitting the dynamics of diverse societies and agents into a single framework. History is traditionally studied in terms of contextual and contingency based research, so simple narratives of evolution from simple societies to more complex ones are poorly suited to this approach. Archeologists have also been skeptical about the feasibility of using one single model to study varying contexts of human social complexity (Tim Kohler et al., 2018; Robert D. Drennan & Peterson, 2012). Some Big Historians have devised the concept of "*multilinear evolution*" to address this difficulty. Building upon the concept devised by Julian Steward, *multilinear evolution* concerns itself with how various societies have evolved in the same general direction yet still doing so in their own peculiar ways (for example: from simple tribal to more complex state-centered social systems) (Carneiro, 2011).

Rather than a linear progression from simple to complex, social complexity can be the study of the processes, evolution, and emergence of different configurations of dynamics as well as connections between varying levels with one another. This approach is compatible with the concept of "*multilinear evolution*" but without the implicit concept of one specific form of social evolution happening across different contexts. Rather the emphasis should be on the comparative study of these varying contexts. There is also the problematic issue of determining when exactly this transition from social simplicity to complexity could be dated and studied. As some researchers have openly admitted: "*Complexity is an unfortunate term because its inversion is simplicity, but no society of Homo sapiens is (or ever has been) simple*" (Tim Kohler et al., 2018). Ongoing research from archeology and other fields has continued to demonstrate that even prehistoric societies have featured considerable levels of complexity (Oxford Archaeologists Discover Monumental Evidence of Prehistoric, 2022).

Issues concerning the presence of both linear and non-linear dynamics in the evolution of social complexity are another reason a more comparative approach might be better suited to Big History. For simplicity purposes, linear and non-linear will be defined as largely analogous to predictable (linear) and unpredictable (non-linear) actions and behaviors. Human behavior on a large scale can incorporate both, although Piero Dominici stresses that human-based complex systems tend more toward being "*uncontrollable, unpredictable, and impossible to break down into their highly dynamic, interconnected, and interdependent parts*" as opposed to more "complicated" mechanical-based systems (Dominici, 2023).

This further underlies the requirement for a greater reliance on comparative research into causal mechanisms, contingencies, and configurations across different historical contexts, levels, and forms of social complexity. This also necessitates the need for scholars to maintain a more multiparadigmatic approach in order to study the multifaceted nature of social complexity across varying historical contexts. Paradigms and methodologies are important, but by nature they are selective of data in order to gain insights of perspective (Kalberg, 2021, p. 37). Different paradigms have significant insights to make, but no one paradigm has all the answers. Synthesis and eclecticism are more the foundational points for multiparadigmatic approaches with the intention that "*each paradigm position is studied and considered in with essential interplay with alternative ones. If an overarching view does take prominence in such synthesis and design, it will only be due to some compelling logic or constellation of ideas that organize in such a way to resonate in [multiple paradigms] to various practitioners*" (Zweibelson, 2025, p. 726). This allows for a more meta-level analysis of the phenomenon of social complexity, and more specifically in this case the interconnections with war (Gioia & Pitre, 1990, p. 598). This is also reflected in the fact mentioned earlier that there is no single consensus definition of what actually constitutes complexity.

The Civilizational field is already familiar with these kinds of debates regarding human socio-cultural macro-dynamics, exemplified by Vlad Alaykin-Izvekov's model of *Civilizational Science* that builds upon the premise of different macro-level sciences and studies "creating a sophisticated web of interconnections" in which they influence one another (Alaykin-Izvekov, 2011). This might prove more helpful than trying to impose a single definition and paradigm upon various contexts where they may not apply. An extremely useful supplement to such an approach is the one proposed by Andrew Latham in his study of late Medieval geopolitics that he terms "constructive holism" that seeks a "theoretically guided, historically sensitive account of the deep socio-political character" of the particular historical and civilizational context under investigation (Latham, 2012, pp. 16, 42). Archeologists Scott Ortman, Lily Blair, and Peter N. Peregrine have proposed a form of "coarse-graining" that has been utilized in the natural sciences (physics and biology) that entail a "stepping back from the details to a level of focus where the behavior of the system becomes more amenable to generalizations." The result of such a framework would be defining human social evolution as the process "where societies still vary in their basic properties but the patterns in this variation appear more regular than they do when viewed in close detail" (Scott Ortman et al., 2018, pp. 187, 188). This example also proves that fruitful cross-disciplinary work between the human and natural sciences is possible without the need to seek to subordinate one form of science to the other. Mutual intellectual enrichment between the different fields and sciences should be the ideal for Big History moving forward.

One major component of social complexity concerns the multi-scale and multi-layers it encompasses, typically divided into *macro*, *meso*, and *micro* levels. This study will use this model to demonstrate examples of these levels in action:

Levels of Social Complexity:

- **Macro-level**: Civilizations, empires, mega-states, world systems, alliance systems, state systems, large-scale networks, supercomplex chiefdoms, etc.
- **Meso-level**: States, kingdoms, complex chiefdoms, complex confederations, etc.
- **Micro-level**: Tribes, villages, small towns, chiefdoms, cities (states), etc.

This further helps to develop the argument concerning the non-linear comparative study of social complex dynamics. Rather than trying to explain the linear progression from tribes to villages to states to world systems for example, Big History (following the example of Civilizational scholarship and Historical Macrosociology) could study the different configurations of how tribes, villages, states, and world systems connected and related to one another across large time scales.

This form of study could also be furthered with the incorporation of the study of *Civilizational Ontology*, which has already been utilized to the study of war and social complexity (Satkiewicz, 2023). This emerging field of study is concerned *with the underlining nature of civilizations and seeks to define their fundamental characteristics.* This conception was formulated through the expansion of Benjamin N. Nelson's model of *civilization complexes* and *inter-civilizational encounters* (Nelson, 1981, p. 90) into a more tripartite model of macro-scale human socio-cultural dynamics.

The major characteristics of Civilizational Ontology are defined as follows:

1. ***Civilizational Complexes or Civilizational Complexity*****:** The most basic underlining principle involved. Civilizations are complex macro-scale sociocultural entities.

2. ***Intra-Civilizational Encounters***: Due to the inherent complex nature of civilizations, many subgroups and sub-entities within civilizations will internally interact with one another on various levels. This can take the form of both cooperative and antagonistic forms.
3. ***Inter-Civilizational Encounters***: Civilizations can and often do interact with one another in external forms as well, but like their internal counterpart can take the form of both cooperative and antagonistic forms (Satkiewicz, 2020b).

One advantage of this model is its attempt to reconcile the process-based and substance-based analysis of macro-scale social complexity that can complicate any worthwhile study in this area.

It is intended to avoid any simplistic "clash of civilizations" interpretations as well. Stephen Kalberg emphasizes that the importance of studying "intra- and inter-civilizational tensions" requires "powerful constructs, modes of analysis, and research strategies..." (Kalberg, 2021, p. 4). It is the intention of *Civilizational Ontology* to provide the overarching framework for such endeavors.

Although previously conceived to analyze contemporary conflicts, it can be applied to many other historical contexts and examine the relationship between war and social complexity. Here are a few possible examples of social complexity and war analyzed with this model:

- ***The Peloponnesian War*** **(431–404 BC):** *Intra-civilizational encounter* between Athens and Sparta; city-states (polis) and alliance systems involved.
- ***Greek-Persian Wars*** **(499–449 BC):** *Inter-civilizational encounter*; city-states (polis), alliance systems, world system, empire involved.
- ***Warring States Period*** **(fifth century BC):** *Intra-civilizational encounter* involving various Chinese kingdoms fighting for dominance.
- ***Nomadic invasions of China*****:** *Inter-civilizational encounters* involving Chinese empire and nomadic complex chiefdoms.

These examples are intended to also compare social complexity and war across two different major ancient historical civilizational contexts: Ancient Greece and Ancient China. One can notice that while both civilizations face both *inter* and *intra* civilizational warfare encounters, it would be wrong to argue that they existed in the exact same form in an evolutionary manner. Ancient Greece during the Hellenic period was based around several independent city-states (*polis*) fighting for dominance and control. By contrast, China's context was based more around competing kingdoms.

This contrast is an interesting point that should be explored in a little further depth, which can help further demonstrate why comparative historical study should be the emphasis as opposed to the linear progression from simple to complex. Even Max Weber, one of the founding fathers of modern social science, noted a strong contrast to how European cities tended to be major arenas of relative autonomous power (depending on which time period one is referencing) as opposed to the tendency for cities in China to be subordinate to a central authority (Kalberg,

2021, pp. 218–223). There has been further recent scholarship establishing the importance of the "city" as a foundational concept in European and later Western political theory (Hörcher, 2021). A simple comparison of Plato and Aristotle utilizing the Greek polis as the foundation for their theorizing about politics as a contrast to Sun Tzu and Confucius using the standard Chinese kingdom as the foundation for their theorizing demonstrates this contrast.

Thus, a rigorous ideational element of comparative history is necessary.

19.3 War and Social Complexity

The connection between war and social complexity is not entirely new. In many ways the idea could even be traced back to Carl von Clausewitz's famous work *On War*, which noted that war is a complex reality of varying factors intermixing with one another. Although often misunderstood, Clausewitz's model of the *Trinity* was meant to capsulate this complexity into three main factors of *passion*, *chance*, and *policy* that operated "like three different codes of law, deep-rooted in their subject and yet variable in their relationship to one another" and to truly understand how these factors operated together required avoiding theories "that ignores any one of them or seeks to fix an arbitrary relationship between them would conflict with reality to such an extent that for this reason alone it would be useless" (Von Clausewitz, 1979, p. 89).

Extensive study and analysis of Clausewitz is beyond the scope of this paper (Satkiewicz, 2020a), but it might be helpful to note that Christopher Bassford and Edward J. Villacresn stipulate that Clausewitz's trinity amounts to an account of *rational*, *irrational*, and *non-rational* factors at play in war (Villacres & Bassford, 1995) and that could also "easily be made to encompass all of human experience" (Bassford, n.d.). Furthermore, in the past couple of decades, scholars have noted significant parallels between Clausewitz's theory of war and contemporary scholarship in complexity theory (D. Beyerchen, 1992). From this foundation, much scholarship has proceeded over the past generation to formulating the connections between war and complexity. The late Colin S. Gray (1943–2020), a leading strategic theorist, made considerable note about the interrelated contexts that war and its planning is influenced within:

1. ***Socio-cultural***
2. ***Economic***
3. ***Technological***
4. ***Military***
5. ***Geographical***
6. ***Historical*** (Gray, 2016, p. 39)

The parallels to the varying elements commonly studied with social complexity are striking and demonstrate that war does not occur in a vacuum but is conditioned upon its context within complex social systems. Hugh Smith has also further stressed this connection:

> *For the social dimension introduces elements into war that military and civilian strategists often find awkward to handle; military historians may prefer campaigns and battles to behavioral science. Increasingly, however, many analysts of war are now coming to emphasize the social factors in modern conflict—culture, sociology, psychology, anthropology, ethnography and the like. . . War now needs social science as it once needed physics and chemistry. (Smith, n.d.)*

One other important factor in consideration for the connections between the evolution of social complexity and war concerns the nature of *norms*, that is, "the patterns of behaviour" that make up the practice of war within certain given socio-cultural contexts (Walker & Roberts, 2023, p. 1). Robert Axelrod also defined *norms* as a "*powerful mechanism for regulating conflict in groups, even where there are more than two people and no central authority. . . A norm exists in a given social setting to the extent that individuals usually act in a certain way and are often punished when seen not to be acting this way*" (Axelrod, 1997, pp. 44, 47).

Peter Roberts and Paddy Walker identify three basic forms of *norms* as follows:

- *Enduring norms:* Patterns of behavior that have existed and persisted on longer timescales.
- *Emerging or evolving norms:* These patterns of behaviors are in the midst of change conditioned by changes in the wider socio-cultural context.
- *New norms:* Sudden changes in behavior that have effect on the wider socio-cultural context (ibid., pp. 2–3).

This addresses a certain paradox that war has a persistent *nature* but often changing *character*. Clausewitz himself made this insight when he famously referred to war as a "true chameleon that slightly adapts its characteristics to the given case" (Von Clausewitz, 1979, p. 89). These norms in war are further influenced by the given configurations of "intersubjective beliefs, social institutions and constitutive norms" in each complex society (Latham, p. 60).

How these *norms* are enforced are through several causal mechanisms. Axelrod identified five major such mechanisms for the enforcement of norms in social complexity that have general resonance with the existence of war and social violence:

- *Metanorms*: That seek to punish those who violate norms but also those who refuse to support the punishment of violators. Rene Girard's conception of the *scapegoat mechanism* is perhaps the most famous example of such metanorms (Andrade, n.d.; Hardy et al., 2022).
- *Dominance*.
- *Internalization*: How a society and culture comes to internally adhere to a norm.
- *Deterrence*.

- *Social Proof:* The more a norm is commonly adhered to within a social system, the more it is perceived as "proof" of its correctness.
- *Membership.*
- *Law.*
- *Reputation.* (Axelrod, 1997, pp. 55-66)

This outlines the mechanisms of how norms are enforced, but the historical-genealogical reality of how norms even arise and evolve over time is through the mechanism of normative complexes that over time produce the process of "*unifying [societies] around a common moral ontology and shared purpose creating a shared life-world within which communicative action becomes possible*" (Phillips, 2011, p. 26). This points to yet another important factor for the proper study of social complexity and war, which is termed the *Military-Ideational Nexus* and its influence on the development of *Strategic Culture* and *History*. First the concept of the *Military-Ideational Nexus* is based on the premise that "*[w]ar and warfare stand at the interface between instinctive behaviour and the driving force of ideas*" (Heuser, 2022, pg.1). Thus the formulation of the practices and ideas of war do not grow out of a vacuum, but rather are reflective of the given "knowledge structures" of the socio-cultural context from which they emerge (Layton, 2018, p. 39).

Perhaps the most famous examples of such "structures of knowledge" are the ones that emerged in the Axial Age during the first millennium B.C.E. out of which emerged the major world religions. The idea of the Axial Age has been studied in great depth in both Big History (Baskin & Bondarenko, 2014) and the Civilizational field (Arjomand & Kalberg, 2021; Eisenstadt, 1986; Rnason et al., 2005), which once again presents another venue for further convergences between the two fields. The main Axial Age civilizations whose legacy continues to influence the world emerged in the ancient Near East, Greece, India, and China. From these emerged the great world religions of Zoroastrianism, Hinduism, Buddhism, Judaism, Greek Philosophy, as well as Christianity and later Islam (Arnason & Hann, 2018, p. 55). All these traditions in some form or another dealt with the subject of war. The famous Chinese strategist Sun Tzu largely lived during the Warring States Period of China, from which also emerged the Axial Age religions of Confucianism and Taoism. Sun Tzu himself was deeply influenced by Taoism (Yuen, 2014). In terms of Western or European Civilization, Beatrice Heuser has identified four main axial points for its tradition of thinking about war:

- Greek philosophy
- Roman Law
- Judaism
- Christianity (Heuser, 2022, pp. 4–6)

These are just a few examples that help to establish the existence of such a *Military-Ideational nexus*. Recent research into civilizations of the ancient Near East has attempted to show the origins of the concept of "just war" that long shaped modern attitudes (Cox, 2023). This means that the Big History of Warfare and the Big History of Ideas have plenty of space to converge on this point.

From the foundations of the *Military-Ideational nexus* emerges the need for such ideas to be put into actual practice. This is where both *Strategic Culture* and *Strategic History* must be addressed. Very much like social complexity, there is no consensus about how to define a *Strategic Culture*, but one helpful definition regards it as *"[T]hat set of shared beliefs, assumptions, and modes of behavior, derived from common experiences and accepted narratives (both oral and written), that shape collective identity and relationships to other groups, and which determine appropriate ends and means for achieving security objectives*" (Kartchner et al., 2023, p. 6).

Thus, how different complex societies and civilizations think and act in war can differ significantly, and this in turn is related to differing strategic cultures built on differing ideational "structures of knowledge." This further emphasizes why a more comparative approach to such strategic cultures is necessary "to address the problem of how best to develop analytical concepts that do not treat the world as an isotropic (uniform) surface, but instead make sense of different military goals and traditions" across differing socio-cultural contexts (Black, 2004, p. 19).

The long-term study of how these *strategic cultures* have operated is the focus of *strategic history*. As opposed to traditional military history, *Strategic History* is more grounded in a theoretical-based study of long-term patterns (Gray, 2013, pp. 1–15). Although not relating directly to *strategic history*, Andrew Latham has also referred to these as "the historical structures of war" that are characterized as being grounded in "macro-level cultural, institutional and epochal contexts and configurations" (Latham, 2012, pp. 10–11, 35–41). So the relevance to the field of Big History is certainly obvious, although Big History would be addressing these factors on much larger time-scales than what is usually standard in strategic history. For example, Gray's study covers only the last two centuries. Latham's study of the Medieval period is better, but can still be expanded even further. This is hopefully the most promising area of growth that both Big History and the Civilizational field can contribute to this scholarly field. An exhaustive study of Strategic history is beyond the scope of this paper, but it is worth citing Colin S. Gray at length as to the foundation for such a field of study:

> *[W] e can be certain that all human communities, in all periods of the past, had and were aware they had, a strategic context within which they were (and are) obliged by necessity to operate. Human beings gather in communities for security. This gathering requires political activity, which inevitably leads to anxiety about rival communities. Thus, the need for strategy is born, always and everywhere. World history includes a ubiquitous strategic theme. We have always had a strategic context.* (Gray, 2018)

So, a vitally important aspect to studying the evolution of complex societies is studying the manner in which they addressed and confronted certain existential threats. This very much aligns with the classic "challenge and response" model developed by Arnold J. Toynbee (1889–1975) in his classic multi-volume work *A Study of History*.

Hopefully, this has served as a comprehensive and useful introduction to the multifaceted scholarship and research into the nature of war and complexity. There is plenty of room for growth for both Big History and Civilizational Sciences in this regard. This also further confirms a major argument of this paper that human social complexity has unique characteristics that need to be considered apart from other natural and even biological forms of complexity, which can present certain challenges for the standard approaches in Big History. Of course, it is also the intention of this paper to assist Big History in surmounting such challenges.

19.3.1 War as a Cause and Effect of Social Complexity: Case Studies Review

A major argument of this study is how war can be both a casual mechanism and effect of social complexity. A short review of several case studies from already existing scholarship will be utilized to help illustrate this point, with the obvious caveat that each case study is the object of intense specialized research. Admittedly, many of these cases will be Euro-centric in orientation. This is not meant to convey any lack of legitimate study in non-European forms of social complexity, but rather simply an honest acknowledgment that the vast array of scholarships available cannot be adequately addressed in one article. The honest acknowledgment of this deficiency should serve as an invitation for further research and scholarship, and plenty of comparative studies in the field of social complexity have already been achieved.

19.3.2 War and the Origins of Social Complexity

If there has been one influential scholar on the study of the rise of social complexity, and especially on the rise of the state as an embodiment of social complexity, over the past half century that would be the late anthropologist Robert L. Carneiro (1927–2020) (Chacon, 2020). In his 1970 article "A Theory of the Origin of the State," Carneiro outlined what he would term his "Circumscription Theory" about the origins of the state. To summarize: communities would face inherent environmental constraints and had to effectively meet and overcome such challenges in order to survive on a long-term basis. One of the most prominent methods used by communities to overcome such constraints was warfare and especially conquest of neighboring communities to exploit their own resources. As Carneiro himself stated: "*Force, and not enlightened self-interest, is the mechanism by which political evolution has led, step by step, from autonomous villages to the state*" (Carneiro, 1970, p. 734).

Although intended as a general theory of the origins of the state and social complexity, Carneiro did specify that his theoretical outline was meant to address ancient societies as varied as the Maya, the Andes, Egypt, and China (Carneiro, 1970, p. 736; 2012). The overall influence of this theory has been immense in the scholarship of social complexity and the origins of the state as we know it. It has been argued to be one of the most cited theories in the field (Zinkina et al., 2016). Throughout the rest of his career and life, Robert L. Carneiro was open to modifying his theory and addressed criticism directly, which proves how vitally important his work has been to the study of war and the origins of social complexity (Carneiro, The Circumscription Theory).

The full implications, nuances, and debates about Cairnero's Circumscription Theory and more specifically the role of war in it are beyond the scope of this chapter. Its main relevance is the demonstration of war being a primary causal mechanism for the evolution of social complexity from simple to complex chiefdoms and finally to states and empires. This growth in complexity was necessitated due to the inherent advantages of the latter over the former formations of social complexity, namely the managing of resources and ability to consolidate power over conquered territories over longer durations. This process can be thoroughly investigated in case studies across the ancient world in Mesopotamia, Egypt, China, the Indus Valley, alongside Mesoamerica and the Andean regions (Malešević, 2017, p. 106). This ability of control through military force was also connected to being able to maintain a certain monopoly of control on the use of violence by these states over the abilities of complex chiefdoms with more decentralized authority structures.

The key similarity between these examples was also the advanced agricultural basis of these early complex societies. More recently, scholars have begun to address the important role of nomadic societies (especially in the Eurasian steppes) in this context of social complexity in the ancient and premodern world, which does also provide an intriguing example of the interrelationship between war and social complexity. Nomadic societies were often fiercely warlike due to hunting being major source of subsistence. As a result, raids and pillaging of more agricultural societies became a regular occurrence and at times this would result in outright conquest of such societies by the nomads (ibid.). This would correspond with the basic foundations of Cairnero's Circumscription Theory, but there is a nuanced twist.

While in the case of agricultural societies, this process leads to the emergence of the state, characterized by Cairnero as operating with more centralized authority (Cairnero, 1970, p. 733), in the case of the nomads it leads not to a centralized authority exactly but more of a confederated structure that has been characterized as forms of "*supercomplex chiefdoms*" that encompassed large swathes of territory as well as people; while also utilizing a "*complex system of titles of chiefs and functionaries, held diplomatic correspondence with neighboring countries, contracted dynastic marriages with agricultural states, neighboring nomadic empires and 'quasi-imperial' polities of nomads*" (Kradin, 2002, pp. 372, 373).

This helps establish that varying forms of social complexity did exist in the premodern world, which relates back to the more theoretical and metatheoretical elements addressed earlier in the chapter, as well as the necessity for the need for a multiparadigmatic approach to studying these varying forms. However, in the case of non-state forms of social complexity (at least as defined by Cairnero in this context), it still demonstrates this intertwined process of the use of social violence and war with the emergence of social complexity of varying forms. This was as much the case in the premodern world as well as the modern world, as will be explained in the next section concerning how this was a critical factor in the rise of the modern state system.

19.3.3 The Rise of the Modern State-System in the West

The noted historical sociologist Charles Tilly (1929–2008) famously summarized his argument that "*war made the state and the state made war*" (Tilly, 1975, p. 42). Although overlapping considerably with elements of Cairnero's theory, it should be noted that Cairnero emphasized more external environmental factors on human agency in the use of war and violence. Charles Tilly emphasized more the centrality of human agency in the connections between war and the emergence of the state.

The modern European state arose largely out of a long-term process that began with a certain "crisis of lordship" that began in the 11th century of the Common Era. Andrew Latham explains that this crisis in the High Medieval period was marked by "large-scale inter-noble violence, insubordination of castellans, dynastic disputes and 'political' assassinations" that was threatening the long-term longevity of the social order of Europe (Latham, 2011, p. 59).

Monarchs sought to bring their kingdoms more definitively under their control to better organize them against external enemies, and end any internal feuds between feudal lords. To help achieve this aim, a more centralized command system was necessary to coordinate the raising and long-term maintenance of armies. As a result, early forms of bureaucracies were designed to help manage these affairs both in military and civil matters. The financing of wars required a more efficient system of tax collecting, to give just one example. Short-term levies were no longer able to fulfill the personnel needs of modern wars, so monarchs needed soldiers on a more permanent basis to help guard their territories and go on foreign adventures. This process would eventually reach its peak in the Thirty Years War (1618–1648) in central Europe that resulted in the Treaty of Westphalia, which firmly entrenched the sovereignty of the modern state. Hence why it is often referred to as the "Westphalian" system (McMillan, 2009).

By the time of the French Revolution (1789–1799), mass politics entered the picture and with it the notion of the patriotic citizen marching forth to defend his homeland. This was demonstrated with the proclamation of mass conscription by the French Republic in August 1793. This would be the first time that the full resources of a nation were brought to bear on the battlefield and with it the further expansion of the powers of the state to organize itself for defense. To this end, the military was conceived as a major institution for building national unity among its recruits from various regions, and educating citizens in patriotic fervor for the cause. This in turn meant that public morale was ever more important for sustaining the war effort, and to this end meant the expansion of certain privileges of modern citizenship (most notably the right to vote) to those who fought and died in the nation's wars.

This would also play a role in giving women the right to vote, since their support for the war effort was vital in World War I (1914–1918); and later in World War II (1939–1945), the large number of women who entered the factories to work on war supplies would later help give impetus to the modern feminist movement in the postwar period. The large number of African-Americans who served in World War II would later inspire the integration of the American military, which in turn would later help give further strength to the emerging Civil Rights Movement. These are merely a few examples to demonstrate that the prominence of military prerogatives in the birth of the modern state and modern politics cannot be overemphasized (Malešević, 2010, pp. 118–145; 2017).

19.3.4 Military Revolutions and Aromorphoses

This close connection between warfare and social evolution can be further addressed when concerning the rise and diffusion of *Military Revolutions*. *Military Revolutions* are large-scale transformations of socio-political systems that have great implications for the restructuring of military organizations as well as the actual waging of war (Macgregor Knox & Murray, 2001, p. 7).

Although military revolutions may demonstrate themselves abruptly on the battlefield, the underlining long-term processes take considerable time to fully develop. These long-term processes and their outcome are not easily discernible before such climatic demonstrations within battlefield conditions. This has historically been the case, as can be demonstrated in several examples.

There have been at least five major Military Revolutions that have shaped the nature of modern Western warfare:

- ***The rise of the modern state (fifteenth–eighteenth centuries)***, and with it the rise of large-scale professional armies.
- ***The French Revolution (1789–1815)***: The birth of mass-politics and with it, mass-warfare.
- ***The Industrial Revolution (nineteenth century)***: The expansion of rationalized control over armies through advanced technological and supply systems.
- ***World War I and II (1914–1945)***: The ultimate culmination of the synthesis of mass-politics, mass-warfare, and industrial efficiency to its full destructive capacity.
- ***Cold War (1945–1991)***: The introduction of nuclear weapons (ibid., 6).

There has been considerable debate since the end of the Cold War, especially over the past couple of decades or so, whether the current stage of warfare can be seen as another military revolution. This is especially true regarding the so-called *New Wars* and later "*Hybrid warfare*" debates that have argued for the possible change in the nature of war over the course of the twenty-first century. The nature of this debate is beyond the scope of this study, but the issue has been addressed at length elsewhere (Satkiewicz, 2020a, 2020b, 2021b).

Several questions arise for further scholarship concerning military revolutions, questions that could be adequately addressed within a Big History perspective. The first question concerns the predominant focus on modern warfare. The concept of military revolutions has rarely been applied to premodern forms of warfare, with the partial exception of the Late Medieval period—mostly to demonstrate the transition to modern warfare. The issue of the accuracy of applying the concept of military revolutions to premodern warfare is an area of scholarly exploration within Big History. One example could be the Neolithic Revolution, which saw the birth of complex social systems and with it the rise of the first military revolution, which would be the birth of organized collective violence. This in itself is the subject of much scholarly debate and research (Arkush & Allen, 2008).

These questions concerning the concept of Military Revolutions could be adequately examined through the prism of social evolution. Military revolutions could be compared to the concept of *social aromorphosis. Social Aromorphosis*, as defined by Lenoid Grinin and Andrey Korotayev, is "*a rare qualitative macrochange that increases in a very significant way complexity, adaptability, and mutual influence of the social systems, that opens new possibilities for social macrodevelopment*" (Grinin et al., 2011).[1] There is a direct parallel here to the concept of military revolutions. So not only the direct relationship of military revolutions to certain stages of social evolution is an area for further study, but also it gives rise to the potential theoretical concept of a "*military aromorphosis*," that is *qualitative macrochanges in the complexity and adaptability of military systems that allow for further macrodevelopments.*

Such a perspective could apply to both premodern and modern forms of war, as well as Western and non-Western forms of war and military systems. This is quite a contrast to the usual Eurocentric and modern centered application of the "military revolution" concept. This need not imply that the traditional concept of "military revolutions" is invalid, but rather an attempt to formulate alternative concepts that can move beyond its inherent limitations.

19.3.5 War as a Casual Mechanism for Cross-Cultural Contact and Exchanges

Another possible theme of exploration within the Big History study of social complexity and war would concern the important role of warfare in cross-cultural contact and exchanges. It is one of the great paradoxes of history that cross-cultural exchanges often serve constructive purposes in the evolution of cultures; yet it is often the ultimate result of more hostile encounters through warfare.

The conquests of Alexander the Great across the Ancient Near East helped in the spread of Greek language and culture to as far as India, which resulted in the

---

[1] See also chapter "Biological and Social Phases of Big History and Complexity Growth" (Grinin et al., 2026) in this volume.—*Editors.*

emergence of the Hellenistic world. Greek became an international language, which helped provide a certain sense of cultural-civilizational unity among peoples of diverse backgrounds. Overtime Greek culture began to syncretize itself with numerous cultures it encountered, which most famously led to the rise of the Mystery Cults involving many Eastern deities. The Cult of Isis originated from the devotion of the Egyptian goddess of motherhood and fertility. The cult of Mithras was a prominent mystery cult devoted to the Persian god of the Sun. Even the common artistic depictions of the Buddha are a result of Hellenistic influence upon Indian art. Hellenistic influences because of Alexander's conquests had great impact upon the evolution of Judaism and Christianity. Although receiving a hostile reaction to Hellenism at first, as depicted in the Biblical book of Maccabees, a Hellenistic form of Judaism did emerge around the first century Before the Common Era. The work of the Jewish philosopher Philo, as well as the Greek translation of the Bible in the Septuagint are important legacies of this process. This Hellenistic tradition of Judaism formed an important foundation for the emerging Christian church, which relied upon the Greek translations of the Bible. This is but a short summary of the ultimate impact the conquests of Alexander the Great had upon cross-cultural exchanges.

Similar arguments also apply concerning the cross-cultural results of Roman conquests across the Mediterranean world. For example, the Mystery Cults and of course Christianity were able to spread throughout that part of the world because of the road system built by Roman legions in order to better secure the Empire and move troops more quickly to areas where needed. Just as Greek was the international language in the East, Latin became the international language in the West, providing a legacy for what would expand later in the Medieval period and continue until the nineteenth century. Another important legacy was the spread of Roman law, which still influences Western legal thinking to this day.

19.3.6 Cliodynamics, Social Solidarity, and the Frontier

An intriguing nuance concerning this relationship between warfare and cross-cultural contacts is the thesis proposed by Peter Turchin. It can shortly be summarized that major empires consistently arise along the so-called meta-ethnic frontiers. Cross-cultural contacts along such frontiers are often marked by constant conflict as well as exchanges. This constancy of conflict helps in the building of stronger bonds of social solidarity ("*Asabiya*" is the term Turchin borrows from Ibn Khaldun). This strengthening of social solidarity will often allow for one of the conflicting parties to eventually overcome the other, and this party will later forge a mighty empire of their own (Turchin, 2007, pp. 4–7).

Turchin provides the example of the interactions between Romans and Germanic tribes to substantiate his thesis. To stand against the stronger Romans, the Germanic tribes first needed to unite into larger tribal federations—which amounted to the strengthening of their social solidarity or "*asabiya*." They also needed to fight more like Romans; achieved through fighting the Romans directly (thus gaining knowledge of Roman fighting methods), but also through the Roman training of Germanic auxiliaries to help impose their rule. One of these auxiliaries was Arminius, who successfully defeated the Romans in the Teutoburg Forest in 9 CE. Eventually the Germanic tribes were successful in their struggle with the Romans, and generations later the previous "meta-ethnic frontier" of Latin vs. German became the core of the future Carolingian Empire led by Charlemagne. A crude summary of Turchin's account of the historical dynamics involved in this case study (ibid., pp. 57–88).

Peter Turchin has further developed his research on the connections between the evolution of human social complexity and war with his thesis of "*ultrasociety*," being based on the "theory of cultural multilevel selection that says that evolution is only possible when societies compete against each other" (Turchin, 2016, p. 19). In order to thrive in such highly competitive environments, social groups must build and maintain their internal cooperation and cohesion to help better preserve themselves more effectively. This is not entirely an original argument, for even Aristotle noted in his *Politics* how "*a common danger unites even the bitterest enemies*" (Aristotle, n.d.). Turchin does stress the argument further, almost also building on the argument made by Charles Tilly about the modern state:

> *It was competition and conflict between human groups that drove the transformation of small hunter-gatherers into huge nation-states. Not to put too fine a point it was war that first created despotic, archaic states and then destroyed them, replacing them with better, more equal societies. War both destroys and creates.* (Turchin, 2016, p. 22)

Turchin then postulates that due to this destructive and creative dialectic, war will eventually "destroy itself and create a world without war" (ibid.). For various reasons, that remains highly speculative and improbable. Peter Turchin's thesis has the advantage of addressing several important themes related to social complexity and war: the relationship of war to cross-cultural exchanges, along with the importance of war in the formation and disintegration of varying forms of social solidarity.

Questions related to Turchin's methodology based upon mathematical modeling ("*Cliodynamics*" as he has termed it) can legitimately be debated, but in the proposed multiparadigmatic method it can legitimately be utilized alongside other methodologies. Although missing a few facts here or there, Turchin's overall points related to these historical dynamics appear to be sound for the most part, which can be demonstrated by relevant parallels in research from archeologists studying early Medieval *ethnogenesis* among the Germanic tribes along the Roman frontier (Curta, 2006).

Further research and debate on questions raised by Turchin's work are necessary. Perhaps the main challenge will be to properly engage such mathematical modeling with the more contextual and rigorous "*mode of analysis*" found within Civilizational scholarship.

19.4 Conclusion

Hopefully, this has served as a decent introduction to the theme related to the study of the interrelationship between human social complexity and war, and especially how the established interdisciplinary field of Civilizational scholarship might be able to assist Big History in further developing its work in this field. There are other potential areas for further research into this form of scholarship—along with the wider awareness and engagement between Big History and Civilizational scholarship. Such a form of scholarship would be important to better understand the wider impact of war upon history. Although war is inherently destructive by nature, it has laid the foundations for creative construction as well. This paradox underlines how war has perhaps been one of the most transformative forces known to humanity, and thus it is important for scholars to be able to accurately assess its nature and long-term effects. Noted historical sociologist Siniša Malešević has probably best summarized this thesis:

> *Nearly every aspect of social life including governance structures, social hierarchies, gender and sexual relations, religious identities, class dynamics, ethno-racial stratification, educational practices, heath systems and administrative apparatuses have all been molded by legacies of specific wars. Whether we like it or not, warfare has historically been and remains a norm that constantly shapes our social order.* (Malešević, 2025)

To this end, the alliance between Big History and Civilizational scholarship could be the proper means of studying those phenomena.

References

Alaykin-Izvekov, V. (2011). Civilizational science: The evolution of a new field. *Comparative Civilizations Review, 64*(64), 103–116.

Andrade, G. (n.d.). René Girard (1923—2015). Internet encyclopedia of philosophy. https://iep.utm.edu/girard/ Retrieved June 27, 2025.

Archer, M. S. (1995). *Realist social theory: The morphogenetic approach*. Cambridge University Press.

Arkush, E. N., & Allen, M. W. (2008). The archaeology of warfare: Prehistories of raiding and conquest.

Aristotle. (n.d.). *Politics*. The Internet classics archive: Retrieved December 1, 2023. https://classics.mit.edu/Aristotle/politics.5.five.html.

Árnason, J. P., & Hann, C. (2018). *Anthropology and civilizational analysis: Eurasian explorations*. State University of New York Press.

Arnason, J. P. ©., Eisenstadt, S. S. N., & Wittrock, B. (2005). *Axial civilizations and world history*. Brill.

Axelrod, R. (1997). *The complexity of cooperation: Agent-based models of competition and collaboration*. Princeton University Press.

Barreiros, D. (2018). Warfare, ethics, ethology: Evolutionary fundamentals for conflict and cooperation in the lineage of man. *Journal of Big History, 2*(2), 19–35.

Barreiros, D., & Ribera Vainfas, D. (2019). Clausewitz, Keegan, and the big history of warfare. *Journal of Big History, 4*(1), 4–12.

Baskin, K., & Bondarenko, D. M. (2014). *The axial ages of world history: Lessons for the 21st century*. ISCE Publishing.

Black, J. (2004). *Rethinking military history*. Psychology Press.

Bassford, C. (n.d.). *Tiptoe through the trinity*. https://clausewitzstudies.org/mobile/trinity8.htm Retrieved February 21, 2024.

Beyerchen D. A. (1992). *Clausewitz, nonlinearity, and the unpredictability of war*. Retrieved December 1, 2023. https://www.clausewitz.com/item/BeyerchenClausewitzNonlinearityAndTheUnpredictabilityOfWar.htm.

Carneiro, R. L. (1970). A theory of the origin of the state. *Science, 169*(3947), 733–738. https://doi.org/10.1126/science.169.3947.733

Carneiro, R. L. (2011). Stellar evolution and social evolution: A study in parallel processes. In R. L. Carneiro, A. Korotayev, L. Grinin, & F. Spier (Eds.), *Evolution: Cosmic, biological, and social* (pp. 66–83). "Uchitel" Publishing House.

Cepelewicz, J. (2023, August 31). Why mathematical proof is a social compact. *Quanta Magazine*. https://www.quantamagazine.org/why-mathematical-proof-is-a-social-compact-20230831/. Retrieved June 27, 2025.

Chacon, R. R. (2020, September). Robert L. *Carneiro (1927–2020). Social Studies*. https://www.sociostudies.org/journal/articles/2759628/ Retrieved September 11, 2025.

Cox, R. (2023). *Origins of the just war: Military ethics and culture in the ancient near east*. Princeton University Press.

Curta, F. (2006). *Borders, barriers, and ethnogenesis: Frontiers in late antiquity and the middle ages*. Brepols Publishers.

Dominici, P. (2023). Democracy is complexity. Social transformation from below. *Società Mutamento Politica, 14*(28), 21–32.

Drennan, R. D., & Peterson, C. E. (2012). Challenges for comparative study of early complex societies. In *The comparative archaeology of complex societies* (pp. 62–87). Cambridge University Press.

Eisenstadt, S. N. (Ed.) (1986). *Origins and Diversity of Axial Civilizations*. State of New York University Press.

Gat, A. (2008). *War in human civilization*. Oxford University Press.

Gibney, E. (2025, July 30). Physicists disagree wildly on what quantum mechanics says about reality, nature survey shows. *Nature News*. https://www.nature.com/articles/d41586-025-02342-y Retrieved September 10, 2025.

Gioia, D. A., & Pitre, E. (1990). Multiparadigm perspectives on theory building. *The Academy of Management Review, 15*(4), 584. https://doi.org/10.2307/258683

Gray, C.S. (2018, August 17). Strategic history—military strategy magazine. Retrieved February 21, 2024. https://www.militarystrategymagazine.com/article/Strategic-History/.

Gray, C. S. (2016). *The strategy bridge: Theory for practice*. Oxford University Press.

Gray, C. S. (2013). *War, peace and international relations: An introduction to strategic history*. Routledge.

Grinin, L., Korotayev, A., Carneiro, R. L., & Spier, F. (2011). Introduction. Evolutionary Megaparadigm: Potential, problems, perspectives. In L. Grinin, A. Korotayev, F. Spier, & R. L. Carneiro (Eds.), *Evolution: Cosmic, biological, and social* (pp. 5–29). Uchitel.

Grinin, L., Markov, A., & Korotayev, A. (2026). Biological and social phases of Big History and complexity growth. In D. J. LePoire, L. Grinin, & A. Korotayev (Eds.), *Complexity in universal evolution—a Big History perspective* (pp. 283–356). Springer Nature. In this volume. https://doi.org/10.1007/978-3-032-20516-2_12

Hardy, Q., Birnbaum, F. L., Maggiori, & Liberation. (2022, September 1). René Girard. Colloquium on Violence & Religion. https://violenceandreligion.com/rene-girard-2/ Retrieved June 27, 2025.

Hartmann, S. (2025, May 22). A theory of everything will never work at all scales: Stephan Hartmann. IAI TV—changing how the world thinks. https://iai.tv/articles/a-theory-of-everything-will-never-work-at-all-scales-auid-3175?_auid=2020. Retrieved June 27, 2025.

Heraclitus. (n.d.). Internet encyclopedia of philosophy. Retrieved February 21, 2024, from https://iep.utm.edu/heraclit/.

Heuser, B. (2022). *War: A genealogy of Western ideas and practices*. Oxford University Press.

Hörcher, F. (2021). *Political philosophy of the European city from polis, through city-state, to megalopolis?* Lexington Books.

Kalberg, S. (2021). *Max weber's sociology of civilizations: A reconstruction*. Routledge.

Kappeler, P. M., Clutton-Brock, T., Shultz, S., & Lukas, D. (2019). Social complexity: Patterns, processes, and evolution. *Behavioral Ecology and Sociobiology, 73*(1). https://doi.org/10.1007/s00265-018-2613-4

Kartchner, K. M., Bowen, B. D., & Johnson, J. L. (2023). *Routledge handbook of strategic culture*. Taylor & Francis.

Kohler, T., Crabtree, S., & Bocinsky, R. K. (2018). Sociopolitical evolution in midrange societies: The Prehispanic Pueblo case. In *The emergence of premodern states: New perspectives on the development of complex societies* (pp. 133–184). Santa Fe Institute Press.

Kradin, N. N. (2002). Nomadism, evolution and world-systems: Pastoral societies in theories of historical development. *Journal of World-Systems Research*, 368–388. https://doi.org/10.5195/jwsr.2002.266

Latham, A. (2012). *Theorizing medieval geopolitics: War and world order in the age of the crusades*. Routledge.

Layton, P. (2018). *Grand Strategy*. Peter Layton.

Macgregor Knox, & Murray, W. (2001). *The dynamics of military revolution, 1300–2050*. Cambridge University Press.

Malešević, S. (2010). *The sociology of war and violence*. Cambridge University Press.

Malešević, S. (2017). *The rise of organised brutality: A historical sociology of violence*. Cambridge University Press.

Malešević, S. (2025, April 30). *Article: War has made our social order*. Thesis Eleven https://thesiseleven.com/2025/04/30/article-war-has-made-our-social-order/. Accessed September 10, 2025.

Mangone, E. (2023). *Pitirim A. Sorokin: Rediscovering a master of sociology*. Vernon Press.

McMillan, A. (2009). *The concise Oxford dictionary of politics*. Oxford University Press.

Nelson, B. (1981). *On the roads to modernity: Conscience, science, and civilizations: Selected writings*. Rowman and Littlefield.

Ortman, S., Blair, L., & Peregrine, P. (2018). The contours of cultural evolution. In *The emergence of premodern states: New perspectives on the development of complex societies* (pp. 187–217). Santa Fe Institute Press.

Overview. (n.d.). New England Complex Systems Institute. Retrieved July 26, 2023, from https://necsi.edu/research-overview.

Oxford Archaeologists Discover Monumental Evidence of Prehistoric Hunting Across Arabian Desert | University of Oxford. (2022, September 2). https://www.ox.ac.uk/news/2022-09-02-oxford-archaeologists-discover-monumental-evidenceprehistoric-hunting-across.

Phillips, A. (2011). *War, religion, and empire: The transformation of international orders*. Cambridge University Press.

Satkiewicz, S. (2020a). War as a complex reality: Comparative analysis of the studies on war and peace by Clausewitz and Pitirim Sorokin. *Biocosmology-Neo-Aristotelism, 10*(3 & 4) https://biocosmology.org/wp-content/uploads/2021/05/Stephen-T.-Satkiewicz_Vol10Nos34.pdf

Satkiewicz, S. (2020b). Civilizational dynamics of hybrid warfare. *Comparative Civilizations Review, 83*(83), 150–163.

Satkiewicz, S. (2021a) *Comparative civilizations review* [Review of Leonid Grinin, *Macrohistory and Globalization,* Russia: Uchitel Publishing House, 2012]. (pp. 134–136).

Satkiewicz, S. (2021b). *Comparative civilizations review* [Review of *the dynamics of military revolution, 1300–2050,* by M. Knox & W. M. Murray]. (pp. 197–199).

Satkiewicz, S. (2022). *Comparative Civilizations Review*. Review of David Christian. *Maps of Time: An Introduction to Big History*. University of California Press, New Edition, 2004 and Leonid E. Grinin, Andrey V. Korotayev, Barry H. Rodrigue eds. *Evolution: A Big History Perspective*. Volgograd: Uchitel Publishing House, 2011. pp. 160–163

Satkiewicz, S. (2023). On civilizational strategic studies. *Comparative Civilizations Review, 88*(88), 18–33.

Satkiewicz, S. T. (2025). Social complexity and war: A civilizational contribution to big history. *Journal of Big History, 8*(1), 154–166.

SFI In Depth | Santa Fe Institute. (n.d.). Www.santafe.edu; Santa Fe Institute. Retrieved July 26, 2023, from https://www.santafe.edu/about/faq.

Smith, H. (n.d.). Clausewitz as sociologist. *Military Strategy Magazine*. Retrieved July 26, 2023, from https://www.militarystrategymagazine.com/article/Clausewitz-as-Sociologist

Sorokin, P. A. (1943). *Sociocultural causality, space, time: A study of referential principles of sociology and social science*. Duke University Press.

Tilly, C. (1975). Reflections on the history of European state making. In C. Tilly (Ed.), *The formation of National States in Western Europe*. Princeton University Press.

Turchin, P. (2007). *War and peace and war: The rise and fall of empires*. Penguin Group.

Turchin, P. (2016). *Ultrasociety: How 10,000 years of war made humans the greatest cooperators on earth*. Beresta Books.

Villacres, E. J., & Bassford, C. (1995). Reclaiming the Clausewitzian trinity. *Parameters, 25*(1).

Von Clausewitz, C. (1979). *On war*. Princeton University Press.

Walker, P., & Roberts, P. (2023). *War's changed landscape?: A primer on conflict's forms and norms*. Howgate Publishing Limited.

Wilson, E. O. (1998). *Consilience the unity of knowledge*. Vintage Books.

Yuen, D. M. C. (2014). *Deciphering Sun Tzu: How to read the Art of War*. Oxford University Press.

Zinkina, J., Korotayev, A., & Andreev, A. (2016). Circumscription theory of the origins of the state: A cross-cultural re-analysis. *Cliodynamics: The Journal of Quantitative History and Cultural Evolution, 7*(2), 187–203. https://doi.org/10.21237/C7clio7230484

Zweibelson, B. (2025). *Reconceptualizing War*. Helion.

Stephen Satkiewicz is an independent scholar whose research areas are related to Civilizational Sciences, Social Complexity, Big History, Historical Sociology, Military History, War Studies, International Relations, Geopolitics, as well as Russian and East European History. He is the Book Review Editor for the academic journal *Comparative Civilizations Review*. He served as the editor for the International Society for the Comparative Study of Civilizations (ISCSC) newsletter from 2016 to 2018. He produces a podcast for the New Book Network on books concerning topics in his research area.

Part VII

Future Complexity

20 Modeling Galactic Evolution and the Potential for Intelligent Life in the Milky Way

Teófilo Vargas, Paolo Musso, and Nicolò Antonietti

Abstract

This study presents a model of the Milky Way's structural and evolutionary properties, focusing on star formation, gas surface density, supernovae rates, and metallicity distribution. The gas surface density follows an exponential decline with distance from the Galactic Center, while the star formation rate is determined using the Schmidt–Kennicutt law. The evolution of stellar populations is modeled through the initial mass function (IMF) and main sequence lifetimes, with implications for supernovae occurrences and heavy element enrichment. Using these factors, we estimate the time since conditions first became favorable for Earth-like planets around G-type stars, providing an age distribution for potential intelligent life across different Galactic radii. The model incorporates uncertainty propagation, highlighting the limitations imposed by measurement uncertainties in key astrophysical parameters. Future refinements will address Galactic asymmetries to better identify habitable regions.

Keywords

Model · Galaxy · Stellar · Life · Complexity · Evolution

This chapter is an extensively re-worked version of the text originally published in Journal of Big History (Vargas et al., 2025).

T. Vargas
Universidad Nacional Mayor, de San Marcos, Peru

P. Musso
University of Insubria, Varese, VA, Italy
e-mail: paolo.musso@uninsubria.it

N. Antonietti (✉)
Institute of Radioastronomy – National Institute of Astrophysics (INAF-IRA), Bologna, Italy

D. J. LePoire et al. (eds.), *Complexity in Universal Evolution*, World-Systems Evolution and Global Futures, https://doi.org/10.1007/978-3-032-20517-9_20

20.1 Model of the Galaxy

20.1.1 Milky Way Size and Age

The Milky Way has an average radius of roughly 53,000 light years, even though, inexplicably, the number of stars drops significantly faster than expected after 40,000 light years (that is around 13 kpc) (Sale et al., 2010). The age of the universe is currently estimated to be around 13.77 billion years, based on the study of the microwave background radiation (Tauber, 2018). Also, the distance of the Solar System from the center of the Milky Way is estimated to be between 7.4 and 8.7 kpc (Francis & Anderson, 2014).

20.1.2 Milky Way Gas Surface Density

The gas surface density in the absence of star formation of the Milky Way, $\Sigma_d\,(r,t)$, decreases exponentially with the distance from its center and is dependent on time, according to the equation (Naab & Ostriker, 2006)

$$\Sigma_d(r,t) = \Sigma_0(t)e^{-r/r_d(t)} \tag{20.1}$$

where r_d is the radial distance from the center of the Galaxy, where the gas surface density is decreased by a factor of e and is called the scale length; Σ_0 is the central surface density of the Galaxy.

The scale length, r_d, linearly increases with time at a rate of 3/11 pcMyr^{-1} and intercept equal to 0. Σ_0 is constant in time, after 2500 My after the formation of the universe ($t = 0$) and equal to around 440 $\mathrm{M_\odot\,pc^{-2}}$ (Sale et al., 2010).

In Fig. 20.1, we plot the gas surface density of the Milky Way from 2.5 Gy (because of the definition of Σ_0) to 13.5 Gy since the formation of the universe and at a distance from 2.5 kpc (to exclude the Galactic Center as it is hard to model it) to 13 kpc to the Galactic Center.

In Fig. 20.2, we plot the gas surface density of the Milky Way from 13.49 Gy to 13.5 Gy since the formation of the universe (recent years) and at a distance from 7.4 kpc to 8.7 kpc to the Galactic Center (round the Solar System).

The gas surface density decreases exponentially along the distance from the Galactic Center at a fixed time, according to Eq. (20.1), but it increases according to the same formula because of the definition of r_d.

20.1.3 Milky Way Stars Total Mass

During the lifetime of the Milky Way, the mass of gas is turned into stars. The Schmidt–Kennicutt law (Schmidt M., 1959; Kennicutt, 1998) predicts the star formation rate ψ starting from the mass of gas M_{gas},

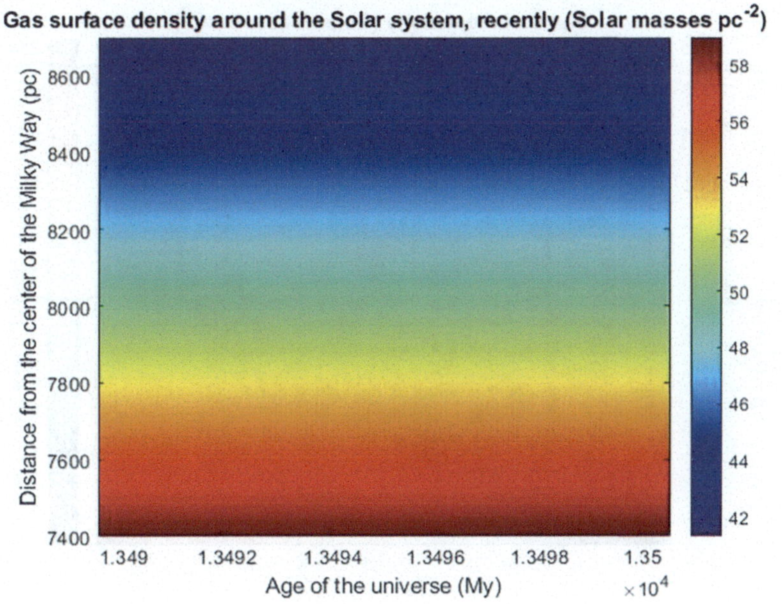

Fig. 20.1 Gas surface density of the Milky Way from 2.5 Gy to 13.5 Gy and from 2.5 kpc to 13 kpc from the Galactic Center

$$\psi = aA\left(\frac{M_{gas}}{a}\right)^n$$

Here, M_{gas} is given by the surface density multiplied by the area around the radial bin a (that is the area of an annulus) in pc^{-2}; $A = 2.5 \cdot 10^{-4}$ $M_{\odot}pc^{-2}Myr^{-1}$. The exponent, n, is estimated as about 1.3. Every time a star is formed, its mass is subtracted to the mass of gas and will not contribute to the star formation any longer.

In Fig. 20.3, the stars' total mass of the Milky Way is plotted for the current age of the universe and around the Sun.

20.1.4 Initial Mass Function

The initial mass function (IMF), ξ(M), returns the distribution of stellar masses in a population of stars.

Here we considered stellar masses to be between $0.08M_{\odot}$ and $100M_{\odot}$, according to Kroupa (2001). Also, there, the IMF is derived as a two-part power law function:

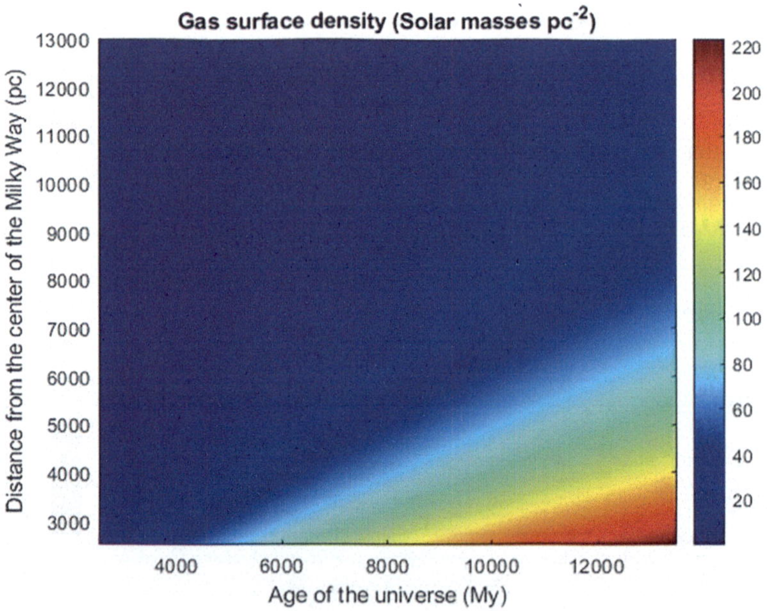

Fig. 20.2 Gas surface density of the Milky Way from 13.49 Gy to 13.5 Gy and from 7.4 kpc to 8.7 kpc from the Galactic Center

$$\xi(M) = \xi_n \left(\frac{m}{M_\odot}\right)^{-\alpha}$$

ξ_n and α are found to be, thanks to normalization conditions:

- $\xi_n = 0.44$, $\alpha = 0.3$, when $0.08 \leq m/M_\odot \leq 0.50$,
- $\xi_n = 0.22$, $\alpha = 1.3$, when $0.50 \leq m/M_\odot$.

In Table 20.1, we show the calculated initial mass function values for different types of stars.

Given the star formation rate and the initial mass function, we know how many stars of each type is formed in the Galaxy, at all times.

20.1.5 Main Sequence Lifetime

A star spends about 90% if its life fusing its core to make heavier elements. When hydrogen is fused to create helium, the star is said to be in its main sequence (MS); the time this process lasts is called main sequence lifetime (MSL), τ_{MS}, and is calculated as (Cosmos, 2024):

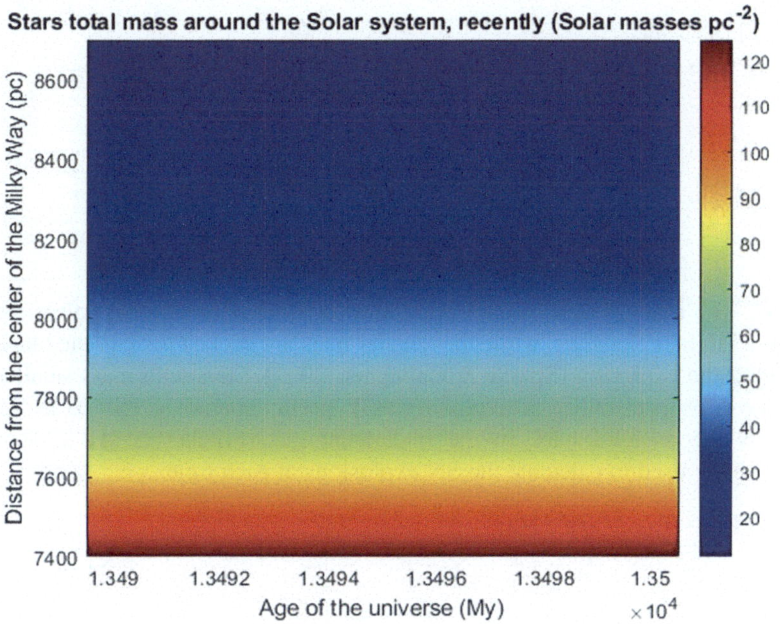

Fig. 20.3 Stars' total mass of the Milky Way from 13.49 Gy to 13.5 Gy and from 7.4 kpc to 8.7 kpc from the Galactic Center

Table 20.1 Initial mass function value for different star types with the initial mass function giving the percentage of the star-forming mass going into producing that star type

Star type	Initial mass function (%)
M-type (0.08–0.5 $M_\odot$)	26.7%
K-type (0.5–0.8 $M_\odot$)	11.6%
G-type (0.8–1.05 $M_\odot$)	6.1%
F-type (1.05–2.0 $M_\odot$)	12.1%
A-type (2.0–4.0 $M_\odot$)	10.6%
B-type (4.0–20.0 $M_\odot$)	13.9%
O-type (20.0–100.0 $M_\odot$)	8.6%

$$\tau_{MS} = \tau_{MS,\odot}\left(\frac{m}{M_\odot}\right)^{-2.5}$$

where $\tau_{MS,\odot} \approx 10$ Gyr is the MSL of the Sun. In Table 20.2, we show the main sequence lifetime for different star types.

20.1.6 Supernovae

At the end of their MS, stars explode as supernovae (SN).

Table 20.2 Main sequence lifetime for different star types

Star type	Main sequence lifetime
M-type (0.08–0.5 $M_\odot$)	221 Gy
K-type (0.5–0.8 $M_\odot$)	30 Gy
G-type (0.8–1.05 $M_\odot$)	12 Gy
F-type (1.05–2.0 $M_\odot$)	3.5 Gy
A-type (2.0–4.0 $M_\odot$)	641 My
B-type (4.0–20.0 $M_\odot$)	20 My
O-type (20.0–100.0 $M_\odot$)	0.3 My

Stars with mass greater than $8M_\odot$ explode as Type II supernovae (SNII), that is, they eject lighter elements, while the heavier ones are sucked back in the black hole born after the explosion (Sparke & Gallagher, 2007). Stars with mass smaller than $8M_\odot$ have 1% probability of exploding as Type Ia supernovae (SNIa) and all the elements are released back to the interstellar gas (Pritchet et al., 2008). Two functions are used to calculate the SN explosions at each time:

$$\int_8^{100} \frac{\xi(M)}{M} SFR(\tau - \tau_{MS}(M))dM, (SNII)$$

$$0.01 \int_{0.08}^{8} \frac{\xi(M)}{M} SFR(\tau - \tau_{MS}(M))dM, (SNIa)$$

The results have been applied to each radial bin, given that an SNII explosion would sterilize planets located up to 8 pc away (Gehrels et al., 2003) and SNIa explosion would sterilize planets up to 19 pc away (Gowanlock et al., 2011).

In Fig. 20.4, the numbers of SNIa explosions per radial bin and in time are given, whereas in Fig. 20.5, the numbers of SNII explosions per radial bin and in time are given.

20.1.7 Metallicity

At the end of their lives, stars eject metals, that is, elements heavier than helium.[1] An object's metallicity [Fe/H] is the ratio of the mass of iron to the mass of hydrogen per unit volume compared with the Sun:

$$[Fe/H] = \log_{10}\left(\frac{M_{Fe}}{M_H}\right)_{star} - \left(\frac{M_{Fe}}{M_H}\right)_{Sun}$$

Estimates of the ratio of the mass of iron to the mass of hydrogen for the stars are given in Raiteri et al. (1996) and Iwamoto et al. (1999). Sun's iron abundance is

[1]For more detail, see chapter "The Evolution of Complexity in the Stelliferous Era: Observations and Patterns" (Korotayev & LePoire, 2026) in this volume.—*Editors*.

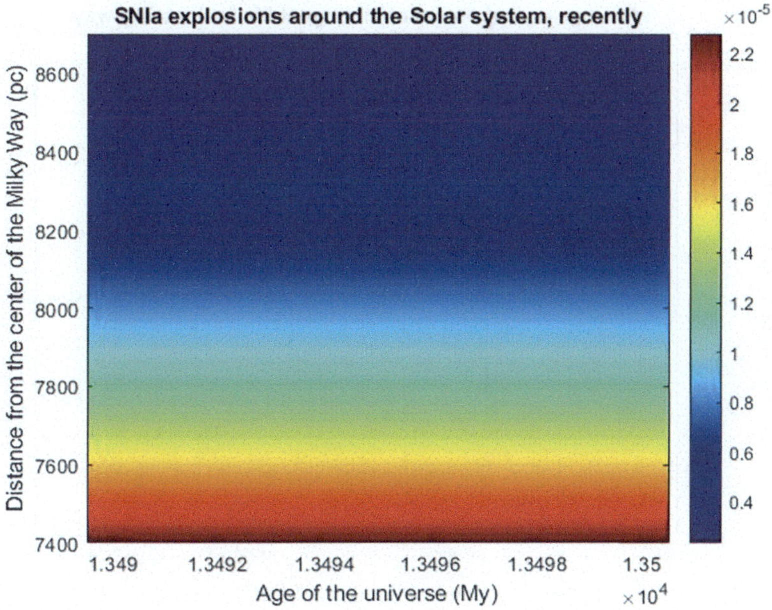

Fig. 20.4 SNIa explosions around the Solar System at around the age of the universe per radial bin and time

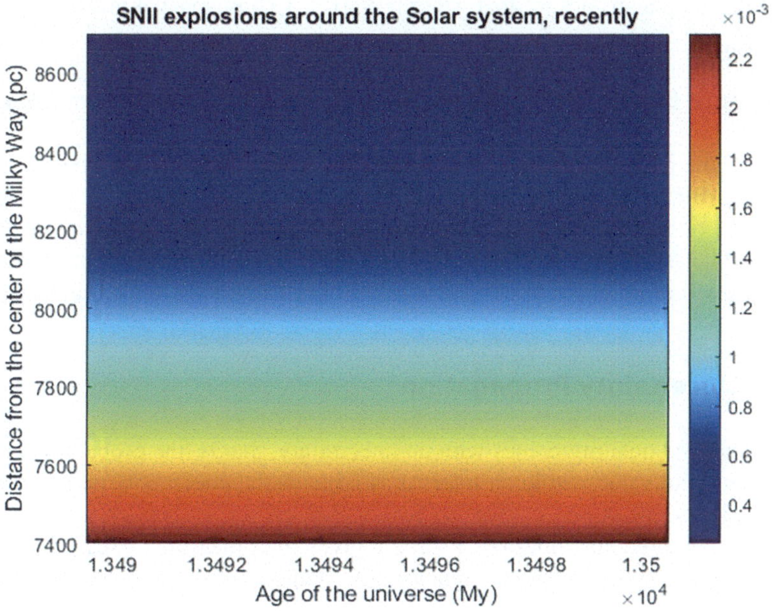

Fig. 20.5 SNII explosions around the Solar System at around the age of the universe per radial bin and time

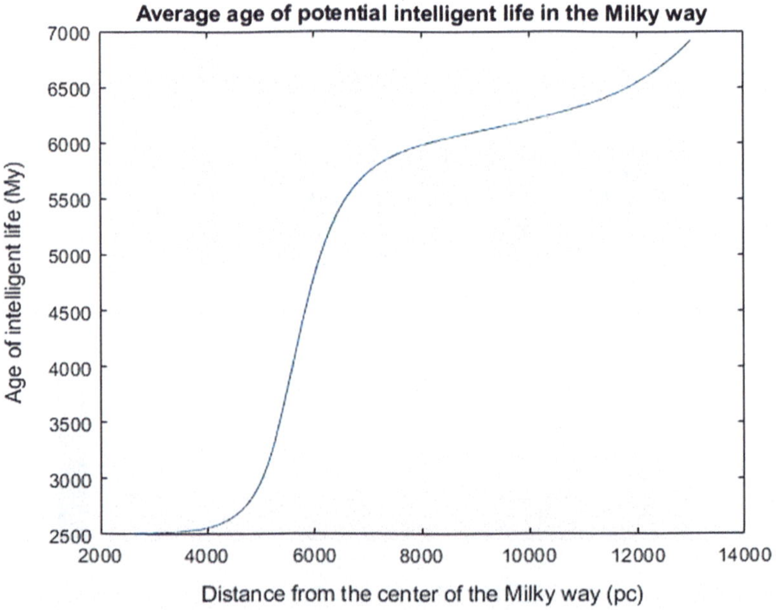

Fig. 20.6 Average age of potential intelligent life in the Milky Way

taken as $1.17 \cdot 10^{-3}$ $M_{\odot}$, oxygen abundance is taken as $9.59 \cdot 10^{-3}$ $M_{\odot}$, and hydrogen abundance is taken as 0.706 $M_{\odot}$.

Earth-sized planets are formed when [Fe/H] > 1 (Johnson & Li, 2012).

20.1.8 Age Distribution of Potential Intelligent Life in Our Galaxy

Mixing all the pieces of information, we can calculate the time since the conditions were first sufficient for an Earth-like planet to form around a G-type star as a function of the Galactic radius, which is referred to as the average age of potential intelligent life in the Galaxy (Fig. 20.6). For example, at 8 kpc, around the Earth, the conditions for the growth of potential intelligent life have been favorable for 6 billion years.

20.2 Uncertainty Propagation

Every measured quantity comes with an uncertainty, that is, the probability to have the quantity real value within an interval as large as the uncertainty and centered at the measured value.

All the measured quantities (U, V, … Z), given their measured values (u, v, … z), are combined to give a calculated quantity (X) with value x. All the uncertainties of the measured values (σ_u, σ_v, ... σ_z) will contribute to the uncertainty of the calculated quantity (σ_x).

If the calculated quantity is the result of multiplications and divisions among the measured quantities with weighting constants, the following uncertainty propagation formula holds true:

$$\frac{\sigma_x^2}{x^2} = \frac{\sigma_u^2}{u^2} + \frac{\sigma_v^2}{v^2} + \cdots \frac{\sigma_z^2}{z^2}$$

We try to estimate the uncertainty on the age distribution of potential intelligent life in our Galaxy, starting from the uncertainty of the measured quantities that contribute to its definition, as they are multiplied together.

The gas surface density is an exponential with respect to r_d. Its relative uncertainty is:

$$\frac{\sigma_\Sigma}{\Sigma} = \frac{\sigma_{r_d}}{r_d}$$

Where r_d is between 2500 and 3500 pc [4]. The variable can thus be thought of as a random variable with a uniform distribution, whose standard deviation is

$$\sigma_{r_d} = \frac{1}{\sqrt{12}}(3500 - 2500) = 288$$

The relative uncertainty on the age distribution of potential intelligent life in the Milky Way, A, due to the contribution of the gas surface density is:

$$\frac{\sigma_A}{A} = \frac{\sigma_\Sigma}{\Sigma} = \frac{\sigma_{r_d}}{r_d} = 288/3000 = 0.1$$

Therefore, because of the uncertainty in the gas surface density measurements, the uncertainty in the age of potential intelligent life is 600 million years around the Earth.

Similarly, we can evaluate the uncertainty due to the initial mass function; we do that for M-type stars. The uncertainty is due to the mass of the stars. The relative uncertainty on the function due to the mass of the star is:

$$\frac{\sigma_{\xi(m)}}{\xi(m)} = \frac{\frac{\xi_n}{aM_\odot}\left(\frac{m}{M_\odot}\right)^{-\alpha-1}}{\xi_n\left(\frac{m}{M_\odot}\right)^{-\alpha}}\sigma_m = \frac{1}{a}\frac{\sigma_m}{m} = \frac{5\%}{0.3} \sim 16\%$$

The relative uncertainty σ_m/m is equal to 5%, according to Tayar et al. (2022). The relative uncertainty on the age of potential intelligent life reaches 19%. We see thus that an estimate on the age of potential intelligent life is limited by the uncertainties on the measurements of the inputs which must be refined.

20.3 Conclusion

We have calculated the average age of potential intelligent life in the Milky Way, as a function of the radius of the Galaxy. Star formation rate, initial mass function, supernovae explosions, and metallicity were taken into account. The model has a radial symmetry and in next improvements, asymmetries will have to be considered so that habitable islands can be isolated.

References

Cosmos. (2024). Main sequence lifetime, Swinburne University of Technology. Accessed July 12, 2024. http://astronomy.swin.edu.au/cosmos/M/Main+Sequence+Lifetime.

Francis, C., & Anderson, E. (2014). Two estimates of the distance to the Galactic Centre. *Monthly Notices of the Royal Astronomical Society, 441*(2), 1105–1114. https://doi.org/10.1093/mnras/stu631

Gehrels, N., Laird, C. M., Jackman, C. H., Cannizzo, J. K., Mattson, B. J., & Chen, W. (2003). Ozone depletion from nearby supernovae. *The Astrophysical Journal, 585*(2), 1169. https://doi.org/10.1086/346127

Gowanlock, M. G., Patton, D. R., & McConnell, S. M. (2011). A model of habitability within the Milky Way Galaxy. *Astrobiology, 11*(9), 855–873. https://doi.org/10.1089/ast.2010.0558

Iwamoto, K., Brachwitz, F., Nomoto, K., Kishimoto, N., Umeda, H., Hix, W. R., & Thielemann, F. (1999). Nucleosynthesis in Chandrasekhar mass models for type Ia supernovae and constraints on progenitor systems and burning-front propagation. *The Astrophysical Journal Supplement Series, 125*(2), 439–462. https://doi.org/10.1086/313271

Johnson, J. L., & Li, H. (2012). The first planets: The critical metallicity for planet formation. *The Astrophysical Journal, 751*(2), 11. https://doi.org/10.1088/0004-637X/751/2/11

Kennicutt, R. C., Jr. (1998). The global Schmidt law in star-forming galaxies. *The Astrophysical Journal, 498*, 541–552. https://doi.org/10.1086/305774

Korotayev, A., & LePoire, D. J. (2026). The evolution of complexity in the Stelliferous era: Observations and patterns. In D. J. LePoire, L. Grinin, & A. Korotayev (Eds.), *Complexity in universal evolution—a big history perspective* (pp. 151–172). Springer Nature. In this volume. https://doi.org/10.1007/978-3-032-20516-2_5

Kroupa, P. (2001). On the variation of the initial mass function. *Monthly Notices of the Royal Astronomical Society, 322*(2), 231–246. https://doi.org/10.1046/j.1365-8711.2001.04022.x

Naab, T., & Ostriker, J. P. (2006). A simple model for the evolution of disc galaxies: The Milky Way. *Monthly Notices of the Royal Astronomical Society, 366*(3), 899–917. https://doi.org/10.1111/j.1365-2966.2005.09919.x

Pritchet, C. J., Howell, A. D., & Sullivan, M. (2008). The progenitors of Type 1a supernovae. *The Astrophysical Journal, 683*(1), 25–28. https://doi.org/10.1086/591314

Raiteri, C. M., Vallata, M., & Navarro, J. F. (1996). Simulations of galactic chemical evolution I. O and Fe abundances in a simple collapse model. *Astronomy and Astrophysics, 315*, 105–115. https://doi.org/10.1111/j.1365-2966.2009.15746.x

Sale, S. E., et al. (2010). The structure of the outer Galactic disc as revealed by IPHAS early A stars. *Monthly Notices of the Royal Astronomical Society, 402*(2), 713–723. https://doi.org/10.1111/j.1365-2966.2009.15746.x

Schmidt, M. (1959). The rate of star formation. *The Astrophysical Journal, 129*, 243–258. https://doi.org/10.1086/146614

Sparke, L. S., & Gallagher, J. S. III (2007). *Galaxies in the universe: An introduction*. 2nd ed. Cambridge University Press.

Tauber, J. (2018). From an almost perfect universe to the best of both worlds. European Space Agency. https://sci.esa.int/web/planck/-/60499-from-an-almost-perfect-universe-to-the-best-of-both-worlds.

Tayar, J., Claytor, R. Z., Huber, D., & van Saders, J. (2022). A guide to realistic uncertainties on the fundamental properties of solar-type exoplanet host stars. *The Astrophysical Journal, 927*, 31. https://doi.org/10.3847/1538-4357/ac4bbc

Vargas, T., Musso, P., & Antonietti, N. (2025). Numerical uncertainty propagation to the number of possible galactic Habitable Islands in the Milky Way. *Journal of Big History, 7*(4), 21–26.

Teófilo Vargas is a researcher with interests in astronomy. His work encompasses various aspects of celestial phenomena and cosmic exploration. Through his research, Vargas contributes to the broader understanding of the universe, collaborating with academic and scientific communities to advance the field of astronomy.

Paolo Musso is a professor at the University de la Insubria in Varese (Italy), where he teaches Science and Science Fiction and Foundations of Modernity courses. Since 2005, he has been a visiting professor of Epistemology at Universidad Católica Sedes Sapientiae.

Nicolò Antonietti is an astronomy and SETI (Search for Extraterrestrial Intelligence) researcher. He specializes in studying the cosmos with a particular focus on the search for signs of extraterrestrial life. Antonietti employs advanced astronomical techniques and data analysis to explore the universe, contributing to our understanding of its vast complexities. His work in SETI involves collaborating with international teams to analyze signals and phenomena that could indicate the presence of intelligent life beyond Earth.

Beyond Intelligence: The Galactic Information Field and the Evolution of Complexity

21

Alexander Panov

Abstract

This study challenges the conventional perspective of Big History, proposing that intelligence, as traditionally defined, may not represent the peak of cosmic evolution. Instead, the increasing complexity of the Universe could give rise to novel forms of matter organization that transcend our current understanding of intelligence. Building on the concept of "superintelligence," this study explores a scenario in which intelligent civilizations across the Galaxy engage in continuous interstellar communication, gradually forming a vast, interconnected Galactic information field. Over time, this distributed network accumulates and influences planetary civilizations, evolving into an over-civilization—an entity shaped by but not reducible to individual intelligences. Due to the finite speed of light and the vastness of space, no single civilization can fully grasp or control this emergent structure. The Galactic information field is thus proposed as a superintelligent entity, distinct from traditional intelligence and beyond the comprehension of any one planetary mind. This perspective challenges the assumption that intelligence is the ultimate evolutionary outcome, suggesting instead that the Universe may be progressing toward forms of complexity that defy human-centric definitions of cognition and agency.

Keywords

Aliens · Superintelligence · Contacts · Risk · Radars · Evolution · Complexity · Big history

This chapter is an extensively re-worked version of the text originally published in Journal of Big History (Panov, 2025).

A. Panov (✉)
Skobeltsyn Institute of Nuclear Physics (MSU SINP), Lomonosov Moscow State University, Moscow, Russia

D. J. LePoire et al. (eds.), *Complexity in Universal Evolution*, World-Systems Evolution and Global Futures, https://doi.org/10.1007/978-3-032-20517-9_21

Big History usually understands our human history as a natural continuation of the history of matter in the Universe, starting from the Big Bang, up to the present day (e.g., Panov et al., 2020). It is not at all obvious that the intelligence in our modern understanding is the top of the evolution of matter. On the contrary, it is likely that the evolution of complexity in the Universe will lead to the emergence of completely new forms of organization of matter, which can hardly even be called intelligence. We cannot say that life is just a very complex chemical process. Life is something qualitatively superior to mere chemistry in its complexity. Similarly, it would be impossible to say that these new, very advanced forms of existence of matter are simply a very highly developed intelligence. It could be something qualitatively different that is not even in principle accessible to our understanding. Intelligence does not have to be the highest point of the evolution in the Universe.

Here is an important example of "superintelligence" scenario (Panov, 2011). Suppose there are many planets in the Galaxy inhabited by intelligent life. Suppose that some of these planetary civilizations tend to transmit messages about themselves into deep space in the hope that it will be accepted and understood on other planets. Suppose that planetary civilizations not only transmit information about themselves, but also retranslate cosmic messages that they themselves receive. Then, once transmitted, the message will not disappear, will be transmitted from one civilization to another; it will be read by many civilizations and will have an impact on many planetary civilizations. Over time, more and more information will be accumulated in the volume of the Galaxy. Planetary civilizations can appear and disappear, but this distributed, cumulative information (the information field) of the Galaxy will exist and develop. An over-civilization object is formed, the carrier of which are individual planetary civilizations, but which itself is not an intelligent civilization. Moreover, no single local civilization (a local mind) can completely understand the processes in the Galactic information field even in principle, since, due to the finiteness of the speed of light, it is simply impossible to collect information about the state of this information field to study it. One can say that the Galaxy information field goes beyond the understanding of any planetary intelligence and it may be understood as a superintelligent object.

We see that one of the paths of the Big History in the Galaxy after the emergence of intelligence on individual planets is evolution, which is the essential result of interaction between individual planetary civilizations. It should be expected that any interaction between individual planetary civilizations, either through communication channels, as in the example with the Galactic information field, or directly, will be a very complex process that is inevitably associated with certain risks. We discuss here the issue of the risks of contacts with other civilizations with the example of our own Earth civilization. Let us start with some patterns about the dangers of contact with alien civilizations that are ingrained in the public mind.

A popular line of reasoning is the following: If aggressive aliens learn of our existence, they will come and destroy us (enslave us). Consequently, we should not transmit messages about ourselves into space, for our location not to be revealed. Suffice it to say that this argument was supported by Stephen Hawking (Devlin, 2016). Even a special scale was invented to assess the danger level of space

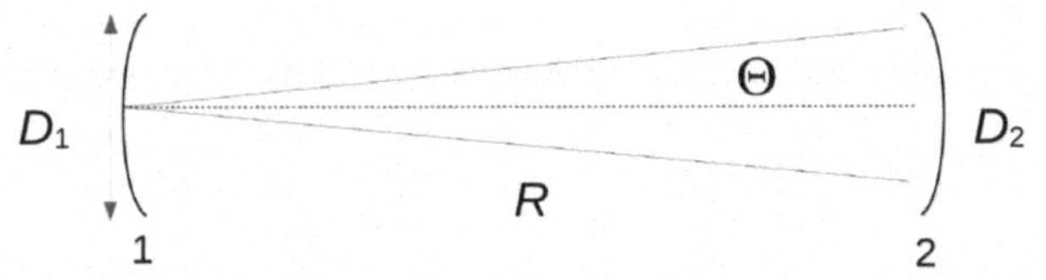

Fig. 21.1 Schematic diagram of the space radio communication channel. 1—transmitter; 2—receiver; D_1 and D_2—diameter of the antenna of the transmitter and the receiver, respectively; R—distance between the transmitter and the receiver; Θ—diffraction angle of divergence of the transmitter radio beam

transmissions (Almár & Shuch, 2007). Actually, there are at least two errors in this popular reasoning.

The first and very simple mistake is to assume that the absence of targeted space transmissions could "hide the Earth." In reality, it is absolutely impossible to hide the Earth from "interstellar aggressors."

Let us review the basics of the theory of interstellar radio communication. Figure 21.1 shows the scheme of organization of interstellar radio communication. There are two civilizations, of which civilization 1 is a transmitter of information, and civilization 2 is a receiver of information. For simplicity, we will consider that civilizations for reception and transmission use round radio antennas (as it is done now on the Earth). Diameter of the transmitter antenna is D_1, diameter of the receiver antenna is D_2, power of the transmitter is P_1, wavelength of the transmitter is λ, and width of the transmitter band is Δf.

Since the radio beam of the transmitter diverges inevitably by an angle $\Theta = \lambda / D_1$ due to diffraction effects, the intensity of the radio beam decreases with distance, which is the reason for the existence of a distance limit at which a radio transmission can still be received or at least detected. In order to determine this limiting range, we must specify the signal-to-noise ratio we want to obtain. Let us denote this ratio by α. For reliable reception of a signal, α must be greater than unity, but only for detection of the carrier frequency of a signal the ratio α can be much less than unity, since the signal can be detected by the accumulation method. For example, quite reasonable for signal detection is a ratio $\alpha \sim 0.01$. Noise is expressed by the value of the so-called antenna temperature T_{ns}, and for all good modern radio telescopes is reduced solely to the intrinsic noise of space, which in the centimeter-wave band of interest is about 10 K.

The radio communication distance limit is expressed by the formula

$$R = \sqrt{\frac{4S_1S_2P_1}{\alpha\pi^2\lambda} \cdot \frac{1}{kT_{ns}\Delta f}}.$$

Here S_1 and S_2 are the antenna squares 1 and 2, k is the Boltzmann constant, and $\pi = 3.14$. Now consider such transmitters like airfield or space radars (which no civilization in the phase of technological explosion can do without). Suppose also that the receiver has a 40 km diameter space antenna (we will explain below why this size seems reasonable). From what distance can our military and airfield radars be detected by a receiver with such an antenna? Here are the parameters that characterize our problem:

$$D_1 \approx 10\ \text{m} => S_1 \approx 10\ \text{m}^2.$$

$$D_2 \approx 40\ \text{km} => S_2 \approx 10^9\ \text{m}^2.$$

$$P_1 = 50\ \text{kW}.$$

$$\lambda = 3\ \text{cm}.$$

$$\Delta f = 1000\ \text{Hz}.$$

$$T_{ns} = 10\ \text{K}.$$

$$\alpha \approx 0.01.$$

Substituting these values into the formula for the limiting R, we obtain $R \approx 5000$ light years.

Suppose there are extraterrestrial civilizations that can be "interstellar aggressors." Consequently, they are capable of "interstellar flights" in some form. Consequently (it is much easier) such extraterrestrial civilizations can build space radio telescopes with size of antenna more than 40 km. As our calculations show, the radiation of our airfield and military radars will be detected by such a civilization from a distance of more than a kiloparsec, and if we take a slightly larger space antenna (which is quite acceptable for civilizations capable of interstellar flights), then across the entire Galaxy. We inevitably will be detected by a potential aggressor. Consequently, targeted space transmissions will not change anything additionally—we will be detected anyway, if there is someone to detect, of course.

The second error is that existence of alien civilizations that could come and destroy us is contradicted by observations. There is two-step reasoning.

Let's first consider the question: Why an alien civilization can want to come and destroy us (or enslave us)?

There are two groups of arguments. The first group of arguments relies on some possible benefit. For example:

1. They will want to enslave us. But slave labor is very unproductive. It is very stupid for a civilization that probably knows how to create even self-reproducing robots (otherwise interstellar flights would be impossible for it).
2. They will want to take over our resources. It looks to be very stupid, because in space they can find any amount of resources on uninhabited planets and asteroids.
3. They will want to take over our planet and settle on it. But it is much easier to settle on a planet without intelligent life: they will not have to fight with anyone. There are plenty of suitable planets without intelligent life in the Galaxy.

And so on. It is very difficult to find a rational reason relying on benefit.

The second group of arguments relies on fear. They may see us as competitors for habitat, they fear aggression from us, so they may want to destroy us. This argument might be correct in principle, so let us assume that it actually is.

We consider only civilizations capable of interstellar flight, since civilizations not capable of interstellar flight pose no physical danger to us. Consider the first civilization in the Galaxy that tends to view other civilizations as potential competitors for habitats and aggressors and is capable of interstellar flight. The primary goal of any extraterrestrial civilization, like all life, is survival. The best way to ensure a civilization's survival is to spread out in space as much as possible. Consequently, already the first civilization capable of interstellar flight will begin to explore other star systems. In 30–50 million years, the civilization must spread over the entire Galaxy, leaving no free places to inhabit. For an aggressive extraterrestrial civilization (tends to view other civilizations as potential aggressors), the easiest way to prevent other civilizations is to prevent intelligence *from appearing* on a habitable planet where intelligence does not yet exist (they can terminate a potentially intelligent species, etc.). Since the first civilizations in the Galaxy could have emerged at least a billion years ago, the possibility of intelligence development on Earth should have been ruled out and Earth should be populated by aliens, since the Solar System is one of the best places in the Galaxy to inhabit (a corotation circle, a very calm place in the Galaxy and a very calm star). But we are here and there are no aggressive aliens, hence there is something wrong in the reasoning above. It is a conclusion from observations, it is not a speculation. What exactly is wrong? Nobody knows. Either interstellar travel is impossible; either exploration of inhabited planets is impossible by some reasons; or flights are possible and exploration of inhabited planets is possible, but some ethical imperatives do not allow aliens to explore inhabited planets (hence aliens may not be aggressive). In any case, physical invasion by aggressive aliens does not threaten us, because if it were possible in principle, it would have already happened.

Turn to a more realistic scenario, for example: The aliens could insert malicious computer code into their space transmission, which we would run, and it would somehow take over our computers for an unknown purpose or force us to solve their problems. Is it possible to imagine a rational reason for such informational invasion? Unfortunately, yes (Turchin, 2018).

Suppose, a civilization, in order to protect its own survival, seeks to settle its colonies near other stars, but *interstellar flights are physically impossible*. However, by means of informational invasion it is possible to try to move the "code" of this civilization to a planet inhabited by another civilization, and by deception to force another civilization to launch this "code" into development. Then it will effectively become possible to move a civilization across stars, but only stars with planetary systems occupied by other civilizations. Such a civilization resembles a viral infection. Viruses cannot spread on their own, but if the right carriers are present, an epidemic of the virus can emerge, allowing viruses to spread over a greater distance.

What should we do to prevent informational aggression? There are two main ways for it.

The first one is the *extreme strategy*: Completely forbidding searching for interstellar alien transmissions, and if one happens occasionally to receive them, forbidding any attempts to decode or study them. But there are problems with this option. The *extreme strategy* is too restrictive because it cuts us off from a huge sector of knowledge about the Universe, but cognition cannot be globally stopped by administrative measures. Moreover, the *extreme strategy* will not work, because there are bound to be people who will search (or decode) alien transmissions in an unspoken, private way. The *extreme strategy* is actually counterproductive because it takes the search for alien signals into the gray sector, taking it out of public scrutiny and making it more dangerous.

The second way is the *soft strategy*: Basic caution must be exercised when examining the alien message materials. Basic caution means measures such as the following. Do not run programs unless their purpose is completely clear. Do not assemble or run devices if their purpose is not completely clear. Do not follow recommendations and instructions if their purpose is in any doubt. Analyze information for implicit "bait"—promises of solutions to some important problems, but with not entirely clear side effects.

Unfortunately, the *soft strategy* has problems as well. The process of examining the alien message material should be as open (i.e., widely discussed) and controlled as possible. But there are doubts that the process of examining the message materials can actually be *completely* open and controlled. The risk of "taking the bait" by an individual or an unsupervised organization cannot be completely eliminated.

The conclusion is that the risk of information aggression cannot be completely eliminated *regardless of our behavior.* The *soft strategy* looks more reasonable in this context.

There are, however, arguments according to which the probability of aggressive behavior of highly advanced technological civilizations is low. Aphoristically speaking, "high knowledge is given only in kind hands." Really, if aggressiveness in any form persists in the later stages of the technological explosion, civilization inevitably self-destructs. Probabilistic estimates based on Drake's formula show that there is likely only one civilization in the Galaxy in the stage of technological explosion (scale of duration is ~200 years), and that is our own Earth civilization. Consequently, any detected signal of an extraterrestrial civilization will belong to a civilization thousands of years older than the technological explosion (e.g., see Nielsen, 2015). If such a civilization has passed the filter of the technological explosion, it seems very unlikely that it is capable of aggression at all. Therefore, although the risk of informational aggression cannot be completely eliminated, the probability of informational aggression seems very small.

However, encountering direct alien information aggression is not the only risk of contacts. The law of Techno-Humanitarian Balance (Nazaretyan, 2017) reads as follows: The level of cultural restrictions of the destructive power of technology grows along the level of technology itself and evolves in parallel with it. The law of Techno-Humanitarian Balance works very well. For example, local social systems that strongly violate the techno-humanitarian balance have repeatedly dropped out of evolution (as, for example, Hitler's Germany or Pol Pot's Kampuchea). A global and profound disruption of the techno-humanitarian balance due to an unexpected new technology could be very dangerous for Earth's civilization as a whole.

There is a possibility of alien messages providing information that we are not ready to receive, or accessing resources that we are not ready to use in a sense that we have not cultural restrictions and cultural adaptation yet adequate to these innovations, because these innovations were not obtained by a natural evolution way. This could mean a deep disruption of the techno-humanitarian balance and have negative consequences for our civilization. We must be prepared to carefully analyze the possible negative consequences of using the information obtained.

Therefore, there are not only one, but multiple sources of risks, but our opinion is that the benefits of interstellar contacts via communication channels greatly outweigh the risks of such contacts.

Finally, note why the discovery of alien civilizations is important just now for us. The Earth civilization goes through a very difficult period of history, often called the singularity zone (Panov, 2011), which is characterized by the overlapping of many existential crises, and it is not certain that there is even a small chance of successfully overcoming these crises with a transition to long-term stable development. Since already the first discovered alien civilization must have a long history after passing the singularity zone, hence this civilization has overcome the problems of the singularity zone and, consequently, the problems arising here are at least in principle solvable. This would also mean that our own Earth civilization has a chance to overcome the existential crises of the singularity zone, and Earth civilization has a chance to go on to a long-term technological development. So far, this is not clear at all. That is why the discovery of the first alien civilization, whether the alien messages are understood or not, will radically change our view of ourselves. Also, the discovery of the first extraterrestrial civilization will mean the transition of our own civilization into the Galactic phase of Big History and complexity growth.

References

Almár, I., & Shuch, H. P. (2007). The San Marino scale: A new analytical tool for assessing transmission risk. *Acta Astronautica, 60*(1), 57–59. https://doi.org/10.1016/j.actaastro.2006.04.040

Devlin, H. (2016). Stephen hawking warns against seeking out aliens in new film. *The Guardian*. https://www.theguardian.com/science/2016/sep/23/stephen-hawking-warns-against-seeking-out-aliens-in-new-film

Nazaretyan, A. P. (2017). Mega-history and the twenty-first century singularity puzzle. *Social Evolution & History, 16*(1), 31–52. https://doi.org/10.30829/2309-6490-2017-1-31-52

Nielsen, J. N. (2015). Addendum on suboptimal civilizations, Grand Strategy: The View from Oregon. https://geopolicraticus.wordpress.com/2015/04/27/addendum-on-suboptimal-civilizations/.

Panov, A. (2011). Post-singular evolution and post-singular civilizations. *Evolution, 2*, 212–231.

Panov, A. (2025). Big history in the galaxy: About risks of contacts with alien civilizations. *Journal of Big History, 8*(1), 27–31.

Panov, A., LePoire, D., & Korotayev, A. (2020). The 21st century singularity in the big history perspective. An overview. In A. Korotayev & D. LePoire (Eds.), *The 21st century singularity and global futures. A big history perspective* (pp. 1–18). Springer Nature. https://doi.org/10.1007/978-3-030-33730-8_1

Turchin, A. (2018). The global catastrophic risks connected with possibility of finding alien AI during SETI. *Journal of British Interplanetary Society, 71*(2), 71–79.

Alexander Panov graduated from Moscow State University, Department of Physics. At present, he is a senior researcher at the Moscow State University Skobeltsyn Institute of Nuclear Physics (MSU SINP), and is CSc (Physics and Mathematics). His major works are devoted to nuclear physics, surface physics, quantum theory of measurement, cosmic rays physics, and problems of universal evolution/Big History. He is the author of about 270 articles in the Russian and international academic press, as well as the author of the monograph *Universal Evolution and Problems of the Search for Extraterrestrial Intelligence* (SETI).

Part VIII

Findings, Summary, Conclusions, and Plans for Future

A Big History Perspective on Complexity in Universal Evolution: Conclusions

22

David J. LePoire, Leonid Grinin, and Andrey Korotayev

Abstract

This collective monograph explores focus on complexity aspects in Big History against the background of complexity growth in the Universe, on our planet, and in biological, social, and cultural systems. Complexity growth is regarded as the connecting thread of evolutionary development and as a leading trend of Big History. The cosmic development chapters examine symmetries in cosmic and terrestrial evolution, energy distribution in cosmic structures, complexity growth patterns during the Stelliferous Era of Big History, modeling planet formation conditions, and risks of extraterrestrial intelligence (ETI) contact. This continues with perspectives on the evolution and complexity development in geology and chemical development. In the realm of biological evolution, the chapters cover the role of viruses in early evolution and ecosystems, infectious diseases in civilization history, and approaches to teaching complexity in undergraduate biology. Human evolution articles discuss human evolution periodization, AI's (artificial intelligence's) potential in understanding brain evolution, and the role of energy flow though social systems in generating complexity. The civilization evolution section explores energy use in empire life cycles, population growth and urban distributions, foundational myths of various civilizations, and war's role in social complexity growth. Notably, some articles span multiple Big

D. J. LePoire
Argonne National Laboratory, Argonne, IL, USA
e-mail: dlepoire@anl.gov

L. Grinin · A. Korotayev (✉)
HSE University, Moscow, Russia

Eurasian Center for Big History and System Forecasting, Institute of Oriental Studies of the Russian Academy of Sciences, Moscow, Russia

D. J. LePoire et al. (eds.), *Complexity in Universal Evolution*, World-Systems Evolution and Global Futures, https://doi.org/10.1007/978-3-032-20517-9_22

History phases, offering valuable insights into cross-disciplinary trends and relationships, such as comparing cosmic and terrestrial evolution or examining energy dissipation across physical and complex adaptive systems. This collection provides a comprehensive look at how complexity manifests and evolves throughout the different phases of Big History and universal evolution, from the cosmic scale to the intricacies of human civilization. By exploring these diverse topics, this book aims to deepen our understanding of the interconnected nature of complexity across various scales and time periods, contributing to the broader field of Big History and universal evolution research.

Keywords

Evolution · Universal evolution · Phases · Research · Complexity · Big history cosmic evolution · Biological evolution · Human evolution · Social evolution

22.1 Introductory Notes

This book, *Complexity in Universal Evolution: A Big History Perspective,* reveals both the richness of the concept of complexity and the many challenges that remain in fully understanding and applying it across the universal evolutionary spectrum—from cosmic origins to civilizational dynamics. Across the five temporal phases of Big History (Cosmic, Life, Human, Civilization, and Future) and five intersecting research paradigms (theory construction, methodological innovation, interdisciplinary synthesis, future projection, and pedagogical transformation), this volume offers a comprehensive synthesis of ideas, models, and questions that push the boundaries of current understanding.

The collected chapters explore how complexity emerges, develops, transforms, and occasionally declines in various domains and scales (often to rise to a new level). While no single model or metric can yet capture the full scope of complexity across these systems, the plurality of approaches presented here reflects the very nature of complexity itself: dynamic, layered, and resistant to reduction. Whether viewed through thermodynamic models, systems thinking, or cosmological simulations, the concept of complexity emerges as both a tool of analysis and a driver of change.

In this Conclusion, we have two tasks.

1. To discuss some issues on the subject of complexity and universal evolution in their interconnectedness.
2. To summarize achievements and values of the volume contributions and to talk about the further research.

22.2 Complexity and Universal Evolution

Complexity growth as a trend is inextricably linked with universal evolution. Not every evolutionary change leads to increase in complexity. On the contrary, only a very small part does—such evolutionary changes can be denoted as aromorphoses.[1]

Such changes may occur only when evolutionary niches are filled, abrupt changes in the environment are observed, and exceptional conditions emerge. But without evolution, evolutionary search, and numerous trials and errors,[2] there can be no movement to new levels and combinations of complexity.

Both characteristics—stability and variability—have great advantages, as well as disadvantages; they are both necessary for the existence of objects, species, and the Universe in general. This can also be observed in social evolution (Grinin & Korotayev, 2023). We argue that such social variability is not rooted in genetic or race-related phenomena (although for the period of anthropogenesis it was quite possible), but rather arises from the particular circumstances of each society—shaped by natural and social environments, the influence of key contingent factors, and even the emergence of exceptional individuals.

22.2.1 Panarchy and Evolution

Now we connect the earlier discussion in previous chapters of the dynamics of complex adaptive systems (CAS) under the panarchy model with an evolutionary perspective. This model offers some insights to the general evolutionary processes and rules that are described in more detail in the following sections. The panarchy model (Gunderson & Holling, 2002; Holling, 2001) describes adaptive cycles of growth, accumulation, crisis, and reorganization within complex systems. Unlike linear or purely progressive models, panarchy emphasizes cycles of expansion and collapse, nested across multiple scales of space and time.

Each adaptive cycle includes four phases:

1. **r (exploitation/growth):** Rapid expansion and innovation, when resources are plentiful.
2. **K (conservation/accumulation):** Slower consolidation, efficiency, and stability, but also increasing rigidity.
3. **Ω (release/collapse):** Crisis and breakdown of accumulated structures, often triggered by disturbance.

[1] See chapter "Biological and Social Phases of Big History and Complexity" (Grinin et al., 2026) in this volume. See also, e.g., Grinin et al. (2009, 2011), Grinin and Grinin (2020, 2023), Grinin (2020), Korotayev (2004, 2020), and Korotayev et al. (2000).

[2] On the rule of paying for evolutionary progress or the cost of complexity principle, see above chapter "Complexity Theory for Big History" (LePoire et al., 2026) in this volume. See also LePoire et al. (2025).

4. **α (reorganization/renewal):** Creative recombination, in some cases, of surviving elements, opening opportunities for new forms.

These cycles are **nested**: smaller, faster systems interact with larger, slower ones, so local disruptions can cascade upward, and broader stability can constrain local dynamics. This multilevel interaction is the essence of *panarchy*.

From an **evolutionary perspective**, panarchy offers several insights:

- **Creative destruction** (Schumpeter, 1942) is built into the cycle: Collapse (Ω) is not an end, but a transition to reorganization (α), enabling novelty (at least in some cases[3]).
- **Rule of preparatory work** (Grinin, 2013, 2017) aligns with the idea that resilience and latent potential accumulate during the conservation phase (K), enabling eventual breakthroughs.
- **Evolutionary inertia** parallels the rigidity of the K phase: Accumulated structures constrain future options.
- **Co-evolution** fits the cross-scale interactions of panarchy, where changes in one domain (e.g., ecosystems, economies, civilizations) trigger responses in others.
- **Block assemblage and memory** resonate with the reorganization phase (α), where surviving components are recombined into new adaptive structures.

In Big History, panarchy can help explain recurring patterns of rise and fall—from ecosystems and species to civilizations and technologies (for the Big History dimension of panarchy, see, e.g., Rodrigue, 2023a, 2023b). It emphasizes that evolution is not a smooth linear ascent, but rather a series of adaptive cycles nested within larger-scale processes. Over the long run, these cycles may generate increasing complexity by reusing, recombining, and reorganizing existing components in new ways.[4]

Thus, the panarchy model complements universal evolutionary rules: it highlights the role of cycles, crises, and cross-scale dynamics in shaping the pathways of universal evolution.

22.2.2 Differences and Similarities Are Two Sides of the Same Coin

We would like to present the following methodological idea. To show the path of "progressive"/arogenic evolution, how it became more complicated and moved to new levels, it is crucial to investigate, figuratively speaking, its vertical development

[3]One should not forget that not every destruction is creative (for example, the destruction of all complex forms of life in the internal parts of the Antarctic Continent around 30–10 million BP was not accompanied by the creation of some new more complex forms of life).

[4]On the cyclical component of social macroevolution, see, e.g., Grinin et al. (2016), Grinin and Korotayev (2018), Korotayev et al. (2006, 2011, 2016), and Chatalova and Korotayev (2022).

(from simple to complex). But if we study it from the general point of view, it is logical to present different levels as different manifestations of changes in the horizontal dimension—in other words, as a multi-line manifestation of general development. In fact, we are talking about changes, transformations in different parts or spheres of the *single* Universe: cosmic web, galaxies, stars, planets, minerals, molecules, living beings, etc. At the same time, it is important to keep in mind that the development of more complex forms is a part of a broader evolution. Thus, abiogenic chemical evolution was actually a lateral line of geochemical evolution, and the latter, in its turn, was a part of geological evolution. And this mere fact determines the similarities. In addition, some types of evolution develop in co-evolution, which imply mutual influence, transformation, and support (see Grinin & Grinin, 2025; Waring & Wood, 2021). Such an approach allows understanding that there are some basic patterns which are differentiated and acquire specific forms related to the peculiarities of the form of matter in which they manifest themselves. It is quite possible to distinguish these common patterns. The more so in the case of evolution on the Earth, where all its forms and levels are very closely connected by a common place of development. Thus, if we consider megaevolution/universal evolution horizontally, that is, in terms of emerging new lines, then we reveal a common basis and if we consider universal evolution/megaevolution vertically as a tree, then we find "genetic" relationship.

The rule of evolutionary inertia (formulated by Ludwig Doderlein and Othenio Abel for biological evolution) can be used for predetermined character of evolution. It deals with the general dependence of subsequent evolution on the previous one, when the past largely determines not only the present but also the future. This is reflected in the significant dependence of subsequent phylogenetic events on the preceding ones, which is interpreted as evidence of the inertial influence of the past evolution on its future. The inertia manifests both in the similarity of development mechanisms and in the fact that every transition to a higher level channels the direction of development more and more. Meanwhile, we are too accustomed to seeing an insurmountable barrier between higher and lower levels of evolution, absolutizing the differences between living and non-living, human and animal. *But one should rather be surprised not by the similarities, but by the differences. The similarities between the levels are more natural, since the birth of a new one does not mean the rejection of the old one.* Until recently, evolution has been mainly additive in nature, so the new did not reject the old, but added to it: elementary particles did not disappear with the emergence of atoms, and the latter with the emergence of molecules; inorganic molecules remained, but organic molecules were added to them, etc. Therefore, the old has a continuous effect on the new, but the new also affects the old where possible. A number of evolutionary rules (see Grinin, 2017, 2020[5]) show that the new is not only different from the old, but also related to

[5] See also chapter "Introduction to Complexity Theory for Big History" (LePoire et al., 2026) in this volume, as well as LePoire et al. (2025).

it, and that it breaks through only in certain directions (in fact, where the old allows it to break through), and that it is formed not in all, but only in some aspects.

22.2.3 The Principle of Creative Destruction

By studying the relationship between catastrophes and evolution, one can formulate the *principle of creative destruction* for phase transitions, transformations, and expansion of diversity if to use Joseph Schumpeter's expression (1994 [1942]). "Creative destruction" is the creation of a new one by destroying or removing the old one from active operation. At the same time, the new is already essential and different from the old. This provides both continuity and space for moving toward the new. However, the destruction itself cannot be creative. It turns out this way only after a great amount of preparatory work. At the same time, first this often leads to regress and only then (i.e., much later) to evolution, as if taken a run-up, starts a new movement forward. In social evolution, one can find many such cases. The most famous examples are the barbarization of Europe after the fall of the Western Roman Empire following the German invasion; as well as destruction of prosperous countries resulting from the Mongol invasion. Both catastrophes would sometimes launch a rise based on a new synthesis which, however, would take much time. Therefore, one can speak about *the rule of preparatory work of evolution*. It means that an evolutionary breakthrough resulting from unique circumstances is never a coincidence, but it is always prepared by a huge and longtime "work" of evolution to advance changes in a certain direction. However, the emergence of unique circumstances in the right place at the right time often depends on chance. At the same time, a phase transition or transformation of an object often needs an impetus or a trigger to start. On the one hand, of course, the latter will not work without the internal readiness of the system; but on the other hand, even a high level of internal readiness by itself cannot launch the transformation process like the gunpowder cannot explode without fire. Without a trigger, a system may remain in a state of potential readiness for transformations for a long time. In this case, the analogues of evolutionary typical/recognized systems are formed.[6]

22.2.4 Why Do We Observe Unity and Similarity in the Mechanisms and Patterns at Different Levels of Universal Evolution?

In this section we will try to present some evolutionary and philosophical ideas that explain the profound similarity in the laws and patterns of evolution at all its levels and phases. What defines this unity? This is one of the most important questions, the

[6]About the analogs/alternatives in social evolution, see Grinin (2003, 2025), Grinin et al. (2004), Grinin and Korotayev (2009), Bondarenko et al. (2002), Korotayev (2003, 2020), Bondarenko (2025), Dementyeva (2025), Kradin (2025), Popov (2025), Shchavelev (2025), Smorchkov (2025), Surikov (2025), and Korotayev et al. (2000). On analogues in cosmic evolution, see Grinin (2013, 2017, 2018), and Grinin and Grinin (2019).

answer to which can significantly change our approach to the study of universal evolution. But it can only be provided by a long and diverse work on the development of *evolutionary studies*. Although a number of researchers left very insightful ideas and assumptions (Jantsch, 1980; Teilhard de Chardin, 1987; Lima-de-Faria, 1988; Nazaretyan, 1991; Snooks, 2005; Spier, 2010, 2011; Christian et al., 2014; Christian, 2018), there is a great task to perform such work in a systemic and consistent manner. In this section, we would like to demonstrate some opportunities and dimensions of such research.

22.2.4.1 The Causes of Evolution

First of all, let us speculate why evolution is possible at all? Some general reasons are: 1) the gradually changing conditions which make it necessary to adjust structure, functions, etc., to the changed conditions; the aspiration for the most harmonious congruence with external environment is caused by *the pursuit to the most favorable energy state*, but the process of this adjustment sometimes leads to an unusual result that can provide some advantages; 2) competition due to limited resources; 3) the desire for self-preservation; and 4) the circulation of matter (see above). But, as already mentioned, in every cycle, this circulation has some differences which tend to accumulate.

22.2.4.2 The Systemic Character, Environment, and the Laws of High Abstraction

There are also quite obvious situations, laws, and patterns that are present at all levels and in all systems. 1) For example, objects or systems exist in the environment and there should be some interaction between them. Despite the variety of environments and situations, there are quite a few basic interaction models; so, they can be quite similar at different levels. 2) The systemic character by itself leads to certain similarities; this was established back in the 1950s and with respect to a number of relations even earlier. 3) The laws of dialectics, formulated by Georg Wilhelm Friedrich Hegel, also have in their abstract form a rather clear mechanism. For example, the law of transition from quantitative to qualitative changes manifests itself because any forces have limits beyond which their impact declines and becomes insignificant, so when the quantitative accumulation reaches this limit, the former structure (order, etc.) must inevitably transform. The law of the unity and struggle of opposites as a part of an even broader pattern of binary (duality, dichotomy) is determined by the fact that any structure or change requires at least a couple of opposing forces, elements, etc. 4) The binary is also related to the universal symmetry, which determines the opposite parts or paired relationship between elements (e.g., of the positively and negatively charged). 5) There are common models and patterns of self-organization and self-regulation in different systems on all levels of evolution.[7]

[7] This is studied, for example, by cybernetics. Cybernetics as a scientific field can be defined as the study of general laws and patterns of receiving, storing, and processing information in complex

22.2.4.3 Parsimony of Evolution

The presence of common laws and patterns is logically explained by the fact that in all aspects it is more advantageous to have a few universal rules than a set of special ones for each case. Here, it is worth mentioning *the rule of rarity of new evolutionary rules*. According to this rule, evolution is wasteful in its "experiments," but rather stingy with respect to mechanisms and patterns and "prefers" to use the already available rather than to invent new ones. Each new rule (or pattern) is related either to the peculiarities of filling evolutionary niches or to the emergence of some new sub-levels, levels, or blocks. This perspective allows us to hope that in the future it will be possible to identify a group of primary (basic) rules and laws of evolution that have already manifested themselves in the first hundreds of millions of years, and then new ones that would appear later. In addition, self-organization does not require a large amount of forces or rules, rather quite a limited number would suffice (Grinin, 2017). One should remember that the diversity of manifestations is based on a limited number of basic rules.

22.2.4.4 More Specific Mechanisms

Much is canalized by rather rigid constraints: energy, efficiency, and previous development. Thus, the choice of the most energetically advantageous regime can occur at different levels, which is accounted for by the same concerns, and, respectively, by the choice of similar forms. But, of course, revealing specific mechanisms united by a common law or rule of universal evolution is of special value. Thus, some things are determined by *the rule of minimization of evolutionary efforts*, when the ready-made solutions are used, and also by *rule of "block assemblage."*

The essence of this rule[8] is that in the course of evolution, there emerge some elementary or more complex units, systems, and constructions which are used in different variations. The elementary particles are the units which form atoms. With the emergence of atoms, there also emerge stellar systems, and in the stellar interior, at certain phases of stellar evolution, new types of atoms including heavy elements are formed from additional elementary particles.[9] Due to the diversity of emerging atoms, one can speak about chemical evolution. Atoms are the universal units and components for the formation of various molecules and this marks the beginning of planetological, geological, and complex molecular organic evolution leading to life. The cells become "bricks" for the formation of living organisms; there progressively emerge whole blocks of organs and systems which are surprisingly similar in

systems. Its main principles are quite suitable for the description of self-regulation (see, e.g., Wiener, 1948; Ashby, 1956; Foerster & Zopf, 1962; Beer, 1994; Umpleby & Dent, 1999; Wieser & Slunecko, 2014). Regulation and self-regulation (as the highest form of regulation) within systems are also major categories in cybernetics. About self-regulation and self-managing of systems in Big History, see Grinin (2016); for technological sphere, see (Grinin et al. (2024).

[8] See chapter "Biological and Social Phases of Big History and Complexity" (Grinin et al., 2026) in this volume; see also Grinin et al. (2009, 2025).

[9] See chapter "The Evolution of Complexity in the Stelliferous Era: Observations and Patterns" (Korotayev & LePoire, 2026) in this volume. See also Korotayev and LePoire (2025).

different classes and even types of living organisms. One can recall genes and chromosomes as standard components and blocks of biological systems. One can insert a gene of a mouse into an elephant DNA, and the gene of a dog into the human DNA! Thus, there is a striking standardization of elements and "components" at all evolutionary levels; and since entirely new objects within evolution are created for 90–99% from the already existing components, the speed of evolution increases dramatically. Let us also add that in human society, the borrowing occurs rather frequently: societies borrow (sometimes to the full extent) religions, and legal, political, and technological systems. Thus, "block assemblage" allows rather fast modernization of societies.

So, the increasing complexity of structure at all levels—from atom to society—is often carried out, figuratively speaking, by *polymerization*, that is, by assembling standard "details." All chemical elements of Mendeleev's periodic table can be represented as gradual complication of the structure of their atoms through adding an atom of hydrogen. The same can be said about complex molecules, multicellular organisms, or expansion of the society by agglomerating small structures (like families, communities).

22.2.4.5 Evolutionary Memory

One can also make some assumptions that evolution has some kind of a code and memory, which are fixed with the help of some imprints, and also function on the basis of the rule of minimization of evolution efforts. Of course, it remains unclear how this memory becomes fixed but there is no doubt that it is based on some rather material things.

For example, everybody knows about the so-called golden ratio. But why does this ratio have such proportion?[10] Why do some patterns become common at all? Probably, because some discoveries of nature and evolution reveal a certain code, a set of ancient and longstanding solutions and combinations, thanks to which, on the one hand, the already available solutions are used to create a new one, while on the other hand, the evolution related to those solutions is canalized and becomes autoevolution, according to Antonio Lima-de-Faria (1988).[11] But this defines certain limits, since the fundamentally new solutions are already made far from easily and only as a result of some rarely occurring breakthrough created by peculiar circumstances.

It is still impossible to reveal how these universal solutions and patterns are encoded, but there probably exists some mechanism. However, if we speak about the "genetic" connection between higher and lower levels of evolution (see above), why should we deny the possibility of "genetic" memory and "genetic" code of

[10] Let us remember that in the rounded percentage value, the golden ratio describes the relationship between two proportions, which is 62% to 38%. This ratio equals 1:1.62 (a common proportion in the construction of objects).

[11] That is in the most general form, the mechanism is similar to that in the genome of living beings, in the form of the so-called genomic ballast combinations of genes of which some are used only in extreme cases.

evolution? Even relatively simple structures have memory. A kind of "memory" can be observed in self-organization and the activation of this "memory" is promoted by the fact that order often turns out to be energetically beneficial.

Another aspect of this assumption is the universal character of information.[12] We learn more and more about different kinds of information, in particular, about chemical signals which even the simplest organisms (bacteria) appear to be able to perceive. In fact, one can observe information already at the level of elementary particles, where it seems to be syncretic with the energy form. But, in any case, it is important that the information interaction can occur only if the properties of objects correspond to each other (Yankovsky, 2000). Also, the electromagnetic and other interactions provide adjustment, as a result of which, for example, negatively and positively charged particles "recognize" each other. In fact, they exchange "codes" and turn out to be complementary, and therefore can create stable structures. It bears repeating that, at this level, the energy and information aspects are inseparable but still different. A greater difference between the energy and information aspects can be observed in catalytic interaction (Ibid.) when one substance (catalyst) changes the rate of chemical reaction between other substances, which are reactive chemicals in this case. Without information that activates the reactive chemicals, the reaction would be much slower or could not take place at all under existing conditions. In other words, information is mostly separated from energy processes; so, the catalysts can only impact the speed but not participate in chemical reactions. But the condition that information between objects is transferred by means of substance or energy exchange is fully fulfilled.

In the general theory of information, the law of information preservation is also formulated: the latter keeps its significance unchanged as long as the information carrier—memory—remains unchanged. The information exchange at the lowest levels, already in the micro-world, evidences the existence of memory (in particular, in the form of recognition). It seems that preservation and transfer of information at different levels and in different systems is not only one of the foundations of interaction between different objects, from particles to galaxies, but also a way to react to environmental changes, and most rules, laws, and patterns manifest themselves just in the interaction with the environment.

Thus, there is a common base, a "common denominator" in the continuity of motion and energy processes, in interactions involving information exchanges, in destruction and new assemblage, and other aspects so that the common may manifest itself in the behavior of different objects. At the same time, it should be implemented not only in standard conditions but also in unusual ones, which are the most interesting for evolutionary studies because it is just the unusual responses to unusual challenges that may give rise to fundamentally new things.

---

[12] For some detail on information and its role in the universal evolution, see chapter "Introduction to Complexity Theory for Big History" (LePoire et al., 2026) in this volume; see also LePoire et al. (2025).

We have also mentioned above the circulation of matter, energy, and information. However, such circulation could not take place without some kind of memory which made possible the new assemblages and new processes of self-organization. Hence, we inevitably return to the fact that there must be some mechanisms of coding, some organizational and system-forming memory. *However, similar to non-specialized stem cells, which can differentiate into different cell types and organs, the matter with such memory in different situations can transform into different types and forms of matter.*

22.3 Synthesis Across the Four Evolutionary Domains

In the **cosmic phase**, contributors such as Andrey Korotayev and David LePoire articulate a compelling symmetry between the Universe's early cooling and Earth's accelerating complexity, offering a "cone model" that visualizes mirrored trajectories of change.[13] This model illustrates how systems can undergo both decelerating and accelerating phases, offering a more dynamic picture of universal change. However, this depiction is shown to oversimplify complexity growth patterns, specifically at where the transitions from the Universe's decelerating physical complexity development and Earths accelerating complex adaptive systems took place. Korotayev and LePoire propose possible proxy measures of complexity that could help us to estimate the universal complexity dynamics in the Stelliferous Era of the Big History in more detail than was done earlier. One such measure is what the astronomers call "metallicity." Astronomers define as "metals" any elements that are heavier than hydrogen and helium (thus, according to them, oxygen, or carbon are also "metals"). Then, the "metallicity" turns out to be a measure indicating the ratio of "metals" (all the elements that are heavier [and, thus, more complex] than helium [let alone hydrogen]) to the "non-metals." Korotayev and LePoire suggest that the mean metallicity of the Universe can be regarded as a measure of universal chemical–physical complexity (reflecting the universal complexity of the element base for the chemical evolution). Their analysis also suggests that, since 150 million years after the Big Bang Singularity, the universal metallicity experienced exponential acceleration with subsequent quasi-logistic acceleration getting to the maximum level around 10–11 billion years BP, which was followed by a new wave of hyperbolic deceleration.[14]

LePoire's complementary analysis of energy gradients and structure formation challenges established views on free energy rate density (FERD), suggesting that the energetic and structural potential of early cosmic environments may have supported

[13] See chapter "The Evolution of Complexity in the Stelliferous Era: Observations and Patterns" (Korotayev & LePoire, 2026) in this volume. See also Korotayev and LePoire (2025), and Korotayev (2024, 2025a).

[14] See chapter "The Evolution of Complexity in the Stelliferous Era: Observations and Patterns" (Korotayev & LePoire, 2026) in this volume. See also Korotayev and LePoire (2025).

more intricate pathways of complexity than previously appreciated.[15] On the other hand, Martin van Duin evaluates Chaisson's energy rate density metric across multiple domains, identifying where it succeeds and where it fails to capture informational and institutional complexity.[16]

Teófilo Vargas, Paolo Musso, and Nicolò Antonietti add precision through modeling uncertainty in galactic habitability, showing how probabilistic reasoning can be applied to cosmic-scale questions.[17] Their work emphasizes how complexity in the form of life-supporting environments may emerge under uncertain but quantifiable conditions. Alexander Panov offers a provocative extension, proposing the Galaxy itself as a superintelligent information field.[18] This redefinition of complexity as an emergent, distributed intelligence challenges conventional anthropocentric assumptions and opens new doors for inquiry within the framework of Big History.

In the **geological and biological phases**, the focus shifts to Earth's deep-time transformations. Leonid Grinin's call to integrate geological evolution more fully into Big History marks an important disciplinary bridge.[19] His work reminds us that the stage upon which biological life emerged—the Earth itself—is a dynamic, evolving system that interacts with life in co-constructive ways. David LePoire follows this thread, tracing how Earth's geophysical processes—including tectonics, atmospheric shifts, and ocean circulation—laid the groundwork for life's complexity.[20] His analysis reaffirms the Gaia hypothesis in a systems science context, showing how feedback loops between biology and geology contribute to Earth's long-term habitability.

Anton Grinin's analysis of viruses uncovers a hidden driver of biological evolution.[21] Far from being marginal actors, viruses emerge as vectors of genetic innovation and ecological regulation. By framing viruses as agents of information flow, Anton Grinin reframes evolutionary narratives around complexity, adaptation, and resilience. Meanwhile, Paul Narguizian translates such ideas into education, demonstrating how complexity science and systems thinking can enhance

---

[15] See chapter "Energy Flow and the Emergence of Complexity in the Universe's Structure" (LePoire, 2026a) in this volume.

[16] See chapter "Scale Dependence of FERD in Big History's Hierarchical Systems" (Van Duin, 2026) in this volume.

[17] See chapter "Modeling Galactic Evolution and the Potential for Intelligent Life in the Milky Way" (Vargas et al., 2026) in this volume.

[18] See chapter "Beyond Intelligence: The Galactic Information Field and the Evolution of Complexity" (Panov, 2026) in this volume.

[19] See chapter "Geological Evolution: A Bridge Between Cosmic and Earthly Development in Big History" (Grinin L., 2026) in this volume.

[20] See chapter "The Earth's Pivotal Role in Big History: From Cosmic Evolution to Life's Emergence" (LePoire, 2026b) in this volume.

[21] See chapter "Viruses in Big History: Evolutionary Roles and Complex Systems" (Grinin A., 2026) in this volume.

undergraduate biology instruction.[22] His proposals offer a path toward a scientifically literate public equipped to engage with the multifaceted nature of biological and ecological systems.

In the **human phase**, complexity takes on cognitive, technological, and cultural dimensions (though biological complexity dimension still remains significant up to the Upper Paleolithic Revolution). Andrey Korotayev undertakes an attempt to propose a periodization of the human evolution on the basis of the most recent scientific data, taking into account all the above-mentioned dimensions.[23] Ekaterina Sazhienko and David LePoire show how artificial intelligence serves both as a future driver of complexity and as a lens for reinterpreting human brain evolution.[24] Their analysis of artificial intelligence (AI) as both a mirror and an amplifier of human capabilities opens new directions for modeling the co-evolution of humans and machines. Nick Nielsen's introduction of "peer complexity" broadens the scope of Big History to include non-human, non-biological emergent systems.[25] By situating Earth's evolution within a broader class of cosmological peers, Nielsen challenges SETI (Search for Extraterrestrial Intelligence) and astrobiology to adopt less anthropocentric models.

In the **civilizational phase**, energy and organization converge in new ways. Mark Ciotola's thermodynamic modeling of dynastic cycles and resource bubbles introduces a compelling "complexity profile" for historical systems, advancing the idea that civilizations can be analyzed as extended dissipative structures.[26] Ken Baskin suggests that, by applying complexity science, we can see that societies throughout history—including during the Axial Age and Modernity—have faced transformational challenges and successfully adapted by rewriting their foundational myths.[27] Antony Harper's research on urbanization reveals how city-size distributions follow patterns of punctuated equilibrium, reflecting deeper systemic reorganizations.[28] His use of power-law distributions and entropy as complexity

[22] See chapter "Biology Education Through Big History's Complex Systems Thinking" (Narguizian, 2026) in this volume.

[23] See chapter "Human Evolution in the Complexity Growth Perspective: Toward Periodization of the Big History Biosocial Era" (Korotayev & LePoire, 2026, this volume). See also Korotayev (2025b).

[24] See chapter "The Role of Artificial Intelligence in Big History: Understanding Human Evolution and Future Scenarios" (Sazhienko & LePoire, 2026) in this volume.

[25] See chapter "Rethinking Complexity in the Universe: Implications for SETI and the Fermi Paradox" (Nielsen, 2026) in this volume.

[26] See chapter "Emergence, Complexity, and Thermodynamics in Resource Bubbles and Historical Dynasties" (Ciotola, 2026) in this volume.

[27] See chapter "Myth, Complexity, and Big History: Rethinking Our Shared Map for the Future" (Baskin, 2026) in this volume.

[28] See chapter "Punctuated Growth and Complexity in Urban Evolution: Insights from Historical City Size Distributions" (Harper, 2026) in this volume.

proxies helps quantify civilizational evolution. Stephen Satkiewicz, returning to the enduring role of conflict, reinterprets war as both a product and catalyst of social complexity.[29] Building on Charles Tilly's dictum that "war made the state and the state made war," Satkiewicz shows how violence and order co-evolve within the arc of human development (on the recent research in this direction see Feinman, 2023; Tymowski, 2023; Jennings & Earle, 2023; Kradin, 2023).

This book synthesizes Big History through multiple analytical frameworks, offering both chronological and thematic lenses to explore universal complexity. While organized around five temporal phases (Cosmic, Life, Humans, Civilization, Future), it equally invites exploration through four interconnected research paradigms: methodological innovation, interdisciplinary synthesis, future projection, and pedagogical transformation. Below is an expanded overview.

Temporal Phases as Foundational Structure:

1. Cosmic Phase: Formation of matter, galaxies, and planetary systems
2. Geological Phase as a base for development of abiotic processes
3. Life Phase: Emergence of biological complexity via evolution
4. Human Phase: Cognitive revolution and symbolic communication
5. Civilization Phase: Agricultural, industrial, and technological revolutions
6. Future Phase: Scenarios for humanity and cosmic complexity

22.4 Thematic Research Paradigms

Research methods develop and apply innovative approaches and metrics to analyze patterns:

- **Astronomical modeling:** Simulating dark matter halos and baryonic feedback in early Galaxy formation
- **Metallicity as complexity proxy:** Tracking heavy element abundance to map the rise of chemical diversity in stars and planets[30]
- **Entropy-driven systems:** Framing human technological advances (e.g., steam engines, semiconductors) as dissipative structures optimizing energy flow
- **Urban scaling laws:** Identifying power-law relationships between city size and innovation rates across historical societies[31]

[29] See chapter "War and the Rise of Social Complexity: A Big History Perspective" (Satkiewicz, 2026) in this volume.

[30] See chapter "The Evolution of Complexity in the Stelliferous Era: Observations and Patterns" (Korotayev & LePoire, 2026); see also Korotayev and LePoire (2025).

[31] See chapter "Punctuated Growth and Complexity in Urban Evolution: Insights from Historical City Size Distributions" (Harper, 2026) in this volume.

Research Integration attempts to bridge disciplines and synthesize cross-domain insights:

- **Geological pivots:** Earth's differentiation (core-mantle-crust) as a gateway for terrestrial complex adaptive systems (CAS)
- **Virology's role:** Viruses as evolutionary drivers through horizontal gene transfer and ecosystem regulation[32]
- **Biosocial evolution:** Co-evolution of biological traits (e.g., lactose tolerance) and cultural practices (dairying)[33]
- **Energy flow contrasts:** Comparing cosmic radiative cooling with Earth's localized trophic energy cascades
- **Warfare paradox:** Analyzing conflict as both a catalyst for technological leaps and a destabilizer of civilizations[34]

Projection Studies envision potential futures by extrapolating trajectories of complexity:

- **Non-terrestrial civilizations:** Classifying potential Type II/III Kardashev-scale societies and their detectability
- **AI as evolutionary inflection:** Examining large language models (LLMs) as potential co-creators of "post-biological" knowledge systems[35]
- **Narrative infrastructure:** Leveraging storytelling frameworks (mythic, scientific, speculative) to align global cooperation[36]

Educational Frontiers offer Rethinking of Pedagogical approaches:

- **Complexity-enhanced biology:** Integrating entropy budgets, scaling laws, and network theory into life sciences curricula[37]
- **University-wide unification:** Designing cross-departmental programs around Big History's grand narrative, bridging astrophysics, anthropology, and sustainability studies

[32] See chapter "Viruses in Big History: Evolutionary Roles and Complex Systems" (Grinin A., 2026) in this volume.

[33] See chapter "Human Evolution in the Complexity Growth Perspective: Toward Periodization of the Big History Biosocial Era" (Korotayev, 2026, this volume). See also Korotayev (2025b).

[34] See chapter "War and the Rise of Social Complexity: A Big History Perspective" (Satkiewicz, 2026) in this volume.

[35] See chapter "The Role of Artificial Intelligence in Big History: Understanding Human Evolution and Future Scenarios" (Sazhienko & LePoire, 2026) in this volume.

[36] See chapter "Myth, Complexity, and Big History: Rethinking Our Shared Map for the Future" (Baskin, 2026) in this volume.

[37] See chapter "Biology Education Through Big History's Complex Systems Thinking" (Narguizian, 2026) in this volume.

Aspects of Complex Adaptive Systems:

- **Scale and hierarchy:** Organizational structures and dynamics from quantum fluctuations to galactic superclusters
- **Energy transduction:** Gravitational and stellar nuclear fusion, photosynthesis, fossil fuels, and computation as phase transitions in energy utilization
- **Information systems:** DNA, writing, printing presses, and digital networks as complexity accelerators

This multidimensional approach enables readers to navigate 13.8 billion years of Big History through both time and concept, revealing universal principles that bind cosmic, biological, and cultural evolution.

22.4.1 Thematic Integration and Emerging Patterns

A few overarching insights emerge from the collection:

1. **Energy Flow as Central Driver:** Across all domains, energy gradients appear as both enabling and constraining factors in the emergence of complexity. Whether in star formation, evolutionary biology, or urban development, energy flow sets the stage for self-organization.
2. **Complexity Growth as a Leading Trend of Big History**.
3. **Emergence, Feedback, and System Transitions:** Complexity often arises at the edges of chaos, where feedback loops, threshold events, and non-linear dynamics drive innovation and reorganization. The studies here illustrate how tipping points—from planetary formation to technological revolutions—are moments of accelerated complexity.
4. **Metric Refinement and Conceptual Flexibility:** The limitations of current metrics like FERD point to the need for multidimensional tools that can incorporate not only energy and structure but also information, adaptability, and resilience.[38] The search for such metrics is a recurring theme.
5. **Pluralism as Strength:** The volume's diversity of methods—from probabilistic modeling to historiographical analysis—reflects the irreducibility of complexity to any single framework. This pluralism is not a weakness but a methodological necessity.
6. **Bridging Scales and Disciplines:** Several contributions cross conventional boundaries, drawing connections between cosmic and terrestrial evolution, or between microbial life and urban systems. These bridges are essential for constructing integrative theories of complexity.

[38] See chapter "Scale Dependence of FERD in Big History's Hierarchical Systems" (Van Duin, 2026) in this volume.

22.5 Future Directions: A Call for Iteration, Collaboration, and Imagination

As Big History continues to develop, the frameworks and hypotheses presented here provide fertile ground for further exploration. Methodologically, new tools from AI, agent-based modeling, and network theory may help capture patterns that remain elusive. Conceptually, scholars must wrestle with the implications of complexity as both a descriptive and normative term—as something we observe and something we aim to foster.

Future research could extend this work by:

- Developing **new complexity metrics** that better capture informational and adaptive dynamics.
- Applying **uncertainty quantification** across historical, biological, and cosmological models.
- Exploring **co-evolutionary feedbacks** among systems more explicitly, particularly between human and environmental domains.
- Investigating **ethical and philosophical implications** of complex systems, especially in the context of AI, ecological collapse, and planetary stewardship.
- Enhancing **pedagogical integration**, ensuring that complexity science becomes a core element of science, history, and policy education.

Ultimately, the questions raised in this volume—about growth, collapse, intelligence, and design—invite not just empirical study but philosophical reflection. What kind of Universe do we inhabit? What forces shape its structure and direction? And what role, if any, do humans play in this vast unfolding?

We extend our deepest thanks to all the contributors and reviewers who made this volume possible. Their intellectual courage, methodological creativity, and commitment to cross-disciplinary dialogue are evident throughout.

This volume is not an endpoint but a waypoint. It provides tools, models, and metaphors for navigating 13.8 billion years of emergence, transformation, and renewal. It invites dialogue between the known and the unknown, between science and story, and between what is and what might be.

As we look ahead, we are reminded that complexity is not merely an object of study—it is a condition of existence. It permeates our atoms, our ecosystems, our cities, and our civilizations. To understand it is to understand ourselves in relation to the Universe. That journey is ongoing, but with each new insight, we chart the terrain a bit more clearly. May this work inspire many more such journeys.

The complexity of everything is not an obstacle, but an invitation—to wonder, to question, and to connect.

22.6 The Next Volume

The next volume will continue to explore these questions through further exploration of phase transitions in Big History. Phase transitions represent pivotal shifts in the structure and dynamics of complex systems throughout universal evolution and Big History. These transitions—whether in cosmological, biological, social, or technological domains—mark major reorganizations in energy flow, information processing, and systemic organization. Models such as aromorphosis and panarchy help understand how transitions are triggered, explored, and stabilized through non-linear dynamics, including creative destruction, cross-scale reverberations, and the interplay of resilience, efficiency, and robustness. Such processes often occur under "Goldilocks" conditions, where ingredients, constraints, and structures align to permit the emergence of new levels of complexity. This understanding helps in the exploration of how the probability and difficulty of transitions vary across time and context, and how these probabilities may evolve with changing environmental and energetic conditions. Measures and models of transition can be proposed to characterize their magnitude, contingency, and likelihood. Finally, this exploration guides consideration of how these transformations contribute to an accelerating overall tempo of evolution—driven by diffusion, internal growth, and combinatorial innovation—and how cycles of collapse and regeneration may structure the long-term dynamics of the Universe's evolving complexity.

Together, these themes motivate the central goal of this collection: to develop a coherent, integrative framework for understanding phase transitions as the primary engines of growing complexity across the history of the Universe. In particular, the collection emphasizes the inseparable link between complexity and phase transitions—the way each new organizational level emerges through discontinuous, non-linear restructuring rather than smooth, incremental change. By examining the general features, regularities, and patterns that recur across cosmological, biological, social, and technological transitions, the collection seeks to clarify how such transformations initiate, propagate, stabilize, or fail.

This cross-domain perspective enables the identification of shared mechanisms—threshold effects, symmetry-breaking, cascading feedbacks, combinatorial innovations—and helps distinguish universal signatures from domain-specific contingencies. The goal is not only to measure and compare transitions, but also to understand the non-linear dynamics that make them both rare and transformative. Ultimately, the collection aims to illuminate how repeated phase transitions collectively generate the long-term growth of complexity; how cycles of collapse and regeneration structure evolutionary tempos; and how these insights can inform our ability to anticipate, model, and perhaps guide future transitions in an increasingly interdependent and rapidly changing world.

References

Ashby, R. (1956). *An introduction to cybernetics*. Chapman and Hall.

Baskin, K. (2026). Myth, complexity, and Big History: Rethinking our shared map for the future. In: D. J. LePoire, L. Grinin, & A. Korotayev (Eds.), *Complexity in Universal Evolution – A Big History Perspective* (469–486). Springer Nature (in this volume). https://doi.org/10.1007/978-3-032-20516-2_16

Bondarenko, D. M. (2025). Evolution and forms of socio-political organization: historical and anthropological approaches (Introductory Remarks). *Social Evolution & History, 24*(2), 118–126. https://doi.org/10.30884/seh/2025.02.06

Bondarenko, D. M., Grinin, L. E., & Korotayev, A. V. (2002). Alternative pathways of social evolution. *Social Evolution & History, 1*(1), 54–79.

von Foerster, H., & Zopf, G. W. (1962). Principles of self-organization. In M. C. Yovits & S. Cameron (Eds.), *Self-organization* (pp. 31–50). Pergamon Press.

Beer, S. (1994). *Decision and control: The meaning of operational research and management cybernetics*. Wiley.

Chatalova, L., & Korotayev, A. (2022). The long cycle perspective on the emerging bio age. *Futures, 141*(2) article # 102991. https://doi.org/10.1016/j.futures.2022.102991

Christian, D. (2018). *Origin story: A big history of everything*. Little, Brown and Company.

Christian, D., Stokes Brown, C., & Benjamin, C. (2014). *Big history: Between nothing and everything*. McGraw Hill.

Ciotola, M. (2026). Emergence, complexity, and thermodynamics in resource bubbles and historical dynasties. In: D. J. LePoire, L. Grinin, & A. Korotayev (Eds.), *Complexity in universal evolution—a Big History perspective* (pp. 487–518). Springer Nature. In this volume. https://doi.org/10.1007/978-3-032-20516-2_17

Dementyeva, V. V. (2025). Etatist and non-etatist approaches to the study of ancient civil community: Discussions on civitas romana. *Social Evolution & History, 24*(2), 178–182. https://doi.org/10.30884/seh/2025.02.16

Feinman, G. M. (2023). Rethinking premodern governance: Passing the torch. *Social Evolution & History, 22*(2), 4–17. https://doi.org/10.30884/seh/2023.02.01

Grinin, A. (2026). Viruses in Big History: Evolutionary roles and complex systems. In: D. J. LePoire, L. Grinin, & A. Korotayev (Eds.), *Complexity in universal evolution—a Big History perspective* (pp. 247–256). Springer Nature. In this volume. https://doi.org/10.1007/978-3-032-20516-2_10

Grinin, L. (2013). *Big history of the world's development: The cosmic evolution*. Uchitel Publishing House. In Russian.

Grinin, L. (2017). *Big history of the world's development: History and evolution of the solar system*. Uchitel Publishing House. In Russian.

Grinin, L. (2018). Evolution of the early solar system in terms of Big History and universal evolution. *Journal of Big History, 2*(1), 15–26. https://doi.org/10.22339/jbh.v2i1.2252

Grinin, L. (2020). *Bol'shaya istoriya razvitiya mira: Planety Solnechnoy sistemy. Ikh istoriya i evolyutsiya. Khimicheskaya evolyutsiya v kosmose i na Zemle*. Moskovskaya redaktsiya izdatel'stva Uchitel'.

Grinin, L. E. (2025). On the stages of the evolution of statehood. *Social Evolution & History, 24*(2), 57–161. https://doi.org/10.30884/seh/2025.02.12

Grinin, L. E. (2003). The early state and its analogues. *Social evolution & history, 2*(1), 131–176.

Grinin, L., & Grinin, A. (2023). Evolutionary phases in big history and complexity. *Journal of Big History, 6*(3), 5–17. https://doi.org/10.22339/jbh.v6i3.6302

Grinin, L., & Grinin, A. (2019). The star-galaxy era in terms of Big History and universal evolution. *Journal of Big History, 3*(4), 69–92.

Grinin, L., & Grinin, A. (2020). Social evolution as an integral part of universal evolution. *Social Evolution & History, 19*(2), 19–45.

Grinin, L., & Korotayev, A. (2023). Social evolution: Theoretical aspects and general outlines. *Social Evolution & History, 22*(2), 111–158. https://doi.org/10.30884/seh/2023.02.07
Grinin, L., & Grinin, A. (2025). Big history periodization. Complexity, directions and phase characteristics. In D. J. LePoire, L. Grinin, & A. Korotayev (Eds.), Navigating complexity in big history—exploring periodization across cosmic and biosocial dimensions (pp. 319–342). Springer Nature. https://doi.org/10.1007/978-3-031-85410-1_10
Grinin, L., Grinin, A., & Korotayev, A. (2024). Cybernetic revolution and global aging: Humankind on the way to cybernetic society, or the next hundred years. Springer Nature https://doi.org/10.1007/978-3-031-56764-3
Grinin, L., & Korotayev, A. (2009). The epoch of the initial politogenesis. *Social Evolution and History, 8*(1), 52–91.
Grinin, L., & Korotayev, A. (2018). The future of the global economy in the light of inflationary and deflationary trends and long cycles theory. *World Futures, 74*(2), 84–103. https://doi.org/10.1080/02604027.2017.1357934
Grinin, L., Korotayev, A., & Tausch, A. (2016). *Economic cycles, crises, and the global periphery*. Springer Nature. https://doi.org/10.1007/978-3-319-41262-7
Grinin, L., Markov, A., & Korotayev, A. (2009). Aromorphoses in biological and social evolution: Some general rules for biological and social forms of macro-evolution. *Social Evolution and History, 8*(2), 6–50.
Grinin, L. E., Markov, A. V., & Korotayev, A. V. (2011). Biological and social aromorphoses: A comparison between two forms of macroevolution. In L. E. Grinin, A. V. Korotayev, R. L. Carneiro, & F. Spier (Eds.), *Evolution: Cosmic, biological, and social* (pp. 162–211). Uchitel.
Grinin, L. E., Markov, A., & Korotayev, A. V. (2025). Biological and social phases of big history and universal evolution: Similarities, differences, and the growth of complexity. *Journal of Big History, 9*(3), 1–51. https://doi.org/10.22339/jbh.v8i3.8301
Grinin, L., Markov, A., & Korotayev, A. (2026). Biological and social phases of big history and complexity. In: D. J. LePoire, L. Grinin, & A. Korotayev (Eds.), *Complexity in universal evolution—a Big History perspective* (pp. 283–356). Springer Nature. In this volume. https://doi.org/10.1007/978-3-032-20516-2_12
Gunderson, L. H., & Holling, C. S. (2002). *Panarchy: Understanding transformations in human and natural systems*. Island Press.
Grinin, L., Carneiro, R., Bondarenko, D., Kradin, N., & Korotayev, A. (Eds.). (2004). *The early state, its alternatives and analogues*. Uchitel.
Harper, A. (2026). Punctuated growth and complexity in urban evolution: Insights from Historical City size distributions. In: D. J. LePoire, L. Grinin, & A. Korotayev (Eds.), *Complexity in universal evolution—a Big History perspective* (pp. 519–536). Springer Nature. In this volume. https://doi.org/10.1007/978-3-032-20516-2_18
Holling, C. S. (2001). Understanding the complexity of economic, ecological, and social systems. *Ecosystems, 4*(5), 390–405.
Jantsch, E. (1980). *The self-organizing universe: Scientific and human implications of the emerging paradigm of evolution*. Pergamon.
Jennings, J., & Earle, T. (2023). Social stratification, genealogical distance, and state formation in early Hawaii and the central Andes: Steps, slopes, and off-ramps in social evolution. *Social Evolution & History, 22*(2), 44–66. https://doi.org/10.30884/seh/2023.02.04
Korotayev, A. (2003). Penetration of nomads to the Arabian south and formation of tribal organization among the north-East Yemen agricultural population. In N. N. Kradin, D. M. Bondarenko, & T. J. Barfield (Eds.), *Nomadic pathways in social evolution* (pp. 114–122). Institute for African Studies.
Korotayev, A. (2004). *World religions and social evolution of the Old World Oikumene civilizations: A cross-cultural perspective*. The Edwin Mellen Press.
Korotayev, A. (2020). Evolution of sociopolitical institutions in north-East Yemen (the 1st millennium BCE–the 2nd millennium CE). In D. M. Bondarenko, S. A. Kowalewski, & D. B. Small

(Eds.), *The evolution of social institutions. Interdisciplinary perspectives* (pp. 161–184). Springer Nature. https://doi.org/10.1007/978-3-030-51437-2_7

Korotayev, A. (2024). Complexity growth patterns in the Big History. Preliminary results of a quantitative analysis. *Journal of Big History, 7*(2), 9–27. https://doi.org/10.22339/jbh.v7i2.7203

Korotayev, A. (2025a). Comparing and contrasting Big History singularity trends of the Big Bang and terrestrial evolution. In D. J. LePoire, L. Grinin, & A. Korotayev (Eds.), *Navigating complexity in Big History. Exploring periodization across cosmic and biosocial dimensions* (pp. 101–157). Springer. https://doi.org/10.1007/978-3-031-85410-1_5

Korotayev, A. (2025b). Toward periodization of the biosocial era of the Big History: A reanalysis. *Journal of Big History, 8*(2), 1–37. https://doi.org/10.22339/jbh.v8i2.8201

Korotayev, A. (2026). Human evolution in the complexity growth perspective. Toward periodization of the big history biosocial era. In: D. J. LePoire, L. Grinin, & A. Korotayev (Eds.), *Complexity in Universal Evolution – A Big History Perspective* (pp. 359–410). Springer Nature (in this volume). https://doi.org/10.1007/978-3-032-20516-2_13

Korotayev, A., Kradin, N., de Munck, V., & Lynsha, V. (2000). Alternativity of social evolution: Introductory notes. In N. N. Kradin, A. V. Korotayev, D. M. Bondarenko, V. de Munck, & P. K. Wason (Eds.), *Alternatives of social evolution* (pp. 12–51). Far Eastern Branch of the Russian Academy of Sciences.

Korotayev, A., & LePoire, D. J. (2025). Complexity growth patterns in the Stelliferous Era of Big History. Preliminary observations. *Journal of Big History, 8*(1), 1–12. https://doi.org/10.22339/jbh.v8i1.8102

Korotayev, A., & LePoire, D. J. (2026). The evolution of complexity in the Stelliferous Era: Observations and patterns. In: D. J. LePoire, L. Grinin, & A. Korotayev (Eds.), *Complexity in universal evolution—a Big History perspective* (pp. 151–172). Springer Nature. In this volume. https://doi.org/10.1007/978-3-032-20516-2_5

Korotayev, A., Malkov, A., & Khaltourina, D. (2006). *Introduction to social macrodynamics: Secular cycles and millennial trends*. KomKniga/URSS.

Korotayev, A., Zinkina, J., & Bogevolnov, J. (2011). Kondratieff waves in global invention activity (1900–2008). *Technological Forecasting and Social Change, 78*(7), 1280–1284. https://doi.org/10.1016/j.techfore.2011.02.011

Korotayev, A., Zinkina, J., & Andreev, A. (2016). Secular cycles and millennial trends. *Cliodynamics: The Journal of Quantitative History and Cultural Evolution, 7*(2), 204–216. https://doi.org/10.21237/C7clio7230484

Kradin, N. N. (2023). Origins of the early Russian State: anthropological perspectives. *Social Evolution & History, 22*(2), 67–93. https://doi.org/10.30884/seh/2023.02.05

Kradin, N. N. (2025). Debates on the origin of the state: Some results and perspectives. *Social Evolution & History, 24*(2), 147–156. https://doi.org/10.30884/seh/2025.02.11

LePoire, D. J. (2026a). Energy flow and the emergence of complexity in the universe's structure. In: D. J. LePoire, L. Grinin, & A. Korotayev (Eds.), *Complexity in universal evolution—a Big History perspective* (pp. 173–190). Springer Nature. In this volume. https://doi.org/10.1007/978-3-032-20516-2_6

LePoire, D. J. (2026b). The earth's pivotal role in Big History: From cosmic evolution to life's emergence. In: D. J. LePoire, L. Grinin, & A. Korotayev (Eds.), *Complexity in universal evolution—a Big History perspective* (pp. 211–220). Springer Nature. In this volume. https://doi.org/10.1007/978-3-032-20516-2_8

LePoire, D. J., Grinin, L., & Korotayev, A. (2025). Complexity in Big History. An introductory exploration. *Journal of Big History, 8*(3), 98–139. https://doi.org/10.22339/jbh.v8i3.8304

LePoire, D. J., Grinin, L., & Korotayev, A. (2026). Introduction to complexity theory for Big History. In: D. J. LePoire, L. Grinin, & A. Korotayev (Eds.), *Complexity in universal evolution—a Big History perspective* (pp. 29–90). Springer Nature. In this volume. https://doi.org/10.1007/978-3-032-20516-2_2

Lima-de-Faria, A. (1988). *Evolution without selection: Form and function by autoevolution*. Elsevier.

Narguizian, P. (2026). Biology education through big history's complex systems thinking. In: D. J. LePoire, L. Grinin, & A. Korotayev (Eds.), *Complexity in universal evolution—a Big History perspective* (pp. 257–282). Springer Nature. In this volume. https://doi.org/10.1007/978-3-032-20516-2_11

Nazaretyan, A. P. (1991). *Intellekt vo Vselennoi: istoki, stanovlenie, perspektivy*. Nedra.

Nielsen, N. (2026). Rethinking complexity in the universe: Implications for SETI and the Fermi paradox. In: D. J. LePoire, L. Grinin, & A. Korotayev (Eds.), *Complexity in universal evolution—a Big History perspective* (pp. 431–466). Springer Nature. In this volume. https://doi.org/10.1007/978-3-032-20516-2_15

Panov, A. (2026). Beyond intelligence: The galactic information field and the evolution of complexity. In: D. J. LePoire, L. Grinin, & A. Korotayev (Eds.), *Complexity in universal evolution—a Big History perspective* (pp. 575–582). Springer Nature. In this volume. https://doi.org/10.1007/978-3-032-20516-2_21

Popov, V. A. (2025). On the discussion of the concepts of 'chiefdom' and 'tribe' in current political anthropology and history of primitive society. *Social Evolution & History, 24*(2), 143–146. https://doi.org/10.30884/seh/2025.02.10

Rodrigue, B. H. (2023a). The bedrock of survival: Humanity, history, and adaptation. *Journal of Globalization Studies, 14*(1), 137–169. https://doi.org/10.30884/jogs/2023.01.09

Rodrigue, B. H. (2023b). The study of all existence: Towards global symbiosis. *Journal of Globalization Studies, 14*(2), 171–185. https://doi.org/10.30884/jogs/2023.02.12

Satkiewicz, S. (2026). War and the rise of social complexity: A big history perspective. In: D. J. LePoire, L. Grinin, & A. Korotayev (Eds.), *Complexity in universal evolution—a Big History perspective* (pp. 537–560). Springer Nature. In this volume. https://doi.org/10.1007/978-3-032-20516-2_19

Sazhienko, E., & LePoire, D. J. (2026). The role of artificial intelligence in big history: Understanding human evolution and future scenarios. In: D. J. LePoire, L. Grinin, & A. Korotayev (Eds.), *Complexity in universal evolution—a Big History perspective* (pp. 411–430). Springer Nature. In this volume. https://doi.org/10.1007/978-3-032-20516-2_14

Schumpeter, J. A. 1994 [1942]. *Capitalism, socialism and democracy*. Routledge.

Shchavelev, A. S. (2025). Why do we need a typology (Classification) and a political anthropological meta-language for describing post-neolithic pre-modern polities. *Social Evolution & History, 24*(2), 127–132. https://doi.org/10.30884/seh/2025.02.07

Spier, F. (2010). *Big history and the future of humanity*. Wiley-Blackwell.

Spier, F. (2011). Complexity in big history. *Cliodynamics: The Journal of Quantitative History and Cultural Evolution, 2*(1), 146–166. https://doi.org/10.21237/C7clio21213

Snooks, G. D. (2005). Big history or big theory? Uncovering the Laws of life. *Social Evolution and History, 4*(1), 160–188.

Smorchkov, A. M. (2025). Historical reality in ancient and modern political terminology: In search of a common. *Social Evolution & History, 24*(2), 133–136. https://doi.org/10.30884/seh/2025.02.08

Surikov, I. E. (2025). Greek polis on the scale of potestary formations: State, 'Proto-State', 'Post-State'? *Social Evolution & History, 24*(2), 174–177. https://doi.org/10.30884/seh/2025.02.15

Teilhard de Chardin, P. (1987). *The phenomenon of man*. Harper.

Tymowski, M. (2023). A five-day week in rwanda in the second half of the nineteenth century and in the early twentieth century. *Social Evolution & History, 22*(2), 32–43. https://doi.org/10.30884/seh/2023.02.03

Umpleby, S. A., & Dent, E. B. (1999). The origins and purposes of several traditions in systems theory and cybernetics. *Cybernetics and Systems, 30*, 79–103.

Van Duin, M. (2026). Complexity pyramids and energy rate density: A quantified system hierarchy for Big History. In: D. J. LePoire, L. Grinin, & A. Korotayev (Eds.), *Complexity in universal evolution—a Big History perspective* (pp. 119–148). Springer Nature. In this volume. https://doi.org/10.1007/978-3-032-20516-2_4

Vargas, T., Musso, P., & Antonietti, N. (2026). Modeling galactic evolution and the potential for intelligent life in the milky way. In: D. J. LePoire, L. Grinin, & A. Korotayev (Eds.), *Complexity in universal evolution—a Big History perspective* (pp. 563–574). Springer Nature. In this volume. https://doi.org/10.1007/978-3-032-20516-2_20

Waring, T. M., & Wood, Z. T. (2021). Long-term gene–culture coevolution and the human evolutionary transition. *Proceedings of the Royal Society B, 288*(1952), 20210538. https://doi.org/10.1098/rspb.2021.0538

Wiener, N. (1948). *Cybernetics, or control and communication in the animal and the machine*. MIT Press.

Wieser, M., & Slunecko, T. (2014). Cybernetics, radical behaviorism, cultural psychology, cultural-historical psychology. In T. Teo (Ed.), *Encyclopedia of critical psychology*. Springer.

Yankovsky, S. Y. (2000). *The concept of the general theory of information*. Beta-Izdat. In Russian.

David LePoire researches, develops, and applies science principles in environmental issues, Big History evolutionary trends, and particle scattering. He has a BS in physics from CalTech, a PhD in computer science from DePaul University, and over 30 years of experience at the Argonne National Laboratory in the development of scientific analyses, software, training, and modeling. His research includes Big History synergistic trends among energy, environment, organization, and information. Currently, he is the President of the International Big History Association (IBHA).

Leonid Grinin is Senior Research Professor at the Center for Stability and Risk Analysis at the HSE University in Moscow, as well as Director of Eurasian Center for Big History & System Forecasting at the Institute of Oriental Studies, Russian Academy of Sciences. He is co-editor of the international journals Social Evolution & History and the Journal of Globalization Studies, as well as co-editor of the international almanacs Evolution, History & Mathematics, and Kondratieff Waves. He is author of more than 600 publications, including 30 monographs in Russian, English, and Chinese. His current research focuses on comparative political studies, theory of revolution, political anthropology, global economy, global history, historical sociology, and futurology. Among his monographs are *The Cybernetic Revolution and the Forthcoming Epoch of Self-Regulating Systems* (Uchitel, 2016, with Anton Grinin), *Big History of Globalization: The Emergence of a Global World System* (Springer 2019, in co-authorship with Julia Zinkina, David Christian, and others), and *Islamism, Arab Spring, and the Future of Democracy: World System and World Values Perspectives* (Springer, 2019; with Andrey Korotayev and Arno Tausch). In 2004, he was awarded as a winner of the best publication of Socis journal during the year for the article: "People of Publicity—a New Social Layer?" In 2012, he was awarded the Gold Kondratieff Medal by the International N. D. Kondratieff Foundation. In 2016, he received the Belyaev award for the book From Bifaces to Nanorobots (the book published in Russian in co-authorship with Anton Grinin). 12, Rozhdestvenka, Moscow, 107031, Russian Federation; email: leonid.grinin@gmail.com; ORCID: 0000-0003-0278-2619.

Andrey Korotayev heads the Center for Stability and Risk Analysis at the HSE University in Moscow, Russia. He is also Senior Research Professor at Eurasian Center for Big History & System Forecasting at the Institute of Oriental Studies, Russian Academy of Sciences and Institute for African Studies, Russian Academy of Sciences, as well as at the Faculty of Global Studies of the Lomonosov Moscow State University. Beginning as a historian of Arabia, he made focused studies, such as seen in Two Social Ecological Crises and Genesis of Tribal Organization in the Yemeni North-East (1996). Korotayev researches topics in quantitative cross-cultural anthropology and seeks to understand biological and social macroevolution, as in A Compact Macromodel of World System Evolution (2005) and Mathematical Modeling of Biological and Social Phases of Big History (2014). He has authored or co-authored over 750 scholarly publications, including such monographs as *Ancient Yemen* (Oxford University Press, 1995), *World Religions and Social Evolution of the Old World Oikumene Civilizations: A Cross-Cultural Perspective* (The Edwin

Mellen Press, 2004), *Introduction to Social Macrodynamics: Compact Macromodels of the World System Growth* (URSS, 2006; with Daria Khaltorina and Artemy Malkov), *Introduction to Social Macrodynamics: Secular Cycles and Millennial Trends* (URSS, 2006; with Daria Khaltorina and Artemy Malkov), *Great Divergence and Great Convergence: A Global Perspective* (Springer Nature, 2015; with Leonid Grinin), Economic Cycles, Crises, and the Global Periphery (Springer Nature, 2016; with Leonid Grinin), and *Islamism, Arab Spring, and* the Future of Democracy: World System and World Values Perspectives (Springer Nature, 2019; with Leonid Grinin and Arno Tausch). He is a founding member of the Eurasian Center for Megahistory & System Forecasting and founder of Evolution, an almanac dedicated to the study of Universal Evolution, and thus is closely linked with Big History. In 2020, he produced, with David LePoire, a collective monograph, *The Twenty-First Century Singularity and Global Futures: A Big History Perspective* (Springer Nature, 2020). Recently, he has produced, with David LePoire and Leonid Grinin, another Big History collective monograph, *Navigating Complexity in Big History: Exploring Periodization Across Cosmic and Biosocial Dimensions* (Springer Nature, 2025). He is a laureate of a Russian Science Support Foundation in "The Best Economists of the Russian Academy of Sciences" Nomination (2006); in 2012, he was awarded the Gold Kondratieff Medal by the International N. D. Kondratieff Foundation. He serves as a board member of the International Big History Association. 20, Myasnitskaya, Moscow 101000, Russian Federation; email: akorotayev@gmail.com; ORCID: 0000-0003-3014-2037.

The manufacturer's authorised representative in the EU is Springer Nature Customer Service Centre GmbH, Europaplatz 3, 69115 Heidelberg, Germany. If you have any concerns regarding our products, please contact ProductSafety@springernature.com

Printed and bound by CPI Group (UK) Ltd, Croydon, CR0 4YY
07/10/2026
02229349-0018